Study smarter and improve your grade

OWL

Online Web Learning

For Organic Chemistry

UMassAmherst

W9-CGK-013

Succeed in Organic Chemistry with **OWL**, a proven online learning system that has already helped hundreds of thousands of students master chemistry concepts, develop problem-solving skills, and improve their grades.

The OWL online learning system:

- **Lets you work at your own pace.** With **OWL**, as soon as you master a concept, you can move on.

- **Allows unlimited practice. OWL** provides instant feedback on every question, along with hints that guide you to the problem's solution. If you still need help, **OWL** will generate a different question of the same type.

- **Includes tools to suit every learning style. OWL** offers tutors, simulations, animations, and **MarvinSketch**, an advanced drawing tool, to ensure that you have everything you need to understand concepts.

- **Provides an interactive 3-dimensional environment** through questions that allow you to rotate molecules and measure bond distances and angles.

- **Is available integrated with a time-saving e-book.** An electronic version of your textbook is linked to the questions in **OWL** to save you hours of study time.

> For this textbook, **OWL** includes end-of-chapter questions marked in the text with a ■

The Chemist's Choice. The Student's Solution.

7e

ORGANIC CHEMISTRY

Enhanced Edition
Volume 2

John McMurry

Cornell University

BROOKS/COLE
CENGAGE Learning

Australia • Brazil • Japan • Korea • Mexico • Singapore • Spain • United Kingdom • United States

BROOKS/COLE
CENGAGE Learning™

Organic Chemistry Enhanced Edition, Seventh Edition
John McMurry

Publisher: Mary Finch

Senior Acquisitions Editor: Lisa Lockwood

Senior Developmental Editor: Sandra Kiselica

Assistant Editor: Elizabeth Woods

Senior Media Editor: Lisa Weber

Marketing Manager: Amee Mosley

Marketing Assistant: Kevin Carroll

Marketing Communications Manager: Linda Yip

Content Project Manager: Teresa L. Trego

Creative Director: Rob Hugel

Art Director: John Walker

Print Buyer: Judy Inouye

Production Service: Graphic World, Inc.

Text Designer: tani hasagawa

Photo Researcher: Marcy Lunetta

Copy Editor: Graphic World, Inc.

Illustrator: ScEYEnce Studios, Patrick Lane; Graphic World, Inc.

OWL Producers: Stephen Battisti, Cindy Stein and David Hart in the Center for Educational Software Development at the University of Massachusetts, Amherst, and Cow Town Productions

Cover Designer: tani hasagawa

Cover Image: Sean Duggan

Compositor: Graphic World, Inc.

For product information and technology assistance, contact us at **Cengage Learning Customer & Sales Support, 1-800-354-9706.**
For permission to use material from this text or product, submit all requests online at **www.cengage.com/permissions.**
Further permissions questions can be e-mailed to **permissionrequest@cengage.com.**

Library of Congress Control Number: 2009926438

ISBN-13: 978-0-495-11258-7
ISBN-10: 0-495-11258-5

Volume 1:
ISBN-13: 978-0-538-73395-3
ISBN-10: 0-538-73395-0

Volume 2:
ISBN-13: 978-1-4390-4931-0
ISBN-10: 1-4390-4931-9

Brooks/Cole
10 Davis Drive
Belmont, CA 94002-3098
USA

Cengage Learning is a leading provider of customized learning solutions with office locations around the globe, including Singapore, the United Kingdom, Australia, Mexico, Brazil, and Japan. Locate your local office at **www.cengage.com/global.**

Cengage Learning products are represented in Canada by Nelson Education, Ltd.

To learn more about Brooks/Cole, visit **www.cengage.com/brookscole**

Purchase any of our products at your local college store or at our preferred online store **www.ichapters.com.**

Printed in Canada
1 2 3 4 5 6 7 13 12 11 10 09

Contents in Brief

This text is available in these student versions:
- Complete text ISBN: 978-0-495-11258-7 • Volume 1 (Chapters 1–15 with Quick Prep and study cards): 978-0-538-73395-3 • Volume 2 (Chapters 16–31 with study cards): 978-1-4390-4931-0

Contents

6 | Alkenes: Structure and Reactivity   172

7 | Alkenes: Reactions and Synthesis   213

© 2006 San Marcos Growers

© Macduff Everton/Corbis

11 | Reactions of Alkyl Halides: Nucleophilic Substitutions and Eliminations   359

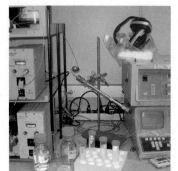

12 | Structure Determination: Mass Spectrometry and Infrared Spectroscopy   408

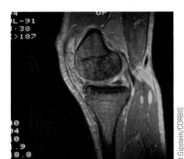

13   Structure Determination: Nuclear Magnetic Resonance
Spectroscopy   440

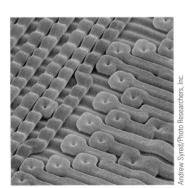

14   Conjugated Compounds and Ultraviolet
Spectroscopy   482

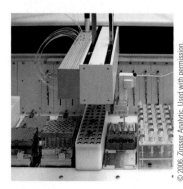

© Karl Weatherly/Getty Images

© Charles O'Rear/Corbis

© Erich Lessing/Art Resource, NY

25 | Biomolecules: Carbohydrates   973

26 | Biomolecules: Amino Acids, Peptides, and Proteins   1016

27 | **Biomolecules: Lipids   1060**

28 | **Biomolecules: Nucleic Acids   1100**

29 | **The Organic Chemistry of Metabolic Pathways   1125**

Preface

I love to write. I get real pleasure from taking a complicated subject, turning it around until I see it clearly, and then explaining it in simple words. I write to explain chemistry to students today the way I wish it had been explained to me years ago.

The enthusiastic response to the six previous editions has been very gratifying and suggests that this book has served students well. I hope you will find that this seventh edition of *Organic Chemistry* builds on the strengths of the first six and serves students even better. I have made every effort to make this new edition as effective, clear, and readable as possible; to show the beauty and logic of organic chemistry; and to make organic chemistry enjoyable to learn.

Organization and Teaching Strategies   This seventh edition, like its predecessors, blends the traditional functional-group approach with a mechanistic approach. The primary organization is by functional group, beginning with the simple (alkenes) and progressing to the more complex. Most faculty will agree that students new to the subject and not yet versed in the subtleties of mechanism do better this way. In other words, the *what* of chemistry is generally easier to grasp than the *why*. Within this primary organization, however, I place heavy emphasis on explaining the fundamental mechanistic similarities of reactions. This emphasis is particularly evident in the chapters on carbonyl-group chemistry (Chapters 19–23), where mechanistically related reactions like the aldol and Claisen condensations are covered together. By the time students reach this material, they have seen all the common mechanisms and the value of mechanisms as an organizing principle has become more evident.

The Lead-Off Reaction: Addition of HBr to Alkenes   Students usually attach great importance to a text's lead-off reaction because it is the first reaction they see and is discussed in such detail. I use the addition of HBr to an alkene as the lead-off to illustrate general principles of organic chemistry for several reasons: the reaction is relatively straightforward; it involves a common but important functional group; no prior knowledge of stereochemistry or kinetics in needed to understand it; and, most important, it is a *polar* reaction. As such, I believe that electrophilic addition reactions represent a much more useful and realistic introduction to functional-group chemistry than a lead-off such as radical alkane chlorination.

Reaction Mechanisms   In the first edition of this book, I introduced an innovative format for explaining reaction mechanisms in which the reaction steps are printed vertically, with the changes taking place in each step described next to the reaction arrow. This format allows a reader to see easily what is occurring at each step without having to flip back and forth between structures and text. Each successive edition has seen an increase in the number and quality of these vertical mechanisms, which are still as fresh and useful as ever.

Organic Synthesis   Organic synthesis is treated in this text as a teaching device to help students organize and deal with a large body of factual information—the same skill so critical in medicine. Two sections, the first in Chapter 8 (Alkynes) and the second in Chapter 16 (Chemistry of Benzene), explain the thought processes involved in working synthesis problems and emphasize the value of starting from what is known and logically working backward. In addition, *Focus On* boxes, including The Art of Organic Synthesis, Combinatorial Chemistry, and Enantioselective Synthesis, further underscore the importance and timeliness of synthesis.

Modular Presentation   Topics are arranged in a roughly modular way. Thus, certain chapters are grouped together: simple hydrocarbons (Chapters 3–8), spectroscopy (Chapters 12–14), carbonyl-group chemistry (Chapters 19–23), and biomolecules (Chapters 25–29). I believe that this organization brings to these subjects a cohesiveness not found in other texts and allows the instructor the flexibility to teach in an order different from that presented in the book.

Basic Learning Aids   In writing and revising this text, I consistently aim for lucid explanations and smooth transitions between paragraphs and between topics. New concepts are introduced only when they are needed, not before, and they are immediately illustrated with concrete examples. Frequent cross-references to earlier material are given, and numerous summaries are provided to draw information together, both within and at the ends of chapters. In addition, the back of this book contains a wealth of material helpful for learning organic chemistry, including a large glossary, an explanation of how to name polyfunctional organic compounds, and answers to all in-text problems. For still further aid, an accompanying *Study Guide and Solutions Manual* gives summaries of name reactions, methods for preparing functional groups, functional-group reactions, and the uses of important reagents.

Changes and Additions for the Seventh Edition

The primary reason for preparing a new edition is to keep the book up to date, both in its scientific coverage and in its pedagogy. My overall aim is always to refine the features that made earlier editions so successful, while adding new ones.

▌ **The writing** has again been revised at the sentence level, streamlining the presentation, improving explanations, and updating a thousand small details. Several little-used reactions have been deleted (the alkali fusion of arenesulfonic acids to give phenols, for instance), and a few new ones have been added (the Sharpless enantioselective epoxidation of alkenes, for instance).

▌ Other notable **content changes** are:

Chapter 2, *Polar Covalent Bonds; Acids and Bases*—A new Section 2.13 on noncovalent interactions has been added.

Chapter 3, *Organic Compounds: Alkanes and Their Stereochemistry*—The chapter has been revised to focus exclusively on open-chain alkanes.

Chapter 4, *Organic Compounds: Cycloalkanes and Their Stereochemistry*—The chapter has been revised to focus exclusively on cycloalkanes.

Chapter 5, *An Overview of Organic Reactions*—A new Section 5.11 comparing biological reactions and laboratory reactions has been added.

Chapter 7, *Alkenes: Reactions and Synthesis*—Alkene epoxidation has been moved to Section 7.8, and Section 7.11 on the biological addition of radicals to alkenes has been substantially expanded.

Chapter 9, *Stereochemistry*—A discussion of chirality at phosphorus and sulfur has been added to Section 9.12, and a discussion of chiral environments has been added to Section 9.14.

Chapter 11, *Reactions of Alkyl Halides: Nucleophilic Substitutions and Eliminations*—A discussion of the E1cB reaction has been added to Section 11.10, and a new Section 11.11 discusses biological elimination reactions.

Chapter 12, *Structure Determination: Mass Spectrometry and Infrared Spectroscopy*—A new Section 12.4 discusses mass spectrometry of biological molecules, focusing on time-of-flight instruments and soft ionization methods such as MALDI.

Chapter 20, *Carboxylic Acids and Nitriles*—A new Section 20.3 discusses biological carboxylic acids and the Henderson–Hasselbalch equation.

Chapter 24, *Amines and Heterocycles*—This chapter now includes a discussion of heterocycles, and a new Section 24.5 on biological amines and the Henderson–Hasselbalch equation has been added.

Chapter 25, *Biomolecules: Carbohydrates*—A new Section 25.7 on the eight essential carbohydrates has been added, and numerous content revisions have been made.

Chapter 26, *Biomolecules: Amino Acids, Peptides, and Proteins*—The chapter has been updated, particularly in its coverage of solid-phase peptide synthesis.

Chapter 27, *Biomolecules: Lipids*—The chapter has been extensively revised, with increased detail on prostaglandins (Section 27.4), terpenoid biosynthesis (Section 27.5), and steroid biosynthesis, (Section 27.7).

Chapter 28, *Biomolecules: Nucleic Acids*—Coverage of heterocyclic chemistry has been moved to Chapter 24.

Chapter 29, *The Organic Chemistry of Metabolic Pathways*—The chapter has been reorganized and extensively revised, with substantially increased detail on important metabolic pathways.

Chapter 30, *Orbitals and Organic Chemistry: Pericyclic Reactions*—All the art in this chapter has been redone.

▌ **The order of topics** remains basically the same but has been changed to devote Chapter 3 entirely to alkanes and Chapter 4 to cycloalkanes. In addition, epoxides are now introduced in Chapter 7 on alkenes, and coverage of heterocyclic chemistry has been moved to Chapter 24.

▌ **The problems** within and at the end of each chapter have been reviewed, and approximately 100 new problems have been added, many of which focus on biological chemistry.

▌ *Focus On* **boxes** at the end of each chapter present interesting applications of organic chemistry relevant to the main chapter subject. Including topics from biology, industry, and day-to-day life, these applications enliven and reinforce the material presented within the chapter. The boxes have been updated, and new ones added, including Where Do Drugs Come From? (Chapter 5),

Green Chemistry (Chapter 11), X-Ray Crystallography (Chapter 22), and Green Chemistry II: Ionic Liquids (Chapter 24).

▌ **Biologically important molecules and mechanisms** have received particular attention in this edition. Many reactions now show biological counterparts to laboratory examples, many new problems illustrate reactions and mechanisms that occur in living organisms, and enhanced detail is given for major metabolic pathways.

More Features

NEW! ▌ Why do we have to learn this? I've been asked this question so many times by students that I thought that it would be appropriate to begin each chapter with the answer. The *Why This Chapter?* section is a short paragraph that appears at the end of the introduction to every chapter and tells students why the material about to be covered is important.

NEW! ▌ Thirteen Key Ideas are highlighted in the book. These include topics pivotal to students' development in organic chemistry, such as Curved Arrows in Reaction Mechanisms (Chapter 5) and Markovnikov's Rule (Chapter 6). These Key Ideas are further reinforced in end-of-chapter problems marked with a ▲ icon. A selection of these problems are also assignable in OWL, denoted by a ■.

▌ Worked Examples are now titled to give students a frame of reference. Each Worked Example includes a Strategy and a worked-out Solution, and then is followed by problems for students to try on their own. This book has more than 1800 in-text and end-of-chapter problems.

▌ An overview chapter, *A Preview of Carbonyl Chemistry,* follows Chapter 18 and highlights the author's belief that studying organic chemistry requires both summarizing and looking ahead.

NEW!

Organic KNOWLEDGE TOOLS

▌ Thorough media integration with Organic Knowledge Tools: CengageNOW for Organic Chemistry and Organic OWL are provided to help students practice and test their knowledge of the concepts. CengageNOW is an online assessment program for self-study with interactive tutorials. Organic OWL is an online homework learning system. Icons throughout the book direct students to CengageNOW at **www.cengage.com/login** A fee-based access code is required for Organic OWL.

NEW! ▌ Approximately 15 to 20 end-of-chapter problems per chapter, denoted with a ■ icon, are assignable in the OWL online learning system. These questions are algorithmically generated, allowing students more practice.

Online Web Learning

UMassAmherst

▌ OWL (Online Web Learning) for Organic Chemistry, developed at the University of Massachusetts, Amherst; class-tested by thousands of students; and used by more than 50,000 students, provides fully class-tested questions and tutors in an easy-to-use format. OWL is also customizable and cross-platform. OWL offers students instant grading and feedback on homework problems, modeling questions, and animations to accompany this text. With parameterization, OWL for Organic Chemistry offers nearly 6000 different questions as well as Marvin-Sketch for viewing and drawing chemical structures, and Jmol, a powerful molecule viewer that helps students visualize stereochemistry.

▌ A number of the figures are animated in CengageNOW. These are designated as **Active Figures** in the figure legends.

▌ The Visualizing Chemistry Problems that begin the exercises at the end of each chapter offer students an opportunity to see chemistry in a different way by visualizing molecules rather than by simply interpreting structural formulas.

▌ Summaries and Key Word lists help students by outlining the key concepts of the chapter.

▌ Summaries of Reactions, at the ends of appropriate chapters, bring together the key reactions from the chapter in one complete list.

Companions to This Text

Supporting instructor materials are available to qualified adopters. Please consult your local Brooks/Cole Cengage Learning representative for details.

Visit **www.cengage.com/chemistry/mcmurry** to:

Find your local representative

Download electronic files of text art and ancillaries from this book's companion site

Request a desk copy

Ancillaries for Students

Study Guide and Solutions Manual, by Susan McMurry, provides answers and explanations to all in-text and end-of-chapter exercises. (0-495-11268-2)

CENGAGENOW™   **CengageNOW** To further student understanding, the text features sensible media integration through **CengageNOW**, a powerful online learning companion that helps students determine their unique study needs and provides them with individualized resources. This dynamic learning companion combines with the text to provide students with a seamless, integrated learning system. The access code required to register for access at **www.cengage.com/login** may be included with a copy of this text or purchased with ISBN 0-495-31869-8 from **www.ichapters.com** .

Online Web Learning

UMassAmherst

OWL for Organic Chemistry, authored by Steve Hixson and Peter Lillya of the University of Massachusetts, Amherst, and William Vining of the State University of New York at Oneonta. Class-tested by thousands of students and used by more than 50,000 students, OWL (Online Web Learning) provides fully class-tested content in an easy-to-use format. OWL is also customizable and cross-platform. OWL offers students instant grading and feedback on homework problems, modeling questions, and animations to accompany this text. With parameterization, OWL for Organic Chemistry offers nearly 6000 questions as well as MarvinSketch, a Java applet for viewing and drawing chemical structures, and Jmol, a powerful molecule viewer that helps students visualize stereochemistry.

This powerful system maximizes the students' learning experience and, at the same time, reduces faculty workload and helps facilitate instruction. New to this edition are 15 to 20 end-of-chapter problems per chapter, denoted by a ■ icon, that are assignable in OWL. A fee-based access code is required for OWL.

Pushing Electrons: A Guide for Students of Organic Chemistry, third edition, by Daniel P. Weeks. A workbook designed to help students learn techniques of electron pushing, its programmed approach emphasizes repetition and active participation. (0-03-020693-6)

NEW!   **Spartan Model Electronic Modeling Kit**, A set of easy-to-use builders allow for the construction and 3-D manipulation of molecules of any size or complexity—from a hydrogen atom to DNA and everything in between. This kit includes the SpartanModel software on CD-ROM, an extensive molecular database, 3-D glasses, and a *Tutorial and Users Guide* that includes a wealth of activities to help you get the most out of your course. (0-495-01793-0)

Ancillaries for Instructors

PowerLecture   A dual-platform digital library and presentation tool on two CD-ROMs that provides art and tables from the main text in a variety of electronic formats that are easily exported into other software packages. PowerLecture also contains simulations, molecular models, and QuickTime movies to supplement lectures as well as electronic files of various print supplements. Instructors can customize the PowerPoint® Lectures by adding their own slides or by deleting or changing existing slides (PowerLecture ISBNs: 0-495-11265-8 and 0-495-55443-X). PowerLecture also includes:

▮ **ExamView Testing**   This easy-to-use software, containing questions and problems authored specifically for the text, allows professors to create, deliver, and customize tests in minutes.

▮ **JoinIn on Turning Point for Organic Chemistry**   Book-specific JoinIn™ content for Response Systems tailored to *Organic Chemistry* allows you to transform your classroom and assess your students' progress with instant in-class quizzes and polls. Our exclusive agreement to offer TurningPoint software lets you pose book-specific questions and display students' answers seamlessly within the Microsoft PowerPoint slides of your own lecture, in conjunction with the "clicker" hardware of your choice. Enhance how your students interact with you, your lecture, and one another. Contact your local Brooks/Cole Cengage Learning representative to learn more.

WebCT/NOW Integration   Instructors and students enter **CengageNOW** through their familiar Blackboard or WebCT environment without the need for a separate user name or password and can access all of the **CengageNOW** assessments and content. Contact your local Brooks/Cole Cengage Learning representative to learn more.

Transparency Acetates   Approximately 200 full-color transparency acetates of key text illustrations, enlarged for use in the classroom and lecture halls. (0-495-11260-7)

Organic Chemistry Laboratory Manuals   Brooks/Cole is pleased to offer a choice of organic chemistry laboratory manuals catered to fit individual needs. Visit www.cengage.com/chemistry. Customizable laboratory manuals also can be assembled—contact your Brooks/Cole Cengage Learning representative to learn more.

Acknowledgments

I thank all the people who helped to shape this book and its message. At Brooks/Cole they include: David Harris, publisher; Sandra Kiselica, senior development editor; Amee Mosley executive marketing manager; Teresa Trego, project manager; Lisa Weber; media editor; and Sylvia Krick, assistant editor, along with Suzanne Kastner and Gwen Gilbert at Graphic World.

I am grateful to colleagues who reviewed the manuscript for this book and participated in a survey about its approach. They include:

Manuscript Reviewers

Arthur W. Bull, Oakland University
Robert Coleman, Ohio State University
Nicholas Drapela, Oregon State University
Christopher Hadad, Ohio State University
Eric J. Kantorowski, California Polytechnic State University
James J. Kiddle, Western Michigan University
Joseph B. Lambert, Northwestern University
Dominic McGrath, University of Arizona
Thomas A. Newton, University of Southern Maine
Michael Rathke, Michigan State University
Laren M. Tolbert, Georgia Institute of Technology

Reviewers of Previous Editions

Wayne Ayers, East Carolina University
Kevin Belfield, University of Central Florida-Orlando
Byron Bennett, University of Las Vegas
Robert A. Benkeser, Purdue University
Donald E. Bergstrom Purdue University
Christine Bilicki, Pasedena City College
Weston J. Borden, University of North Texas
Steven Branz, San Jose State University
Larry Bray, Miami-Dade Community College
James Canary, New York University
Ronald Caple, University of Minnesota-Duluth
John Cawley, Villanova University
George Clemans, Bowling Green State University
Bob Coleman, Ohio State University
Paul L. Cook, Albion College
Douglas Dyckes, University of Colorado-Denver
Kenneth S. Feldman, Pennsylvania State University
Martin Feldman, Howard University

Kent Gates, University of Missouri-Columbia
Warren Gierring, Boston University
Daniel Gregory, St. Cloud State University
David Hart, Ohio State University
David Harpp, McGill University
Norbert Hepfinger, Rensselaer Polytechnic Institute
Werner Herz, Florida State University
John Hogg, Texas A&M University
Paul Hopkins, University of Washington
John Huffman, Clemson University
Jack Kampmeier, University of Rochester
Thomas Katz, Columbia University
Glen Kauffman, Eastern Mennonite College
Andrew S. Kendle, University of North Carolina- Wilmington
Paul E. Klinedinst, Jr., California State University- Northridge
Joseph Lamber, Northwestern University
John T. Landrum, Florida International University

Peter Lillya, University of Massachusetts

Thomas Livinghouse, Montana State University

James Long, University of Oregon

Todd Lowary, University of Alberta

Luis Martinez, University of Texas, El Paso

Eugene A. Mash, University of Arizona

Guy Matson, University of Central Florida

Fred Matthews, Austin Peay State University

Keith Mead, Mississippi State University

Michael Montague-Smith, University of Maryland

Andrew Morehead, East Carolina University

Harry Morrison, Purdue University

Cary Morrow, University of New Mexico

Clarence Murphy, East Stroudsburg University

Roger Murray, St. Joseph's University

Oliver Muscio, Murray State University

Ed Neeland, University of British Columbia

Jacqueline Nikles, University of Alabama

Mike Oglioruso, Virginia Polytechnic Institute and State University

Wesley A. Pearson, St. Olaf College

Robert Phillips, University of Georgia

Carmelo Rizzo, Vanderbilt University

William E. Russey, Juniata College

Neil E. Schore, University of California-Davis

Gerald Selter, California State University- San Jose

Eric Simanek, Texas A&M University

Jan Simek, California Polytechnic State University

Ernest Simpson, California State Polytechnic University- Pomona

Peter W. Slade, University College of Fraser Valley

Gary Snyder, University of Massachusetts

Ronald Starkey, University of Wisconsin- Green Bay

J. William Suggs, Brown University

Michelle Sulikowski, Vanderbilt University

Douglas Taber, University of Delaware

Dennis Taylor, University of Adelaide

Marcus W. Thomsen, Franklin & Marshall College

Walter Trahanovsky, Iowa State University

Harry Ungar, Cabrillo College

Joseph J. Villafranca, Pennsylvania State University

Barbara J. Whitlock, University of Wisconsin-Madison

Vera Zalkow, Kennesaw College

16

Chemistry of Benzene: Electrophilic Aromatic Substitution

Organic KNOWLEDGE TOOLS

CENGAGENOW™ Throughout this chapter, sign in at **www.cengage.com/login** for online self-study and interactive tutorials based on your level of understanding.

OWL Online homework for this chapter may be assigned in Organic OWL.

In the preceding chapter, we looked at *aromaticity*—the stability associated with benzene and related compounds that contain a cyclic conjugated system of $4n + 2$ π electrons. In this chapter, we'll look at some of the unique reactions that aromatic molecules undergo.

The most common reaction of aromatic compounds is **electrophilic aromatic substitution**. That is, an electrophile reacts with an aromatic ring and substitutes for one of the hydrogens. The reaction is characteristic of all aromatic rings, not just benzene and substituted benzenes. In fact, the ability of a compound to undergo electrophilic substitution is a good test of aromaticity.

Many different substituents can be introduced onto an aromatic ring through electrophilic substitution reactions. To list some possibilities, an aromatic ring can be substituted by a halogen ($-Cl$, $-Br$, I), a nitro group ($-NO_2$), a sulfonic acid group ($-SO_3H$), a hydroxyl group ($-OH$), an alkyl group ($-R$), or an acyl group ($-COR$). Starting from only a few simple materials, it's possible to prepare many thousands of substituted aromatic compounds.

Halogenation

Acylation

Nitration

Alkylation

Aromatic ring

Sulfonation

Hydroxylation

Sean Duggan

WHY THIS CHAPTER?

This chapter generally continues the coverage of aromatic molecules begun in the preceding chapter, but we'll shift focus to concentrate on reactions, looking at the relationship between aromatic structure and reactivity. This relationship is critical to an understanding of how many biological molecules and pharmaceutical agents are synthesized and why they behave as they do.

16.1 | Electrophilic Aromatic Substitution Reactions: Bromination

CENGAGENOW™ Click *Organic Process* to **view an animation of the bromination of aromatic rings**.

Before seeing how electrophilic aromatic substitutions occur, let's briefly recall what we said in Chapter 6 about electrophilic alkene additions. When a reagent such as HCl adds to an alkene, the electrophilic hydrogen approaches the *p* orbitals of the double bond and forms a bond to one carbon, leaving a positive charge at the other carbon. This carbocation intermediate then reacts with the nucleophilic Cl⁻ ion to yield the addition product.

Alkene   **Carbocation**   **Addition product**
                    intermediate

CENGAGENOW™ Click *Organic Interactive* to **practice your problem-solving skills on the mechanism of electrophilic aromatic substitution**.

An electrophilic aromatic substitution reaction begins in a similar way, but there are a number of differences. One difference is that aromatic rings are less reactive toward electrophiles than alkenes are. For example, Br_2 in CH_2Cl_2 solution reacts instantly with most alkenes but does not react with benzene at room temperature. For bromination of benzene to take place, a catalyst such as $FeBr_3$ is needed. The catalyst makes the Br_2 molecule more electrophilic by polarizing it to give an $FeBr_4^- Br^+$ species that reacts as if it were Br^+. The polarized Br_2 molecule then reacts with the nucleophilic benzene ring to yield a nonaromatic carbocation intermediate that is doubly allylic (Section 11.5) and has three resonance forms.

$$Br-Br \; + \; FeBr_3 \longrightarrow Br^+ \; {}^-FeBr_4$$

Although more stable than a typical alkyl carbocation because of resonance, the intermediate in electrophilic aromatic substitution is nevertheless much less stable than the starting benzene ring itself, with its 150 kJ/mol (36 kcal/mol) of aromatic stability. Thus, the reaction of an electrophile with a benzene ring is endergonic, has a substantial activation energy, and is rather slow. Figure 16.1 shows an energy diagram comparing the reaction of an electrophile with an alkene and with benzene. The benzene reaction is slower (higher $\Delta G^{\ddagger}$) because the starting material is more stable.

Figure 16.1 A comparison of the reactions of an electrophile (E^+) with an alkene and with benzene: $\Delta G^{\ddagger}_{alkene} < \Delta G^{\ddagger}_{benzene}$.

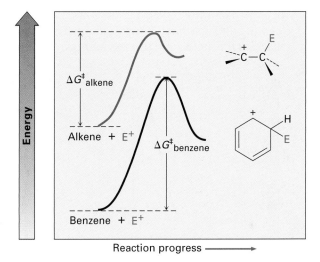

A second difference between alkene addition and aromatic substitution occurs after the carbocation intermediate has formed. Instead of adding Br^- to give an addition product, the carbocation intermediate loses H^+ from the bromine-bearing carbon to give a substitution product. Note that this loss of H^+ is similar to what occurs in the second step of an E1 reaction (Section 11.10). The net effect of reaction of Br_2 with benzene is the substitution of H^+ by Br^+ by the overall mechanism shown in Figure 16.2.

Figure 16.2 MECHANISM: The mechanism of the electrophilic bromination of benzene. The reaction occurs in two steps and involves a resonance-stabilized carbocation intermediate.

1 An electron pair from the benzene ring attacks the positively polarized bromine, forming a new C–Br bond and leaving a nonaromatic carbocation intermediate.

2 A base removes H^+ from the carbocation intermediate, and the neutral substitution product forms as two electrons from the C–H bond move to re-form the aromatic ring.

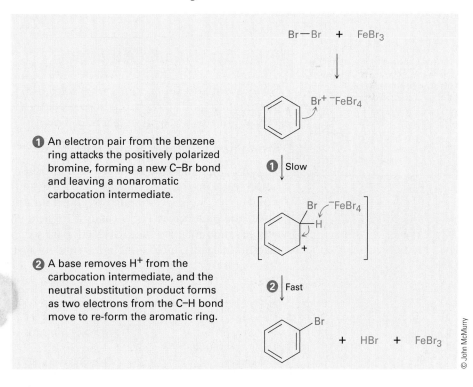

Why does the reaction of Br_2 with benzene take a different course than its reaction with an alkene? The answer is straightforward. If addition occurred, the 150 kJ/mol stabilization energy of the aromatic ring would be lost and the overall reaction would be endergonic. When substitution occurs, though, the stability of the aromatic ring is retained and the reaction is exergonic. An energy diagram for the overall process is shown in Figure 16.3.

Figure 16.3 An energy diagram for the electrophilic bromination of benzene. The overall process is exergonic.

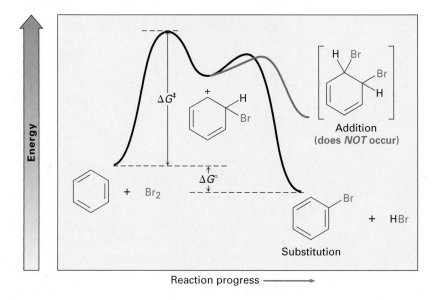

Problem 16.1 Monobromination of toluene gives a mixture of three bromotoluene products. Draw and name them.

16.2 | Other Aromatic Substitutions

There are many other kinds of electrophilic aromatic substitutions besides bromination, and all are thought to occur by the same general mechanism. Let's look at some of these other reactions briefly.

Aromatic Chlorination and Iodination

Chlorine and iodine can be introduced into aromatic rings by electrophilic substitution reactions, but fluorine is too reactive and only poor yields of monofluoroaromatic products are obtained by direct fluorination. Aromatic rings react with Cl_2 in the presence of $FeCl_3$ catalyst to yield chlorobenzenes, just as they react with Br_2 and $FeBr_3$. This kind of reaction is used in the synthesis of numerous pharmaceutical agents, including the antianxiety agent diazepam, marketed as Valium.

Benzene + Cl_2 → (FeCl₃ catalyst) **Chlorobenzene (86%)** + HCl

Diazepam

Iodine itself is unreactive toward aromatic rings, and an oxidizing agent such as hydrogen peroxide or a copper salt such as $CuCl_2$ must be added to the reaction. These substances accelerate the iodination reaction by oxidizing I_2 to a more powerful electrophilic species that reacts as if it were I^+. The aromatic ring then reacts with I^+ in the typical way, yielding a substitution product.

$$I_2 \ + \ 2\,Cu^{2+} \ \longrightarrow \ 2\,I^+ \ + \ 2\,Cu^+$$

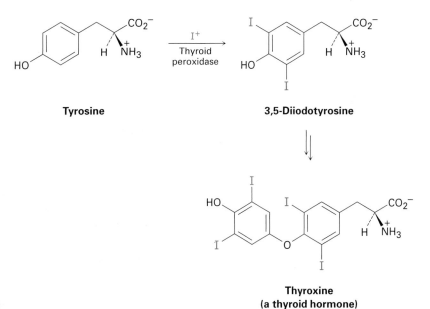

Benzene   **Iodobenzene (65%)**

Electrophilic aromatic halogenations occur in the biosynthesis of numerous naturally occurring molecules, particularly those produced by marine organisms. In humans, the best-known example occurs in the thyroid gland during the biosynthesis of thyroxine, a thyroid hormone involved in regulating growth and metabolism. The amino acid tyrosine is first iodinated by thyroid peroxidase, and two of the iodinated tyrosine molecules then couple. The electrophilic iodinating agent is an I^+ species, perhaps hypoiodous acid (HIO), that is formed from iodide ion by oxidation with H_2O_2.

Tyrosine   **3,5-Diiodotyrosine**

Thyroxine
(a thyroid hormone)

Aromatic Nitration

Aromatic rings can be nitrated by reaction with a mixture of concentrated nitric and sulfuric acids. The electrophile is the nitronium ion, NO_2^+, which is generated from HNO_3 by protonation and loss of water. The nitronium ion reacts with benzene to yield a carbocation intermediate, and loss of H^+ from this intermediate gives the neutral substitution product, nitrobenzene (Figure 16.4).

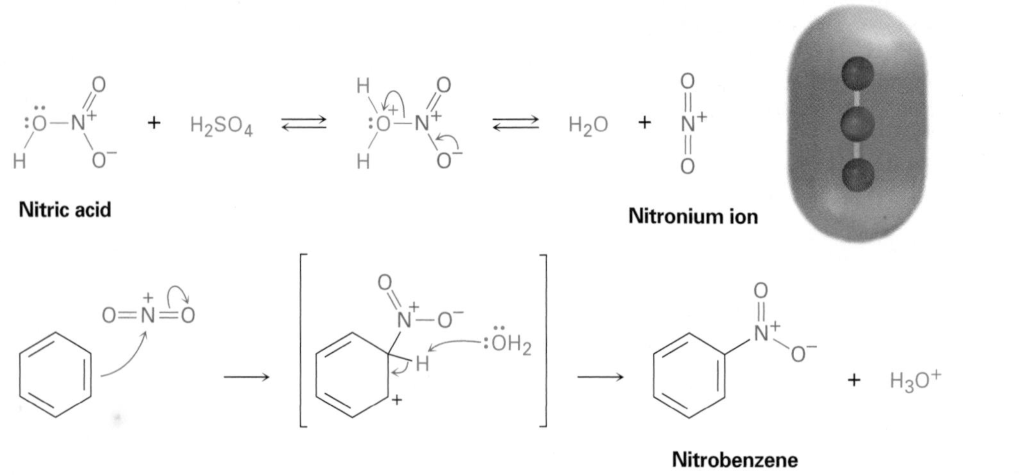

Figure 16.4 The mechanism of electrophilic nitration of an aromatic ring. An electrostatic potential map of the reactive electrophile NO_2^+ shows that the nitrogen atom is most positive (blue).

Nitric acid

Nitronium ion

Nitrobenzene

Nitration of an aromatic ring does not occur in nature but is particularly important in the laboratory because the nitro-substituted product can be reduced by reagents such as iron, tin, or $SnCl_2$ to yield an *arylamine,* $ArNH_2$. Attachment of an amino group to an aromatic ring by the two-step nitration/reduction sequence is a key part of the industrial synthesis of many dyes and pharmaceutical agents. We'll discuss this reduction and other reactions of aromatic nitrogen compounds in Chapter 24.

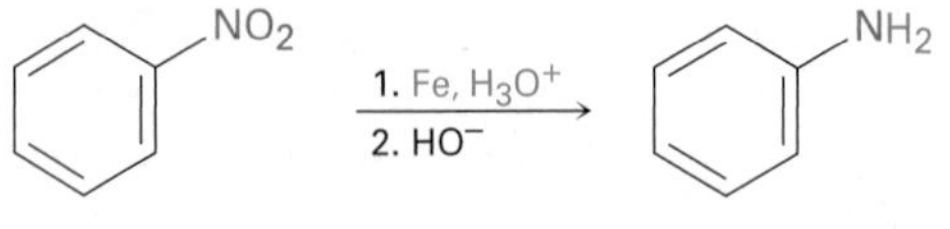

Nitrobenzene   Aniline (95%)

Aromatic Sulfonation

Aromatic rings can be sulfonated by reaction with fuming sulfuric acid, a mixture of H_2SO_4 and SO_3. The reactive electrophile is either HSO_3^+ or neutral SO_3, depending on reaction conditions, and substitution occurs by the same two-step mechanism seen previously for bromination and nitration (Figure 16.5). Note, however, that the sulfonation reaction is readily reversible; it can occur either forward or backward, depending on the reaction conditions. Sulfonation is favored in strong acid, but desulfonation is favored in hot, dilute aqueous acid.

Like nitration, aromatic sulfonation does not occur naturally but is widely used in the preparation of dyes and pharmaceutical agents. For example, the sulfa drugs, such as sulfanilamide, were among the first clinically useful antibiotics. Although largely replaced today by more effective agents, sulfa drugs are still used in the treatment of meningitis and urinary tract infections. These drugs are prepared commercially by a process that involves aromatic sulfonation as the key step.

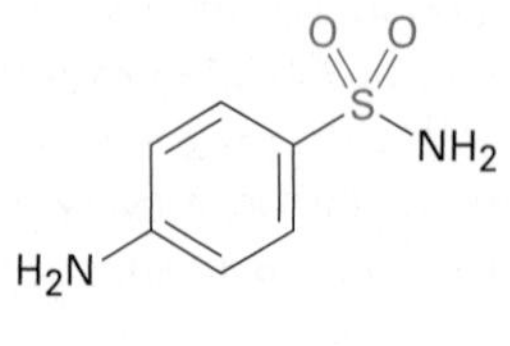

Sulfanilamide (an antibiotic)

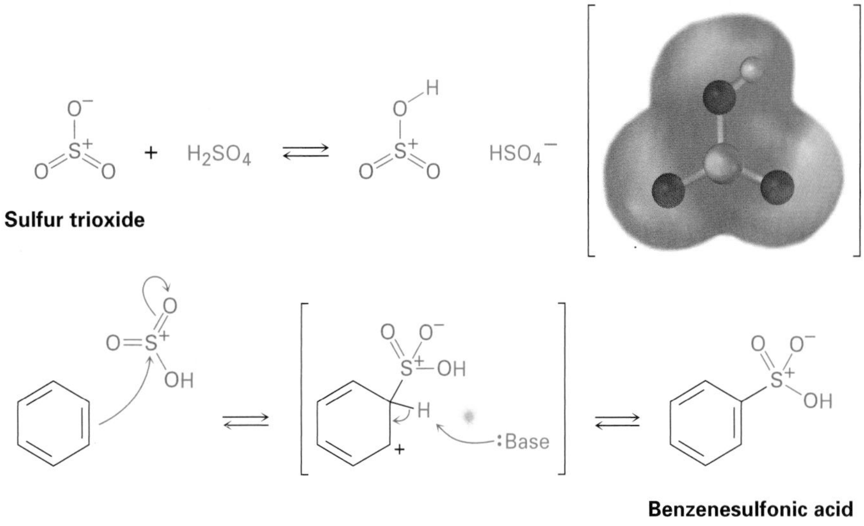

Figure 16.5 The mechanism of electrophilic sulfonation of an aromatic ring. An electrostatic potential map of the reactive electrophile $HOSO_2^+$ shows that sulfur and hydrogen are the most positive atoms (blue).

Sulfur trioxide

Benzenesulfonic acid

Aromatic Hydroxylation

Direct hydroxylation of an aromatic ring to yield a hydroxybenzene (a *phenol*) is difficult and rarely done in the laboratory, but occurs much more frequently in biological pathways. An example is the hydroxylation of *p*-hydroxyphenyl acetate to give 3,4-dihydroxyphenyl acetate. The reaction is catalyzed by *p*-hydroxyphenylacetate-3-hydroxylase and requires molecular oxygen plus the coenzyme reduced flavin adenine dinucleotide, abbreviated $FADH_2$.

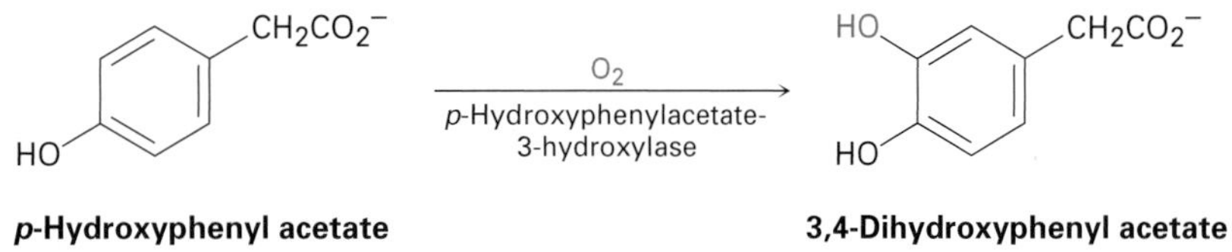

p-Hydroxyphenyl acetate   **3,4-Dihydroxyphenyl acetate**

By analogy with other electrophilic aromatic substitutions, you might expect that an electrophilic oxygen species acting as an "OH^+ equivalent" is needed for the hydroxylation reaction. That is exactly what happens, with the electrophilic oxygen arising by protonation of FAD hydroperoxide, RO—OH (Figure 16.6); that is, $RO-OH + H^+ \rightarrow ROH + OH^+$. The FAD hydroperoxide is itself formed by reaction of $FADH_2$ with O_2.

Problem 16.2 How many products might be formed on chlorination of *o*-xylene (*o*-dimethyl-benzene), *m*-xylene, and *p*-xylene?

Problem 16.3 When benzene is treated with D_2SO_4, deuterium slowly replaces all six hydrogens in the aromatic ring. Explain.

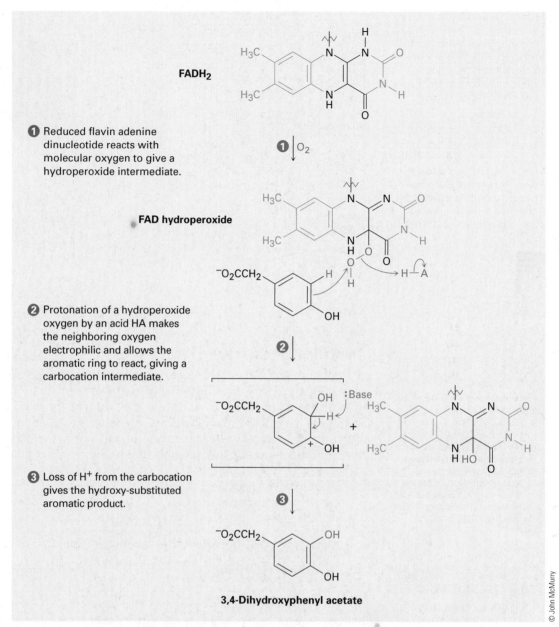

1 Reduced flavin adenine dinucleotide reacts with molecular oxygen to give a hydroperoxide intermediate.

2 Protonation of a hydroperoxide oxygen by an acid HA makes the neighboring oxygen electrophilic and allows the aromatic ring to react, giving a carbocation intermediate.

3 Loss of H$^+$ from the carbocation gives the hydroxy-substituted aromatic product.

3,4-Dihydroxyphenyl acetate

Figure 16.6 MECHANISM: Mechanism of the electrophilic hydroxylation of *p*-hydroxyphenyl acetate, by reaction with FAD hydroperoxide. The hydroxylating species is an "OH$^+$ equivalent" that arises by protonation of FAD hydroperoxide, RO$-$OH + H$^+$ $\rightarrow$ ROH + OH$^+$.

16.3 | Alkylation and Acylation of Aromatic Rings: The Friedel–Crafts Reaction

Among the most useful electrophilic aromatic substitution reactions in the laboratory is **alkylation**—the introduction of an alkyl group onto the benzene ring. Called the **Friedel–Crafts reaction** after its discoverers, the reaction is carried out by treating the aromatic compound with an alkyl chloride, RCl, in the presence of AlCl₃ to generate a carbocation electrophile, R⁺. Aluminum chloride catalyzes the reaction by helping the alkyl halide to dissociate in much the same way that FeBr₃ catalyzes aromatic brominations by polarizing Br₂ (Section 16.1). Loss of H⁺ then completes the reaction (Figure 16.7).

Figure 16.7 MECHANISM:
Mechanism of the Friedel–Crafts alkylation reaction of benzene with 2-chloropropane to yield isopropylbenzene (cumene). The electrophile is a carbocation, generated by AlCl₃-assisted dissociation of an alkyl halide.

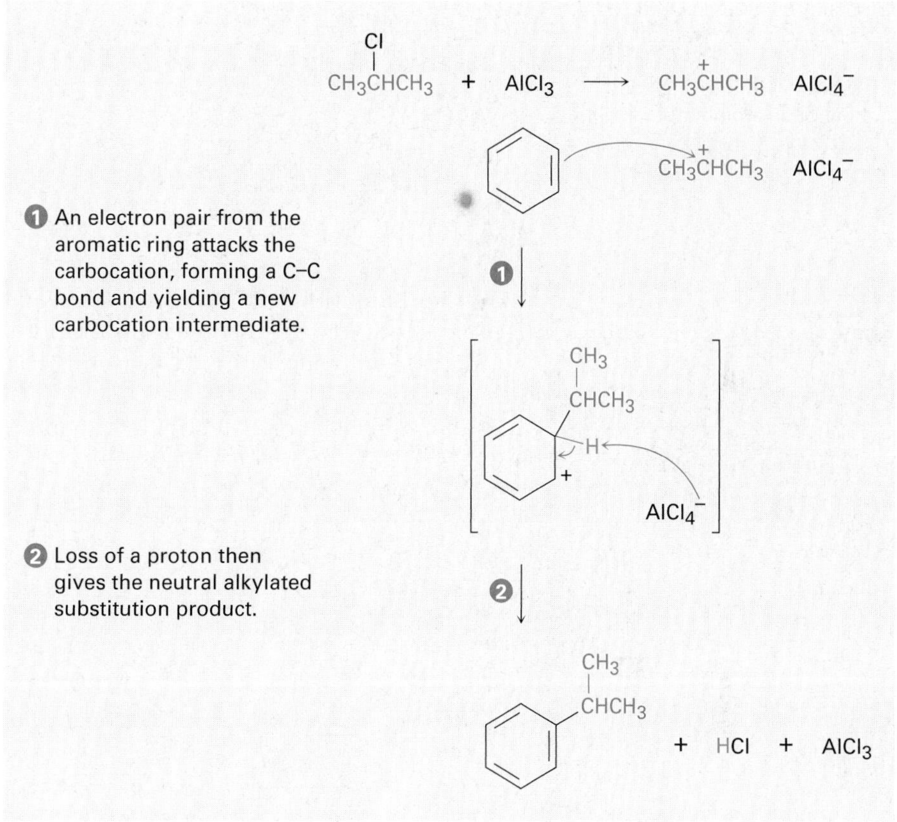

① An electron pair from the aromatic ring attacks the carbocation, forming a C–C bond and yielding a new carbocation intermediate.

② Loss of a proton then gives the neutral alkylated substitution product.

© John McMurry

Charles Friedel

Charles Friedel (1832–1899) was born in Strasbourg, France, and studied at the Sorbonne in Paris. Trained as both a mineralogist and a chemist, he was among the first to attempt to manufacture synthetic diamonds. He was professor of mineralogy at the School of Mines before becoming professor of chemistry at the Sorbonne (1884–1899).

James Mason Crafts

James Mason Crafts (1839–1917) was born in Boston, Massachusetts, and graduated from Harvard in 1858. Although he did not receive a Ph.D., he studied with eminent chemists in Europe for several years and was appointed in 1868 as the first professor of chemistry at the newly founded Cornell University in Ithaca, New York. Ithaca winters proved too severe, however, and he soon moved to the Massachusetts Institute of Technology, where he served as president from 1897 to 1900.

Despite its utility, the Friedel–Crafts alkylation has several limitations. For one thing, only *alkyl* halides can be used. Aromatic *(aryl)* halides and vinylic halides do not react because aryl and vinylic carbocations are too high in energy to form under Friedel–Crafts conditions.

An aryl halide   **A vinylic halide**

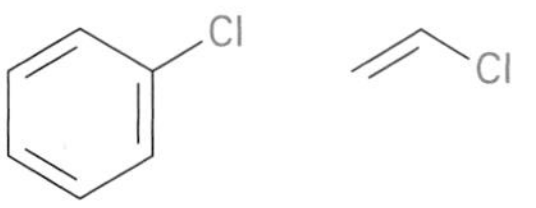

NOT reactive

Another limitation is that Friedel–Crafts reactions don't succeed on aromatic rings that are substituted either by a strongly electron-withdrawing group such as carbonyl (C=O) or by an amino group (−NH$_2$, NHR, −NR$_2$). We'll see in the next section that the presence of a substituent group already on a ring can have a dramatic effect on that ring's subsequent reactivity toward further electrophilic substitution. Rings that contain any of the substituents listed in Figure 16.8 do not undergo Friedel–Crafts alkylation.

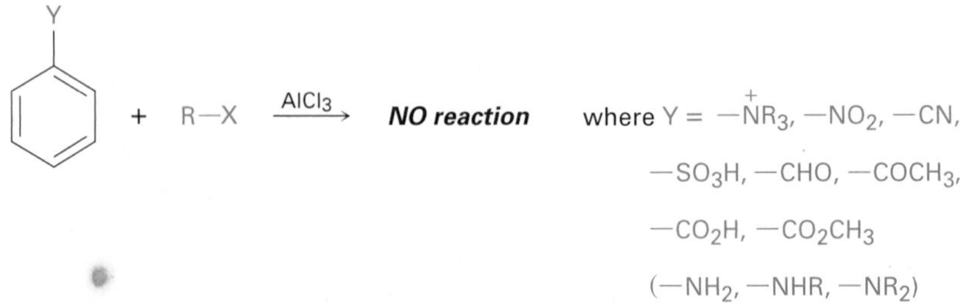

Figure 16.8 Limitations on the aromatic substrate in Friedel–Crafts reactions. No reaction occurs if the substrate has either an electron-withdrawing substituent or an amino group.

A third limitation to the Friedel–Crafts alkylation is that it's often difficult to stop the reaction after a single substitution. Once the first alkyl group is on the ring, a second substitution reaction is facilitated for reasons we'll discuss in the next section. Thus, we often observe *polyalkylation*. Reaction of benzene with 1 mol equivalent of 2-chloro-2-methylpropane, for example, yields *p-di-tert-*butylbenzene as the major product, along with small amounts of *tert*-butyl-benzene and unreacted benzene. A high yield of monoalkylation product is obtained only when a large excess of benzene is used.

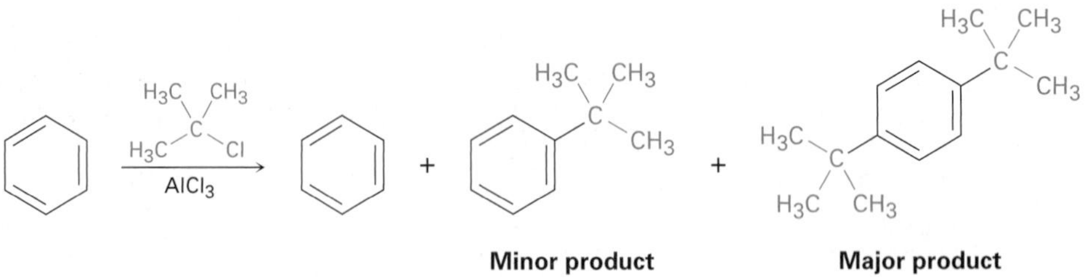

Minor product   **Major product**

Yet a final limitation to the Friedel–Crafts reaction is that a skeletal rearrangement of the alkyl carbocation electrophile sometimes occurs during reaction, particularly when a primary alkyl halide is used. Treatment of benzene with 1-chlorobutane at 0 °C, for instance, gives an approximately 2:1 ratio of rearranged (*sec*-butyl) to unrearranged (butyl) products.

The carbocation rearrangements that accompany Friedel–Crafts reactions are like those that accompany electrophilic additions to alkenes (Section 6.11) and occur either by hydride shift or alkyl shift. For example, the relatively unstable primary butyl carbocation produced by reaction of 1-chlorobutane with AlCl$_3$ rearranges to the more stable secondary butyl carbocation by shift of a hydrogen atom and its electron pair (a hydride ion, H:$^-$) from C2 to C1. Similarly, alkylation of benzene with 1-chloro-2,2-dimethylpropane yields (1,1-dimethylpropyl)benzene. The initially formed primary carbocation rearranges to a tertiary carbocation by shift of a methyl group and its electron pair from C2 to C1.

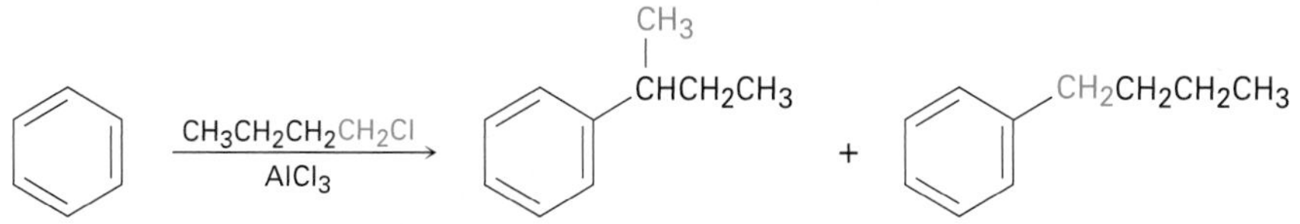

Benzene → sec-Butylbenzene (65%) + Butylbenzene (35%)

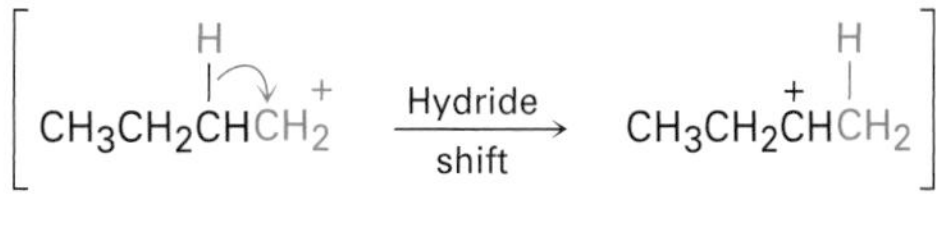

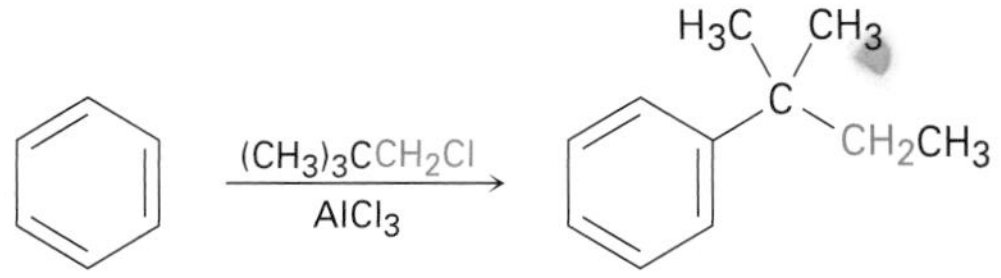

$$CH_3CH_2CH\overset{H}{\overset{|}{C}}H_2^+ \xrightarrow[\text{shift}]{\text{Hydride}} CH_3CH_2\overset{+}{C}\overset{H}{\overset{|}{C}}H_2$$

Benzene → (1,1-Dimethylpropyl)benzene

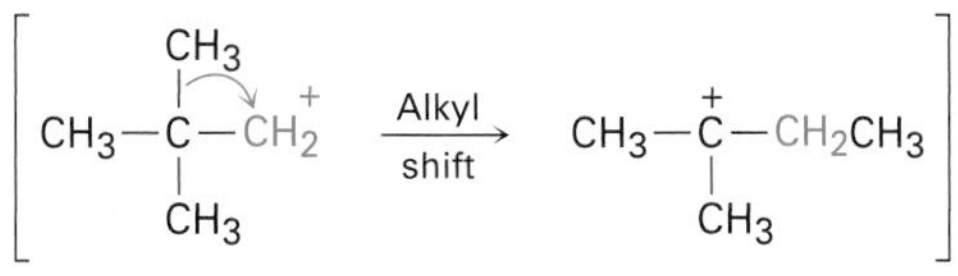

$$CH_3-\overset{CH_3}{\underset{CH_3}{\overset{|}{\underset{|}{C}}}}-\overset{+}{C}H_2 \xrightarrow[\text{shift}]{\text{Alkyl}} CH_3-\overset{+}{C}-CH_2CH_3$$

Just as an aromatic ring is alkylated by reaction with an alkyl chloride, it is **acylated** by reaction with a carboxylic acid chloride, RCOCl, in the presence of AlCl$_3$. That is, an **acyl group** (—COR; pronounced **a**-sil) is substituted onto the aromatic ring. For example, reaction of benzene with acetyl chloride yields the ketone, acetophenone.

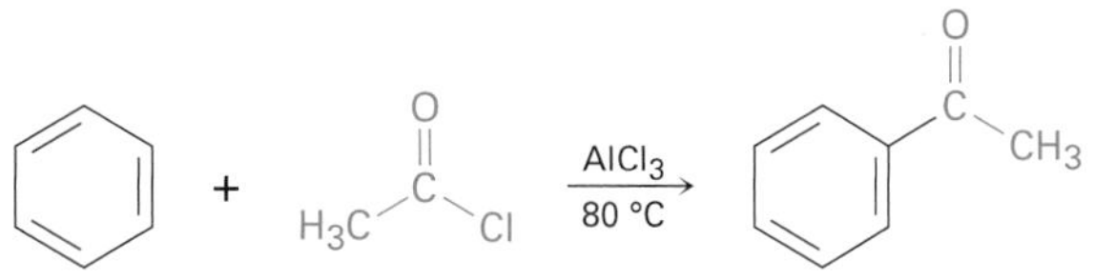

Benzene + Acetyl chloride → Acetophenone (95%)

The mechanism of the Friedel–Crafts acylation reaction is similar to that of Friedel–Crafts alkylation, and the same limitations on the aromatic substrate noted previously in Figure 16.8 for alkylation also apply to acylation. The reactive electrophile is a resonance-stabilized acyl cation, generated by reaction between the acyl chloride and AlCl$_3$ (Figure 16.9). As the resonance structures in the figure indicate, an acyl cation is stabilized by interaction of the vacant orbital on carbon with lone-pair electrons on the neighboring oxygen. Because of this stabilization, no carbocation rearrangement occurs during acylation.

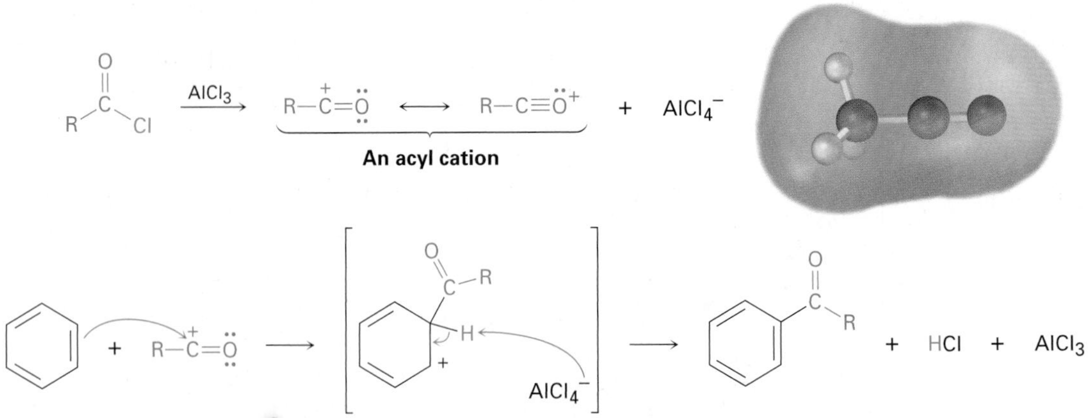

Figure 16.9 Mechanism of the Friedel–Crafts acylation reaction. The electrophile is a resonance-stabilized acyl cation, whose electrostatic potential map indicates that carbon is the most positive atom (blue).

Unlike the multiple substitutions that often occur in Friedel–Crafts alkylations, acylations never occur more than once on a ring because the product acylbenzene is less reactive than the nonacylated starting material. We'll account for this reactivity difference in the next section.

Aromatic alkylations occur in numerous biological pathways, although there is of course no $AlCl_3$ present in living systems to catalyze the reaction. Instead, the carbocation electrophile is usually formed by dissociation of an organodiphosphate, as we saw in Section 11.6. The dissociation is typically assisted by complexation to a divalent metal cation such as Mg^{2+} to help neutralize charge.

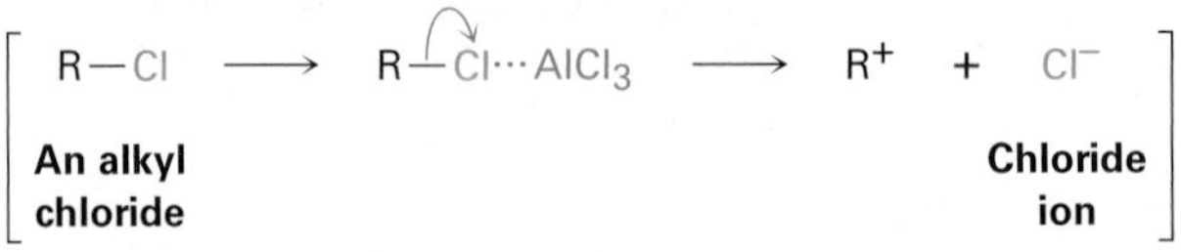

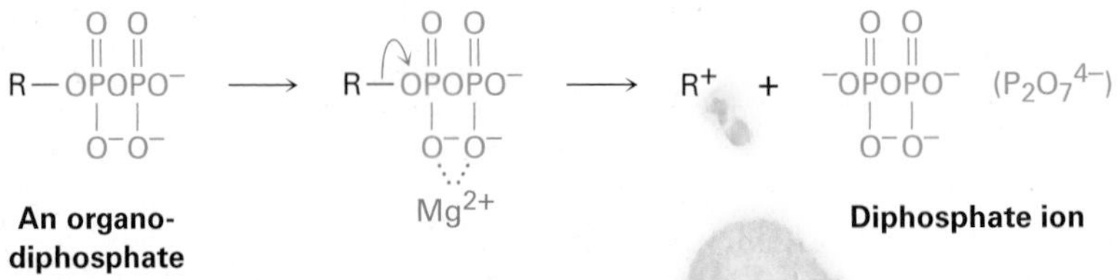

An example of a biological Friedel–Crafts reaction occurs during the biosynthesis of phylloquinone, or vitamin K_1, the human blood-clotting factor. Phylloquinone is formed by reaction of 1,4-dihydroxynaphthoic acid with phytyl diphosphate. Phytyl diphosphate first dissociates to a resonance-stabilized allylic carbocation, which then substitutes onto the aromatic ring in the typical way. Several further transformations lead to phylloquinone (Figure 16.10).

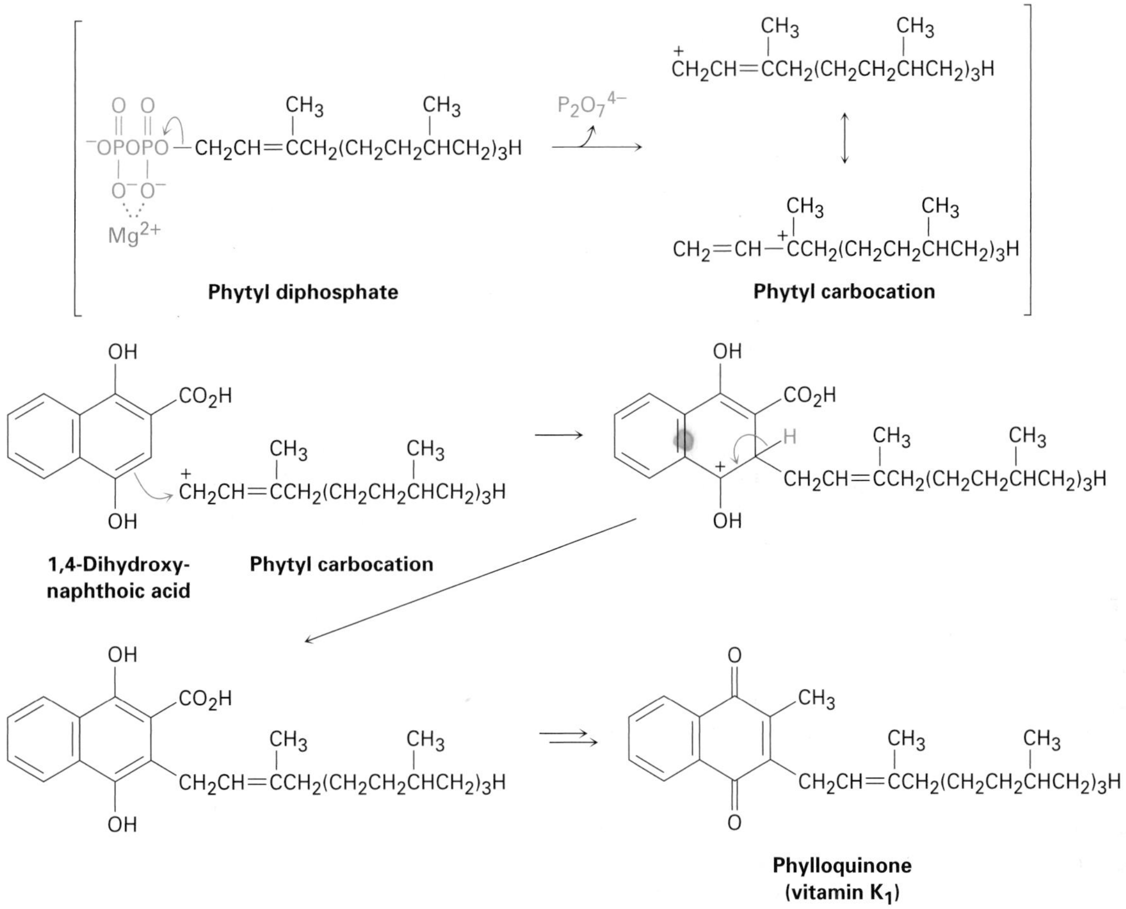

Figure 16.10 Biosynthesis of phylloquinone (vitamin K_1) from 1,4-dihydroxynaphthoic acid. The key step that joins the 20-carbon phytyl side chain to the aromatic ring is a Friedel–Crafts-like electrophilic substitution reaction.

WORKED EXAMPLE 16.1

Predicting the Product of a Carbocation Rearrangement

The Friedel–Crafts reaction of benzene with 2-chloro-3-methylbutane in the presence of $AlCl_3$ occurs with a carbocation rearrangement. What is the structure of the product?

Strategy A Friedel–Crafts reaction involves initial formation of a carbocation, which can rearrange by either a hydride shift or an alkyl shift to give a more stable carbocation. Draw the initial carbocation, assess its stability, and see if the shift of a hydride ion or an alkyl group from a neighboring carbon will result in increased stability. In the present instance, the initial carbocation is a secondary one that can rearrange to a more stable tertiary one by a hydride shift.

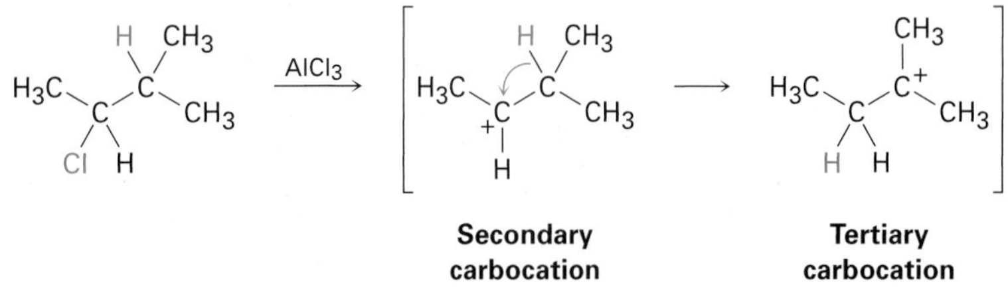

Use this more stable tertiary carbocation to complete the Friedel–Crafts reaction.

Solution

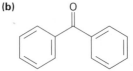

Problem 16.4 Which of the following alkyl halides would you expect to undergo Friedel–Crafts reaction *without* rearrangement? Explain.
(a) CH_3CH_2Cl (b) $CH_3CH_2CH(Cl)CH_3$ (c) $CH_3CH_2CH_2Cl$
(d) $(CH_3)_3CCH_2Cl$ (e) Chlorocyclohexane

Problem 16.5 What is the major monosubstitution product from the Friedel–Crafts reaction of benzene with 1-chloro-2-methylpropane in the presence of $AlCl_3$?

Problem 16.6 Identify the carboxylic acid chloride that might be used in a Friedel–Crafts acylation reaction to prepare each of the following acylbenzenes:

(a)

(b)

16.4 | Substituent Effects in Substituted Aromatic Rings

Only one product can form when an electrophilic substitution occurs on benzene, but what would happen if we were to carry out a reaction on an aromatic ring that already has a substituent? A substituent already present on the ring has two effects.

▮ **Substituents affect the *reactivity* of the aromatic ring.** Some substituents activate the ring, making it more reactive than benzene, and some deactivate the ring, making it less reactive than benzene. In aromatic nitration, for instance, an $-OH$ substituent makes the ring 1000 times more reactive than benzene, while an $-NO_2$ substituent makes the ring more than 10 million times less reactive.

Relative rate of nitration: 6×10^{-8} 0.033   1   1000

Reactivity

▮ **Substituents affect the *orientation* of the reaction.** The three possible disubstituted products—ortho, meta, and para—are usually not formed in equal amounts. Instead, the nature of the substituent already present on the benzene ring determines the position of the second substitution. Table 16.1 lists experimental results for the nitration of some substituted benzenes and shows that some groups direct substitution primarily to the ortho and para positions, while other groups direct substitution primarily to the meta position.

Table 16.1   Orientation of Nitration in Substituted Benzenes

	Product (%)					Product (%)		
	Ortho	Meta	Para			Ortho	Meta	Para
Meta-directing deactivators					Ortho- and para-directing deactivators			
$-\overset{+}{N}(CH_3)_3$	2	87	11		$-F$	13	1	86
$-NO_2$	7	91	2		$-Cl$	35	1	64
$-CO_2H$	22	76	2		$-Br$	43	1	56
$-CN$	17	81	2		$-I$	45	1	54
$-CO_2CH_3$	28	66	6		Ortho- and para-directing activators			
$-COCH_3$	26	72	2		$-CH_3$	63	3	34
$-CHO$	19	72	9		$-OH$	50	0	50
					$-NHCOCH_3$	19	2	79

Substituents can be classified into three groups, as shown in Figure 16.11: *ortho- and para-directing activators, ortho- and para-directing deactivators,* and *meta-directing deactivators.* There are no meta-directing activators. Notice how the directing effects of the groups correlate with their reactivities. All meta-directing groups are deactivating, and most ortho- and para-directing groups are activating. The halogens are unique in being ortho- and para-directing but weakly deactivating.

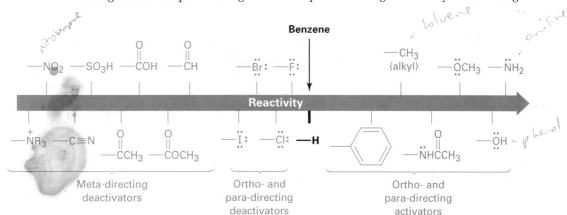

Active Figure 16.11 Classification of substituent effects in electrophilic aromatic substitution. All activating groups are ortho- and para-directing, and all deactivating groups other than halogen are meta-directing. The halogens are unique in being deactivating but ortho- and para-directing. *Sign in at* **www.cengage.com/login** *to see a simulation based on this figure and to take a short quiz.*

Reactivity and orientation in electrophilic aromatic substitutions are controlled by an interplay of inductive effects and resonance effects. As we saw in Sections 2.1 and 6.9, an **inductive effect** is the withdrawal or donation of electrons through a σ bond due to electronegativity. Halogens, hydroxyl groups, carbonyl groups, cyano groups, and nitro groups inductively *withdraw* electrons through the σ bond linking the substituent to a benzene ring. The effect is most pronounced in halobenzenes and phenols, in which the electronegative atom is directly attached to the ring, but is also significant in carbonyl compounds, nitriles, and nitro compounds, in which the electronegative atom is farther removed. Alkyl groups, on the other hand, inductively *donate* electrons. This is the same hyperconjugative donating effect that causes alkyl substituents to stabilize alkenes (Section 6.6) and carbocations (Section 6.9).

Inductive electron withdrawal

Inductive electron donation

A **resonance effect** is the withdrawal or donation of electrons through a π bond due to the overlap of a *p* orbital on the substituent with a *p* orbital on the aromatic ring. Carbonyl, cyano, and nitro substituents, for example, *withdraw* electrons from the aromatic ring by resonance. Pi electrons flow from the rings to the substituents, leaving a positive charge in the ring. Note that substituents with an electron-withdrawing resonance effect have the general structure $-Y=Z$, where the Z atom is more electronegative than Y.

**Resonance electron-
withdrawing group**

Conversely, halogen, hydroxyl, alkoxyl (−OR), and amino substituents *donate* electrons to the aromatic ring by resonance. Lone-pair electrons flow from the substituents to the ring, placing a negative charge in the ring. Substituents with an electron-donating resonance effect have the general structure −Ÿ, where the Y atom has a lone pair of electrons available for donation to the ring.

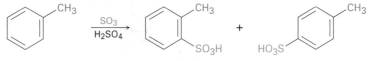

Resonance electron-donating group

One further point: inductive effects and resonance effects don't necessarily act in the same direction. Halogen, hydroxyl, alkoxyl, and amino substituents, for instance, have electron-*withdrawing* inductive effects because of the electronegativity of the −X, −O, or −N atom bonded to the aromatic ring but have electron-*donating* resonance effects because of the lone-pair electrons on those same −X, −O, or −N atoms. When the two effects act in opposite directions, the stronger of the two dominates.

WORKED EXAMPLE 16.2

Predicting the Product of an Electrophilic Aromatic Substitution Reaction

Predict the major product of the sulfonation of toluene.

Strategy

Identify the substituent present on the ring, and decide whether it is ortho- and para-directing or meta-directing. According to Figure 16.11, an alkyl substituent is ortho- and para-directing, so sulfonation of toluene will give primarily a mixture of *o*-toluenesulfonic acid and *p*-toluenesulfonic acid.

Solution

| Toluene | *o*-Toluenesulfonic acid | *p*-Toluenesulfonic acid |

Problem 16.7 | Write resonance structures for nitrobenzene to show the electron-withdrawing resonance effect of the nitro group.

Problem 16.8 | Write resonance structures for chlorobenzene to show the electron-donating resonance effect of the chloro group.

Problem 16.9 | Predict the major products of the following reactions:
(a) Nitration of bromobenzene   **(b)** Bromination of nitrobenzene
(c) Chlorination of phenol   **(d)** Bromination of aniline

16.5 | An Explanation of Substituent Effects

Activation and Deactivation of Aromatic Rings

What makes a group either activating or deactivating? The common characteristic of all activating groups is that they *donate* electrons to the ring, thereby making the ring more electron-rich, stabilizing the carbocation intermediate, and lowering the activation energy for its formation. Hydroxyl, alkoxyl, and amino groups are activating because their stronger electron-donating resonance effect outweighs their weaker electron-withdrawing inductive effect. Alkyl groups are activating because of their electron-donating inductive effect.

Conversely, the common characteristic of all deactivating groups is that they *withdraw* electrons from the ring, thereby making the ring more electron-poor, destabilizing the carbocation intermediate, and raising the activation energy for its formation. Carbonyl, cyano, and nitro groups are deactivating because of both electron-withdrawing resonance and inductive effects. Halogens are deactivating because their stronger electron-withdrawing inductive effect outweighs their weaker electron-donating resonance effect.

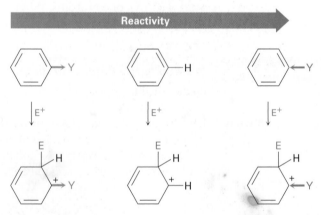

Y withdraws electrons; carbocation intermediate is less stable, and ring is less reactive.

Y donates electrons; carbocation intermediate is more stable, and ring is more reactive.

Figure 16.12 compares electrostatic potential maps of benzaldehyde (deactivated), chlorobenzene (weakly deactivated), and phenol (activated) with that of benzene. The ring is more positive (yellow-green) when an electron-withdrawing group such as −CHO or −Cl is present and more negative (red) when an electron-donating group such as −OH is present.

Figure 16.12 Electrostatic potential maps of benzene and several substituted benzenes show that an electron-withdrawing group (−CHO or −Cl) makes the ring more electron-poor (yellow-green), while an electron-donating group (−OH) makes the ring more electron-rich (red).

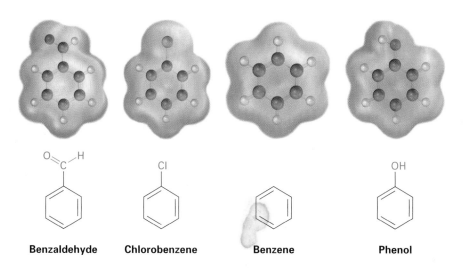

Benzaldehyde   **Chlorobenzene**   **Benzene**   **Phenol**

Problem 16.10   Rank the compounds in each group in order of their reactivity to electrophilic substitution: NO₂ OH CH₃
(a) Nitrobenzene, phenol, toluene, benzene
(b) Phenol, benzene, chlorobenzene, benzoic acid
(c) Benzene, bromobenzene, benzaldehyde, aniline NH₂

Problem 16.11   Use Figure 16.11 to explain why Friedel–Crafts alkylations often give polysubstitution but Friedel–Crafts acylations do not.

Problem 16.12   An electrostatic potential map of (trifluoromethyl)benzene, $C_6H_5CF_3$, is shown. Would you expect (trifluoromethyl)benzene to be more reactive or less reactive than toluene toward electrophilic substitution? Explain.

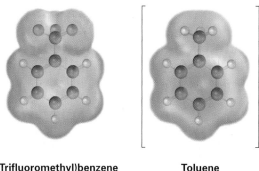

(Trifluoromethyl)benzene   **Toluene**

Ortho- and Para-Directing Activators: Alkyl Groups

Inductive and resonance effects account for the directing effects of substituents as well as for their activating or deactivating effects. Take alkyl groups, for instance, which have an electron-donating inductive effect and are ortho and para directors. The results of toluene nitration are shown in Figure 16.13.

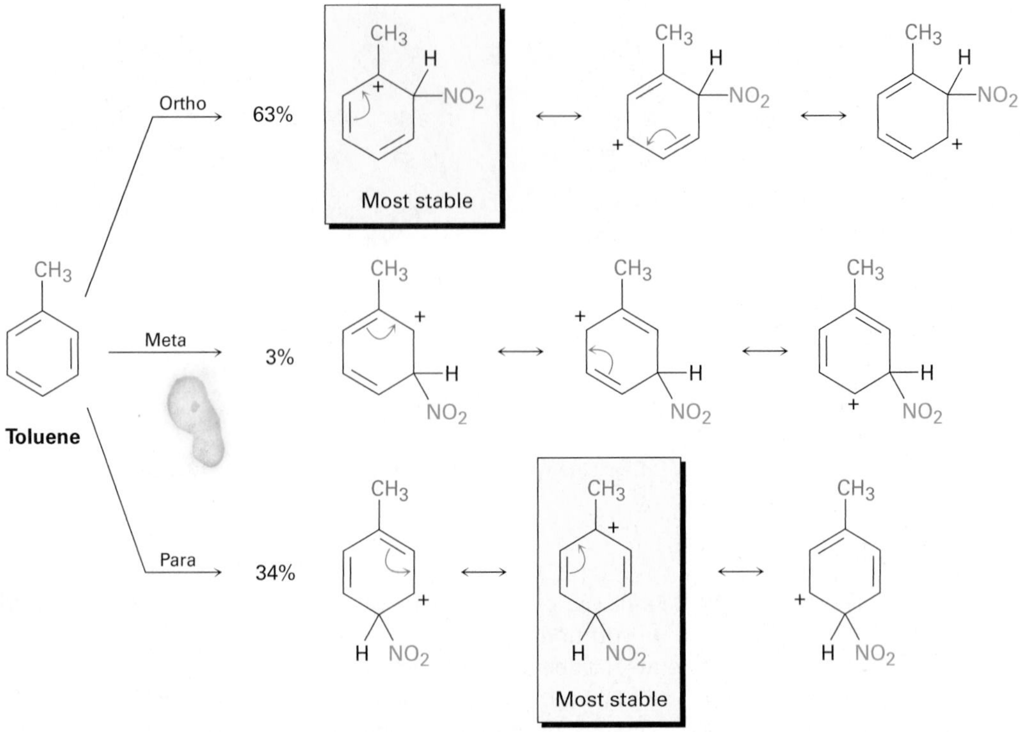

Figure 16.13 Carbocation intermediates in the nitration of toluene. Ortho and para intermediates are more stable than the meta intermediate because the positive charge is on a tertiary carbon rather than a secondary carbon.

Nitration of toluene might occur either ortho, meta, or para to the methyl group, giving the three carbocation intermediates shown in Figure 16.13. All three intermediates are resonance-stabilized, but *the ortho and para intermediates are more stabilized than the meta intermediate.* For both the ortho and para reactions, but not for the meta reaction, a resonance form places the positive charge directly on the methyl-substituted carbon, where it is in a tertiary position and can best be stabilized by the electron-donating inductive effect of the methyl group. The ortho and para intermediates are thus lower in energy than the meta intermediate and form faster.

Ortho- and Para-Directing Activators: OH and NH$_2$

Hydroxyl, alkoxyl, and amino groups are also ortho–para activators, but for a different reason than for alkyl groups. As described in the previous section, hydroxyl, alkoxyl, and amino groups have a strong, electron-donating resonance effect that outweighs a weaker electron-withdrawing inductive effect. When phenol is nitrated, for instance, only ortho and para reaction is observed. As shown in Figure 16.14, all three possible carbocation intermediates are stabilized by resonance, but the intermediates from ortho and para reaction are stabilized most. Only the ortho and para intermediates have resonance forms in which the positive charge is stabilized by donation of an electron pair from oxygen. The intermediate from meta reaction has no such stabilization.

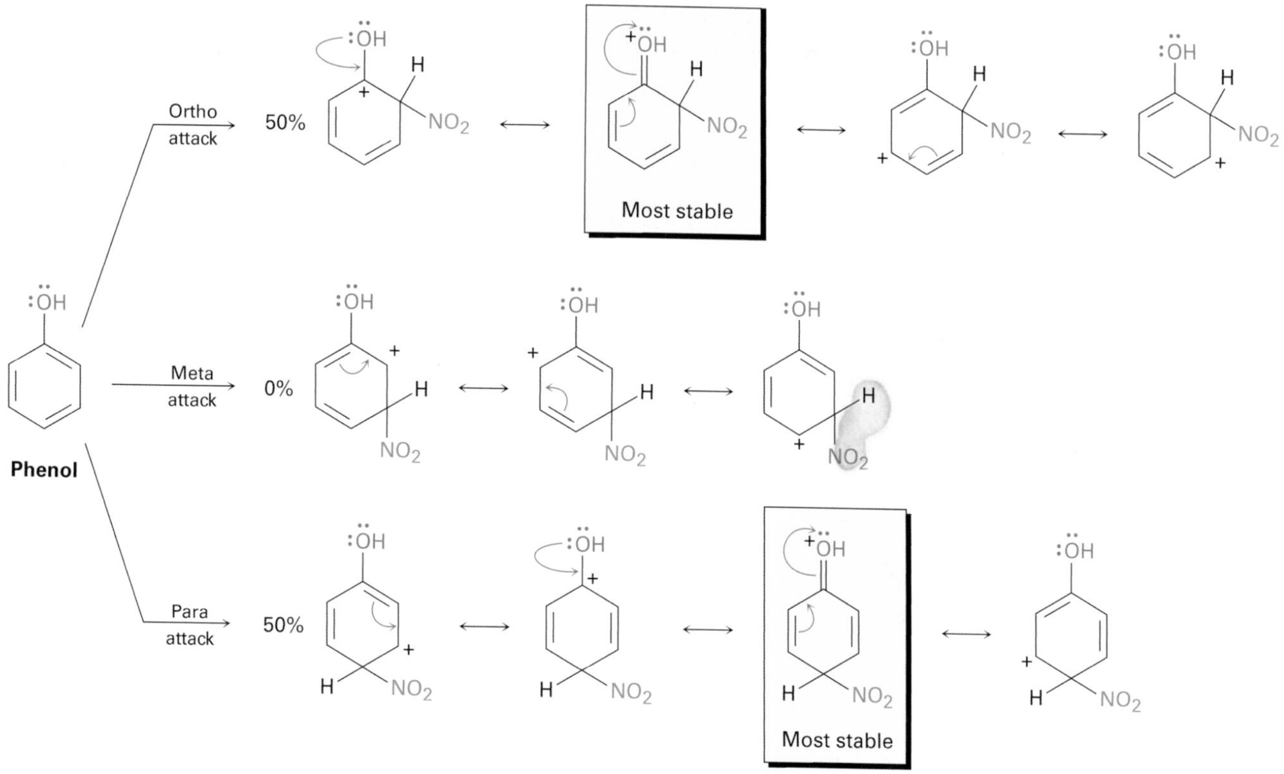

Figure 16.14 Carbocation intermediates in the nitration of phenol. The ortho and para intermediates are more stable than the meta intermediate because of resonance donation of electrons from oxygen.

Problem 16.13   Acetanilide is less reactive than aniline toward electrophilic substitution. Explain.

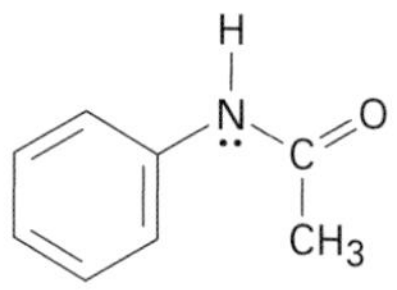

Acetanilide

Ortho- and Para-Directing Deactivators: Halogens

Halogens are deactivating because their stronger electron-withdrawing inductive effect outweighs their weaker electron-donating resonance effect. Although weak, that electron-donating resonance effect is felt only at the ortho and para positions (Figure 16.15). Thus, a halogen substituent can stabilize the positive charge of the carbocation intermediates from ortho and para reaction in the same way that hydroxyl and amino substituents can. The meta intermediate, however, has no such stabilization and is therefore formed more slowly.

Note again that halogens, hydroxyl, alkoxyl, and amino groups *withdraw* electrons inductively and *donate* electrons by resonance. Halogens have a

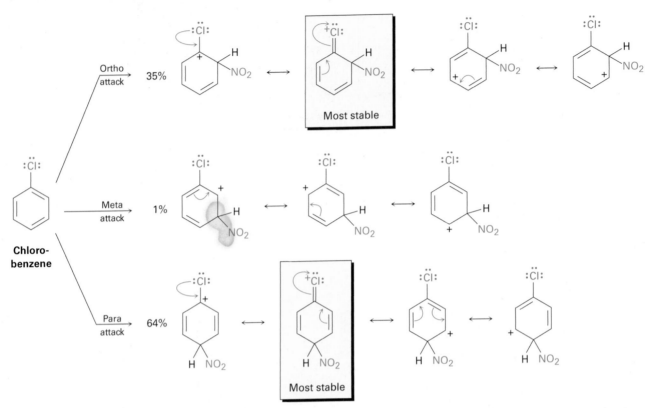

Figure 16.15 Carbocation intermediates in the nitration of chlorobenzene. The ortho and para intermediates are more stable than the meta intermediate because of electron donation of the halogen lone-pair electrons.

stronger electron-withdrawing inductive effect but a weaker electron-donating resonance effect and are thus deactivators. Hydroxyl, alkoxyl, and amino groups have a weaker electron-withdrawing inductive effect but a stronger electron-donating resonance effect and are thus activators. All are ortho and para directors, however, because of the lone pair of electrons on the atom bonded to the aromatic ring.

Meta-Directing Deactivators

Meta-directing deactivators, such as −CHO, act through a combination of electron-withdrawing inductive and resonance effects that reinforce each other and are felt most strongly at the ortho and para positions. As a result, the ortho and para intermediates are less stable so reaction with an electrophile occurs at the meta position (Figure 16.16).

Problem 16.14 Draw resonance structures for the intermediates from reaction of an electrophile at the ortho, meta, and para positions of nitrobenzene. Which intermediates are most stable?

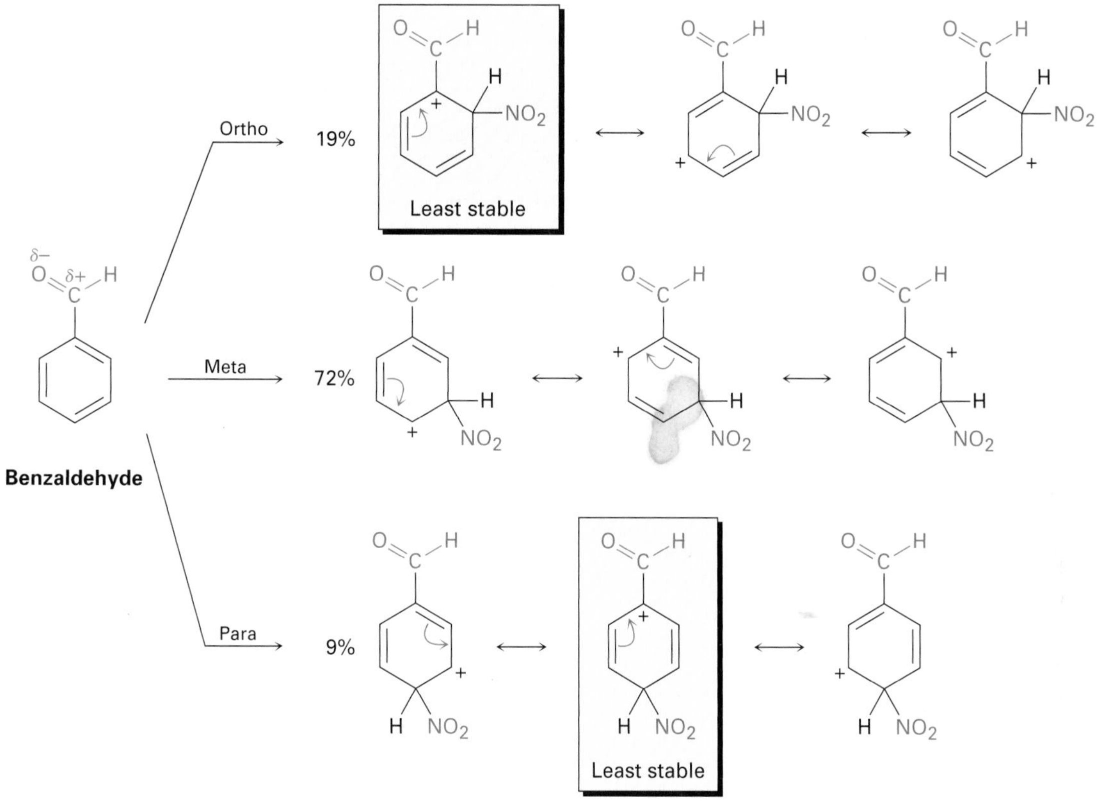

Figure 16.16 Carbocation intermediates in the chlorination of benzaldehyde. The ortho and para intermediates are less stable than the meta intermediate.

Key IDEAS

Test your knowledge of Key Ideas by using resources in CengageNOW or by answering end-of-chapter problems marked with ▲.

A Summary of Substituent Effects in Aromatic Substitution   A summary of the activating and directing effects of substituents in electrophilic aromatic substitution is shown in Table 16.2.

Table 16.2   Substituent Effects in Electrophilic Aromatic Substitution

Substituent	Reactivity	Orienting effect	Inductive effect	Resonance effect
–CH₃	Activating	Ortho, para	Weak donating	—
–OH, –NH₂	Activating	Ortho, para	Weak withdrawing	Strong donating
–F, –Cl –Br, –I	Deactivating	Ortho, para	Strong withdrawing	Weak donating
–NO₂, –CN, –CHO, –CO₂R –COR, –CO₂H	Deactivating	Meta	Strong withdrawing	Strong withdrawing

16.6 | Trisubstituted Benzenes: Additivity of Effects

Electrophilic substitution of a disubstituted benzene ring is governed by the same resonance and inductive effects that affect monosubstituted rings. The only difference is that it's now necessary to consider the additive effects of two different groups. In practice, this isn't as difficult as it sounds; three rules are usually sufficient.

1. If the directing effects of the two groups reinforce each other, the situation is straightforward. In *p*-nitrotoluene, for example, both the methyl and the nitro group direct further substitution to the same position (ortho to the methyl = meta to the nitro). A single product is thus formed on electrophilic substitution.

p-Nitrotoluene   2-Bromo-4-nitrotoluene

2. If the directing effects of the two groups oppose each other, the more powerful activating group has the dominant influence, but mixtures of products often result. For example, bromination of *p*-methylphenol yields primarily 2-bromo-4-methylphenol because −OH is a more powerful activator than −CH$_3$.

p-Methylphenol   4-Methyl-2-nitrophenol

3. Further substitution rarely occurs between the two groups in a meta-disubstituted compound because this site is too hindered. Aromatic rings with three adjacent substituents must therefore be prepared by some other route, usually by substitution of an ortho-disubstituted compound.

m-Chlorotoluene   3,4-Dichlorotoluene   2,5-Dichlorotoluene   *NOT* formed

But:

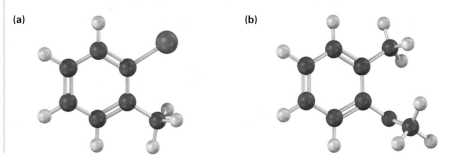

o-Nitrotoluene   2-Chloro-6-nitrotoluene   4-Chloro-2-nitrotoluene

---

WORKED EXAMPLE 16.3   ***Predicting the Product of Substitution on a Disubstituted Benzene***

What product would you expect from bromination of *p*-methylbenzoic acid?

Strategy   Identify the two substituents present on the ring, decide the directing effect of each and, if necessary, decide which substituent is the stronger activator. In the present case, the carboxyl group ($-CO_2H$) is a meta director and the methyl group is an ortho and para director. Both groups direct bromination to the position next to the methyl group, yielding 3-bromo-4-methylbenzoic acid.

Solution

p-Methylbenzoic acid   →   3-Bromo-4-methylbenzoic acid

Problem 16.15   At what position would you expect electrophilic substitution to occur in each of the following substances?

(a) OCH$_3$ ... Br

(b) NH$_2$ ... Br

(c) NO$_2$ ... Cl

Problem 16.16   Show the major product(s) from reaction of the following substances with (i) CH_3CH_2Cl, $AlCl_3$ and (ii) HNO_3, H_2SO_4.

(a)

(b)

16.7 | Nucleophilic Aromatic Substitution

CENGAGENOW Click *Organic Process* to **view an animation showing a nucleophilic aromatic substitution reaction**.

As we've seen, aromatic substitution reactions usually occur by an *electrophilic* mechanism. Aryl halides that have electron-withdrawing substituents, however, can also undergo **nucleophilic aromatic substitution**. For example, 2,4,6-trinitrochlorobenzene reacts with aqueous NaOH at room temperature to give 2,4,6-trinitrophenol. The nucleophile OH^- has substituted for Cl^-.

2,4,6-Trinitrochlorobenzene   **2,4,6-Trinitrophenol (100%)**

Nucleophilic aromatic substitution is much less common than electrophilic substitution but nevertheless does have certain uses. One such use is the reaction of proteins with 2,4-dinitrofluorobenzene, known as *Sanger's reagent,* to attach a "label" to the terminal NH_2 group of the amino acid at one end of the protein chain.

2,4-Dinitro-fluorobenzene   **A protein**   **A labeled protein**

How does this reaction take place? Although it appears superficially similar to the S_N1 and S_N2 nucleophilic substitution reactions of alkyl halides discussed in Chapter 11, it must be different because aryl halides are inert to both S_N1 and S_N2 conditions. S_N1 reactions don't occur with aryl halides because dissociation of the halide is energetically unfavorable due to the instability of the potential aryl cation product. S_N2 reactions don't occur with aryl halides because the halo-substituted carbon of the aromatic ring is sterically shielded from backside approach. For a nucleophile to react with an aryl halide, it would have to approach directly through the aromatic ring and invert the stereochemistry of the aromatic ring carbon—a geometric impossibility.

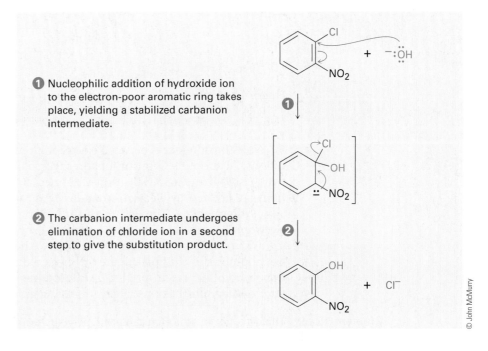

Dissociation reaction does not occur because the aryl cation is unstable; therefore, no S_N1 reaction.

Backside displacement is sterically blocked; therefore, no S_N2 reaction.

Jacob Meisenheimer

Jacob Meisenheimer (1876–1934) was born in Greisheim, Germany, and received his Ph.D. at Munich. He was professor of chemistry at the universities of Berlin and Tübingen.

Figure 16.17 MECHANISM: Mechanism of nucleophilic aromatic substitution. The reaction occurs in two steps and involves a resonance-stabilized carbanion intermediate.

Nucleophilic substitutions on an aromatic ring proceed by the mechanism shown in Figure 16.17. The nucleophile first adds to the electron-deficient aryl halide, forming a resonance-stabilized negatively charged intermediate called a *Meisenheimer complex.* Halide ion is then eliminated in the second step.

① Nucleophilic addition of hydroxide ion to the electron-poor aromatic ring takes place, yielding a stabilized carbanion intermediate.

② The carbanion intermediate undergoes elimination of chloride ion in a second step to give the substitution product.

Nucleophilic aromatic substitution occurs only if the aromatic ring has an electron-withdrawing substituent in a position ortho or para to the leaving group. The more such substituents there are, the faster the reaction. As shown in Figure 16.18, only ortho and para electron-withdrawing substituents stabilize the anion intermediate through resonance; a meta substituent offers no such resonance stabilization. Thus, *p*-chloronitrobenzene and *o*-chloronitrobenzene react with hydroxide ion at 130 °C to yield substitution products, but *m*-chloronitrobenzene is inert to OH⁻.

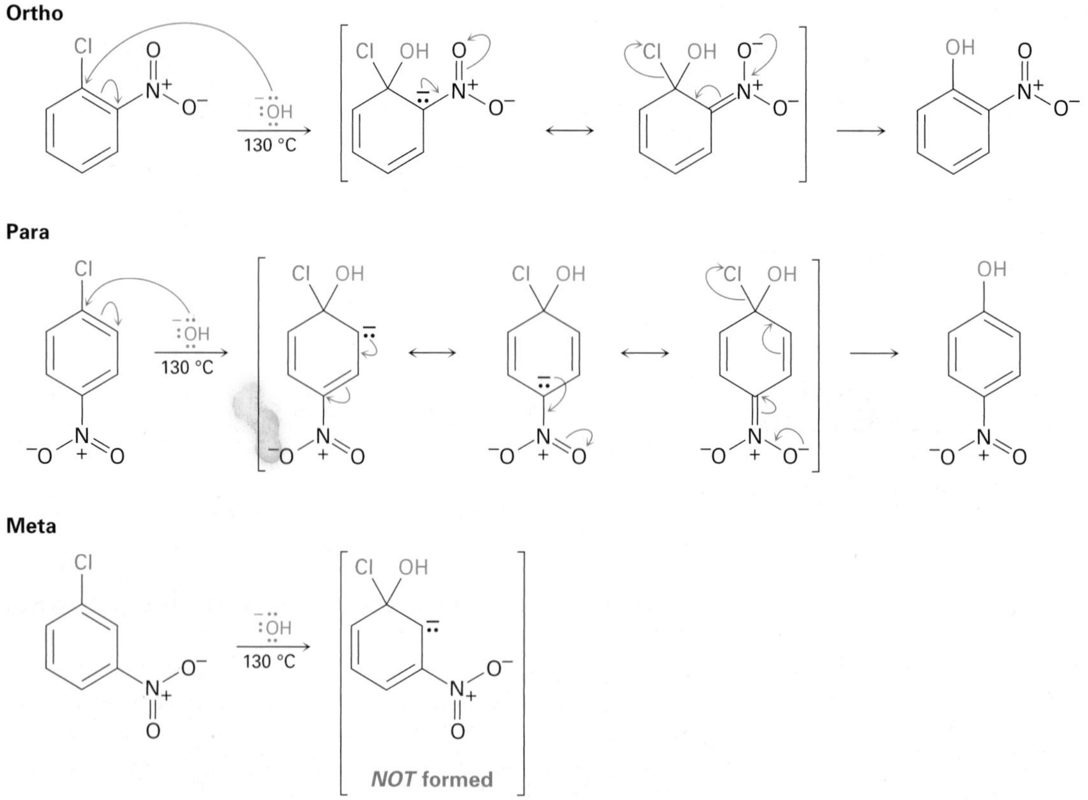

Ortho

Para

Meta

Figure 16.18 Nucleophilic aromatic substitution on nitrochlorobenzenes. Only in the ortho and para intermediates is the negative charge stabilized by a resonance interaction with the nitro group, so only the ortho and para isomers undergo reaction.

Note the differences between electrophilic and nucleophilic aromatic substitutions. Electrophilic substitutions are favored by electron-*donating* substituents, which stabilize the carbocation intermediate, while nucleophilic substitutions are favored by electron-*withdrawing* substituents, which stabilize a carbanion intermediate. The electron-withdrawing groups that *deactivate* rings for electrophilic substitution (nitro, carbonyl, cyano, and so on) *activate* them for nucleophilic substitution. What's more, these groups are meta directors in electrophilic substitution but are ortho–para directors in nucleophilic substitution. In addition, electrophilic substitutions replace hydrogen on the ring, while nucleophilic substitutions replace a leaving group, usually halide ion.

Problem 16.17 | The herbicide oxyfluorfen can be prepared by reaction between a phenol and an aryl fluoride. Propose a mechanism.

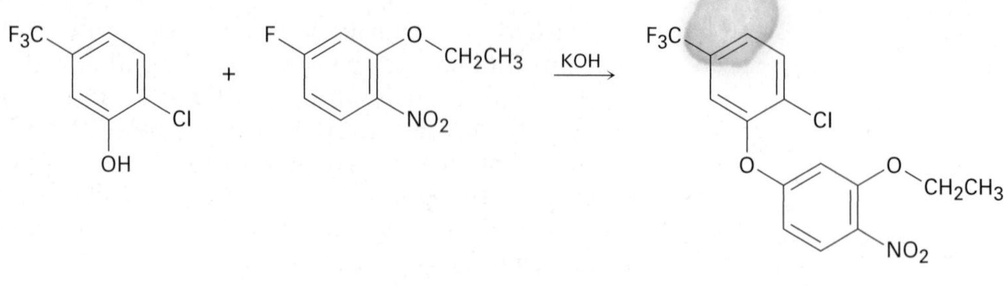

Oxyfluorfen

16.8 | Benzyne

Halobenzenes without electron-withdrawing substituents don't react with nucleophiles under most conditions. At high temperature and pressure, however, even chlorobenzene can be forced to react. Chemists at the Dow Chemical Company discovered in 1928 that phenol could be prepared on a large industrial scale by treatment of chlorobenzene with dilute aqueous NaOH at 340 °C under 170 atm pressure.

Chlorobenzene 1. NaOH, H_2O, 340 °C, 170 atm 2. H_3O^+ → **Phenol** + NaCl

A similar substitution reaction occurs with other strong bases. Treatment of bromobenzene with potassium amide (KNH_2) in liquid NH_3 solvent, for instance, gives aniline. Curiously, though, when bromobenzene labeled with radioactive ^{14}C at the C1 position is used, the substitution product has equal amounts of the label at both C1 and C2, implying the presence of a symmetrical reaction intermediate in which C1 and C2 are equivalent.

Bromobenzene → (K^+ $^-NH_2$ / NH_3) → products

50 : 50

Aniline

Further mechanistic evidence comes from trapping experiments. When bromobenzene is treated with KNH_2 in the presence of a diene such as furan, a Diels–Alder reaction (Section 14.5) occurs, implying that the symmetrical intermediate is a **benzyne**, formed by elimination of HBr from bromobenzene. Benzyne is too reactive to be isolated as a pure compound but, in the presence of water, addition occurs to give the phenol. In the presence of a diene, Diels–Alder cycloaddition takes place.

Chlorobenzene → (−HCl, elimination) → **Benzyne** → (H_2O, addition) → **Phenol**

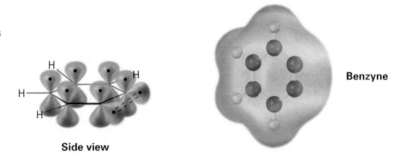

Bromobenzene   **Benzyne**   50%   50%
                         (symmetrical)   **Aniline**

The electronic structure of benzyne, shown in Figure 16.19, is that of a highly distorted alkyne. Although a typical alkyne triple bond uses *sp*-hybridized carbon atoms, the benzyne triple bond uses sp^2-hybridized carbons. Furthermore, a typical alkyne triple bond has two mutually perpendicular π bonds formed by *p–p* overlap, but the benzyne triple bond has one π bond formed by *p–p* overlap and one π bond formed by sp^2–sp^2 overlap. The latter π bond is in the plane of the ring and is very weak.

Figure 16.19 An orbital picture and electrostatic potential map of benzyne. The benzyne carbons are sp^2-hybridized, and the "third" bond results from weak overlap of two adjacent sp^2 orbitals.

Side view

Benzyne

Problem 16.18 Treatment of *p*-bromotoluene with NaOH at 300 °C yields a mixture of *two* products, but treatment of *m*-bromotoluene with NaOH yields a mixture of *three* products. Explain.

16.9 | Oxidation of Aromatic Compounds

Oxidation of Alkylbenzene Side Chains

Despite its unsaturation, the benzene ring is inert to strong oxidizing agents such as $KMnO_4$ and $Na_2Cr_2O_7$, reagents that will cleave alkene carbon–carbon bonds (Section 7.9). It turns out, however, that the presence of the aromatic ring has a dramatic effect on alkyl side chains. Alkyl side chains react rapidly with oxidizing agents and are converted into carboxyl groups, $-CO_2H$. The net effect is conversion of an alkylbenzene into a benzoic acid, $Ar-R \rightarrow Ar-CO_2H$. As an example, butylbenzene is oxidized by aqueous $KMnO_4$ in high yield to give benzoic acid.

$CH_2CH_2CH_2CH_3$

$\xrightarrow[H_2O]{KMnO_4}$

Butylbenzene   **Benzoic acid (85%)**

A similar oxidation is employed industrially for the preparation of the terephthalic acid used in the production of polyester fibers. Approximately 5 million tons per year of *p*-xylene are oxidized, using air as the oxidant and Co(III) salts as catalyst.

p-Xylene   **Terephthalic acid**

The mechanism of side-chain oxidation is complex and involves reaction of C−H bonds at the position next to the aromatic ring to form intermediate benzylic radicals. *tert*-Butylbenzene has no benzylic hydrogens, however, and is therefore inert.

tert-Butylbenzene

Analogous side-chain oxidations occur in various biosynthetic pathways. The neurotransmitter norepinephrine, for instance, is biosynthesized from dopamine by a benzylic hydroxylation reaction. The process is catalyzed by the copper-containing enzyme dopamine β-monooxygenase and occurs by a radical mechanism. A copper–oxygen species in the enzyme first abstracts the *pro-R* benzylic hydrogen to give a radical, and a hydroxyl is then transferred from copper to carbon.

Dopamine   **Norepinephrine**

Problem 16.19 | What aromatic products would you obtain from the KMnO$_4$ oxidation of the following substances?

(a) O$_2$N—⟨ring⟩—CH(CH$_3$)$_2$ **(b)** H$_3$C—⟨ring⟩—C(CH$_3$)$_3$

Bromination of Alkylbenzene Side Chains

Side-chain bromination at the benzylic position occurs when an alkylbenzene is treated with *N*-bromosuccinimide (NBS). For example, propylbenzene gives (1-bromopropyl)benzene in 97% yield on reaction with NBS in the presence of benzoyl peroxide, $(PhCO_2)_2$, as a radical initiator. Bromination occurs exclusively in the benzylic position and does not give a mixture of products.

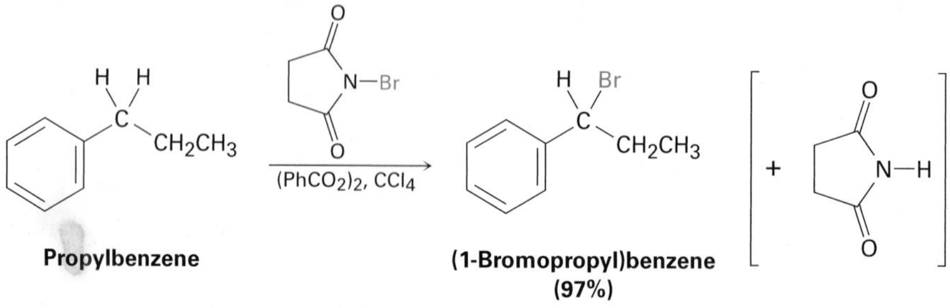

Propylbenzene

$(PhCO_2)_2$, CCl_4

(1-Bromopropyl)benzene (97%)

The mechanism of benzylic bromination is similar to that discussed in Section 10.4 for allylic bromination of alkenes. Abstraction of a benzylic hydrogen atom generates an intermediate benzylic radical, which reacts with Br_2 to yield product and a Br· radical that cycles back into the reaction to carry on the chain. The Br_2 necessary for reaction with the benzylic radical is produced by a concurrent reaction of HBr with NBS.

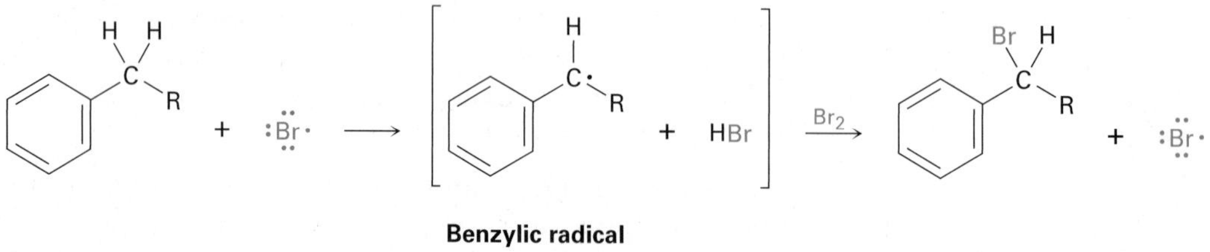

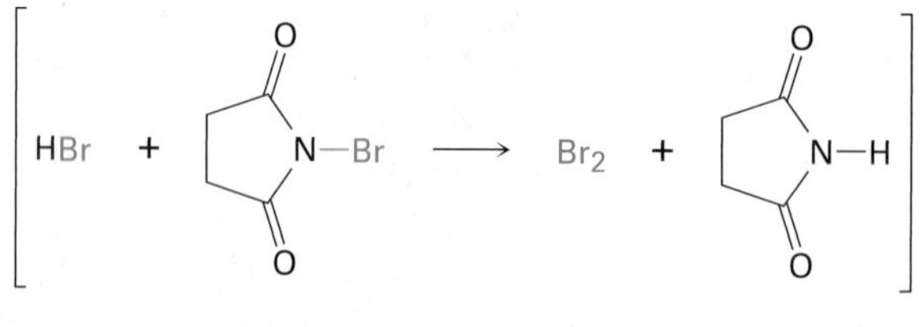

Benzylic radical

Reaction occurs exclusively at the benzylic position because the benzylic radical intermediate is stabilized by resonance. Figure 16.20 shows how the benzyl radical is stabilized by overlap of its *p* orbital with the ring π electron system.

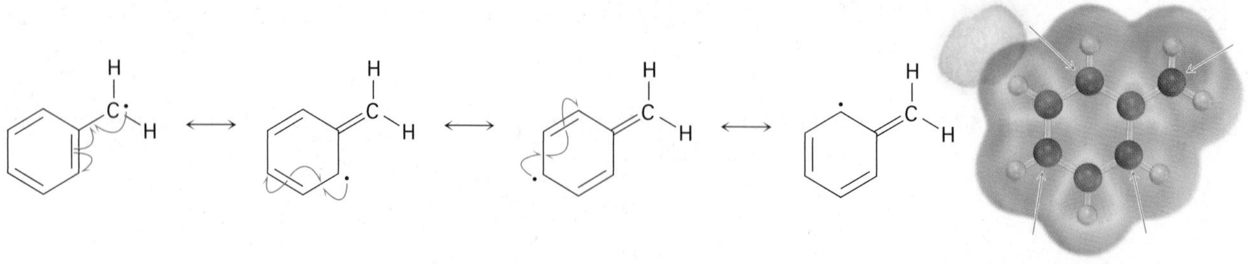

Figure 16.20 A resonance-stabilized benzylic radical. The spin-density surface shows that the unpaired electron (blue) is shared by the ortho and para carbons of the ring.

Problem 16.20   Refer to Table 5.3 on page 156 for a quantitative idea of the stability of a benzyl radical. How much more stable (in kJ/mol) is the benzyl radical than a primary alkyl radical? How does a benzyl radical compare in stability to an allyl radical?

Problem 16.21   Styrene, the simplest alkenylbenzene, is prepared commercially for use in plastics manufacture by catalytic dehydrogenation of ethylbenzene. How might you prepare styrene from benzene using reactions you've studied?

Styrene

16.10 | Reduction of Aromatic Compounds

Catalytic Hydrogenation of Aromatic Rings

Just as aromatic rings are generally inert to oxidation, they're also inert to catalytic hydrogenation under conditions that reduce typical alkene double bonds. As a result, it's possible to reduce an alkene double bond selectively in the presence of an aromatic ring. For example, 4-phenyl-3-buten-2-one is reduced to 4-phenyl-2-butanone at room temperature and atmospheric pressure using a palladium catalyst. Neither the benzene ring nor the ketone carbonyl group is affected.

$$\xrightarrow[\text{Ethanol}]{\text{H}_2, \text{Pd}}$$

4-Phenylbut-3-en-2-one   **4-Phenylbutan-2-one**
                                              (100%)

To hydrogenate an aromatic ring, it's necessary either to use a platinum catalyst with hydrogen gas at several hundred atmospheres pressure or to use a more effective catalyst such as rhodium on carbon. Under these conditions, aromatic rings are converted into cyclohexanes. For example, o-xylene yields 1,2-dimethylcyclohexane, and 4-tert-butylphenol gives 4-tert-butylcyclohexanol.

$$\xrightarrow[\text{130 atm, 25 °C}]{\text{H}_2, \text{Pt; ethanol}}$$

o-Xylene   **cis-1,2-Dimethyl-
                            cyclohexane**

4-*tert*-Butylphenol

cis-4-*tert*-Butyl-
cyclohexane

Reduction of Aryl Alkyl Ketones

Just as an aromatic ring activates a neighboring (benzylic) hydrogen toward oxidation, it also activates a neighboring carbonyl group toward reduction. Thus, an aryl alkyl ketone prepared by Friedel–Crafts acylation of an aromatic ring can be converted into an alkylbenzene by catalytic hydrogenation over a palladium catalyst. Propiophenone, for instance, is reduced to propylbenzene by catalytic hydrogenation. Since the net effect of Friedel–Crafts acylation followed by reduction is the preparation of a primary alkylbenzene, this two-step sequence of reactions makes it possible to circumvent the carbocation rearrangement problems associated with direct Friedel–Crafts alkylation using a primary alkyl halide (Section 16.3).

Propiophenone (95%)

Propylbenzene (100%)

Propylbenzene

Isopropylbenzene

Mixture of two products

Note that the conversion of a carbonyl group into a methylene group ($C=O \rightarrow CH_2$) by catalytic hydrogenation is limited to *aryl* alkyl ketones; dialkyl ketones are not reduced under these conditions. Furthermore, the catalytic reduction of aryl alkyl ketones is not compatible with the presence of a nitro substituent on the aromatic ring because a nitro group is reduced to an amino group under the reaction conditions. We'll see a more general method for reducing all ketone carbonyl groups to yield alkanes in Section 19.9.

m-Nitroacetophenone

m-Ethylaniline

Problem 16.22 | How would you prepare diphenylmethane, $(Ph)_2CH_2$, from benzene and an acid chloride?

16.11 | Synthesis of Trisubstituted Benzenes

One of the surest ways to learn organic chemistry is to work synthesis problems. The ability to plan a successful multistep synthesis of a complex molecule requires a working knowledge of the uses and limitations of a great many organic reactions. Not only must you know *which* reactions to use, you must also know *when* to use them because the order in which reactions are carried out is often critical to the success of the overall scheme.

The ability to plan a sequence of reactions in the right order is particularly valuable in the synthesis of substituted aromatic rings, where the introduction of a new substituent is strongly affected by the directing effects of other substituents. Planning syntheses of substituted aromatic compounds is therefore an excellent way to gain confidence using the many reactions learned in the past few chapters.

During our previous discussion of strategies for working synthesis problems in Section 8.9, we said that it's usually best to work a problem backward, or *retrosynthetically*. Look at the target molecule and ask yourself, "What is an immediate precursor of this compound?" Choose a likely answer and continue working backward, one step at a time, until you arrive at a simple starting material. Let's try some examples.

WORKED EXAMPLE 16.4 | ***Synthesizing a Polysubstituted Benzene***

Synthesize 4-bromo-2-nitrotoluene from benzene.

Strategy | Draw the target molecule, identify the substituents, and recall how each group can be introduced separately. Then plan retrosynthetically.

4-Bromo-2-nitrotoluene

The three substituents on the ring are a bromine, a methyl group, and a nitro group. A bromine can be introduced by bromination with $Br_2/FeBr_3$, a methyl group can be introduced by Friedel–Crafts alkylation with $CH_3Cl/AlCl_3$, and a nitro group can be introduced by nitration with HNO_3/H_2SO_4.

Solution | "What is an immediate precursor of the target?" The final step will involve introduction of one of three groups—bromine, methyl, or nitro—so we have to consider three possibilities. Of the three, the bromination of *o*-nitrotoluene could be used because the activating methyl group would dominate the deactivating nitro group and direct bromination to the right position. Unfortunately, a mixture of product isomers would be formed. A Friedel–Crafts reaction can't be used as the final step because this reaction doesn't work on a nitro-substituted (strongly deactivated) benzene. The best precursor of the desired product is probably *p*-bromotoluene, which can be nitrated ortho to the activating methyl group to give a single product.

o-Nitrotoluene

This ring will give a mixture of isomers on bromination.

m-Bromonitrobenzene

This deactivated ring will not undergo a Friedel–Crafts reaction.

p-Bromotoluene

This ring will give only the desired isomer on nitration.

Br_2 FeBr$_3$

HNO_3 H$_2$SO$_4$

4-Bromo-2-nitrotoluene

Next ask yourself, "What is an immediate precursor of *p*-bromotoluene?" Perhaps toluene is an immediate precursor because the methyl group would direct bromination to the ortho and para positions. Alternatively, bromobenzene might be an immediate precursor because we could carry out a Friedel–Crafts methylation and obtain a mixture of ortho and para products. Both answers are satisfactory, although both would also lead unavoidably to a product mixture that would have to be separated.

Toluene

Br_2 FeBr$_3$

p-Bromotoluene (+ ortho isomer)

CH_3Cl AlCl$_3$

Bromobenzene

"What is an immediate precursor of toluene?" Benzene, which could be methylated in a Friedel–Crafts reaction. Alternatively, "What is an immediate precursor of bromobenzene?" Benzene, which could be brominated.

The retrosynthetic analysis has provided two valid routes from benzene to 4-bromo-2-nitrotoluene.

Benzene

CH_3Cl AlCl$_3$

Toluene

Br_2 FeBr$_3$

Br_2 FeBr$_3$

Bromobenzene

CH_3Cl AlCl$_3$

p-Bromotoluene

HNO_3 H$_2$SO$_4$

4-Bromo-2-nitrotoluene

| WORKED EXAMPLE 16.5 | *Synthesizing a Polysubstituted Benzene* |

Synthesize 4-chloro-2-propylbenzenesulfonic acid from benzene.

Strategy Draw the target molecule, identify its substituents, and recall how each of the three can be introduced. Then plan retrosynthetically.

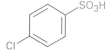

4-Chloro-2-propylbenzenesulfonic acid

The three substituents on the ring are a chlorine, a propyl group, and a sulfonic acid group. A chlorine can be introduced by chlorination with $Cl_2/FeCl_3$, a propyl group can be introduced by Friedel–Crafts acylation with $CH_3CH_2COCl/AlCl_3$ followed by reduction with H_2/Pd, and a sulfonic acid group can be introduced by sulfonation with SO_3/H_2SO_4.

Solution "What is an immediate precursor of the target?" The final step will involve introduction of one of three groups—chlorine, propyl, or sulfonic acid—so we have to consider three possibilities. Of the three, the chlorination of *o*-propylbenzenesulfonic acid can't be used because the reaction would occur at the wrong position. Similarly, a Friedel–Crafts reaction can't be used as the final step because this reaction doesn't work on sulfonic acid-substituted (strongly deactivated) benzenes. Thus, the immediate precursor of the desired product is probably *m*-chloropropylbenzene, which can be sulfonated to give a mixture of product isomers that must then be separated.

**o-Propylbenzene-
sulfonic acid**

This ring will give the wrong
isomer on chlorination.

**p-Chlorobenzene-
sulfonic acid**

This deactivated ring will not
undergo a Friedel–Crafts reaction.

m-Chloropropylbenzene

This ring will give the desired
product on sulfonation.

SO_3

H_2SO_4

4-Chloro-2-propylbenzenesulfonic acid

"What is an immediate precursor of *m*-chloropropylbenzene?" Because the two substituents have a meta relationship, the first substituent placed on the ring must be a meta director so that the second substitution will take place at the proper position. Furthermore, because primary alkyl groups such as propyl can't be introduced directly by Friedel–Crafts alkylation, the precursor of *m*-chloropropylbenzene is probably *m*-chloropropiophenone, which could be catalytically reduced.

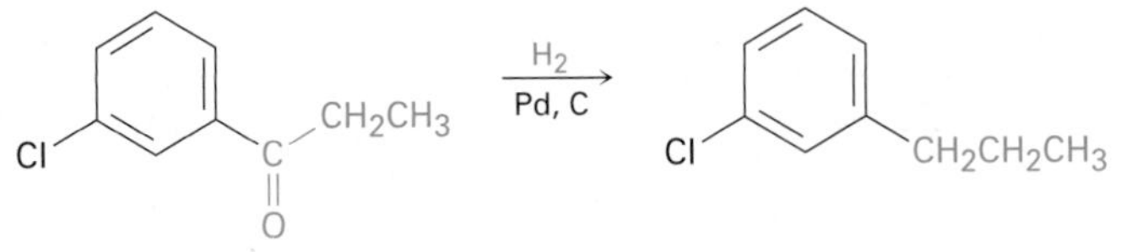

_m_-Chloropropiophenone   **_m_-Chloropropylbenzene**

"What is an immediate precursor of *m*-chloropropiophenone?" Propiophenone, which could be chlorinated in the meta position.

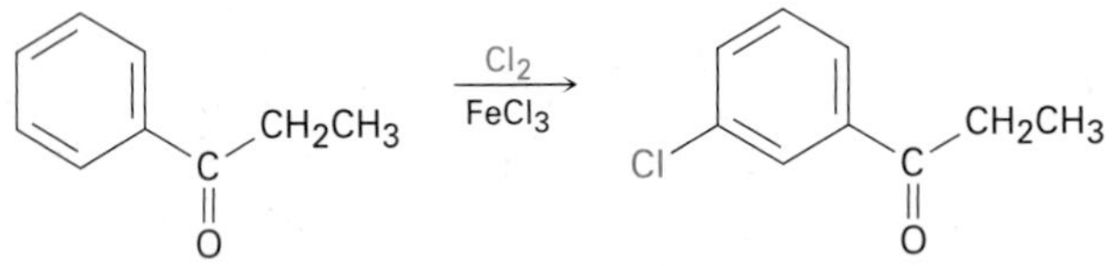

Propiophenone   **_m_-Chloropropiophenone**

"What is an immediate precursor of propiophenone?" Benzene, which could undergo Friedel–Crafts acylation with propanoyl chloride and AlCl$_3$.

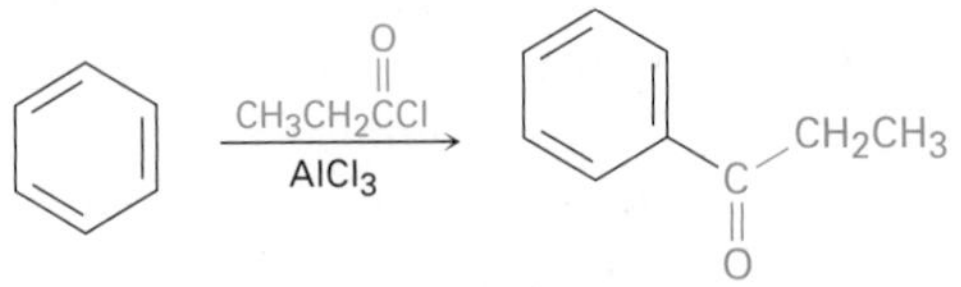

Benzene   **Propiophenone**

The final synthesis is a four-step route from benzene:

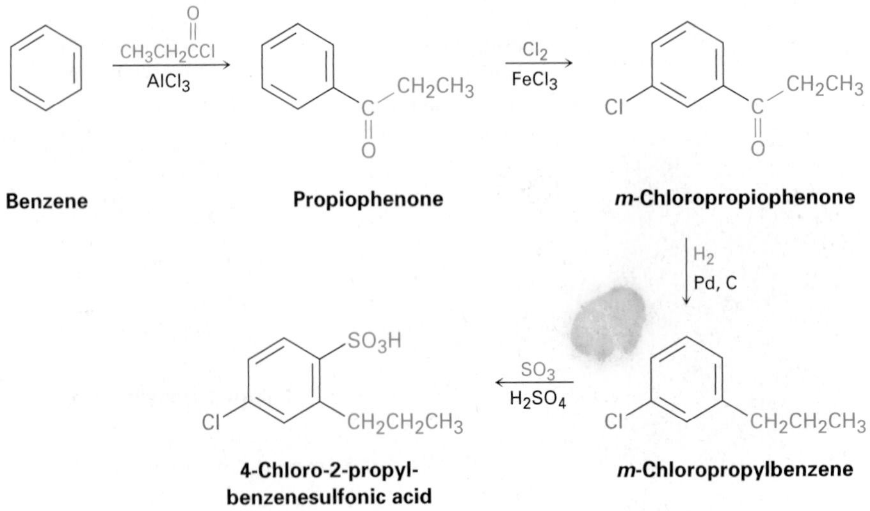

Benzene   **Propiophenone**   **_m_-Chloropropiophenone**

4-Chloro-2-propyl-benzenesulfonic acid   **_m_-Chloropropylbenzene**

Planning organic syntheses has been compared with playing chess. There are no tricks; all that's required is a knowledge of the allowable moves (the organic reactions) and the discipline to plan ahead, carefully evaluating the consequences of each move. Practicing is not always easy, but there is no surer way to learn organic chemistry.

Problem 16.23 | Propose syntheses of the following substances from benzene:
(a) *m*-Chloronitrobenzene   (b) *m*-Chloroethylbenzene
(c) 4-Chloro-1-nitro-2-propylbenzene   (d) 3-Bromo-2-methylbenzenesulfonic acid

Problem 16.24 | In planning a synthesis, it's as important to know what not to do as to know what to do. As written, the following reaction schemes have flaws in them. What is wrong with each?

(a)

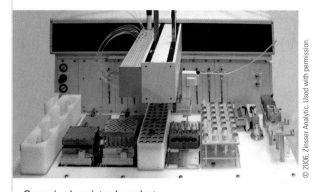

Organic chemistry by robot
means no spilled flasks!

Focus On . . .

Combinatorial Chemistry

Traditionally, organic compounds have been synthesized one at a time. This works well for preparing large amounts of a few substances, but it doesn't work so well for preparing small amounts of a great many substances. This latter goal is particularly important in the pharmaceutical industry, where vast numbers of structurally similar compounds must be screened to find the optimum drug candidate.

(continued)

To speed the process of drug discovery, *combinatorial chemistry* has been developed to prepare what are called *combinatorial libraries,* in which anywhere from a few dozen to several hundred thousand substances are prepared simultaneously. Among the early successes of combinatorial chemistry is the development of a benzodiazepine library, a class of aromatic compounds much used as antianxiety agents.

Benzodiazepine library
(R_1–R_4 are various organic substituents)

Two main approaches to combinatorial chemistry are used—*parallel synthesis* and *split synthesis.* In parallel synthesis, each compound is prepared independently. Typically, a reactant is first linked to the surface of polymer beads, which are then placed into small wells on a 96-well glass plate. Programmable robotic instruments add different sequences of building blocks to the different wells, thereby making 96 different products. When the reaction sequences are complete, the polymer beads are washed and their products are released.

In split synthesis, the initial reactant is again linked to the surface of polymer beads, which are then divided into several groups. A different building block is added to each group of beads, the different groups are combined, and the reassembled mix is again split to form new groups. Another building block is added to each group, the groups are again combined and redivided, and the process continues. If, for example, the beads are divided into four groups at each step, the number of compounds increases in the progression $4 \rightarrow 16 \rightarrow 64 \rightarrow 256$. After 10 steps, more than 1 million compounds have been prepared (Figure 16.21).

Figure 16.21 The results of split combinatorial synthesis. Assuming that 4 different building blocks are used at each step, 64 compounds result after 3 steps, and more than 1 million compounds result after 10 steps.

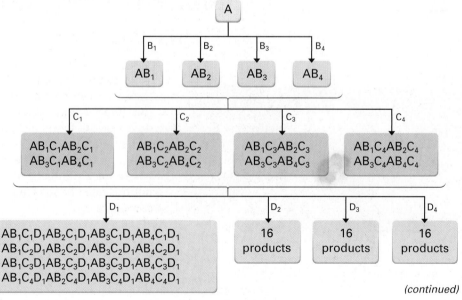

(continued)

Of course, with so many different final products mixed together, the problem is to identify them. What structure is linked to what bead? Several approaches to this problem have been developed, all of which involve the attachment of encoding labels to each polymer bead to keep track of the chemistry each has undergone. Encoding labels used thus far have included proteins, nucleic acids, halogenated aromatic compounds, and even computer chips.

SUMMARY AND KEY WORDS

acyl group, 557

acylation, 557

alkylation, 554

benzyne, 575

electrophilic aromatic substitution, 547

Friedel–Crafts reaction, 554

inductive effect, 562

nucleophilic aromatic substitution, 572

resonance effect, 562

An **electrophilic aromatic substitution reaction** takes place in two steps—initial reaction of an electrophile, E^+, with the aromatic ring, followed by loss of H^+ from the resonance-stabilized carbocation intermediate to regenerate the aromatic ring.

Many variations of the reaction can be carried out, including halogenation, nitration, and sulfonation. **Friedel–Crafts alkylation** and **acylation** reactions, which involve reaction of an aromatic ring with carbocation electrophiles, are particularly useful. They are limited, however, by the fact that the aromatic ring must be at least as reactive as a halobenzene. In addition, polyalkylation and carbocation rearrangements often occur in Friedel–Crafts alkylation.

Substituents on the benzene ring affect both the reactivity of the ring toward further substitution and the orientation of that substitution. Groups can be classified as *ortho- and para-directing activators, ortho- and para-directing deactivators,* or *meta-directing deactivators.* Substituents influence aromatic rings by a combination of resonance and inductive effects. **Resonance effects** are transmitted through π bonds; **inductive effects** are transmitted through σ bonds.

Halobenzenes undergo **nucleophilic aromatic substitution** through either of two mechanisms. If the halobenzene has a strongly electron-withdrawing substituent in the ortho or para position, substitution occurs by addition of a nucleophile to the ring, followed by elimination of halide from the intermediate anion. If the halobenzene is not activated by an electron-withdrawing substituent, substitution can occur by elimination of HX to give a **benzyne,** followed by addition of a nucleophile.

The benzylic position of an alkylbenzene can be brominated by reaction with *N*-bromosuccinimide, and the entire side chain can be degraded to a carboxyl group by oxidation with aqueous $KMnO_4$. Although aromatic rings are less reactive than isolated alkene double bonds, they can be reduced to cyclohexanes by hydrogenation over a platinum or rhodium catalyst. In addition, aryl alkyl ketones are reduced to alkylbenzenes by hydrogenation over a platinum catalyst.

SUMMARY OF REACTIONS

1. Electrophilic aromatic substitution
   (a) Bromination (Section 16.1)

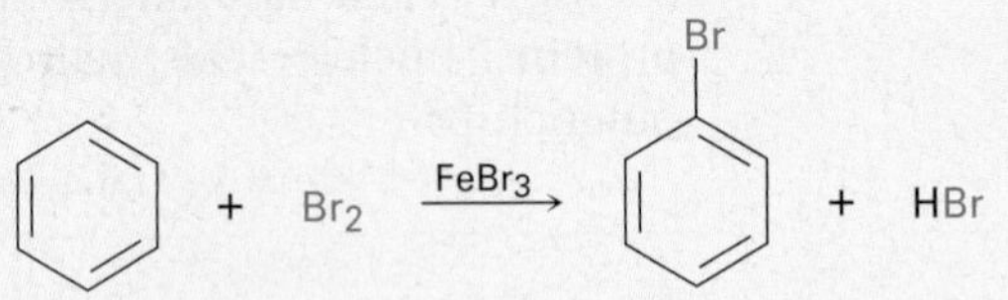

   (b) Chlorination (Section 16.2)

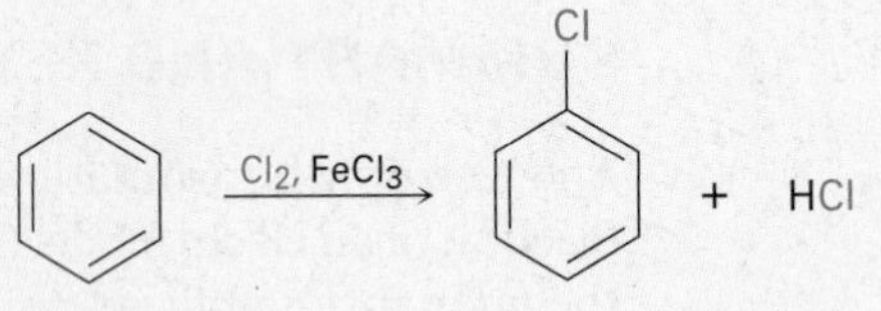

   (c) Iodination (Section 16.2)

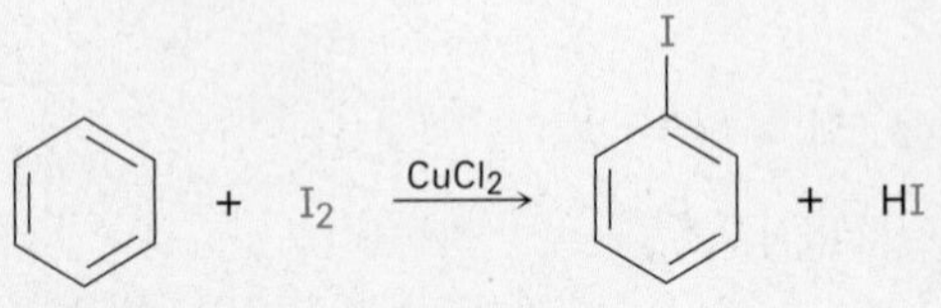

   (d) Nitration (Section 16.2)

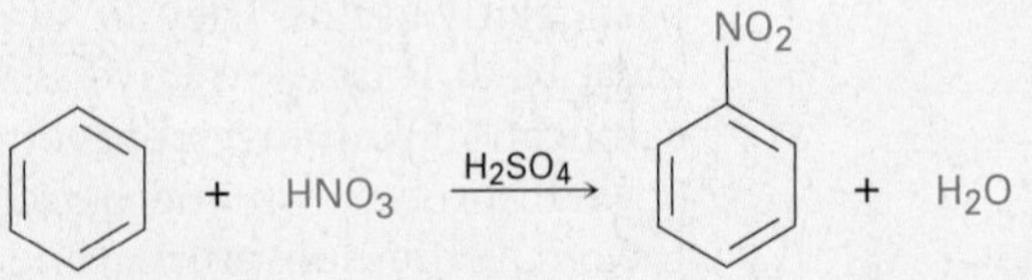

   (e) Sulfonation (Section 16.2)

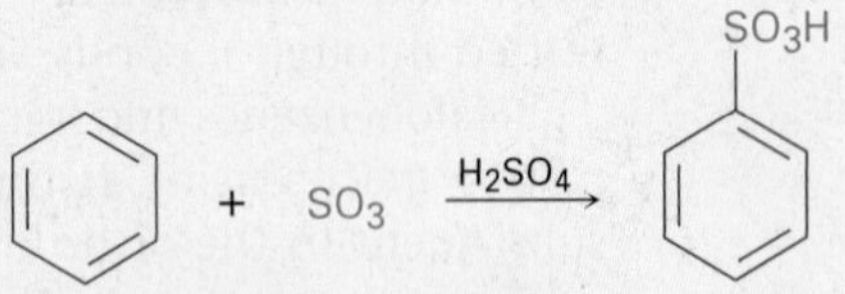

   (f) Friedel–Crafts alkylation (Section 16.3)

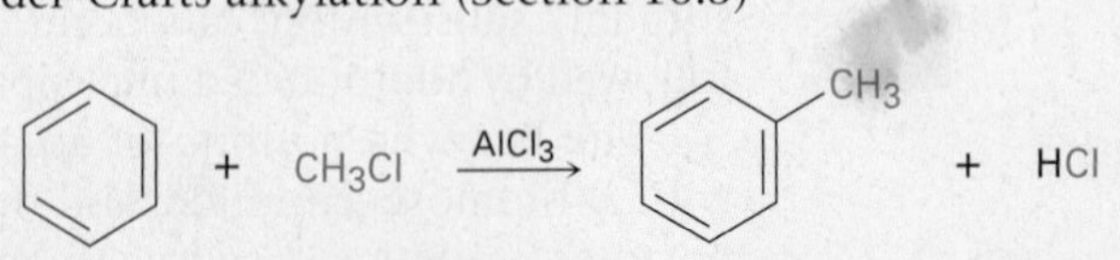

   Aromatic ring. Must be at least as reactive as a halobenzene.
   Alkyl halide. Primary alkyl halides undergo carbocation rearrangement.

(g) Friedel–Crafts acylation (Section 16.3)

2. Reduction of aromatic nitro groups (Section 16.2)

3. Nucleophilic aromatic substitution
   (a) By addition to activated aryl halides (Section 16.7)

   (b) By formation of benzyne intermediate from unactivated aryl halide
       (Section 16.8)

4. Oxidation of alkylbenzene side chain (Section 16.9)

5. Benzylic bromination of alkylbenzene side chain (Section 16.9)

6. Catalytic hydrogenation of aromatic ring (Section 16.10)

7. Reduction of aryl alkyl ketones (Section 16.10)

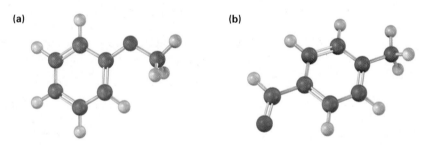

EXERCISES

Organic KNOWLEDGE TOOLS

CENGAGENOW™ Sign in at **www.cengage.com/login** to assess your knowledge of this chapter's topics by taking a pre-test. The pre-test will link you to interactive organic chemistry resources based on your score in each concept area.

OWL Online homework for this chapter may be assigned in Organic OWL.

■ indicates problems assignable in Organic OWL.

▲ denotes problems linked to Key Ideas of this chapter and testable in CengageNOW.

VISUALIZING CHEMISTRY

(Problems 16.1–16.24 appear within the chapter.)

16.25 ■ Draw the product from reaction of each of the following substances with (i) Br_2, $FeBr_3$ and (ii) CH_3COCl, $AlCl_3$.

(a) **(b)**

16.26 The following molecular model of a dimethyl-substituted biphenyl represents the lowest-energy conformation of the molecule. Why are the two benzene rings tilted at a 63° angle to each other rather than being in the same plane so that their *p* orbitals can overlap? Why doesn't complete rotation around the single bond joining the two rings occur?

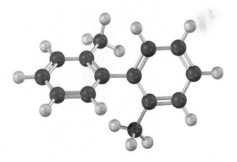

16.27 ■ How would you synthesize the following compound starting from benzene? More than one step is needed.

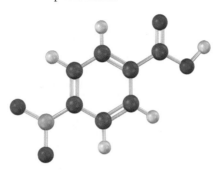

16.28 The following compound can't be synthesized using the methods discussed in this chapter. Why not?

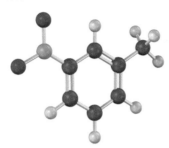

ADDITIONAL PROBLEMS

16.29 ■ Identify each of the following groups as an activator or deactivator and as an *o,p*-director or *m*-director:

(a) $-N(CH_3)_2$ (b)   (c) $-OCH_2CH_3$ (d)

16.30 ■ Predict the major product(s) of nitration of the following substances. Which react faster than benzene, and which slower?
(a) Bromobenzene   (b) Benzonitrile   (c) Benzoic acid
(d) Nitrobenzene   (e) Benzenesulfonic acid   (f) Methoxybenzene

16.31 ■ ▲ Rank the compounds in each group according to their reactivity toward electrophilic substitution.
(a) Chlorobenzene, *o*-dichlorobenzene, benzene
(b) *p*-Bromonitrobenzene, nitrobenzene, phenol
(c) Fluorobenzene, benzaldehyde, *o*-xylene
(d) Benzonitrile, *p*-methylbenzonitrile, *p*-methoxybenzonitrile

16.32 ■ ▲ Predict the major monoalkylation products you would expect to obtain from reaction of the following substances with chloromethane and AlCl₃:
(a) Bromobenzene   (b) *m*-Bromophenol
(c) *p*-Chloroaniline   (d) 2,4-Dichloronitrobenzene
(e) 2,4-Dichlorophenol   (f) Benzoic acid
(g) *p*-Methylbenzenesulfonic acid   (h) 2,5-Dibromotoluene

■ Assignable in OWL   ▲ Key Idea Problems

16.33 ■ Name and draw the major product(s) of electrophilic chlorination of the following compounds:
(a) *m*-Nitrophenol   (b) *o*-Xylene
(c) *p*-Nitrobenzoic acid   (d) *p*-Bromobenzenesulfonic acid

16.34 Predict the major product(s) you would obtain from sulfonation of the following compounds:
(a) Fluorobenzene   (b) *m*-Bromophenol
(c) *m*-Dichlorobenzene   (d) 2,4-Dibromophenol

16.35 Rank the following aromatic compounds in the expected order of their reactivity toward Friedel–Crafts alkylation. Which compounds are unreactive?
(a) Bromobenzene   (b) Toluene   (c) Phenol
(d) Aniline   (e) Nitrobenzene   (f) *p*-Bromotoluene

16.36 ■ What product(s) would you expect to obtain from the following reactions?

(a)

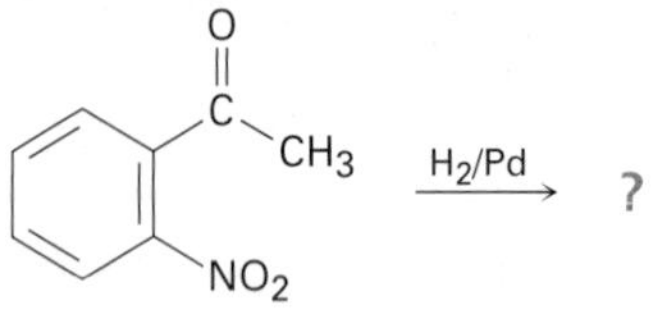

(b)

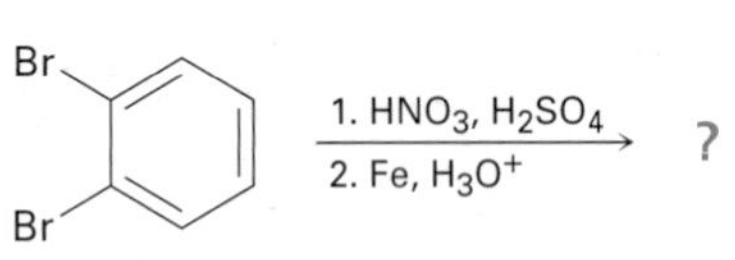

(c)

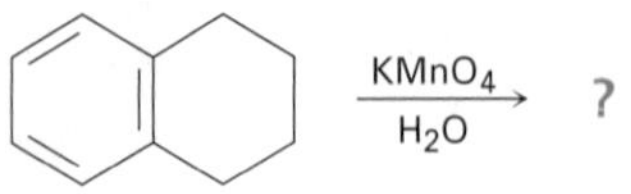

(d)

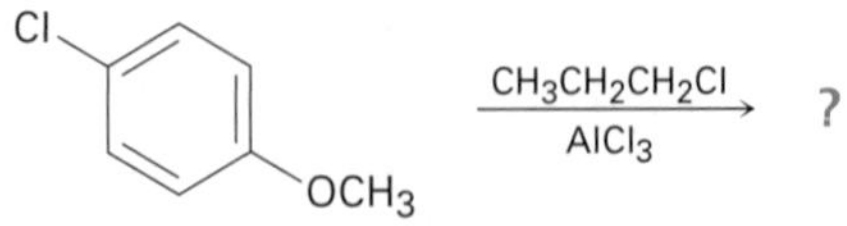

16.37 ■ Predict the major product(s) of the following reactions:

(a)

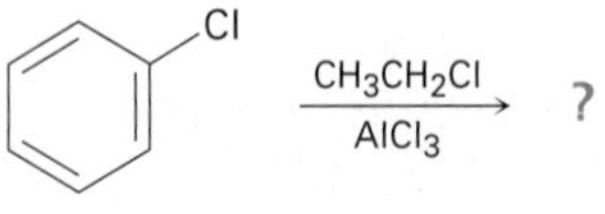

(b)

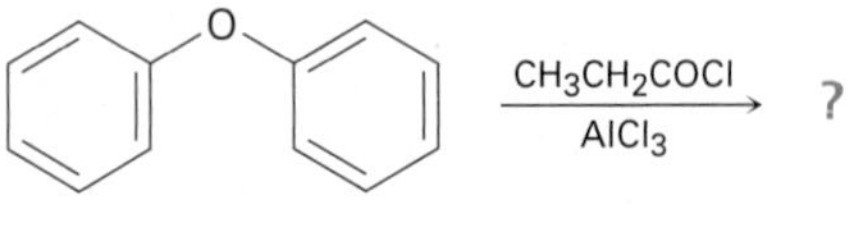

(c)

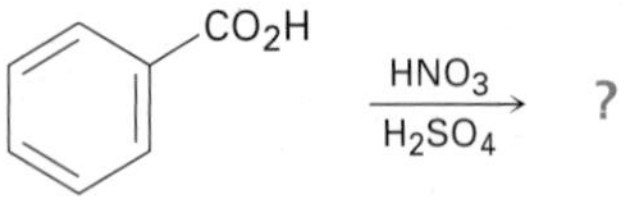

(d)

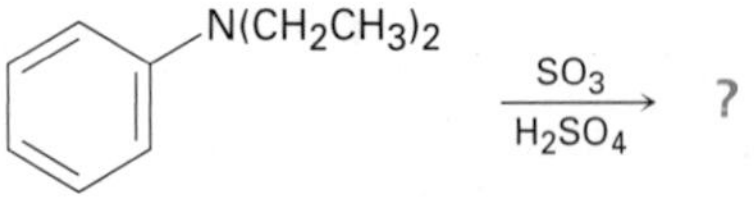

16.38 Aromatic iodination can be carried out with a number of reagents, including iodine monochloride, ICl. What is the direction of polarization of ICl? Propose a mechanism for the iodination of an aromatic ring with ICl.

16.39 ■ The sulfonation of an aromatic ring with SO_3 and H_2SO_4 is reversible. That is, heating benzenesulfonic acid with H_2SO_4 yields benzene. Show the mechanism of the desulfonation reaction. What is the electrophile?

16.40 ■ The carbocation electrophile in a Friedel–Crafts reaction can be generated in ways other than by reaction of an alkyl chloride with $AlCl_3$. For example, reaction of benzene with 2-methylpropene in the presence of H_3PO_4 yields *tert*-butylbenzene. Propose a mechanism for this reaction.

16.41 The *N,N,N*-trimethylammonium group, $-\overset{+}{N}(CH_3)_3$, is one of the few groups that is a meta-directing deactivator yet has no electron-withdrawing resonance effect. Explain.

16.42 The nitroso group, $-N=O$, is one of the few nonhalogens that is an ortho- and para-directing deactivator. Explain by drawing resonance structures of the carbocation intermediates in ortho, meta, and para electrophilic reaction on nitrosobenzene, $C_6H_5N=O$.

16.43 Using resonance structures of the intermediates, explain why bromination of biphenyl occurs at ortho and para positions rather than at meta.

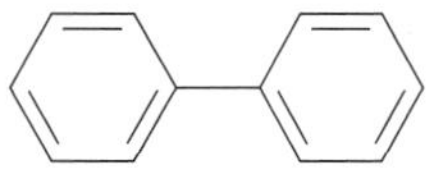

Biphenyl

16.44 ■ ▲ At what position and on what ring do you expect nitration of 4-bromo-biphenyl to occur? Explain, using resonance structures of the potential intermediates.

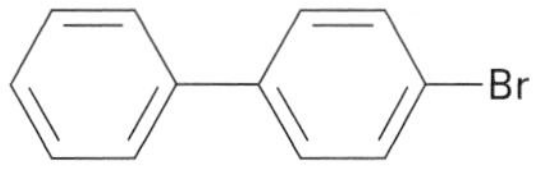

Br   **4-Bromobiphenyl**

16.45 ▲ Electrophilic substitution on 3-phenylpropanenitrile occurs at the ortho and para positions, but reaction with 3-phenylpropenenitrile occurs at the meta position. Explain, using resonance structures of the intermediates.

CH₂CH₂CN   CN

3-Phenylpropanenitrile   **3-Phenylpropenenitrile**

16.46 Addition of HBr to 1-phenylpropene yields only (1-bromopropyl)benzene. Propose a mechanism for the reaction, and explain why none of the other regioisomer is produced.

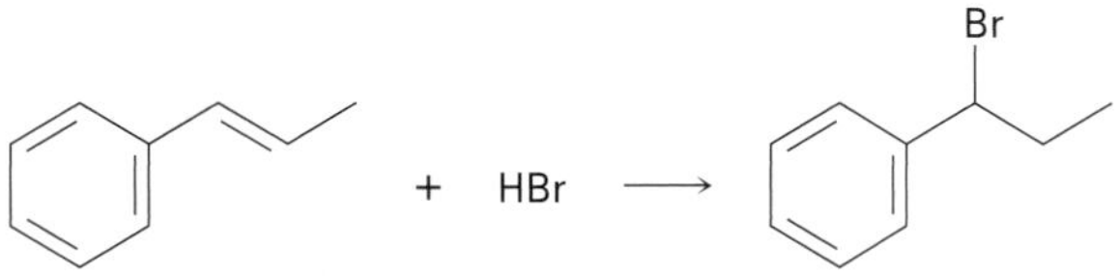

Br

+   HBr $\longrightarrow$

16.47 Triphenylmethane can be prepared by reaction of benzene and chloroform in the presence of $AlCl_3$. Propose a mechanism for the reaction.

+   CHCl₃ $\xrightarrow{AlCl_3}$

H

C

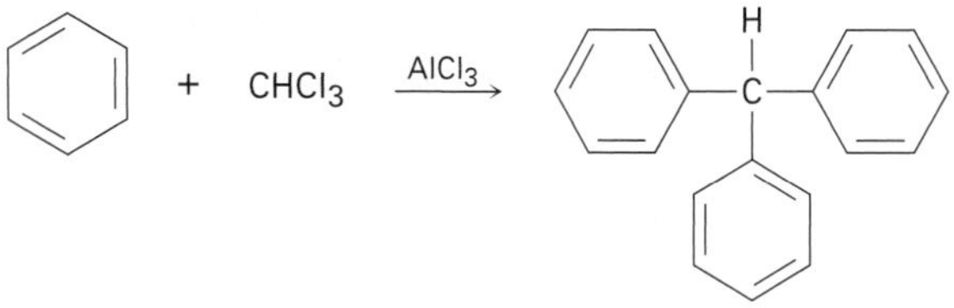

16.48 ■ At what position, and on what ring, would you expect the following substances to undergo electrophilic substitution?

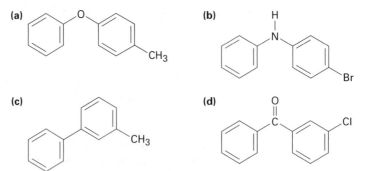

(a)

(b)

(c)

(d)

16.49 ■ At what position, and on what ring, would you expect bromination of benzanilide to occur? Explain by drawing resonance structures of the intermediates.

Benzanilide

16.50 Would you expect the Friedel–Crafts reaction of benzene with (R)-2-chlorobutane to yield optically active or racemic product? Explain.

16.51 ■ How would you synthesize the following substances starting from benzene or phenol? Assume that ortho- and para-substitution products can be separated.
(a) o-Bromobenzoic acid   (b) p-Methoxytoluene
(c) 2,4,6-Trinitrobenzoic acid   (d) m-Bromoaniline

16.52 ■ Starting with benzene as your only source of aromatic compounds, how would you synthesize the following substances? Assume that you can separate ortho and para isomers if necessary.
(a) p-Chloroacetophenone   (b) m-Bromonitrobenzene
(c) o-Bromobenzenesulfonic acid   (d) m-Chlorobenzenesulfonic acid

16.53 ■ Starting with either benzene or toluene, how would you synthesize the following substances? Assume that ortho and para isomers can be separated.
(a) 2-Bromo-4-nitrotoluene   (b) 1,3,5-Trinitrobenzene
(c) 2,4,6-Tribromoaniline

16.54 As written, the following syntheses have flaws. What is wrong with each?

(a) CH$_3$
→ [1. Cl$_2$, FeCl$_3$; 2. KMnO$_4$] → CO$_2$H, Cl

(b) Cl
→ [1. HNO$_3$, H$_2$SO$_4$; 2. CH$_3$Cl, AlCl$_3$; 3. Fe, H$_3$O$^+$; 4. NaOH, H$_2$O] → Cl, CH$_3$, NH$_2$

(c) CH$_3$
→ [1. CH$_3$CCl, AlCl$_3$; 2. HNO$_3$, H$_2$SO$_4$; 3. H$_2$/Pd; ethanol] → CH$_3$, NO$_2$, CH$_2$CH$_3$

16.55 ■ How would you synthesize the following substances starting from benzene?

(a) 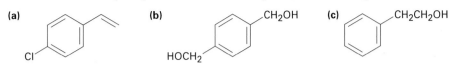 (b) (c)

16.56 The compound MON-0585 is a nontoxic, biodegradable larvicide that is highly selective against mosquito larvae. Synthesize MON-0585 using either benzene or phenol as a source of the aromatic rings.

MON-0585

16.57 ■ Hexachlorophene, a substance used in the manufacture of germicidal soaps, is prepared by reaction of 2,4,5-trichlorophenol with formaldehyde in the presence of concentrated sulfuric acid. Propose a mechanism for the reaction.

$$\xrightarrow[H_2SO_4]{CH_2O}$$

Hexachlorophene

16.58 ■ Benzenediazonium carboxylate decomposes when heated to yield N_2, CO_2, and a reactive substance that can't be isolated. When benzenediazonium carboxylate is heated in the presence of furan, the following reaction is observed:

$$+ \xrightarrow{Heat} + CO_2 + N_2$$

What intermediate is involved in this reaction? Propose a mechanism for its formation.

16.59 Phenylboronic acid, $C_6H_5B(OH)_2$, is nitrated to give 15% ortho-substitution product and 85% meta. Explain the meta-directing effect of the $-B(OH)_2$ group.

16.60 Draw resonance structures of the intermediate carbocations in the bromination of naphthalene, and account for the fact that naphthalene undergoes electrophilic substitution at C1 rather than C2.

$$\xrightarrow{Br_2}$$

■ Assignable in OWL   ▲ Key Idea Problems

16.61 ■ Propose a mechanism for the reaction of 1-chloroanthraquinone with methoxide ion to give the substitution product 1-methoxyanthraquinone. Use curved arrows to show the electron flow in each step.

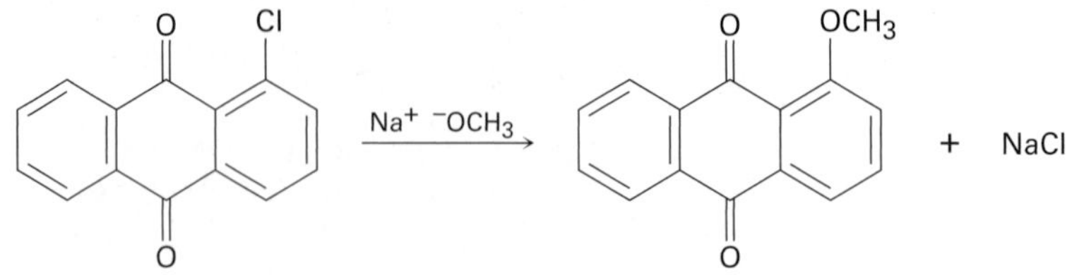

1-Chloroanthraquinone   **1-Methoxyanthraquinone**

16.62 ■ 4-Chloropyridine undergoes reaction with dimethylamine to yield 4-dimethyl-aminopyridine. Propose a mechanism for the reaction.

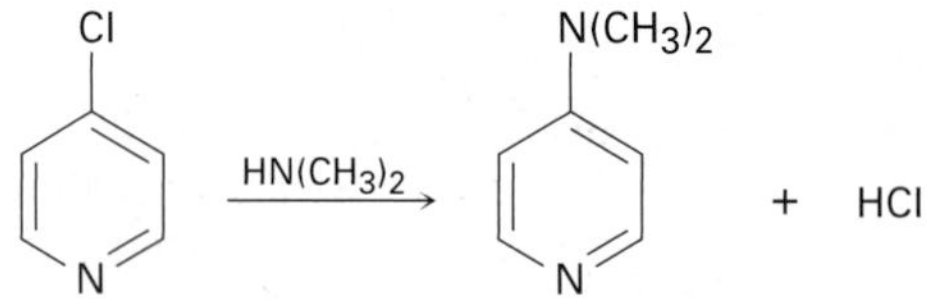

16.63 ■ *p*-Bromotoluene reacts with potassium amide to give a mixture of *m*- and *p*-methylaniline. Explain.

16.64 ■ Propose a mechanism to account for the reaction of benzene with 2,2,5,5-tetramethyltetrahydrofuran.

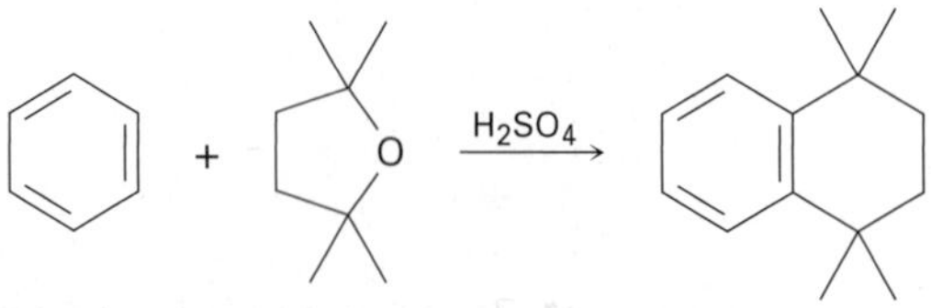

16.65 Propose a mechanism to account for the following reaction:

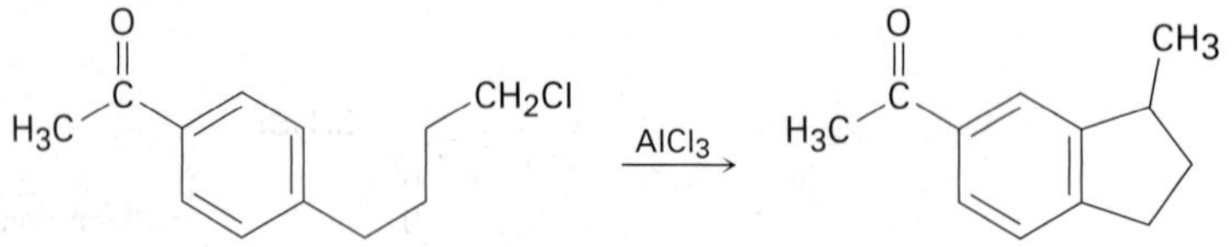

16.66 ■ In the *Gatterman–Koch reaction*, a formyl group (−CHO) is introduced directly onto a benzene ring. For example, reaction of toluene with CO and HCl in the presence of mixed CuCl/AlCl₃ gives *p*-methylbenzaldehyde. Propose a mechanism.

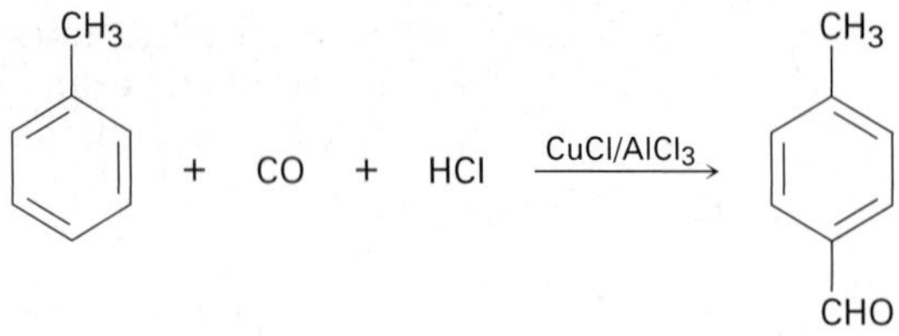

■ Assignable in OWL   ▲ Key Idea Problems

16.67 ■ Treatment of *p-tert*-butylphenol with a strong acid such as H_2SO_4 yields phenol and 2-methylpropene. Propose a mechanism.

16.68 ■ Benzene and alkyl-substituted benzenes can be hydroxylated by reaction with H_2O_2 in the presence of an acidic catalyst. What is the structure of the reactive electrophile? Propose a mechanism for the reaction.

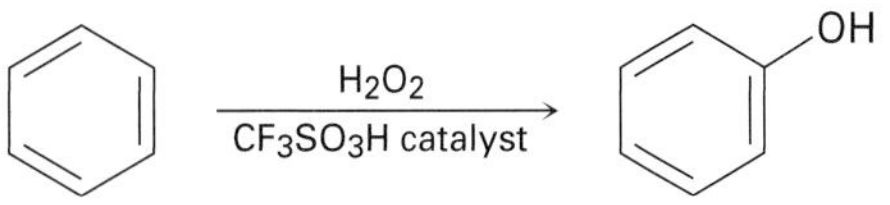

16.69 How would you synthesize the following compounds from benzene? Assume that ortho and para isomers can be separated.

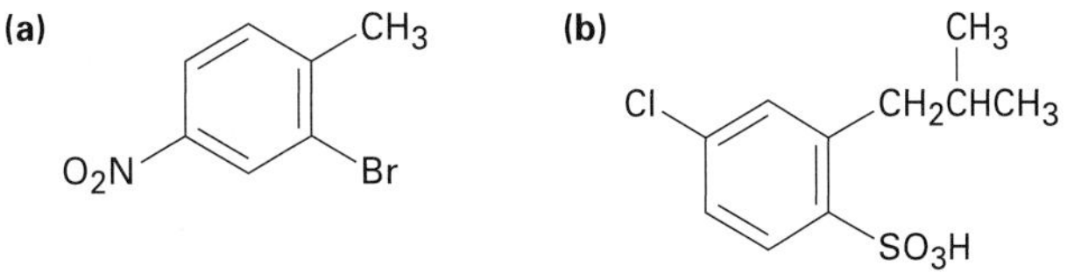

(a)

(b)

16.70 You know the mechanism of HBr addition to alkenes, and you know the effects of various substituent groups on aromatic substitution. Use this knowledge to predict which of the following two alkenes reacts faster with HBr. Explain your answer by drawing resonance structures of the carbocation intermediates.

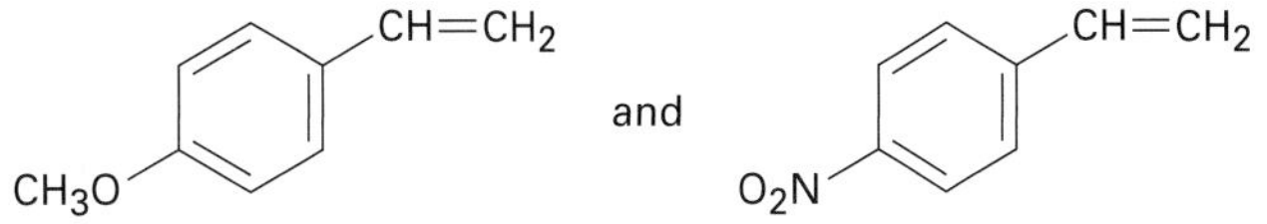

and

16.71 Benzyl bromide is converted into benzaldehyde by heating in dimethyl sulfoxide. Propose a structure for the intermediate, and show the mechanisms of the two steps in the reaction.

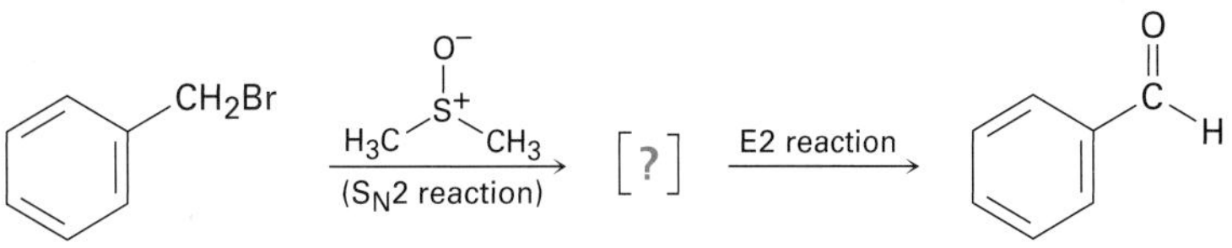

16.72 Use your knowledge of directing effects, along with the following data, to deduce the directions of the dipole moments in aniline and bromobenzene.

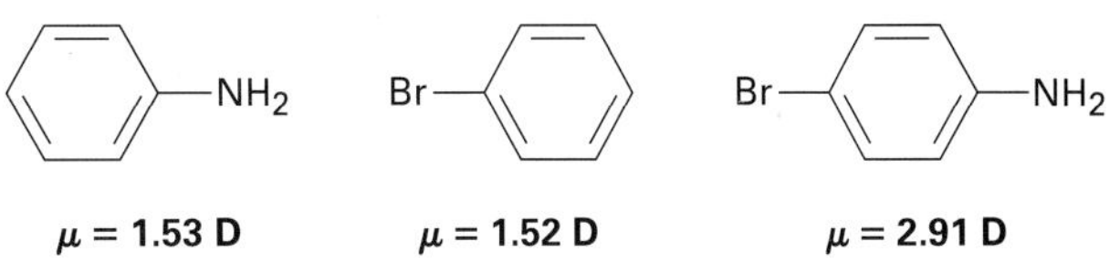

$\mu = 1.53$ D $\mu = 1.52$ D $\mu = 2.91$ D

16.73 Identify the reagents represented by the letters **a–e** in the following scheme:

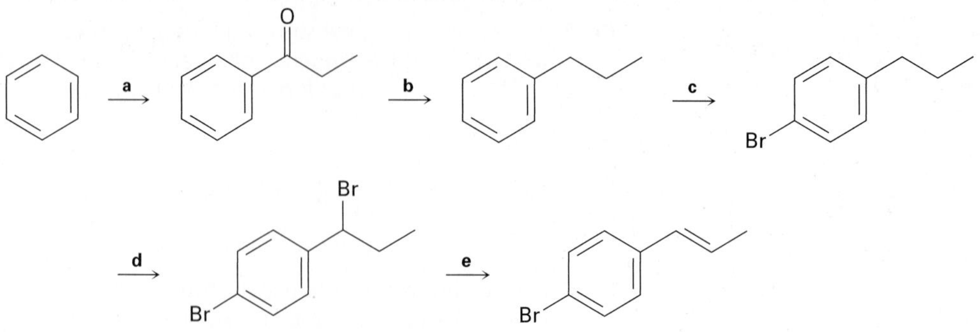

16.74 Phenols (ArOH) are relatively acidic, and the presence of a substituent group on the aromatic ring has a large effect. The pK_a of unsubstituted phenol, for example, is 9.89, while that of p-nitrophenol is 7.15. Draw resonance structures of the corresponding phenoxide anions and explain the data.

16.75 Would you expect p-methylphenol to be more acidic or less acidic than unsubstituted phenol? Explain. (See Problem 16.74.)

17

Alcohols and Phenols

Organic KNOWLEDGE TOOLS

CENGAGENOW™ Throughout this chapter, sign in at **www.cengage.com/login** for online self-study and interactive tutorials based on your level of understanding.

⭐WL Online homework for this chapter may be assigned in Organic OWL.

Alcohols and **phenols** can be thought of as organic derivatives of water in which one of the water hydrogens is replaced by an organic group: $H-O-H$ versus $R-O-H$ and $Ar-O-H$. In practice, the group name *alcohol* is restricted to compounds that have their $-OH$ group bonded to a saturated, sp^3-hybridized carbon atom, while compounds with their $-OH$ group bonded to a vinylic, sp^2-hybridized carbon are called *enols*. We'll look at enols in Chapter 22.

An alcohol   **A phenol**   **An enol**

Alcohols occur widely in nature and have many industrial and pharmaceutical applications. Methanol, for instance, is one of the most important of all industrial chemicals. Historically, methanol was prepared by heating wood in the absence of air and thus came to be called *wood alcohol*. Today, approximately 1.3 billion gallons of methanol is manufactured each year in the United States by catalytic reduction of carbon monoxide with hydrogen gas. Methanol is toxic to humans, causing blindness in small doses (15 mL) and death in larger amounts (100–250 mL). Industrially, it is used both as a solvent and as a starting material for production of formaldehyde (CH_2O) and acetic acid (CH_3CO_2H).

$$CO \ + \ 2\,H_2 \ \xrightarrow[\substack{\text{Zinc oxide/chromia} \\ \text{catalyst}}]{400\,°C} \ CH_2OH$$

Ethanol was one of the first organic chemicals to be prepared and purified. Its production by fermentation of grains and sugars has been carried out for perhaps 9000 years, and its purification by distillation goes back at least as far as the 12th century. Today, approximately 4 billion gallons of ethanol is produced annually in the United States by fermentation of corn, barley, and sorghum, and production is expected to double by 2012. Essentially the entire amount is used to make E85 automobile fuel, a blend of 85% ethanol and 15% gasoline.

Ethanol for nonbeverage use is obtained by acid-catalyzed hydration of ethylene. Approximately 110 million gallons of ethanol a year is produced in the United States for use as a solvent or as a chemical intermediate in other industrial reactions.

$$H_2C{=}CH_2 \xrightarrow[\substack{H_3PO_4 \\ 250\ °C}]{H_2O} CH_3CH_2OH$$

Phenols occur widely throughout nature and also serve as intermediates in the industrial synthesis of products as diverse as adhesives and antiseptics. Phenol itself is a general disinfectant found in coal tar; methyl salicylate is a flavoring agent found in oil of wintergreen; and the urushiols are the allergenic constituents of poison oak and poison ivy. Note that the word *phenol* is the name both of the specific compound hydroxybenzene and of a class of compounds.

Phenol
(also known as
carbolic acid)

Methyl salicylate

Urushiols
(R = different C_{15} alkyl
and alkenyl chains)

WHY THIS CHAPTER?

Up to this point, we've focused on developing some general ideas of organic reactivity, on looking at the chemistry of hydrocarbons, and on seeing some of the tools used in structural studies. With that background, it's now time to begin a study of the oxygen-containing functional groups that lie at the heart of biological chemistry. We'll look at alcohols in this chapter and then move on to carbonyl compounds in Chapters 19 through 23.

17.1 | Naming Alcohols and Phenols

CENGAGENOW™ Click *Organic Interactive* to **use a web-based palette to draw structures of alcohols based on their IUPAC names.**

Alcohols are classified as primary (1°), secondary (2°), or tertiary (3°), depending on the number of organic groups bonded to the hydroxyl-bearing carbon.

A primary (1°) alcohol   A secondary (2°) alcohol   A tertiary (3°) alcohol

Simple alcohols are named by the IUPAC system as derivatives of the parent alkane, using the suffix -*ol*.

Rule 1   Select the longest carbon chain containing the hydroxyl group, and derive the parent name by replacing the -*e* ending of the corresponding alkane with -*ol*. The -*e* is deleted to prevent the occurrence of two adjacent vowels: propanol rather than propaneol, for example.

Rule 2   Number the alkane chain beginning at the end nearer the hydroxyl group.

Rule 3   Number the substituents according to their position on the chain, and write the name listing the substituents in alphabetical order and identifying the position to which the —OH is bonded. Note that in naming *cis*-1,4-cyclohexanediol, the final -*e* of cyclohexane is not deleted because the next letter, *d*, is not a vowel, that is, cyclohexanediol rather than cyclohexandiol. Also, as with alkenes (Section 6.3), newer IUPAC naming recommendations place the locant immediately before the suffix rather than before the parent.

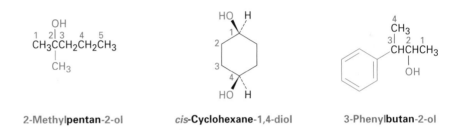

2-Methylpentan-2-ol   *cis*-**Cyclohexane-1,4-diol**   **3-Phenylbutan-2-ol**

Some simple and widely occurring alcohols have common names that are accepted by IUPAC. For example:

Benzyl alcohol   **Allyl alcohol**   ***tert*-Butyl alcohol**   **Ethylene glycol**   **Glycerol**
(phenylmethanol)   **(prop-2-en-1-ol)**   **(2-methylpropan-2-ol)**   **(ethane-1,2-diol)**   **(propane-1,2,3-triol)**

Phenols are named as described previously for aromatic compounds according to the rules discussed in Section 15.1. Note that -*phenol* is used as the parent name rather than -*benzene*.

***m*-Methylphenol**   **2,4-Dinitrophenol**
(*m*-Cresol)

Problem 17.1 | Give IUPAC names for the following compounds:

(a)
$$\underset{\underset{\displaystyle CH_3}{|}}{CH_3CHCH_2CHCHCH_3}$$
OH   OH

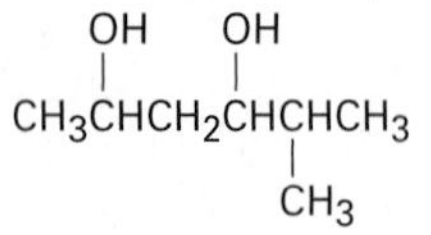

(b)
$$\underset{\underset{\displaystyle CH_3}{|}}{-CH_2CH_2CCH_3}$$
OH

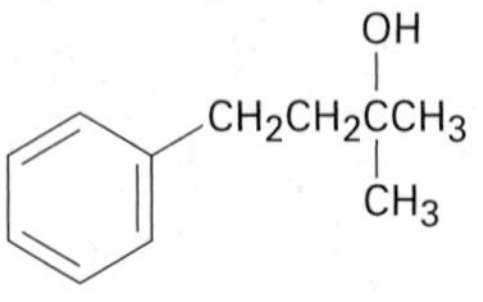

(c) HO—⬡—CH₃, CH₃

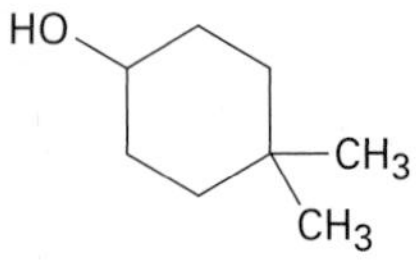

(d) H, Br / H, OH (cyclopentane)

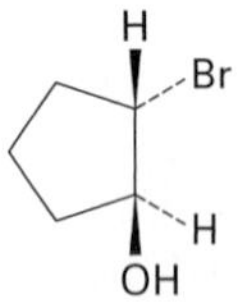

(e) H₃C—⬡—OH, Br

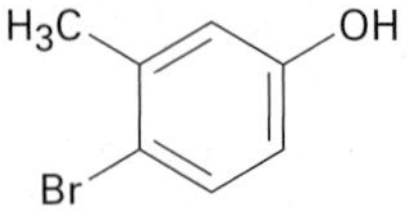

(f) ⬠—OH (cyclopentene)

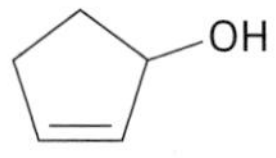

Problem 17.2 | Draw structures corresponding to the following IUPAC names:
(a) (Z)-2-Ethyl-2-buten-1-ol   (b) 3-Cyclohexen-1-ol
(c) *trans*-3-Chlorocycloheptanol   (d) 1,4-Pentanediol
(e) 2,6-Dimethylphenol   (f) *o*-(2-Hydroxyethyl)phenol

17.2 | Properties of Alcohols and Phenols

Alcohols and phenols have nearly the same geometry around the oxygen atom as water. The R−O−H bond angle has an approximately tetrahedral value (109° in methanol, for example), and the oxygen atom is sp^3-hybridized.

Also like water, alcohols and phenols have higher boiling points than might be expected because of hydrogen-bonding (Section 2.13). A positively polarized −OH hydrogen atom from one molecule is attracted to a lone pair of electrons on the electronegative oxygen atom of another molecule, resulting in a weak force that holds the molecules together (Figure 17.1). These inter-molecular attractions must be overcome for a molecule to break free from the liquid and enter the vapor state, so the boiling temperature is raised. For example, 1-propanol (MW = 60), butane (MW = 58), and chloroethane (MW = 65) have similar molecular weights, yet 1-propanol boils at 97 °C, compared with −0.5 °C for the alkane and 12.5 °C for the chloroalkane.

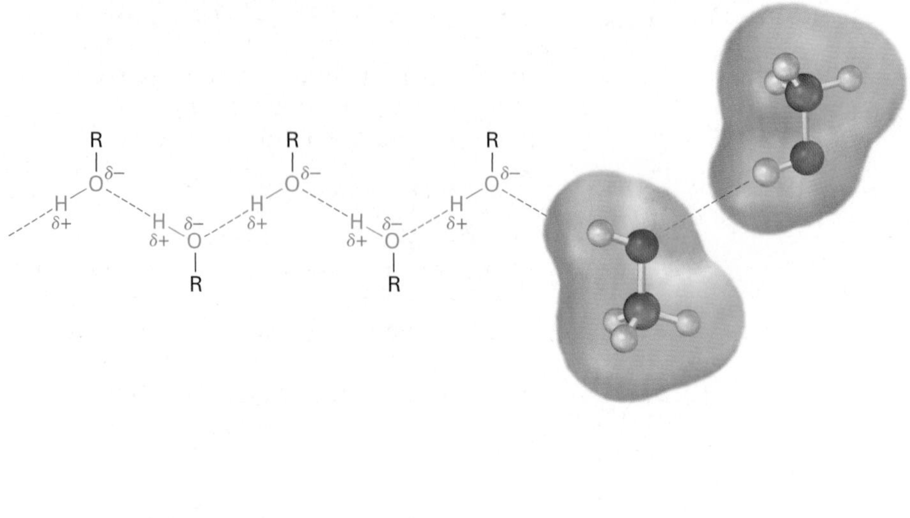

Figure 17.1 Hydrogen-bonding in alcohols and phenols. A weak attraction between a positively polarized OH hydrogen and a negatively polarized oxygen holds molecules together. The electrostatic potential map of methanol shows the positively polarized O−H hydrogen (blue) and the negatively polarized oxygen (red).

Another similarity with water is that alcohols and phenols are both weakly basic and weakly acidic. As weak bases, they are reversibly protonated by strong acids to yield oxonium ions, ROH_2^+.

An alcohol   **An oxonium ion**

$$\left[\text{or}\quad \text{ArOH} \quad + \quad \text{HX} \quad \rightleftharpoons \quad \overset{+}{\text{ArOH}}_2 \ \ \text{X}^-\right]$$

As weak acids, they dissociate slightly in dilute aqueous solution by donating a proton to water, generating H_3O^+ and an **alkoxide ion, RO⁻**, or a **phenoxide ion, ArO⁻**.

An alcohol   **An alkoxide ion**

A phenol   **A phenoxide ion**

Recall from the earlier discussion of acidity in Sections 2.7 through 2.11 that the strength of any acid HA in water can be expressed by an acidity constant, K_a.

$$K_a = \frac{[\text{A}^-]\,[\text{H}_3\text{O}^+]}{[\text{HA}]} \qquad pK_a = -\log K_a$$

Compounds with a smaller K_a and larger pK_a are less acidic, whereas compounds with a larger K_a and smaller pK_a are more acidic. As shown by the data in Table 17.1, simple alcohols like methanol and ethanol are about as acidic as water but substituent groups can have a significant effect. *tert*-Butyl alcohol is a weaker acid, for instance, and 2,2,2-trifluoroethanol is stronger. Phenols and *thiols,* the sulfur analogs of alcohols, are substantially more acidic than water.

The effect of alkyl substitution on alcohol acidity is due primarily to solvation of the alkoxide ion that results from dissociation. The more readily the alkoxide ion is solvated by water, the more stable it is, the more its formation is energetically favored, and the greater the acidity of the parent alcohol. For example, the oxygen atom of an unhindered alkoxide ion, such as that from methanol, is sterically accessible and is easily solvated by water. The oxygen

Table 17.1 | **Acidity Constants of Some Alcohols and Phenols**

Compound	pK_a	
$(CH_3)_3COH$	18.00	Weaker acid
CH_3CH_2OH	16.00	
H_2O	15.74	
CH_3OH	15.54	
CF_3CH_2OH	12.43	
p-Aminophenol	10.46	
CH_3SH	10.3	
p-Methylphenol	10.17	
Phenol	9.89	
p-Chlorophenol	9.38	
p-Nitrophenol	7.15	Stronger acid

atom of a hindered alkoxide ion, however, such as that from *tert*-butyl alcohol, is less easily solvated and is therefore less stabilized.

Sterically accessible; less hindered and more easily solvated.

Sterically less accessible; more hindered and less easily solvated.

Methoxide ion, CH₃O⁻
(p*K*ₐ = 15.54)

***tert*-Butoxide ion, (CH₃)₃CO⁻**
(p*K*ₐ = 18.00)

   Inductive effects (Section 16.4) are also important in determining alcohol acidities. Electron-withdrawing halogen substituents, for example, stabilize an alkoxide ion by spreading out the charge over a larger volume, thus making the alcohol more acidic. Compare, for example, the acidities of ethanol (pK_a = 16.00) and 2,2,2-trifluoroethanol (pK_a = 12.43), or of *tert*-butyl alcohol (pK_a = 18.0) and nonafluoro-*tert*-butyl alcohol (pK_a = 5.4).

Electron-withdrawing groups stabilize the alkoxide ion and lower the *p*Ka.

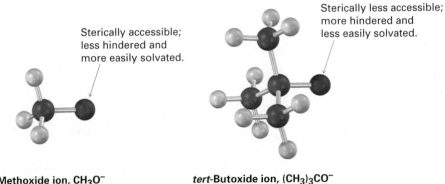

pKa = 5.4   versus   pKa = 18.0

Because alcohols are weak acids, they don't react with weak bases such as amines or bicarbonate ion, and they react to only a limited extent with metal hydroxides such as NaOH. Alcohols do, however, react with alkali metals and with strong bases such as sodium hydride (NaH), sodium amide ($NaNH_2$), and Grignard reagents (RMgX). Alkoxides are themselves bases that are frequently used as reagents in organic chemistry. They are named systematically by adding the *-ate* suffix to the name of the alcohol. Methanol becomes methanolate, for instance.

tert-Butyl alcohol
(2-methylpropan-2-ol)

Potassium *tert*-butoxide
(potassium 2-methyl-propan-2-olate)

$$CH_3OH \; + \; NaH \longrightarrow CH_3O^- \; Na^+ \; + \; H_2$$

Methanol

Sodium methoxide
(sodium methanolate)

$$CH_3CH_2OH \; + \; NaNH_2 \longrightarrow CH_3CH_2O^- \; Na^+ \; + \; NH_3$$

Ethanol

Sodium ethoxide
(sodium ethanolate)

Cyclohexanol

Bromomagnesium cyclohexanolate

Phenols are about a million times more acidic than alcohols (Table 17.1). They are therefore soluble in dilute aqueous NaOH and can often be separated from a mixture simply by basic extraction into aqueous solution, followed by reacidification.

Phenol

Sodium phenoxide
(sodium phenolate)

Phenols are more acidic than alcohols because the phenoxide anion is resonance-stabilized. Delocalization of the negative charge over the ortho and para positions of the aromatic ring results in increased stability of the phenoxide anion relative to undissociated phenol and in a consequently lower $\Delta G°$ for dissociation. Figure 17.2 compares electrostatic potential maps of an alkoxide ion (CH_3O^-) with phenoxide ion and shows how the negative charge in phenoxide ion is delocalized from oxygen to the ring.

Figure 17.2 The resonance-stabilized phenoxide ion is more stable than an alkoxide ion. Electrostatic potential maps show how the negative charge is concentrated on oxygen in the methoxide ion but is spread over the aromatic ring in the phenoxide ion.

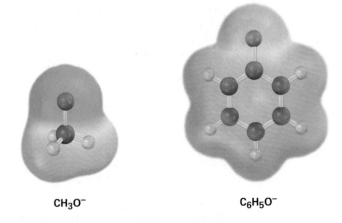

CH₃O⁻   C₆H₅O⁻

Substituted phenols can be either more acidic or less acidic than phenol itself, depending on whether the substituent is electron-withdrawing or electron-donating (Section 16.4). Phenols with an electron-withdrawing substituent are more acidic because these substituents delocalize the negative charge; phenols with an electron-donating substituent are less acidic because these substituents concentrate the charge. The acidifying effect of an electron-withdrawing substituent is particularly noticeable in phenols with a nitro group at the ortho or para position.

| WORKED EXAMPLE 17.1 | *Predicting the Relative Acidity of a Substituted Phenol* |

Is *p*-hydroxybenzaldehyde more acidic or less acidic than phenol?

Strategy   Identify the substituent on the aromatic ring, and decide whether it is electron-donating or electron-withdrawing. Electron-withdrawing substituents make the phenol more acidic by stabilizing the phenoxide anion, and electron-donating substituents make the phenol less acidic by destabilizing the anion.

Solution   We saw in Section 16.4 that a carbonyl group is electron-withdrawing. Thus, *p*-hydroxybenzaldehyde is more acidic (pK_a = 7.9) than phenol (pK_a = 9.89).

p-Hydroxybenzaldehyde
(pK_a = 7.9)

Problem 17.3 | The following data for isomeric four-carbon alcohols show that there is a decrease in boiling point with increasing substitution of the OH-bearing carbon. How might you account for this trend?

1-Butanol, bp 117.5 °C

2-Butanol, bp 99.5 °C

2-Methyl-2-propanol, bp 82.2 °C

Problem 17.4 | Rank the following substances in order of increasing acidity:
(a) $(CH_3)_2CHOH$, $HC \equiv CH$, $(CF_3)_2CHOH$, CH_3OH
(b) Phenol, *p*-methylphenol, *p*-(trifluoromethyl)phenol
(c) Benzyl alcohol, phenol, *p*-hydroxybenzoic acid

Problem 17.5 | *p*-Nitrobenzyl alcohol is more acidic than benzyl alcohol but *p*-methoxybenzyl alcohol is less acidic. Explain.

17.3 | Preparation of Alcohols: A Review

Alcohols occupy a central position in organic chemistry. They can be prepared from many other kinds of compounds (alkenes, alkyl halides, ketones, esters, and aldehydes, among others), and they can be transformed into an equally wide assortment of compounds (Figure 17.3).

Figure 17.3 The central position of alcohols in organic chemistry. Alcohols can be prepared from, and converted into, many other kinds of compounds.

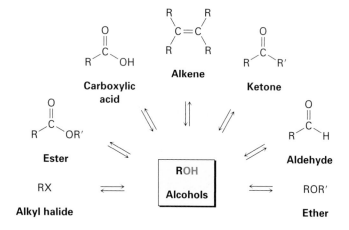

We've already seen several methods of alcohol synthesis:

▌ Alcohols can be prepared by hydration of alkenes. Because the direct hydration of alkenes with aqueous acid is generally a poor reaction in the laboratory, two indirect methods are commonly used. Hydroboration/oxidation yields the product of syn, non-Markovnikov hydration (Section 7.5), whereas oxymercuration/reduction yields the product of Markovnikov hydration (Section 7.4).

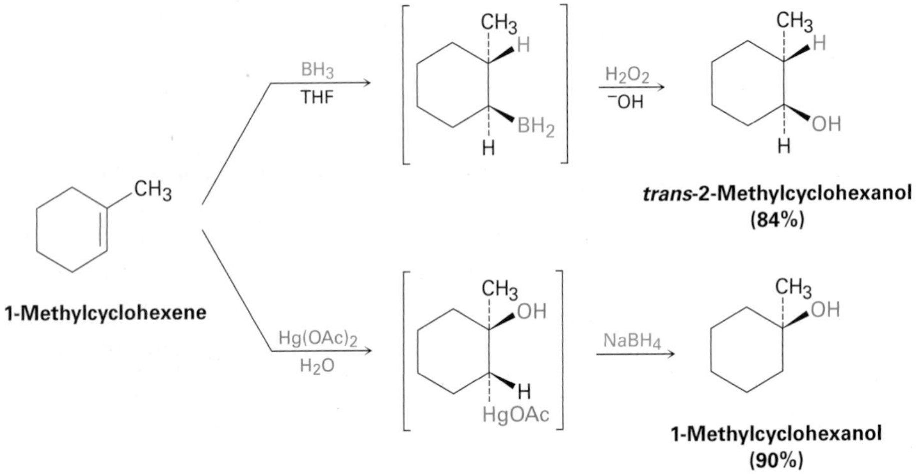

trans-2-Methylcyclohexanol (84%)

1-Methylcyclohexene

1-Methylcyclohexanol (90%)

▌ 1,2-Diols can be prepared either by direct hydroxylation of an alkene with OsO$_4$ followed by reduction with NaHSO$_3$ or by acid-catalyzed hydrolysis of an epoxide (Section 7.8). The OsO$_4$ reaction occurs with syn stereochemistry to give a cis diol, and epoxide opening occurs with anti stereochemistry to give a trans diol.

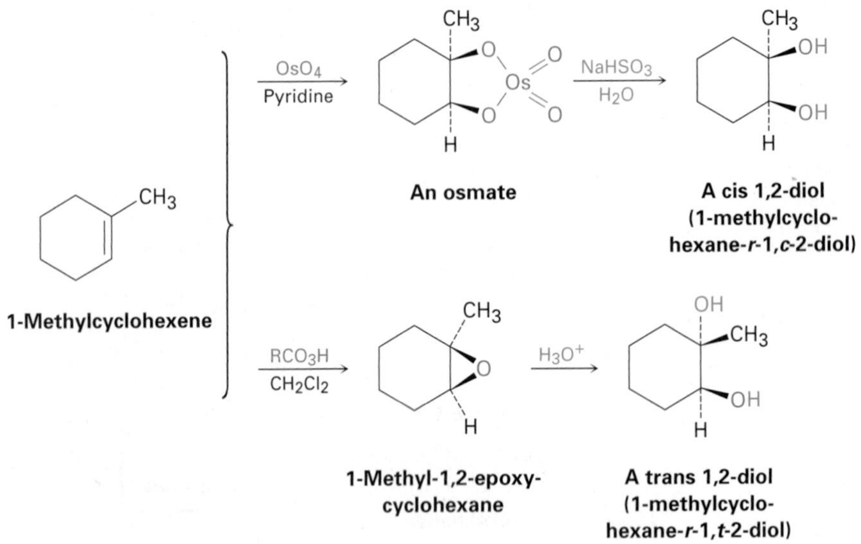

An osmate

A cis 1,2-diol (1-methylcyclo- hexane-r-1,c-2-diol)

1-Methylcyclohexene

1-Methyl-1,2-epoxy- cyclohexane

A trans 1,2-diol (1-methylcyclo- hexane-r-1,t-2-diol)

As noted at the end of Section 7.8, the prefixes *cis-* and *trans-* would be ambiguous when naming the diols derived from 1-methylcyclohexene because the ring has three substituents. Instead, a reference substituent *r* is chosen and other substituents are either cis (*c*) or trans (*t*) to that reference. For the two 1-methyl-1,2-cyclohexanediol isomers, the −OH group at C1 is the reference (*r*-1), and the −OH at C2 is either cis (*c*-2) or trans (*t*-2) to that reference. Thus, the diol isomer derived by cis hydroxylation is named 1-methyl-*r*-1,*c*-2-cyclohexanediol, and the isomer derived by trans hydroxylation is named 1-methyl-*r*-1,*t*-2-cyclohexanediol.

Problem 17.6 | Predict the products of the following reactions:

(a)

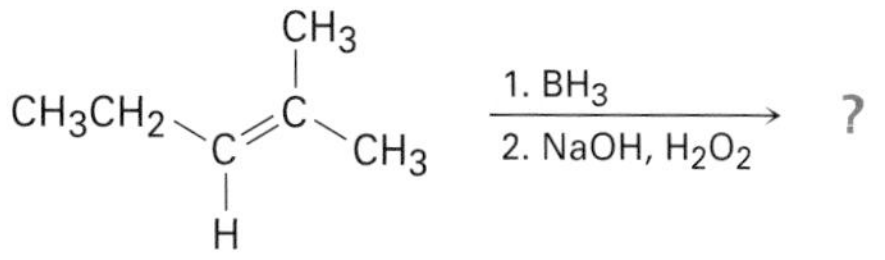

$$\underset{\text{CH}_3\text{CH}_2}{}\overset{\text{CH}_3}{\underset{\text{H}}{\text{C}=\text{C}}}\overset{}{\underset{}{\text{CH}_3}} \quad \xrightarrow[\text{2. NaOH, H}_2\text{O}_2]{\text{1. BH}_3} \quad ?$$

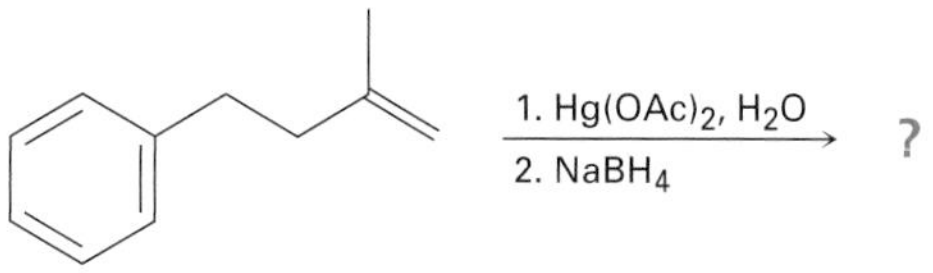

(b)

$$\xrightarrow[\text{2. NaBH}_4]{\text{1. Hg(OAc)}_2, \text{H}_2\text{O}} \quad ?$$

(c) $\text{CH}_3\text{CH}_2\text{CH}_2\text{CH}_2 \quad \text{CH}_2\text{CH}_2\text{CH}_2\text{CH}_3$

$$\underset{\text{H}}{}\overset{}{\text{C}=\text{C}}\underset{\text{H}}{} \quad \xrightarrow[\text{2. NaHSO}_3, \text{H}_2\text{O}]{\text{1. OsO}_4} \quad ?$$

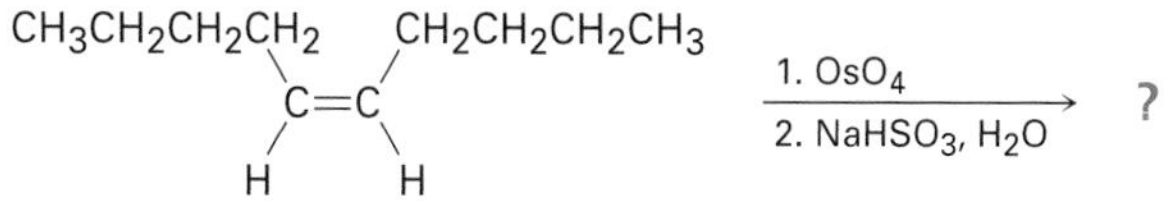

17.4 | Alcohols from Reduction of Carbonyl Compounds

The most general method for preparing alcohols, both in the laboratory and in living organisms, is by the reduction of a carbonyl compound. Just as reduction of an alkene adds hydrogen to a C=C bond to give an alkane (Section 7.7), reduction of a carbonyl compound adds hydrogen to a C=O bond to give an alcohol. All kinds of carbonyl compounds can be reduced, including aldehydes, ketones, carboxylic acids, and esters.

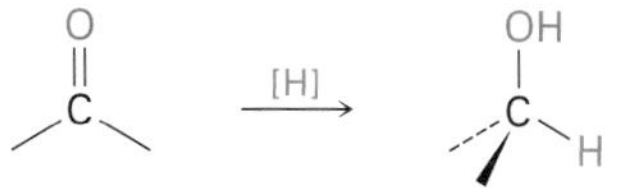

where [H] is a reducing agent

A carbonyl compound   **An alcohol**

Reduction of Aldehydes and Ketones

Aldehydes are easily reduced to give primary alcohols, and ketones are reduced to give secondary alcohols.

An aldehyde   **A primary alcohol**   **A ketone**   **A secondary alcohol**

Literally dozens of reagents are used in the laboratory to reduce aldehydes and ketones, depending on the circumstances, but sodium borohydride, NaBH_4, is usually chosen because of its safety and ease of handling. Sodium borohydride is a white, crystalline solid that can be weighed in the open atmosphere and used in either water or alcohol solution to give high yields of products.

Aldehyde reduction

$$CH_3CH_2CH_2\overset{O}{\underset{}{\overset{\|}{C}}}H \quad \xrightarrow[\text{2. } H_3O^+]{\text{1. } NaBH_4, \text{ ethanol}} \quad CH_3CH_2CH_2\overset{OH}{\underset{H}{\overset{|}{C}}}H$$

Butanal   **Butan-1-ol (85%)**
                               (a 1° alcohol)

Ketone reduction

                    $$\xrightarrow[\text{2. } H_3O^+]{\text{1. } NaBH_4, \text{ ethanol}}$$

Dicyclohexyl ketone   **Dicyclohexylmethanol (88%)**
                                   (a 2° alcohol)

Lithium aluminum hydride, $LiAlH_4$, is another reducing agent often used for reduction of aldehydes and ketones. A grayish powder that is soluble in ether and tetrahydrofuran, $LiAlH_4$ is much more reactive than $NaBH_4$ but also more dangerous. It reacts violently with water and decomposes explosively when heated above 120 °C.

                    $$\xrightarrow[\text{2. } H_3O^+]{\text{1. } LiAlH_4, \text{ ether}}$$

Cyclohex-2-enone   **Cyclohex-2-enol (94%)**

We'll defer a detailed discussion of the mechanisms of these reductions until Chapter 19. For the moment, we'll simply note that they involve the addition of a nucleophilic hydride ion (:H⁻) to the positively polarized, electrophilic carbon atom of the carbonyl group. The initial product is an alkoxide ion, which is protonated by addition of H_3O^+ in a second step to yield the alcohol product.

A carbonyl   **An alkoxide ion**   **An alcohol**
compound   **intermediate**

In living organisms, aldehyde and ketone reductions are carried out by either of the coenzymes NADH (reduced nicotinamide adenine dinucleotide) or NADPH (reduced nicotinamide adenine dinucleotide phosphate). Although these biological "reagents" are much more complex structurally than $NaBH_4$ or $LiAlH_4$, the mechanisms of laboratory and biological reactions are similar. The coenzyme acts as a hydride-ion donor, and the intermediate anion is then protonated by acid. An example is the reduction of acetoacetyl ACP to β-hydroxybutyryl ACP, a step in the biological synthesis of fats (Figure 17.4). Note that the *pro-R* hydrogen of NADPH is the one transferred in this example. Enzyme-catalyzed reactions usually occur with high specificity, although it's not usually possible to predict the stereochemical result before the fact.

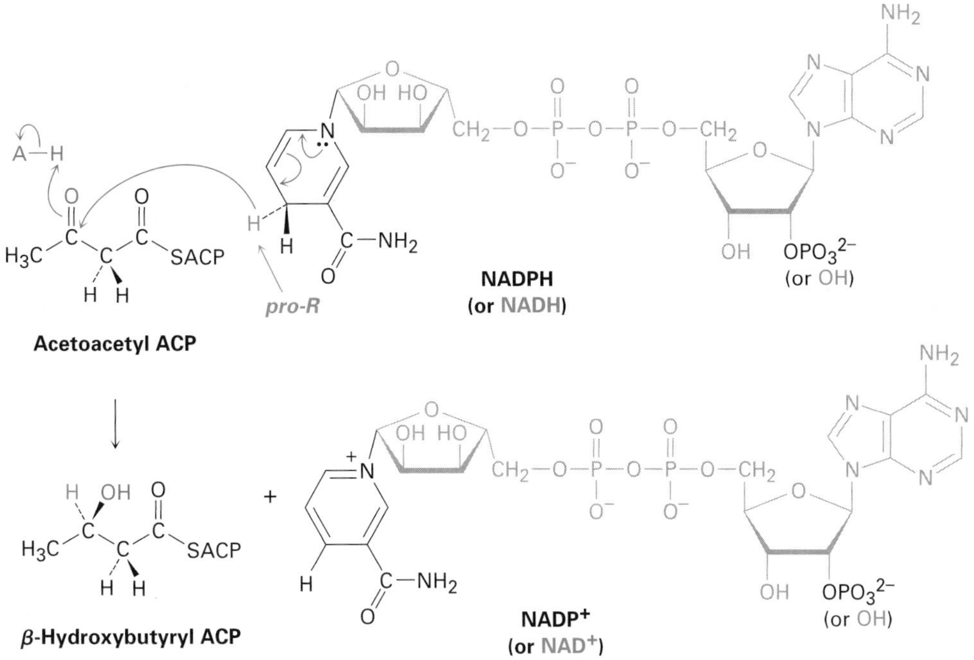

Figure 17.4 The biological reduction of a ketone (acetoacetyl ACP) to an alcohol (β-hydroxybutyryl ACP) by NADPH.

Reduction of Carboxylic Acids and Esters

Carboxylic acids and esters are reduced to give primary alcohols.

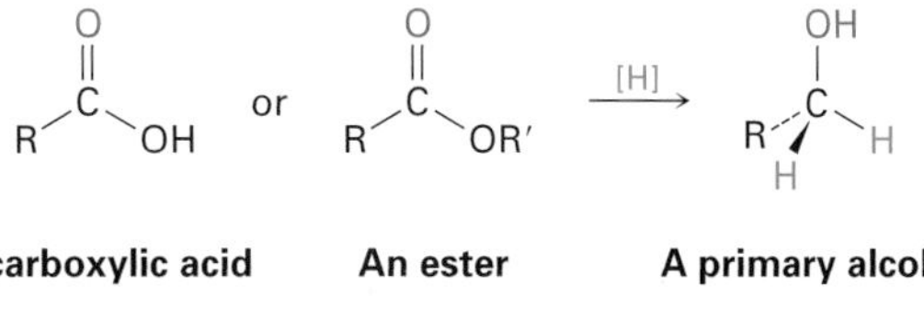

These reactions aren't as rapid as the reductions of aldehydes and ketones. $NaBH_4$ reduces esters very slowly and does not reduce carboxylic acids at all. Instead, carboxylic acid and ester reductions are usually carried out with the more reactive reducing agent $LiAlH_4$. All carbonyl groups, including acids, esters, ketones, and aldehydes, are reduced by $LiAlH_4$. Note that one hydrogen atom is delivered to the carbonyl carbon atom during aldehyde and ketone reductions but that two hydrogens become bonded to the former carbonyl carbon during carboxylic acid and ester reductions. We'll defer a discussion of the mechanisms of these reactions until Chapter 21.

Carboxylic acid reduction

$$CH_3(CH_2)_7CH=CH(CH_2)_7\overset{\overset{\displaystyle O}{\|}}{C}OH \xrightarrow[\text{2. H}_3O^+]{\text{1. LiAlH}_4, \text{ ether}} CH_3(CH_2)_7CH=CH(CH_2)_7CH_2OH$$

Octadec-9-enoic acid   Octadec-9-en-1-ol (87%)
(oleic acid)

Ester reduction

$$CH_3CH_2CH=CH\overset{\overset{\displaystyle O}{\|}}{C}OCH_3 \xrightarrow[\text{2. H}_3O^+]{\text{1. LiAlH}_4, \text{ ether}} CH_3CH_2CH=CHCH_2OH \ + \ CH_3OH$$

Methyl pent-2-enoate   Pent-2-en-1-ol (91%)

WORKED EXAMPLE 17.2	*Predicting the Structure of a Reactant, Given the Product*

What carbonyl compounds would you reduce to obtain the following alcohols?

(a)
$$\overset{\qquad CH_3 \quad OH}{\underset{CH_3CH_2CHCH_2CHCH_3}{\quad | \qquad |}}$$

(b)

Strategy   Identify the target alcohol as primary, secondary, or tertiary. A primary alcohol can be prepared by reduction of an aldehyde, an ester, or a carboxylic acid; a secondary alcohol can be prepared by reduction of a ketone; and a tertiary alcohol can't be prepared by reduction.

Solution   (a) The target molecule is a secondary alcohol, which can be prepared only by reduction of a ketone. Either NaBH₄ or LiAlH₄ can be used.

$$\overset{\quad CH_3 \quad O}{\underset{CH_3CH_2CHCH_2CCH_3}{\quad | \qquad \|}} \xrightarrow[\text{2. H}_3O^+]{\text{1. NaBH}_4 \text{ or LiAlH}_4} \overset{\quad CH_3 \quad OH}{\underset{CH_3CH_2CHCH_2CHCH_3}{\quad | \qquad |}}$$

(b) The target molecule is a primary alcohol, which can be prepared by reduction of an aldehyde, an ester, or a carboxylic acid. LiAlH₄ is needed for the ester and carboxylic acid reductions.

Problem 17.7 What reagent would you use to accomplish each of the following reactions?

(a)

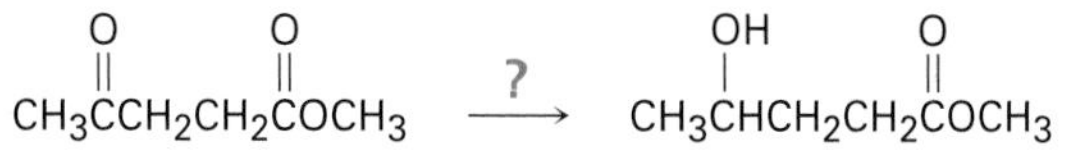

CH3CCH2CH2COCH3
?
CH3CHCH2CH2COCH3

(b)

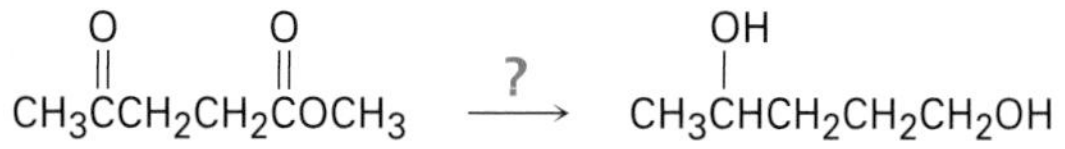

CH3CCH2CH2COCH3
?
CH3CHCH2CH2CH2OH

(c)

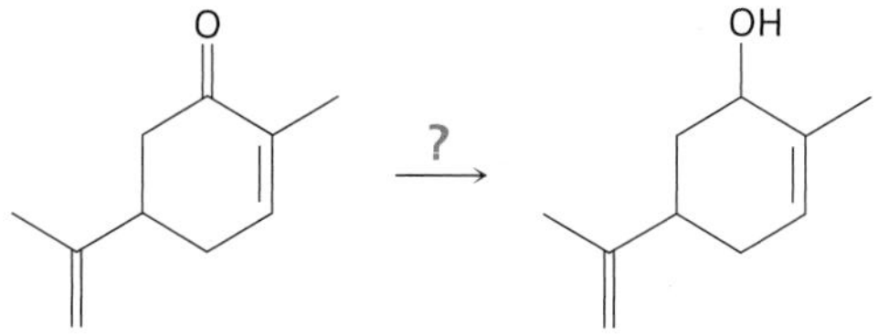

?
OH

Problem 17.8 What carbonyl compounds give the following alcohols on reduction with $LiAlH_4$? Show all possibilities.

(a)

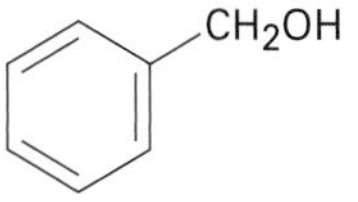

CH2OH

(b)

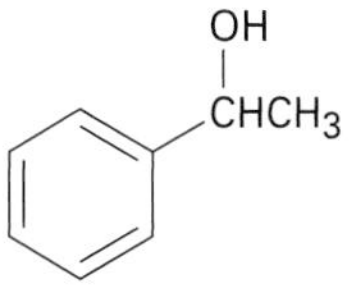

OH
CHCH3

(c)

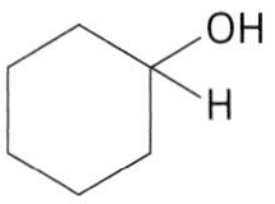

OH
H

(d) $(CH_3)_2CHCH_2OH$

17.5 Alcohols from Reaction of Carbonyl Compounds with Grignard Reagents

CENGAGENOW™ Click *Organic Interactive* to **find supplemental problems and stepwise solutions to the design of Grignard syntheses.**

We saw in Section 10.7 that alkyl, aryl, and vinylic halides react with magnesium in ether or tetrahydrofuran to generate Grignard reagents, RMgX, which act as carbon-based nucleophiles. These Grignard reagents react with carbonyl compounds to yield alcohols in much the same way that hydride reducing agents do.

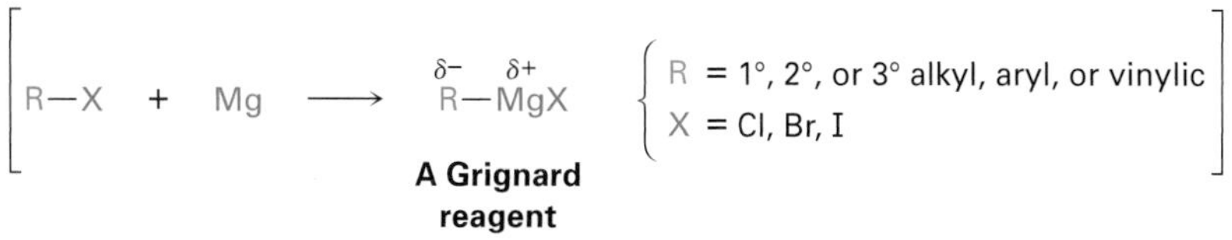

R—X + Mg
δ− δ+
R—MgX
A Grignard reagent
R = 1°, 2°, or 3° alkyl, aryl, or vinylic
X = Cl, Br, I

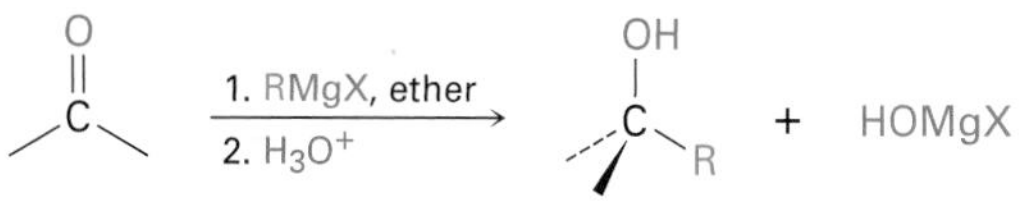

1. RMgX, ether
2. H3O+
OH
R
+ HOMgX

The reaction of Grignard reagents with carbonyl compounds has no direct biological counterpart, because organomagnesium compounds are too strongly basic to exist in an aqueous medium. The reaction *does* have an indirect biological counterpart, however, for we'll see in Chapter 23 that the addition of stabilized carbon nucleophiles to carbonyl compounds is used in almost all metabolic pathways as the major process for forming carbon–carbon bonds.

As examples of their addition to carbonyl compounds, Grignard reagents react with formaldehyde, $H_2C=O$, to give primary alcohols, with aldehydes to give secondary alcohols, and with ketones to give tertiary alcohols.

Formaldehyde reaction

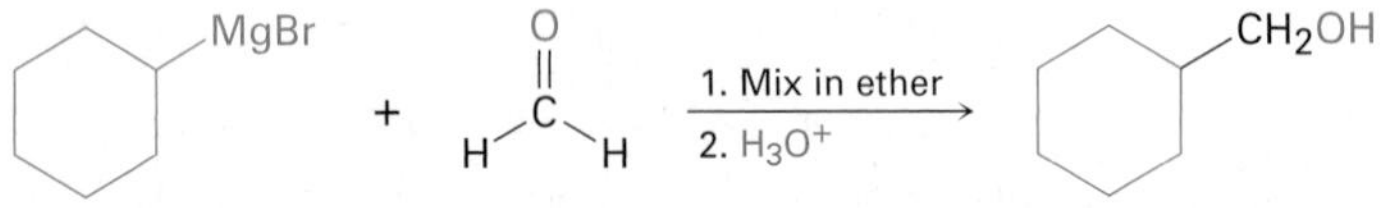

| Cyclohexyl-magnesium bromide | Formaldehyde | Cyclohexylmethanol (65%) (a 1° alcohol) |

Aldehyde reaction

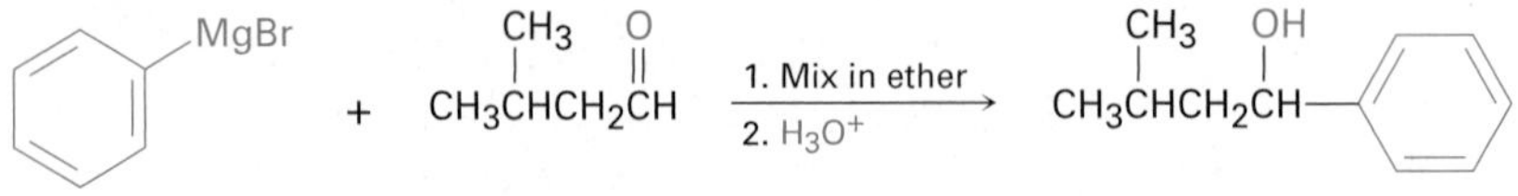

| Phenylmagnesium bromide | 3-Methylbutanal | 3-Methyl-1-phenyl-butan-1-ol (73%) (a 2° alcohol) |

Ketone reaction

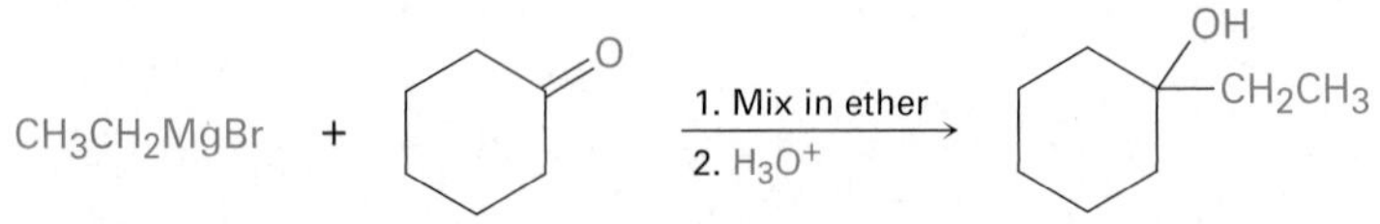

| Ethylmagnesium bromide | Cyclohexanone | 1-Ethylcyclohexanol (89%) (a 3° alcohol) |

Esters react with Grignard reagents to yield tertiary alcohols in which two of the substituents bonded to the hydroxyl-bearing carbon have come from the Grignard reagent, just as $LiAlH_4$ reduction of an ester adds two hydrogens.

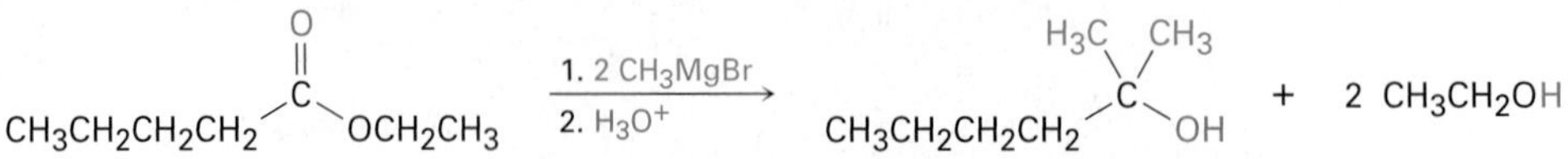

| Ethyl pentanoate | | 2-Methylhexan-2-ol (85%) (a 2° alcohol) |

Carboxylic acids don't give addition products with Grignard reagents because the acidic carboxyl hydrogen reacts with the basic Grignard reagent to yield a hydrocarbon and the magnesium salt of the acid. We saw this reaction in Section 10.7 as a means of reducing an alkyl halide to an alkane.

$$\left[R-X \quad + \quad Mg \quad \longrightarrow \quad R-MgX \right]$$

| **A carboxylic acid** | **A carboxylic acid salt** |

The Grignard reaction, although useful, also has limitations. One major problem is that a Grignard reagent can't be prepared from an organohalide if other reactive functional groups are in the same molecule. For example, a compound that is both an alkyl halide and a ketone can't form a Grignard reagent because it would react with itself. Similarly, a compound that is both an alkyl halide and a carboxylic acid, an alcohol, or an amine can't form a Grignard reagent because the acidic RCO_2H, ROH, or RNH_2 hydrogen present in the same molecule would react with the basic Grignard reagent as rapidly as it forms. In general, Grignard reagents can't be prepared from alkyl halides that contain the following functional groups (FG):

Br — Molecule — FG

where FG = —OH, —NH, —SH, —CO₂H   The Grignard reagent is protonated by these groups.

$$FG = \begin{matrix} O \\ \| \\ -CH, \end{matrix} \quad \begin{matrix} O \\ \| \\ -CR, \end{matrix} \quad \begin{matrix} O \\ \| \\ -CNR_2 \end{matrix}$$

—C≡N, —NO₂, —SO₂R   The Grignard reagent adds to these groups.

As with the reduction of carbonyl compounds discussed in the previous section, we'll defer a detailed treatment of the mechanism of Grignard reactions until Chapter 19. For the moment, it's sufficient to note that Grignard reagents act as nucleophilic carbon anions, or *carbanions* ($:R^-$), and that the addition of a Grignard reagent to a carbonyl compound is analogous to the addition of hydride ion. The intermediate is an alkoxide ion, which is protonated by addition of H_3O^+ in a second step.

| **A carbonyl compound** | **An alkoxide ion intermediate** | **An alcohol** |

| WORKED EXAMPLE 17.3 | *Using a Grignard Reaction to Synthesize an Alcohol* |

How could you use the addition of a Grignard reagent to a ketone to synthesize 2-phenyl-2-butanol?

Strategy Draw the product, and identify the three groups bonded to the alcohol carbon atom. One of the three will have come from the Grignard reagent, and the remaining two will have come from the ketone.

Solution 2-Phenyl-2-butanol has a methyl group, an ethyl group, and a phenyl group ($-C_6H_5$) attached to the alcohol carbon atom. Thus, the possibilities are addition of ethylmagnesium bromide to acetophenone, addition of methylmagnesium bromide to propiophenone, and addition of phenylmagnesium bromide to 2-butanone.

Acetophenone

1. CH_3CH_2MgBr
2. H_3O^+

Propiophenone

1. CH_3MgBr
2. H_3O^+

2-Phenylbutan-2-ol

Butan-2-one

1. C_6H_5MgBr
2. H_3O^+

| WORKED EXAMPLE 17.4 | *Using a Grignard Reaction to Synthesize an Alcohol* |

How could you use the reaction of a Grignard reagent with a carbonyl compound to synthesize 2-methyl-2-pentanol?

Strategy Draw the product, and identify the three groups bonded to the alcohol carbon atom. If the three groups are all different, the starting carbonyl compound must be a ketone. If two of the three groups are identical, the starting carbonyl compound might be either a ketone or an ester.

Solution In the present instance, the product is a tertiary alcohol with two methyl groups and one propyl group. Starting from a ketone, the possibilities are addition of methylmagnesium bromide to 2-pentanone and addition of propylmagnesium bromide to acetone.

$$CH_3CH_2CH_2 \overset{\overset{\displaystyle O}{\|}}{C} CH_3$$

2-Pentanone

1. CH_3MgBr
2. H_3O^+

$$CH_3CH_2CH_2 \overset{\overset{\displaystyle H_3C \quad CH_3}{|}}{\underset{OH}{C}}$$

2-Methyl-2-pentanol

$$H_3C \overset{\overset{\displaystyle O}{\|}}{C} CH_3$$

Acetone

1. $CH_3CH_2CH_2MgBr$
2. H_3O^+

Starting from an ester, the only possibility is addition of methylmagnesium bromide to an ester of butanoic acid, such as methyl butanoate.

$$CH_3CH_2CH_2 \overset{\overset{\displaystyle O}{\|}}{C} OCH_3$$

Methyl butanoate

1. $2\ CH_3MgBr$
2. H_3O^+

$$CH_3CH_2CH_2 \overset{\overset{\displaystyle H_3C \quad CH_3}{|}}{\underset{OH}{C}} \quad + \quad CH_3OH$$

2-Methyl-2-pentanol

Problem 17.9 | Show the products obtained from addition of methylmagnesium bromide to the following compounds:
(a) Cyclopentanone (b) Benzophenone (diphenyl ketone)
(c) 3-Hexanone

Problem 17.10 | Use a Grignard reaction to prepare the following alcohols:
(a) 2-Methyl-2-propanol (b) 1-Methylcyclohexanol (c) 3-Methyl-3-pentanol
(d) 2-Phenyl-2-butanol (e) Benzyl alcohol (f) 4-Methyl-1-pentanol

Problem 17.11 | Use the reaction of a Grignard reagent with a carbonyl compound to synthesize the following compound:

17.6 | Reactions of Alcohols

CENGAGENOW™ Click *Organic Interactive* to **use a web-based palette to predict products from a variety of reactions involving alcohols**.

We've already seen several reactions of alcohols—their conversion into alkyl halides and tosylates in Section 10.6 and their dehydration to give alkenes in Section 7.1—although without mechanistic details. Let's now look at those details.

Conversion of Alcohols into Alkyl Halides

Tertiary alcohols react with either HCl or HBr at 0 °C by an S_N1 mechanism through a carbocation intermediate. Primary and secondary alcohols are much more resistant to acid, however, and are best converted into halides by treatment with either $SOCl_2$ or PBr_3 through an S_N2 mechanism.

The reaction of a tertiary alcohol with HX takes place by an S_N1 mechanism when acid protonates the hydroxyl oxygen atom, water is expelled to generate a carbocation, and the cation reacts with nucleophilic halide ion to give the alkyl halide product.

The reactions of primary and secondary alcohols with $SOCl_2$ and PBr_3 take place by S_N2 mechanisms. Hydroxide ion itself is too poor a leaving group to be displaced by nucleophiles in S_N2 reactions, but reaction of an alcohol with $SOCl_2$ or PBr_3 converts the −OH into a much better leaving group, either a chlorosulfite (−OSOCl) or a dibromophosphite (−OPBr₂), that is readily expelled by backside nucleophilic substitution.

Conversion of Alcohols into Tosylates

Alcohols react with *p*-toluenesulfonyl chloride (tosyl chloride, *p*-TosCl) in pyridine solution to yield alkyl tosylates, ROTos (Section 11.1). Only the O−H bond of the alcohol is broken in this reaction; the C−O bond remains intact, so no change of configuration occurs if the oxygen is attached to a chirality center. The resultant alkyl tosylates behave much like alkyl halides, undergoing both S_N1 and S_N2 substitution reactions.

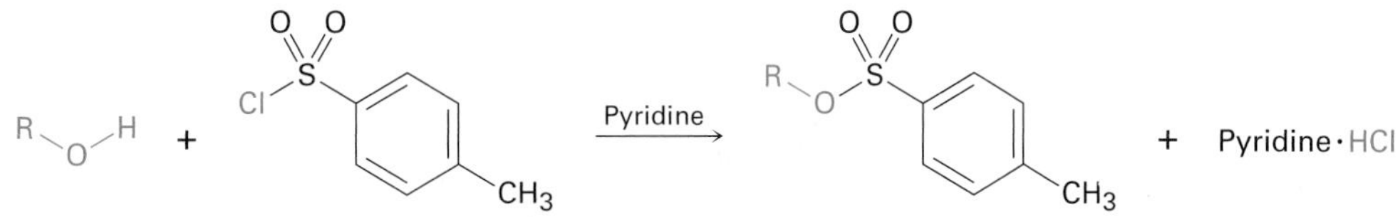

An alcohol   *p*-Toluenesulfonyl chloride   A tosylate (ROTos)

One of the most important reasons for using tosylates in S_N2 reactions is stereochemical. The S_N2 reaction of an alcohol via an alkyl halide proceeds with *two* inversions of configuration—one to make the halide from the alcohol and one to substitute the halide—and yields a product with the same stereochemistry as the starting alcohol. The S_N2 reaction of an alcohol via a tosylate, however, proceeds with only *one* inversion and yields a product of opposite stereochemistry to the starting alcohol. Figure 17.5 shows a series of reactions on the *R* enantiomer of 2-octanol that illustrates these stereochemical relationships.

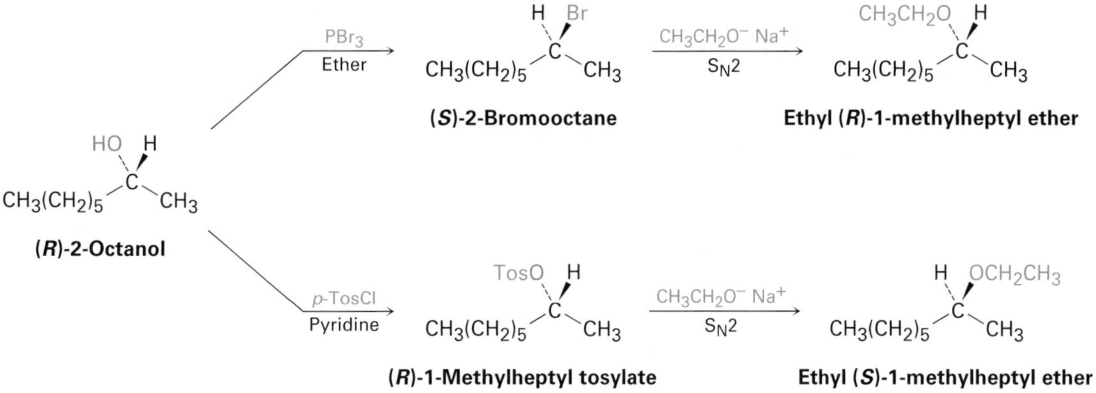

Active Figure 17.5 Stereochemical consequences of S_N2 reactions on derivatives of (*R*)-2-octanol. Substitution through the halide gives a product with the same stereochemistry as the starting alcohol; substitution through the tosylate gives a product with opposite stereochemistry to the starting alcohol. *Sign in at* **www.cengage.com/login** *to see a simulation based on this figure and to take a short quiz.*

Problem 17.12 | How would you carry out the following transformation, a step used in the commercial synthesis of (*S*)-ibuprofen?

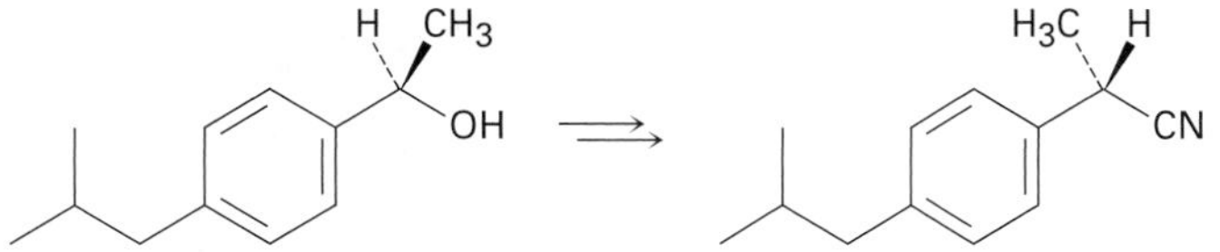

Dehydration of Alcohols to Yield Alkenes

A third important reaction of alcohols, both in the laboratory and in biological pathways, is their dehydration to give alkenes. The C—O bond and a neighboring C—H are broken, and an alkene π bond is formed.

A dehydration reaction

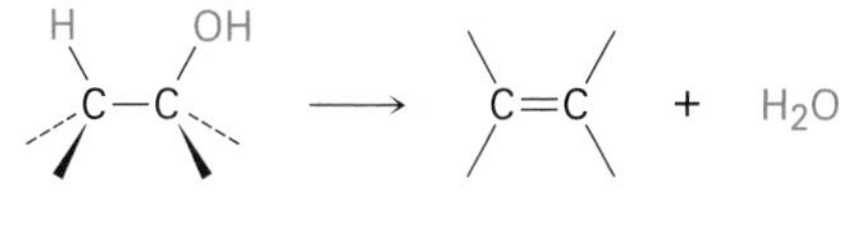

Because of the usefulness of the reaction, a number of ways have been devised for carrying out dehydrations. One method that works particularly well for tertiary alcohols is the acid-catalyzed reaction discussed in Section 7.1. For example, treatment of 1-methylcyclohexanol with warm aqueous sulfuric acid in a solvent such as tetrahydrofuran results in loss of water and formation of 1-methylcyclohexene.

1-Methylcyclohexanol   **1-Methylcyclohexene (91%)**

Acid-catalyzed dehydrations usually follow Zaitsev's rule (Section 11.7) and yield the more stable alkene as the major product. Thus, 2-methyl-2-butanol gives primarily 2-methyl-2-butene (trisubstituted double bond) rather than 2-methyl-1-butene (disubstituted double bond).

2-Methylbutan-2-ol

2-Methylbut-2-ene **2-Methylbut-1-ene**
(trisubstituted) **(disubstituted)**

Major product   Minor product

The reaction is an E1 process and occurs through the three-step mechanism shown in Figure 17.6). As usual for E1 reactions (Section 11.10), only tertiary alcohols are readily dehydrated with acid. Secondary alcohols can be made to react, but the conditions are severe (75% H_2SO_4, 100 °C) and sensitive molecules don't survive. Primary alcohols are even less reactive than secondary ones, and very harsh conditions are necessary to cause dehydration (95% H_2SO_4, 150 °C). Thus, the reactivity order for acid-catalyzed dehydrations is

Primary < **Secondary** < **Tertiary**

Reactivity

To circumvent the need for strong acid and allow the dehydration of secondary alcohols, reagents have been developed that are effective under mild, basic conditions. One such reagent, phosphorus oxychloride ($POCl_3$) in the basic amine solvent pyridine, is often able to effect the dehydration of secondary and tertiary alcohols at 0 °C.

1-Methylcyclohexanol   **1-Methylcyclohexene (96%)**

Figure 17.6 MECHANISM:
Mechanism of the acid-catalyzed dehydration of an alcohol to yield an alkene. The process is an E1 reaction and involves a carbocation intermediate.

CENGAGENOW™ Click *Organic Process* to **view animations showing the E1 acid-catalyzed dehydration of an alcohol**.

① Two electrons from the oxygen atom bond to H⁺, yielding a protonated alcohol intermediate.

Protonated alcohol

② The carbon–oxygen bond breaks, and the two electrons from the bond stay with oxygen, leaving a carbocation intermediate.

Carbocation

③ Two electrons from a neighboring carbon–hydrogen bond form the alkene π bond, and H⁺ (a proton) is eliminated.

© John McMurry

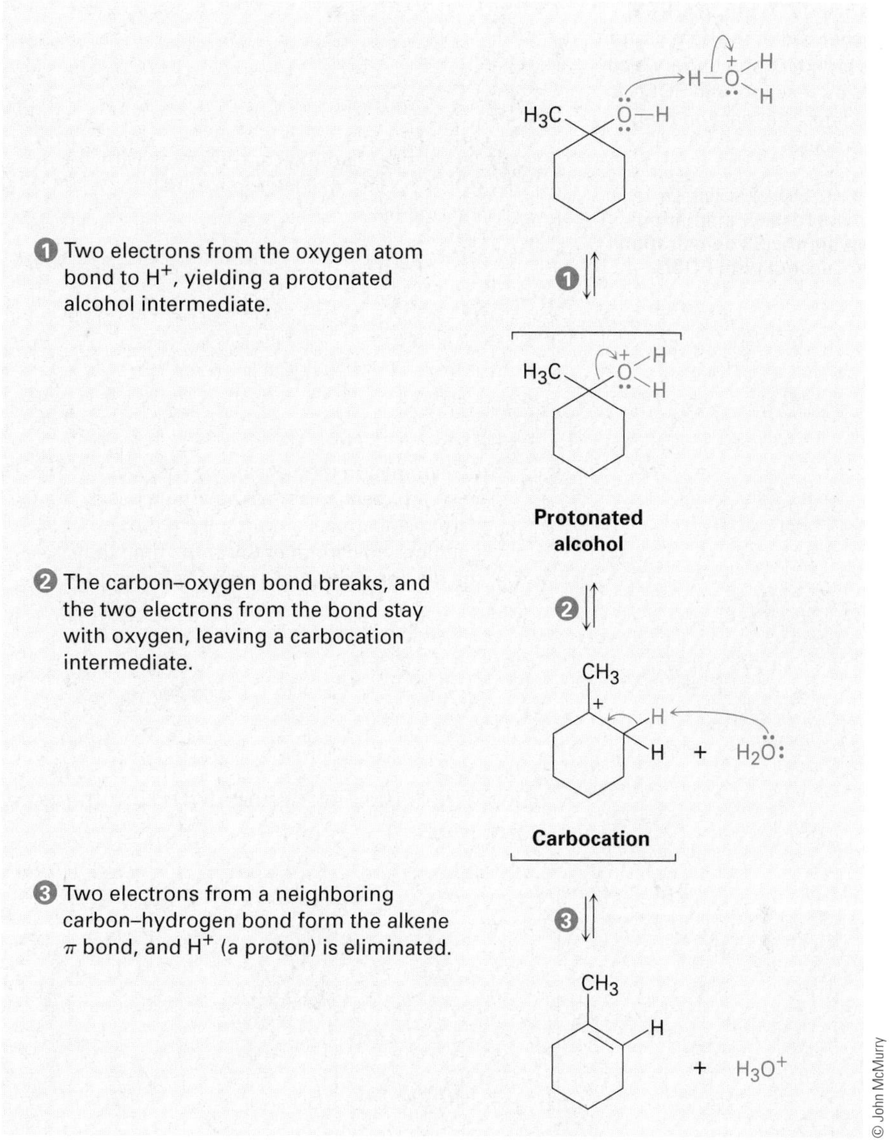

Alcohol dehydrations carried out with $POCl_3$ in pyridine take place by an E2 mechanism, as shown in Figure 17.7. Because hydroxide ion is a poor leaving group, direct E2 elimination of water from an alcohol does not occur. On reaction with $POCl_3$, however, the −OH group is converted into a dichlorophosphate ($-OPOCl_2$), which is a good leaving group and is readily eliminated. Pyridine is both the reaction solvent and the base that removes a neighboring proton in the E2 elimination step.

As noted previously in Section 11.10, biological dehydrations are also common and usually occur by an E1cB mechanism on a substrate in which the −OH group is two carbons away from a carbonyl group. An example occurs in the biosynthesis of the aromatic amino acid tyrosine. A base first abstracts a proton from the carbon adjacent to the carbonyl group, and the anion intermediate

Figure 17.7 MECHANISM:
Mechanism of the dehydration of secondary and tertiary alcohols by reaction with POCl$_3$ in pyridine. The reaction is an E2 process.

CENGAGENOW™ Click *Organic Process* to **view animations showing the E2 dehydration of an alcohol with POCl$_3$.**

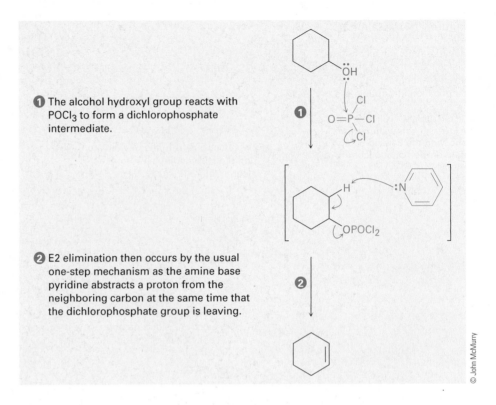

1 The alcohol hydroxyl group reacts with POCl$_3$ to form a dichlorophosphate intermediate.

2 E2 elimination then occurs by the usual one-step mechanism as the amine base pyridine abstracts a proton from the neighboring carbon at the same time that the dichlorophosphate group is leaving.

© John McMurry

then expels the −OH group with simultaneous protonation by an acid (HA) to form water.

| 5-Dehydroquinate | Anion intermediate | 5-Dehydroshikimate | Tyrosine |

Problem 17.13 | What product(s) would you expect from dehydration of the following alcohols with POCl$_3$ in pyridine? Indicate the major product in each case.

(a)   CH$_3$CH$_2$CHCHCH$_3$ with OH and CH$_3$ substituents

(b) cyclohexane with H, OH, H, CH$_3$

(c) cyclohexane with H, OH, CH$_3$, H

(d) H$_3$C OH
    CH$_3$CHCCH$_2$CH$_3$
         CH$_3$

(e)   OH
    CH$_3$CH$_2$CH$_2$CCH$_3$
         CH$_3$

Conversion of Alcohols into Esters

Alcohols react with carboxylic acids to give esters, a reaction that is common in both the laboratory and living organisms. In the laboratory, the reaction can be carried out in a single step if a strong acid is used as catalyst. More frequently, though, the reactivity of the carboxylic acid is enhanced by first converting it into a carboxylic acid chloride, which then reacts with the alcohol. We'll look in detail at the mechanisms of these reactions in Chapter 21.

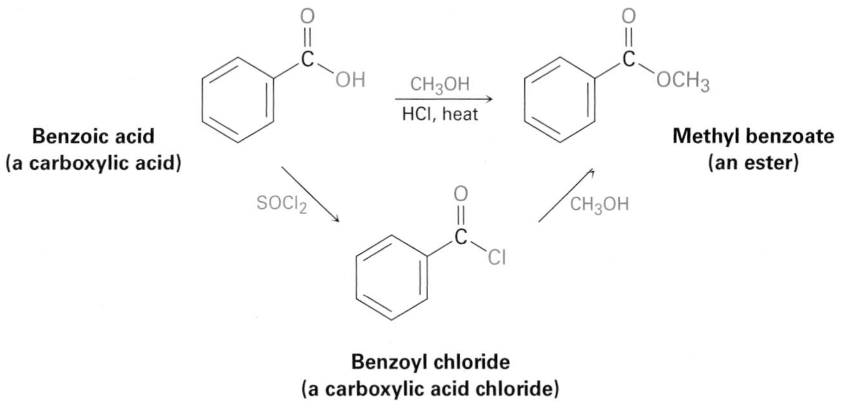

Benzoic acid
(a carboxylic acid)

CH_3OH
HCl, heat

$SOCl_2$

CH_3OH

Methyl benzoate
(an ester)

Benzoyl chloride
(a carboxylic acid chloride)

In living organisms, a similar process occurs, although a thioester or acyl adenosyl phosphate is the substrate rather than a carboxylic acid chloride.

A thioester

ROH

An ester

An acyl adenosyl phosphate

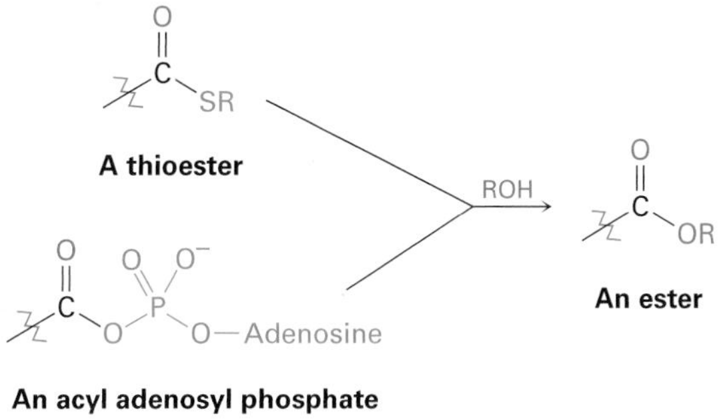

17.7 | Oxidation of Alcohols

Perhaps the most valuable reaction of alcohols is their oxidation to yield carbonyl compounds—the opposite of the reduction of carbonyl compounds to yield alcohols. Primary alcohols yield aldehydes or carboxylic acids, secondary alcohols yield ketones, but tertiary alcohols don't normally react with most oxidizing agents.

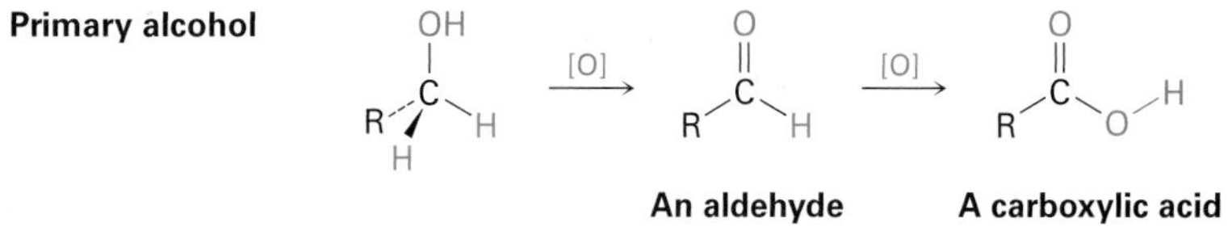

Primary alcohol

[O]

[O]

An aldehyde

A carboxylic acid

Secondary alcohol

A ketone

Tertiary alcohol

NO reaction

The oxidation of a primary or secondary alcohol can be accomplished by any of a large number of reagents, including $KMnO_4$, CrO_3, and $Na_2Cr_2O_7$. Which reagent is used in a specific case depends on such factors as cost, convenience, reaction yield, and alcohol sensitivity. For example, the large-scale oxidation of a simple, inexpensive alcohol such as cyclohexanol might best be done with a cheap oxidant such as $Na_2Cr_2O_7$. On the other hand, the small-scale oxidation of a delicate and expensive polyfunctional alcohol might best be done with one of several mild and high-yielding reagents, regardless of cost.

Primary alcohols are oxidized to either aldehydes or carboxylic acids, depending on the reagents chosen and the conditions used. One of the best methods for preparing an aldehyde from a primary alcohol on a small laboratory scale, as opposed to a large industrial scale, is to use pyridinium chlorochromate (PCC, $C_5H_6NCrO_3Cl$) in dichloromethane solvent.

Citronellol (from rose oil)   **Citronellal (82%)**

Most other oxidizing agents, such as chromium trioxide (CrO_3) in aqueous acid, oxidize primary alcohols directly to carboxylic acids. An aldehyde is involved as an intermediate in this reaction but can't usually be isolated because it is further oxidized too rapidly.

Decan-1-ol   **Decanoic acid (93%)**

Secondary alcohols are oxidized easily and in high yield to give ketones. For large-scale oxidations, an inexpensive reagent such as $Na_2Cr_2O_7$ in aqueous acetic acid might be used. For a more sensitive or costly alcohol, however, pyridinium chlorochromate is often used because the reaction is milder and occurs at lower temperatures.

Testosterone
(male sex hormone)

Androst-4-ene-3,17-dione (82%)

All these oxidations occur by a pathway that is closely related to the E2 reaction (Section 11.8). The first step involves reaction between the alcohol and a Cr(VI) reagent to form a *chromate* intermediate, followed by expulsion of chromium as the leaving group to yield the carbonyl product. Although we usually think of the E2 reaction as a means of generating a carbon–*carbon* double bond by elimination of a halide leaving group, the reaction is also useful for generating a carbon–*oxygen* double bond by elimination of a reduced metal as the leaving group.

An alcohol

A chromate intermediate

Carbonyl product

Biological alcohol oxidations are the exact opposite of biological carbonyl reductions and are carried by the coenzymes NAD⁺ and NADP⁺. A base removes the −OH proton, and the alkoxide ion transfers a hydride ion to the coenzyme. An example is the oxidation of *sn*-glycerol 3-phosphate to dihydroxyacetone phosphate, a step in the biological metabolism of fats (Figure 17.8). Note that addition occurs exclusively on the *Re* face of the NAD⁺ ring, adding a hydrogen with *pro-R* stereochemistry.

Problem 17.14 | What alcohols would give the following products on oxidation?

(a)

(b)

$$CH_3CHCHO$$
(with CH₃ on the middle carbon)

CH₃CHCHO

(c)

Problem 17.15 | What products would you expect from oxidation of the following compounds with CrO₃ in aqueous acid? With pyridinium chlorochromate?
(a) 1-Hexanol   (b) 2-Hexanol   (c) Hexanal

Figure 17.8 The biological oxidation of an alcohol (*sn*-glycerol 3-phosphate) to give a ketone (dihydroxyacetone phosphate). This mechanism is the exact opposite of the ketone reduction shown previously in Figure 17.4.

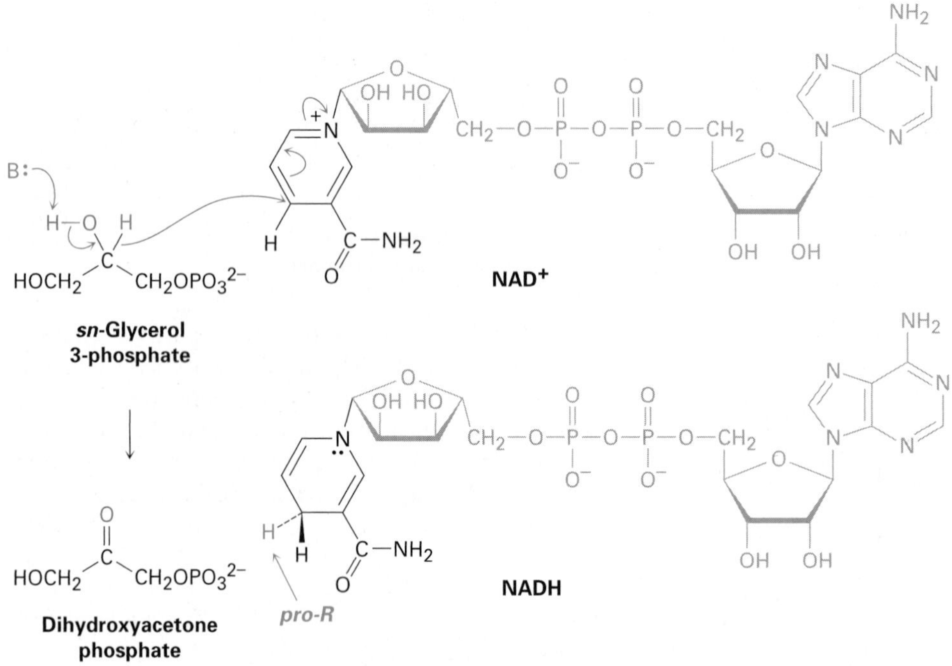

17.8 Protection of Alcohols

It often happens, particularly during the synthesis of complex molecules, that one functional group in a molecule interferes with an intended reaction on a second functional group elsewhere in the same molecule. For example, we saw earlier in this chapter that a Grignard reagent can't be prepared from a halo alcohol because the C–Mg bond is not compatible with the presence of an acidic –OH group in the same molecule.

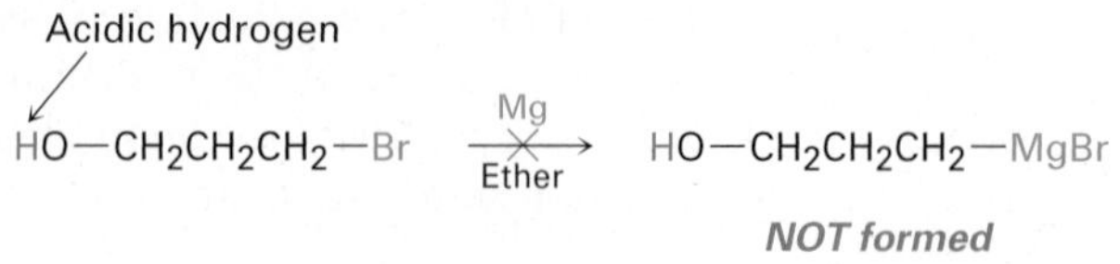

When this kind of incompatibility arises, it's sometimes possible to circumvent the problem by *protecting* the interfering functional group. Protection involves three steps: (1) introducing a **protecting group** to block the interfering function, (2) carrying out the desired reaction, and (3) removing the protecting group.

One of the more common methods of alcohol protection is by reaction with a chlorotrialkylsilane, $Cl–SiR_3$, to yield a trialkylsilyl ether, $R'–O–SiR_3$. Chlorotrimethylsilane is often used, and the reaction is carried out in the presence of a base, such as triethylamine, to help form the alkoxide anion from the alcohol and to remove the HCl by-product from the reaction.

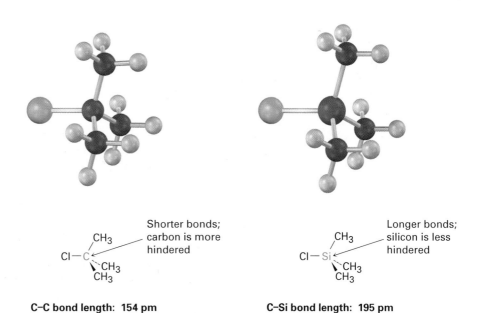

An alcohol   Chlorotrimethyl-silane   A trimethylsilyl (TMS) ether

For example:

Cyclohexanol   Cyclohexyl trimethylsilyl ether (94%)

The ether-forming step is an S_N2-like reaction of the alkoxide ion on the silicon atom, with concurrent loss of the leaving chloride anion. Unlike most S_N2 reactions, though, this reaction takes place at a *tertiary* center—a trialkyl-substituted silicon atom. The reaction occurs because silicon, a third-row atom, is larger than carbon and forms longer bonds. The three methyl substituents attached to silicon thus offer less steric hindrance to reaction than they do in the analogous *tert*-butyl chloride.

Shorter bonds; carbon is more hindered

Longer bonds; silicon is less hindered

C–C bond length: 154 pm   **C–Si bond length: 195 pm**

Like most other ethers, which we'll study in the next chapter, TMS ethers are relatively unreactive. They have no acidic hydrogens and don't react with oxidizing agents, reducing agents, or Grignard reagents. They do, however, react with aqueous acid or with fluoride ion to regenerate the alcohol.

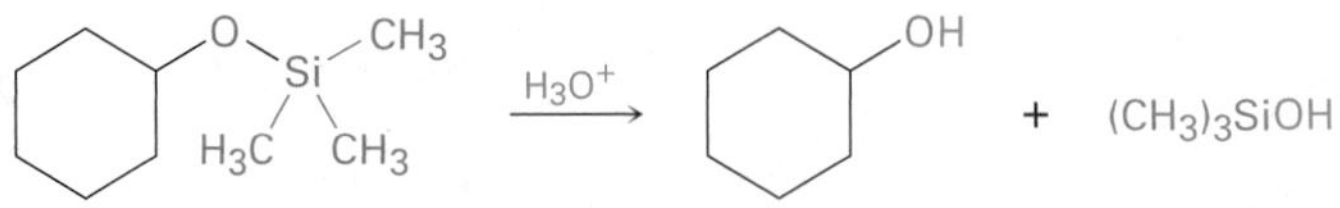

Cyclohexyl trimethylsilyl ether → **Cyclohexanol** + $(CH_3)_3SiOH$

To now solve the problem posed at the beginning of this section, it's possible to use a halo alcohol in a Grignard reaction by employing a protection sequence. For example, we can add 3-bromo-1-propanol to acetaldehyde by the route shown in Figure 17.9.

Figure 17.9 Use of a TMS-protected alcohol during a Grignard reaction.

Step 1 Protect alcohol:

$HOCH_2CH_2CH_2Br$ + $(CH_3)_3SiCl$ $\xrightarrow{(CH_3CH_2)_3N}$ $(CH_3)_3SiOCH_2CH_2CH_2Br$

Step 2a Form Grignard reagent:

$(CH_3)_3SiOCH_2CH_2CH_2Br$ $\xrightarrow[\text{Ether}]{\text{Mg}}$ $(CH_3)_3SiOCH_2CH_2CH_2MgBr$

Step 2b Do Grignard reaction:

$(CH_3)_3SiOCH_2CH_2CH_2MgBr$ $\xrightarrow[\text{2. } H_3O^+]{\text{1. } CH_3CH(=O)}$ $(CH_3)_3SiOCH_2CH_2CH_2CHCH_3$ (OH)

Step 3 Remove protecting group:

$(CH_3)_3SiOCH_2CH_2CH_2CHCH_3$ (OH) $\xrightarrow{H_3O^+}$ $HOCH_2CH_2CH_2CHCH_3$ (OH) + $(CH_3)_3SiOH$

Problem 17.16 TMS ethers can be removed by treatment with fluoride ion as well as by acid-catalyzed hydrolysis. Propose a mechanism for the reaction of cyclohexyl TMS ether with LiF. Fluorotrimethylsilane is a product.

17.9 | Phenols and Their Uses

Historically, the outbreak of the first World War provided a stimulus for the industrial preparation of large amounts of synthetic phenol, which was needed as a raw material to manufacture the explosive picric acid (2,4,6-trinitrophenol). Today, more than 2 million tons of phenol is manufactured each year in the United States for use in such products as Bakelite resin and adhesives for binding plywood.

Phenol was manufactured for many years by the Dow process, in which chlorobenzene reacts with NaOH at high temperature and pressure (Section 16.8). Now, however, an alternative synthesis from isopropylbenzene, commonly called *cumene,* is used. Cumene reacts with air at high temperature by benzylic oxidation through a radical mechanism to form cumene hydroperoxide, which is converted into phenol and acetone by treatment with acid. This is a particularly efficient process because two valuable chemicals are prepared at the same time.

Cumene
(isopropylbenzene)

Cumene
hydroperoxide

Phenol

Acetone

The reaction occurs by protonation of oxygen, followed by rearrangement of the phenyl group from carbon to oxygen with simultaneous loss of water. Readdition of water then yields an intermediate called a *hemiacetal*—a compound that contains one −OR group and one −OH group bonded to the same carbon atom—which breaks down to phenol and acetone (Figure 17.10).

In addition to its use in making resins and adhesives, phenol is also the starting material for the synthesis of chlorinated phenols and the food preservatives BHT (butylated hydroxytoluene) and BHA (butylated hydroxyanisole). Pentachlorophenol, a widely used wood preservative, is prepared by reaction of phenol with excess Cl_2. The herbicide 2,4-D (2,4-dichlorophenoxyacetic acid) is prepared from 2,4-dichlorophenol, and the hospital antiseptic agent hexachlorophene is prepared from 2,4,5-trichlorophenol.

Pentachlorophenol
(wood preservative)

2,4-Dichlorophenoxyacetic acid,
2,4-D (herbicide)

Hexachlorophene
(antiseptic)

The food preservative BHT is prepared by Friedel–Crafts alkylation of *p*-methylphenol (*p*-cresol) with 2-methylpropene in the presence of acid; BHA is prepared similarly by alkylation of *p*-methoxyphenol.

BHT

BHA

Problem 17.17 | Show the mechanism of the reaction of *p*-methylphenol with 2-methylpropene and H_3PO_4 catalyst to yield the food additive BHT.

Figure 17.10 MECHANISM:
Mechanism of the formation of phenol by acid-catalyzed rearrangement of cumene hydroperoxide.

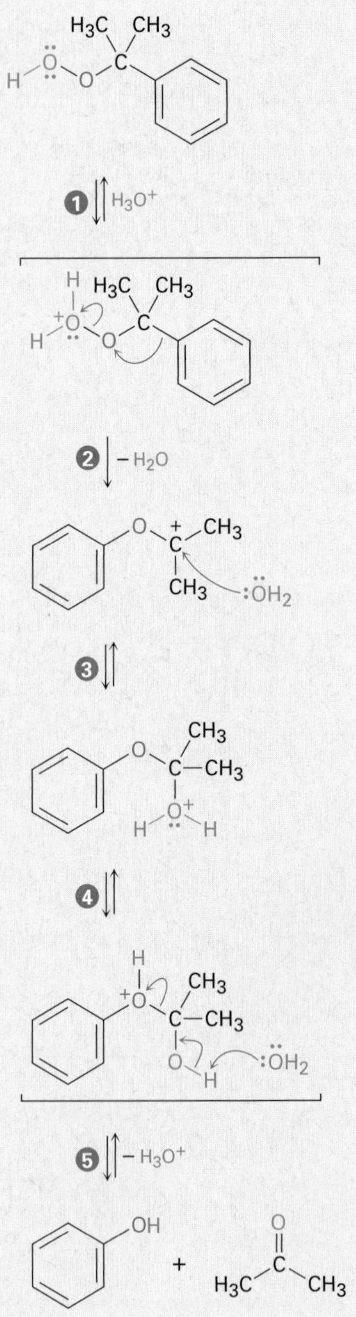

① Protonation of the hydroperoxy group on the terminal oxygen atom gives an oxonium ion . . .

① ‖ H_3O^+

② . . . which undergoes rearrangement by migration of the phenyl ring from carbon to oxygen, expelling water as the leaving group and giving a carbocation.

② ↓ $-H_2O$

③ Nucleophilic addition of water to the carbocation yields another oxonium ion . . .

③ ‖

④ . . . which rearranges by a proton shift from one oxygen to another.

④ ‖

⑤ Elimination of phenol gives acetone as co-product and regenerates the acid catalyst.

⑤ ‖ $-H_3O^+$

© John McMurry

17.10 | Reactions of Phenols

Electrophilic Aromatic Substitution Reactions

The hydroxyl group is a strongly activating, ortho- and para-directing substituent in electrophilic aromatic substitution reactions (Section 16.4). As a result, phenols are highly reactive substrates for electrophilic halogenation, nitration, sulfonation, and Friedel–Crafts reactions.

Oxidation of Phenols: Quinones

Phenols don't undergo oxidation in the same way that alcohols do because they don't have a hydrogen atom on the hydroxyl-bearing carbon. Instead, reaction of a phenol with a strong oxidizing agent yields a 2,5-cyclohexadiene-1,4-dione, or **quinone**. Older procedures employed $Na_2Cr_2O_7$ as oxidant, but Fremy's salt [potassium nitrosodisulfonate, $(KSO_3)_2NO$] is now preferred. The reaction takes place under mild conditions through a radical mechanism.

Phenol   **Benzoquinone (79%)**

Quinones are an interesting and valuable class of compounds because of their oxidation–reduction, or *redox*, properties. They can be easily reduced to **hydroquinones** (*p*-dihydroxybenzenes) by reagents such as $NaBH_4$ and $SnCl_2$, and hydroquinones can be easily reoxidized back to quinones by Fremy's salt.

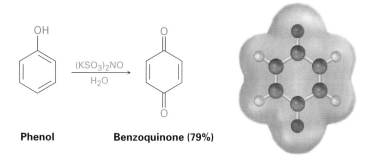

Benzoquinone   **Hydroquinone**

The redox properties of quinones are crucial to the functioning of living cells, where compounds called *ubiquinones* act as biochemical oxidizing agents to mediate the electron-transfer processes involved in energy production. Ubiquinones, also called *coenzymes Q*, are components of the cells of all aerobic organisms, from the simplest bacterium to humans. They are so named because of their ubiquitous occurrence in nature.

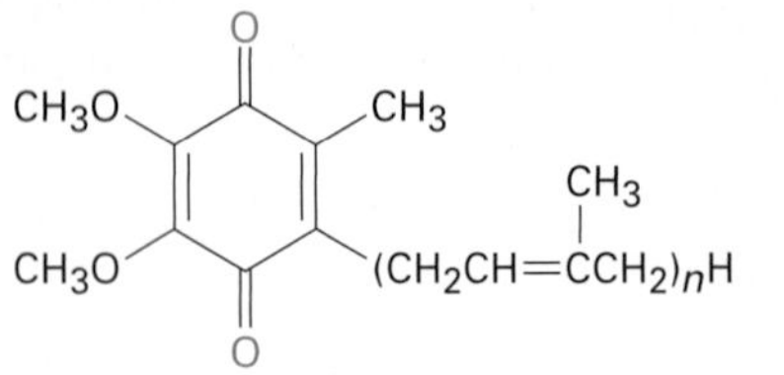

Ubiquinones (n = 1–10)

Ubiquinones function within the mitochondria of cells to mediate the respiration process in which electrons are transported from the biological reducing agent NADH to molecular oxygen. Through a complex series of steps, the ultimate result is a cycle whereby NADH is oxidized to NAD^+, O_2 is reduced to water, and energy is produced. Ubiquinone acts only as an intermediary and is itself unchanged.

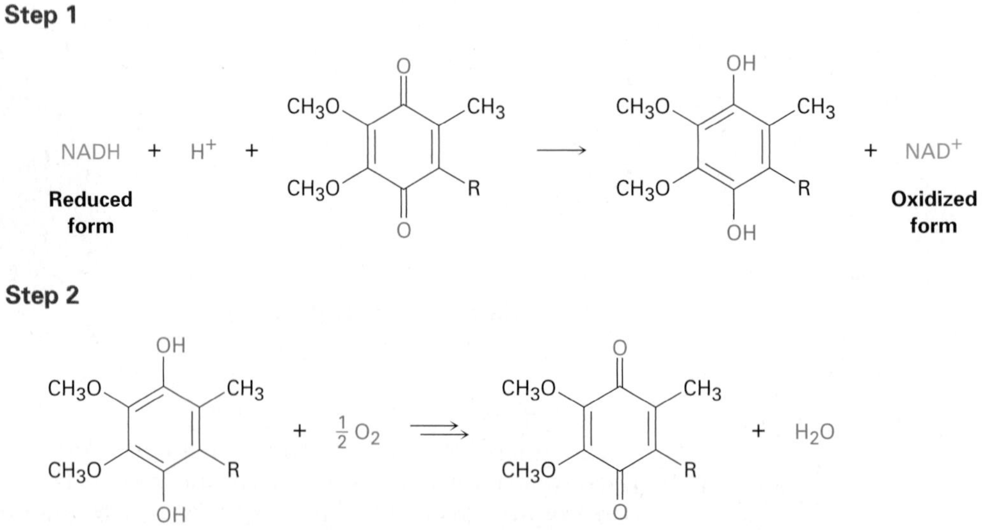

Step 1

Step 2

Net change: NADH   + $\frac{1}{2}O_2$ + H^+ $\longrightarrow$ NAD^+ + H_2O

17.11 | Spectroscopy of Alcohols and Phenols

Infrared Spectroscopy

Alcohols have a strong C–O stretching absorption near 1050 cm^{-1} and a characteristic O–H stretching absorption at 3300 to 3600 cm^{-1}. The exact position of the O–H stretch depends on the extent of hydrogen bonding in the molecule.

Unassociated alcohols show a fairly sharp absorption near 3600 cm^{-1}, whereas hydrogen-bonded alcohols show a broader absorption in the 3300 to 3400 cm^{-1} range. The hydrogen-bonded hydroxyl absorption appears at 3350 cm^{-1} in the IR spectrum of cyclohexanol (Figure 17.11).

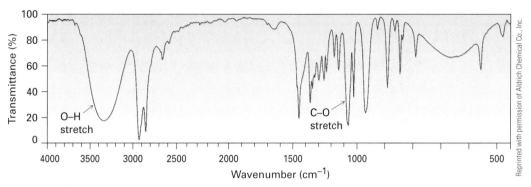

Figure 17.11 Infrared spectrum of cyclohexanol. Characteristic O−H and C−O stretching absorptions are indicated.

Phenols also show a characteristic broad IR absorption at 3500 cm^{-1} due to the −OH group, as well as the usual 1500 and 1600 cm^{-1} aromatic bands (Figure 17.12). In phenol itself, the monosubstituted aromatic-ring peaks at 690 and 760 cm^{-1} are visible.

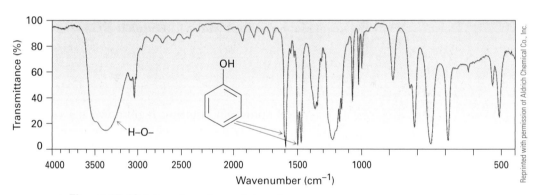

Figure 17.12 Infrared spectrum of phenol.

Problem 17.18 Assume that you need to prepare 5-cholesten-3-one from cholesterol. How could you use IR spectroscopy to tell whether the reaction was successful? What differences would you look for in the IR spectra of starting material and product?

Cholesterol   **Cholest-5-ene-3-one**

Nuclear Magnetic Resonance Spectroscopy

Carbon atoms bonded to electron-withdrawing −OH groups are deshielded and absorb at a lower field in the ^{13}C NMR spectrum than do typical alkane carbons. Most alcohol carbon absorptions fall in the range 50 to 80 δ, as the following data illustrate for cyclohexanol:

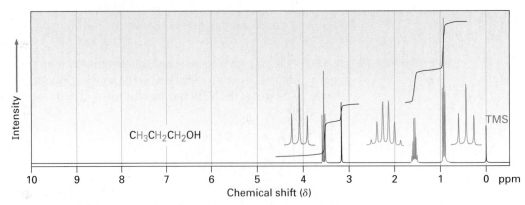

Alcohols also show characteristic absorptions in the ^{1}H NMR spectrum. Hydrogens on the oxygen-bearing carbon atom are deshielded by the electron-withdrawing effect of the nearby oxygen, and their absorptions occur in the range 3.4 to 4.5 δ. Spin–spin splitting, however, is not usually observed between the O−H proton of an alcohol and the neighboring protons on carbon. Most samples contain small amounts of acidic impurities, which catalyze an exchange of the O−H proton on a timescale so rapid that the effect of spin–spin splitting is removed. It's often possible to take advantage of this rapid proton exchange to identify the position of the O−H absorption. If a small amount of deuterated water, D_2O, is added to the NMR sample tube, the O−H proton is rapidly exchanged for deuterium, and the hydroxyl absorption disappears from the spectrum.

Typical spin–spin splitting *is* observed between protons on the oxygen-bearing carbon and other neighbors. For example, the signal of the two −CH₂O− protons in 1-propanol is split into a triplet by coupling with the neighboring −CH₂− protons (Figure 17.13).

Figure 17.13 ^{1}H NMR spectrum of 1-propanol. The protons on the oxygen-bearing carbon are split into a triplet at 3.58 δ.

Phenols, like all aromatic compounds, show ^{1}H NMR absorptions near 7 to 8 δ, the expected position for aromatic-ring protons (Section 15.8). In addition, phenol O−H protons absorb at 3 to 8 δ. In neither case are these absorptions uniquely diagnostic for phenols, since other kinds of protons absorb in the same range.

Problem 17.19 When the ^1H NMR spectrum of an alcohol is run in dimethyl sulfoxide (DMSO) solvent rather than in chloroform, exchange of the O−H proton is slow and spin–spin splitting is seen between the O−H proton and C−H protons on the adjacent carbon. What spin multiplicities would you expect for the hydroxyl protons in the following alcohols?

(a) 2-Methyl-2-propanol   (b) Cyclohexanol   (c) Ethanol
(d) 2-Propanol   (e) Cholesterol   (f) 1-Methylcyclohexanol

Mass Spectrometry

As noted previously in Section 12.3, alcohols undergo fragmentation in the mass spectrometer by two characteristic pathways, *alpha cleavage* and *dehydration*. In the alpha-cleavage pathway, a C−C bond nearest the hydroxyl group is broken, yielding a neutral radical plus a resonance-stabilized, oxygen-containing cation.

In the dehydration pathway, water is eliminated, yielding an alkene radical cation.

Both fragmentation modes are apparent in the mass spectrum of 1-butanol (Figure 17.14). The peak at $m/z = 56$ is due to loss of water from the molecular ion, and the peak at $m/z = 31$ is due to an alpha cleavage.

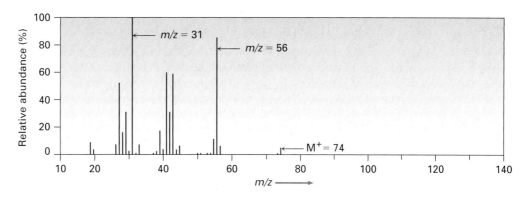

Figure 17.14 Mass spectrum of 1-butanol ($M^+ = 74$). Dehydration gives a peak at $m/z = 56$, and fragmentation by alpha cleavage gives a peak at $m/z = 31$.

Focus On . . .

Ethanol: Chemical, Drug, and Poison

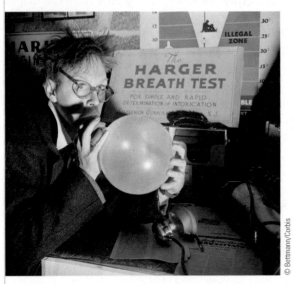

The Harger Drunkometer was introduced in 1938 to help convict drunk drivers.

The production of ethanol by fermentation of grains and sugars is one of the oldest known organic reactions, going back at least 8000 years in the Middle East and perhaps as many as 9000 years in China. Fermentation is carried out by adding yeast to an aqueous sugar solution, where enzymes break down carbohydrates into ethanol and CO_2. As noted in the chapter introduction, approximately 4 billion gallons of ethanol is produced each year in the United States by fermentation, with essentially the entire amount used to make E85 automobile fuel.

$$C_6H_{12}O_6 \xrightarrow{\text{Yeast}} 2\ CH_3CH_2OH\ +\ 2\ CO_2$$

A carbohydrate

Ethanol is classified for medical purposes as a central nervous system (CNS) depressant. Its effects—that is, being drunk—resemble the human response to anesthetics. There is an initial excitability and increase in sociable behavior, but this results from depression of inhibition rather than from stimulation. At a blood alcohol concentration of 0.1% to 0.3%, motor coordination is affected, accompanied by loss of balance, slurred speech, and amnesia. When blood alcohol concentration rises to 0.3% to 0.4%, nausea and loss of consciousness occur. Above 0.6%, spontaneous respiration and cardiovascular regulation are affected, ultimately leading to death. The LD_{50} of ethanol is 10.6 g/kg (Chapter 1 *Focus On*).

The passage of ethanol through the body begins with its absorption in the stomach and small intestine, followed by rapid distribution to all body fluids and organs. In the pituitary gland, ethanol inhibits the production of a hormone that regulates urine flow, causing increased urine production and dehydration. In the stomach, ethanol stimulates production of acid. Throughout the body, ethanol causes blood vessels to dilate, resulting in flushing of the skin and a sensation of warmth as blood moves into capillaries beneath the surface. The result is not a warming of the body, but an increased loss of heat at the surface.

Ethanol metabolism occurs mainly in the liver and proceeds by oxidation in two steps, first to acetaldehyde (CH_3CHO) and then to acetic acid (CH_3CO_2H). When continuously present in the body, ethanol and acetaldehyde are toxic, leading to the devastating physical and metabolic deterioration

© Bettmann/Corbis

(continued)

seen in chronic alcoholics. The liver usually suffers the worst damage since it is the major site of alcohol metabolism.

Approximately 17,000 people are killed each year in the United States in alcohol-related automobile accidents. Thus, all 50 states—Massachusetts was the last holdout—have made it illegal to drive with a blood alcohol concentration (BAC) above 0.08%. Fortunately, simple tests have been devised for measuring blood alcohol concentration. The *Breathalyzer test* measures alcohol concentration in expired air by the color change that occurs when the bright orange oxidizing agent potassium dichromate ($K_2Cr_2O_7$) is reduced to blue-green chromium(III). The *Intoxilyzer* test uses IR spectroscopy to measure blood alcohol levels in expired air. Just breathe into the machine, and let the spectrum tell the tale.

SUMMARY AND KEY WORDS

alcohol (ROH), 599

alkoxide ion (RO$^-$), 603

hydroquinone, 631

phenol (ArOH), 599

phenoxide ion (ArO$^-$), 603

protecting group, 626

quinone, 631

Alcohols are among the most versatile of all organic compounds. They occur widely in nature, are important industrially, and have an unusually rich chemistry. The most widely used methods of alcohol synthesis start with carbonyl compounds. Aldehydes, ketones, esters, and carboxylic acids are reduced by reaction with $LiAlH_4$. Aldehydes, esters, and carboxylic acids yield primary alcohols (RCH_2OH) on reduction; ketones yield secondary alcohols (R_2CHOH).

Alcohols are also prepared by reaction of carbonyl compounds with Grignard reagents, RMgX. Addition of a Grignard reagent to formaldehyde yields a primary alcohol, addition to an aldehyde yields a secondary alcohol, and addition to a ketone or an ester yields a tertiary alcohol. The Grignard reaction is limited by the fact that Grignard reagents can't be prepared from alkyl halides that contain reactive functional groups in the same molecule. This problem can sometimes be avoided by **protecting** the interfering functional group. Alcohols are often protected by formation of trimethylsilyl (TMS) ethers.

Alcohols undergo many reactions and can be converted into many other functional groups. They can be dehydrated to give alkenes by treatment with $POCl_3$ and can be transformed into alkyl halides by treatment with PBr_3 or $SOCl_2$. Furthermore, alcohols are weakly acidic ($pK_a \approx 16$–18) and react with strong bases and with alkali metals to form **alkoxide anions**, which are used frequently in organic synthesis.

Perhaps the most important reaction of alcohols is their oxidation to carbonyl compounds. Primary alcohols yield either aldehydes or carboxylic acids, secondary alcohols yield ketones, but tertiary alcohols are not normally oxidized. Pyridinium chlorochromate (PCC) in dichloromethane is often used for oxidizing primary alcohols to aldehydes and secondary alcohols to ketones. A solution of CrO_3 in aqueous acid is frequently used for oxidizing primary alcohols to carboxylic acids and secondary alcohols to ketones.

Phenols are aromatic counterparts of alcohols but are more acidic ($pK_a \approx 10$) because the corresponding **phenoxide anions** are resonance stabilized by delocalization of the negative charge into the aromatic ring. Substitution of the aromatic ring by an electron-withdrawing group increases phenol acidity, and substitution by an electron-donating group decreases acidity. Phenols can be oxidized to **quinones** by reaction with Fremy's salt (potassium nitrosodisulfonate), and quinones can be reduced to **hydroquinones** by reaction with $NaBH_4$.

SUMMARY OF REACTIONS

1. Synthesis of alcohols
   (a) Reduction of carbonyl compounds (Section 17.4)
      (1) Aldehydes

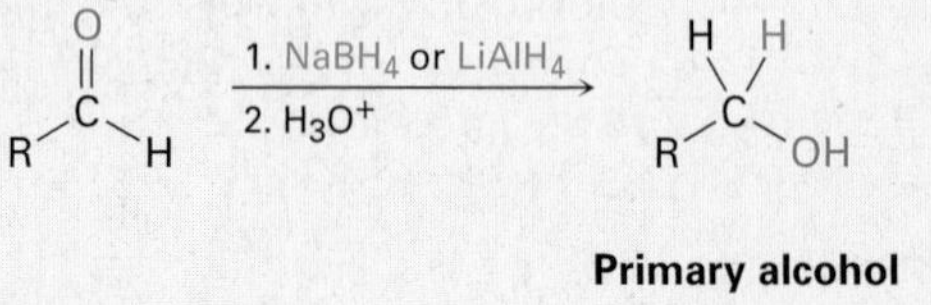

Primary alcohol

      (2) Ketones

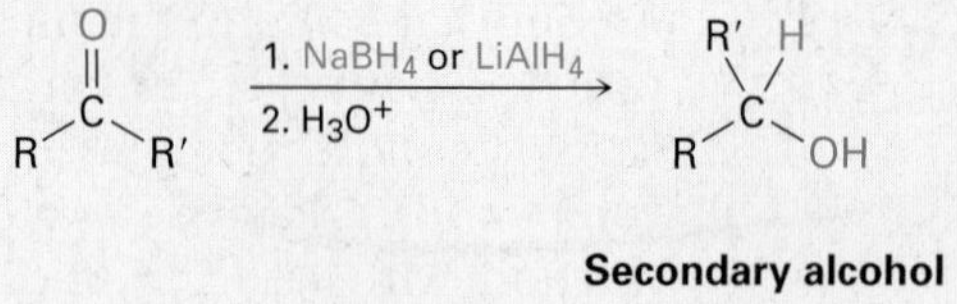

Secondary alcohol

      (3) Esters

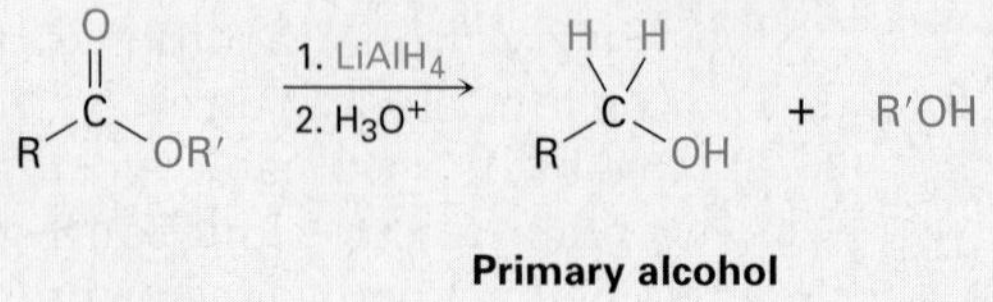

Primary alcohol

      (4) Carboxylic acids

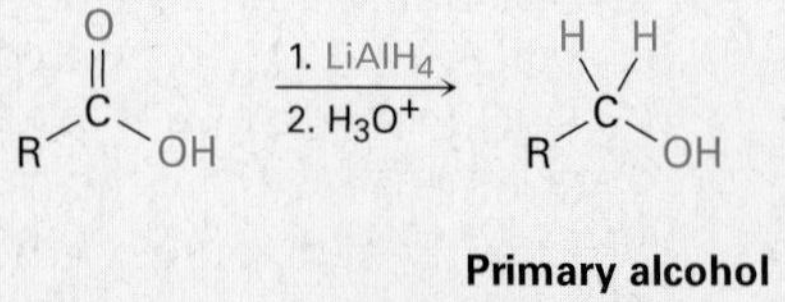

Primary alcohol

   (b) Grignard addition to carbonyl compounds (Section 17.5)
      (1) Formaldehyde

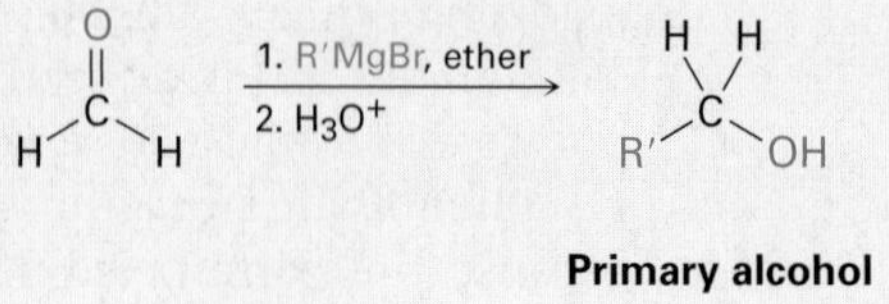

Primary alcohol

(2) Aldehydes

Secondary alcohol

(3) Ketones

Tertiary alcohol

(4) Esters

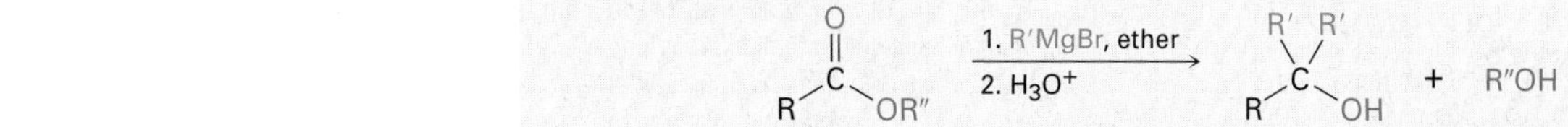

Tertiary alcohol

2. Reactions of alcohols
   (a) Dehydration (Section 17.6)
      (1) Tertiary alcohols

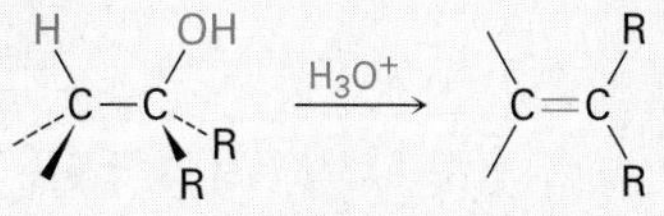

   (2) Secondary and tertiary alcohols

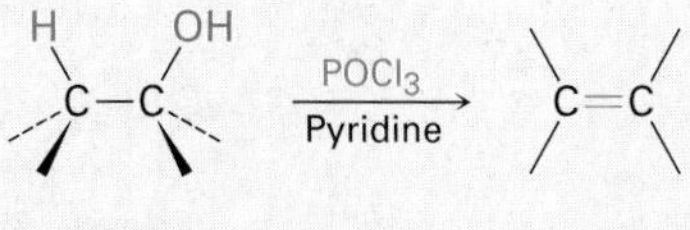

   (b) Oxidation (Section 17.7)
      (1) Primary alcohols

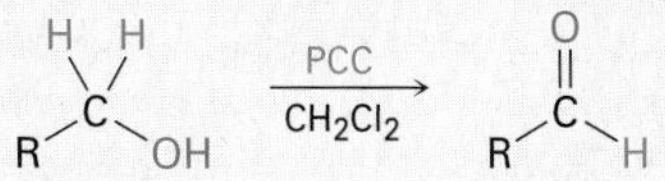

Aldehyde

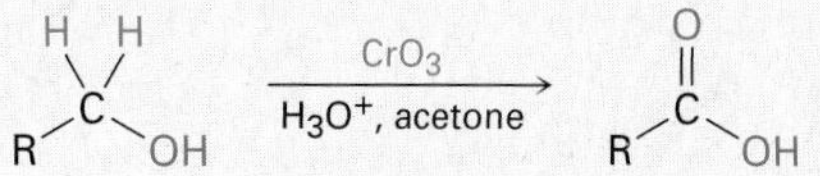

Carboxylic acid

(2) Secondary alcohols

$$\underset{\substack{\text{R} \quad \text{R}'}}{\overset{\substack{\text{H} \quad \text{OH}}}{\text{C}}} \xrightarrow[\text{CH}_2\text{Cl}_2]{\text{PCC}} \underset{\substack{\text{R} \quad \text{R}'}}{\overset{\text{O}}{\text{C}}}$$

Ketone

3. Oxidation of phenols to quinones (Section 17.10)

$$\text{(phenol)} \xrightarrow[\text{H}_2\text{O}]{(\text{KSO}_3)_2\text{NO}} \text{(quinone)}$$

EXERCISES

Organic KNOWLEDGE TOOLS

CENGAGENOW™ Sign in at **www.cengage.com/login** to assess your knowledge of this chapter's topics by taking a pre-test. The pre-test will link you to interactive organic chemistry resources based on your score in each concept area.

ⓦWL Online homework for this chapter may be assigned in Organic OWL.

■ indicates problems assignable in Organic OWL.

VISUALIZING CHEMISTRY

(Problems 17.1–17.19 appear within the chapter.)

17.20 ■ Give IUPAC names for the following compounds:

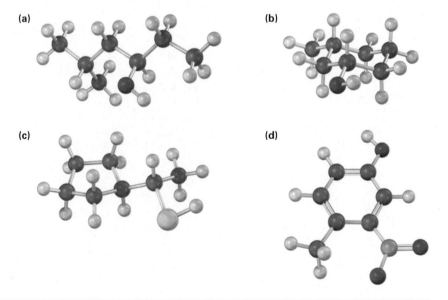

(a)

(b)

(c)

(d)

17.21 ■ Draw the structure of the carbonyl compound(s) from which each of the following alcohols might have been prepared, and show the products you would obtain by treatment of each alcohol with (i) Na metal, (ii) SOCl$_2$, and (iii) pyridinium chlorochromate.

(a)   **(b)**

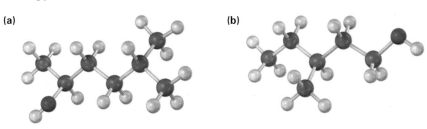

17.22 ■ Predict the product from reaction of the following substance (reddish brown = Br) with:
(a) PBr$_3$ **(b)** Aqueous H$_2$SO$_4$ **(c)** SOCl$_2$
(d) PCC   **(e)** Br$_2$, FeBr$_3$

17.23 ■ Predict the product from reaction of the following substance with:
(a) NaBH$_4$; then H$_3$O$^+$ **(b)** LiAlH$_4$; then H$_3$O$^+$
(c) CH$_3$CH$_2$MgBr; then H$_3$O$^+$

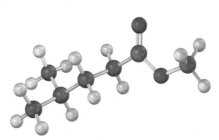

17.24 Name and assign *R* or *S* stereochemistry to the product(s) you would obtain by reaction of the following substance with ethylmagnesium bromide. Is the product chiral? Is it optically active? Explain.

ADDITIONAL PROBLEMS

17.25 ■ Give IUPAC names for the following compounds:

(a)

$$\underset{\underset{HOCH_2CH_2CHCH_2OH}{|}}{\overset{CH_3}{}}$$

(b)

$$\underset{\underset{CH_2CH_2CH_3}{|}}{\overset{OH}{\underset{}{}}}\; CH_3CHCHCH_2CH_3$$

(c)

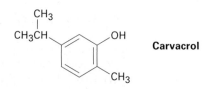

(d)

(e) Ph

(f)

17.26 Draw and name the eight isomeric alcohols with formula $C_5H_{12}O$.

17.27 Which of the eight alcohols you identified in Problem 17.26 react with CrO_3 in aqueous acid? Show the products you would expect from each reaction.

17.28 Named *bombykol*, the sex pheromone secreted by the female silkworm moth has the formula $C_{16}H_{28}O$ and the systematic name (10*E*,12*Z*)-10,12-hexa-decadien-1-ol. Draw bombykol showing correct geometry for the two double bonds.

17.29 *Carvacrol* is a naturally occurring substance isolated from oregano, thyme, and marjoram. What is its IUPAC name?

$$\underset{CH_3CH}{\overset{\overset{CH_3}{|}}{}}$$

Carvacrol

17.30 ■ What products would you obtain from reaction of 1-pentanol with the following reagents?
(a) PBr_3 (b) $SOCl_2$ (c) CrO_3, H_2O, H_2SO_4 (d) PCC

17.31 ■ How would you prepare the following compounds from 2-phenylethanol? More than one step may be required.
(a) Styrene ($PhCH{=}CH_2$)   (b) Phenylacetaldehyde ($PhCH_2CHO$)
(c) Phenylacetic acid ($PhCH_2CO_2H$)   (d) Benzoic acid
(e) Ethylbenzene   (f) Benzaldehyde
(g) 1-Phenylethanol   (h) 1-Bromo-2-phenylethane

17.32 ■ How would you prepare the following compounds from 1-phenylethanol? More than one step may be required.
(a) Acetophenone ($PhCOCH_3$)   (b) Benzyl alcohol
(c) *m*-Bromobenzoic acid   (d) 2-Phenyl-2-propanol

17.33 ■ What Grignard reagent and what carbonyl compound might you start with to prepare the following alcohols?

(a)

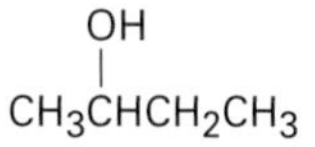

OH
|
CH₃CHCH₂CH₃

(b)

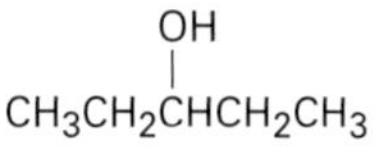

OH
|
CH₃CH₂CHCH₂CH₃

(c)

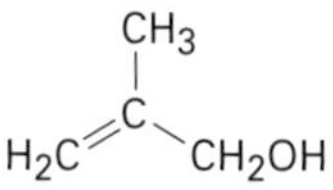

CH₃
|
H₂C=C
      \
       CH₂OH

(d)

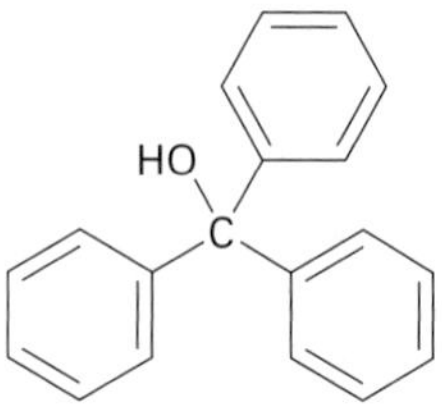

HO
|
C(three phenyl groups)

(e)

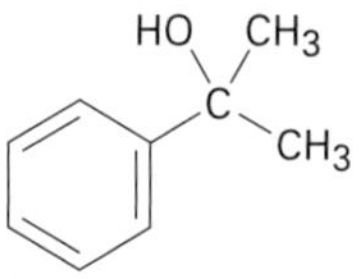

HO CH₃
  \ /
   C
   |
   CH₃

(f)

(cyclohexene)—CH₂OH

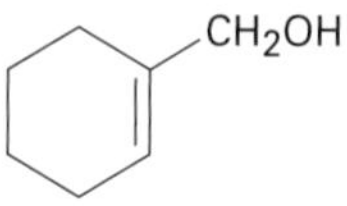

17.34 What carbonyl compounds would you reduce to prepare the following alcohols? List all possibilities.

(a)

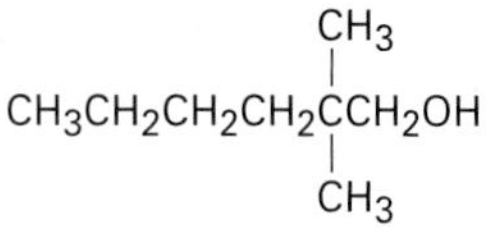

                 CH₃
                 |
CH₃CH₂CH₂CH₂CCH₂OH
                 |
                 CH₃

(b)

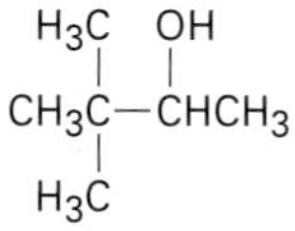

H₃C OH
  | |
CH₃C—CHCH₃
  |
  H₃C

(c)

OH
|
(cyclohexane)—CHCH₂CH₃

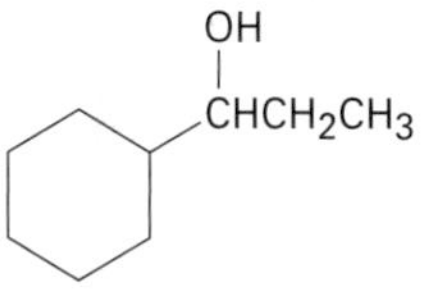

17.35 ■ How would you carry out the following transformations?

(a)  (phenyl)—CH=CH—CO₂H   →?   (phenyl)—CH₂CH₂—CO₂H

(b) 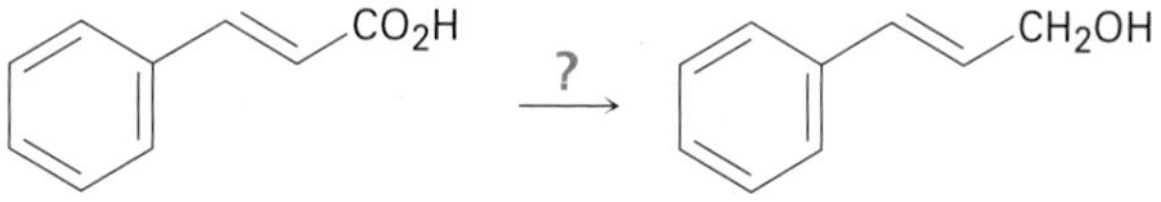 (phenyl)—CH=CH—CO₂H   →?   (phenyl)—CH=CH—CH₂OH

(c) 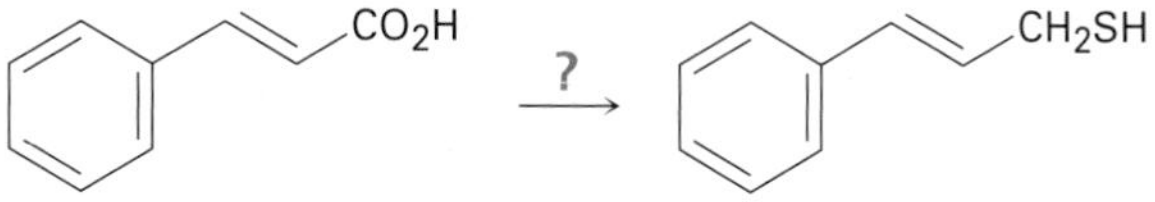 (phenyl)—CH=CH—CO₂H   →?   (phenyl)—CH=CH—CH₂SH

17.36 ■ What carbonyl compounds might you start with to prepare the following compounds by Grignard reaction? List all possibilities.

(a) 2-Methyl-2-propanol   **(b)** 1-Ethylcyclohexanol
(c) 3-Phenyl-3-pentanol   **(d)** 2-Phenyl-2-pentanol

(e)

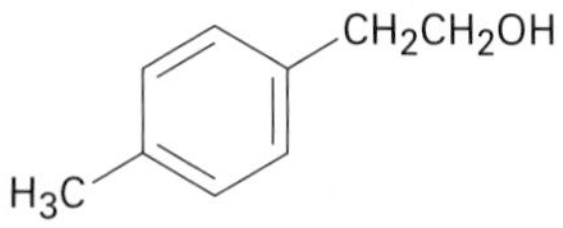

(phenyl with H₃C)—CH₂CH₂OH

(f)

              OH
              |
(cyclopentane)—CH₂CCH₃
              |
              CH₃

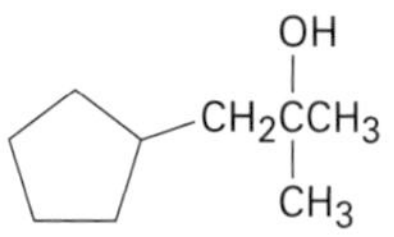

17.37 ■ Evidence for the intermediate carbocations in the acid-catalyzed dehydration of alcohols comes from the observation that rearrangements sometimes occur. Propose a mechanism to account for the formation of 2,3-dimethyl-2-butene from 3,3-dimethyl-2-butanol.

17.38 ■ Acid-catalyzed dehydration of 2,2-dimethylcyclohexanol yields a mixture of 1,2-dimethylcyclohexene and isopropylidenecyclopentane. Propose a mechanism to account for the formation of both products.

Isopropylidenecyclopentane

17.39 Epoxides react with Grignard reagents to yield alcohols. Propose a mechanism.

17.40 ■ How would you prepare the following substances from cyclopentanol? More than one step may be required.
(a) Cyclopentanone   (b) Cyclopentene
(c) 1-Methylcyclopentanol (d) *trans*-2-Methylcyclopentanol

17.41 ■ What products would you expect to obtain from reaction of 1-methylcyclohexanol with the following reagents?
(a) HBr   (b) NaH   (c) H$_2$SO$_4$ (d) Na$_2$Cr$_2$O$_7$

17.42 Treatment of the following epoxide with aqueous acid produces a carbocation intermediate that reacts with water to give a diol product. Show the structure of the carbocation, and propose a mechanism for the second step.

17.43 Benzoquinone is an excellent dienophile in the Diels–Alder reaction. What product would you expect from reaction of benzoquinone with 1 equivalent of 1,3-butadiene? From reaction with 2 equivalents of 1,3-butadiene?

17.44 ■ Rank the following substituted phenols in order of increasing acidity, and explain your answer:

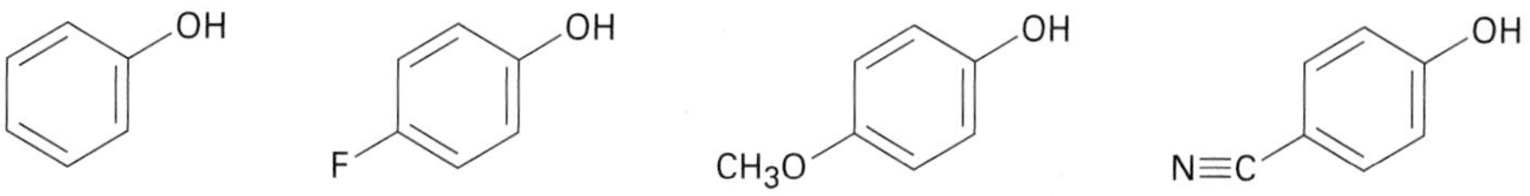

17.45 Benzyl chloride can be converted into benzaldehyde by treatment with nitromethane and base. The reaction involves initial conversion of nitromethane into its anion, followed by S_N2 reaction of the anion with benzyl chloride and subsequent E2 reaction. Write the mechanism in detail, using curved arrows to indicate the electron flow in each step.

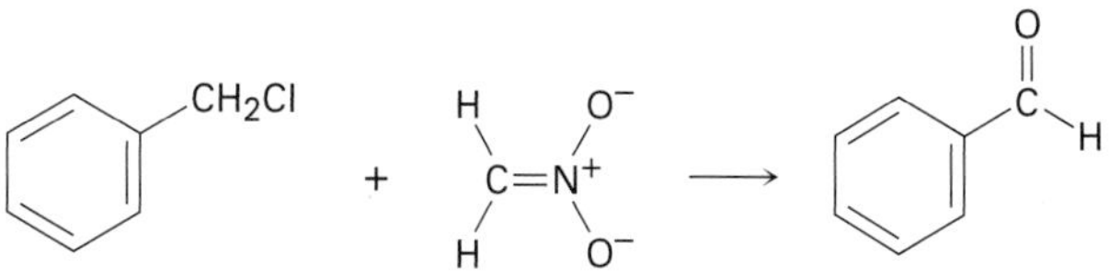

Benzyl chloride   **Nitromethane anion**   **Benzaldehyde**

17.46 Reduction of 2-butanone with $NaBH_4$ yields 2-butanol. Is the product chiral? Is it optically active? Explain.

17.47 Reaction of (*S*)-3-methyl-2-pentanone with methylmagnesium bromide followed by acidification yields 2,3-dimethyl-2-pentanol. What is the stereochemistry of the product? Is the product optically active?

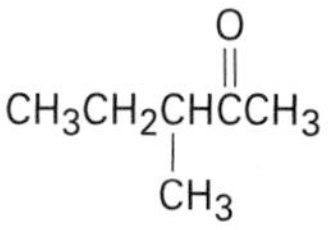

$$CH_3CH_2\overset{\displaystyle CH}{\underset{\displaystyle CH_3}{|}}\overset{\displaystyle O}{\overset{\|}{C}}CH_3 \qquad \textbf{3-Methylpentan-2-one}$$

17.48 ■ Testosterone is one of the most important male steroid hormones. When testosterone is dehydrated by treatment with acid, rearrangement occurs to yield the product shown. Propose a mechanism to account for this reaction.

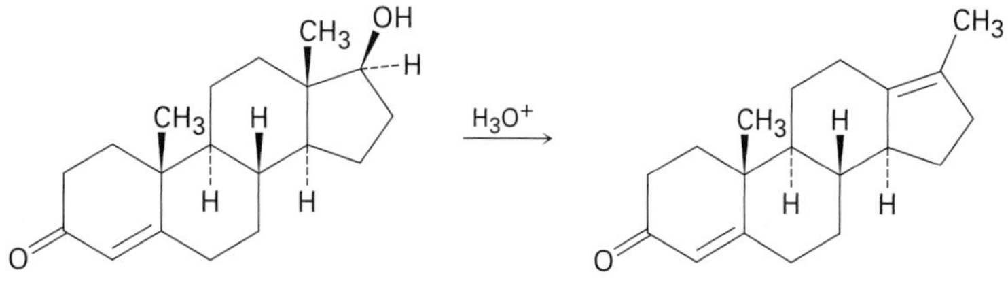

Testosterone

17.49 Starting from testosterone (Problem 17.48), how would you prepare the following substances?

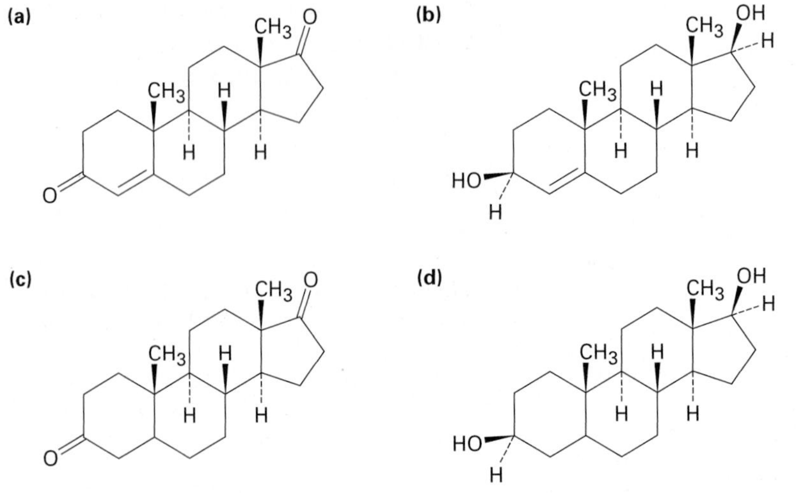

(a)

(b)

(c)

(d)

17.50 Compound **A**, $C_{10}H_{18}O$, undergoes reaction with dilute H_2SO_4 at 25 °C to yield a mixture of two alkenes, $C_{10}H_{16}$. The major alkene product, **B**, gives only cyclopentanone after ozone treatment followed by reduction with zinc in acetic acid. Write the reactions involved, and identify **A** and **B**.

17.51 Dehydration of *trans*-2-methylcyclopentanol with $POCl_3$ in pyridine yields predominantly 3-methylcyclopentene. Is the stereochemistry of this dehydration syn or anti? Can you suggest a reason for formation of the observed product? (Make molecular models!)

17.52 How would you synthesize the following alcohols, starting with benzene and other alcohols of six or fewer carbons as your only organic reagents?

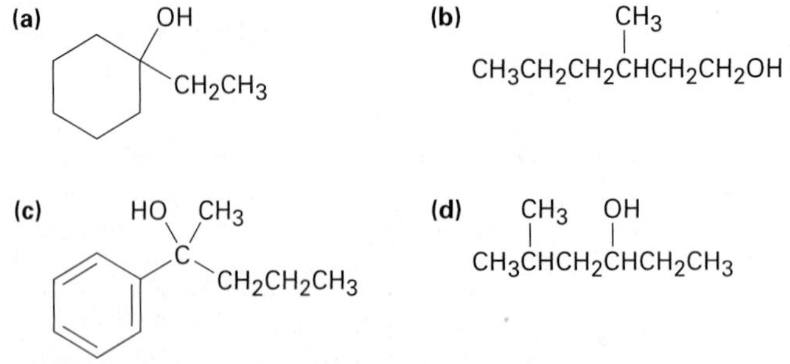

(a)

(b)

$$\underset{\underset{CH_3}{|}}{CH_3CH_2CH_2CHCH_2CH_2OH}$$

(c)

(d)

$$\underset{\underset{CH_3}{|}\underset{OH}{|}}{CH_3CHCH_2CHCH_2CH_3}$$

17.53 ■ 2,3-Dimethyl-2,3-butanediol has the common name *pinacol*. On heating with aqueous acid, pinacol rearranges to *pinacolone*, 3,3-dimethyl-2-butanone. Suggest a mechanism for this reaction.

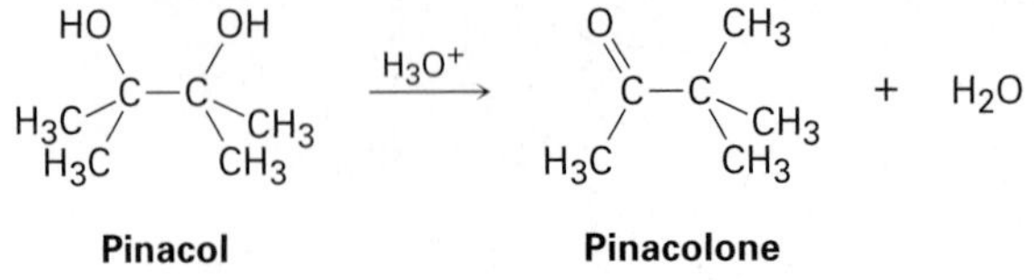

Pinacol   Pinacolone

17.54 As a rule, axial alcohols oxidize somewhat faster than equatorial alcohols. Which would you expect to oxidize faster, *cis*-4-*tert*-butylcyclohexanol or *trans*-4-*tert*-butylcyclohexanol? Draw the more stable chair conformation of each molecule.

17.55 Propose a synthesis of bicyclohexylidene, starting from cyclohexanone as the only source of carbon.

Bicyclohexylidene

17.56 A problem often encountered in the oxidation of primary alcohols to acids is that esters are sometimes produced as by-products. For example, oxidation of ethanol yields acetic acid and ethyl acetate:

$$CH_3CH_2OH \xrightarrow{CrO_3} CH_3\overset{O}{\underset{||}{C}}OH \ + \ CH_3\overset{O}{\underset{||}{C}}OCH_2CH_3$$

Propose a mechanism to account for the formation of ethyl acetate. Take into account the reversible reaction between aldehydes and alcohols:

$$R\overset{O}{\underset{||}{C}}H \ + \ R'OH \ \rightleftharpoons \ R\overset{HO\ \ OR'}{\underset{H}{C}}$$

17.57 Identify the reagents **a–f** in the following scheme:

17.58 Galactose, a constituent of the disaccharide lactose found in dairy products, is metabolized by a pathway that includes the isomerization of UDP-galactose to UDP-glucose, where UDP = uridylyl diphosphate. The enzyme responsible for the transformation uses NAD^+ as cofactor. Propose a mechanism.

UDP-galactose

UDP-glucose

17.59 ■ Propose a structure consistent with the following spectral data for a compound $C_8H_{18}O_2$:

IR: 3350 cm^{-1}

^{1}H NMR: 1.24 δ (12 H, singlet); 1.56 δ (4 H, singlet); 1.95 δ (2 H, singlet)

17.60 The ^{1}H NMR spectrum shown is that of 3-methyl-3-buten-1-ol. Assign all the observed resonance peaks to specific protons, and account for the splitting patterns.

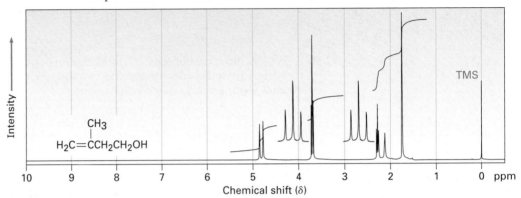

17.61 ■ Compound A, $C_5H_{10}O$, is one of the basic building blocks of nature. All steroids and many other naturally occurring compounds are built from compound A. Spectroscopic analysis of A yields the following information:

IR: 3400 cm^{-1}; 1640 cm^{-1}

^{1}H NMR: 1.63 δ (3 H, singlet); 1.70 δ (3 H, singlet); 3.83 δ (1 H, broad singlet); 4.15 δ (2 H, doublet, J = 7 Hz); 5.70 δ (1 H, triplet, J = 7 Hz)

(a) How many double bonds and/or rings does **A** have?

(b) From the IR spectrum, what is the identity of the oxygen-containing functional group?

(c) What kinds of protons are responsible for the NMR absorptions listed?

(d) Propose a structure for **A**.

17.62 ■ A compound of unknown structure gave the following spectroscopic data:

Mass spectrum: M$^+$ = 88.1

IR: 3600 cm^{-1}

^{1}H NMR: 1.4 δ (2 H, quartet, J = 7 Hz); 1.2 δ (6 H, singlet); 1.0 δ (1 H, singlet); 0.9 δ (3 H, triplet, J = 7 Hz)

^{13}C NMR: 74, 35, 27, 25 δ

(a) Assuming that the compound contains C and H but may or may not contain O, give three possible molecular formulas.

(b) How many protons (H) does the compound contain?

(c) What functional group(s) does the compound contain?

(d) How many carbons does the compound contain?

(e) What is the molecular formula of the compound?

(f) What is the structure of the compound?

(g) Assign the peaks in the ^{1}H NMR spectrum of the molecule to specific protons.

17.63 ■ The following ^{1}H NMR spectrum is that of an alcohol, $C_8H_{10}O$. Propose a structure.

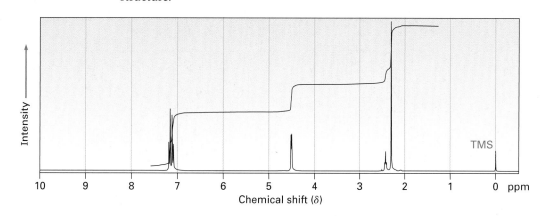

17.64 ■ Propose structures for alcohols that have the following ^{1}H NMR spectra:

(a) $C_5H_{12}O$

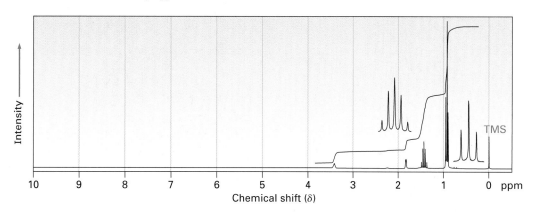

(b) $C_8H_{10}O$

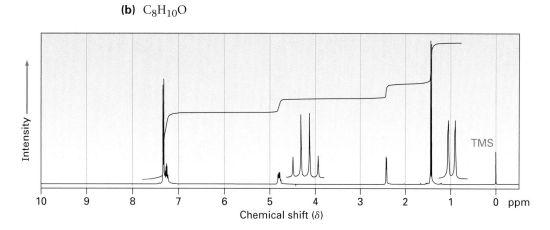

17.65 ■ Propose structures for alcohols that have the following ^{1}H NMR spectra:

(a) $C_9H_{12}O$

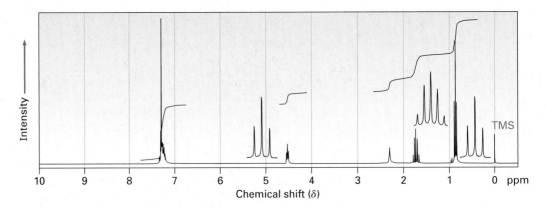

(b) $C_8H_{10}O_2$

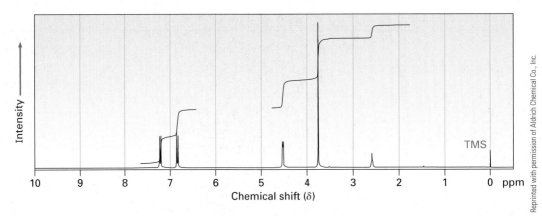

17.66 ■ Compound A, $C_8H_{10}O$, has the IR and ^{1}H NMR spectra shown. Propose a structure consistent with the observed spectra, and assign each peak in the NMR spectrum. Note that the absorption at 5.5 δ disappears when D_2O is added.

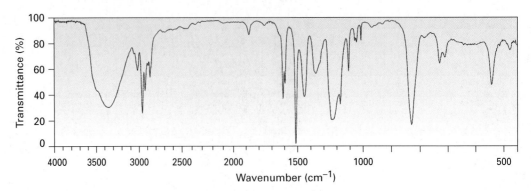

■ Assignable in OWL

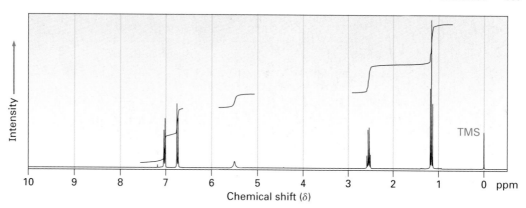

17.67 ■ Propose a structure for a compound $C_{15}H_{24}O$ that has the following ¹H NMR spectrum. The peak marked by an asterisk disappears when D_2O is added to the sample.

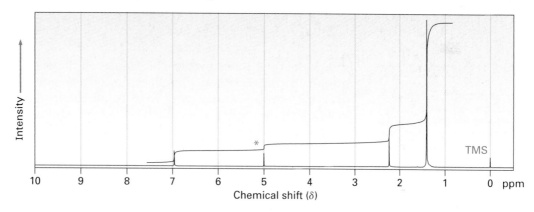

17.68 The reduction of carbonyl compounds by reaction with hydride reagents (H:⁻) and the Grignard addition by reaction with organomagnesium halides (R:⁻ ⁺MgBr) are examples of *nucleophilic carbonyl addition reactions*. What analogous product do you think might result from reaction of cyanide ion with a ketone?

$$\overset{\displaystyle O}{\underset{\displaystyle C}{\parallel}} \quad \xrightarrow[\text{H}_3\text{O}^+]{\text{CN}^-} \quad ?$$

17.69 Ethers can be prepared by reaction of an alkoxide or phenoxide ion with a primary alkyl halide. Anisole, for instance, results from reaction of sodium phenoxide with iodomethane. What kind of reaction is occurring? Show the mechanism.

O⁻ Na⁺

\+   CH₃I   ⟶   OCH₃

Sodium phenoxide   **Anisole**

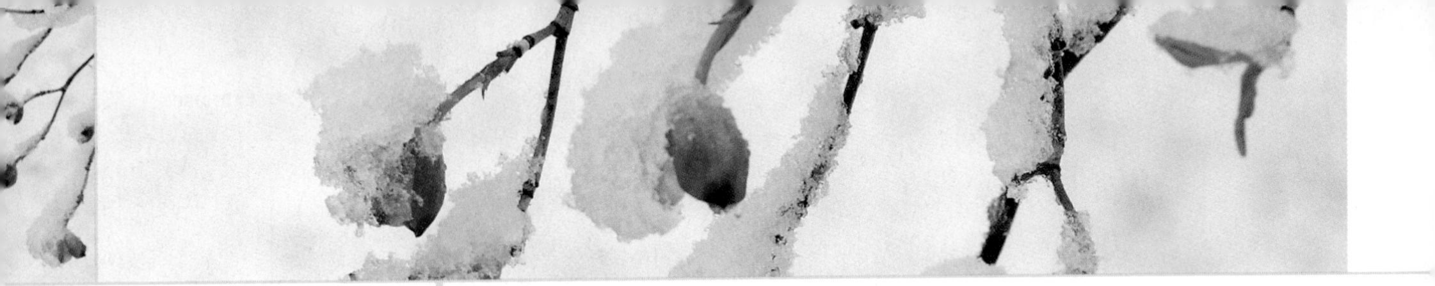

18

Ethers and Epoxides; Thiols and Sulfides

Organic **KNOWLEDGE TOOLS**

CENGAGENOW™ Throughout this chapter, sign in at **www.cengage.com/login** for online self-study and interactive tutorials based on your level of understanding.

ⓦWL Online homework for this chapter may be assigned in Organic OWL.

Ethers (R—O—R'), like the alcohols we saw in the preceding chapter, are also organic derivatives of water but have two organic groups bonded to the same oxygen atom rather than one. The organic groups might be alkyl, aryl, or vinylic, and the oxygen atom might be in an open chain or a ring. Perhaps the most well-known ether is diethyl ether, which has a long history of medicinal use as an anesthetic and industrial use as a solvent. Other useful ethers include anisole, a pleasant-smelling aromatic ether used in perfumery, and tetrahydrofuran (THF), a cyclic ether often used as a solvent.

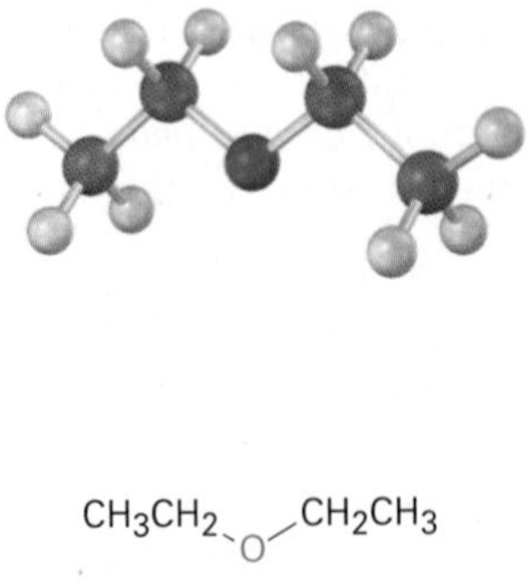

CH_3CH_2 CH_2CH_3
         O

Diethyl ether

O—CH_3

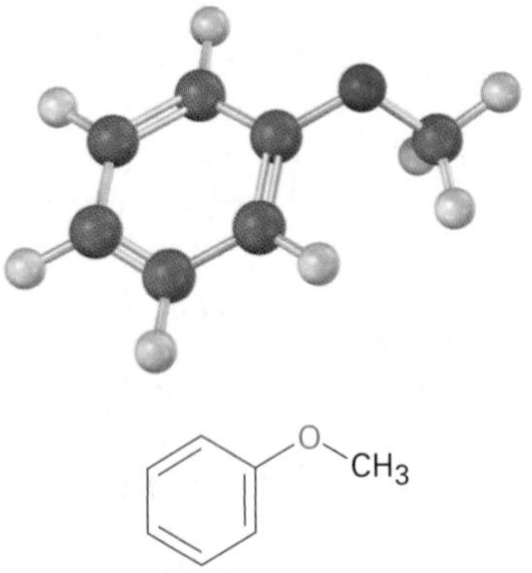

Anisole
(methyl phenyl ether)

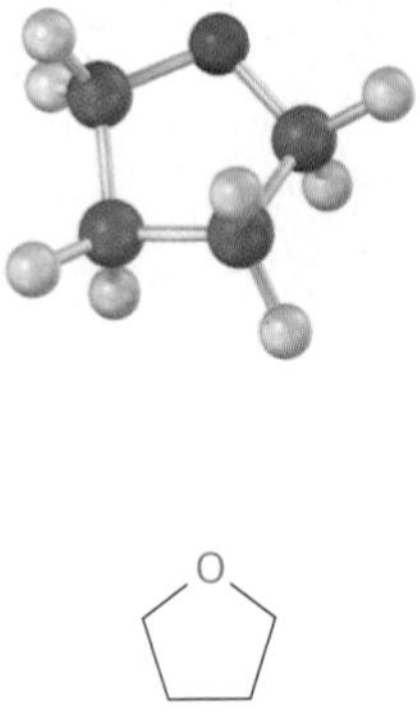

Tetrahydrofuran

Thiols (R—S—H) and **sulfides (R—S—R')** are sulfur analogs of alcohols and ethers, respectively. Both functional groups are found in various biomolecules, although not as commonly as their oxygen-containing relatives.

Sean Duggan

WHY THIS CHAPTER?

This chapter finishes the coverage of functional groups with C—O and C—S single bonds that was begun in Chapter 17. We'll focus primarily on ethers and take only a brief look at thiols and sulfides before going on to an extensive coverage of compounds with C=O bonds in Chapters 19 through 23.

18.1 | Names and Properties of Ethers

Simple ethers with no other functional groups are named by identifying the two organic substituents and adding the word *ether.*

Isopropyl methyl ether   **Ethyl phenyl ether**

If other functional groups are present, the ether part is considered an *alkoxy* substituent. For example:

p-Dimethoxy**benzene**   **4-*tert*-Butoxy-1-cyclohexene**

CENGAGENOW™ Click *Organic Interactive* to **use a web-based palette to draw ether structures based on their IUPAC names**.

Like alcohols, ethers have nearly the same geometry as water. The R—O—R bonds have an approximately tetrahedral bond angle (112° in dimethyl ether), and the oxygen atom is sp^3-hybridized.

The electronegative oxygen atom gives ethers a slight dipole moment, and the boiling points of ethers are often slightly higher than the boiling points of comparable alkanes. Table 18.1 compares the boiling points of some common ethers and the corresponding hydrocarbons.

Ethers are relatively stable and unreactive in many respects, but some ethers react slowly with the oxygen in air to give *peroxides,* compounds that contain an O—O bond. The peroxides from low-molecular-weight ethers such as diisopropyl ether and tetrahydrofuran are explosive and extremely dangerous, even in tiny amounts. Ethers are very useful as solvents in the laboratory, but they must always be used cautiously and should not be stored for long periods of time.

Table 18.1 | **Comparison of Boiling Points of Ethers and Hydrocarbons**

Ether	Boiling point °C	Hydrocarbon	Boiling point °C
CH_3OCH_3	−25	$CH_3CH_2CH_3$	−45
$CH_3CH_2OCH_2CH_3$	34.6	$CH_3CH_2CH_2CH_2CH_3$	36
	65	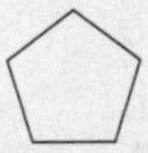	49
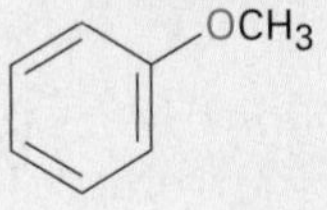	158	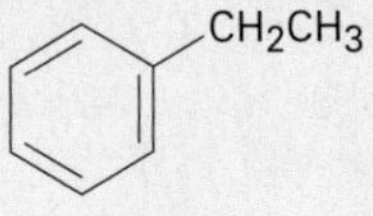	136

Problem 18.1 | Name the following ethers:

(a)
$$\underset{CH_3CHOCHCH_3}{\overset{CH_3\quad CH_3}{|\qquad\ |}}$$

(b)

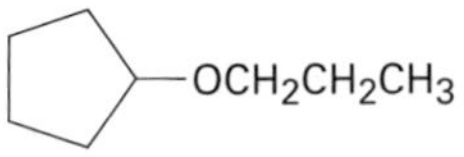

(c)

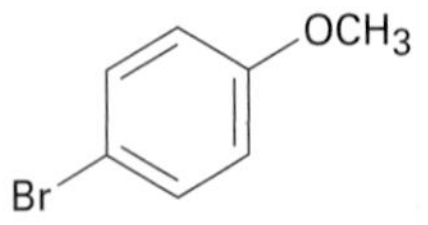

(d)

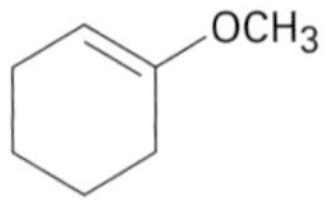

(e)
$$\underset{CH_3CHCH_2OCH_2CH_3}{\overset{CH_3}{|}}$$

(f) $H_2C{=}CHCH_2OCH{=}CH_2$

18.2 | Synthesis of Ethers

Diethyl ether and other simple symmetrical ethers are prepared industrially by the sulfuric acid–catalyzed dehydration of alcohols. The reaction occurs by S_N2 displacement of water from a protonated ethanol molecule by the oxygen atom of a second ethanol. Unfortunately, the method is limited to use with primary alcohols because secondary and tertiary alcohols dehydrate by an E1 mechanism to yield alkenes (Section 17.6).

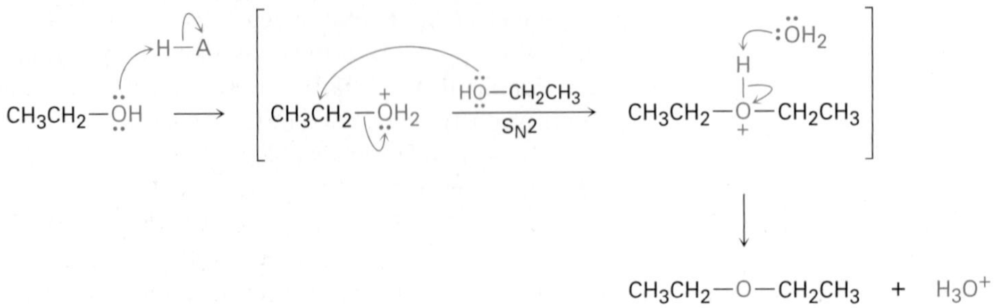

$$CH_3CH_2{-}O{-}CH_2CH_3 \quad + \quad H_3O^+$$

The Williamson Ether Synthesis

The most generally useful method of preparing ethers is by the *Williamson ether synthesis,* in which an alkoxide ion reacts with a primary alkyl halide or tosylate in an S_N2 reaction. As we saw earlier in Section 17.2, the alkoxide ion is normally prepared by reaction of an alcohol with a strong base such as sodium hydride, NaH.

| Cyclopentanol | Alkoxide ion | Cyclopentyl methyl ether (74%) |

A useful variation of the Williamson synthesis involves silver oxide, Ag_2O, as a mild base rather than NaH. Under these conditions, the free alcohol reacts directly with alkyl halide, so there is no need to preform the metal alkoxide intermediate. Sugars react particularly well; glucose, for example, reacts with excess iodomethane in the presence of Ag_2O to generate a pentaether in 85% yield.

α-D-Glucose   α-D-Glucose pentamethyl ether (85%)

Because the Williamson synthesis is an S_N2 reaction, it is subject to all the usual constraints, as discussed in Section 11.2. Primary halides and tosylates work best because competitive E2 elimination can occur with more hindered substrates. Unsymmetrical ethers should therefore be synthesized by reaction between the more hindered alkoxide partner and less hindered halide partner rather than vice versa. For example, *tert*-butyl methyl ether, a substance used in the 1990s as an octane booster in gasoline, is best prepared by reaction of *tert*-butoxide ion with iodomethane rather than by reaction of methoxide ion with 2-chloro-2-methylpropane.

tert-Butoxide   Iodomethane   *tert*-Butyl methyl ether

2-Chloro-2-methylpropane   2-Methylpropene

Problem 18.2 Why do you suppose only symmetrical ethers are prepared by the sulfuric acid–catalyzed dehydration procedure? What product(s) would you expect if ethanol and 1-propanol were allowed to react together? In what ratio would the products be formed if the two alcohols were of equal reactivity?

Problem 18.3 How would you prepare the following ethers using a Williamson synthesis?
(a) Methyl propyl ether   (b) Anisole (methyl phenyl ether)
(c) Benzyl isopropyl ether   (d) Ethyl 2,2-dimethylpropyl ether

Alkoxymercuration of Alkenes

We saw in Section 7.4 that alkenes react with water in the presence of mercuric acetate to yield a hydroxymercuration product. Subsequent treatment with $NaBH_4$ breaks the C–Hg bond and yields the alcohol. A similar **alkoxymercuration** reaction occurs when an alkene is treated with an *alcohol* in the presence of mercuric acetate or, even better, mercuric trifluoroacetate, $(CF_3CO_2)_2Hg$. Demercuration by reaction with $NaBH_4$ then yields an ether. The net result is Markovnikov addition of the alcohol to the alkene.

Styrene   **1-Methoxy-1-phenylethane (97%)**

Cyclohexene   **Cyclohexyl ethyl ether (100%)**

CENGAGENOW™ Click *Organic Interactive* to **practice your problem-solving skills designing syntheses of ethers**.

The mechanism of the alkoxymercuration reaction is similar to that described in Section 7.4 for hydroxymercuration. The reaction is initiated by electrophilic addition of Hg^{2+} to the alkene, followed by reaction of the intermediate cation with alcohol and reduction of the C–Hg bond by $NaBH_4$. A variety of alcohols and alkenes can be used in the alkoxymercuration reaction. Primary, secondary, and even tertiary alcohols react well, but ditertiary ethers can't be prepared because of steric hindrance to reaction.

WORKED EXAMPLE 18.1   *Synthesizing an Ether*

How would you prepare ethyl phenyl ether? Use whichever method you think is more appropriate, the Williamson synthesis or the alkoxymercuration reaction.

Strategy   Draw the target ether, identify the two groups attached to oxygen, and recall the limitations of the two methods for preparing ethers. The Williamson synthesis uses an S_N2 reaction and requires that one of the two groups attached to oxygen be either secondary or (preferably) primary. The alkoxymercuration reaction requires that one of the two groups come from an alkene precursor. Ethyl phenyl ether could be made by either method.

Ethyl phenyl ether

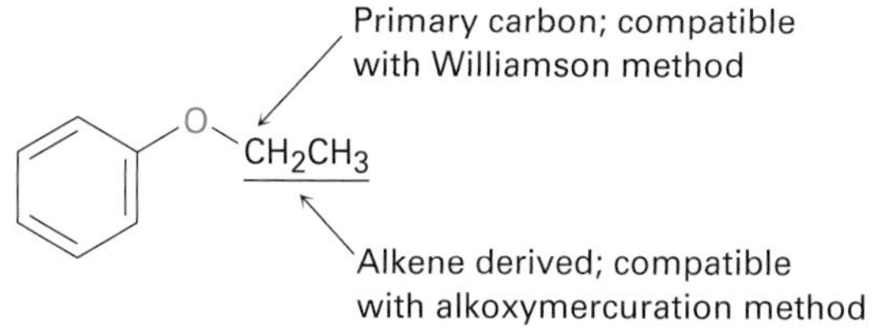

Solution

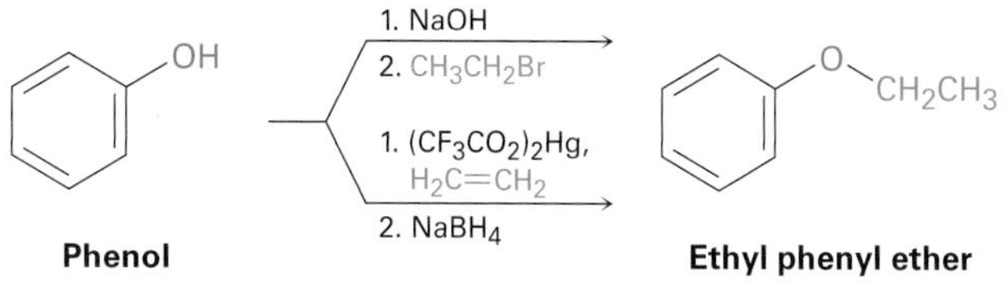

Phenol   Ethyl phenyl ether

Problem 18.4 Review the mechanism of oxymercuration shown in Figure 7.4 (p. 225), and then write the mechanism of the alkoxymercuration reaction of 1-methylcyclopentene with ethanol. Use curved arrows to show the electron flow in each step.

Problem 18.5 How would you prepare the following ethers? Use whichever method you think is more appropriate, the Williamson synthesis or the alkoxymercuration reaction.

(a) Butyl cyclohexyl ether **(b)** Benzyl ethyl ether ($C_6H_5CH_2OCH_2CH_3$)
(c) *sec*-Butyl *tert*-butyl ether **(d)** Tetrahydrofuran

Problem 18.6 Rank the following halides in order of their reactivity in the Williamson synthesis:

(a) Bromoethane, 2-bromopropane, bromobenzene
(b) Chloroethane, bromoethane, 1-iodopropene

18.3 Reactions of Ethers: Acidic Cleavage

Ethers are unreactive to many reagents used in organic chemistry, a property that accounts for their wide use as reaction solvents. Halogens, dilute acids, bases, and nucleophiles have no effect on most ethers. In fact, ethers undergo only one reaction of general use—they are cleaved by strong acids. Aqueous HBr and HI both work well, but HCl does not cleave ethers.

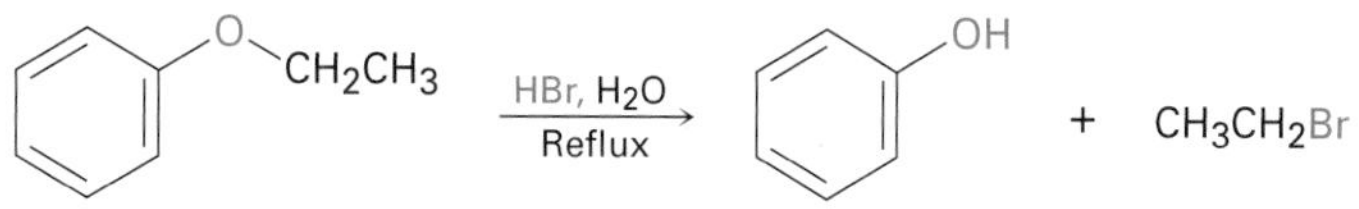

Ethyl phenyl ether   Phenol   Bromoethane

Acidic ether cleavages are typical nucleophilic substitution reactions, either S_N1 or S_N2 depending on the structure of the substrate. Ethers with only primary and secondary alkyl groups react by an S_N2 mechanism, in which I^- or Br^- attacks the protonated ether at the less hindered site. This usually results in a selective cleavage into a single alcohol and a single alkyl halide. For example, ethyl isopropyl ether yields exclusively isopropyl alcohol and iodoethane on cleavage by HI because nucleophilic attack by iodide ion occurs at the less hindered primary site rather than at the more hindered secondary site.

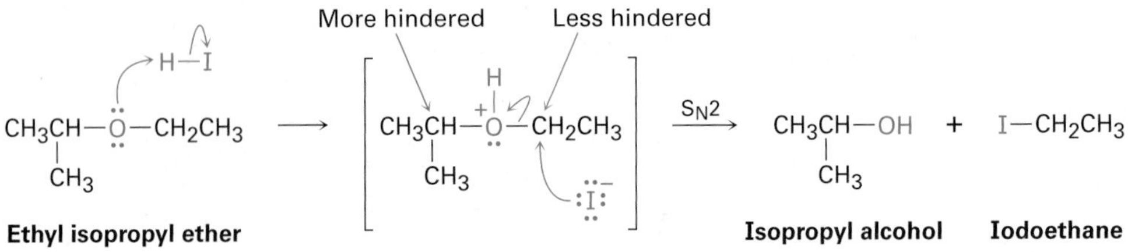

Ethyl isopropyl ether   **Isopropyl alcohol**   **Iodoethane**

Ethers with a tertiary, benzylic, or allylic group cleave by an S_N1 or E1 mechanism because these substrates can produce stable intermediate carbocations. These reactions are often fast and take place at moderate temperatures. *tert*-Butyl ethers, for example, react by an E1 mechanism on treatment with trifluoroacetic acid at 0 °C. We'll see in Section 26.7 that the reaction is often used in the laboratory synthesis of peptides.

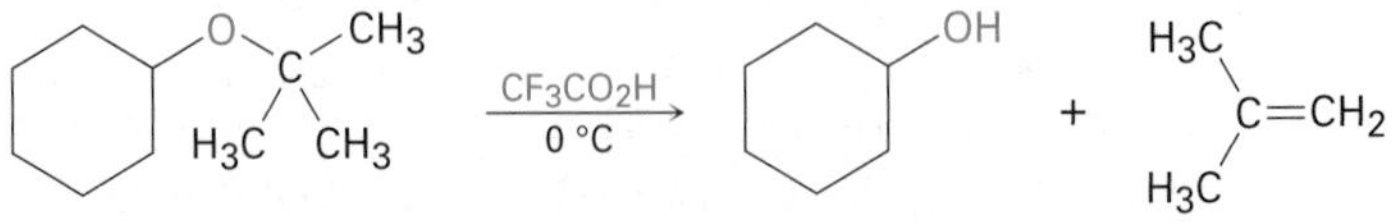

tert-**Butyl cyclohexyl ether**   **Cyclohexanol**   **2-Methylpropene**
                                            (90%)

WORKED EXAMPLE 18.2   ***Predicting the Product of an Ether Cleavage Reaction***

Predict the products of the following reaction:

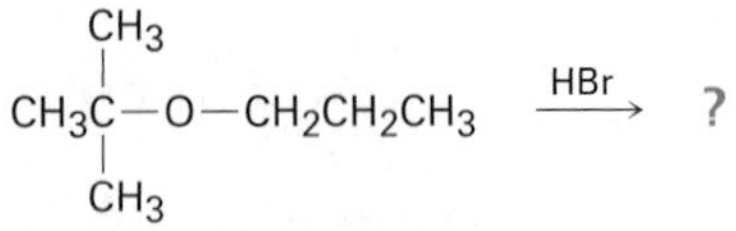

Strategy   Identify the substitution pattern of the two groups attached to oxygen—in this case a tertiary alkyl group and a primary alkyl group. Then recall the guidelines for ether cleavages. An ether with only primary and secondary alkyl groups usually undergoes cleavage by S_N2 attack of a nucleophile on the less hindered alkyl group, but an ether with a tertiary alkyl group usually undergoes cleavage by an S_N1 mechanism. In this case, an S_N1 cleavage of the tertiary C—O bond will occur, giving 1-propanol and a tertiary alkyl bromide.

Solution

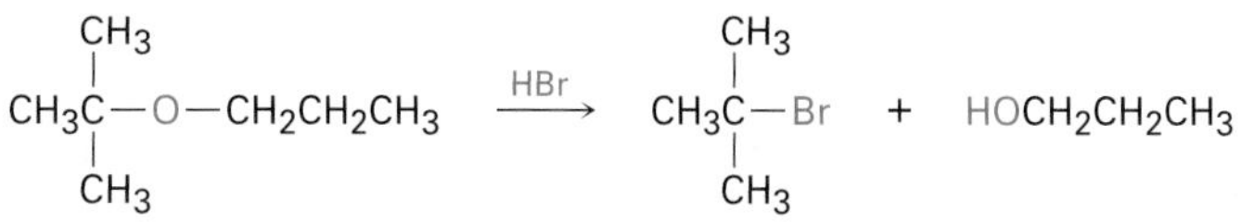

***tert*-Butyl propyl ether** | **2-Bromo-2-methylpropane** | **Propan-1-ol**

Problem 18.7 Predict the products of the following reactions:

(a)

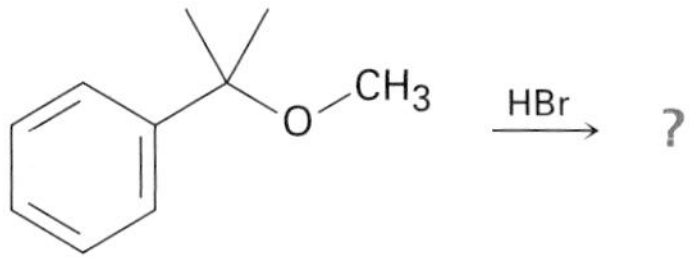

(b)

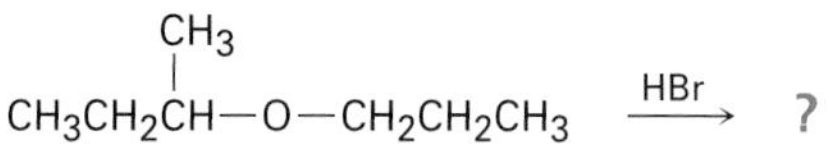

Problem 18.8 Write the mechanism of the acid-catalyzed cleavage of *tert*-butyl cyclohexyl ether to yield cyclohexanol and 2-methylpropene.

Problem 18.9 Why are HI and HBr more effective than HCl in cleaving ethers? (See Section 11.3.)

18.4 Reactions of Ethers: Claisen Rearrangement

> **Ludwig Claisen**
>
> **Ludwig Claisen** (1851–1930) was born in Cologne, Germany, and received his Ph.D. at the University of Bonn, studying under August Kekulé. He never married, but devoted himself throughout his life to organic chemistry. Among his positions, he was professor at the University of Bonn, Owens College (Manchester), and the universities of Munich, Aachen, Kiel, and Berlin.

Unlike the acid-catalyzed ether cleavage reaction discussed in the previous section, which is general to all ethers, the **Claisen rearrangement** is specific to allyl aryl ethers, Ar—O—$CH_2CH{=}CH_2$. Treatment of a phenoxide ion with 3-bromopropene (allyl bromide) results in a Williamson ether synthesis and formation of an allyl aryl ether. Heating the allyl aryl ether to 200 to 250 °C then effects Claisen rearrangement, leading to an *o*-allylphenol. The net result is alkylation of the phenol in an ortho position.

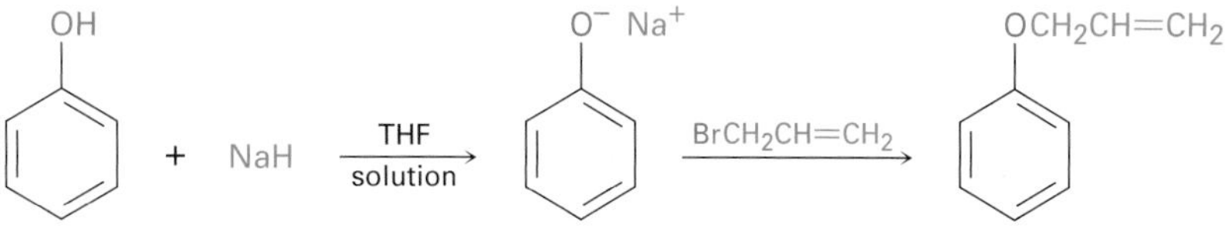

Phenol | **Sodium phenoxide** | **Allyl phenyl ether**

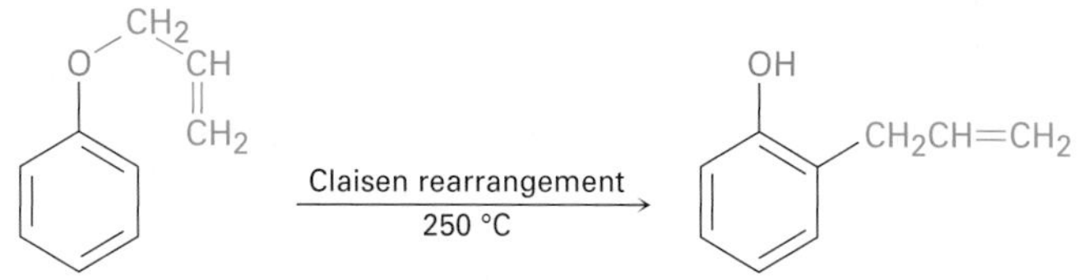

Allyl phenyl ether | ***o*-Allylphenol**

Like the Diels–Alder reaction discussed in Sections 14.4 and 14.5, the Claisen rearrangement reaction takes place through a pericyclic mechanism in which a concerted reorganization of bonding electrons occurs through a six-membered, cyclic transition state. The 6-allyl-2,4-cyclohexadienone intermediate then isomerizes to *o*-allylphenol (Figure 18.1).

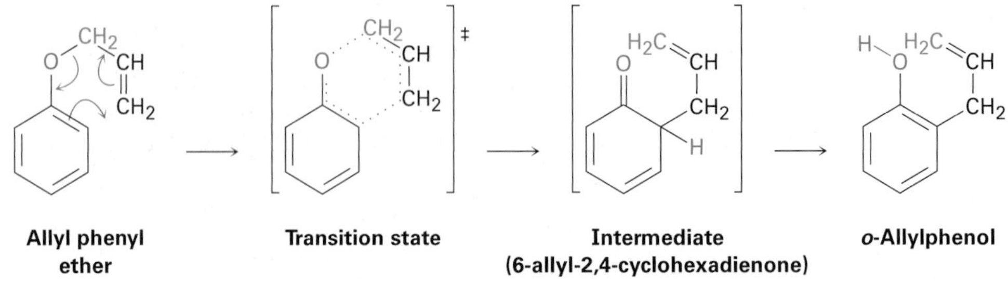

Allyl phenyl ether **Transition state** **Intermediate (6-allyl-2,4-cyclohexadienone)** ***o*-Allylphenol**

Active Figure 18.1 The mechanism of the Claisen rearrangement. The C–O bond-breaking and C–C bond-making occur simultaneously. *Sign in at* **www.cengage.com/login** *to see a simulation based on this figure and to take a short quiz.*

Evidence for this mechanism comes from the observation that the rearrangement takes place with an inversion of the allyl group. That is, allyl phenyl ether containing a ^{14}C label on the allyl *ether* carbon atom yields *o*-allylphenol in which the label is on the *terminal* vinylic carbon (green in Figure 18.1). It would be very difficult to explain this result by any mechanism other than a pericyclic one. We'll look at the reaction in more detail in Section 30.8.

Problem 18.10 | What product would you expect from Claisen rearrangement of 2-butenyl phenyl ether?

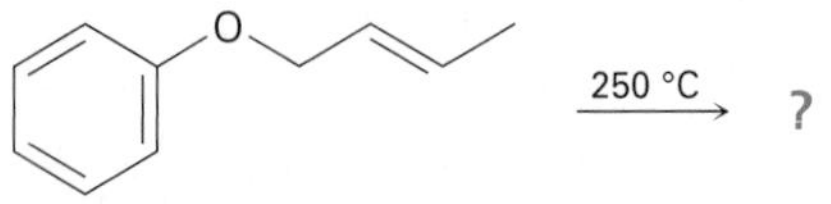

2-Butenyl phenyl ether

18.5 | Cyclic Ethers: Epoxides

For the most part, cyclic ethers behave like acyclic ethers. The chemistry of the ether functional group is the same, whether it's in an open chain or in a ring. Common cyclic ethers such as tetrahydrofuran and dioxane, for example, are often used as solvents because of their inertness, yet they can be cleaved by strong acids.

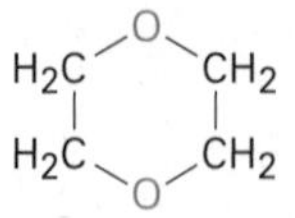

1,4-Dioxane 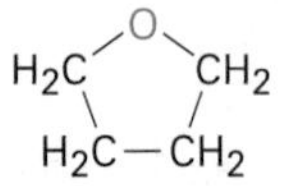 **Tetrahydrofuran**

The one group of cyclic ethers that behaves differently from open-chain ethers contains the three-membered-ring compounds called *epoxides,* or *oxiranes,* which we saw in Section 7.8. The strain of the three-membered ring gives epoxides unique chemical reactivity.

Ethylene oxide, the simplest epoxide, is an intermediate in the manufacture of both ethylene glycol, used for automobile antifreeze, and polyester polymers. More than 4 million tons of ethylene oxide is produced each year in the United States by air oxidation of ethylene over a silver oxide catalyst at 300 °C. This process is not useful for other epoxides, however, and is of little value in the laboratory. Note that the name *ethylene oxide* is not a systematic one because the *-ene* ending implies the presence of a double bond in the molecule. The name is frequently used, however, because ethylene oxide is derived *from* ethylene by addition of an oxygen atom. Other simple epoxides are named similarly. The systematic name for ethylene oxide is 1,2-epoxyethane.

$H_2C=CH_2$ $\xrightarrow[\text{Ag}_2\text{O},\ 300\ °C]{O_2}$ H_2C-CH_2 (epoxide)

Ethylene　　　　　　　**Ethylene oxide**

In the laboratory, as we saw in Section 7.8, epoxides are prepared by treatment of an alkene with a peroxyacid (RCO_3H), typically *m*-chloroperoxybenzoic acid.

Cycloheptene　　**meta-Chloroperoxybenzoic acid**　　**1,2-Epoxycycloheptane**　　**meta-Chlorobenzoic acid**

Another method for the synthesis of epoxides is through the use of halohydrins, prepared by electrophilic addition of HO—X to alkenes (Section 7.3). When halohydrins are treated with base, HX is eliminated and an epoxide is produced by an *intramolecular* Williamson ether synthesis. That is, the nucleophilic alkoxide ion and the electrophilic alkyl halide are in the same molecule.

Cyclohexene　　**trans-2-Chlorocyclohexanol**　　　　**1,2-Epoxycyclohexane**

Problem 18.11　Reaction of *cis*-2-butene with *m*-chloroperoxybenzoic acid yields an epoxide different from that obtained by reaction of the trans isomer. Explain.

18.6 | Reactions of Epoxides: Ring-Opening

Acid-Catalyzed Epoxide Opening

Epoxides are cleaved by treatment with acid just as other ethers are, but under much milder conditions because of ring strain. As we saw in Section 7.8, dilute aqueous acid at room temperature is sufficient to cause the hydrolysis of epoxides to 1,2-diols, also called *vicinal glycols*. (The word *vicinal* means "adjacent," and a *glycol* is a diol.) The epoxide cleavage takes place by S_N2-like backside attack of a nucleophile on the protonated epoxide, giving a *trans*-1,2-diol as product.

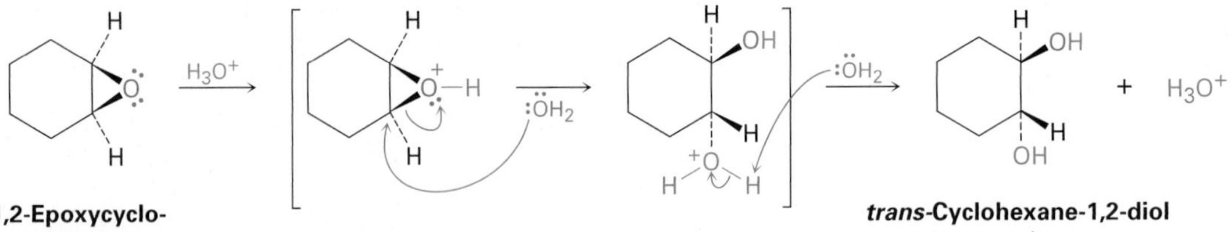

1,2-Epoxycyclo-hexane

trans-Cyclohexane-1,2-diol
(86%)

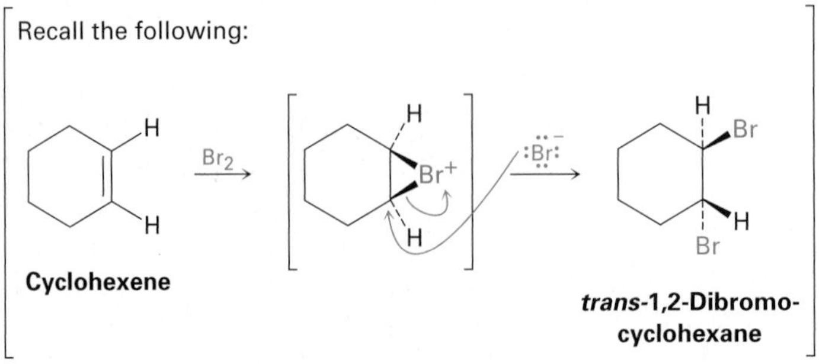

Recall the following:

Cyclohexene

trans-1,2-Dibromo-cyclohexane

Epoxides can also be opened by reaction with acids other than H_3O^+. If anhydrous HX is used, for instance, an epoxide is converted into a trans halohydrin.

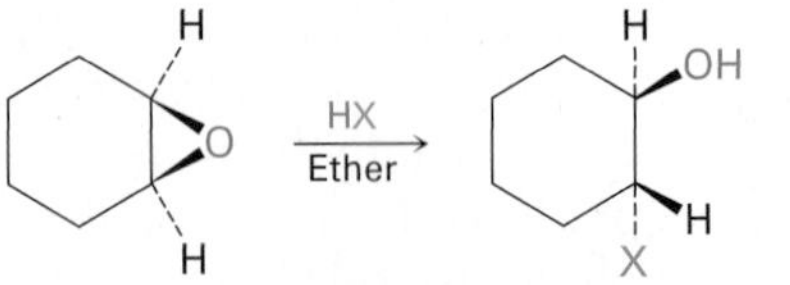

A trans 2-halocyclohexanol

where   X = F, Br, Cl, or I

The regiochemistry of acid-catalyzed ring-opening depends on the epoxide's structure, and a mixture of products is often formed. When both epoxide carbon atoms are either primary or secondary, attack of the nucleophile occurs primarily at the *less* highly substituted site—an S_N2-like result. When one of the epoxide carbon atoms is tertiary, however, nucleophilic attack occurs primarily at the *more* highly substituted site—an S_N1-like result. Thus, 1,2-epoxypropane reacts with HCl to give primarily 1-chloro-2-propanol, but 2-methyl-1,2-epoxypropane gives 2-chloro-2-methyl-1-propanol as the major product.

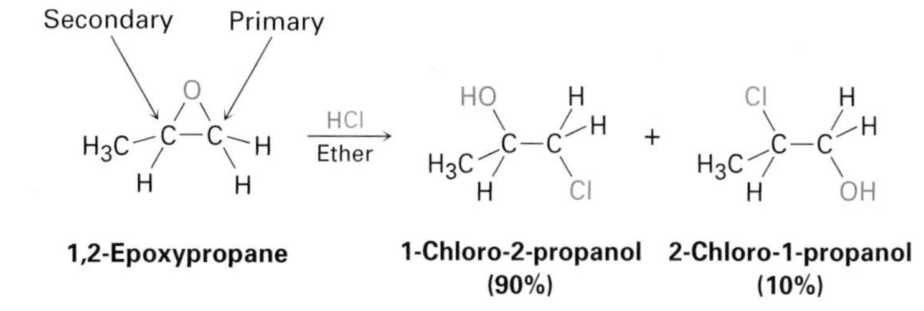

Secondary   Primary

1,2-Epoxypropane   →(HCl / Ether)→   **1-Chloro-2-propanol (90%)**   +   **2-Chloro-1-propanol (10%)**

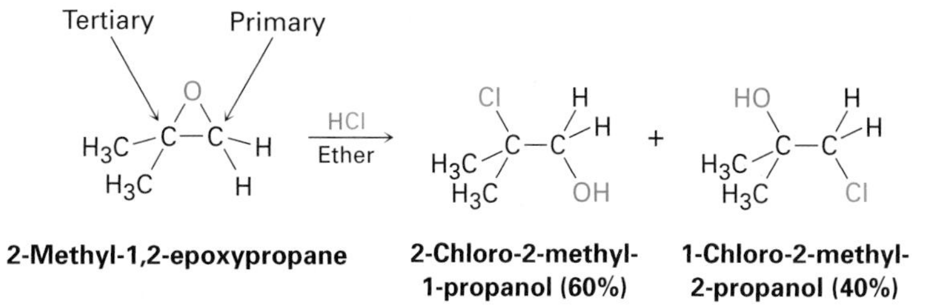

Tertiary   Primary

2-Methyl-1,2-epoxypropane   →(HCl / Ether)→   **2-Chloro-2-methyl-1-propanol (60%)**   +   **1-Chloro-2-methyl-2-propanol (40%)**

The mechanisms of these acid-catalyzed epoxide openings are more complex than they at first appear. They seem to be neither purely S_N1 nor S_N2 but instead to be midway between the two extremes and to have characteristics of both. Take the reaction of 1,2-epoxy-1-methylcyclohexane with HBr shown in Figure 18.2, for instance. The reaction yields only a single stereoisomer of 2-bromo-2-methyl-cyclohexanol in which the −Br and −OH groups are trans, an S_N2-like result caused by backside displacement of the epoxide oxygen. But the fact that Br⁻ attacks the more hindered tertiary side of the epoxide rather than the less hindered secondary side is an S_N1-like result in which the more stable, tertiary carbocation is involved.

Evidently, the transition state for acid-catalyzed epoxide opening has an S_N2-like geometry but also has a large amount of S_N1-like carbocationic character. Since the positive charge in the protonated epoxide is shared by the more highly substituted carbon atom, backside attack of Br⁻ occurs at the more highly substituted site.

Active Figure 18.2 Acid-induced ring-opening of 1,2-epoxy-1-methylcyclohexane with HBr. There is a high degree of S_N1-like carbocation character in the transition state, which leads to backside attack of the nucleophile at the tertiary center and to formation of the isomer of 2-bromo-2-methyl-cyclohexanol that has −Br and −OH groups trans. (Naming of trisubstituted cyclohexanes was explained in Section 7.8.) *Sign in at* **www.cengage.com/login** *to see a simulation based on this figure and to take a short quiz.*

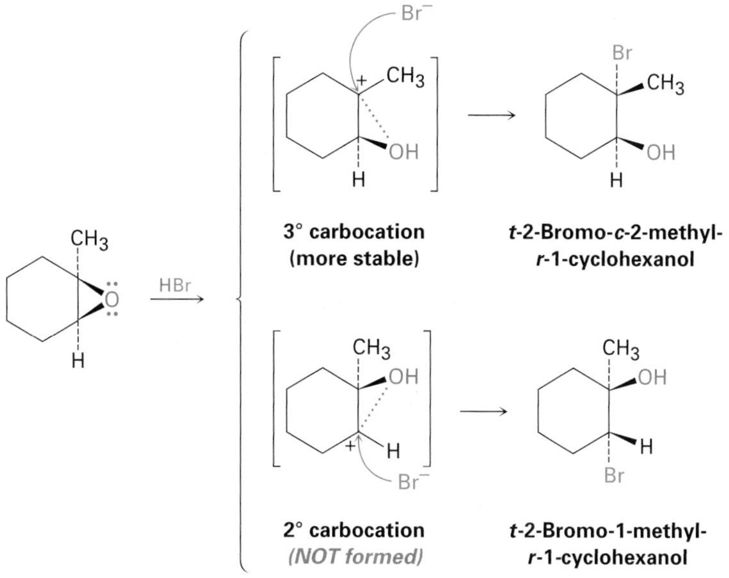

3° carbocation (more stable)   →   t-2-Bromo-c-2-methyl-r-1-cyclohexanol

2° carbocation (NOT formed)   →   t-2-Bromo-1-methyl-r-1-cyclohexanol

WORKED EXAMPLE 18.3 | ***Predicting the Product of Epoxide Ring-Opening***

Predict the major product of the following reaction:

Strategy Identify the substitution pattern of the two epoxide carbon atoms—in this case, one carbon is secondary and one is primary. Then recall the guidelines for epoxide cleavages. An epoxide with only primary and secondary carbons usually undergoes cleavage by S_N2-like attack of a nucleophile on the less hindered carbon, but an epoxide with a tertiary carbon atom usually undergoes cleavage by backside attack on the more hindered carbon. In this case, an S_N2 cleavage of the primary C–O epoxide bond will occur.

Solution

Problem 18.12 | Predict the major product of each of the following reactions:

(a)

HCl / Ether → ?

(b)

HCl / Ether → ?

Problem 18.13 | How would you prepare the following diols?

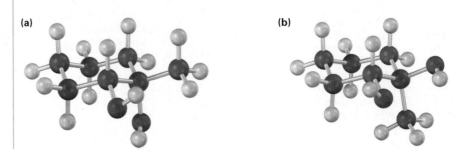

(a)

(b)

Base-Catalyzed Epoxide Opening

Unlike other ethers, epoxide rings can be cleaved by base as well as by acid. Although an ether oxygen is normally a poor leaving group in an S_N2 reaction (Section 11.3), the strain of the three-membered ring causes epoxides to react with hydroxide ion at elevated temperatures.

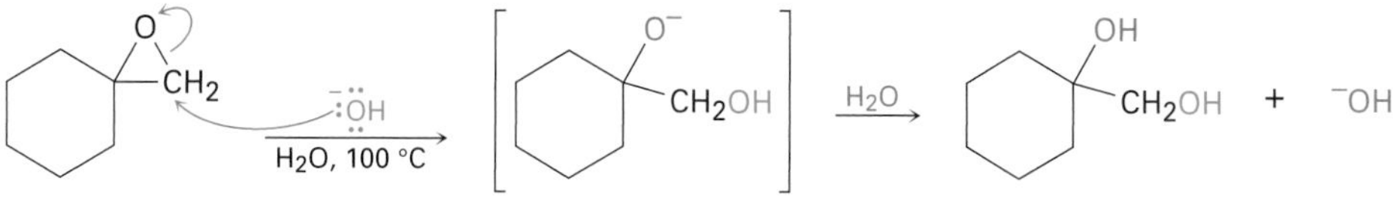

Methylenecyclohexane oxide

1-Hydroxymethyl-cyclohexanol (70%)

A similar nucleophilic ring-opening occurs when epoxides are treated with Grignard reagents. Ethylene oxide is frequently used, thereby allowing the conversion of a Grignard reagent into a primary alcohol having two more carbons than the starting alkyl halide. 1-Bromobutane, for example, is converted into 1-hexanol by reaction of its Grignard reagent with ethylene oxide.

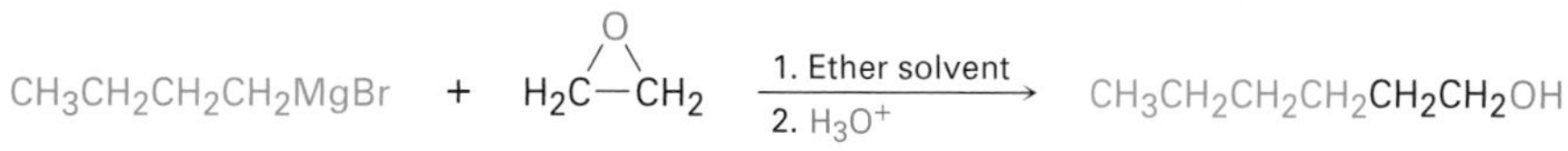

$$CH_3CH_2CH_2CH_2MgBr \ + \ H_2C{-}CH_2 \xrightarrow[\text{2. } H_3O^+]{\text{1. Ether solvent}} CH_3CH_2CH_2CH_2CH_2CH_2OH$$

Butylmagnesium bromide   **Ethylene oxide**   **1-Hexanol (62%)**

CENGAGENOW™ Click *Organic Interactive* to **use a web-based palette to predict products from a variety of reactions involving ethers and epoxides**.

Base-catalyzed epoxide opening is a typical S_N2 reaction in which attack of the nucleophile takes place at the less hindered epoxide carbon. For example, 1,2-epoxypropane reacts with ethoxide ion exclusively at the less highly substituted, primary, carbon to give 1-ethoxy-2-propanol.

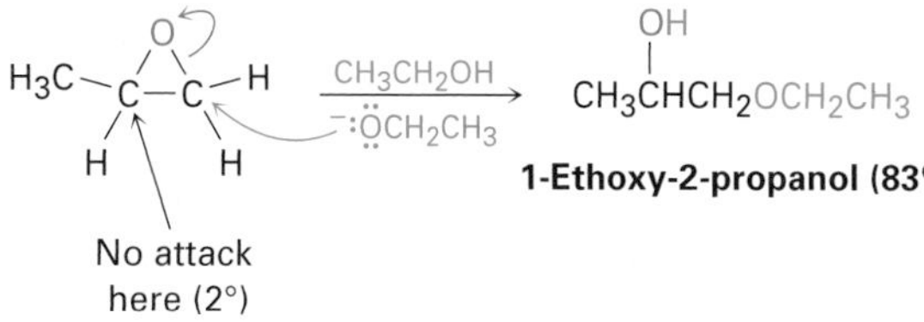

1-Ethoxy-2-propanol (83%)

No attack here (2°)

Problem 18.14 | Predict the major product of the following reactions:

(a)

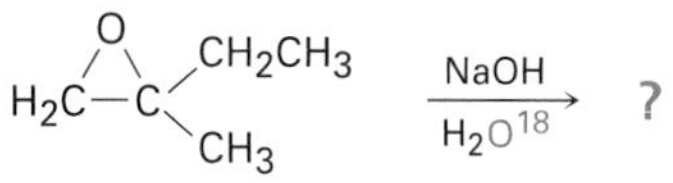

(b)

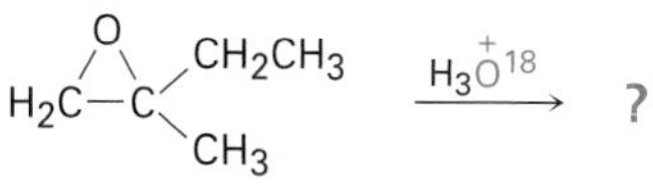

(c)

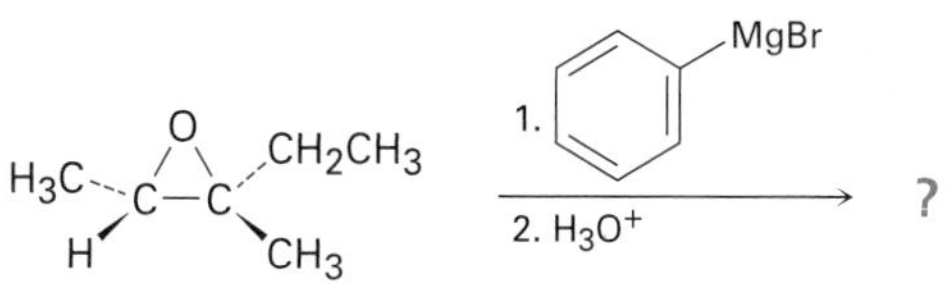

18.7 | Crown Ethers

Crown ethers, discovered in the early 1960s by Charles Pedersen at the DuPont Company, are a relatively recent addition to the ether family. Crown ethers are named according to the general format *x*-crown-*y*, where *x* is the total number of atoms in the ring and *y* is the number of oxygen atoms. Thus, 18-crown-6 ether is an 18-membered ring containing 6 ether oxygen atoms. Note the size and negative (red) character of the crown ether cavity in the following electrostatic potential map.

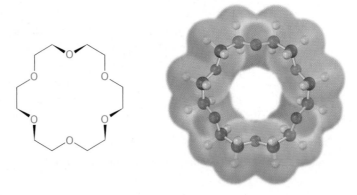

18-Crown-6 ether

The importance of crown ethers derives from their extraordinary ability to solvate metal cations by sequestering the metal in the center of the polyether cavity. For example, 18-crown-6 complexes strongly with potassium ion. Complexes between crown ethers and ionic salts are soluble in nonpolar organic solvents, thus allowing many reactions to be carried out under aprotic conditions that would otherwise have to be carried out in aqueous solution. Potassium permanganate, $KMnO_4$, dissolves in toluene in the presence of 18-crown-6, for instance, and the resulting solution is a valuable reagent for oxidizing alkenes.

Many other inorganic salts, including KF, KCN, and NaN_3, also dissolve in organic solvents with the help of crown ethers. The effect of using a crown ether to dissolve a salt in a hydrocarbon or ether solvent is similar to the effect of dissolving the salt in a polar aprotic solvent such as DMSO, DMF, or HMPA (Section 11.3). In both cases, the metal cation is strongly solvated, leaving the anion bare. Thus, the S_N2 reactivity of an anion is tremendously enhanced in the presence of a crown ether.

Problem 18.15   15-Crown-5 and 12-crown-4 ethers complex Na^+ and Li^+, respectively. Make models of these crown ethers, and compare the sizes of the cavities.

18.8 | Thiols and Sulfides

Thiols

Thiols, sometimes called *mercaptans,* are sulfur analogs of alcohols. They are named by the same system used for alcohols, with the suffix *-thiol* used in place of *-ol.* The —SH group itself is referred to as a **mercapto group**.

CH₃CH₂SH		
Ethanethiol	**Cyclohexane**thiol	*m*-**Mercapto**benzoic acid

The most striking characteristic of thiols is their appalling odor. Skunk scent, for instance, is caused primarily by the simple thiols 3-methyl-1-butanethiol and 2-butene-1-thiol. Volatile thiols such as ethanethiol are also added to natural gas and liquefied propane to serve as an easily detectable warning in case of leaks.

Thiols are usually prepared from alkyl halides by S_N2 displacement with a sulfur nucleophile such as hydrosulfide anion, ⁻SH.

CH₃CH₂CH₂CH₂CH₂CH₂CH₂CH₂—Br + :ṠH ⟶ CH₃CH₂CH₂CH₂CH₂CH₂CH₂CH₂—SH + Br⁻

1-Bromooctane   **Octane-1-thiol (83%)**

The reaction often works poorly unless an excess of the nucleophile is used because the product thiol can undergo a second S_N2 reaction with alkyl halide to give a sulfide as a by-product. To circumvent this problem, thiourea, $(NH_2)_2C=S$, is often used as the nucleophile in the preparation of a thiol from an alkyl halide. The reaction occurs by displacement of the halide ion to yield an intermediate alkyl isothiourea salt, which is hydrolyzed by subsequent reaction with aqueous base.

CH₃CH₂CH₂CH₂CH₂CH₂CH₂CH₂—Br + thiourea ⟶ [alkyl isothiourea salt]

1-Bromooctane   **Thiourea**

↓ H₂O, NaOH

CH₃CH₂CH₂CH₂CH₂CH₂CH₂CH₂—SH + Urea

Octane-1-thiol (83%)   **Urea**

Thiols can be oxidized by Br_2 or I_2 to yield **disulfides (RSSR′)**. The reaction is easily reversed, and a disulfide can be reduced back to a thiol by treatment with zinc and acid.

$$2\,R-SH \underset{Zn,\,H^+}{\overset{I_2}{\rightleftharpoons}} R-S-S-R \;+\; 2\,HI$$

A thiol   **A disulfide**

This thiol–disulfide interconversion is a key part of numerous biological processes. We'll see in Chapter 26, for instance, that disulfide formation is involved in defining the structure and three-dimensional conformations of proteins, where disulfide "bridges" often form cross-links between cysteine amino acid units in the protein chains. Disulfide formation is also involved in the process by which cells protect themselves from oxidative degradation. A cellular component called *glutathione* removes potentially harmful oxidants and is itself oxidized to glutathione disulfide in the process. Reduction back to the thiol requires the coenzyme flavin adenine dinucleotide (reduced), abbreviated $FADH_2$.

Glutathione (GSH)   **Glutathione disulfide (GSSG)**

Sulfides

Sulfides are the sulfur analogs of ethers just as thiols are the sulfur analogs of alcohols. Sulfides are named by following the same rules used for ethers, with *sulfide* used in place of *ether* for simple compounds and *alkylthio* used in place of *alkoxy* for more complex substances.

Dimethyl sulfide   **Methyl phenyl sulfide**   **3-(Methylthio)cyclohexene**

Treatment of a thiol with a base, such as NaH, gives the corresponding **thiolate ion (RS⁻)**, which undergoes reaction with a primary or secondary alkyl halide to give a sulfide. The reaction occurs by an S_N2 mechanism, analogous to the Williamson synthesis of ethers (Section 18.2). Thiolate anions are among the best nucleophiles known, and product yields are usually high in these S_N2 reactions.

Sodium benzenethiolate   **Methyl phenyl sulfide (96%)**

Perhaps surprisingly in light of their close structural similarity, disulfides and ethers differ substantially in their chemistry. Because the valence electrons on sulfur are farther from the nucleus and are less tightly held than those on oxygen ($3p$ electrons versus $2p$ electrons), sulfur compounds are more nucleophilic than their oxygen analogs. Unlike dialkyl ethers, dialkyl sulfides are good nucleophiles that react rapidly with primary alkyl halides by an S_N2 mechanism to give **sulfonium ions (R_3S^+)**.

Dimethyl sulfide   Iodomethane   Trimethylsulfonium iodide

The most common example of this process in living organisms is the reaction of the amino acid methionine with adenosine triphosphate (ATP; Section 5.8) to give *S*-adenosylmethionine. The reaction is somewhat unusual in that the biological leaving group in this S_N2 process is the *triphosphate* ion rather than the more frequently seen *diphosphate* ion (Section 11.6).

Methionine   **Triphosphate ion**

Adenosine triphosphate (ATP)   ***S*-Adenosylmethionine**

Sulfonium ions are themselves useful alkylating agents because a nucleophile can attack one of the groups bonded to the positively charged sulfur, displacing a neutral sulfide as leaving group. We saw an example in Section 11.6

(Figure 11.16) in which *S*-adenosylmethionine transferred a methyl group to norepinephrine to give adrenaline.

Another difference between sulfides and ethers is that sulfides are easily oxidized. Treatment of a sulfide with hydrogen peroxide, H_2O_2, at room temperature yields the corresponding **sulfoxide (R_2SO)**, and further oxidation of the sulfoxide with a peroxyacid yields a **sulfone (R_2SO_2)**.

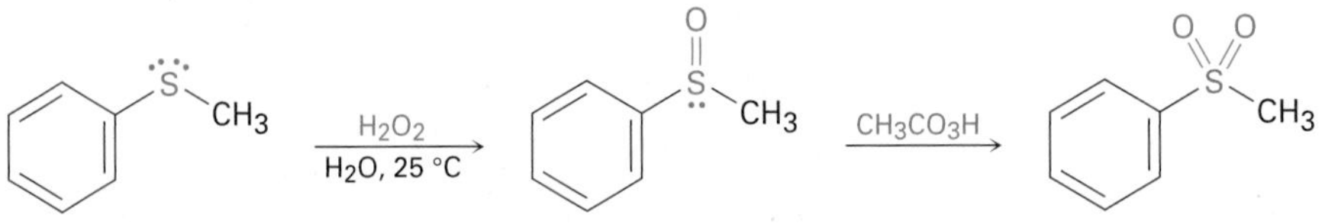
CH3
H2O2
H2O, 25 °C
CH3
CH3CO3H
CH3

Methyl phenyl sulfide **Methyl phenyl sulfoxide** **Methyl phenyl sulfone**

Dimethyl sulfoxide (DMSO) is a particularly well-known sulfoxide that is often used as a polar aprotic solvent. It must be handled with care, however, because it has a remarkable ability to penetrate the skin, carrying along whatever is dissolved in it.

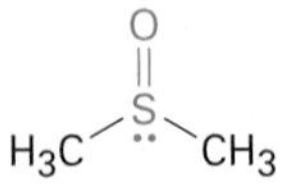
H3C
CH3

Dimethyl sulfoxide (a polar aprotic solvent)

Problem 18.16 Name the following compounds:

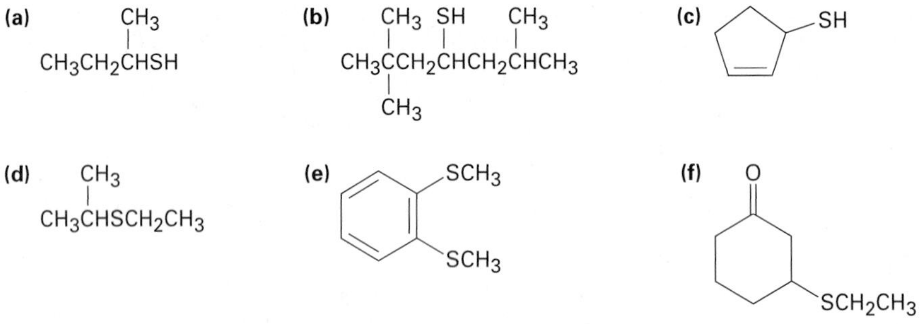
(a)
CH3
CH3CH2CHSH
(b)
CH3 SH CH3
CH3CCH2CHCH2CHCH3
CH3
(c)
SH
(d)
CH3
CH3CHSCH2CH3
(e)
SCH3
SCH3
(f)
SCH2CH3

Problem 18.17 2-Butene-1-thiol is one component of skunk spray. How would you synthesize this substance from methyl 2-butenoate? From 1,3-butadiene?

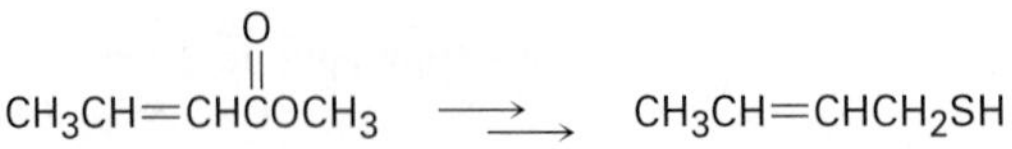
CH3CH=CHCOCH3
CH3CH=CHCH2SH

Methyl but-2-enoate **But-2-ene-1-thiol**

18.9 | Spectroscopy of Ethers

Infrared Spectroscopy

Ethers are difficult to identify by IR spectroscopy. Although they show an absorption due to $C-O$ single-bond stretching in the range 1050 to 1150 cm^{-1}, many other kinds of absorptions occur in the same range. Figure 18.3 shows the IR spectrum of diethyl ether and identifies the $C-O$ stretch.

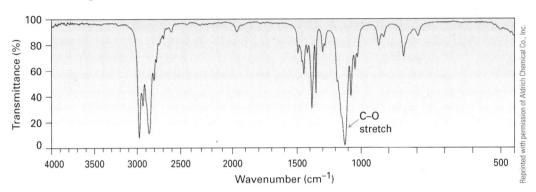

Figure 18.3 The infrared spectrum of diethyl ether, $CH_3CH_2OCH_2CH_3$.

Nuclear Magnetic Resonance Spectroscopy

Hydrogens on carbon next to an ether oxygen are shifted downfield from the normal alkane resonance and show ^{1}H NMR absorptions in the region 3.4 to 4.5 δ. This downfield shift is clearly seen in the spectrum of dipropyl ether shown in Figure 18.4.

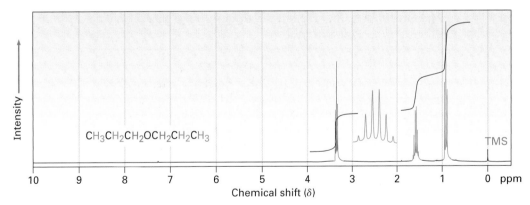

Figure 18.4 The ^{1}H NMR spectrum of dipropyl ether. Protons on carbon next to oxygen are shifted downfield to 3.4 δ.

Epoxides absorb at a slightly higher field than other ethers and show characteristic resonances at 2.5 to 3.5 δ in their ^{1}H NMR spectra, as indicated for 1,2-epoxypropane in Figure 18.5.

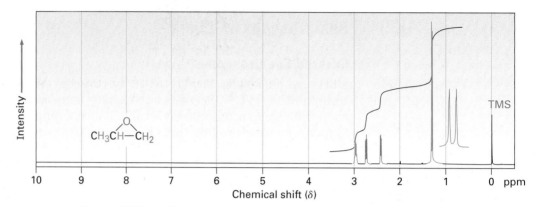

Figure 18.5 The ^{1}H NMR spectrum of 1,2-epoxypropane.

Ether carbon atoms also exhibit a downfield shift in the ^{13}C NMR spectrum, where they usually absorb in the 50 to 80 δ range. For example, the carbon atoms next to oxygen in methyl propyl ether absorb at 58.5 and 74.8 δ. Similarly, the methyl carbon in anisole absorbs at 54.8 δ.

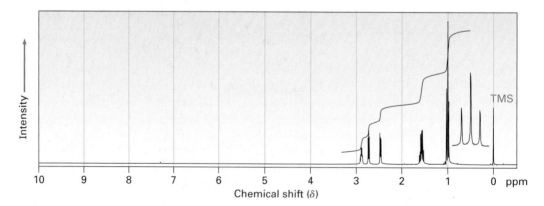

Problem 18.18 | The ^{1}H NMR spectrum shown is that of an ether with the formula C_4H_8O. Propose a structure.

Focus On . . .

Epoxy Resins and Adhesives

Kayaks are often made of a high-strength polymer coated with epoxy resin.

Few nonchemists know exactly what an epoxide is, but practically everyone has used an "epoxy glue" for household repairs or an epoxy resin for a protective coating. Epoxy resins and adhesives generally consist of two components that are mixed just prior to use. One component is a liquid "prepolymer," and the second is a "curing agent" that reacts with the prepolymer and causes it to solidify.

The most widely used epoxy resins and adhesives are based on a prepolymer made from bisphenol A and epichlorohydrin. On treatment with base, bisphenol A is converted into its anion, which acts as a nucleophile in an S_N2 reaction with epichlorohydrin. Each epichlorohydrin molecule can react with two molecules of bisphenol A, once by S_N2 displacement of chloride ion and once by nucleophilic opening of the epoxide ring. At the same time, each bisphenol A molecule can react with two epichlorohydrins, leading to a long polymer chain. Each end of a prepolymer chain has an unreacted epoxy group, and each chain has numerous secondary alcohol groups spaced regularly along its midsection.

Bisphenol A   **Epichlorohydrin**

"Prepolymer"

When the epoxide is to be used, a basic curing agent such as a tertiary amine, R_3N, is added to cause the individual prepolymer chains to link together. This "cross-linking" of chains is simply a base-catalyzed epoxide

(continued)

ring-opening of an −OH group in the middle of one chain with an epoxide group on the end of another chain. The result of such cross-linking is formation of a vast, three-dimensional tangle that has enormous strength and chemical resistance.

SUMMARY AND KEY WORDS

alkoxymercuration, 656

Claisen rearrangement, 659

crown ether, 666

disulfide (RSSR′), 668

ether (ROR′), 652

mercapto group (−SH), 667

sulfide (RSR′), 652

sulfone (R$_2$SO$_2$), 670

sulfonium ion (R$_3$S$^+$), 669

sulfoxide (R$_2$SO), 670

thiol (RSH), 652

thiolate ion (RS$^-$), 668

Ethers are compounds that have two organic groups bonded to the same oxygen atom, ROR′. The organic groups can be alkyl, vinylic, or aryl, and the oxygen atom can be in a ring or in an open chain. Ethers are prepared by either the Williamson ether synthesis, which involves S$_N$2 reaction of an alkoxide ion with a primary alkyl halide, or the **alkoxymercuration** reaction, which involves Markovnikov addition of an alcohol to an alkene.

Ethers are inert to most reagents but react with strong acids to give cleavage products. Both HI and HBr are often used. The cleavage reaction takes place by an S$_N$2 mechanism at the less highly substituted site if only primary and secondary alkyl groups are bonded to the ether oxygen, but by an S$_N$1 or E1 mechanism if one of the alkyl groups bonded to oxygen is tertiary. Aryl allyl ethers undergo **Claisen rearrangement** to give o-allylphenols.

Epoxides are cyclic ethers with a three-membered, oxygen-containing ring. Because of the strain in the ring, epoxides undergo a cleavage reaction with both acids and bases. Acid-induced ring-opening occurs with a regiochemistry that depends on the structure of the epoxide. Cleavage of the C−O bond at the less highly substituted site occurs if both epoxide carbons are primary or secondary, but cleavage of the C−O bond to the more highly substituted site occurs if one of the epoxide carbons is tertiary. Base-catalyzed epoxide ring-opening occurs by S$_N$2 reaction of a nucleophile at the less hindered epoxide carbon.

Thiols, the sulfur analogs of alcohols, are usually prepared by S$_N$2 reaction of an alkyl halide with thiourea. Mild oxidation of a thiol yields a **disulfide,** and mild reduction of a disulfide gives back the thiol. **Sulfides,** the sulfur analogs of ethers, are prepared by an S$_N$2 reaction between a thiolate anion and a primary or secondary alkyl halide. Sulfides are much more nucleophilic than ethers and can be oxidized to **sulfoxides** and to **sulfones.** Sulfides can also be alkylated by reaction with a primary alkyl halide to yield **sulfonium ions.**

SUMMARY OF REACTIONS

1. Synthesis of ethers (Section 18.2)
   (a) Williamson ether synthesis

$$RO^- \ + \ R'CH_2X \ \longrightarrow \ ROCH_2R' \ + \ X^-$$

   (b) Alkoxymercuration/demercuration

2. Reactions of ethers
   (a) Cleavage by HBr or HI (Section 18.3)

$$R-O-R' \ \xrightarrow[H_2O]{HX} \ RX \ + \ R'OH$$

   (b) Claisen rearrangement (Section 18.4)

   (c) Acid-catalyzed epoxide opening (Section 18.6)

   (d) Base-catalyzed epoxide opening (Section 18.6)

3. Synthesis of thiols (Section 18.8)

$$RCH_2Br \xrightarrow[\text{2. } H_2O, \text{ NaOH}]{\text{1. } (H_2N)_2C{=}S} RCH_2SH$$

4. Oxidation of thiols to disulfides (Section 18.8)

$$2\,RSH \xrightarrow{I_2,\ H_2O} RS{-}SR$$

5. Synthesis of sulfides (Section 18.8)

$$RS^- \ + \ R'CH_2Br \longrightarrow RSCH_2R' \ + \ Br^-$$

6. Oxidation of sulfides to sulfoxides and sulfones (Section 18.8)

$$R\overset{\cdot\cdot}{\underset{\cdot\cdot}{S}}R' \xrightarrow{H_2O_2} R\overset{O}{\underset{\cdot\cdot}{S}}R'$$

$$R\overset{O}{\underset{\cdot\cdot}{S}}R' \xrightarrow{RCO_3H} R\overset{O\ \ O}{\underset{\cdot\cdot}{S}}R'$$

EXERCISES

VISUALIZING CHEMISTRY

(Problems 18.1–18.18 appear within the chapter)

18.19 ■ Give IUPAC names for the following compounds (reddish brown = Br):

(a)   (b)   (c)

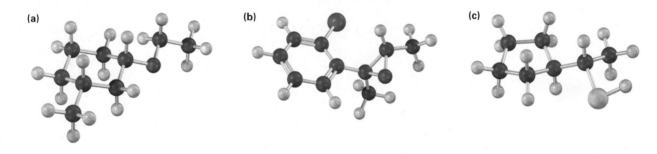

18.20 ■ Show the product, including stereochemistry, that would result from reaction of the following epoxide with HBr:

18.21 Show the product, including stereochemistry, of the following reaction:

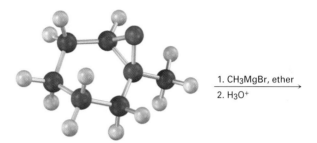

1. CH₃MgBr, ether
2. H₃O⁺

18.22 Treatment of the following alkene with a peroxyacid yields an epoxide different from that obtained by reaction with aqueous Br₂ followed by base treatment. Propose structures for the two epoxides, and explain the result.

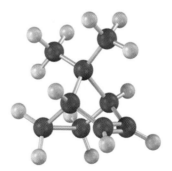

ADDITIONAL PROBLEMS

18.23 ■ Draw structures corresponding to the following IUPAC names:
    (a) Ethyl 1-ethylpropyl ether   **(b)** Di(*p*-chlorophenyl) ether
    (c) 3,4-Dimethoxybenzoic acid   **(d)** Cyclopentyloxycyclohexane
    (e) 4-Allyl-2-methoxyphenol (eugenol; from oil of cloves)

18.24 ■ Give IUPAC names for the following structures:

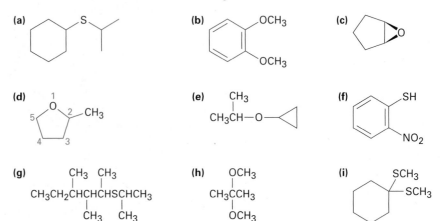

18.25 ■ Predict the products of the following ether cleavage reactions:

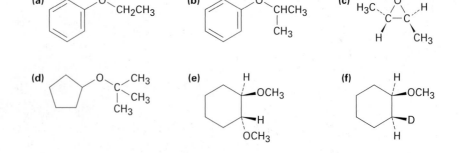

(a) [cyclohexyl-O-CH₂CH₃] $\xrightarrow[\text{H}_2\text{O}]{\text{HI}}$?

(b) [phenyl-O-C(CH₃)₂CH₃ with H₃C CH₃] $\xrightarrow{\text{CF}_3\text{CO}_2\text{H}}$?

(c) H₂C=CH—O—CH₂CH₃ $\xrightarrow[\text{H}_2\text{O}]{\text{HI}}$?

(d)
$$\begin{array}{c} \text{CH}_3 \\ | \\ \text{CH}_3\text{CCH}_2\text{—O—CH}_2\text{CH}_3 \\ | \\ \text{CH}_3 \end{array} \xrightarrow[\text{H}_2\text{O}]{\text{HI}} \text{ ?}$$

18.26 ■ How would you prepare the following ethers?

(a) [phenyl-O-CH₂CH₃]

(b) [phenyl-O-CHCH₃ with CH₃]

(c) [epoxide: H₃C—C—C—H with H and CH₃]

(d) [cyclopentyl-O-C(CH₃)₃]

(e) [cyclohexane with H, OCH₃, H, OCH₃]

(f) [cyclohexane with H, OCH₃, D, H]

18.27 ■ How would you prepare the following compounds from 1-phenylethanol?
(a) Methyl 1-phenylethyl ether   (b) Phenylepoxyethane
(c) *tert*-Butyl 1-phenylethyl ether   (d) 1-Phenylethanethiol

---

18.28 ■ Predict the products of the following reactions:

(a)

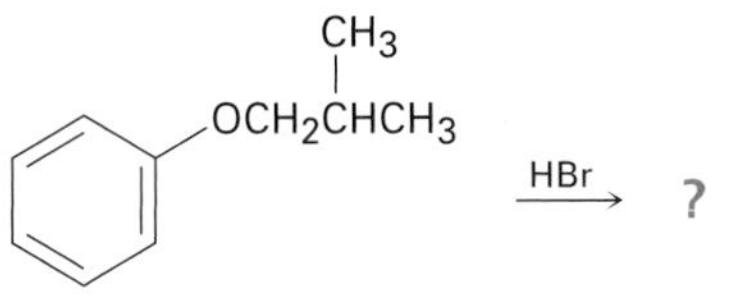

OCH₂CHCH₃ with CH₃ substituent, on benzene ring

$\xrightarrow{\text{HBr}}$?

(b)

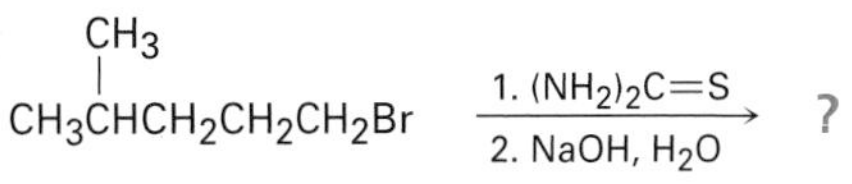

CH₃
CH₃CHCH₂CH₂CH₂Br $\xrightarrow[\text{2. NaOH, H}_2\text{O}]{\text{1. (NH}_2)_2\text{C}=\text{S}}$?

(c)

cyclopentane—SH $\xrightarrow{\text{Br}_2}$?

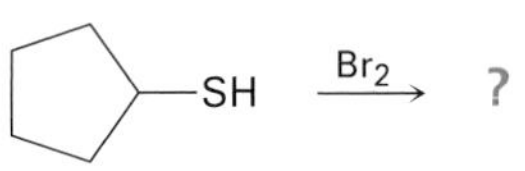

(d)

cyclohexene with SCH₂CH₃ $\xrightarrow{\text{H}_2\text{O}_2, \text{H}_2\text{O}}$?

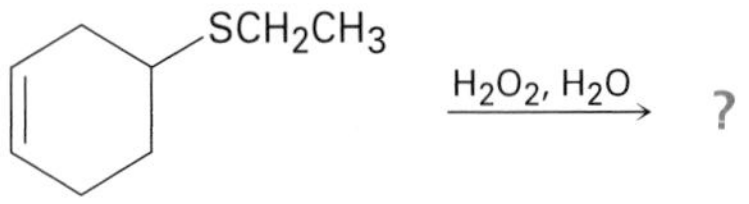

18.29 ■ How would you carry out the following transformations? More than one step may be required.

(a)

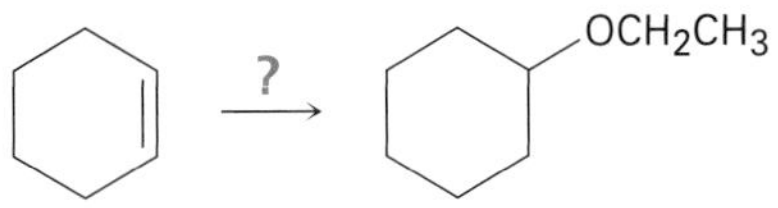

cyclohexene $\xrightarrow{?}$ cyclohexane—OCH₂CH₃

(b)

$\xrightarrow{?}$

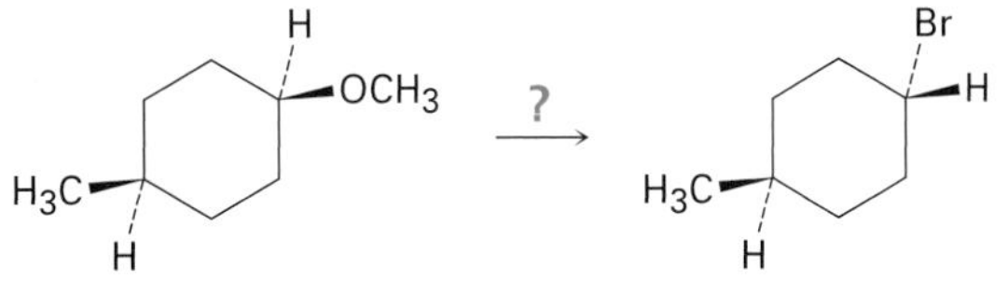

(c)

$\xrightarrow{?}$

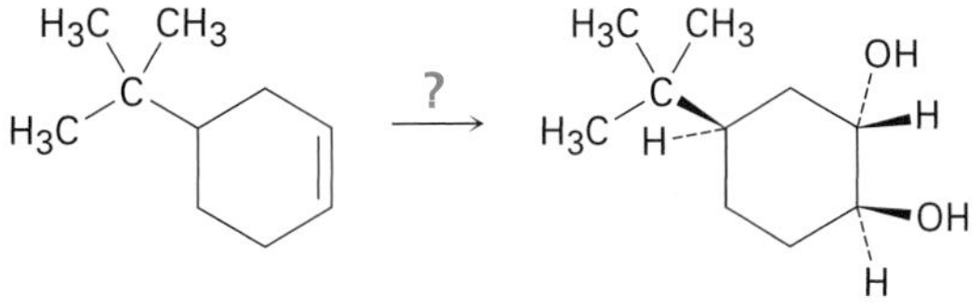

(d) CH₃CH₂CH₂CH₂C≡CH $\xrightarrow{?}$ CH₃CH₂CH₂CH₂CH₂CH₂OCH₃

(e)

CH₃CH₂CH₂CH₂C≡CH $\xrightarrow{?}$

OCH₃
CH₃CH₂CH₂CH₂CHCH₃

18.30 What product would you expect from cleavage of tetrahydrofuran with HI?

18.31 How could you prepare benzyl phenyl ether from benzene and phenol? More than one step is required.

18.32 ■ When 2-methyl-2,5-pentanediol is treated with sulfuric acid, dehydration occurs and 2,2-dimethyltetrahydrofuran is formed. Suggest a mechanism for this reaction. Which of the two oxygen atoms is most likely to be eliminated, and why?

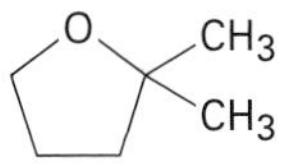

2,2-Dimethyltetrahydrofuran

18.33 ■ Write the mechanism of the hydrolysis of *cis*-5,6-epoxydecane by reaction with aqueous acid. What is the stereochemistry of the product, assuming normal backside S_N2 attack?

18.34 What is the stereochemistry of the product from acid-catalyzed hydrolysis of *trans*-5,6-epoxydecane? How does the product differ from that formed in Problem 18.33?

18.35 Methyl aryl ethers, such as anisole, are cleaved to iodomethane and a phenoxide ion by treatment with LiI in hot DMF. Propose a mechanism for this reaction.

18.36 ■ *tert*-Butyl ethers can be prepared by the reaction of an alcohol with 2-methylpropene in the presence of an acid catalyst. Propose a mechanism for this reaction.

18.37 *Meerwein's reagent,* triethyloxonium tetrafluoroborate, is a powerful ethylating agent that converts alcohols into ethyl ethers at neutral pH. Show the reaction of Meerwein's reagent with cyclohexanol, and account for the fact that trialkyloxonium salts are much more reactive alkylating agents than alkyl iodides.

$$(CH_3CH_2)_3O^+ \ BF_4^- \qquad \textbf{Meerwein's reagent}$$

18.38 Safrole, a substance isolated from oil of sassafras, is used as a perfumery agent. Propose a synthesis of safrole from catechol (1,2-benzenediol).

Safrole

18.39 Epoxides are reduced by treatment with lithium aluminum hydride to yield alcohols. Propose a mechanism for this reaction.

18.40 Show the structure and stereochemistry of the alcohol that would result if 1,2-epoxycyclohexane (Problem 18.39) were reduced with lithium aluminum deuteride, $LiAlD_4$.

18.41 Acid-catalyzed hydrolysis of a 1,2-epoxycyclohexane produces a trans-diaxial 1,2-diol. What product would you expect to obtain from acidic hydrolysis of *cis*-3-*tert*-butyl-1,2-epoxycyclohexane? (Recall that the bulky *tert*-butyl group locks the cyclohexane ring into a specific conformation.)

18.42 Grignard reagents react with oxetane, a four-membered cyclic ether, to yield primary alcohols, but the reaction is much slower than the corresponding reaction with ethylene oxide. Suggest a reason for the difference in reactivity between oxetane and ethylene oxide.

Oxetane

■ Assignable in OWL

18.43 Treatment of *trans*-2-chlorocyclohexanol with NaOH yields 1,2-epoxycyclo-hexane, but reaction of the cis isomer under the same conditions yields cyclo-hexanone. Propose mechanisms for both reactions, and explain why the different results are obtained.

18.44 Ethers undergo an acid-catalyzed cleavage reaction when treated with the Lewis acid BBr_3 at room temperature. Propose a mechanism for the reaction.

18.45 The *Zeisel method* is an analytical procedure for determining the number of methoxyl groups in a compound. A weighed amount of the compound is heated with concentrated HI, ether cleavage occurs, and the iodomethane product is distilled off and passed into an alcohol solution of $AgNO_3$, where it reacts to form a precipitate of silver iodide. The AgI is then collected and weighed, and the percentage of methoxyl groups in the sample is thereby determined. For example, 1.06 g of vanillin, the material responsible for the characteristic odor of vanilla, yields 1.60 g of AgI. If vanillin has a molecular weight of 152, how many methoxyl groups does it contain?

18.46 Disparlure, $C_{19}H_{38}O$, is a sex attractant released by the female gypsy moth, *Lymantria dispar*. The ^1H NMR spectrum of disparlure shows a large absorption in the alkane region, 1 to 2 δ, and a triplet at 2.8 δ. Treatment of disparlure, first with aqueous acid and then with $KMnO_4$, yields two carboxylic acids identified as undecanoic acid and 6-methylheptanoic acid. ($KMnO_4$ cleaves 1,2-diols to yield carboxylic acids.) Neglecting stereochemistry, propose a structure for disparlure. The actual compound is a chiral molecule with 7*R*,8*S* stereochemistry. Draw disparlure, showing the correct stereochemistry.

18.47 How would you synthesize racemic disparlure (Problem 18.46) from compounds having ten or fewer carbons?

18.48 ■ Treatment of 1,1-diphenyl-1,2-epoxyethane with aqueous acid yields diphenylacetaldehyde as the major product. Propose a mechanism for the reaction.

■ Assignable in OWL

18.49 How would you prepare *o*-hydroxyphenylacetaldehyde from phenol? More than one step is required.

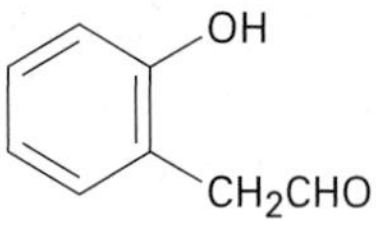

o-Hydroxyphenylacetaldehyde

18.50 ■ Imagine that you have treated (2R,3R)-2,3-epoxy-3-methylpentane with aqueous acid to carry out a ring-opening reaction.

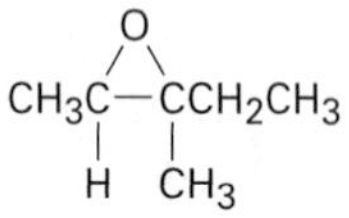

2,3-Epoxy-3-methylpentane
(no stereochemistry implied)

(a) Draw the epoxide, showing stereochemistry.
(b) Draw and name the product, showing stereochemistry.
(c) Is the product chiral? Explain.
(d) Is the product optically active? Explain.

18.51 Identify the reagents **a–e** in the following scheme:

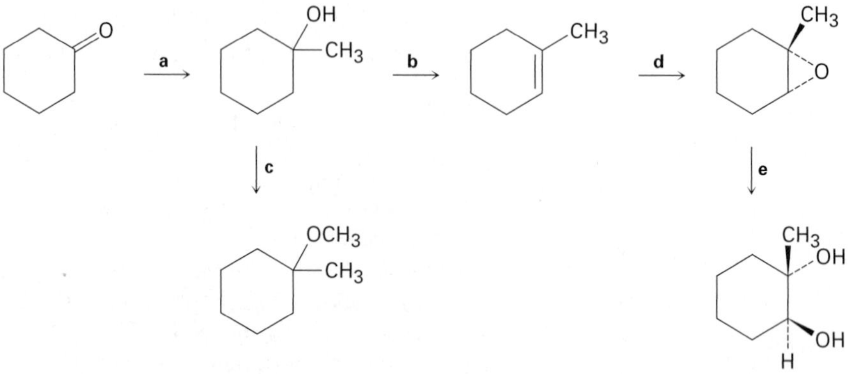

18.52 Fluoxetine, a heavily prescribed antidepressant marketed under the name Prozac, can be prepared by a route that begins with reaction between a phenol and an alkyl chloride.

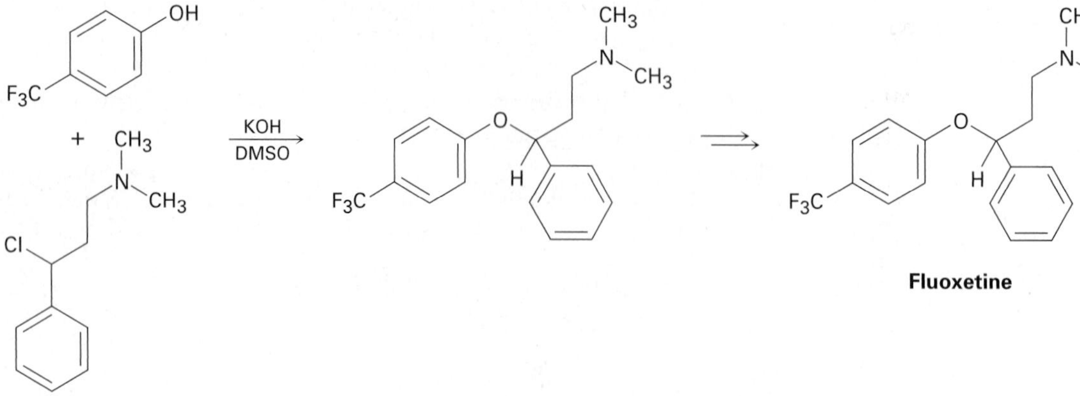

Fluoxetine

(a) The rate of the reaction depends on both phenol and alkyl halide. Is this an S_N1 or an S_N2 reaction? Show the mechanism.

(b) The physiologically active enantiomer of fluoxetine has (S) stereochemistry. Based on your answer in part (a), draw the structure of the alkyl chloride you would need, showing the correct stereochemistry.

18.53 The herbicide acifluorfen can be prepared by a route that begins with reaction between a phenol and an aryl fluoride. Propose a mechanism.

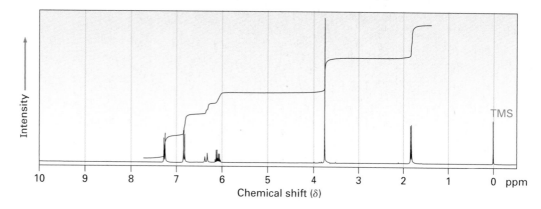

Acifluorfen

18.54 ■ The red fox *(Vulpes vulpes)* uses a chemical communication system based on scent marks in urine. Recent work has shown one component of fox urine to be a sulfide. Mass spectral analysis of the pure scent-mark component shows $M^+ = 116$. IR spectroscopy shows an intense band at 890 cm^{-1}, and ^1H NMR spectroscopy reveals the following peaks:

1.74 δ (3 H, singlet); 2.11 δ (3 H, singlet); 2.27 δ (2 H, triplet, $J = 4.2$ Hz); 2.57 δ (2 H, triplet, $J = 4.2$ Hz); 4.73 δ (2 H, broad)

Propose a structure consistent with these data. [Note: $(CH_3)_2S$ absorbs at 2.1 δ.]

18.55 ■ Anethole, $C_{10}H_{12}O$, a major constituent of the oil of anise, has the ^1H NMR spectrum shown. On oxidation with $Na_2Cr_2O_7$, anethole yields *p*-methoxybenzoic acid. What is the structure of anethole? Assign all peaks in the NMR spectrum, and account for the observed splitting patterns.

18.56 How would you synthesize anethole (Problem 18.55) from phenol?

18.57 ■ Propose structures for compounds that have the following ^1H NMR spectra:

(a) $C_5H_{12}S$

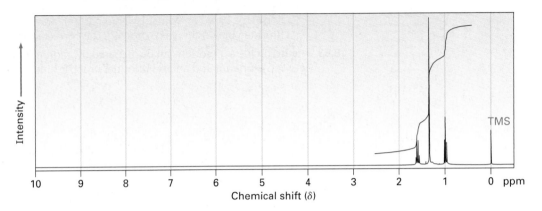

(b) $C_9H_{11}BrO$

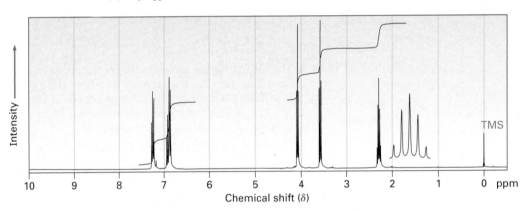

(c) $C_4H_{10}O_2$

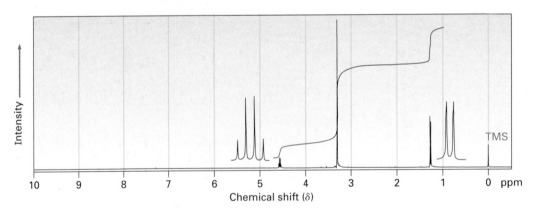

(d) $C_9H_{10}O$

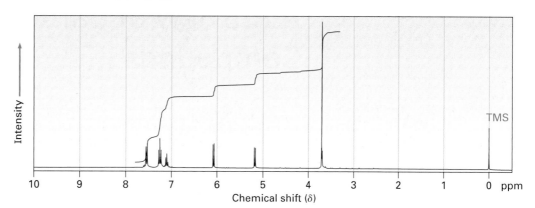

18.58 Aldehydes and ketones undergo acid-catalyzed reaction with alcohols to yield *hemiacetals,* compounds that have one alcohol-like oxygen and one ether-like oxygen bonded to the same carbon. Further reaction of a hemiacetal with alcohol then yields an *acetal,* a compound that has two ether-like oxygens bonded to the same carbon.

A hemiacetal   **An acetal**

(a) Show the structures of the hemiacetal and acetal you would obtain by reaction of cyclohexanone with ethanol.
(b) Propose a mechanism for the conversion of a hemiacetal into an acetal.

18.59 We saw in Section 17.4 that ketones react with $NaBH_4$ to yield alcohols. We'll also see in Section 22.3 that ketones react with Br_2 to yield α-bromo ketones. Perhaps surprisingly, treatment with $NaBH_4$ of the α-bromo ketone from acetophenone yields an epoxide rather than a bromo alcohol. Show the structure of the epoxide, and explain its formation.

Acetophenone   **An α-bromo ketone**

A Preview of Carbonyl Compounds

Carbonyl compounds are everywhere. Most biological molecules contain carbonyl groups, as do most pharmaceutical agents and many of the synthetic chemicals that touch our everyday lives. Citric acid, found in lemons and oranges; acetaminophen, the active ingredient in many over-the-counter headache remedies; and Dacron, the polyester material used in clothing, all contain different kinds of carbonyl groups.

Citric acid	Acetaminophen	Dacron
(a carboxylic acid)	(an amide)	(a polyester)

In the next five chapters, we'll discuss the chemistry of the **carbonyl group**, **C=O** (pronounced car-bo-**neel**). Although there are many different kinds of carbonyl compounds and many different reactions, there are only a few fundamental principles that tie the entire field together. The purpose of this brief preview is not to show details of specific reactions but rather to provide a framework for learning carbonyl-group chemistry. Read through this preview now, and return to it on occasion to remind yourself of the larger picture.

I. Kinds of Carbonyl Compounds

Table 1 shows some of the many different kinds of carbonyl compounds. All contain an **acyl group** (R−C=O) bonded to another substituent. The R part of the acyl group can be practically any organic part-structure, and the other substituent to which the acyl group is bonded can be a carbon, hydrogen, oxygen, halogen, nitrogen, or sulfur.

It's useful to classify carbonyl compounds into two categories based on the kinds of chemistry they undergo. In one category are aldehydes and ketones; in the other are carboxylic acids and their derivatives. The acyl group in an aldehyde or ketone is bonded to an atom (H or C, respectively) that can't stabilize a negative charge and therefore can't act as a leaving group in a nucleophilic substitution reaction. The acyl group in a carboxylic acid or its derivative, however, is bonded to an atom (oxygen, halogen, sulfur, nitrogen) that *can* stabilize a

Table 1 | Some Types of Carbonyl Compounds

Name	General formula	Name ending	Name	General formula	Name ending
Aldehyde	R–C(=O)–H	-al	Ester	R–C(=O)–O–R′	-oate
Ketone	R–C(=O)–R′	-one	Lactone (cyclic ester)	cyclic C–C(=O)–O	None
Carboxylic acid	R–C(=O)–O–H	-oic acid	Thioester	R–C(=O)–S–R′	-thioate
Acid halide	R–C(=O)–X	-yl or -oyl halide	Amide	R–C(=O)–N	-amide
Acid anhydride	R–C(=O)–O–C(=O)–R′	-oic anhydride	Lactam (cyclic amide)	cyclic C–C(=O)–N	None
Acyl phosphate	R–C(=O)–O–P(=O)(O⁻)(O⁻)	-yl phosphate			

negative charge and therefore *can* act as a leaving group in a nucleophilic substitution reaction.

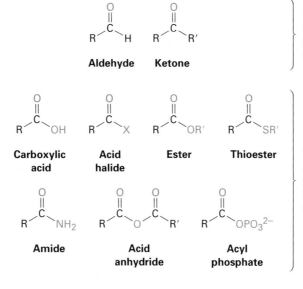

Aldehyde **Ketone**

The –R′ and –H in these compounds *can't* act as leaving groups in nucleophilic substitution reactions.

Carboxylic acid **Acid halide** **Ester** **Thioester**

Amide **Acid anhydride** **Acyl phosphate**

The $-OH$, $-X$, $-OR'$, $-SR$, $-NH_2$, $-OCOR'$, and $-OPO_3^{2-}$ in these compounds *can* act as leaving groups in nucleophilic substitution reactions.

II.   Nature of the Carbonyl Group

The carbon–oxygen double bond of a carbonyl group is similar in many respects to the carbon–carbon double bond of an alkene. The carbonyl carbon atom is sp^2-hybridized and forms three σ bonds. The fourth valence electron remains in a carbon p orbital and forms a π bond to oxygen by overlap with an oxygen p orbital. The oxygen atom also has two nonbonding pairs of electrons, which occupy its remaining two orbitals.

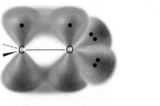

Carbonyl group   **Alkene**

Like alkenes, carbonyl compounds are planar about the double bond and have bond angles of approximately 120°. Figure 1 shows the structure of acetaldehyde and indicates its bond lengths and angles. As you might expect, the carbon–oxygen double bond is both shorter (122 pm versus 143 pm) and stronger [732 kJ/mol (175 kcal/mol) versus 385 kJ/mol (92 kcal/mol)] than a C—O single bond.

Figure 1 Structure of acetaldehyde.

Bond angle	(°)	Bond length	(pm)
H—C—C	118	C=O	122
C—C=O	121	C—C	150
H—C=O	121	OC—H	109

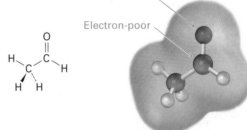

Electron-rich

Electron-poor

As indicated by the electrostatic potential map in Figure 1, the carbon–oxygen double bond is strongly polarized because of the high electronegativity of oxygen relative to carbon. Thus, the carbonyl carbon atom carries a partial positive charge, is an electrophilic (Lewis acidic) site, and reacts with nucleophiles. Conversely, the carbonyl oxygen atom carries a partial negative charge, is a nucleophilic (Lewis basic) site, and reacts with electrophiles. We'll see in the next five chapters that the majority of carbonyl-group reactions can be rationalized by simple polarity arguments.

III.   General Reactions of Carbonyl Compounds

Both in the laboratory and in living organisms, the reactions of carbonyl compounds take place by one of four general mechanisms: *nucleophilic addition, nucleophilic acyl substitution, alpha substitution,* and *carbonyl condensation.* These mechanisms have many variations, just as alkene electrophilic addition reactions and S_N2 reactions do, but the variations are much easier to learn when the fundamental features of the mechanisms are made clear. Let's see what the four mechanisms are and what kinds of chemistry carbonyl compounds undergo.

Nucleophilic Addition Reactions of Aldehydes and Ketones (Chapter 19)

The most common reaction of aldehydes and ketones is the **nucleophilic addition reaction**, in which a nucleophile, $:Nu^-$, adds to the electrophilic carbon of the carbonyl group. Since the nucleophile uses an electron pair to form a new bond to carbon, two electrons from the carbon–oxygen double bond must move toward the electronegative oxygen atom to give an alkoxide anion. The carbonyl carbon rehybridizes from sp^2 to sp^3 during the reaction, and the alkoxide ion product therefore has tetrahedral geometry.

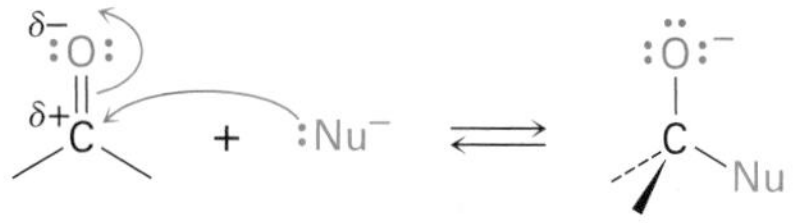

A carbonyl compound
(**sp^2**-hybridized carbon)

A tetrahedral intermediate
(**sp^3**-hybridized carbon)

Once formed, and depending on the nature of the nucleophile, the tetrahedral alkoxide intermediate can undergo either of two further reactions, as shown in Figure 2. Often, the tetrahedral alkoxide intermediate is simply protonated by water or acid to form an alcohol product. Alternatively, the tetrahedral intermediate can be protonated and expel the oxygen to form a new double bond between the carbonyl carbon and the nucleophile. We'll study both processes in detail in Chapter 19.

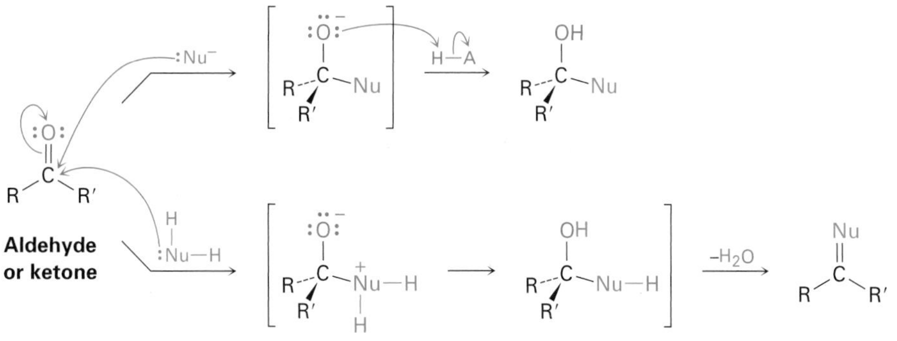

Figure 2 The addition reaction of an aldehyde or a ketone with a nucleophile. Depending on the nucleophile, either an alcohol or a compound with a C=Nu bond is formed.

Formation of an Alcohol The simplest reaction of a tetrahedral alkoxide intermediate is protonation to yield an alcohol. We've already seen two examples of this kind of process during reduction of aldehydes and ketones with hydride reagents such as $NaBH_4$ and $LiAlH_4$ (Section 17.4) and during Grignard reactions (Section 17.5). During a reduction, the nucleophile that adds to the carbonyl group is a hydride ion, H:⁻, while during a Grignard reaction, the nucleophile is a carbanion, $R_3C:^-$.

Reduction

Grignard reaction

Formation of C=Nu The second mode of nucleophilic addition, which often occurs with amine nucleophiles, involves elimination of oxygen and formation of a C=Nu bond. For example, aldehydes and ketones react with primary amines, RNH_2, to form *imines*, $R_2C=NR'$. These reactions proceed through exactly the same kind of tetrahedral intermediate as that formed during hydride reduction and Grignard reaction, but the initially formed alkoxide ion is not isolated. Instead, it is protonated and then loses water to form an imine, as shown in Figure 3.

Figure 3 MECHANISM:
Formation of an imine, $R_2C=NR'$, by reaction of an amine with an aldehyde or a ketone.

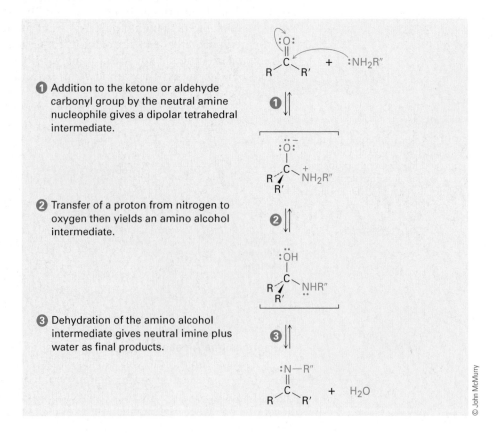

① Addition to the ketone or aldehyde carbonyl group by the neutral amine nucleophile gives a dipolar tetrahedral intermediate.

② Transfer of a proton from nitrogen to oxygen then yields an amino alcohol intermediate.

③ Dehydration of the amino alcohol intermediate gives neutral imine plus water as final products.

Nucleophilic Acyl Substitution Reactions of Carboxylic Acid Derivatives (Chapter 21)

The second fundamental reaction of carbonyl compounds, **nucleophilic acyl substitution**, is related to the nucleophilic addition reaction just discussed but occurs only with carboxylic acid derivatives rather than with aldehydes and ketones. When the carbonyl group of a carboxylic acid derivative reacts with a nucleophile, addition occurs in the usual way, but the initially formed tetrahedral alkoxide intermediate is not isolated. Because carboxylic acid derivatives have a leaving group bonded to the carbonyl-group carbon, the tetrahedral intermediate can react further by expelling the leaving group and forming a new carbonyl compound:

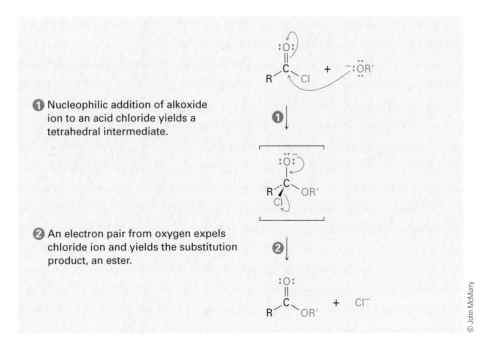

The net effect of nucleophilic acyl substitution is the replacement of the leaving group by the entering nucleophile. We'll see in Chapter 21, for instance, that acid chlorides are rapidly converted into esters by treatment with alkoxide ions (Figure 4).

Figure 4 MECHANISM: The nucleophilic acyl substitution reaction of an acid chloride with an alkoxide ion yields an ester.

1. Nucleophilic addition of alkoxide ion to an acid chloride yields a tetrahedral intermediate.

2. An electron pair from oxygen expels chloride ion and yields the substitution product, an ester.

© John McMurry

Alpha-Substitution Reactions (Chapter 22)

The third major reaction of carbonyl compounds, **alpha substitution**, occurs at the position *next to* the carbonyl group—the alpha (α) position. This reaction, which takes place with all carbonyl compounds regardless of structure, results in the substitution of an α hydrogen by an electrophile through the formation of an intermediate *enol* or *enolate ion:*

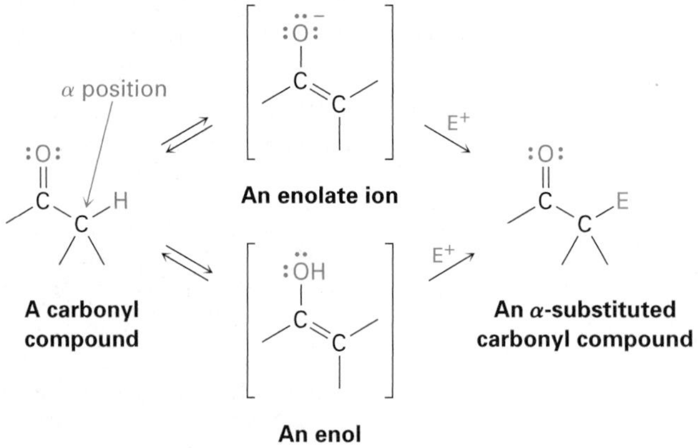

For reasons that we'll explore in Chapter 22, the presence of a carbonyl group renders the hydrogens on the α carbon acidic. Carbonyl compounds therefore react with strong base to yield enolate ions.

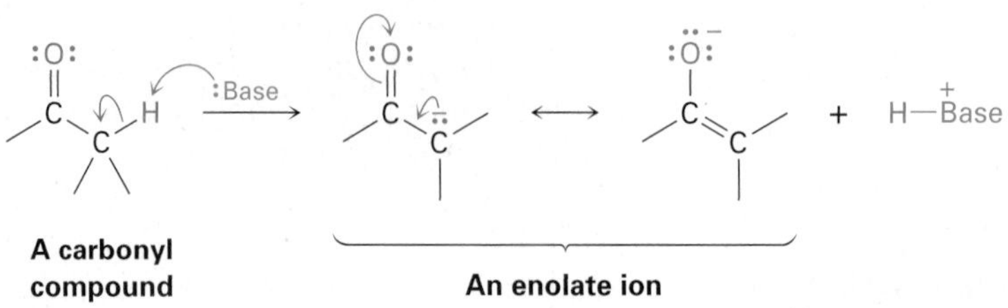

Because they're negatively charged, enolate ions act as nucleophiles and undergo many of the reactions we've already studied. For example, enolates react with primary alkyl halides in the S_N2 reaction. The nucleophilic enolate ion displaces halide ion, and a new C—C bond forms:

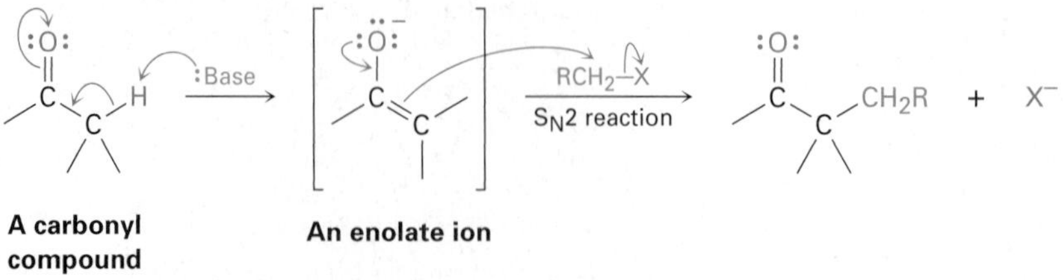

The S_N2 alkylation reaction between an enolate ion and an alkyl halide is a powerful method for making C—C bonds, thereby building up larger molecules from smaller precursors. We'll study the alkylation of many kinds of carbonyl compounds in Chapter 22.

Carbonyl Condensation Reactions (Chapter 23)

The fourth and last fundamental reaction of carbonyl groups, **carbonyl condensation**, takes place when two carbonyl compounds react with each other. When acetaldehyde is treated with base, for instance, two molecules combine to yield the hydroxy aldehyde product known as *aldol* (*ald*ehyde + alcoh*ol*):

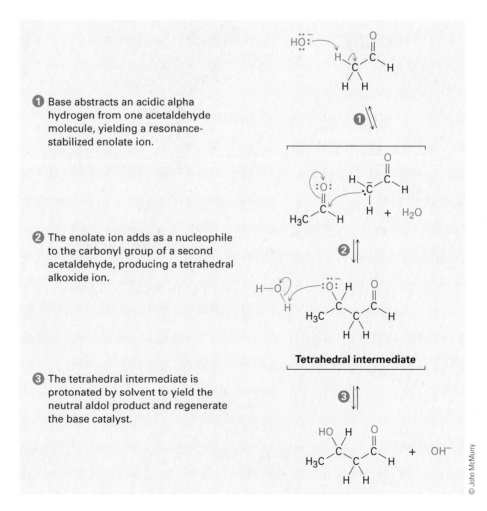

Two acetaldehydes   **Aldol**

Although the carbonyl condensation reaction appears different from the three processes already discussed, it's actually quite similar. A carbonyl condensation reaction is simply a *combination* of a nucleophilic addition step and an α-substitution step. The initially formed enolate ion of one acetaldehyde molecule acts as a nucleophile and adds to the carbonyl group of another acetaldehyde molecule, as shown in Figure 5.

Figure 5 MECHANISM:
A carbonyl condensation reaction between two molecules of acetaldehyde yields a hydroxy aldehyde product.

1 Base abstracts an acidic alpha hydrogen from one acetaldehyde molecule, yielding a resonance-stabilized enolate ion.

2 The enolate ion adds as a nucleophile to the carbonyl group of a second acetaldehyde, producing a tetrahedral alkoxide ion.

Tetrahedral intermediate

3 The tetrahedral intermediate is protonated by solvent to yield the neutral aldol product and regenerate the base catalyst.

© John McMurry

693

IV. Summary

The purpose of this short preview is not to show details of specific reactions but rather to lay the groundwork for the next five chapters. All the carbonyl-group reactions we'll be studying in Chapters 19 through 23 fall into one of the four fundamental categories discussed in this preview. Knowing where we'll be heading should help you to keep matters straight in understanding this most important of all functional groups.

PROBLEMS

1. Judging from the following electrostatic potential maps, which kind of carbonyl compound has the more electrophilic carbonyl carbon atom, a ketone or an acid chloride? Which has the more nucleophilic carbonyl oxygen atom? Explain.

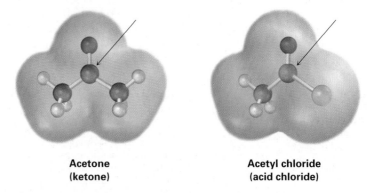

Acetone
(ketone)

Acetyl chloride
(acid chloride)

2. Predict the product formed by nucleophilic addition of cyanide ion (CN^-) to the carbonyl group of acetone, followed by protonation to give an alcohol:

$$\underset{\textbf{Acetone}}{H_3C-\overset{\overset{\textstyle O}{\|}}{C}-CH_3} \quad \xrightarrow[\text{2. } H_3O^+]{\text{1. } CN^-} \quad ?$$

3. Identify each of the following reactions as a nucleophilic addition, nucleophilic acyl substitution, an α substitution, or a carbonyl condensation:

(a)

$$H_3C-\overset{\overset{\textstyle O}{\|}}{C}-Cl \quad \xrightarrow{NH_3} \quad H_3C-\overset{\overset{\textstyle O}{\|}}{C}-NH_2$$

(b)

$$H_3C-\overset{\overset{\textstyle O}{\|}}{C}-H \quad \xrightarrow{NH_2OH} \quad H_3C-\overset{\overset{\textstyle NOH}{\|}}{C}-H$$

(c)

$$2 \; \overset{\bigcirc}{=}O \quad \xrightarrow{NaOH} \quad$$

19

Aldehydes and Ketones: Nucleophilic Addition Reactions

Aldehydes (RCHO) and **ketones (R_2CO)** are among the most widely occurring of all compounds. In nature, many substances required by living organisms are aldehydes or ketones. The aldehyde pyridoxal phosphate, for instance, is a coenzyme involved in a large number of metabolic reactions; the ketone hydrocortisone is a steroid hormone secreted by the adrenal glands to regulate fat, protein, and carbohydrate metabolism.

Pyridoxal phosphate (PLP)

Hydrocortisone

In the chemical industry, simple aldehydes and ketones are produced in large quantities for use as solvents and as starting materials to prepare a host of other compounds. For example, more than 1.9 million tons per year of formaldehyde, $H_2C{=}O$, is produced in the United States for use in building insulation materials and in the adhesive resins that bind particle board and plywood. Acetone, $(CH_3)_2C{=}O$, is widely used as an industrial solvent; approximately 1.2 million tons per year is produced in the United States. Formaldehyde is synthesized industrially by catalytic oxidation of methanol, and one method of acetone preparation involves oxidation of 2-propanol.

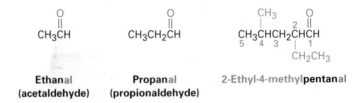

Methanol → Formaldehyde (Catalyst, Heat)

2-Propanol → Acetone (ZnO, 380 °C)

WHY THIS CHAPTER?

Much of organic chemistry is simply the chemistry of carbonyl compounds. Aldehydes and ketones, in particular, are intermediates in the synthesis of many pharmaceutical agents, in almost all biological pathways, and in numerous industrial processes, so an understanding of their properties and reactions is essential. We'll look in this chapter at some of their most important reactions.

19.1 | Naming Aldehydes and Ketones

CENGAGENOW™ Click *Organic Interactive* to **use a web-based palette to draw structures of aldehydes and ketones based on their IUPAC names.**

Aldehydes are named by replacing the terminal *-e* of the corresponding alkane name with *-al*. The parent chain must contain the −CHO group, and the −CHO carbon is numbered as carbon 1. For example:

Ethanal
(acetaldehyde)

Propanal
(propionaldehyde)

2-Ethyl-4-methylpentanal

Note that the longest chain in 2-ethyl-4-methylpentanal is a hexane, but this chain does not include the −CHO group and thus is not considered the parent.

For cyclic aldehydes in which the −CHO group is directly attached to a ring, the suffix *-carbaldehyde* is used.

Cyclohexanecarbaldehyde

2-Naphthalenecarbaldehyde

A few simple and well-known aldehydes have common names that are recognized by IUPAC. Several that you might encounter are listed in Table 19.1.

Table 19.1 | **Common Names of Some Simple Aldehydes**

Formula	Common name	Systematic name
HCHO	Formaldehyde	Methanal
CH_3CHO	Acetaldehyde	Ethanal
H_2C=CHCHO	Acrolein	Propenal
CH_3CH=CHCHO	Crotonaldehyde	2-Butenal
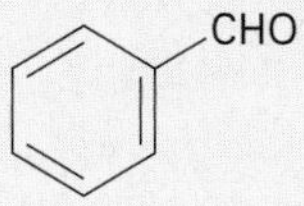 CHO	Benzaldehyde	Benzenecarbaldehyde

Ketones are named by replacing the terminal *-e* of the corresponding alkane name with *-one*. The parent chain is the longest one that contains the ketone group, and the numbering begins at the end nearer the carbonyl carbon. As with alkenes (Section 6.3) and alcohols (Section 17.1), the locant is placed before the parent name in older rules but before the suffix in newer IUPAC recommendations. For example:

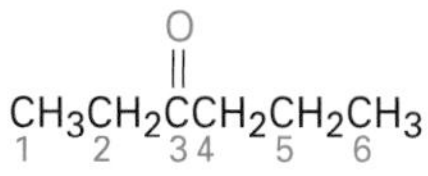

Hexan-3-one 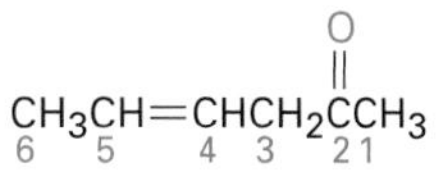   **Hex-4-en-2-one** 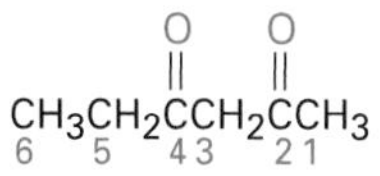   **Hexane-2,4-dione**

A few ketones are allowed by IUPAC to retain their common names.

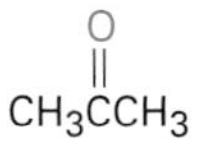

Acetone 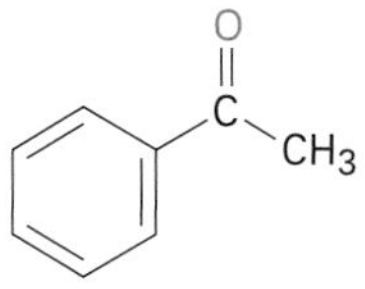   **Acetophenone** 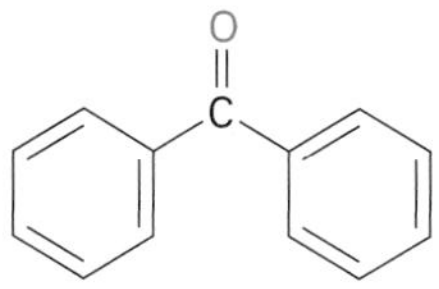   **Benzophenone**

When it's necessary to refer to the R—C=O as a substituent, the name **acyl** (a-sil) **group** is used and the name ending *-yl* is attached. Thus, CH_3CO is an *acetyl* group, CHO is a *formyl* group, and C_6H_5CO is a *benzoyl* group.

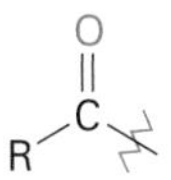

An acyl group 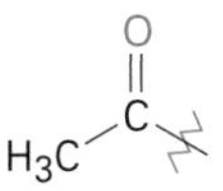   **Acetyl** 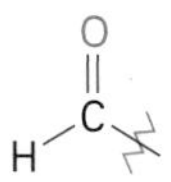   **Formyl** 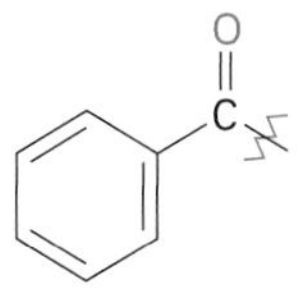   **Benzoyl**

If other functional groups are present and the doubly bonded oxygen is considered a substituent on a parent chain, the prefix *oxo-* is used. For example:

$$\underset{6}{CH_3}\underset{5}{CH_2}\underset{4}{CH_2}\underset{3}{\overset{O}{\overset{\|}{C}}}\underset{2}{CH_2}\underset{1}{\overset{O}{\overset{\|}{C}}}OCH_3 \qquad \textbf{Methyl 3-oxohexanoate}$$

Problem 19.1 | Name the following aldehydes and ketones according to IUPAC rules:

(a)
$$CH_3CH_2\overset{O}{\overset{\|}{C}}\underset{\underset{CH_3}{|}}{CH}CH_3$$

(b) (benzene ring)—CH_2CH_2CHO

(c)
$$CH_3\overset{O}{\overset{\|}{C}}CH_2CH_2CH_2\overset{O}{\overset{\|}{C}}CH_2CH_3$$

(d) (cyclohexane ring with CH_3 and H and CHO substituents)

(e)
$$CH_3CH{=}CHCH_2CH_2\overset{O}{\overset{\|}{C}}H$$

(f) (cyclohexanone ring with H_3C, H, H, CH_3 substituents)

Problem 19.2 | Draw structures corresponding to the following names:
(a) 3-Methylbutanal   (b) 4-Chloro-2-pentanone
(c) Phenylacetaldehyde   (d) *cis*-3-*tert*-Butylcyclohexanecarbaldehyde
(e) 3-Methyl-3-butenal   (f) 2-(1-Chloroethyl)-5-methylheptanal

19.2 | Preparation of Aldehydes and Ketones

Preparing Aldehydes

We've already discussed two methods of aldehyde synthesis: oxidation of primary alcohols and oxidative cleavage of alkenes.

▌ Primary alcohols can be oxidized to give aldehydes (Section 17.7). The reaction is often carried out using pyridinium chlorochromate (PCC) in dichloromethane solvent at room temperature.

(structure) —CH_2OH $\xrightarrow[\text{CH}_2\text{Cl}_2]{\text{PCC}}$ (structure) —CHO

Citronellol   **Citronellal (82%)**

▌ Alkenes with at least one vinylic hydrogen undergo oxidative cleavage when treated with ozone, yielding aldehydes (Section 7.9). If the ozonolysis reaction is carried out on a cyclic alkene, a dicarbonyl compound results.

(cyclohexene ring with CH_3 and H) $\xrightarrow[\text{2. Zn, CH}_3\text{CO}_2\text{H}]{\text{1. O}_3}$ $CH_3\overset{O}{\overset{\|}{C}}CH_2CH_2CH_2CH_2\overset{O}{\overset{\|}{C}}H$

1-Methylcyclohexene   **6-Oxoheptanal (86%)**

A third method of aldehyde synthesis is one that we'll mention here just briefly and then return to in Section 21.6. Certain carboxylic acid derivatives can be *partially* reduced to yield aldehydes. The partial reduction of an ester by diisobutylaluminum hydride (DIBAH), for instance, is an important laboratory-scale method of aldehyde synthesis, and mechanistically related processes also occur in biological pathways. The reaction is normally carried out at $-78\ °C$ (dry-ice temperature) in toluene solution.

$$CH_3(CH_2)_{10}\overset{\overset{\displaystyle O}{\|}}{C}OCH_3 \quad \xrightarrow[\text{2. H}_3\text{O}^+]{\text{1. DIBAH, toluene, –78 °C}} \quad CH_3(CH_2)_{10}\overset{\overset{\displaystyle O}{\|}}{C}H$$

Methyl dodecanoate   **Dodecanal (88%)**

where DIBAH = $CH_3CHCH_2-\underset{\underset{\displaystyle CH_3}{|}}{\overset{\overset{\displaystyle H}{|}}{Al}}-CH_2\underset{\underset{\displaystyle CH_3}{|}}{CH}CH_3$

Problem 19.3 How would you prepare pentanal from the following starting materials?
(a) $CH_3CH_2CH_2CH_2CH_2OH$ (b) $CH_3CH_2CH_2CH_2CH=CH_2$
(c) $CH_3CH_2CH_2CH_2CO_2CH_3$

Preparing Ketones

For the most part, methods of ketone synthesis are similar to those for aldehydes.

▌ Secondary alcohols are oxidized by a variety of reagents to give ketones (Section 17.8). The choice of oxidant depends on such factors as reaction scale, cost, and acid or base sensitivity of the alcohol.

4-*tert*-Butylcyclohexanol   **4-*tert*-Butylcyclohexanone (90%)**

▌ Ozonolysis of alkenes yields ketones if one of the unsaturated carbon atoms is disubstituted (Section 7.9).

70%

▮ Aryl ketones are prepared by Friedel–Crafts acylation of an aromatic ring with an acid chloride in the presence of $AlCl_3$ catalyst (Section 16.3).

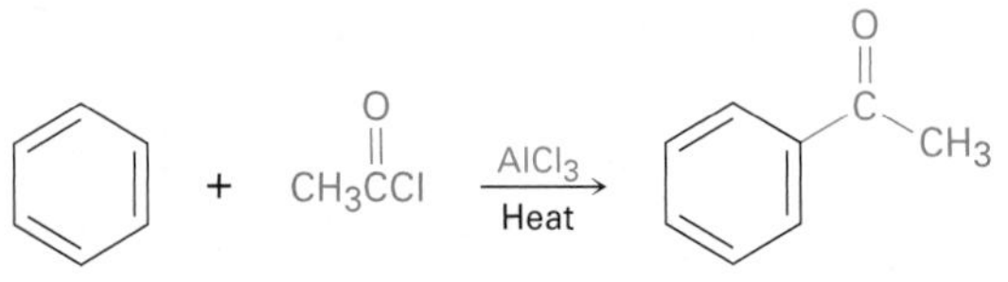

Benzene + CH_3CCl (Acetyl chloride) $\xrightarrow[\text{Heat}]{AlCl_3}$ Acetophenone (95%)

▮ Methyl ketones are prepared by hydration of terminal alkynes in the presence of Hg^{2+} catalyst (Section 8.4).

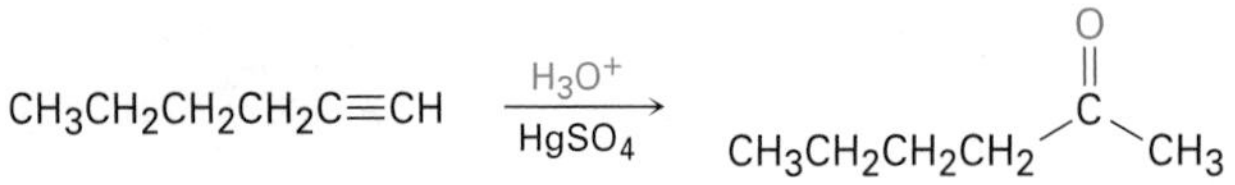

$CH_3CH_2CH_2CH_2C\equiv CH$ $\xrightarrow[\text{HgSO}_4]{H_3O^+}$ $CH_3CH_2CH_2CH_2$ $\overset{O}{\underset{}{C}}$ CH_3

1-Hexyne   **2-Hexanone (78%)**

In addition to those methods already discussed, ketones can also be prepared from certain carboxylic acid derivatives, just as aldehydes can. Among the most useful reactions of this type is that between an acid chloride and a Gilman diorganocopper reagent such as we saw in Section 10.8. We'll discuss this subject in more detail in Section 21.4.

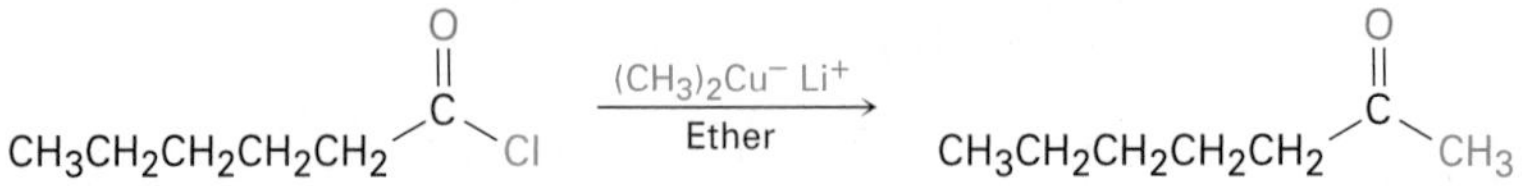

$CH_3CH_2CH_2CH_2CH_2$ $\overset{O}{\underset{}{C}}$ Cl $\xrightarrow[\text{Ether}]{(CH_3)_2Cu^- Li^+}$ $CH_3CH_2CH_2CH_2CH_2$ $\overset{O}{\underset{}{C}}$ CH_3

Hexanoyl chloride   **2-Heptanone (81%)**

Problem 19.4 How would you carry out the following reactions? More than one step may be required.
(a) 3-Hexyne → 3-Hexanone
(b) Benzene → *m*-Bromoacetophenone
(c) Bromobenzene → Acetophenone
(d) 1-Methylcyclohexene → 2-Methylcyclohexanone

19.3 | Oxidation of Aldehydes and Ketones

CENGAGENOW Click *Organic Interactive* to **use a web-based palette to predict products from a variety of oxidation reactions involving aldehydes and ketones.**

Aldehydes are easily oxidized to yield carboxylic acids, but ketones are generally inert toward oxidation. The difference is a consequence of structure: aldehydes have a −CHO proton that can be abstracted during oxidation, but ketones do not.

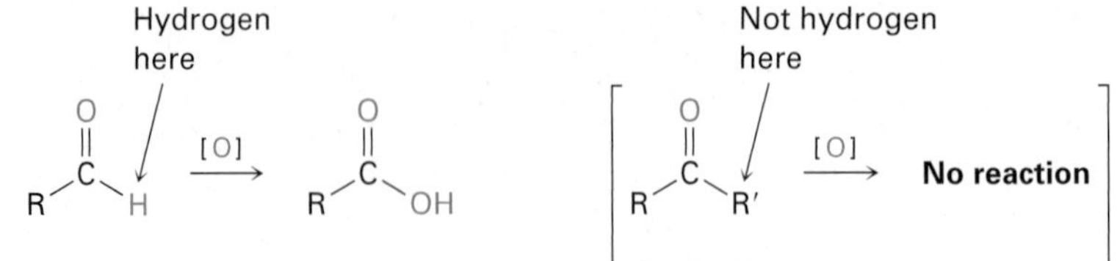

Hydrogen here

R $\overset{O}{\underset{H}{C}}$ $\xrightarrow{[O]}$ R $\overset{O}{\underset{OH}{C}}$

An aldehyde   **A carboxylic acid**

Not hydrogen here

$\left[R \overset{O}{\underset{R'}{C}} \xrightarrow{[O]} \textbf{No reaction} \right]$

A ketone

Many oxidizing agents, including $KMnO_4$ and hot HNO_3, convert aldehydes into carboxylic acids, but CrO_3 in aqueous acid is a more common choice. The oxidation occurs rapidly at room temperature and generally results in good yields.

$$CH_3CH_2CH_2CH_2CH_2\overset{\displaystyle O}{\overset{\|}{C}}H \xrightarrow[\text{Acetone, 0 °C}]{CrO_3,\ H_3O^+} CH_3CH_2CH_2CH_2CH_2\overset{\displaystyle O}{\overset{\|}{C}}OH$$

Hexanal   **Hexanoic acid (85%)**

One drawback to this CrO_3 oxidation is that it takes place under acidic conditions, and sensitive molecules sometimes undergo side reactions. In such cases, the laboratory oxidation of an aldehyde can be carried out using a solution of silver oxide, Ag_2O, in aqueous ammonia, the so-called Tollens' reagent. Aldehydes are oxidized by Tollens' reagent in high yield without harming carbon–carbon double bonds or other acid-sensitive functional groups in a molecule.

Benzaldehyde   **Benzoic acid**

Aldehyde oxidations occur through intermediate 1,1-diols, or *hydrates,* which are formed by a reversible nucleophilic addition of water to the carbonyl group. Even though formed to only a small extent at equilibrium, the hydrate reacts like any typical primary or secondary alcohol and is oxidized to a carbonyl compound (Section 17.7).

An aldehyde   **A hydrate**   **A carboxylic acid**

Ketones are inert to most oxidizing agents but undergo a slow cleavage reaction when treated with hot alkaline $KMnO_4$. The C–C bond next to the carbonyl group is broken, and carboxylic acids are produced. The reaction is useful primarily for symmetrical ketones such as cyclohexanone because product mixtures are formed from unsymmetrical ketones.

Cyclohexanone   **Hexanedioic acid (79%)**

19.4 | Nucleophilic Addition Reactions of Aldehydes and Ketones

As we saw in *A Preview of Carbonyl Compounds,* the most general reaction of aldehydes and ketones is the **nucleophilic addition reaction**. A nucleophile, :Nu⁻, approaches along the C=O bond from an angle of about 75° to the plane of the carbonyl group and adds to the electrophilic C=O carbon atom. At the same time, rehybridization of the carbonyl carbon from sp^2 to sp^3 occurs, an electron pair from the C=O bond moves toward the electronegative oxygen atom, and a tetrahedral alkoxide ion intermediate is produced (Figure 19.1).

Figure 19.1 MECHANISM:
A nucleophilic addition reaction to an aldehyde or ketone. The nucleophile approaches the carbonyl group from an angle of approximately 75° to the plane of the sp^2 orbitals, the carbonyl carbon rehybridizes from sp^2 to sp^3, and an alkoxide ion is formed.

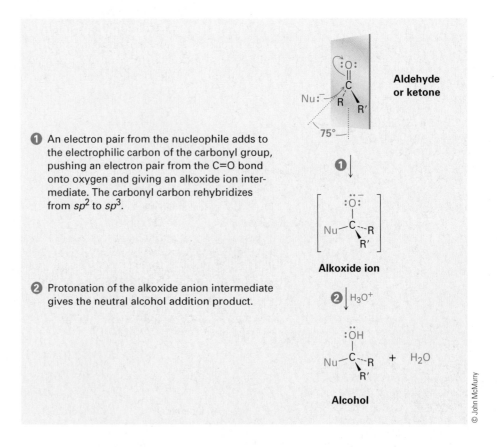

① An electron pair from the nucleophile adds to the electrophilic carbon of the carbonyl group, pushing an electron pair from the C=O bond onto oxygen and giving an alkoxide ion intermediate. The carbonyl carbon rehybridizes from sp^2 to sp^3.

② Protonation of the alkoxide anion intermediate gives the neutral alcohol addition product.

© John McMurry

The nucleophile can be either negatively charged (:Nu⁻) or neutral (:Nu). If it's neutral, however, it usually carries a hydrogen atom that can subsequently be eliminated, :Nu−H. For example:

Some negatively charged nucleophiles

HÖ:⁻ (hydroxide ion)

H:⁻ (hydride ion)

R₃C:⁻ (a carbanion)

RÖ:⁻ (an alkoxide ion)

N≡C:⁻ (cyanide ion)

$$\text{Some neutral nucleophiles} \begin{cases} \text{H}\ddot{\text{O}}\text{H (water)} \\ \text{R}\ddot{\text{O}}\text{H (an alcohol)} \\ \text{H}_3\text{N}\text{: (ammonia)} \\ \text{R}\ddot{\text{N}}\text{H}_2 \text{ (an amine)} \end{cases}$$

Nucleophilic additions to aldehydes and ketones have two general variations, as shown in Figure 19.2. In one variation, the tetrahedral intermediate is protonated by water or acid to give an alcohol as the final product; in the second variation, the carbonyl oxygen atom is protonated and then eliminated as HO⁻ or H₂O to give a product with a C=Nu bond.

Active Figure 19.2 Two general reaction pathways following addition of a nucleophile to an aldehyde or ketone. The top pathway leads to an alcohol product; the bottom pathway leads to a product with a C=Nu bond. *Sign in at* **www.cengage.com/login** *to see a simulation based on this figure and to take a short quiz.*

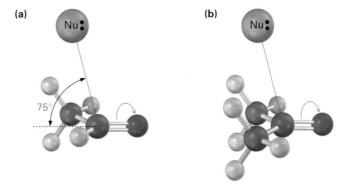

Aldehydes are generally more reactive than ketones in nucleophilic addition reactions for both steric and electronic reasons. Sterically, the presence of only one large substituent bonded to the C=O carbon in an aldehyde versus two large substituents in a ketone means that a nucleophile is able to approach an aldehyde more readily. Thus, the transition state leading to the tetrahedral intermediate is less crowded and lower in energy for an aldehyde than for a ketone (Figure 19.3).

Figure 19.3 (a) Nucleophilic addition to an aldehyde is sterically less hindered because only one relatively large substituent is attached to the carbonyl-group carbon. **(b)** A ketone, however, has two large substituents and is more hindered.

(a) Nu:

75°

(b) Nu:

Electronically, aldehydes are more reactive than ketones because of the greater polarization of aldehyde carbonyl groups. To see this polarity difference, recall the stability order of carbocations (Section 6.9). A primary carbocation is higher in energy and thus more reactive than a secondary carbocation because it has only one alkyl group inductively stabilizing the positive charge rather than two. In the same way, an aldehyde has only one alkyl group inductively stabilizing the partial positive charge on the carbonyl carbon rather than two, is a bit more electrophilic, and is therefore more reactive than a ketone.

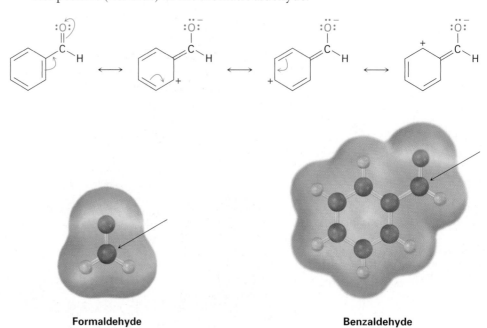

1° carbocation
(less stable, more reactive)

2° carbocation
(more stable, less reactive)

Aldehyde
(less stabilization of δ+, more reactive)

Ketone
(more stabilization of δ+, less reactive)

One further comparison: aromatic aldehydes, such as benzaldehyde, are less reactive in nucleophilic addition reactions than aliphatic aldehydes because the electron-donating resonance effect of the aromatic ring makes the carbonyl group less electrophilic. Comparing electrostatic potential maps of formaldehyde and benzaldehyde, for example, shows that the carbonyl carbon atom is less positive (less blue) in the aromatic aldehyde.

Formaldehyde

Benzaldehyde

Problem 19.5 | Treatment of an aldehyde or ketone with cyanide ion ($^-$:C≡N), followed by protonation of the tetrahedral alkoxide ion intermediate, gives a *cyanohydrin*. Show the structure of the cyanohydrin obtained from cyclohexanone.

Problem 19.6 | *p*-Nitrobenzaldehyde is more reactive toward nucleophilic additions than *p*-methoxybenzaldehyde. Explain.

19.5 | Nucleophilic Addition of H₂O: Hydration

Aldehydes and ketones react with water to yield 1,1-diols, or *geminal (gem)* diols. The hydration reaction is reversible, and a gem diol can eliminate water to regenerate an aldehyde or ketone.

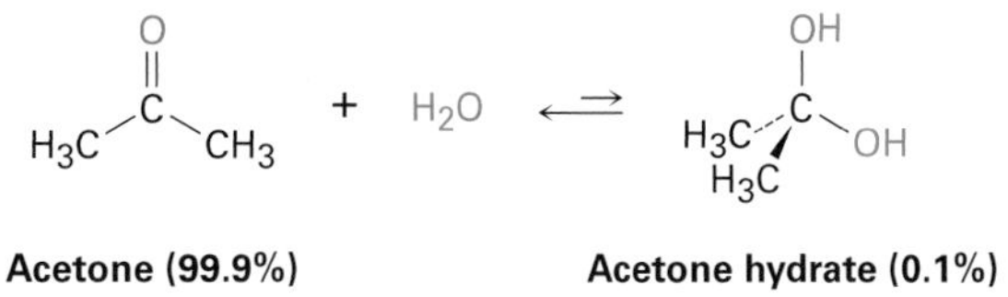

Acetone (99.9%)   **Acetone hydrate (0.1%)**

The position of the equilibrium between a gem diol and an aldehyde or ketone depends on the structure of the carbonyl compound. The equilibrium generally favors the carbonyl compound for steric reasons, but the gem diol is favored for a few simple aldehydes. For example, an aqueous solution of formaldehyde consists of 99.9% gem diol and 0.1% aldehyde, whereas an aqueous solution of acetone consists of only about 0.1% gem diol and 99.9% ketone.

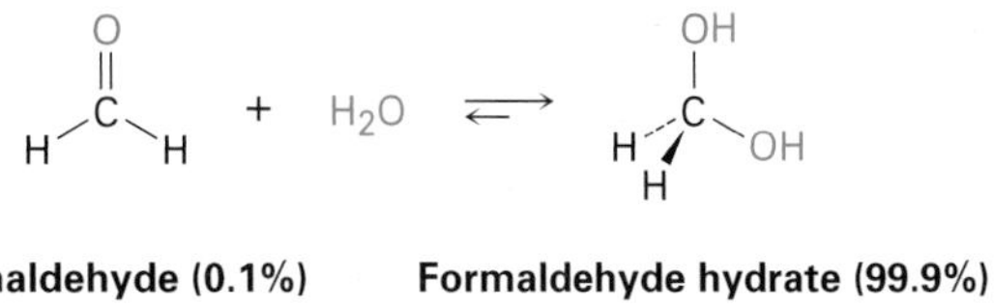

Formaldehyde (0.1%)   **Formaldehyde hydrate (99.9%)**

The nucleophilic addition of water to an aldehyde or ketone is slow under neutral conditions but is catalyzed by both base and acid. The base-catalyzed hydration reaction takes place as shown in Figure 19.4. The nucleophile is the

Figure 19.4 MECHANISM:
The mechanism of base-catalyzed hydration of an aldehyde or ketone. Hydroxide ion is a more reactive nucleophile than neutral water.

CENGAGENOW Click *Organic Process* to **view an animation of the base-catalyzed hydration of a carbonyl**.

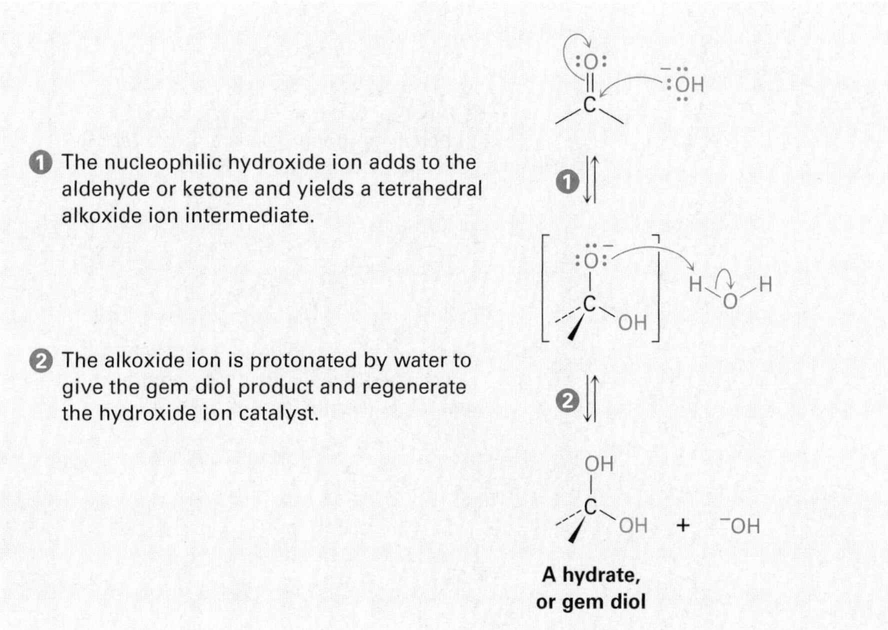

① The nucleophilic hydroxide ion adds to the aldehyde or ketone and yields a tetrahedral alkoxide ion intermediate.

② The alkoxide ion is protonated by water to give the gem diol product and regenerate the hydroxide ion catalyst.

A hydrate, or gem diol

© John McMurry

hydroxide ion, which is much more reactive than neutral water because of its negative charge.

The acid-catalyzed hydration reaction begins with protonation of the carbonyl oxygen atom, which places a positive charge on oxygen and makes the carbonyl group more electrophilic. Subsequent nucleophilic addition of water to the protonated aldehyde or ketone then yields a protonated gem diol, which loses H$^+$ to give the neutral product (Figure 19.5).

Note the key difference between the base-catalyzed and acid-catalyzed reactions. The base-catalyzed reaction takes place rapidly because water is converted into hydroxide ion, a much better *nucleophile*. The acid-catalyzed reaction takes place rapidly because the carbonyl compound is converted by protonation into a much better *electrophile*.

The hydration reaction just described is typical of what happens when an aldehyde or ketone is treated with a nucleophile of the type H–Y, where the Y atom is electronegative and can stabilize a negative charge (oxygen, halogen, or sulfur, for instance). In such reactions, the nucleophilic addition is reversible, with the equilibrium generally favoring the carbonyl reactant rather than the tetrahedral addition product. In other words, treatment of an aldehyde or

Figure 19.5 MECHANISM:
The mechanism of acid-catalyzed hydration of an aldehyde or ketone. Acid protonates the carbonyl group, making it more electrophilic and more reactive.

CENGAGENOW™ Click *Organic Process* to **view an animation of the acid-catalyzed hydration of a carbonyl**.

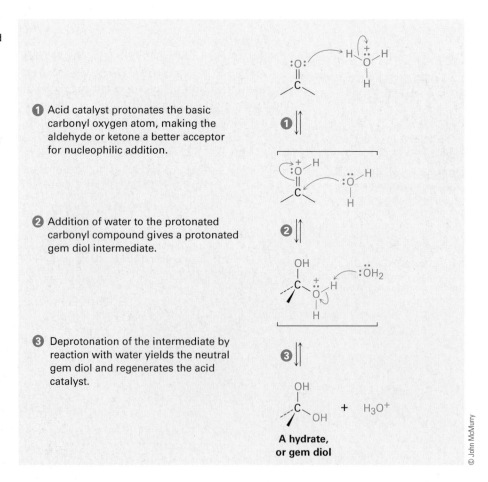

❶ Acid catalyst protonates the basic carbonyl oxygen atom, making the aldehyde or ketone a better acceptor for nucleophilic addition.

❷ Addition of water to the protonated carbonyl compound gives a protonated gem diol intermediate.

❸ Deprotonation of the intermediate by reaction with water yields the neutral gem diol and regenerates the acid catalyst.

A hydrate, or gem diol

© John McMurry

ketone with CH_3OH, H_2O, HCl, HBr, or H_2SO_4 does not normally lead to a stable alcohol addition product.

Favored when
$Y = -OCH_3, -OH, -Br, -Cl, HSO_4^-$

Problem 19.7 | When dissolved in water, trichloroacetaldehyde (chloral, CCl_3CHO) exists primarily as chloral hydrate, $CCl_3CH(OH)_2$, better known as "knockout drops." Show the structure of chloral hydrate.

Problem 19.8 | The oxygen in water is primarily (99.8%) ^{16}O, but water enriched with the heavy isotope ^{18}O is also available. When an aldehyde or ketone is dissolved in ^{18}O-enriched water, the isotopic label becomes incorporated into the carbonyl group. Explain.

$$R_2C=O + H_2O \rightleftharpoons R_2C=O + H_2O \qquad \text{where } O = {}^{18}O$$

19.6 | Nucleophilic Addition of HCN: Cyanohydrin Formation

Arthur Lapworth

Arthur Lapworth (1872–1941) was born in Galashiels, Scotland, and received a D.Sc. at the City and Guilds Institute, London. He was professor of chemistry at the University of Manchester from 1909 until his retirement in 1937.

Aldehydes and unhindered ketones undergo a nucleophilic addition reaction with HCN to yield **cyanohydrins**, $RCH(OH)C\equiv N$. Studies carried out in the early 1900s by Arthur Lapworth showed that cyanohydrin formation is reversible and base-catalyzed. Reaction occurs slowly when pure HCN is used but rapidly when a small amount of base is added to generate the nucleophilic cyanide ion, CN^-. Alternatively, a small amount of KCN can be added to HCN to catalyze the reaction. Addition of CN^- takes place by a typical nucleophilic addition pathway, yielding a tetrahedral intermediate that is protonated by HCN to give cyanohydrin product plus regenerated CN^-.

| Benzaldehyde | Tetrahedral intermediate | Mandelonitrile (88%) |

Cyanohydrin formation is somewhat unusual because it is one of the few examples of the addition of a protic acid (H−Y) to a carbonyl group. As noted in the previous section, protic acids such as H_2O, HBr, HCl, and H_2SO_4 don't normally yield carbonyl addition products because the equilibrium constants are unfavorable. With HCN, however, the equilibrium favors the cyanohydrin adduct.

Cyanohydrin formation is useful because of the further chemistry that can be carried out on the product. For example, a nitrile ($R-C\equiv N$) can be reduced with $LiAlH_4$ to yield a primary amine (RCH_2NH_2) and can be hydrolyzed by hot aqueous acid to yield a carboxylic acid. Thus, cyanohydrin formation provides a method for transforming an aldehyde or ketone into a different functional group.

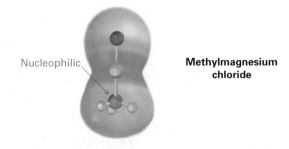

Problem 19.9 | Cyclohexanone forms a cyanohydrin in good yield but 2,2,6-trimethylcyclo-hexanone does not. Explain.

19.7 | Nucleophilic Addition of Grignard and Hydride Reagents: Alcohol Formation

We saw in Section 17.5 that treatment of an aldehyde or ketone with a Grignard reagent, RMgX, yields an alcohol by nucleophilic addition of a carbon anion, or **carbanion**. A carbon–magnesium bond is strongly polarized, so a Grignard reagent reacts for all practical purposes as R:$^-$ $^+$MgX.

Nucleophilic

Methylmagnesium chloride

A Grignard reaction begins with an acid–base complexation of Mg^{2+} to the carbonyl oxygen atom of the aldehyde or ketone, thereby making the carbonyl group a better electrophile. Nucleophilic addition of R:$^-$ then produces a tetra-hedral magnesium alkoxide intermediate, and protonation by addition of water or dilute aqueous acid in a separate step yields the neutral alcohol (Figure 19.6). Unlike the nucleophilic additions of water and HCN, Grignard additions are effectively irreversible because a carbanion is too poor a leaving group to be expelled in a reversal step.

Figure 19.6 MECHANISM:
Mechanism of the Grignard reaction. Nucleophilic addition of a carbanion to an aldehyde or ketone, followed by protonation of the alkoxide intermediate, yields an alcohol.

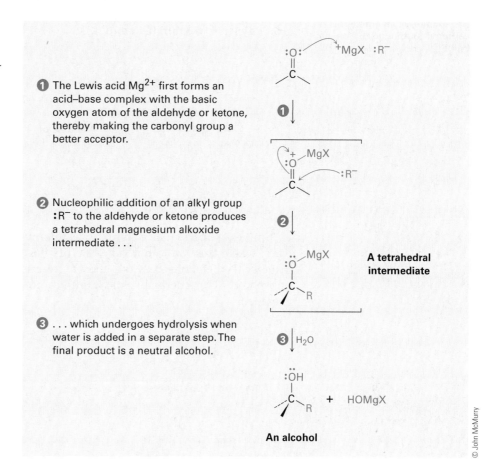

1 The Lewis acid Mg^{2+} first forms an acid–base complex with the basic oxygen atom of the aldehyde or ketone, thereby making the carbonyl group a better acceptor.

2 Nucleophilic addition of an alkyl group **:R⁻** to the aldehyde or ketone produces a tetrahedral magnesium alkoxide intermediate . . .

3 . . . which undergoes hydrolysis when water is added in a separate step. The final product is a neutral alcohol.

A tetrahedral intermediate

An alcohol

© John McMurry

Just as addition of a Grignard reagent to an aldehyde or ketone yields an alcohol, so does addition of hydride ion, **:H⁻** (Section 17.4). Although the details of carbonyl-group reductions are complex, $LiAlH_4$ and $NaBH_4$ act as if they were donors of hydride ion in a nucleophilic addition reaction (Figure 19.7). Addition of water or aqueous acid after the hydride addition step protonates the tetrahedral alkoxide intermediate and gives the alcohol product.

Figure 19.7 Mechanism of carbonyl-group reduction by nucleophilic addition of "hydride ion" from $NaBH_4$ or $LiAlH_4$.

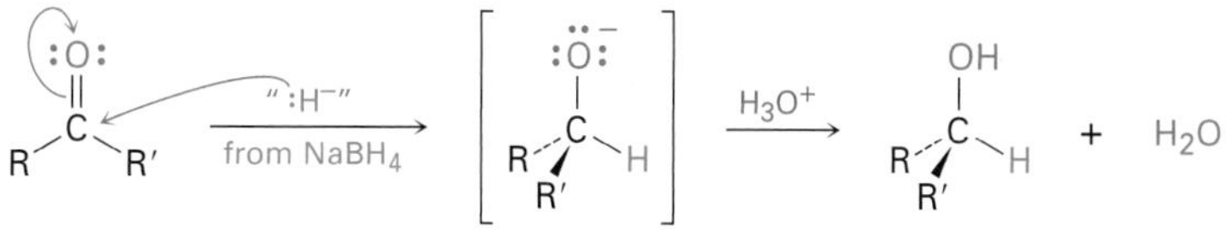

19.8 | Nucleophilic Addition of Amines: Imine and Enamine Formation

Primary amines, RNH_2, add to aldehydes and ketones to yield **imines**, $R_2C=NR$. Secondary amines, R_2NH, add similarly to yield **enamines**, $R_2N-CR=CR_2$ (*ene* + *amine* = unsaturated amine).

A ketone or an aldehyde

An imine   **An enamine**

Imines are particularly common as intermediates in many biological pathways, where they are often called **Schiff bases**. The amino acid alanine, for instance, is metabolized in the body by reaction with the aldehyde pyridoxal phosphate (PLP), a derivative of vitamin B_6, to yield a Schiff base that is further degraded.

Pyridoxal phosphate   **Alanine**   **An imine**

Imine formation and enamine formation appear different because one leads to a product with a $C=N$ bond and the other leads to a product with a $C=C$ bond. Actually, though, the reactions are quite similar. Both are typical examples of nucleophilic addition reactions in which water is eliminated from the initially formed tetrahedral intermediate and a new $C=Nu$ bond is formed.

Imines are formed in a reversible, acid-catalyzed process that begins with nucleophilic addition of the primary amine to the carbonyl group, followed by transfer of a proton from nitrogen to oxygen to yield a neutral amino alcohol, or *carbinolamine*. Protonation of the carbinolamine oxygen by an acid catalyst then converts the $-OH$ into a better leaving group ($-OH_2^+$), and E1-like loss of water produces an iminium ion. Loss of a proton from nitrogen gives the final product and regenerates the acid catalyst (Figure 19.8).

Figure 19.8 MECHANISM:
Mechanism of imine formation
by reaction of an aldehyde or
ketone with a primary amine.
The key step is nucleophilic addition to yield a carbinolamine
intermediate, which then loses
water to give the imine.

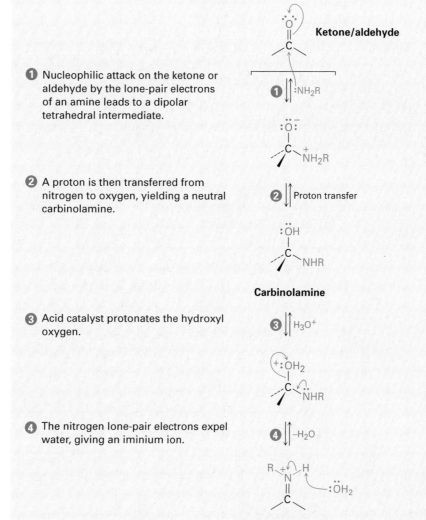

1 Nucleophilic attack on the ketone or
aldehyde by the lone-pair electrons
of an amine leads to a dipolar
tetrahedral intermediate.

2 A proton is then transferred from
nitrogen to oxygen, yielding a neutral
carbinolamine.

3 Acid catalyst protonates the hydroxyl
oxygen.

4 The nitrogen lone-pair electrons expel
water, giving an iminium ion.

5 Loss of H⁺ from nitrogen then gives
the neutral imine product.

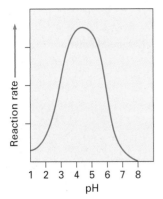

Figure 19.9 Dependence on pH of the rate of reaction between acetone and hydroxylamine: $(CH_3)_2C{=}O + NH_2OH \rightarrow$ $(CH_3)_2C{=}NOH + H_2O$.

Imine and enamine formation are slow at both high pH and low pH but reach a maximum rate at a weakly acidic pH around 4 to 5. For example, the profile of pH versus rate shown in Figure 19.9 for the reaction between acetone and hydroxylamine, NH_2OH, indicates that the maximum reaction rate is obtained at pH 4.5.

We can explain the observed pH dependence of imine formation by looking at the individual steps in the mechanism. As indicated in Figure 19.8, an acid catalyst is required in step 3 to protonate the intermediate carbinolamine, thereby converting the −OH into a better leaving group. Thus, reaction will be slow if not enough acid is present (that is, at high pH). On the other hand, if too much acid is present (low pH), the basic amine nucleophile is completely protonated, so the initial nucleophilic addition step can't occur.

Evidently, a pH of 4.5 represents a compromise between the need for *some* acid to catalyze the rate-limiting dehydration step but *not too much* acid so as to avoid complete protonation of the amine. Each individual nucleophilic addition reaction has its own requirements, and reaction conditions must be optimized to obtain maximum reaction rates.

Imine formation from such reagents as hydroxylamine and 2,4-dinitrophenylhydrazine is sometimes useful because the products of these reactions—*oximes* and *2,4-dinitrophenylhydrazones (2,4-DNPs)*, respectively—are often crystalline and easy to handle. Such crystalline derivatives are occasionally prepared as a means of purifying and characterizing liquid ketones or aldehydes.

Oxime

Cyclohexanone   Hydroxylamine   Cyclohexanone oxime (mp 90 °C)

2,4-Dinitrophenyl-hydrazone

Acetone   2,4-Dinitrophenyl-hydrazine   Acetone 2,4-dinitrophenyl-hydrazone (mp 126 °C)

Reaction of an aldehyde or ketone with a secondary amine, R_2NH, rather than a primary amine yields an enamine. The process is identical to imine formation up to the iminium ion stage, but at this point there is no proton on nitrogen that can be lost to form a neutral imine product. Instead, a proton is lost from the *neighboring* carbon (the α carbon), yielding an enamine (Figure 19.10).

Figure 19.10 MECHANISM: Mechanism of enamine formation by reaction of an aldehyde or ketone with a secondary amine, R_2NH. The iminium ion intermediate has no hydrogen attached to N and so must lose H^+ from the carbon two atoms away.

❶ Nucleophilic addition of a secondary amine to the ketone or aldehyde, followed by proton transfer from nitrogen to oxygen, yields an intermediate carbinolamine in the normal way.

❷ Protonation of the hydroxyl by acid catalyst converts it into a better leaving group.

❸ Elimination of water by the lone-pair electrons on nitrogen then yields an intermediate iminium ion.

❹ Loss of a proton from the alpha carbon atom yields the enamine product and regenerates the acid catalyst.

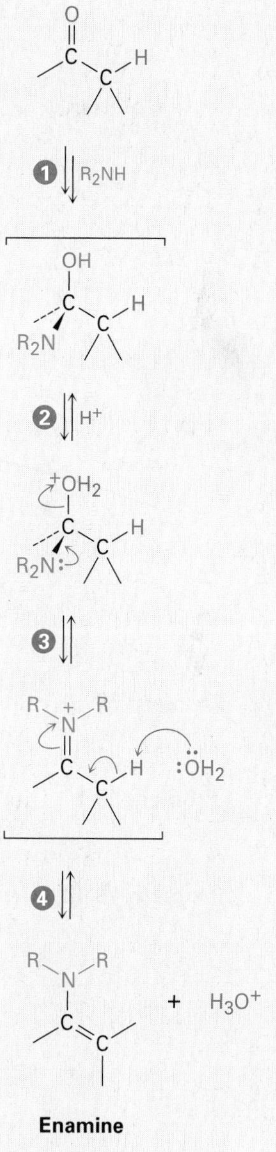

Enamine

© John McMurry

WORKED EXAMPLE 19.1   *Predicting the Product of Reaction between a Ketone and an Amine*

Show the products you would obtain by acid-catalyzed reaction of 3-pentanone with methylamine, CH_3NH_2, and with dimethylamine, $(CH_3)_2NH$.

Strategy   An aldehyde or ketone reacts with a primary amine, RNH_2, to yield an imine, in which the carbonyl oxygen atom has been replaced by the $=N-R$ group of the amine. Reaction of the same aldehyde or ketone with a secondary amine, R_2NH, yields an enamine, in which the oxygen atom has been replaced by the $-NR_2$ group of the amine and the double bond has moved to a position between the former carbonyl carbon and the neighboring carbon.

Solution

An imine

3-Pentanone

An enamine

Problem 19.10   Show the products you would obtain by acid-catalyzed reaction of cyclohexanone with ethylamine, $CH_3CH_2NH_2$, and with diethylamine, $(CH_3CH_2)_2NH$.

Problem 19.11   Imine formation is reversible. Show all the steps involved in the acid-catalyzed reaction of an imine with water (hydrolysis) to yield an aldehyde or ketone plus primary amine.

Problem 19.12   Draw the following molecule as a line-bond structure, and show how it can be prepared from a ketone and an amine.

19.9 Nucleophilic Addition of Hydrazine: The Wolff–Kishner Reaction

Ludwig Wolff

Ludwig Wolff (1857–1919) was born in Neustadt/Hardt, Germany, and received his Ph.D. from the University of Strasbourg working with Rudolf Fittig. He was professor of chemistry at the University of Jena.

A useful variant of the imine-forming reaction just discussed involves the treatment of an aldehyde or ketone with hydrazine, H_2NNH_2, in the presence of KOH. Called the **Wolff–Kishner reaction**, the process is a useful and general method for converting an aldehyde or ketone into an alkane, $R_2C=O \rightarrow R_2CH_2$.

Propiophenone → **Propylbenzene (82%)** $+ N_2 + H_2O$

Cyclopropane-carbaldehyde → **Methylcyclo-propane (72%)** $+ N_2 + H_2O$

N. M. Kishner

N. M. Kishner (1867–1935) was born in Moscow and received his Ph.D. at the University of Moscow working with Vladimir Markovnikov. He became professor, first at the University of Tomsk and then at the University of Moscow.

The Wolff–Kishner reaction involves formation of a *hydrazone* intermediate, $R_2C=NNH_2$, followed by base-catalyzed double-bond migration, loss of N_2 gas, and protonation to give the alkane product (Figure 19.11). The double-bond migration takes place when base removes one of the weakly acidic NH protons to generate a hydrazone anion, which has an allylic resonance structure that places the double bond between nitrogens and the negative charge on carbon. Reprotonation then occurs on carbon to generate the double-bond rearrangement product. The next step—loss of nitrogen and formation of an alkyl anion—is driven by the large thermodynamic stability of the N_2 molecule.

Note that the Wolff–Kishner reduction accomplishes the same overall transformation as the catalytic hydrogenation of an acylbenzene to yield an alkylbenzene (Section 16.10). The Wolff–Kishner reduction is more general and more useful than catalytic hydrogenation, however, because it works well with both alkyl and aryl ketones.

Problem 19.13 Show how you could prepare the following compounds from 4-methyl-3-penten-2-one, $(CH_3)_2C=CHCOCH_3$.

(a)
$$\begin{array}{cc} CH_3 & O \\ | & || \\ CH_3CHCH_2CCH_3 \end{array}$$

(b)
$$\begin{array}{c} CH_3 \\ | \\ CH_3C=CHCH_2CH_3 \end{array}$$

(c)
$$\begin{array}{c} CH_3 \\ | \\ CH_3CHCH_2CH_2CH_3 \end{array}$$

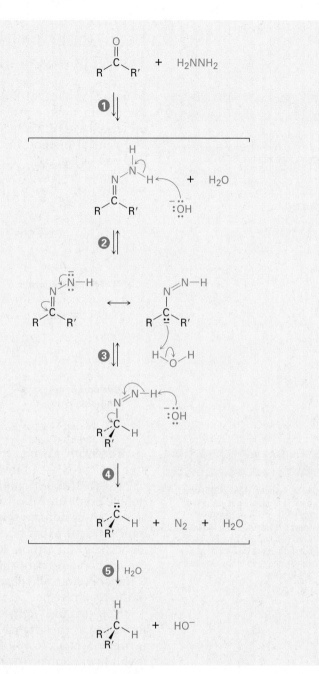

① Reaction of the aldehyde or ketone with hydrazine yields a hydrazone in the normal way.

② Base abstracts a weakly acidic N–H proton, yielding a hydrazone anion. This anion has a resonance form that places the negative charge on carbon and the double bond between nitrogens.

③ Protonation of the hydrazone anion takes place on carbon to yield a neutral intermediate.

④ Deprotonation of the remaining weakly acidic N–H occurs with simultaneous loss of nitrogen to give a carbanion . . .

⑤ . . . which is protonated to give the alkane product.

Figure 19.11 MECHANISM: Mechanism of the Wolff–Kishner reduction of an aldehyde or ketone to yield an alkane.

19.10 | Nucleophilic Addition of Alcohols: Acetal Formation

Aldehydes and ketones react reversibly with 2 equivalents of an alcohol in the presence of an acid catalyst to yield **acetals, $R_2C(OR')_2$**, sometimes called *ketals* if derived from a ketone. Cyclohexanone, for instance, reacts with methanol in the presence of HCl to give the corresponding dimethyl acetal.

Cyclohexanone   **Cyclohexanone dimethyl acetal**

Acetal formation is similar to the hydration reaction discussed in Section 19.5. Like water, alcohols are weak nucleophiles that add to aldehydes and ketones only slowly under neutral conditions. Under acidic conditions, however, the reactivity of the carbonyl group is increased by protonation, so addition of an alcohol occurs rapidly.

A neutral carbonyl group is moderately electrophilic because of the polarity of the C–O bond.

A protonated carbonyl group is strongly electrophilic because of the positive charge on carbon.

Nucleophilic addition of an alcohol to the carbonyl group initially yields a hydroxy ether called a **hemiacetal**, analogous to the gem diol formed by addition of water. Hemiacetals are formed reversibly, with the equilibrium normally favoring the carbonyl compound. In the presence of acid, however, a further reaction occurs. Protonation of the −OH group, followed by an E1-like loss of water, leads to an oxonium ion, $R_2C=OR^+$, which undergoes a second nucleophilic addition of alcohol to yield the acetal. The mechanism is shown in Figure 19.12.

Because all the steps in acetal formation are reversible, the reaction can be driven either forward (from carbonyl compound to acetal) or backward (from acetal to carbonyl compound), depending on the conditions. The forward reaction is favored by conditions that remove water from the medium and thus drive the equilibrium to the right. In practice, this is often done by distilling off water as it forms. The reverse reaction is favored by treating the acetal with a large excess of aqueous acid to drive the equilibrium to the left.

Acetals are useful because they can act as protecting groups for aldehydes and ketones in the same way that trimethylsilyl ethers act as protecting groups for alcohols (Section 17.8). As we saw previously, it sometimes happens that one functional group interferes with intended chemistry elsewhere

Figure 19.12 MECHANISM:
Mechanism of acid-catalyzed acetal formation by reaction of an aldehyde or ketone with an alcohol.

CENGAGENOW™ Click *Organic Process* to **view an animation of acetal formation from an alcohol and an aldehyde**.

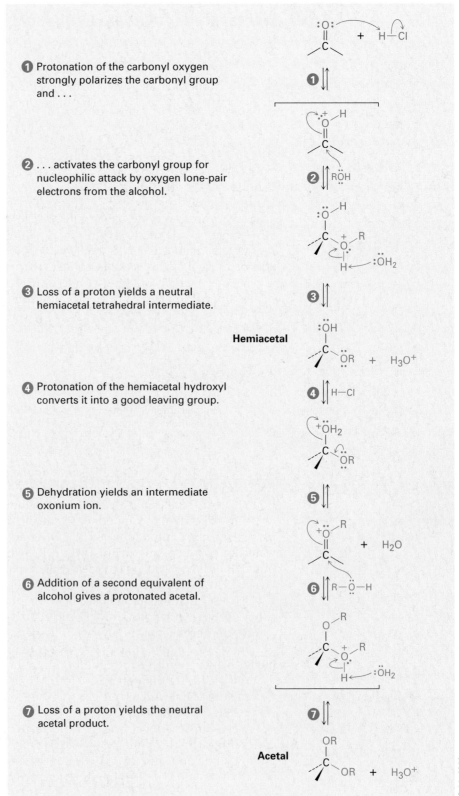

① Protonation of the carbonyl oxygen strongly polarizes the carbonyl group and . . .

② . . . activates the carbonyl group for nucleophilic attack by oxygen lone-pair electrons from the alcohol.

③ Loss of a proton yields a neutral hemiacetal tetrahedral intermediate.

Hemiacetal

④ Protonation of the hemiacetal hydroxyl converts it into a good leaving group.

⑤ Dehydration yields an intermediate oxonium ion.

⑥ Addition of a second equivalent of alcohol gives a protonated acetal.

⑦ Loss of a proton yields the neutral acetal product.

Acetal

© John McMurry

in a complex molecule. For example, if we wanted to reduce only the ester group of ethyl 4-oxopentanoate, the ketone would interfere. Treatment of the starting keto ester with LiAlH$_4$ would reduce both the keto and the ester groups to give a diol product.

$$CH_3CCH_2CH_2COCH_2CH_3 \xrightarrow{?} CH_3CCH_2CH_2CH_2OH$$

Ethyl 4-oxopentanoate   **5-Hydroxy-2-pentanone**

By protecting the keto group as an acetal, however, the problem can be circumvented. Like other ethers, acetals are unreactive to bases, hydride reducing agents, Grignard reagents, and catalytic reducing conditions, but are cleaved by acid. Thus, we can accomplish the selective reduction of the ester group in ethyl 4-oxopentanoate by first converting the keto group to an acetal, then reducing the ester with LiAlH$_4$, and then removing the acetal by treatment with aqueous acid.

Ethyl 4-oxopentanoate

5-Hydroxy-2-pentanone

In practice, it's convenient to use 1 equivalent of a diol such as ethylene glycol as the alcohol and to form a *cyclic* acetal. The mechanism of cyclic acetal formation using ethylene glycol is exactly the same as that using 2 equivalents of methanol or other monoalcohol. The only difference is that both −OH groups are in the same molecule.

Acetal and hemiacetal groups are particularly common in carbohydrate chemistry. Glucose, for instance, is a polyhydroxy aldehyde that undergoes an *internal* nucleophilic addition reaction and exists primarily as a cyclic hemiacetal.

Glucose—open chain   **Glucose—cyclic hemiacetal**

WORKED EXAMPLE 19.2   *Predicting the Product of Reaction between a Ketone and an Alcohol*

Show the structure of the acetal you would obtain by acid-catalyzed reaction of 2-pentanone with 1,3-propanediol.

Strategy   Acid-catalyzed reaction of an aldehyde or ketone with 2 equivalents of a mono-alcohol or 1 equivalent of a diol yields an acetal, in which the carbonyl oxygen atom is replaced by two −OR groups from the alcohol.

Solution

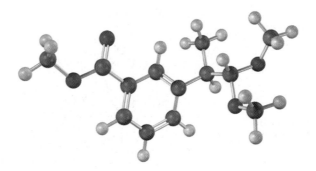

2-Pentanone

Problem 19.14 | Show all the steps in the acid-catalyzed formation of a cyclic acetal from ethylene glycol and an aldehyde or ketone.

Problem 19.15 | Identify the carbonyl compound and the alcohol that were used to prepare the following acetal:

19.11 | Nucleophilic Addition of Phosphorus Ylides: The Wittig Reaction

Georg F. K. Wittig

Georg F. K. Wittig (1897–1987) was born in Berlin, Germany, and received his Ph.D. at the University of Marburg in 1926, working with von Auwers. He then became professor of chemistry, first at the University of Braunschweig and later in Freiburg, Tübingen, and Heidelberg. In 1979, he received the Nobel Prize in chemistry for his work on phosphorus-containing organic compounds.

Aldehydes and ketones are converted into alkenes by means of a nucleophilic addition called the **Wittig reaction**. The reaction has no direct biological counterpart but is important both because of its wide use in the laboratory and drug manufacture and because of its mechanistic similarity to reactions of the coenzyme thiamin diphosphate, which we'll see in Section 29.6.

In the Wittig reaction, a phosphorus *ylide*, $R_2\bar{C}—\overset{+}{P}(C_6H_5)_3$, also called a *phosphorane* and sometimes written in the resonance form $R_2C=P(C_6H_5)_3$, adds to an aldehyde or ketone to yield a dipolar intermediate called a *betaine*. (An **ylide**—pronounced **ill**-id—is a neutral, dipolar compound with adjacent plus and minus charges. A **betaine**—pronounced **bay**-ta-een—is a neutral, dipolar compound with nonadjacent charges.)

The betaine intermediate is not isolated; rather, it spontaneously decomposes through a four-membered ring to yield alkene plus triphenylphosphine

Active Figure 19.13
MECHANISM: The mechanism of the Wittig reaction between a phosphorus ylide and an aldehyde or ketone to yield an alkene. *Sign in at* **www.cengage.com/login** *to see a simulation based on this figure and to take a short quiz.*

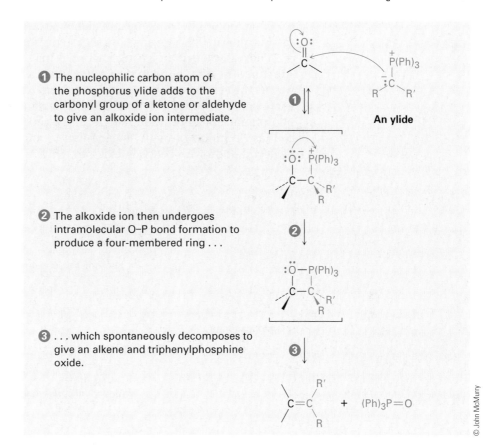

① The nucleophilic carbon atom of the phosphorus ylide adds to the carbonyl group of a ketone or aldehyde to give an alkoxide ion intermediate.

② The alkoxide ion then undergoes intramolecular O–P bond formation to produce a four-membered ring . . .

③ . . . which spontaneously decomposes to give an alkene and triphenylphosphine oxide.

An ylide

© John McMurry

CENGAGENOW™ Click *Organic Interactive* to **use your problem-solving skills to design syntheses involving Wittig reactions.**

oxide, $(Ph)_3P{=}O$. The net result is replacement of the carbonyl oxygen atom by the $R_2C{=}$ group originally bonded to phosphorus (Figure 19.13).

The phosphorus ylides necessary for Wittig reaction are easily prepared by S_N2 reaction of primary (and some secondary) alkyl halides with triphenylphosphine, followed by treatment with base. Triphenylphosphine, $(Ph)_3P$, is a good nucleophile in S_N2 reactions, and yields of the resultant alkyltriphenylphosphonium salts are high. Because of the positive charge on phosphorus, the hydrogen on the neighboring carbon is weakly acidic and can be removed by a strong base such as butyllithium (BuLi) to generate the neutral ylide. For example:

Triphenylphosphine   **Methyltriphenyl-phosphonium bromide**   **Methylenetriphenyl-phosphorane**

Bromo-methane

The Wittig reaction is extremely general, and a great many monosubstituted, disubstituted, and trisubstituted alkenes can be prepared from the appropriate combination of phosphorane and aldehyde or ketone. Tetrasubstituted alkenes can't be prepared, however, because of steric hindrance during the reaction.

The real value of the Wittig reaction is that it yields a pure alkene of defined structure. The C=C bond in the product is always exactly where the C=O group was in the reactant, and no alkene isomers (except *E,Z* isomers) are formed. For example, Wittig reaction of cyclohexanone with methylenetriphenylphosphorane yields only the single alkene product methylenecyclohexane. By contrast, addition of methylmagnesium bromide to cyclohexanone, followed by dehydration with POCl$_3$, yields a roughly 9:1 mixture of two alkenes.

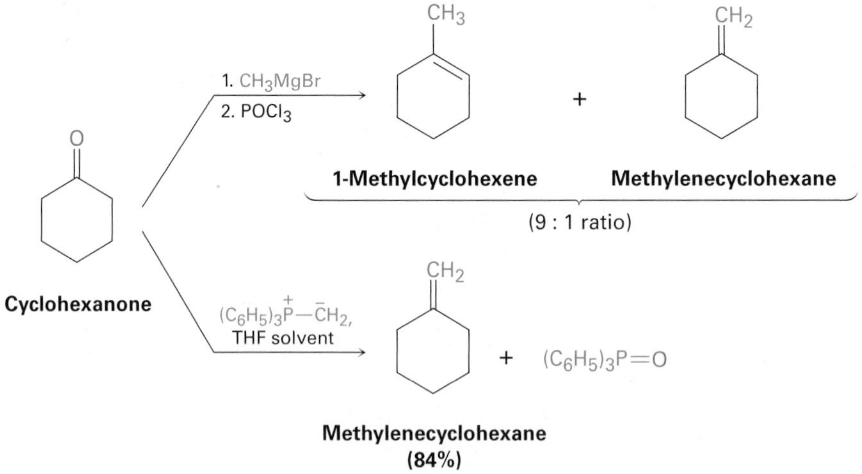

1-Methylcyclohexene   **Methylenecyclohexane**

(9 : 1 ratio)

Cyclohexanone

Methylenecyclohexane
(84%)

Wittig reactions are used commercially in the synthesis of numerous pharmaceutical agents. For example, the German chemical company BASF prepares vitamin A by Wittig reaction between a 15-carbon ylide and a 5-carbon aldehyde.

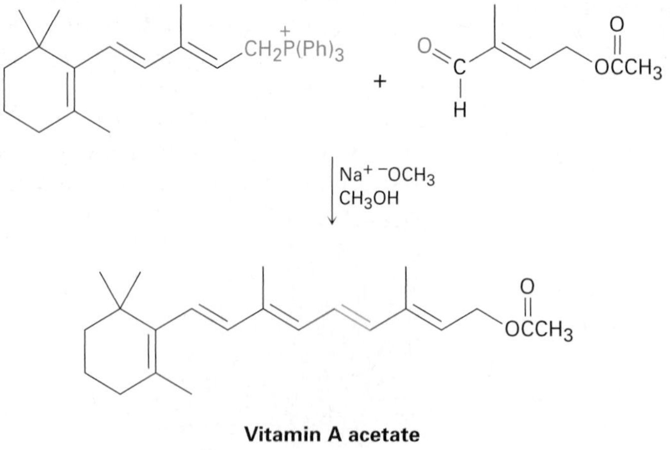

Vitamin A acetate

WORKED EXAMPLE 19.3   *Synthesizing an Alkene Using a Wittig Reaction*

What carbonyl compound and what phosphorus ylide might you use to prepare 3-ethyl-2-pentene?

Strategy An aldehyde or ketone reacts with a phosphorus ylide to yield an alkene in which the oxygen atom of the carbonyl reactant is replaced by the $=CR_2$ of the ylide. Preparation of the phosphorus ylide itself usually involves S_N2 reaction of a primary alkyl halide with triphenylphosphine, so the ylide is typically primary, $RCH=P(Ph)_3$. This means that the disubstituted alkene carbon in the product comes from the carbonyl reactant, while the monosubstituted alkene carbon comes from the ylide.

Solution

Disubstituted; from ketone
Monosubstituted; from ylide

$$CH_3CH_2\overset{\underset{\displaystyle CH_2CH_3}{|}}{C}=O \quad \xrightarrow[\text{THF}]{(Ph)_3\overset{+}{P}-\overset{-}{C}HCH_3} \quad CH_3CH_2\overset{\underset{\displaystyle CH_2CH_3}{|}}{C}=CHCH_3$$

3-Pentanone   **3-Ethyl-2-pentene**

Problem 19.16 What carbonyl compound and what phosphorus ylide might you use to prepare each of the following compounds?

(a)

(b)

(c)

(d)

(e)

(f)

Problem 19.17 β-Carotene, a yellow food-coloring agent and dietary source of vitamin A, can be prepared by a *double* Wittig reaction between 2 equivalents of β-ionylideneacetaldehyde and a *diylide*. Show the structure of the β-carotene product.

β-Ionylideneacetaldehyde   **A diylide**

19.12 Biological Reductions

As a general rule, nucleophilic addition reactions are characteristic only of aldehydes and ketones, not of carboxylic acid derivatives. The reason for the difference is structural. As discussed previously in *A Preview of Carbonyl Compounds* and shown in Figure 19.14, the tetrahedral intermediate produced by addition of a nucleophile to a carboxylic acid derivative can eliminate a leaving group, leading to a net nucleophilic acyl substitution reaction. The tetrahedral intermediate produced by addition of a nucleophile to an aldehyde or ketone, however, has only alkyl or hydrogen substituents and thus can't usually expel a leaving group. One exception to this rule, however, is the **Cannizzaro reaction**, discovered in 1853.

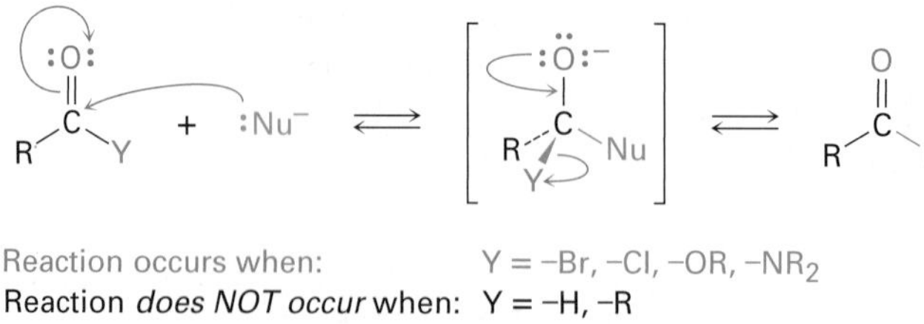

Reaction occurs when: $Y = -Br, -Cl, -OR, -NR_2$
Reaction *does NOT occur* when: $Y = -H, -R$

Figure 19.14 Carboxylic acid derivatives have an electronegative substituent $Y = -Br$, $-Cl$, $-OR$, $-NR_2$ that can be expelled as a leaving group from the tetrahedral intermediate formed by nucleophilic addition. Aldehydes and ketones have no such leaving group and thus do not usually undergo this reaction.

The Cannizzaro reaction takes place by nucleophilic addition of OH^- to an aldehyde to give a tetrahedral intermediate, *which expels hydride ion as a leaving group* and is thereby oxidized. A second aldehyde molecule accepts the hydride ion in another nucleophilic addition step and is thereby reduced. Benzaldehyde, for instance, yields benzyl alcohol plus benzoic acid when heated with aqueous NaOH.

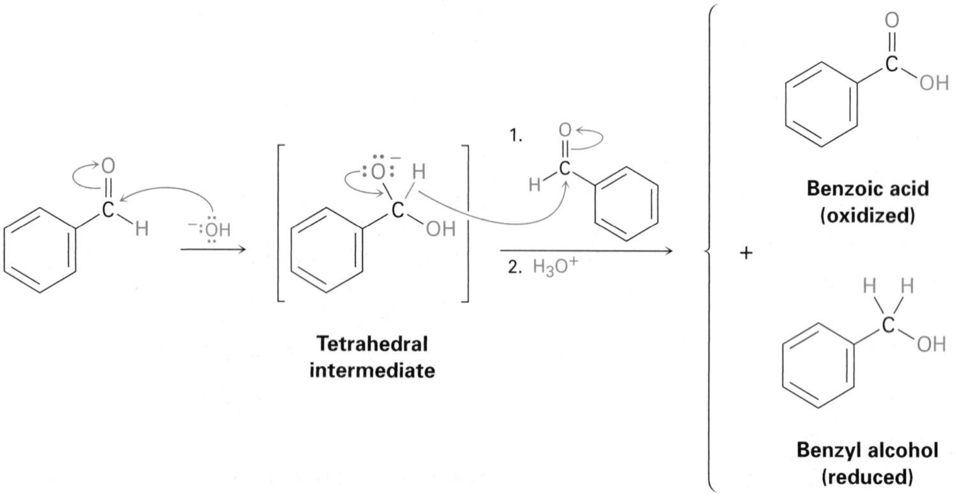

The Cannizzaro reaction has little use but is interesting mechanistically because it is a simple laboratory analogy for the primary biological pathway by which carbonyl reductions occur in living organisms. In nature, as we saw in Section 17.4, one of the most important reducing agents is NADH, reduced nicotinamide adenine dinucleotide. NADH donates H^- to aldehydes and ketones, thereby reducing them, in much the same way that the tetrahedral alkoxide intermediate in a Cannizzaro reaction does. The electron lone pair on a nitrogen atom of NADH expels H^- as leaving group, which adds to a carbonyl group in another molecule (Figure 19.15). As an example, pyruvate is converted during intense muscle activity to (*S*)-lactate, a reaction catalyzed by lactate dehydrogenase.

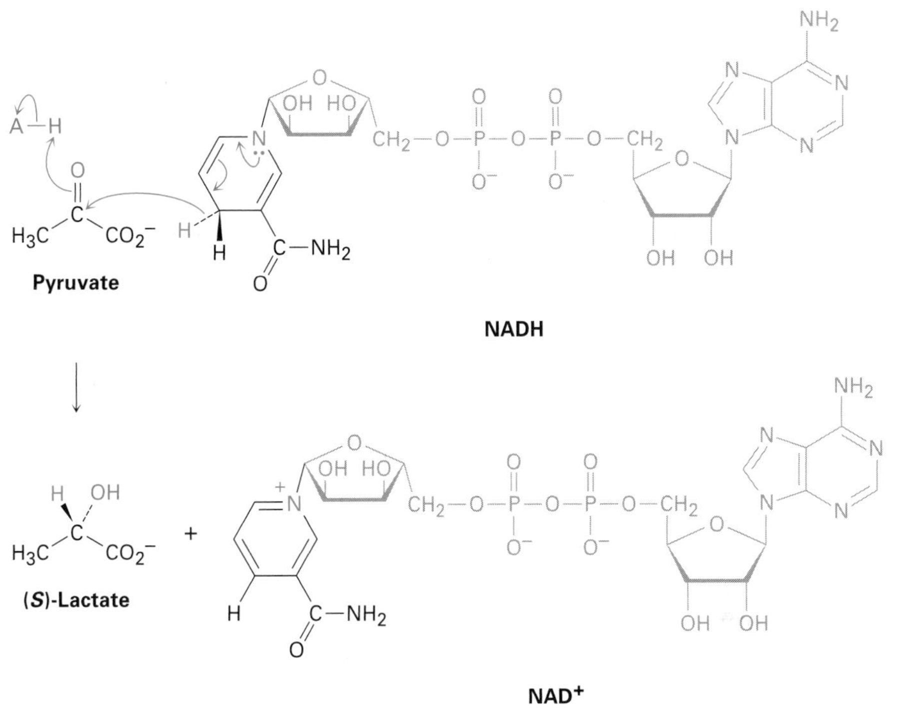

Figure 19.15 Mechanism of biological aldehyde and ketone reductions by the coenzyme NADH.

Problem 19.18   When *o*-phthalaldehyde is treated with base, *o*-(hydroxymethyl)benzoic acid is formed. Show the mechanism of this reaction.

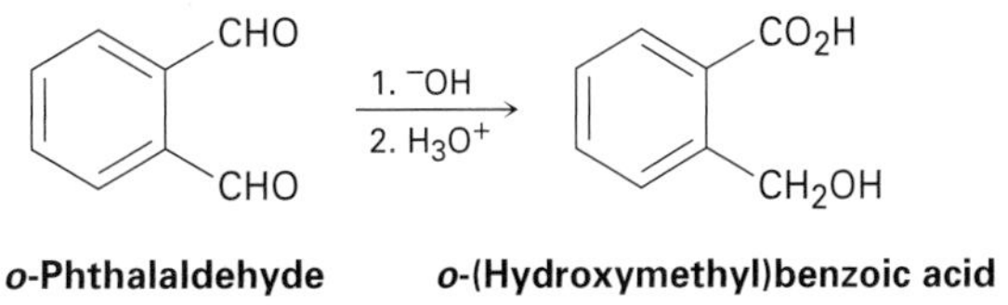

o-Phthalaldehyde   *o*-(Hydroxymethyl)benzoic acid

Problem 19.19   What is the stereochemistry of the pyruvate reduction shown in Figure 19.15? Does NADH lose its *pro-R* or *pro-S* hydrogen? Does addition occur to the *Si* face or *Re* face of pyruvate?

19.13 | Conjugate Nucleophilic Addition to α,β-Unsaturated Aldehydes and Ketones

CENGAGENOW™ Click *Organic Interactive* to **use a web-based palette to predict products from a variety of conjugate addition reactions**.

All the reactions we've been discussing to this point have involved the addition of a nucleophile directly to the carbonyl group, a so-called **1,2-addition**. Closely related to this direct addition is the **conjugate addition**, or **1,4-addition**, of a nucleophile to the C=C bond of an α,β-unsaturated aldehyde or ketone. (The carbon atom next to a carbonyl group is often called the α *carbon,* the next carbon is the β *carbon,* and so on. Thus, an α,β-*unsaturated* aldehyde or ketone has a double bond conjugated with the carbonyl group.) The initial product of conjugate addition is a resonance-stabilized *enolate ion,* which typically undergoes protonation on the α carbon to give a saturated aldehyde or ketone product (Figure 19.16).

Figure 19.16 A comparison of direct (1,2) and conjugate (1,4) nucleophilic addition reactions.

Direct (1,2) addition

Conjugate (1,4) addition

α,β-Unsaturated aldehyde/ketone

Enolate ion

Saturated aldehyde/ketone

The conjugate addition of a nucleophile to an α,β-unsaturated aldehyde or ketone is caused by the same electronic factors that are responsible for direct addition. The electronegative oxygen atom of the α,β-unsaturated carbonyl compound withdraws electrons from the β carbon, thereby making it electron-poor and more electrophilic than a typical alkene carbon.

Electrophilic   Electrophilic

As noted previously, conjugate addition of a nucleophile to the β carbon of an α,β-unsaturated aldehyde or ketone leads to an enolate ion intermediate, which is protonated on the α carbon to give the saturated product (Figure 19.16). The net effect is addition of the nucleophile to the C=C bond, with the carbonyl group itself unchanged. In fact, of course, the carbonyl group is crucial to the success of the reaction. The C=C bond would not be activated for addition, and no reaction would occur, without the carbonyl group.

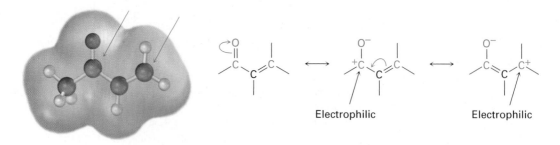

Activated double bond

1. :Nu⁻
2. H₃O⁺

Unactivated double bond → NO reaction

Conjugate Addition of Amines

Both primary and secondary amines add to α,β-unsaturated aldehydes and ketones to yield β-amino aldehydes and ketones rather than the alternative imines. Under typical reaction conditions, both modes of addition occur rapidly. But because the reactions are reversible, they generally proceed with thermodynamic control rather than kinetic control (Section 14.3), so the more stable conjugate addition product is often obtained to the complete exclusion of the less stable direct addition product.

Conjugate Addition of Water

Water can add reversibly to α,β-unsaturated aldehydes and ketones to yield β-hydroxy aldehydes and ketones, although the position of the equilibrium generally favors unsaturated reactant rather than saturated adduct. A related addition to an α,β-unsaturated carboxylic acid occurs in numerous biological pathways, such as the citric acid cycle of food metabolism where *cis*-aconitate is converted into isocitrate by conjugate addition of water to a double bond.

cis-Aconitate   Isocitrate

Problem 19.20 | Assign *R* or *S* stereochemistry to the two chirality centers in isocitrate, and tell whether OH and H add to the *Si* face or the *Re* face of the double bond.

Conjugate Addition of Alkyl Groups: Organocopper Reactions

Conjugate addition of an alkyl group to an α,β-unsaturated ketone (but not aldehyde) is one of the more useful 1,4-addition reactions, just as direct addition of a Grignard reagent is one of the more useful 1,2-additions.

α,β-Unsaturated ketone

Conjugate addition of an alkyl group is carried out by treating the α,β-unsaturated ketone with a lithium diorganocopper reagent. As we saw in Section 10.8, diorganocopper (Gilman) reagents can be prepared by reaction between 1 equivalent of cuprous iodide and 2 equivalents of organolithium.

$$RX \xrightarrow[\text{Pentane}]{2 \text{ Li}} RLi + Li^+ X^-$$

$$2 \text{ RLi} \xrightarrow[\text{Ether}]{\text{CuI}} Li^+(R\bar{C}uR) + Li^+ I^-$$

**A lithium
diorganocopper
(Gilman reagent)**

Primary, secondary, and even tertiary alkyl groups undergo the addition reaction, as do aryl and alkenyl groups. Alkynyl groups, however, react poorly in the conjugate addition process. Diorganocopper reagents are unique in their ability to give conjugate addition products. Other organometallic reagents, such as Grignard reagents and organolithiums, normally give direct carbonyl addition on reaction with α,β-unsaturated ketones.

1-Methyl-2-cyclohexen-1-ol
(95%)

2-Cyclohexenone

3-Methylcyclohexanone
(97%)

The mechanism of the reaction is thought to involve conjugate nucleophilic addition of the diorganocopper anion, R_2Cu^-, to the enone to give a copper-containing intermediate. Transfer of an R group and elimination of a neutral organocopper species, RCu, gives the final product.

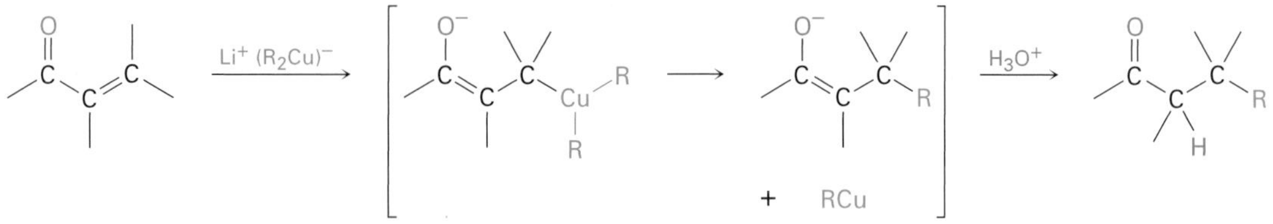

| **WORKED EXAMPLE 19.4** | **Synthesis Using Conjugate Addition Reactions** |

How might you use a conjugate addition reaction to prepare 2-methyl-3-propyl-cyclopentanone?

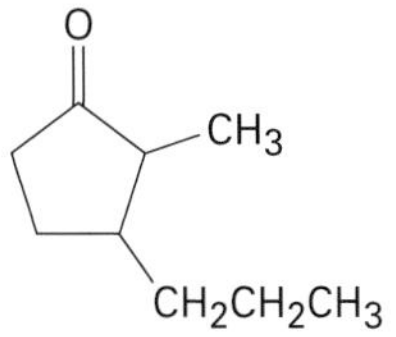

2-Methyl-3-propylcyclopentanone

Strategy   A ketone with a substituent group in its β position might be prepared by a conjugate addition of that group to an α,β-unsaturated ketone. In the present instance, the target molecule has a propyl substituent on the β carbon and might therefore be prepared from 2-methyl-2-cyclopentenone by reaction with lithium dipropylcopper.

Solution

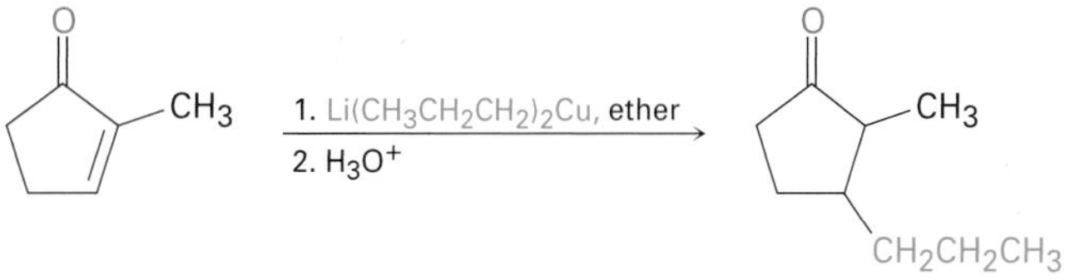

1. Li(CH$_3$CH$_2$CH$_2$)$_2$Cu, ether
2. H$_3$O$^+$

2-Methyl-2-cyclopentenone   **2-Methyl-3-propylcyclopentanone**

Problem 19.21   Treatment of 2-cyclohexenone with HCN/KCN yields a saturated keto nitrile rather than an unsaturated cyanohydrin. Show the structure of the product, and propose a mechanism for the reaction.

Problem 19.22   How might conjugate addition reactions of lithium diorganocopper reagents be used to synthesize the following compounds?

(a)

$$CH_3CH_2CH_2CH_2CH_2CCH_3$$

(with O double bonded to the C)

(b)

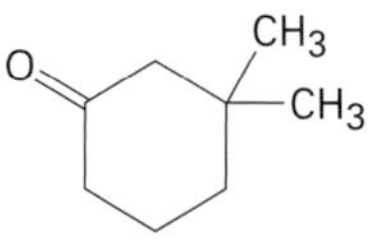

(c)

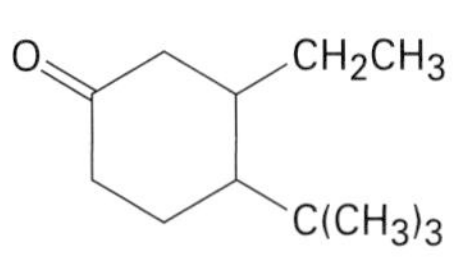

(d)

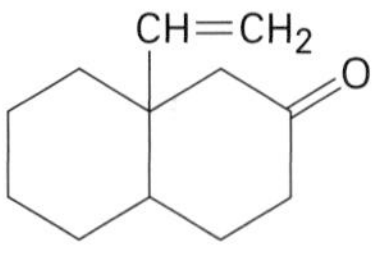

19.14 | Spectroscopy of Aldehydes and Ketones

Infrared Spectroscopy

Aldehydes and ketones show a strong C=O bond absorption in the IR region from 1660 to 1770 cm^{-1}, as the spectra of benzaldehyde and cyclohexanone demonstrate (Figure 19.17). In addition, aldehydes show two characteristic C−H absorptions in the range 2720 to 2820 cm^{-1}.

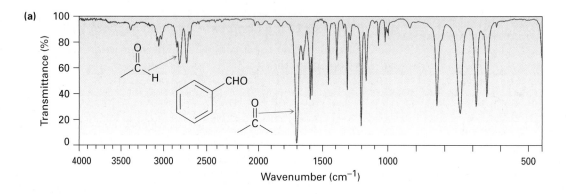

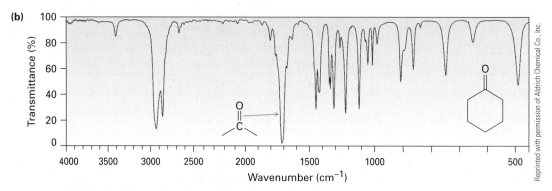

Figure 19.17 Infrared spectra of **(a)** benzaldehyde and **(b)** cyclohexanone.

The exact position of the C=O absorption is diagnostic of the nature of the carbonyl group. As the data in Table 19.2 indicate, saturated aldehydes usually show carbonyl absorptions near 1730 cm^{-1} in the IR spectrum, but conjugation of the aldehyde to an aromatic ring or a double bond lowers the absorption by 25 cm^{-1} to near 1705 cm^{-1}. Saturated aliphatic ketones and cyclohexanones both absorb near 1715 cm^{-1}, and conjugation with a double bond or an aromatic ring again lowers the absorption by 30 cm^{-1} to 1685 to 1690 cm^{-1}. Angle strain in the carbonyl group caused by reducing the ring size of cyclic ketones to four or five raises the absorption position.

The values given in Table 19.2 are remarkably constant from one aldehyde or ketone to another. As a result, IR spectroscopy is a powerful tool for identifying the kind of a carbonyl group in a molecule of unknown structure. An unknown that shows an IR absorption at 1730 cm^{-1} is almost certainly an aldehyde rather than a ketone; an unknown that shows an IR absorption at 1750 cm^{-1} is almost certainly a cyclopentanone, and so on.

Table 19.2 | Infrared Absorptions of Some Aldehydes and Ketones

Carbonyl type	Example	Absorption (cm^{-1})
Saturated aldehyde	CH_3CHO	1730
Aromatic aldehyde	PhCHO	1705
α,β-Unsaturated aldehyde	$H_2C=CHCHO$	1705
Saturated ketone	CH_3COCH_3	1715
Cyclohexanone		1715
Cyclopentanone		1750
Cyclobutanone		1785
Aromatic ketone	$PhCOCH_3$	1690
α,β-Unsaturated ketone	$H_2C=CHCOCH_3$	1705

Problem 19.23 How might you use IR spectroscopy to determine whether reaction between 2-cyclo-hexenone and lithium dimethylcopper gives the direct addition product or the conjugate addition product?

Problem 19.24 Where would you expect each of the following compounds to absorb in the IR spectrum?
(a) 4-Penten-2-one
(b) 3-Penten-2-one
(c) 2,2-Dimethylcyclopentanone
(d) *m*-Chlorobenzaldehyde
(e) 3-Cyclohexenone
(f) 2-Hexenal

Nuclear Magnetic Resonance Spectroscopy

Aldehyde protons (RCHO) absorb near 10 δ in the ^{1}H NMR spectrum and are very distinctive because no other absorptions occur in this region. The aldehyde proton shows spin–spin coupling with protons on the neighboring carbon, with coupling constant $J \approx 3$ Hz. Acetaldehyde, for example, shows a quartet at 9.8 δ for the aldehyde proton, indicating that there are three protons neighboring the −CHO group (Figure 19.18).

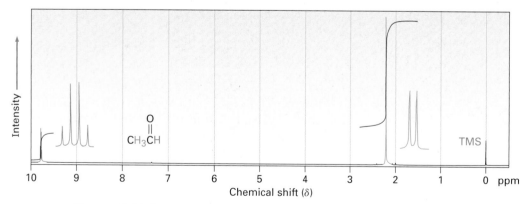

Figure 19.18 ^{1}H NMR spectrum of acetaldehyde. The absorption of the aldehyde proton appears at 9.8 δ and is split into a quartet.

Hydrogens on the carbon next to a carbonyl group are slightly deshielded and normally absorb near 2.0 to 2.3 δ. The acetaldehyde methyl group in Figure 19.18, for instance, absorbs at 2.20 δ. Methyl ketones are particularly distinctive because they always show a sharp three-proton singlet near 2.1 δ.

The carbonyl-group carbon atoms of aldehydes and ketones have characteristic ^{13}C NMR resonances in the range 190 to 215 δ. Since no other kinds of carbons absorb in this range, the presence of an NMR absorption near 200 δ is clear evidence for a carbonyl group. Saturated aldehyde or ketone carbons usually absorb in the region from 200 to 215 δ, while aromatic and α,β-unsaturated carbonyl carbons absorb in the 190 to 200 δ region.

Mass Spectrometry

Aliphatic aldehydes and ketones that have hydrogens on their gamma (γ) carbon atoms undergo a characteristic mass spectral cleavage called the **McLafferty rearrangement**. A hydrogen atom is transferred from the γ carbon to the carbonyl oxygen, the bond between the α and β carbons is broken, and a neutral alkene fragment is produced. The charge remains with the oxygen-containing fragment.

In addition to fragmentation by the McLafferty rearrangement, aldehydes and ketones also undergo cleavage of the bond between the carbonyl group and the α carbon, a so-called α cleavage. Alpha cleavage yields a neutral radical and a resonance-stabilized acyl cation.

Fragment ions from both McLafferty rearrangement and α cleavage are visible in the mass spectrum of 5-methyl-2-hexanone shown in Figure 19.19. McLafferty rearrangement with loss of 2-methylpropene yields a fragment with $m/z = 58$. Alpha cleavage occurs primarily at the more substituted side of the carbonyl group, leading to a $[CH_3CO]^+$ fragment with $m/z = 43$.

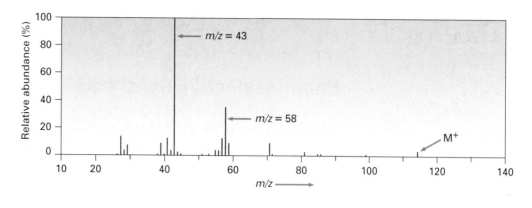

Figure 19.19 Mass spectrum of 5-methyl-2-hexanone. The peak at $m/z = 58$ is due to McLafferty rearrangement. The abundant peak at $m/z = 43$ is due to α cleavage at the more highly substituted side of the carbonyl group. Note that the peak due to the molecular ion is very small.

Problem 19.25 How might you use mass spectrometry to distinguish between the following pairs of isomers?
(a) 3-Methyl-2-hexanone and 4-methyl-2-hexanone
(b) 3-Heptanone and 4-heptanone
(c) 2-Methylpentanal and 3-methylpentanal

Problem 19.26 Tell the prominent IR absorptions and mass spectral peaks you would expect for the following compound:

Focus On . . .

Enantioselective Synthesis

Whenever a chiral product is formed by reaction between achiral reagents, the product is racemic; that is, both enantiomers of the product are formed in equal amounts. The epoxidation reaction of geraniol with *m*-chloroperoxybenzoic acid, for instance, gives a racemic mixture of (2*R*,3*S*) and (2*S*,3*R*) epoxides.

Geraniol

50%

+

50%

Unfortunately, it's usually the case that only a *single* enantiomer of a given drug or other important substance has the desired biological properties. The other enantiomer might be inactive or even dangerous. Thus, much work is currently being done on developing *enantioselective* methods of synthesis, which yield only one of two possible enantiomers. So important has enantioselective synthesis become that the 2001 Nobel Prize in chemistry was awarded to three pioneers in the field: William S. Knowles, K. Barry Sharpless, and Ryoji Noyori.

Several approaches to enantioselective synthesis have been taken, but the most efficient are those that use chiral catalysts to temporarily hold a substrate molecule in an unsymmetrical environment—exactly the same strategy that nature uses when catalyzing reactions with chiral enzymes. While in that unsymmetrical environment, the substrate may be more open to reaction on one side than on another, leading to an excess of one enantiomeric product over another. As an analogy, think about picking up a coffee mug in your

(continued)

A substance made from the tartaric acid found at the bottom of this wine vat catalyzes enantioselective reactions.

right hand to take a drink. The mug by itself is achiral, but as soon as you pick it up by the handle, it becomes chiral. One side of the mug now faces toward you so you can drink from it, but the other side faces away. The two sides are different, with one side much more accessible to you than the other.

Among the many enantioselective reactions now known, one of the most general is the so-called Sharpless epoxidation, in which an allylic alcohol, such as geraniol, is treated with *tert*-butyl hydroperoxide, $(CH_3)_3C\!-\!OOH$, in the presence of titanium tetraisopropoxide and diethyl tartrate (DET) as a chiral auxiliary reagent. When the (*R,R*) tartrate is used, geraniol is converted into its 2*R*,3*S* epoxide with 98% selectivity, whereas use of the (*S,S*) tartrate gives the 2*S*,3*R* epoxide enantiomer. We say that the major product in each case is formed with an *enantiomeric excess* of 96%, meaning that 4% of the product is racemic (2% 2*R*,3*S* plus 2% 2*S*,3*R*) and an extra 96% of a single enantiomer is formed. The mechanistic details by which the chiral catalyst works are a bit complex, although it appears that a chiral complex of two tartrate molecules with one titanium is involved.

SUMMARY AND KEY WORDS

Aldehydes and ketones are among the most important of all compounds, both in biochemistry and in the chemical industry. Aldehydes are normally prepared in the laboratory by oxidation of primary alcohols or by partial reduction of esters. Ketones are similarly prepared by oxidation of secondary alcohols or by addition of diorganocopper reagents to acid chlorides.

The **nucleophilic addition reaction** is the most common reaction of aldehydes and ketones. Many different kinds of products can be prepared by nucleophilic additions. Aldehydes and ketones are reduced by $NaBH_4$ or $LiAlH_4$ to yield secondary and primary alcohols, respectively. Addition of Grignard reagents to aldehydes and ketones also gives alcohols (tertiary and secondary, respectively), and addition of HCN yields **cyanohydrins**. Primary amines add to carbonyl compounds yielding **imines**, and secondary amines yield **enamines**. Reaction of an aldehyde or ketone with hydrazine and base gives an alkane (the **Wolff–Kishner reaction**). Alcohols add to carbonyl groups to yield **acetals**, which are valuable as protecting groups. Phosphoranes add to aldehydes and ketones to give alkenes (the **Wittig reaction**) in which the new C=C bond in the product is exactly where the C=O bond was in the starting material.

α,β-Unsaturated aldehydes and ketones often react with nucleophiles to give the product of **conjugate addition**, or **1,4-addition**. Particularly useful is the reaction with a diorganocopper reagent, which results in the addition of an alkyl, aryl, or alkenyl group to the double bond.

IR spectroscopy is helpful for identifying aldehydes and ketones. Carbonyl groups absorb in the IR range 1660 to 1770 cm^{-1}, with the exact position highly diagnostic of the kind of carbonyl group present in the molecule. ^{13}C NMR spectroscopy is also useful for aldehydes and ketones because their carbonyl carbons show resonances in the 190 to 215 δ range. ^{1}H NMR is useful for aldehyde —CHO protons, which absorb near 10 δ. Aldehydes and ketones undergo two characteristic kinds of fragmentation in the mass spectrometer: α cleavage and **McLafferty rearrangement**.

SUMMARY OF REACTIONS

1. Preparation of aldehydes (Section 19.2)
   (a) Oxidation of primary alcohols (Section 17.7)

   (b) Partial reduction of esters (Section 19.2)

2. Preparation of ketones (Section 19.2)

Diorganocopper reaction with acid chlorides

3. Reactions of aldehydes (Section 19.3)

Oxidation to give carboxylic acids

4. Nucleophilic addition reactions of aldehydes and ketones

(a) Addition of hydride: alcohols (Section 19.7)

(b) Addition of Grignard reagents: alcohols (Section 19.7)

(c) Addition of HCN: cyanohydrins (Section 19.6)

(d) Addition of primary amines: imines (Section 19.8)

(e) Addition of secondary amines: enamines (Section 19.8)

(f) Wolff–Kishner reaction (Section 19.9)

(g) Addition of alcohols: acetals (Section 19.10)

(h) Addition of phosphorus ylides: Wittig reaction (Section 19.11)

5. Conjugate additions to α,β-unsaturated aldehydes and ketones (Section 19.13)
   (a) Conjugate addition of amines

(b) Conjugate addition of water

(c) Conjugate addition of alkyl groups: diorganocopper reaction

EXERCISES

Organic KNOWLEDGE TOOLS

CENGAGENOW™ Sign in at **www.cengage.com/login** to assess your knowledge of this chapter's topics by taking a pre-test. The pre-test will link you to interactive organic chemistry resources based on your score in each concept area.

⊗WL Online homework for this chapter may be assigned in Organic OWL.

■ indicates problems assignable in Organic OWL.

VISUALIZING CHEMISTRY

(Problems 19.1–19.26 appear within the chapter.)

19.27 ■ Each of the following substances can be prepared by a nucleophilic addition reaction between an aldehyde or ketone and a nucleophile. Identify the reactants from which each was prepared. If the substance is an acetal, identify the carbonyl compound and the alcohol; if it is an imine, identify the carbonyl compound and the amine; and so forth.

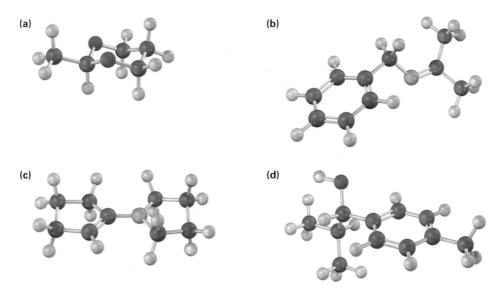

(a)　　　　　　　　　　　　　　　(b)

(c)　　　　　　　　　　　　　　　(d)

19.28 ■ The following molecular model represents a tetrahedral intermediate resulting from addition of a nucleophile to an aldehyde or ketone. Identify the reactants, and write the structure of the final product when the nucleophilic addition reaction is complete.

19.29 The enamine prepared from acetone and dimethylamine is shown here in its lowest-energy form.
  (a) What is the geometry and hybridization of the nitrogen atom?
  (b) What orbital on nitrogen holds the lone pair of electrons?
  (c) What is the geometric relationship between the *p* orbitals of the double bond and the nitrogen orbital that holds the lone pair? Why do you think this geometry represents the minimum energy?

ADDITIONAL PROBLEMS

19.30 ■ Draw structures corresponding to the following names:
  (a) Bromoacetone
  (b) (*S*)-2-Hydroxypropanal
  (c) 2-Methyl-3-heptanone
  (d) (2*S*,3*R*)-2,3,4-Trihydroxybutanal
  (e) 2,2,4,4-Tetramethyl-3-pentanone
  (f) 4-Methyl-3-penten-2-one
  (g) Butanedial
  (h) 3-Phenyl-2-propenal
  (i) 6,6-Dimethyl-2,4-cyclohexadienone
  (j) *p*-Nitroacetophenone

19.31 ■ Draw and name the seven aldehydes and ketones with the formula $C_5H_{10}O$. Which are chiral?

19.32 Give IUPAC names for the following structures:

(a)

(b)   CHO
      H—C—OH
      CH_2OH

(c)

(d)   O
       ‖
   $CH_3CHCCH_2CH_3$
       |
       CH_3

(e)   OH   O
       | ‖
   CH_3CHCH_2CH

(f)   CHO

   OHC

19.33 Give structures that fit the following descriptions:
  (a) An α,β-unsaturated ketone, C_6H_8O **(b)** An α-diketone
  (c) An aromatic ketone, $C_9H_{10}O$ **(d)** A diene aldehyde, C_7H_8O

19.34 ■ Predict the products of the reaction of (i) phenylacetaldehyde and (ii) aceto-phenone with the following reagents:

(a) NaBH$_4$, then H$_3$O$^+$ (b) Tollens' reagent
(c) NH$_2$OH, HCl catalyst   (d) CH$_3$MgBr, then H$_3$O$^+$
(e) 2 CH$_3$OH, HCl catalyst   (f) H$_2$NNH$_2$, KOH
(g) (C$_6$H$_5$)$_3$P=CH$_2$ (h) HCN, KCN

19.35 ■ How would you prepare the following substances from 2-cyclohexenone? More than one step may be required.

19.36 ■ Show how the Wittig reaction might be used to prepare the following alkenes. Identify the alkyl halide and the carbonyl components that would be used.

(a)   (b)

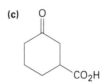

19.37 ■ How would you use a Grignard reaction on an aldehyde or ketone to synthesize the following compounds?

(a) 2-Pentanol   (b) 1-Butanol
(c) 1-Phenylcyclohexanol   (d) Diphenylmethanol

19.38 Aldehydes can be prepared by the Wittig reaction using (methoxymethylene)-triphenylphosphorane as the Wittig reagent and then hydrolyzing the product with acid. For example,

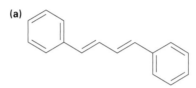

(a) How would you prepare the necessary phosphorane?
(b) Propose a mechanism for the hydrolysis step.

19.39 When 4-hydroxybutanal is treated with methanol in the presence of an acid catalyst, 2-methoxytetrahydrofuran is formed. Explain.

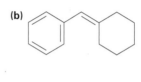

19.40 ■ How might you carry out the following selective transformations? One of the two schemes requires a protection step. (Recall from Section 19.5 that aldehydes are more reactive than ketones toward nucleophilic addition.)

(a)

$$CH_3CCH_2CH_2CH_2CH \longrightarrow CH_3CCH_2CH_2CH_2CH_2OH$$

(b)

$$CH_3CCH_2CH_2CH_2CH \longrightarrow CH_3CHCH_2CH_2CH_2CH$$

19.41 ■ How would you synthesize the following substances from benzaldehyde and any other reagents needed?

(a) C_6H_5—CH_2CHO   (b)   (c)

19.42 ■ Carvone is the major constituent of spearmint oil. What products would you expect from reaction of carvone with the following reagents?

Carvone

(a) $(CH_3)_2Cu^- Li^+$, then H_3O^+ (b) $LiAlH_4$, then H_3O^+
(c) CH_3NH_2 (d) C_6H_5MgBr, then H_3O^+
(e) H_2/Pd (f) CrO_3, H_3O^+
(g) $(C_6H_5)_3\overset{+}{P}\overset{-}{C}HCH_3$ (h) $HOCH_2CH_2OH$, HCl

19.43 The S_N2 reaction of (dibromomethyl)benzene, $C_6H_5CHBr_2$, with NaOH yields benzaldehyde rather than (dihydroxymethyl)benzene, $C_6H_5CH(OH)_2$. Explain.

19.44 Reaction of 2-butanone with HCN yields a chiral product. What stereochemistry does the product have? Is it optically active?

19.45 ■ How would you synthesize the following compounds from cyclohexanone?
(a) 1-Methylcyclohexene   (b) 2-Phenylcyclohexanone
(c) cis-1,2-Cyclohexanediol   (d) 1-Cyclohexylcyclohexanol

19.46 One of the steps in the metabolism of fats is the reaction of an unsaturated acyl CoA with water to give a β-hydroxyacyl CoA. Propose a mechanism.

$$RCH_2CH_2CH=CHCSCoA \xrightarrow{\text{H}_2\text{O}} RCH_2CH_2CH-CH_2CSCoA$$

Unsaturated acyl CoA   **β-Hydroxyacyl CoA**

19.47 The amino acid methionine is biosynthesized by a multistep route that includes reaction of an imine of pyridoxal phosphate (PLP) to give an unsaturated imine, which then reacts with cysteine. What kinds of reactions are occurring in the two steps?

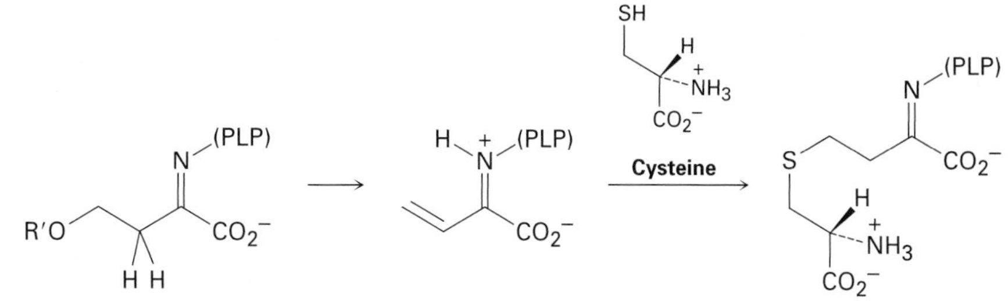

O-Succinylhomoserine–
PLP imine   **Unsaturated**
                                  imine

19.48 Each of the following reaction schemes contains one or more flaws. What is wrong in each case? How would you correct each scheme?

(a)

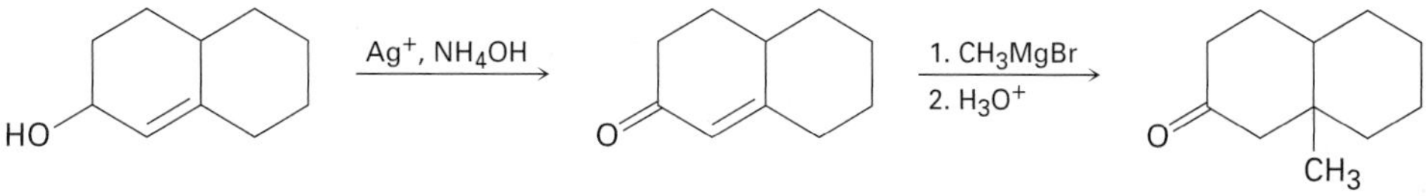

(b) $C_6H_5CH=CHCH_2OH$ $\xrightarrow[\text{H}_3\text{O}^+]{\text{CrO}_3}$ $C_6H_5CH=CHCHO$ $\xrightarrow{\text{H}^+,\ \text{CH}_3\text{OH}}$ $C_6H_5CH=CHCH(OCH_3)_2$

(c)

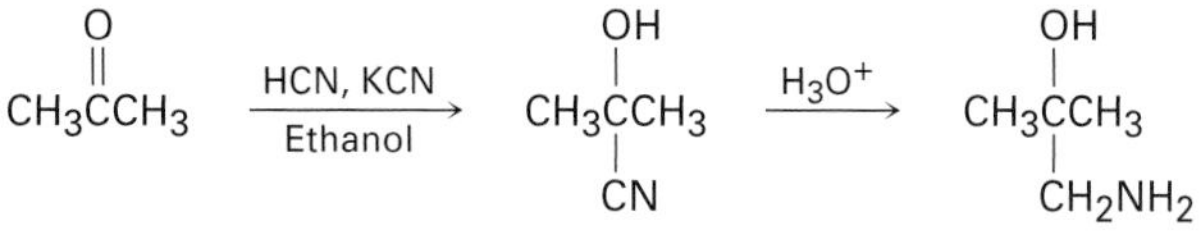

19.49 6-Methyl-5-hepten-2-one is a constituent of lemongrass oil. How could you synthesize this substance from methyl 4-oxopentanoate?

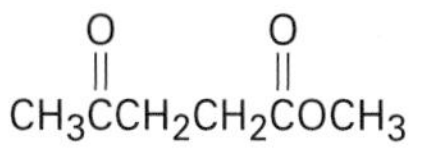

$$\overset{\text{O}}{\overset{\|}{CH_3C}}CH_2CH_2\overset{\text{O}}{\overset{\|}{C}}OCH_3 \quad \textbf{Methyl 4-oxopentanoate}$$

19.50 Aldehydes and ketones react with thiols to yield *thioacetals* just as they react with alcohols to yield acetals. Predict the product of the following reaction, and propose a mechanism:

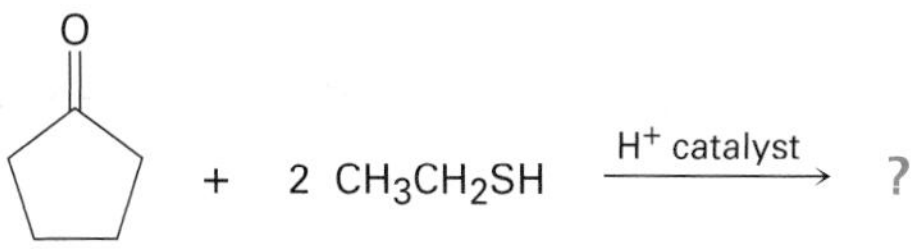

■ Assignable in OWL

19.51 ■ Ketones react with dimethylsulfonium methylide to yield epoxides. Suggest a mechanism for the reaction.

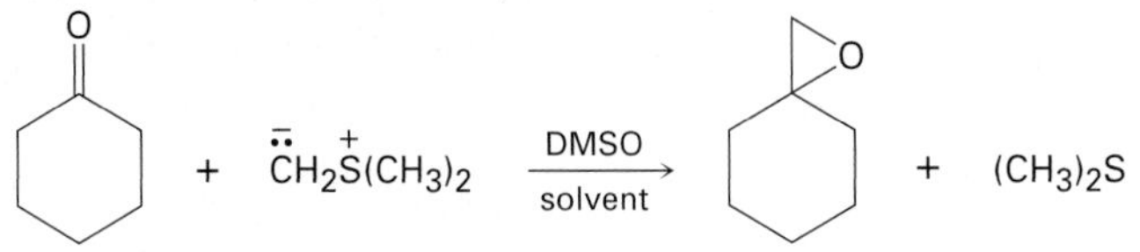

**Dimethylsulfonium
methylide**

19.52 ■ When cyclohexanone is heated in the presence of a large amount of acetone cyanohydrin and a small amount of base, cyclohexanone cyanohydrin and acetone are formed. Propose a mechanism.

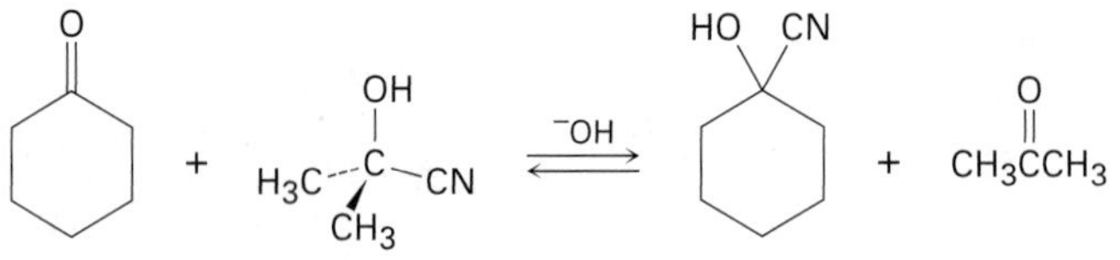

19.53 Tamoxifen is a drug used in the treatment of breast cancer. How would you prepare tamoxifen from benzene, the following ketone, and any other reagents needed?

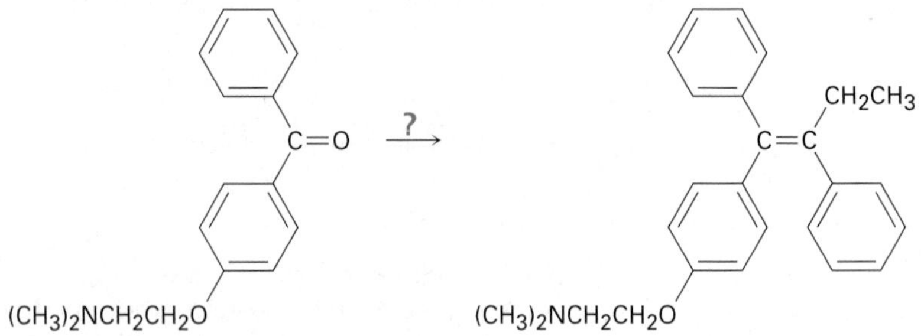

Tamoxifen

19.54 Paraldehyde, a sedative and hypnotic agent, is prepared by treatment of acetaldehyde with an acidic catalyst. Propose a mechanism for the reaction.

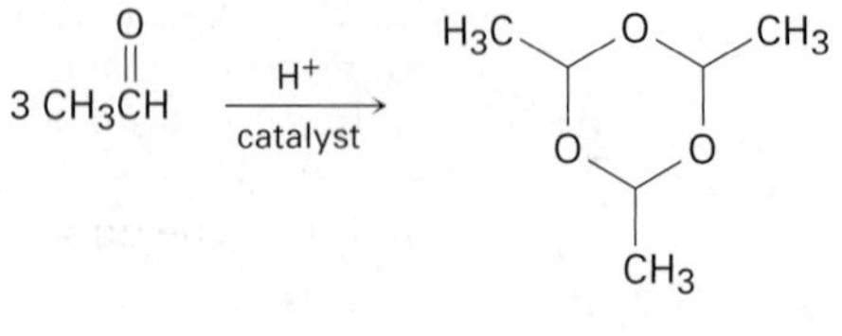

Paraldehyde

19.55 The Meerwein–Ponndorf–Verley reaction involves reduction of a ketone by treatment with an excess of aluminum triisopropoxide. The mechanism of the process is closely related to the Cannizzaro reaction in that a hydride ion acts as a leaving group. Propose a mechanism.

19.56 Propose a mechanism to account for the formation of 3,5-dimethylpyrazole from hydrazine and 2,4-pentanedione. Look carefully to see what has happened to each carbonyl carbon in going from starting material to product.

2,4-Pentanedione   **3,5-Dimethylpyrazole**

19.57 In light of your answer to Problem 19.56, propose a mechanism for the formation of 3,5-dimethylisoxazole from hydroxylamine and 2,4-pentanedione.

3,5-Dimethylisoxazole

19.58 ■ Trans alkenes are converted into their cis isomers and vice versa on epoxidation followed by treatment of the epoxide with triphenylphosphine. Propose a mechanism for the epoxide → alkene reaction.

19.59 ■ Treatment of an α,β-unsaturated ketone with basic aqueous hydrogen peroxide yields an epoxy ketone. The reaction is specific to unsaturated ketones; isolated alkene double bonds do not react. Propose a mechanism.

19.60 ■ One of the biological pathways by which an amine is converted to a ketone involves two steps: (1) oxidation of the amine by NAD^+ to give an imine, and (2) hydrolysis of the imine to give a ketone plus ammonia. Glutamate, for instance, is converted by this process into α-ketoglutarate. Show the structure of the imine intermediate, and propose mechanisms for both steps.

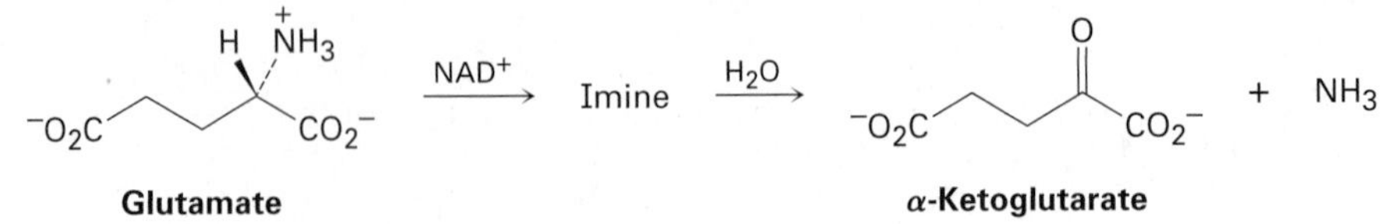

Glutamate   α-Ketoglutarate

19.61 At what position would you expect to observe IR absorptions for the following molecules?

(a)

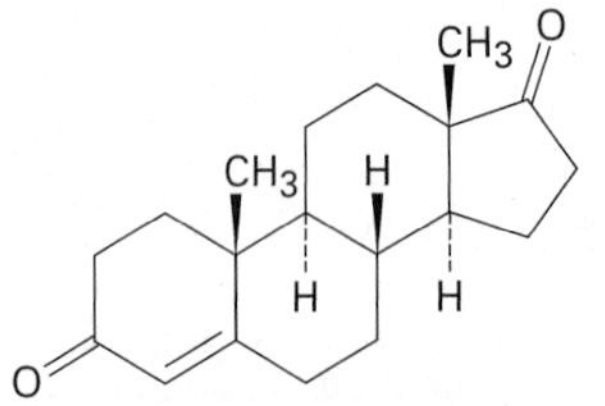

4-Androstene-3,17-dione

(b)

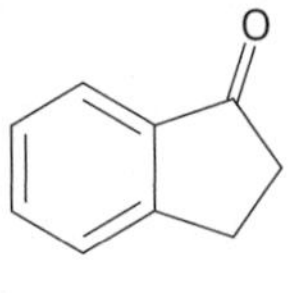

1-Indanone

(c)

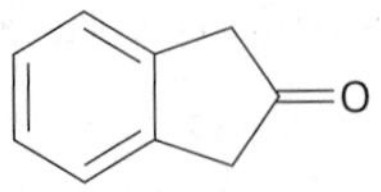

(d)

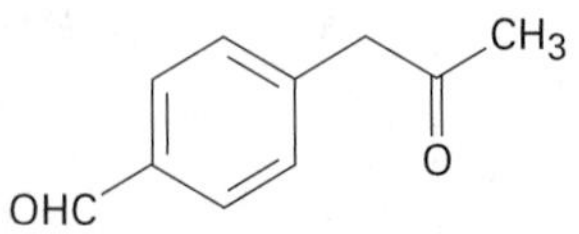

19.62 Acid-catalyzed dehydration of 3-hydroxy-3-phenylcyclohexanone leads to an unsaturated ketone. What possible structures are there for the product? At what position in the IR spectrum would you expect each to absorb? If the actual product has an absorption at 1670 cm^{-1}, what is its structure?

19.63 ■ Compound **A**, MW = 86, shows an IR absorption at 1730 cm^{-1} and a very simple ^{1}H NMR spectrum with peaks at 9.7 δ (1 H, singlet) and 1.2 δ (9 H, singlet). Propose a structure for **A**.

19.64 ■ Compound **B** is isomeric with **A** (Problem 19.63) and shows an IR peak at 1715 cm^{-1}. The ^{1}H NMR spectrum of **B** has peaks at 2.4 δ (1 H, septet, $J = 7$ Hz), 2.1 δ (3 H, singlet), and 1.2 δ (6 H, doublet, $J = 7$ Hz). What is the structure of **B**?

19.65 ■ The ^{1}H NMR spectrum shown is that of a compound with formula $C_9H_{10}O$. How many double bonds and/or rings does this compound contain? If the unknown has an IR absorption at 1690 cm^{-1}, what is a likely structure?

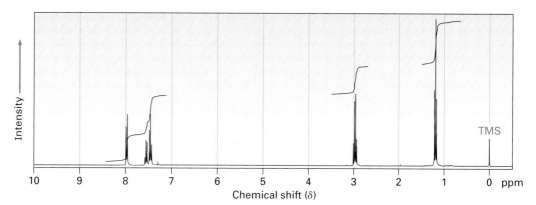

19.66 The ^{1}H NMR spectrum shown is that of a compound isomeric with the one in Problem 19.65. This isomer has an IR absorption at 1730 cm^{-1}. Propose a structure. [Note: Aldehyde protons (CHO) often show low coupling constants to adjacent hydrogens, so the splitting of aldehyde signals is not always apparent.]

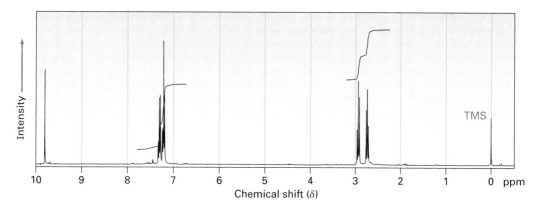

19.67 ■ Propose structures for molecules that meet the following descriptions. Assume that the kinds of carbons (1°, 2°, 3°, or 4°) have been assigned by DEPT-NMR.

(a) $C_6H_{12}O$; IR: 1715 cm^{-1}; ^{13}C NMR: 8.0 δ (1°), 18.5 δ (1°), 33.5 δ (2°), 40.6 δ (3°), 214.0 δ (4°)

(b) $C_5H_{10}O$; IR: 1730 cm^{-1}; ^{13}C NMR: 22.6 δ (1°), 23.6 δ (3°), 52.8 δ (2°), 202.4 δ (3°)

(c) C_6H_8O; IR: 1680 cm^{-1}; ^{13}C NMR: 22.9 δ (2°), 25.8 δ (2°), 38.2 δ (2°), 129.8 δ (3°), 150.6 δ (3°), 198.7 δ (4°)

19.68 Compound **A**, $C_8H_{10}O_2$, has an intense IR absorption at 1750 cm^{-1} and gives the ^{13}C NMR spectrum shown. Propose a structure for **A**.

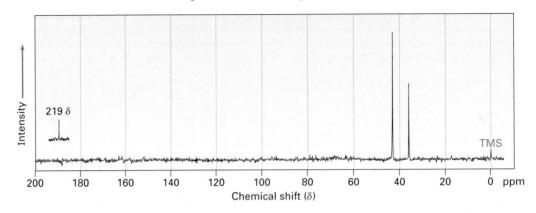

19.69 Propose structures for ketones or aldehydes that have the following ^1H NMR spectra:

(a) C_4H_7ClO
   IR: 1715 cm^{-1}

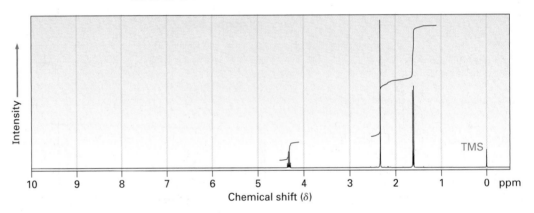

(b) $C_7H_{14}O$
   IR: 1710 cm^{-1}

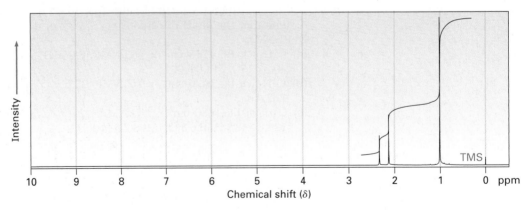

(c) $C_9H_{10}O_2$
IR: 1695 cm^{-1}

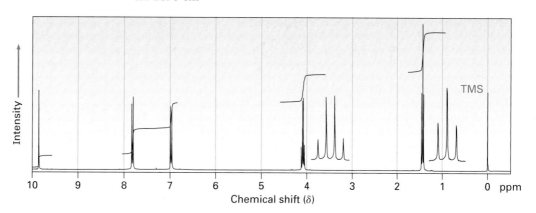

19.70 Propose structures for ketones or aldehydes that have the following ^{1}H NMR spectra.

(a) $C_{10}H_{12}O$
IR: 1710 cm^{-1}

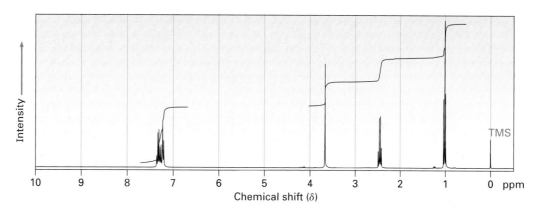

(b) $C_6H_{12}O_3$
IR: 1715 cm^{-1}

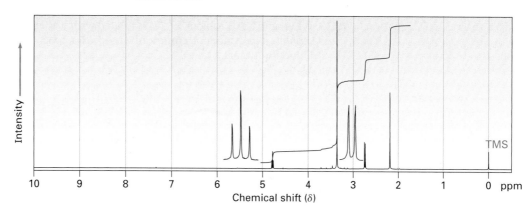

(c) C_4H_6O
IR: 1690 cm^{-1}

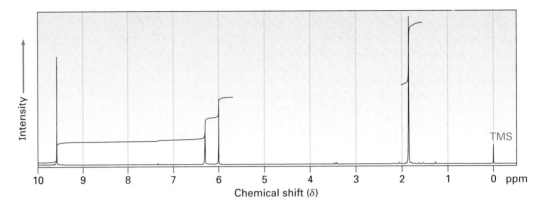

19.71 Primary amines react with esters to yield amides: $RCO_2R' + R''NH_2 \rightarrow$
$RCONHR'' + R'OH$. Propose a mechanism for the following reaction of an
α,β-unsaturated ester.

19.72 When crystals of pure α-glucose are dissolved in water, isomerization slowly
occurs to produce β-glucose. Propose a mechanism for the isomerization.

α-Glucose   **β-Glucose**

19.73 When glucose (Problem 19.72) is treated with $NaBH_4$, reaction occurs to yield
sorbitol, a polyalcohol commonly used as a food additive. Show how this
reduction occurs.

Glucose   **Sorbitol**

■ Assignable in OWL

20

Carboxylic Acids and Nitriles

Carboxylic acids, RCO$_2$H, occupy a central place among carbonyl compounds. Not only are they valuable in themselves, they also serve as starting materials for preparing numerous *acyl derivatives* such as acid chlorides, esters, amides, and thioesters. In addition, carboxylic acids are present in the majority of biological pathways. We'll look both at acids and at their close relatives, *nitriles (RC≡N),* in this chapter and at acyl derivatives in the next chapter.

A carboxylic acid

An acid chloride   **An ester**   **An amide**   **A thioester**

A great many carboxylic acids are found in nature: acetic acid, CH$_3$CO$_2$H, is the chief organic component of vinegar; butanoic acid, CH$_3$CH$_2$CH$_2$CO$_2$H, is responsible for the rancid odor of sour butter; and hexanoic acid (caproic acid), CH$_3$(CH$_2$)$_4$CO$_2$H, is responsible for the unmistakable aroma of goats and dirty gym socks (the name comes from the Latin *caper,* meaning "goat"). Other examples are cholic acid, a major component of human bile, and long-chain aliphatic acids such as palmitic acid, CH$_3$(CH$_2$)$_{14}$CO$_2$H, a biological precursor of fats and vegetable oils.

Sean Duggan

Cholic acid

Approximately 2.5 million tons of acetic acid is produced each year in the United States for a variety of purposes, including preparation of the vinyl acetate polymer used in paints and adhesives. About 20% of the acetic acid synthesized industrially is obtained by oxidation of acetaldehyde. Much of the remaining 80% is prepared by the rhodium-catalyzed reaction of methanol with carbon monoxide.

WHY THIS CHAPTER?

Carboxylic acids are present in many industrial processes and most biological pathways and are the starting materials from which other acyl derivatives are made. Thus, an understanding of their properties and reactions is fundamental to understanding organic chemistry. In this chapter, we'll look both at acids and at their close relatives, *nitriles (RC≡N)*. In the next chapter, we'll look at acyl derivatives.

20.1 | Naming Carboxylic Acids and Nitriles

Carboxylic Acids, RCO₂H

CENGAGENOW™ Click *Organic Interactive* to **use a web-based palette to draw structures of carboxylic acids based on their IUPAC names**.

Simple carboxylic acids derived from open-chain alkanes are systematically named by replacing the terminal *-e* of the corresponding alkane name with *-oic acid*. The −CO₂H carbon atom is numbered C1.

Propanoic acid   4-Methyl**pentan**oic acid   3-Ethyl-6-methyl**octane**dioic acid

Compounds that have a −CO₂H group bonded to a ring are named using the suffix *-carboxylic acid*. The CO₂H carbon is attached to C1 in this system and is not itself numbered. As a substituent, the CO₂H group is called a **carboxyl group**.

trans-4-**Hydroxycyclohexane**carboxylic acid   1-**Cyclopent**enecarboxylic acid

Because many carboxylic acids were among the first organic compounds to be isolated and purified, a large number of common names exist (Table 20.1). Biological chemists, in particular, make frequent use of these names. We'll use systematic names in this book, with a few exceptions such as formic (methanoic) acid and acetic (ethanoic) acid, whose names are accepted by IUPAC and are so well known that it makes little sense to refer to them any other way. Also listed in Table 20.1 are the common names used for acyl groups derived from the parent acids. Except for the small handful at the top of Table 20.1, acyl groups are named by changing the *-ic acid* or *-oic acid* ending to *-oyl*.

Table 20.1 | **Common Names of Some Carboxylic Acids and Acyl Groups**

Structure	Name	Acyl group
HCO_2H	Formic	Formyl
CH_3CO_2H	Acetic	Acetyl
$CH_3CH_2CO_2H$	Propionic	Propionyl
$CH_3CH_2CH_2CO_2H$	Butyric	Butyryl
HO_2CCO_2H	Oxalic	Oxalyl
$HO_2CCH_2CO_2H$	Malonic	Malonyl
$HO_2CCH_2CH_2CO_2H$	Succinic	Succinyl
$HO_2CCH_2CH_2CO_2H$	Glutaric	Glutaryl
$HO_2CCH_2CH_2CH_2CH_2CO_2H$	Adipic	Adipoyl
$H_2C\!=\!CHCO_2H$	Acrylic	Acryloyl
$HO_2CCH\!=\!CHCO_2H$	Maleic (cis)	Maleoyl
	Fumaric (trans)	Fumaroyl
$HOCH_2CO_2H$	Glycolic	Glycoloyl
$\overset{\displaystyle OH}{\underset{\displaystyle \vert}{CH_3CHCO_2H}}$	Lactic	Lactoyl
$\overset{\displaystyle O}{\underset{\displaystyle \parallel}{CH_3CCO_2H}}$	Pyruvic	Pyruvoyl
$\overset{\displaystyle OH}{\underset{\displaystyle \vert}{HOCH_2CHCO_2H}}$	Glyceric	Gylceroyl
$\overset{\displaystyle OH}{\underset{\displaystyle \vert}{HO_2CCHCH_2CO_2H}}$	Malic	Maloyl
$\overset{\displaystyle O}{\underset{\displaystyle \parallel}{HO_2CCCH_2CO_2H}}$	Oxaloacetic	Oxaloacetyl
benzene ring with CO_2H	Benzoic	Benzoyl
benzene ring with two CO_2H (ortho)	Phthalic	Phthaloyl

Nitriles, RC≡N

Compounds containing the −C≡N functional group are called **nitriles** and undergo some chemistry similar to that of carboxylic acids. Simple open-chain nitriles are named by adding -*nitrile* as a suffix to the alkane name, with the nitrile carbon numbered C1.

$$\underset{5}{CH_3}\underset{4}{CH}\underset{3}{CH_2}\underset{2}{CH_2}\underset{1}{CN} \qquad \text{4-Methyl\textbf{pentane}nitrile}$$

with CH₃ substituent on C4.

Nitriles can also be named as derivatives of carboxylic acids by replacing the -*ic acid* or -*oic acid* ending with -*onitrile*, or by replacing the -*carboxylic acid* ending with -*carbonitrile*. The nitrile carbon atom is attached to C1 but is not itself numbered.

CH₃C≡N

Acetonitrile
(from acetic acid)

Benzene ring with C≡N

Benzonitrile
(from benzoic acid)

Cyclohexane ring (numbered 1–6) with CN on C1 and two CH₃ on C2

2,2-Dimethylcyclohexanecarbonitrile
(from 2,2-dimethylcyclohexane-carboxylic acid)

Problem 20.1 | Give IUPAC names for the following compounds:

(a)
$$\underset{}{CH_3CHCH_2COH}$$
with CH₃ and O (=O) substituents

(b)
$$CH_3CHCH_2CH_2COH$$
with Br and O (=O) substituents

(c)
$$CH_3CH_2CHCH_2CH_2CH_3$$
with CO₂H substituent

(d)
C=C double bond with H, H, H₃C, and CH₂CH₂COH (=O) groups

(e)
$$CH_3CHCH_2CHCH_3$$
with CH₃ and CN substituents

(f)
Cyclopentane ring with H, H, HO₂C and CO₂H groups

Problem 20.2 | Draw structures corresponding to the following IUPAC names:
(a) 2,3-Dimethylhexanoic acid   (b) 4-Methylpentanoic acid
(c) *trans*-1,2-Cyclobutanedicarboxylic acid   (d) *o*-Hydroxybenzoic acid
(e) (9Z,12Z)-9,12-Octadecadienoic acid   (f) 2-Pentenenitrile

20.2 | Structure and Properties of Carboxylic Acids

Carboxylic acids are similar in some respects to both ketones and alcohols. Like ketones, the carboxyl carbon is sp^2-hybridized, and carboxylic acid groups are therefore planar with C−C=O and O=C−O bond angles of approximately 120° (Table 20.2).

Like alcohols, carboxylic acids are strongly associated because of hydrogen bonding. Most carboxylic acids exist as cyclic dimers held together by two hydrogen bonds. This strong hydrogen bonding has a noticeable effect on boiling points, making carboxylic acids much higher boiling than the corresponding

Table 20.2 | **Physical Parameters for Acetic Acid**

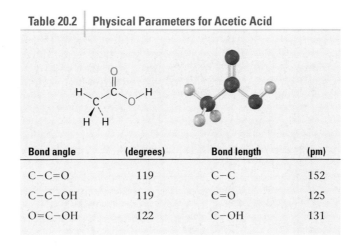

Bond angle	(degrees)	Bond length	(pm)
C–C=O	119	C–C	152
C–C–OH	119	C=O	125
O=C–OH	122	C–OH	131

alcohols. Acetic acid, for instance, has a boiling point of 117.9 °C, versus 78.3 °C for ethanol, even though both compounds have two carbons.

Acetic acid dimer

The most obvious property of carboxylic acids is implied by their name: carboxylic acids are *acidic*. They therefore react with bases such as NaOH and NaHCO$_3$ to give metal carboxylate salts, RCO$_2^-$ M$^+$. Carboxylic acids with more than six carbons are only slightly soluble in water, but the alkali metal salts of carboxylic acids are often highly water-soluble. In fact, it's often possible to purify an acid by extracting its salt into aqueous base, then reacidifying and extracting the pure acid back into an organic solvent.

A carboxylic acid
(water-insoluble)

A carboxylic acid salt
(water-soluble)

Like other Brønsted–Lowry acids discussed in Section 2.7, carboxylic acids dissociate slightly in dilute aqueous solution to give H$_3$O$^+$ and the corresponding carboxylate anions, RCO$_2^-$. The extent of dissociation is given by an acidity constant, K_a.

$$K_a = \frac{[\text{RCO}_2^-][\text{H}_3\text{O}^+]}{[\text{RCO}_2\text{H}]} \quad \text{and} \quad \text{p}K_a = -\log K_a$$

A list of K_a values for various carboxylic acids is given in Table 20.3. For most, K_a is approximately 10^{-4} to 10^{-5}. Acetic acid, for instance, has $K_a = 1.75 \times 10^{-5}$, which corresponds to a pK_a of 4.76. In practical terms, a K_a value near 10^{-5} means that only about 0.1% of the molecules in a 0.1 M solution are dissociated, as opposed to the 100% dissociation found with strong mineral acids like HCl.

Table 20.3 | **Acidity of Some Carboxylic Acids**

Structure	K_a	pK_a	
CF_3CO_2H	0.59	0.23	Stronger acid
HCO_2H	1.77×10^{-4}	3.75	
$HOCH_2CO_2H$	1.5×10^{-4}	3.84	
$C_6H_5CO_2H$	6.46×10^{-5}	4.19	
$H_2C{=}CHCO_2H$	5.6×10^{-5}	4.25	
CH_3CO_2H	1.75×10^{-5}	4.76	
$CH_3CH_2CO_2H$	1.34×10^{-5}	4.87	
CH_3CH_2OH (ethanol)	(1.00×10^{-16})	(16.00)	Weaker acid

Although much weaker than mineral acids, carboxylic acids are nevertheless much stronger acids than alcohols and phenols. The K_a of ethanol, for example, is approximately 10^{-16}, making ethanol a weaker acid than acetic acid by a factor of 10^{11}.

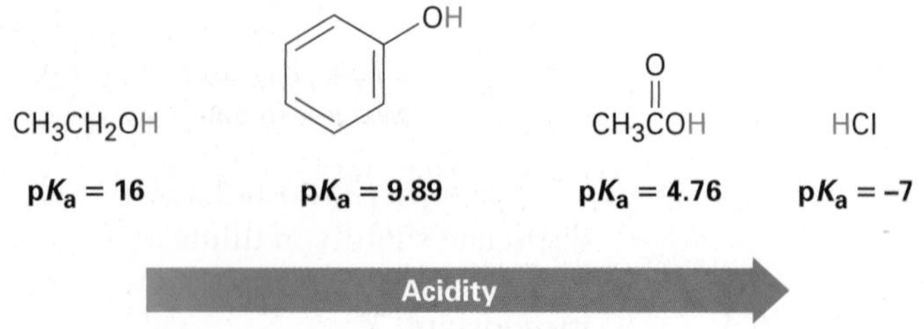

CH_3CH_2OH	OH (phenol)	$CH_3\overset{O}{\overset{\|}{C}}OH$	HCl
$pK_a = 16$	$pK_a = 9.89$	$pK_a = 4.76$	$pK_a = -7$

Acidity →

Why are carboxylic acids so much more acidic than alcohols, even though both contain −OH groups? An alcohol dissociates to give an alkoxide ion, in which the negative charge is localized on a single electronegative atom. A carboxylic acid, however, gives a carboxylate ion, in which the negative charge is delocalized over *two* equivalent oxygen atoms (Figure 20.1). In resonance terms (Section 2.4), a carboxylate ion is a stabilized resonance hybrid of two equivalent structures. Since a carboxylate ion is more stable than an alkoxide ion, it is lower in energy and more favored in the dissociation equilibrium.

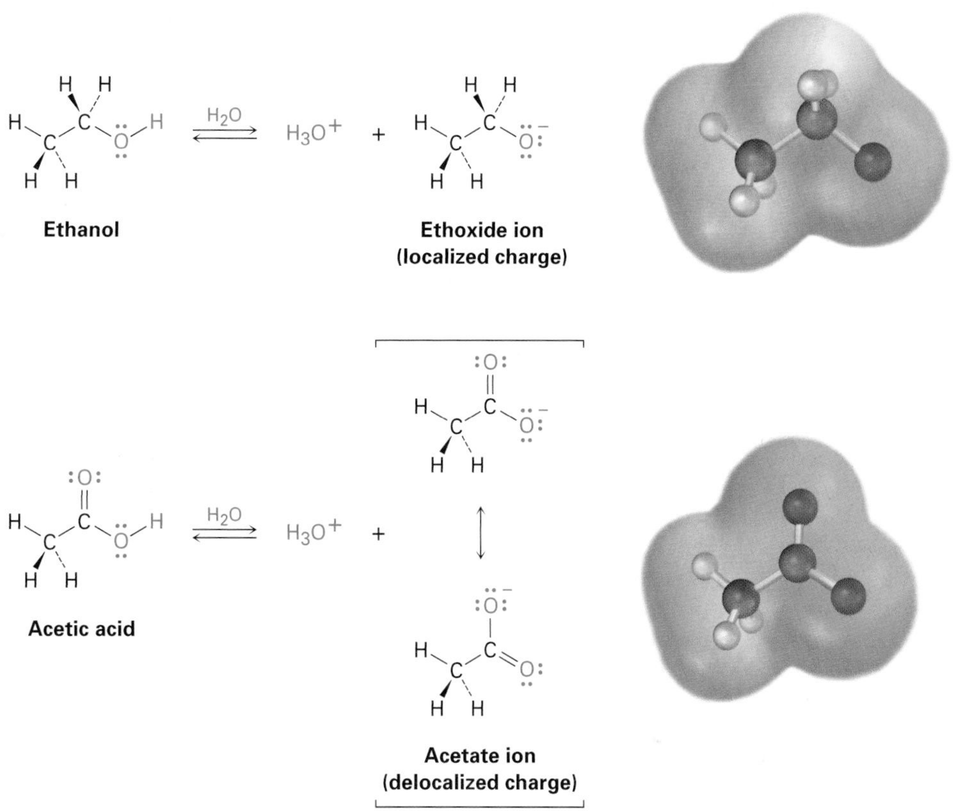

Ethanol

Ethoxide ion
(localized charge)

Acetic acid

Acetate ion
(delocalized charge)

Active Figure 20.1 An alkoxide ion has its charge localized on one oxygen atom and is less stable, while a carboxylate ion has the charge spread equally over both oxygens and is therefore more stable. *Sign in at* **www.cengage.com/login** *to see a simulation based on this figure and to take a short quiz.*

Experimental evidence for the equivalence of the two carboxylate oxygens comes from X-ray crystallographic studies on sodium formate. Both carbon–oxygen bonds are 127 pm in length, midway between the C=O bond (120 pm) and C−O bond (134 pm) of formic acid. An electrostatic potential map of the formate ion also shows how the negative charge (red) is dispersed equally over both oxygens.

127 pm

$H-C$ — Na$^+$

Sodium formate

120 pm
134 pm

$H-C$

Formic acid

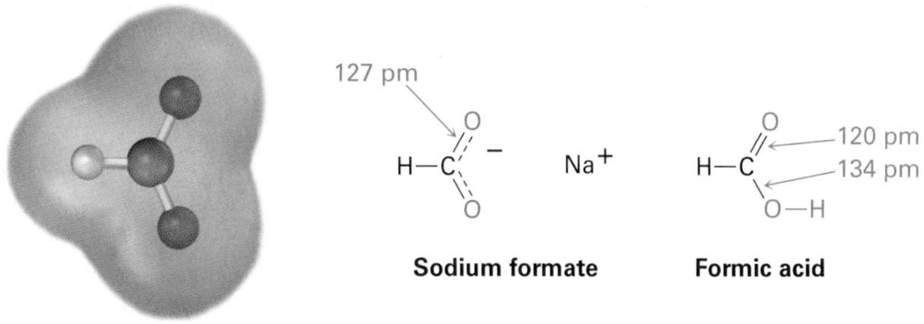

Problem 20.3 | Assume you have a mixture of naphthalene and benzoic acid that you want to separate. How might you take advantage of the acidity of one component in the mixture to effect a separation?

Problem 20.4 | The K_a for dichloroacetic acid is 3.32×10^{-2}. Approximately what percentage of the acid is dissociated in a 0.10 M aqueous solution?

20.3 | Biological Acids and the Henderson–Hasselbalch Equation

In acidic solution at low pH, a carboxylic acid is completely undissociated and exists entirely as RCO_2H. In basic solution at high pH, a carboxylic acid is completely dissociated and exists entirely as RCO_2^-. Inside living cells, however, the pH is neither acidic nor basic but is instead buffered to nearly neutral pH—in humans, to pH = 7.3, a value often referred to as *physiological pH*. In what form, then, do carboxylic acids exist inside cells? The question is an important one for understanding the acid catalysts so often found in biological reactions.

If the pK_a value of a given acid and the pH of the medium are known, the percentages of dissociated and undissociated forms can be calculated using what is called the **Henderson–Hasselbalch equation**.

For any acid HA, we have

$$pK_a = -\log\frac{[H_3O^+]\,[A^-]}{[HA]} = -\log[H_3O^+] - \log\frac{[A^-]}{[HA]}$$

$$= pH - \log\frac{[A^-]}{[HA]}$$

which can be rearranged to give

$$pH = pK_a + \log\frac{[A^-]}{[HA]} \qquad \text{Henderson–Hasselbalch equation}$$

$$\text{so} \quad \log\frac{[A^-]}{[HA]} = pH - pK_a$$

This equation says that the logarithm of the concentration of dissociated acid $[A^-]$ divided by the concentration of undissociated acid $[HA]$ is equal to the pH of the solution minus the pK_a of the acid. Thus, if we know both the pH of the solution and the pK_a of the acid, we can calculate the ratio of $[A^-]$ to $[HA]$. Furthermore, when pH = pK_a, the two forms HA and A^- are present in equal amounts because log 1 = 0.

As an example of how to use the Henderson–Hasselbalch equation, let's find out what species are present in a 0.0010 M solution of acetic acid at pH = 7.3. According to Table 20.3, the pK_a of acetic acid is 4.76. From the Henderson–Hasselbalch equation, we have

$$\log\frac{[A^-]}{[HA]} = pH - pK_a = 7.3 - 4.76 = 2.54$$

$$\frac{[A^-]}{[HA]} = \text{antilog } 2.54 = 3.5 \times 10^2 \qquad \text{so} \qquad [A^-] = 3.5 \times 10^2\,[HA]$$

In addition, we know that

$$[A^-] + [HA] = 0.0010 \text{ M}$$

Solving the two simultaneous equations gives $[A^-] = 0.0010$ M and $[HA] = 3 \times 10^{-6}$ M. In other words, at a physiological pH of 7.3, essentially 100% of acetic acid molecules in a 0.0010 M solution are dissociated to the acetate ion.

What is true for acetic acid is also true for other carboxylic acids: at the physiological pH that exists inside cells, carboxylic acids are almost entirely dissociated. To reflect this fact, we always refer to cellular carboxylic acids by the name of their anion—acetate, lactate, citrate, and so forth, rather than acetic acid, lactic acid, and citric acid.

Problem 20.5 | Calculate the percentages of dissociated and undissociated forms present in the following solutions:
(a) 0.0010 M glycolic acid ($HOCH_2CO_2H$; $pK_a = 3.83$) at pH = 4.50
(b) 0.0020 M propanoic acid ($pK_a = 4.87$) at pH = 5.30

20.4 | Substituent Effects on Acidity

The listing of pK_a values shown previously in Table 20.3 indicates that there are substantial differences in acidity from one carboxylic acid to another. For example, trifluoroacetic acid ($K_a = 0.59$) is 33,000 times as strong as acetic acid ($K_a = 1.75 \times 10^{-5}$). How can we account for such differences?

Because the dissociation of a carboxylic acid is an equilibrium process, any factor that stabilizes the carboxylate anion relative to undissociated carboxylic acid will drive the equilibrium toward increased dissociation and result in increased acidity. An electron-withdrawing chlorine atom, for instance, makes chloroacetic acid ($K_a = 1.4 \times 10^{-3}$) approximately 80 times as strong as acetic acid; introduction of two chlorines makes dichloroacetic acid 3000 times as strong as acetic acid, and introduction of three chlorines makes trichloroacetic acid more than 12,000 times as strong.

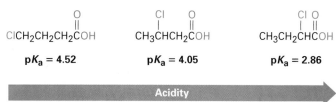

Because inductive effects operate through σ bonds and are dependent on distance, the effect of halogen substitution decreases as the substituent moves farther from the carboxyl. Thus, 2-chlorobutanoic acid has $pK_a = 2.86$, 3-chlorobutanoic acid has $pK_a = 4.05$, and 4-chlorobutanoic acid has $pK_a = 4.52$, similar to that of butanoic acid itself.

$$ClCH_2CH_2CH_2COH \qquad CH_3CHCH_2COH \qquad CH_3CH_2CHCOH$$

$pK_a = 4.52$ $pK_a = 4.05$ $pK_a = 2.86$

Acidity →

Substituent effects on acidity are also found in substituted benzoic acids. We saw during the discussion of electrophilic aromatic substitution in Section 16.4 that substituents on the aromatic ring dramatically affect reactivity. Aromatic rings with electron-donating groups are activated toward further electrophilic substitution, and aromatic rings with electron-withdrawing groups are deactivated. Exactly the same effects are noticed on the acidity of substituted benzoic acids (Table 20.4).

Table 20.4 | **Substituent Effects on Acidity of *p*-Substituted Benzoic Acids**

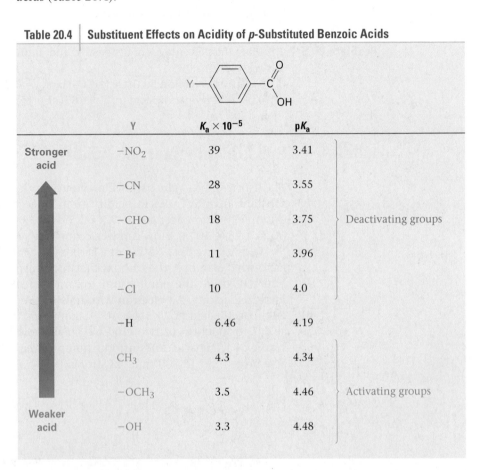

	Y	$K_a \times 10^{-5}$	pK_a	
Stronger acid	$-NO_2$	39	3.41	
	$-CN$	28	3.55	
	$-CHO$	18	3.75	Deactivating groups
	$-Br$	11	3.96	
	$-Cl$	10	4.0	
	$-H$	6.46	4.19	
	CH_3	4.3	4.34	
	$-OCH_3$	3.5	4.46	Activating groups
Weaker acid	$-OH$	3.3	4.48	

As Table 20.4 shows, an electron-withdrawing (deactivating) group such as nitro increases acidity by stabilizing the carboxylate anion, and an electron-donating (activating) group such as methoxy decreases acidity by destabilizing the carboxylate anion.

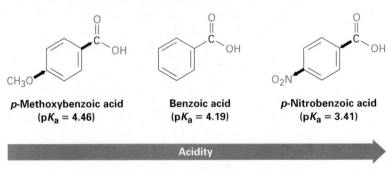

p-Methoxybenzoic acid ($pK_a = 4.46$)   Benzoic acid ($pK_a = 4.19$)   *p*-Nitrobenzoic acid ($pK_a = 3.41$)

Acidity

Because it's much easier to measure the acidity of a substituted benzoic acid than it is to determine the relative reactivity of an aromatic ring toward electrophilic substitution, the correlation between the two effects is useful for predicting reactivity. If we want to know the effect of a certain substituent on electrophilic reactivity, we can simply find the acidity of the corresponding benzoic acid. Worked Example 20.1 gives an example.

Finding the K_a
of this acid . . .

. . . lets us predict
the reactivity of this
substituted benzene
to electrophilic attack.

WORKED EXAMPLE 20.1   *Predicting the Effect of a Substituent on the Reactivity of an Aromatic Ring toward Electrophilic Substitution*

The pK_a of *p*-(trifluoromethyl)benzoic acid is 3.6. Is the trifluoromethyl substituent an activating or deactivating group in electrophilic aromatic substitution?

Strategy   Decide whether *p*-(trifluoromethyl)benzoic acid is stronger or weaker than benzoic acid. A substituent that strengthens the acid is a deactivating group because it withdraws electrons, and a substituent that weakens the acid is an activating group because it donates electrons.

Solution   A pK_a of 3.6 means that *p*-(trifluoromethyl)benzoic acid is stronger than benzoic acid, whose pK_a is 4.19. Thus, the trifluoromethyl substituent favors dissociation by helping stabilize the negative charge. Trifluoromethyl must therefore be an electron-withdrawing, deactivating group.

Problem 20.6   Which would you expect to be a stronger acid, the lactic acid found in tired muscles or acetic acid? Explain.

HO O
| ||
CH₃CHCOH   **Lactic acid**

Problem 20.7   Dicarboxylic acids have two dissociation constants, one for the initial dissociation into a monoanion and one for the second dissociation into a dianion. For oxalic acid, HO_2C-CO_2H, the first ionization constant has pK_{a1} = 1.2 and the second ionization constant has pK_{a2} = 4.2. Why is the second carboxyl group so much less acidic than the first?

Problem 20.8   The pK_a of *p*-cyclopropylbenzoic acid is 4.45. Is cyclopropylbenzene likely to be more reactive or less reactive than benzene toward electrophilic bromination? Explain.

Problem 20.9   Rank the following compounds in order of increasing acidity. Don't look at a table of pK_a data to help with your answer.
(a) Benzoic acid, *p*-methylbenzoic acid, *p*-chlorobenzoic acid
(b) *p*-Nitrobenzoic acid, acetic acid, benzoic acid

20.5 | Preparation of Carboxylic Acids

CENGAGENOW™ Click *Organic Interactive* to **use your problem-solving skills to design syntheses of carboxylic acids**.

Let's review briefly some of the methods for preparing carboxylic acids that we've seen in past chapters.

▌ Oxidation of a substituted alkylbenzene with $KMnO_4$ or $Na_2Cr_2O_7$ gives a substituted benzoic acid (Section 16.9). Both primary and secondary alkyl groups can be oxidized, but tertiary groups are not affected.

p-Nitrotoluene   *p*-Nitrobenzoic acid (88%)

▌ Oxidative cleavage of an alkene with $KMnO_4$ gives a carboxylic acid if the alkene has at least one vinylic hydrogen (Section 7.9).

Oleic acid   Nonanoic acid   Nonanedioic acid

▌ Oxidation of a primary alcohol or an aldehyde yields a carboxylic acid (Sections 17.7 and 19.3). Primary alcohols are often oxidized with CrO_3 in aqueous acid, and aldehydes are oxidized with either acidic CrO_3 or basic silver oxide (Tollens' reagent).

4-Methyl-1-pentanol   4-Methylpentanoic acid

Hexanal   Hexanoic acid

Hydrolysis of Nitriles

Carboxylic acids can be prepared from nitriles by reaction with hot aqueous acid or base by a mechanism that we'll see in Section 20.9. Since nitriles themselves are usually made by S_N2 reaction of a primary or secondary alkyl halide with CN^-, the two-step sequence of cyanide displacement followed by nitrile hydrolysis is a good way to make a carboxylic acid from an alkyl halide ($RBr \rightarrow RC\equiv N \rightarrow RCO_2H$).

Note that the product acid has one more carbon than the starting alkyl halide. An example occurs in the commercial synthesis of fenoprofen, a nonsteroidal anti-inflammatory drug, or NSAID, marketed under the trade name Nalfon. (See Chapter 15 *Focus On.*)

Fenoprofen
(an antiarthritis agent)

Carboxylation of Grignard Reagents

Another method for preparing carboxylic acids is by reaction of a Grignard reagent with CO_2 to yield a metal carboxylate, followed by protonation to give the carboxylic acid. This **carboxylation** reaction is usually carried out by bubbling a stream of dry CO_2 gas through a solution of the Grignard reagent. The organomagnesium halide adds to a C=O bond of carbon dioxide in a typical nucleophilic carbonyl addition reaction, and protonation of the carboxylate by addition of aqueous HCl in a separate step then gives the free carboxylic acid. For example

Phenylmagnesium
bromide

Benzoic acid

There are, of course, no Grignard reagents inside living cells, but there are other types of stabilized carbanions that are often carboxylated. One of the initial steps in fatty-acid biosynthesis, for instance, involves formation of a carbanion from acetyl CoA, followed by carboxylation to yield malonyl CoA.

Acetyl CoA

Malonyl CoA

WORKED EXAMPLE 20.2 | *Devising a Synthesis Route for a Carboxylic Acid*

How would you prepare phenylacetic acid (PhCH$_2$CO$_2$H) from benzyl bromide (PhCH$_2$Br)?

Strategy   We've seen two methods for preparing carboxylic acids from alkyl halides: (1) cyanide ion displacement followed by hydrolysis and (2) formation of a Grignard reagent followed by carboxylation. The first method involves an S$_N$2 reaction and is therefore limited to use with primary and some secondary alkyl halides. The second method involves formation of a Grignard reagent and is therefore limited to use with organic halides that have no acidic hydrogens or reactive functional groups elsewhere in the molecule. In the present instance, either method would work well.

Solution

Problem 20.10 | How would you prepare the following carboxylic acids?
(a) (CH$_3$)$_3$CCO$_2$H from (CH$_3$)$_3$CCl
(b) CH$_3$CH$_2$CH$_2$CO$_2$H from CH$_3$CH$_2$CH$_2$Br

20.6 | Reactions of Carboxylic Acids: An Overview

We commented earlier in this chapter that carboxylic acids are similar in some respects to both alcohols and ketones. Like alcohols, carboxylic acids can be deprotonated to give anions, which are good nucleophiles in S$_N$2 reactions. Like ketones, carboxylic acids undergo addition of nucleophiles to the carbonyl group. In addition, carboxylic acids undergo other reactions characteristic of neither alcohols nor ketones. Figure 20.2 shows some of the general reactions of carboxylic acids.

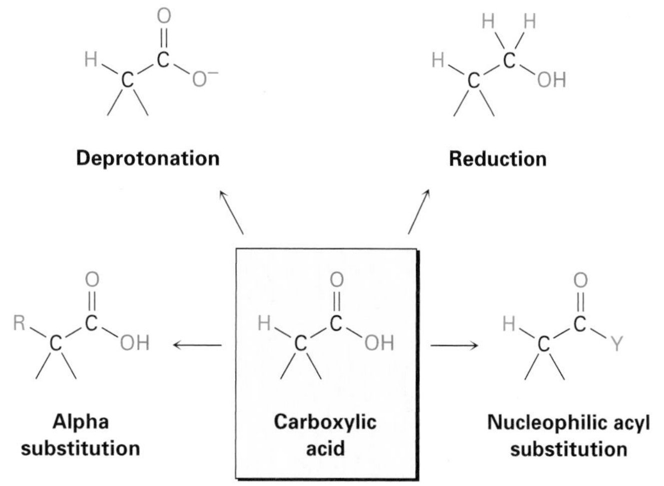

Figure 20.2 Some general reactions of carboxylic acids.

Deprotonation

Reduction

Alpha substitution

Carboxylic acid

Nucleophilic acyl substitution

Reactions of carboxylic acids can be grouped into the four categories indicated in Figure 20.2. Of the four, we've already discussed the acidic behavior of carboxylic acids in Sections 20.2 through 20.4, and we mentioned reduction by treatment of the acid with LiAlH$_4$ in Section 17.4. The remaining two categories are examples of fundamental carbonyl-group reaction mechanisms—nucleophilic acyl substitution and α substitution—that will be discussed in detail in Chapters 21 and 22.

Problem 20.11 | How might you prepare 2-phenylethanol from benzyl bromide? More than one step is needed.

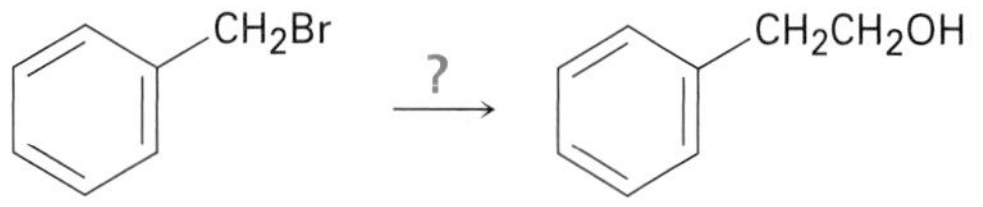

Problem 20.12 | How might you carry out the following transformation? More than one step is needed.

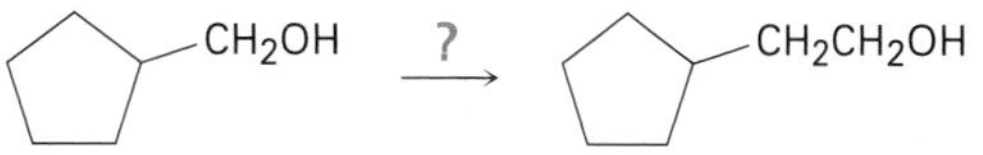

20.7 | Chemistry of Nitriles

Nitriles are analogous to carboxylic acids in that both have a carbon atom with three bonds to an electronegative atom, and both contain a π bond. Thus, some reactions of nitriles and carboxylic acids are similar. Both kinds of compounds are electrophiles, for instance, and both undergo nucleophilic addition reactions.

$$R-C\equiv N \qquad\qquad R-\overset{\displaystyle O}{\underset{\displaystyle OH}{C}}$$

**A nitrile—three
bonds to nitrogen**

**An acid—three
bonds to two oxygens**

Nitriles occur infrequently in living organisms, although several hundred examples of their occurrence are known. Cyanocycline A, for instance, has been isolated from the bacterium *Streptomyces lavendulae* and found to have both antimicrobial and antitumor activity. In addition, more than 1000 compounds called *cyanogenic glycosides* are known. Derived primarily from plants, cyanogenic glycosides contain a sugar with an acetal carbon, one oxygen of which is bonded to a nitrile-bearing carbon (sugar$-$O$-$C$-$CN). On hydrolysis with aqueous acid, the acetal is cleaved (Section 19.10), generating a cyanohydrin (HO$-$C$-$CN), which releases hydrogen cyanide. It's thought that the primary function of cyanogenic glycosides is to protect the plant by poisoning any animal foolish enough to eat it. Lotaustralin from the cassava plant is an example.

Cyanocycline A

**Lotaustralin
(a cyanogenic glycoside)**

Preparation of Nitriles

The simplest method of nitrile preparation is the S_N2 reaction of CN^- with a primary or secondary alkyl halide, as discussed in Section 20.5. Another method for preparing nitriles is by dehydration of a primary amide, $RCONH_2$. Thionyl chloride is often used for the reaction, although other dehydrating agents such as $POCl_3$ also work.

2-Ethylhexanamide

2-Ethylhexanenitrile (94%)

The dehydration occurs by initial reaction of $SOCl_2$ on the nucleophilic amide oxygen atom, followed by deprotonation and a subsequent E2-like elimination reaction.

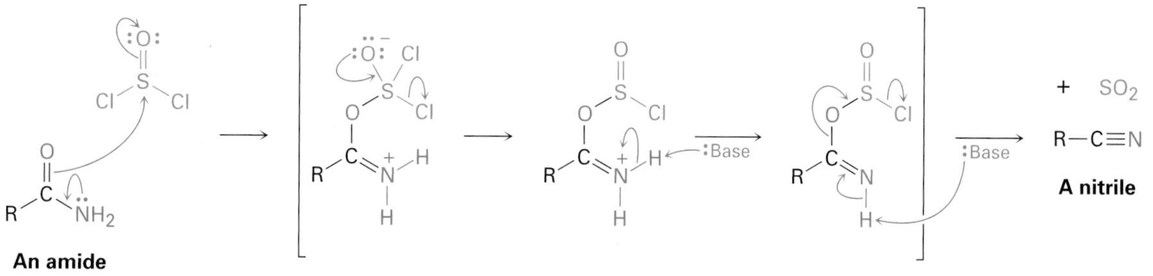

Both methods of nitrile synthesis—S_N2 displacement by CN^- on an alkyl halide and amide dehydration—are useful, but the synthesis from amides is more general because it is not limited by steric hindrance.

Reactions of Nitriles

Like a carbonyl group, a nitrile group is strongly polarized and has an electrophilic carbon atom. Nitriles therefore react with nucleophiles to yield sp^2-hybridized imine anions in a reaction analogous to the formation of an sp^3-hybridized alkoxide ion by nucleophilic addition to a carbonyl group.

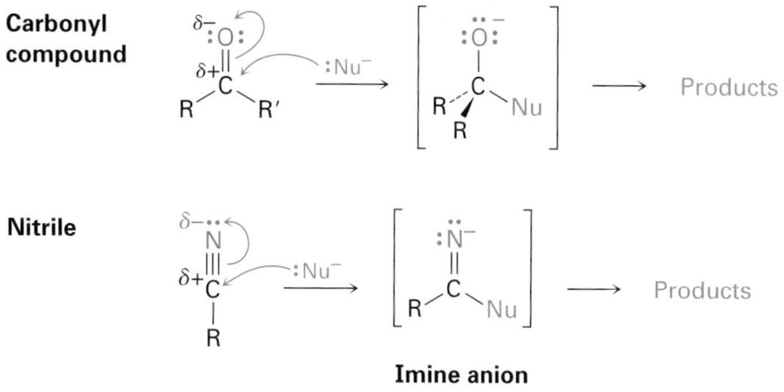

Among the most useful reactions of nitriles are hydrolysis to yield first an amide and then a carboxylic acid plus ammonia, reduction to yield an amine, and Grignard reaction to yield a ketone (Figure 20.3).

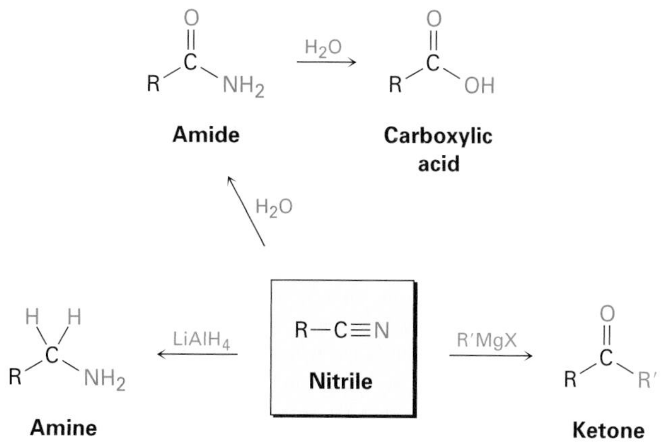

Figure 20.3 Some reactions of nitriles.

Hydrolysis: Conversion of Nitriles into Carboxylic Acids   A nitrile is hydrolyzed in either basic or acidic aqueous solution to yield a carboxylic acid plus ammonia or an amine.

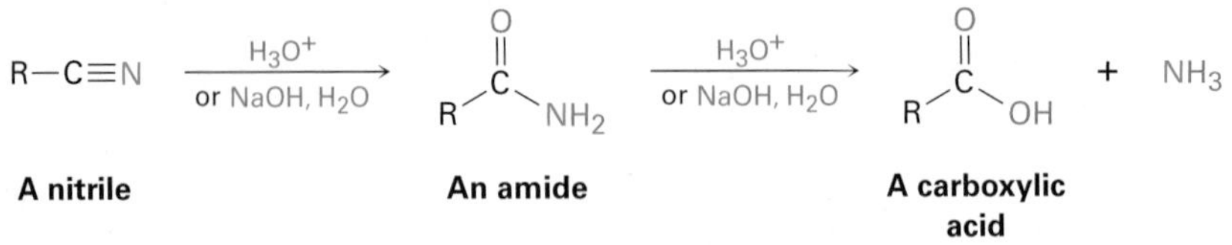

Base catalyzed nitrile hydrolysis involves nucleophilic addition of hydroxide ion to the polar C≡N bond to give an imine anion in a process similar to nucleophilic addition to a polar C=O bond to give an alkoxide anion. Protonation then gives a hydroxy imine, which tautomerizes (Section 8.4) to an amide in a step similar to the tautomerization of an enol to a ketone. The mechanism is shown in Figure 20.4.

Active Figure 20.4 MECHANISM: Mechanism of the basic hydrolysis of a nitrile to yield an amide, which is subsequently hydrolyzed further to a carboxylic acid anion. *Sign in at* **www.cengage.com/login** *to see a simulation based on this figure and to take a short quiz.*

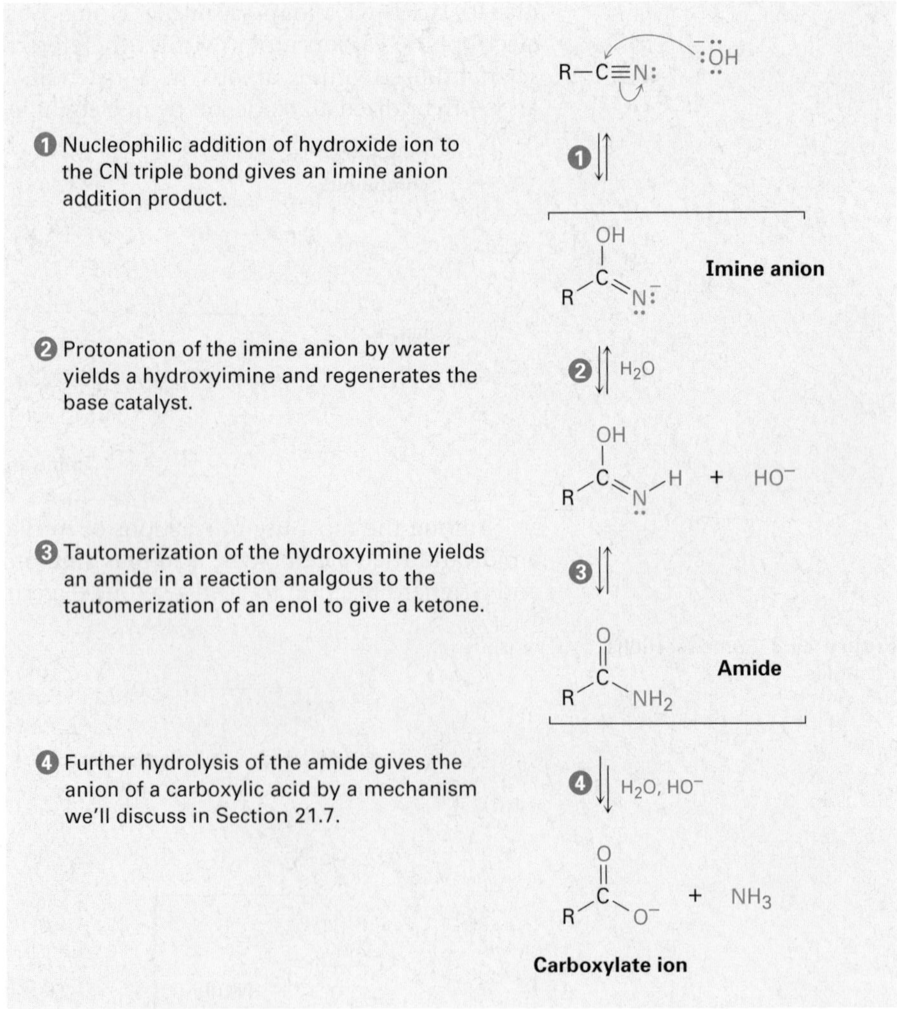

① Nucleophilic addition of hydroxide ion to the CN triple bond gives an imine anion addition product.

② Protonation of the imine anion by water yields a hydroxyimine and regenerates the base catalyst.

③ Tautomerization of the hydroxyimine yields an amide in a reaction analgous to the tautomerization of an enol to give a ketone.

④ Further hydrolysis of the amide gives the anion of a carboxylic acid by a mechanism we'll discuss in Section 21.7.

© John McMurry

Following formation of the amide intermediate, a second nucleophilic addition of hydroxide ion to the amide carbonyl group then yields a tetrahedral alkoxide ion, which expels amide ion, NH_2^-, as leaving group and gives the carboxylate ion, thereby driving the reaction toward products. Subsequent acidification in a separate step yields the carboxylic acid. We'll look at this process in more detail in Section 21.7.

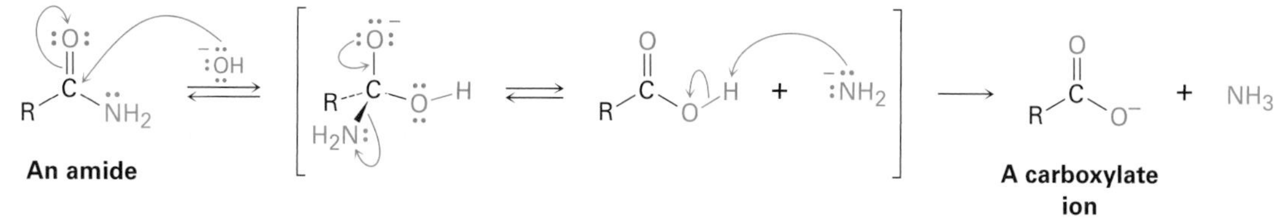

An amide   **A carboxylate ion**

Reduction: Conversion of Nitriles into Amines   Reduction of a nitrile with $LiAlH_4$ gives a primary amine, RNH_2. The reaction occurs by nucleophilic addition of hydride ion to the polar C≡N bond, yielding an imine anion, which still contains a C=N bond and therefore undergoes a second nucleophilic addition of hydride to give a *dianion*. Both monoanion and dianion intermediates are undoubtedly stabilized by Lewis acid–base complexation to an aluminum species, facilitating the second addition that would otherwise be difficult. Protonation of the dianion by addition of water in a subsequent step gives the amine.

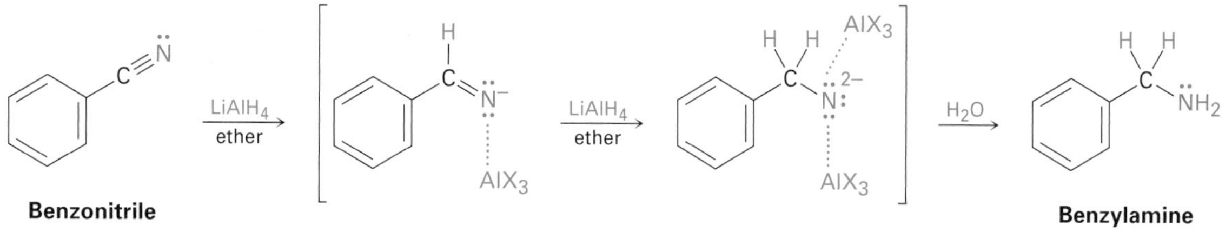

Benzonitrile   **Benzylamine**

Reaction of Nitriles with Organometallic Reagents   Grignard reagents add to a nitrile to give an intermediate imine anion that is hydrolyzed by addition of water to yield a ketone.

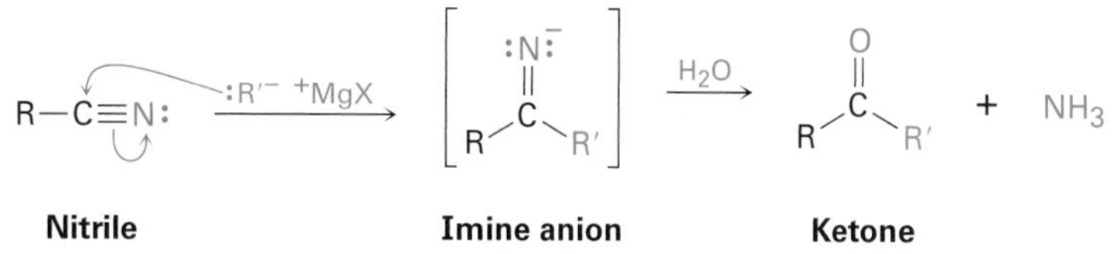

Nitrile   **Imine anion**   **Ketone**

The reaction is similar to the reduction of a nitrile to an amine, except that only one nucleophilic addition occurs rather than two, and the attacking nucleophile is a carbanion ($R:^-$) rather than a hydride ion. For example:

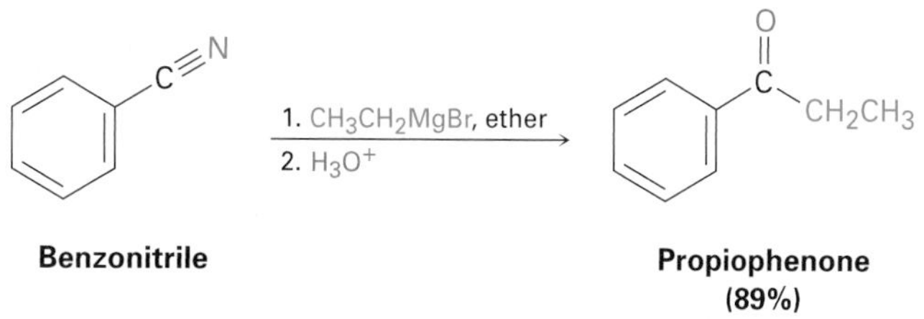

Benzonitrile   **Propiophenone (89%)**

| WORKED EXAMPLE 20.3 | ***Synthesizing a Ketone from a Nitrile*** |

How would you prepare 2-methyl-3-pentanone from a nitrile?

$$CH_3CH_2\overset{\overset{\displaystyle O}{\|}}{C}\underset{\underset{\displaystyle CH_3}{|}}{C}HCH_3 \quad \textbf{2-Methylpentan-3-one}$$

Strategy A ketone results from the reaction between a Grignard reagent and a nitrile, with the C≡N carbon of the nitrile becoming the carbonyl carbon. Identify the two groups attached to the carbonyl carbon atom in the product. One will come from the Grignard reagent, and the other will come from the nitrile.

Solution There are two possibilities.

$$\left\{\begin{array}{c}CH_3CH_2C\equiv N \\ + \\ (CH_3)_2CHMgBr\end{array}\right\} \xrightarrow[\text{2. H}_3O^+]{\text{1. Grignard}} CH_3CH_2\overset{\overset{\displaystyle O}{\|}}{C}\underset{\underset{\displaystyle CH_3}{|}}{C}HCH_3 \xleftarrow[\text{2. H}_3O^+]{\text{1. Grignard}} \left\{\begin{array}{c}\overset{\overset{\displaystyle CH_3}{|}}{C}H_3CHC\equiv N \\ + \\ CH_3CH_2MgBr\end{array}\right.$$

2-Methylpentan-3-one

Problem 20.13 How would you prepare the following carbonyl compounds from a nitrile?

(a)

$$CH_3CH_2\overset{\overset{\displaystyle O}{\|}}{C}CH_2CH_3$$

(b)

$$O_2N\text{—}\bigcirc\text{—}\overset{\overset{\displaystyle O}{\|}}{C}\diagdown CH_3$$

Problem 20.14 How would you prepare 1-phenyl-2-butanone, $C_6H_5CH_2COCH_2CH_3$, from benzyl bromide, $C_6H_5CH_2Br$? More than one step is required.

20.8 | Spectroscopy of Carboxylic Acids and Nitriles

Infrared Spectroscopy

Carboxylic acids have two characteristic IR absorptions that make the $-CO_2H$ group easily identifiable. The O–H bond of the carboxyl group gives rise to a very broad absorption over the range 2500 to 3300 cm^{-1}, and the C=O bond shows an absorption between 1710 and 1760 cm^{-1}. The exact position of C=O absorption depends both on the structure of the molecule and on whether the acid is free (monomeric) or hydrogen-bonded (dimeric). Free carboxyl groups absorb at 1760 cm^{-1}, but the more commonly encountered dimeric carboxyl groups absorb in a broad band centered around 1710 cm^{-1}.

Free carboxyl (uncommon), 1760 cm^{-1}

$$R\text{—}C\overset{\overset{\displaystyle O}{\diagup}}{\diagdown}O\text{—}H$$

Associated carboxyl (usual case), 1710 cm^{-1}

$$R\text{—}C\overset{O\cdots\cdots H\text{—}O}{\underset{O\text{—}H\cdots\cdots O}{}}C\text{—}R$$

Both the broad O–H absorption and the C=O absorption at 1710 cm^{-1} (dimeric) are identified in the IR spectrum of butanoic acid shown in Figure 20.5.

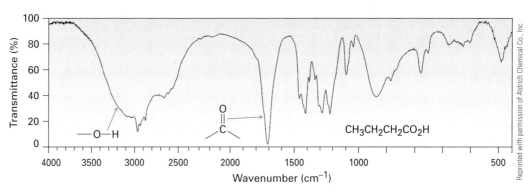

Figure 20.5 IR spectrum of butanoic acid, $CH_3CH_2CH_2CO_2H$.

Nitriles show an intense and easily recognizable C≡N bond absorption near 2250 cm^{-1} for saturated compounds and 2230 cm^{-1} for aromatic and conjugated molecules. Since few other functional groups absorb in this region, IR spectroscopy is highly diagnostic for nitriles.

Problem 20.15 Cyclopentanecarboxylic acid and 4-hydroxycyclohexanone have the same formula ($C_6H_{10}O_2$), and both contain an −OH and a C=O group. How could you distinguish between them by IR spectroscopy?

Nuclear Magnetic Resonance Spectroscopy

Carboxylic acid groups can be detected by both ^{1}H and ^{13}C NMR spectroscopy. Carboxyl carbon atoms absorb in the range 165 to 185 δ in the ^{13}C NMR spectrum, with aromatic and α,β-unsaturated acids near the upfield end of the range (~165 δ) and saturated aliphatic acids near the downfield end (~185 δ). Nitrile carbons absorb in the range 115 to 130 δ.

In the ^{1}H NMR spectrum, the acidic −CO$_2$H proton normally absorbs as a singlet near 12 δ. As with alcohols (Section 17.11), the −CO$_2$H proton can be replaced by deuterium when D$_2$O is added to the sample tube, causing the absorption to disappear from the NMR spectrum. Figure 20.6 shows the ^{1}H NMR spectrum of phenylacetic acid. Note that the carboxyl proton absorption occurs at 12.0 δ.

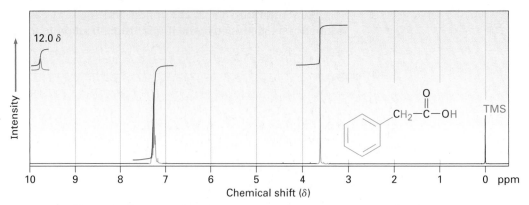

Figure 20.6 Proton NMR spectrum of phenylacetic acid, $C_6H_5CH_2CO_2H$.

Problem 20.16 | How could you distinguish between the isomers cyclopentanecarboxylic acid and 4-hydroxycyclohexanone by ^1H and ^{13}C NMR spectroscopy? (See Problem 20.14.)

Focus On . . .

Vitamin C

In addition to the hazards of weather, participants in early polar expeditions often suffered from scurvy, caused by a dietary vitamin C deficiency.

Vitamin C, or ascorbic acid, is surely the best known of all vitamins. It was the first vitamin to be discovered (1928), the first to be structurally characterized (1933), and the first to be synthesized in the laboratory (1933). Over 200 million pounds of vitamin C are now synthesized worldwide each year, more than the total amount of all other vitamins combined. In addition to its use as a vitamin supplement, vitamin C is used as a food preservative, a "flour improver" in bakeries, and an animal food additive.

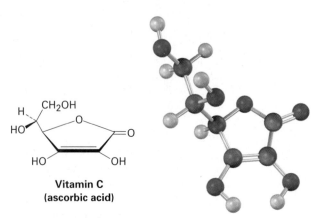

**Vitamin C
(ascorbic acid)**

(continued)

Vitamin C is perhaps most famous for its antiscorbutic properties, meaning that it prevents the onset of scurvy, a bleeding disease affecting those with a deficiency of fresh vegetables and citrus fruits in their diet. Sailors in the Age of Exploration were particularly susceptible to scurvy, and the death toll was high. The Portuguese explorer Vasco da Gama lost more than half his crew to scurvy during his 2-year voyage around the Cape of Good Hope in 1497–1499.

In more recent times, large doses of vitamin C have been claimed to prevent the common cold, cure infertility, delay the onset of symptoms in acquired immunodeficiency syndrome (AIDS), and inhibit the development of gastric and cervical cancers. None of these claims have been backed by medical evidence, however. In the largest study yet done of the effect of vitamin C on the common cold, a meta-analysis of more than 100 separate trials covering 40,000 people found no difference in the incidence of colds between those who took supplemental vitamin C regularly and those who did not. When taken during a cold, however, vitamin C does appear to decrease the cold's duration by 8%.

The industrial preparation of vitamin C involves an unusual blend of biological and laboratory organic chemistry. The Hoffmann-La Roche company synthesizes ascorbic acid from glucose through the five-step route shown in Figure 20.7. Glucose, a pentahydroxy aldehyde, is first reduced to sorbitol, which is then oxidized by the microorganism *Acetobacter suboxydans*. No chemical reagent is known that is selective enough to oxidize only one of the six alcohol groups in sorbitol, so an enzymatic reaction is used. Treatment with acetone and an acid catalyst then protects four of the remaining hydroxyl groups in acetal linkages, and the unprotected hydroxyl group is chemically oxidized to the carboxylic acid by reaction with aqueous NaOCl (household bleach). Hydrolysis with acid then removes the two acetal groups and causes an internal ester-forming reaction to take place to give ascorbic acid. Each of the five steps takes place in better than 90% yield.

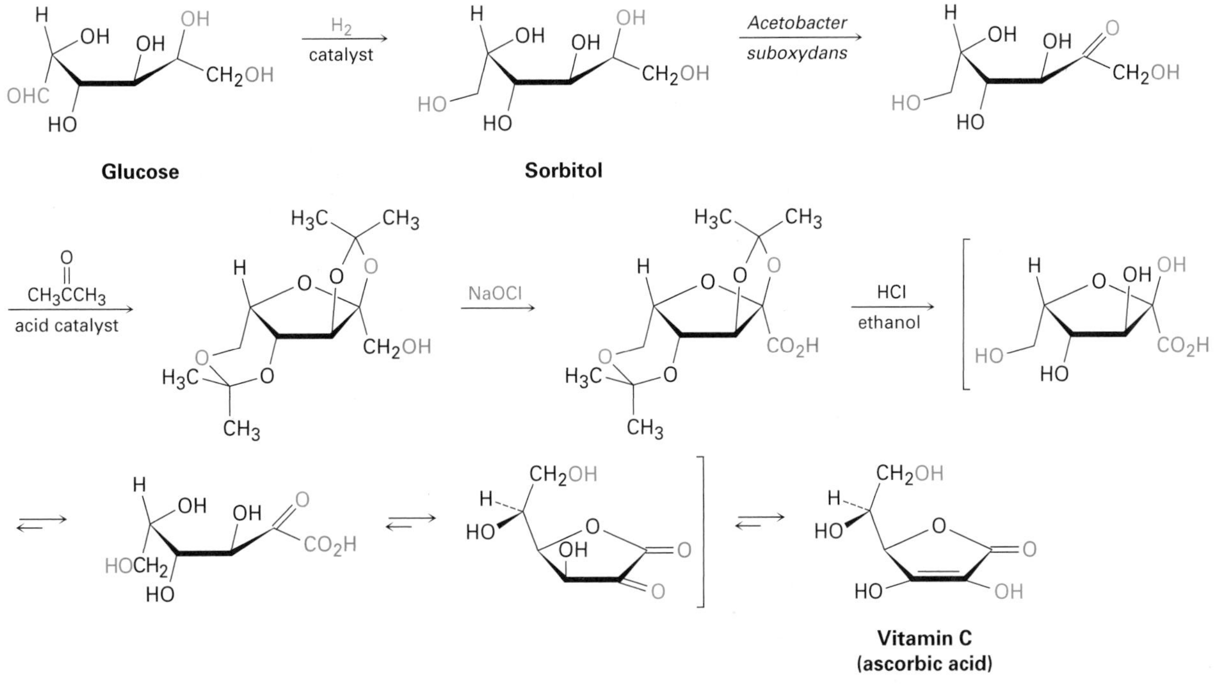

Figure 20.7 The industrial synthesis of ascorbic acid from glucose.

SUMMARY AND KEY WORDS

carboxyl group, 752

carboxylation, 763

carboxylic acid (RCO_2H), 751

Henderson–Hasselbalch equation, 758

nitrile ($RC{\equiv}N$), 754

Carboxylic acids are among the most useful building blocks for synthesizing other molecules, both in nature and in the chemical laboratory. They are named systematically by replacing the terminal -*e* of the corresponding alkane name with -*oic acid*. Like aldehydes and ketones, the carbonyl carbon atom is sp^2-hybridized; like alcohols, carboxylic acids are associated through hydrogen-bonding and therefore have high boiling points.

The distinguishing characteristic of carboxylic acids is their acidity. Although weaker than mineral acids such as HCl, carboxylic acids dissociate much more readily than alcohols because the resultant carboxylate ions are stabilized by resonance between two equivalent forms.

Most carboxylic acids have pK_a values near 5, but the exact pK_a of a given acid depends on structure. Carboxylic acids substituted by electron-withdrawing groups are more acidic (have a lower pK_a) because their carboxylate ions are stabilized. Carboxylic acids substituted by electron-donating groups are less acidic (have a higher pK_a) because their carboxylate ions are destabilized. The extent of dissociation of a carboxylic acid in a buffered solution of a given pH can be calculated with the **Henderson–Hasselbalch equation**. Inside living cells, where the physiological pH = 7.3, carboxylic acids are entirely dissociated and exist as their carboxylate anions.

Methods of synthesis for carboxylic acids include (1) oxidation of alkylbenzenes, (2) oxidative cleavage of alkenes, (3) oxidation of primary alcohols or aldehydes, (4) hydrolysis of nitriles, and (5) reaction of Grignard reagents with CO_2 (**carboxylation**). General reactions of carboxylic acids include (1) loss of the acidic proton, (2) nucleophilic acyl substitution at the carbonyl group, (3) substitution on the α carbon, and (4) reduction.

Nitriles are similar in some respects to carboxylic acids and are prepared either by S_N2 reaction of an alkyl halide with cyanide ion or by dehydration of an amide. Nitriles undergo nucleophilic addition to the polar $C{\equiv}N$ bond in the same way that carbonyl compounds do. The most important reactions of nitriles are their hydrolysis to carboxylic acids, reduction to primary amines, and reaction with organometallic reagents to yield ketones.

Carboxylic acids and nitriles are easily distinguished spectroscopically. Acids show a characteristic IR absorption at 2500 to 3300 cm^{-1} due to the O–H and another at 1710 to 1760 cm^{-1} due to the C=O; nitriles have an absorption at 2250 cm^{-1}. Acids also show ^{13}C NMR absorptions at 165 to 185 δ and ^1H NMR absorptions near 12 δ; nitriles have a ^{13}C NMR absorption in the range 115 to 130 δ.

SUMMARY OF REACTIONS

1. Preparation of carboxylic acids (Section 20.5)
   (a) Carboxylation of Grignard reagents

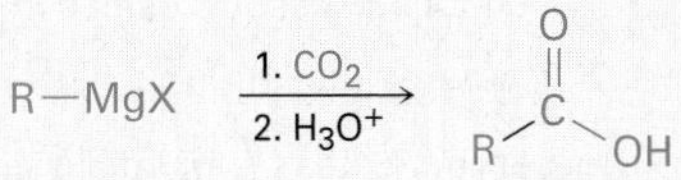

$$R-MgX \xrightarrow[\text{2. } H_3O^+]{\text{1. } CO_2} \underset{R}{\overset{O}{\underset{\,}{\|}}} C \diagdown_{OH}$$

   (b) Hydrolysis of nitriles

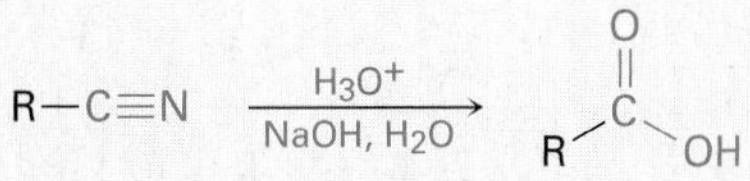

$$R-C\equiv N \xrightarrow[\text{NaOH, } H_2O]{H_3O^+} \underset{R}{\overset{O}{\|}} C \diagdown_{OH}$$

2. Preparation of nitriles (Section 20.7)
   (a) S_N2 reaction of alkyl halides

$$RCH_2Br \xrightarrow{NaCN} RCH_2C\equiv N$$

   (b) Dehydration of amides

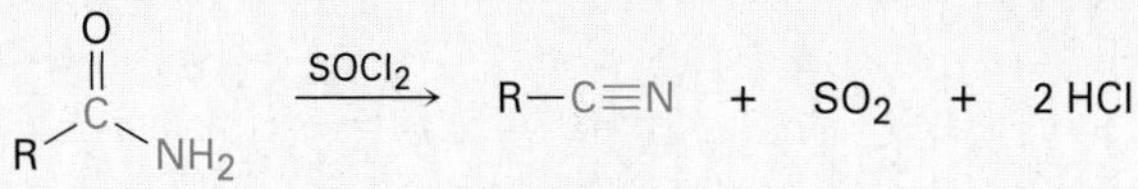

$$\underset{R}{\overset{O}{\|}} C \diagdown_{NH_2} \xrightarrow{SOCl_2} R-C\equiv N \ + \ SO_2 \ + \ 2\,HCl$$

3. Reactions of nitriles (Section 20.7)
   (a) Hydrolysis to yield carboxylic acids

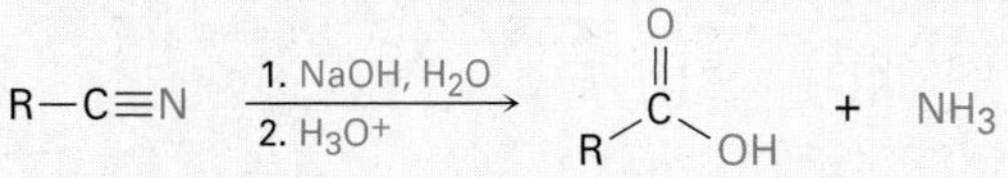

$$R-C\equiv N \xrightarrow[\text{2. } H_3O^+]{\text{1. NaOH, } H_2O} \underset{R}{\overset{O}{\|}} C \diagdown_{OH} \ + \ NH_3$$

   (b) Reduction to yield primary amines

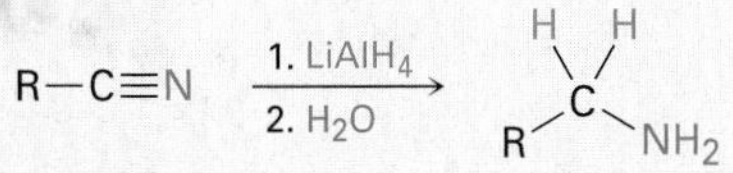

$$R-C\equiv N \xrightarrow[\text{2. } H_2O]{\text{1. } LiAlH_4} \underset{R}{\overset{H \ \ H}{\underset{\,}{C}}} \diagdown_{NH_2}$$

   (c) Reaction with Grignard reagents to yield ketones

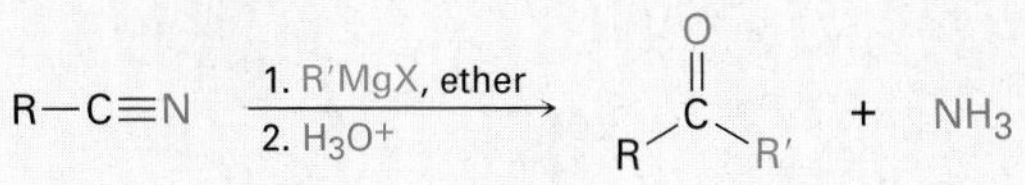

$$R-C\equiv N \xrightarrow[\text{2. } H_3O^+]{\text{1. } R'MgX, \text{ ether}} \underset{R}{\overset{O}{\|}} C \diagdown_{R'} \ + \ NH_3$$

EXERCISES

VISUALIZING CHEMISTRY

(Problems 20.1–20.16 appear within the chapter.)

20.17 ■ Give IUPAC names for the following carboxylic acids (reddish brown = Br):

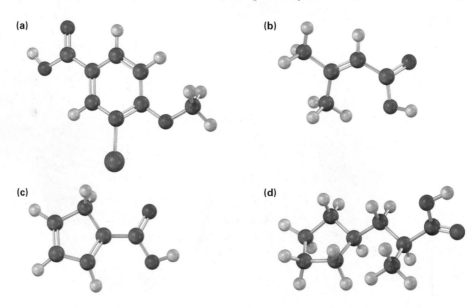

(a)   **(b)**

(c)   **(d)**

20.18 Would you expect the following carboxylic acids to be more acidic or less acidic than benzoic acid? Explain. (Reddish brown = Br.)

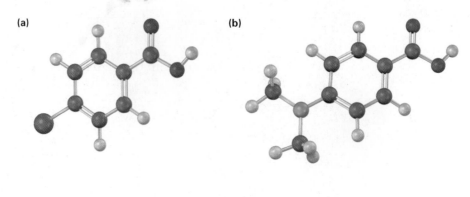

(a)   **(b)**

20.19 The following carboxylic acid can't be prepared from an alkyl halide by either the nitrile hydrolysis route or the Grignard carboxylation route. Explain.

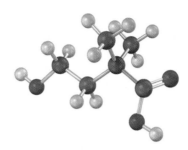

20.20 Electrostatic potential maps of anisole and thioanisole are shown. Which do you think is the stronger acid, *p*-methoxybenzoic acid or *p*-(methylthio)benzoic acid? Explain.

Anisole (C₆H₅OCH₃)   Thioanisole (C₆H₅SCH₃)

ADDITIONAL PROBLEMS

20.21 ■ Give IUPAC names for the following compounds:

(a)   CO₂H   CO₂H
CH₃CHCH₂CH₂CHCH₃

(b)   CH₃
CH₃CCO₂H
CH₃

(c) NC ⟨benzene ring⟩ CO₂H

(d) ⟨cyclodecene ring⟩ CO₂H

(e)   CH₃
CH₃CCN
CH₃

(f)   CH₂CO₂H
CH₃CH₂CH₂CHCH₂CH₃

(g)   Br
BrCH₂CHCH₂CH₂CO₂H

(h) ⟨cyclopentene ring⟩ CN

20.22 ■ Draw structures corresponding to the following IUPAC names:
(a) *cis*-1,2-Cyclohexanedicarboxylic acid   **(b)** Heptanedioic acid
(c) 2-Hexen-4-ynoic acid   **(d)** 4-Ethyl-2-propyloctanoic acid
(e) 3-Chlorophthalic acid   **(f)** Triphenylacetic acid
(g) 2-Cyclobutenecarbonitrile   **(h)** *m*-Benzoylbenzonitrile

20.23 Draw and name the following:
   (a) The eight carboxylic acids with the formula $C_6H_{12}O_2$
   (b) Three nitriles with the formula C_5H_7N

20.24 Isocitric acid, an intermediate in the citric acid cycle of food metabolism, has the systematic name (2R,3S)-3-carboxy-2-hydroxypentanedioic acid. Draw the structure.

20.25 ■ Order the compounds in each of the following sets with respect to increasing acidity:
   (a) Acetic acid, oxalic acid, formic acid
   (b) p-Bromobenzoic acid, p-nitrobenzoic acid, 2,4-dinitrobenzoic acid
   (c) Fluoroacetic acid, 3-fluoropropanoic acid, iodoacetic acid

20.26 ■ Arrange the compounds in each of the following sets in order of increasing basicity:
   (a) Magnesium acetate, magnesium hydroxide, methylmagnesium bromide
   (b) Sodium benzoate, sodium p-nitrobenzoate, sodium acetylide
   (c) Lithium hydroxide, lithium ethoxide, lithium formate

20.27 ■ How could you convert butanoic acid into the following compounds? Write each step showing the reagents needed.
   (a) 1-Butanol   (b) 1-Bromobutane   (c) Pentanoic acid
   (d) 1-Butene   (e) Octane

20.28 ■ How could you convert each of the following compounds into butanoic acid? Write each step showing all reagents.
   (a) 1-Butanol   (b) 1-Bromobutane   (c) 1-Butene
   (d) 1-Bromopropane   (e) 4-Octene

20.29 ■ How could you convert butanenitrile into the following compounds? Write each step showing the reagents needed.
   (a) 1-Butanol   (b) Butylamine   (c) 2-Methyl-3-hexanone

20.30 ■ How would you prepare the following compounds from benzene? More than one step is required in each case.
   (a) m-Chlorobenzoic acid   (b) p-Bromobenzoic acid
   (c) Phenylacetic acid, $C_6H_5CH_2CO_2H$

20.31 ■ Calculate pK_a's for the following acids:
   (a) Lactic acid, $K_a = 8.4 \times 10^{-4}$ (b) Acrylic acid, $K_a = 5.6 \times 10^{-6}$

20.32 ■ Calculate K_a's for the following acids:
   (a) Citric acid, p$K_a = 3.14$ (b) Tartaric acid, p$K_a = 2.98$

20.33 ■ Thioglycolic acid, $HSCH_2CO_2H$, a substance used in depilatory agents (hair removers) has p$K_a = 3.42$. What is the percent dissociation of thioglycolic acid in a buffer solution at pH = 3.0?

20.34 ■ In humans, the final product of purine degradation from DNA is uric acid, p$K_a = 5.61$, which is excreted in the urine. What is the percent dissociation of uric acid in urine at a typical pH = 6.0? Why do you think uric acid is acidic even though it does not have a CO_2H group?

Uric acid

20.35 Shown here are some pK_a data for simple dibasic acids. How can you account for the fact that the difference between the first and second ionization constants decreases with increasing distance between the carboxyl groups?

Name	Structure	pK_1	pK_2
Oxalic	HO_2CCO_2H	1.2	4.2
Succinic	$HO_2CCH_2CH_2CO_2H$	4.2	5.6
Adipic	$HO_2C(CH_2)_4CO_2H$	4.4	5.4

20.36 ■ Predict the product of the reaction of *p*-methylbenzoic acid with each of the following:
(a) $LiAlH_4$, then H_3O^+
(b) *N*-Bromosuccinimide in CCl_4
(c) CH_3MgBr in ether, then H_3O^+
(d) $KMnO_4$, H_3O^+

20.37 Using $^{13}CO_2$ as your only source of labeled carbon, along with any other compounds needed, how would you synthesize the following compounds?
(a) $CH_3CH_2{}^{13}CO_2H$
(b) $CH_3{}^{13}CH_2CO_2H$

20.38 How would you carry out the following transformations?

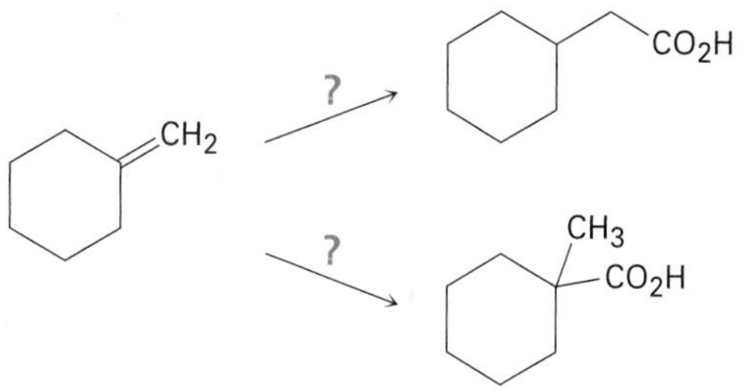

20.39 Which method—Grignard carboxylation or nitrile hydrolysis—would you use for each of the following reactions? Explain.

(a)

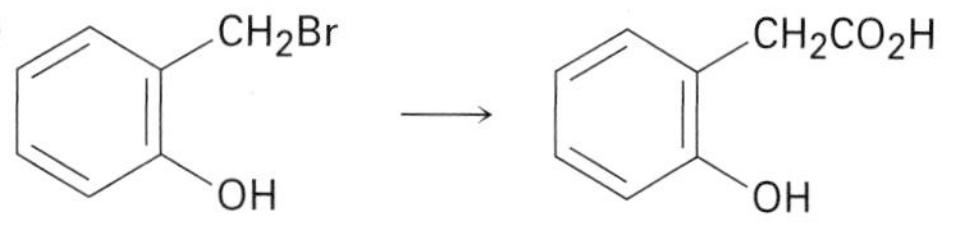

(b)

$$\underset{CH_3CH_2CHCH_3}{\overset{Br}{|}} \longrightarrow \underset{CH_3CH_2CHCO_2H}{\overset{CH_3}{|}}$$

(c)

$$\underset{CH_3CCH_2CH_2CH_2I}{\overset{O}{\|}} \longrightarrow \underset{CH_3CCH_2CH_2CH_2CO_2H}{\overset{O}{\|}}$$

(d) $HOCH_2CH_2CH_2Br \longrightarrow HOCH_2CH_2CH_2CO_2H$

20.40 1,6-Hexanediamine, a starting material needed for making nylon, can be made from 1,3-butadiene. How would you accomplish this synthesis?

$$H_2C=CHCH=CH_2 \overset{?}{\longrightarrow} H_2NCH_2CH_2CH_2CH_2CH_2CH_2NH_2$$

20.41 A chemist in need of 2,2-dimethylpentanoic acid decided to synthesize some by reaction of 2-chloro-2-methylpentane with NaCN, followed by hydrolysis of the product. After the reaction sequence was carried out, however, none of the desired product could be found. What do you suppose went wrong?

20.42 Show how you might prepare the anti-inflammatory agent ibuprofen starting from isobutylbenzene. More than one step is needed.

Isobutylbenzene   **Ibuprofen**

20.43 The following synthetic schemes all have at least one flaw in them. What is wrong with each?

(a)

$$CH_3CH_2\overset{\overset{\displaystyle Br}{|}}{C}HCH_2CH_3 \xrightarrow[\substack{2.\ NaCN \\ 3.\ H_3O^+}]{1.\ Mg} CH_3CH_2\overset{\overset{\displaystyle CO_2H}{|}}{C}HCH_2CH_3$$

(b)

(c)

$$CH_3\overset{\overset{\displaystyle OH}{|}}{\underset{\underset{\displaystyle CH_3}{|}}{C}}CH_2CH_2Cl \xrightarrow[\substack{2.\ H_3O^+}]{1.\ NaCN} CH_3\overset{\overset{\displaystyle OH}{|}}{\underset{\underset{\displaystyle CH_3}{|}}{C}}CH_2CH_2\overset{\overset{\displaystyle O}{||}}{C}OH$$

20.44 Naturally occurring compounds called *cyanogenic glycosides,* such as lotaustralin, release hydrogen cyanide, HCN, when treated with aqueous acid. The reaction occurs by hydrolysis of the acetal linkage to form a cyanohydrin, which then expels HCN and gives a carbonyl compound.
(a) Show the mechanism of the acetal hydrolysis and the structure of the cyanohydrin that results.
(b) Propose a mechanism for the loss of HCN, and show the structure of the carbonyl compound that forms.

Lotaustralin

20.45 Acid-catalyzed hydrolysis of a nitrile to give a carboxylic acid occurs by initial protonation of the nitrogen atom, followed by nucleophilic addition of water. Review the mechanism of base-catalyzed nitrile hydrolysis in Section 20.7, and then write all the steps involved in the acid-catalyzed reaction, using curved arrows to represent electron flow in each step.

20.46 *p*-Aminobenzoic acid (PABA) is widely used as a sunscreen agent. Propose a synthesis of PABA starting from toluene.

20.47 Propose a synthesis of the anti-inflammatory drug Fenclorac from phenyl-cyclohexane.

Fenclorac

■ Assignable in OWL

20.48 The pK_a's of five p-substituted benzoic acids ($YC_6H_4CO_2H$) follow. Rank the corresponding substituted benzenes (YC_6H_5) in order of their increasing reactivity toward electrophilic aromatic substitution. If benzoic acid has $pK_a = 4.19$, which of the substituents are activators and which are deactivators?

Substituent Y	pK_a of 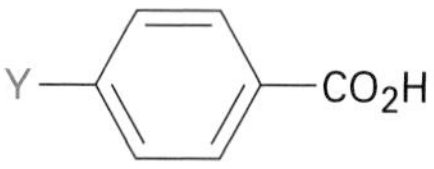 $Y-\!\!\bigcirc\!\!-CO_2H$
$-Si(CH_3)_3$	4.27
$-CH=CHC\equiv N$	4.03
$-HgCH_3$	4.10
$-OSO_2CH_3$	3.84
$-PCl_2$	3.59

20.49 How would you carry out the following transformations? More than one step is required in each case.

(a)

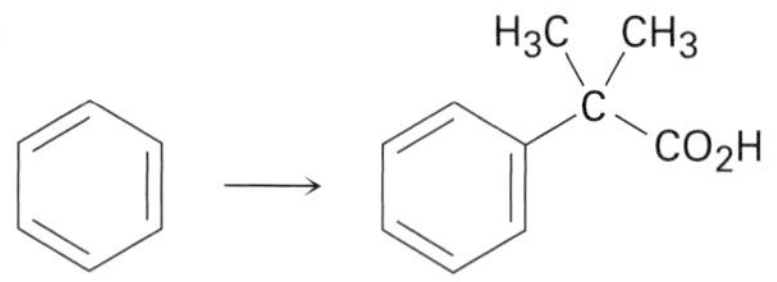

(b)

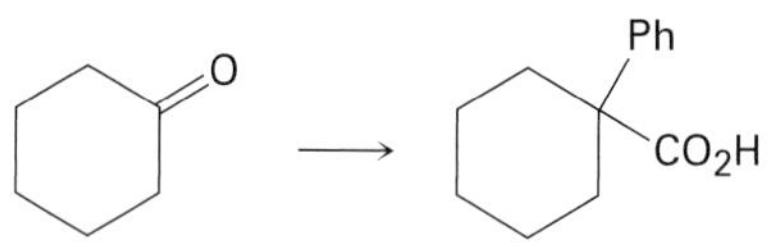

20.50 The following pK_a values have been measured. Explain why a hydroxyl group in the para position decreases the acidity while a hydroxyl group in the meta position increases the acidity.

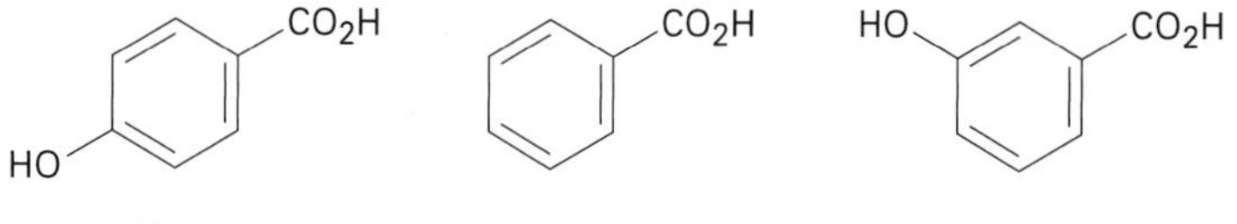

$pK_a = 4.48$ $pK_a = 4.19$ $pK_a = 4.07$

20.51 3-Methyl-2-hexenoic acid (mixture of E and Z isomers) has been identified as the substance responsible for the odor of human sweat. Synthesize the compound from starting materials having five or fewer carbons.

20.52 Identify the missing reagents **a–f** in the following scheme:

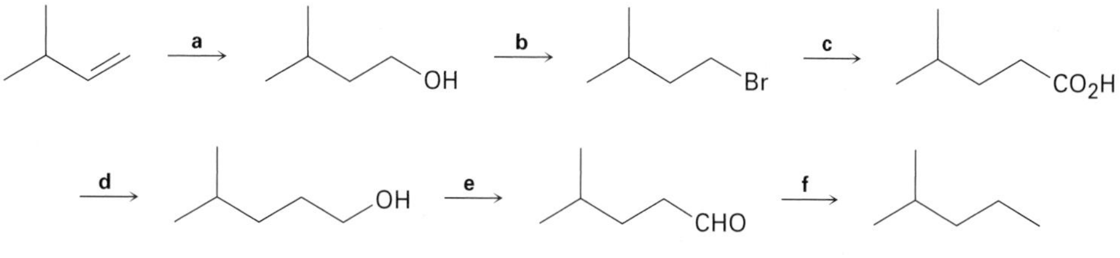

20.53 ■ 2-Bromo-6,6-dimethylcyclohexanone gives 2,2-dimethylcyclopentane-carboxylic acid on treatment with aqueous NaOH followed by acidification, a process called the *Favorskii reaction*. Propose a mechanism.

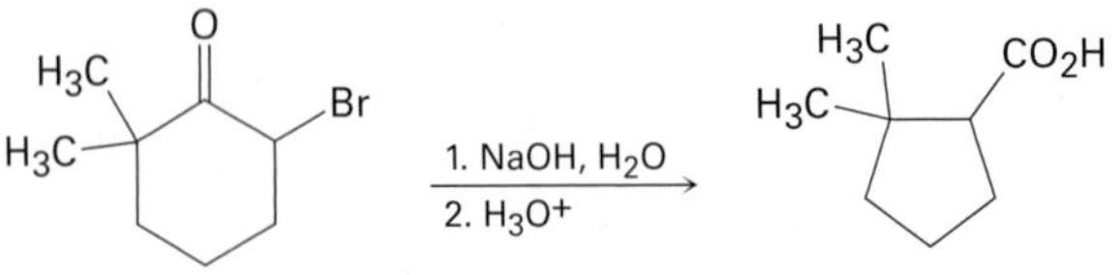

20.54 In plants, terpenes (see Chapter 6 *Focus On*) are biosynthesized by a pathway that involves loss of CO_2 from 3-phosphomevalonate 5-diphosphate to yield isopentenyl diphosphate. Use curved arrows to show the mechanism of this reaction.

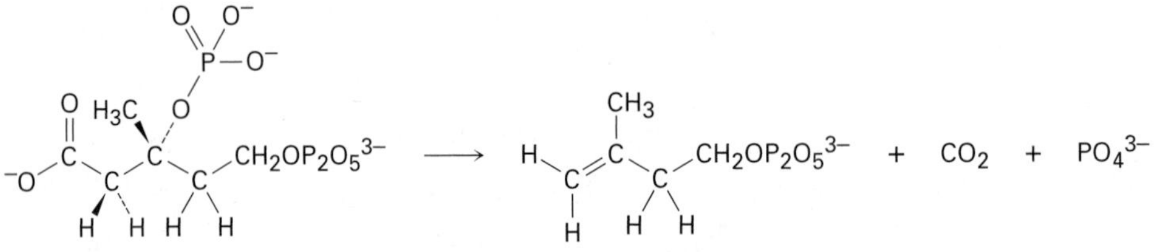

3-Phosphomevalonate 5-diphosphate

Isopentenyl diphosphate

20.55 ■ Propose a structure for a compound $C_6H_{12}O_2$ that dissolves in dilute NaOH and shows the following ^1H NMR spectrum: 1.08 δ (9 H, singlet), 2.2 δ (2 H, singlet), and 11.2 δ (1 H, singlet).

20.56 What spectroscopic method could you use to distinguish among the following three isomeric acids? Tell what characteristic features you would expect for each acid.

$$CH_3(CH_2)_3CO_2H \qquad (CH_3)_2CHCH_2CO_2H \qquad (CH_3)_3CCO_2H$$

Pentanoic acid   **3-Methylbutanoic acid**   **2,2-Dimethylpropanoic acid**

20.57 How would you use NMR (either ^{13}C or ^1H) to distinguish between the following pairs of isomers?

(a)

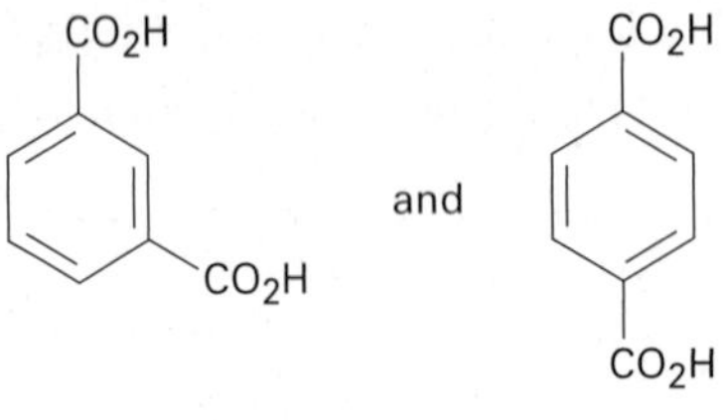

and

(b) $HO_2CCH_2CH_2CO_2H$ and $CH_3CH(CO_2H)_2$

(c) $CH_3CH_2CH_2CO_2H$ and $HOCH_2CH_2CH_2CHO$

(d) $(CH_3)_2C{=}CHCH_2CO_2H$ and

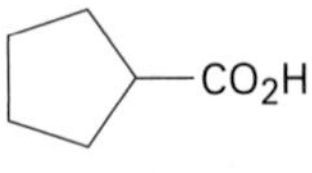

20.58 ■ Compound **A**, $C_4H_8O_3$, has infrared absorptions at 1710 and 2500 to 3100 cm^{-1} and has the ^{1}H NMR spectrum shown. Propose a structure for **A**.

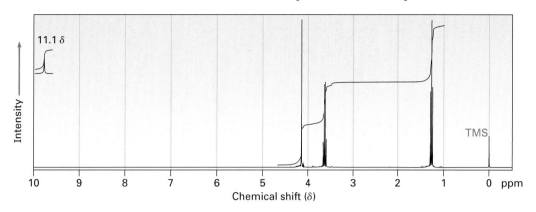

20.59 ■ Propose a structure for a compound, C_4H_7N, that has the following IR and ^{1}H NMR spectra:

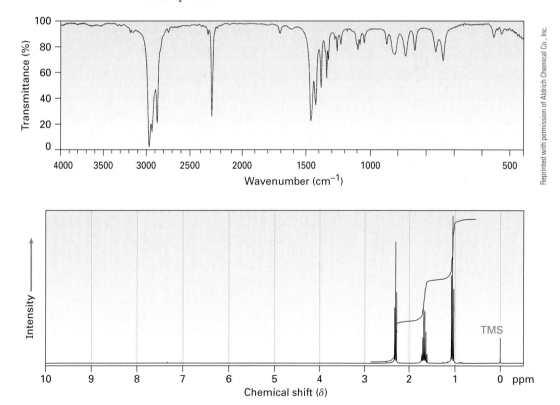

20.60 ■ The two ^1H NMR spectra shown here belong to crotonic acid (*trans*-$CH_3CH=CHCO_2H$) and methacrylic acid [$H_2C=C(CH_3)CO_2H$]. Which spectrum corresponds to which acid? Explain.

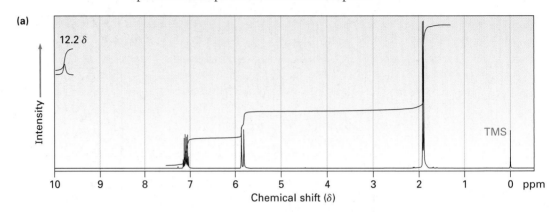

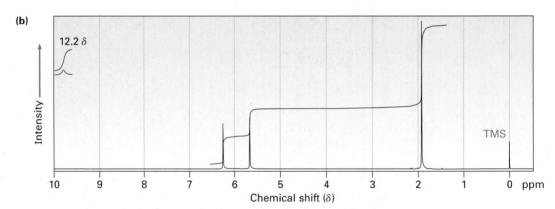

20.61 ■ Propose structures for carboxylic acids that show the following peaks in their ^{13}C NMR spectra. Assume that the kinds of carbons (1°, 2°, 3°, or 4°) have been assigned by DEPT-NMR.
 (a) $C_7H_{12}O_2$: 25.5 δ (2°), 25.9 δ (2°), 29.0 δ (2°), 43.1 δ (3°), 183.0 δ (4°)
 (b) $C_8H_8O_2$: 21.4 δ (1°), 128.3 δ (4°), 129.0 δ (3°), 129.7 δ (3°), 143.1 δ (4°), 168.2 δ (4°)

20.62 Carboxylic acids having a second carbonyl group two atoms away lose CO_2 (*decarboxylate*) through an intermediate enolate ion when treated with base. Write the mechanism of this decarboxylation reaction using curved arrows to show the electron flow in each step.

$$CH_3\overset{O}{\overset{||}{C}}CH_2\overset{O}{\overset{||}{C}}OH \xrightarrow[\text{H}_2\text{O}]{\text{NaOH}} \left[CH_3\overset{O^-}{\overset{|}{C}}=CH_2 \right] + CO_2 \xrightarrow{\text{H}_2\text{O}} CH_3\overset{O}{\overset{||}{C}}CH_3$$

An enolate ion

21

Carboxylic Acid Derivatives: Nucleophilic Acyl Substitution Reactions

Organic **KNOWLEDGE TOOLS**

CENGAGENOW Throughout this chapter, sign in at **www.cengage.com/login** for online self-study and interactive tutorials based on your level of understanding.

OWL Online homework for this chapter may be assigned in Organic OWL.

Closely related to the carboxylic acids and nitriles discussed in the previous chapter are the **carboxylic acid derivatives**, compounds in which an acyl group is bonded to an electronegative atom or substituent that can act as a leaving group in a substitution reaction. Many kinds of acid derivatives are known, but we'll be concerned primarily with four of the more common ones: **acid halides, acid anhydrides, esters,** and **amides**. Esters and amides are common in both laboratory and biological chemistry, while acid halides and acid anhydrides are used only in the laboratory. **Thioesters** and **acyl phosphates** are encountered primarily in biological chemistry. Note the structural similarity between acid anhydrides and acyl phosphates.

Carboxylic acid	**Acid halide** (X = Cl, Br)	**Acid anhydride**	**Ester**
Amide	**Thioester**	**Acyl phosphate**	

The chemistry of all acid derivatives is similar and is dominated by a single reaction—the nucleophilic acyl substitution reaction that we saw briefly in *A Preview of Carbonyl Compounds.*

$$
R-\underset{\underset{Y}{}}{\overset{\overset{O}{\|}}{C}} \ + \ :Nu^- \ \longrightarrow \ R-\underset{\underset{Nu}{}}{\overset{\overset{O}{\|}}{C}} \ + \ :Y^-
$$

WHY THIS CHAPTER?

Carboxylic acid derivatives are among the most widespread of all molecules, both in laboratory chemistry and in biological pathways. Thus, a study of them and their primary reaction—nucleophilic acyl substitution—is fundamental to understanding organic chemistry. We'll begin this chapter by first learning about carboxylic acid derivatives, and then we'll explore the chemistry of acyl substitution reactions.

21.1 | Naming Carboxylic Acid Derivatives

Acid Halides, RCOX

CENGAGENOW™ Click *Organic Interactive* to **use a web-based palette to draw structures of acyl derivatives based on their IUPAC names.**

Acid halides are named by identifying first the acyl group and then the halide. The acyl group name is derived from the carboxylic acid name by replacing the *-ic acid* ending with *-yl* or the *-carboxylic acid* ending with *-carbonyl,* as described previously in Section 20.1 and shown in Table 20.1 on page 753. For example:

| **Acetyl chloride** | **Benzoyl bromide** | **Cyclohexanecarbonyl chloride** |

Acid Anhydrides, RCO₂COR′

Symmetrical anhydrides of unsubstituted monocarboxylic acids and cyclic anhydrides of dicarboxylic acids are named by replacing the word *acid* with *anhydride.*

| **Acetic anhydride** | **Benzoic anhydride** | **Succinic anhydride** |

Unsymmetrical anhydrides—those prepared from two different carboxylic acids—are named by citing the two acids alphabetically and then adding *anhydride.*

Acetic benzoic anhydride

Amides, RCONH$_2$

Amides with an unsubstituted −NH$_2$ group are named by replacing the -*oic acid* or -*ic acid* ending with -*amide*, or by replacing the -*carboxylic acid* ending with -*carboxamide*.

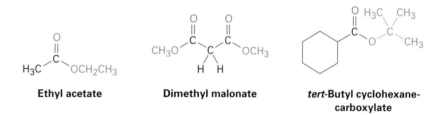

Acetamide   **Hexanamide**   **Cyclopentane-
                                                              carboxamide**

If the nitrogen atom is further substituted, the compound is named by first identifying the substituent groups and then the parent amide. The substituents are preceded by the letter *N* to identify them as being directly attached to nitrogen.

N-Methyl**propanamide**   *N,N*-Diethyl**cyclohexanecarboxamide**

Esters, RCO$_2$R′

Esters are named by first identifying the alkyl group attached to oxygen and then the carboxylic acid, with the -*ic acid* ending replaced by -*ate*.

Ethyl acetate   **Dimethyl malonate**   ***tert*-Butyl cyclohexane-
                                                              carboxylate**

Thioesters, RCOSR′

Thioesters are named like the corresponding esters. If the related ester has a common name, the prefix *thio*- is added to the name of the carboxylate; acetate becomes thioacetate, for instance. If the related ester has a systematic name, the -*oate* or -*carboxylate* ending is replaced by -*thioate* or -*carbothioate*; butanoate becomes butanethioate and cyclohexanecarboxylate becomes cyclohexanecarbothioate, for instance.

Methyl **thioacetate**   Ethyl **butanethioate**   Methyl **cyclohexane-
                                                              carbothioate**

Acyl Phosphates, $RCO_2PO_3^{2-}$ and $RCO_2PO_3R'^-$

Acyl phosphates are named by citing the acyl group and adding the word *phosphate*. If an alkyl group is attached to one of the phosphate oxygens, it is identified after the name of the acyl group. In biological chemistry, acyl adenosyl phosphates are particularly common.

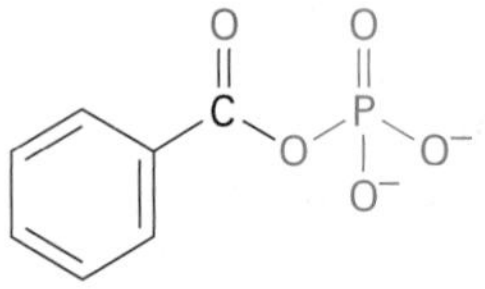

Benzoyl phosphate

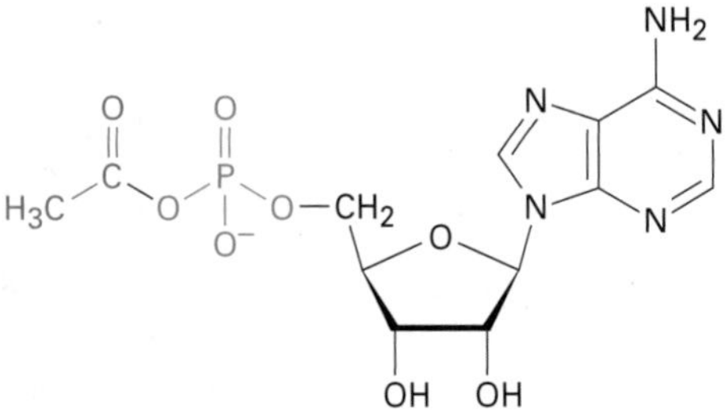

Acetyl adenosyl phosphate

A summary of nomenclature rules for carboxylic acid derivatives is given in Table 21.1.

Table 21.1 | **Nomenclature of Carboxylic Acid Derivatives**

Functional group	Structure	Name ending
Carboxylic acid	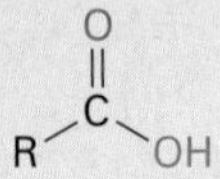	*-ic acid* (*-carboxylic acid*)
Acid halide	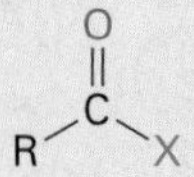	*-oyl halide* (*-carbonyl halide*)
Acid anhydride	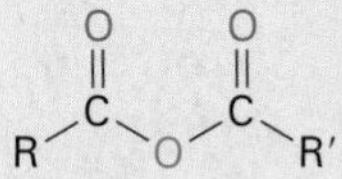	*anhydride*
Amide	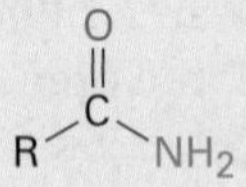	*-amide* (*-carboxamide*)
Ester	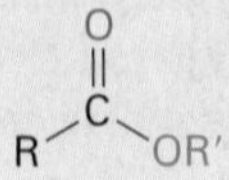	*-ate* (*-carboxylate*)
Thioester	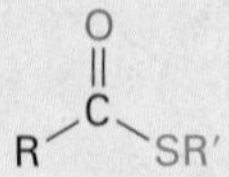	*-thioate* (*-carbothioate*)
Acyl phosphate	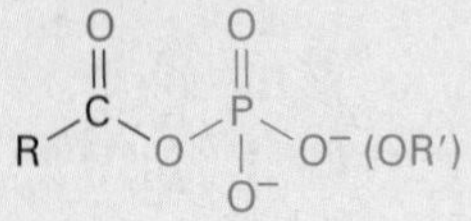	*-yl phosphate*

Problem 21.1 | Give IUPAC names for the following substances:

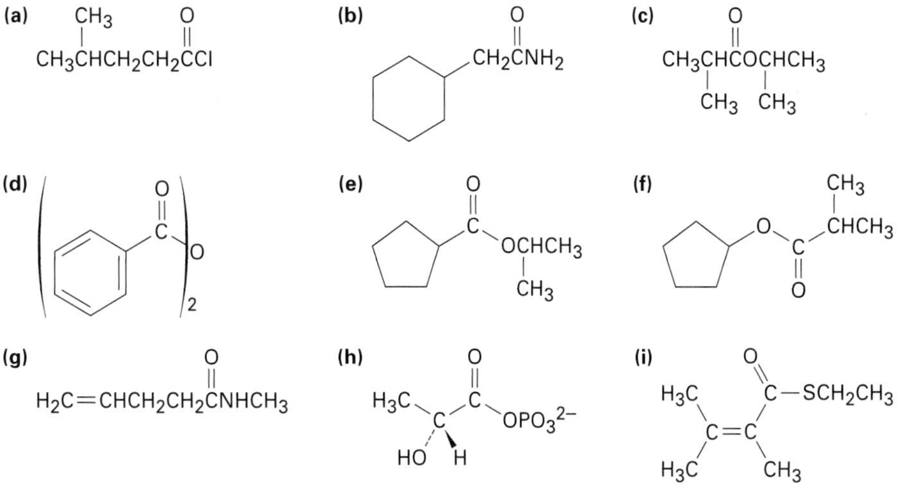

(a)

$$CH_3CHCH_2CH_2CCl$$

with CH_3 and O labeled

(b)

CH_2CNH_2 attached to cyclohexane, with O

(c)

$$CH_3CHCOCHCH_3$$
$$CH_3 \quad CH_3$$

with O

(d)

benzoate structure with O, C, O in brackets, subscript 2

(e)

cyclopentane with C, O, $OCHCH_3$ and CH_3

(f)

cyclopentane with O, C, $CHCH_3$, CH_3, O

(g)

$$H_2C=CHCH_2CH_2CNHCH_3$$

with O

(h)

$$H_3C \quad C \quad C \quad OPO_3^{2-}$$
$$HO \quad H$$

with O

(i)

$$H_3C \quad C-SCH_2CH_3$$
$$C=C$$
$$H_3C \quad CH_3$$

with O

Problem 21.2 | Draw structures corresponding to the following names:
(a) Phenyl benzoate
(b) N-Ethyl-N-methylbutanamide
(c) 2,4-Dimethylpentanoyl chloride
(d) Methyl 1-methylcyclohexanecarboxylate
(e) Ethyl 3-oxopentanoate
(f) Methyl p-bromobenzenethioate
(g) Formic propanoic anhydride
(h) cis-2-Methylcyclopentanecarbonyl bromide

21.2 | Nucleophilic Acyl Substitution Reactions

CENGAGENOW˙ Click *Organic Interactive* to **learn to predict the course of an acyl transfer reaction by examining reactants and leaving groups.**

The addition of a nucleophile to a polar C=O bond is the key step in three of the four major carbonyl-group reactions. We saw in Chapter 19 that when a nucleophile adds to an aldehyde or ketone, the initially formed tetrahedral intermediate either can be protonated to yield an alcohol or can eliminate the carbonyl oxygen, leading to a new C=Nu bond. When a nucleophile adds to a carboxylic acid derivative, however, a different reaction course is followed. The initially formed tetrahedral intermediate eliminates one of the two substituents originally bonded to the carbonyl carbon, leading to a net **nucleophilic acyl substitution reaction** (Figure 21.1).

The difference in behavior between aldehydes/ketones and carboxylic acid derivatives is a consequence of structure. Carboxylic acid derivatives have an acyl carbon bonded to a group −Y that can leave as a stable anion. As soon as the tetrahedral intermediate is formed, the leaving group is expelled to generate a new carbonyl compound. Aldehydes and ketones have no such leaving group, however, and therefore don't undergo substitution.

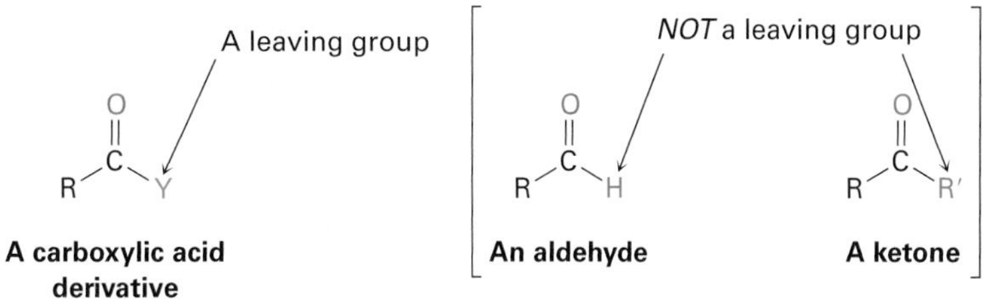

A leaving group

$$\begin{array}{c} O \\ \| \\ R-C-Y \end{array}$$

A carboxylic acid derivative

NOT a leaving group

$$\left[\begin{array}{cc} O & O \\ \| & \| \\ R-C-H & R-C-R' \end{array} \right]$$

An aldehyde   **A ketone**

Figure 21.1 MECHANISM:
General mechanism of a nucleophilic acyl substitution reaction.

CENGAGENOW™ Click *Organic Process* to **view animations showing chemistry of the acyl transfer process.**

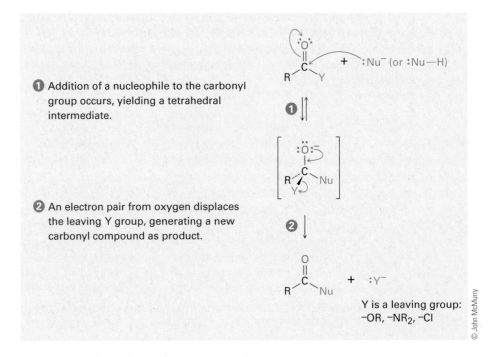

① Addition of a nucleophile to the carbonyl group occurs, yielding a tetrahedral intermediate.

② An electron pair from oxygen displaces the leaving Y group, generating a new carbonyl compound as product.

Y is a leaving group: −OR, −NR$_2$, −Cl

© John McMurry

The net effect of the addition/elimination sequence is a substitution of the nucleophile for the −Y group originally bonded to the acyl carbon. Thus, the overall reaction is superficially similar to the kind of nucleophilic substitution that occurs during an S_N2 reaction (Section 11.3), but the *mechanisms* of the two reactions are completely different. An S_N2 reaction occurs in a single step by backside displacement of the leaving group; a nucleophilic acyl substitution takes place in two steps and involves a tetrahedral intermediate.

Problem 21.3 | Show the mechanism of the following nucleophilic acyl substitution reaction, using curved arrows to indicate the electron flow in each step:

$$\text{PhC(=O)Cl} \xrightarrow[\text{CH}_3\text{OH}]{\text{Na}^+ \ ^-\text{OCH}_3} \text{PhC(=O)OCH}_3$$

Relative Reactivity of Carboxylic Acid Derivatives

Both the initial addition step and the subsequent elimination step can affect the overall rate of a nucleophilic acyl substitution reaction, but the addition step is generally the rate-limiting one. Thus, any factor that makes the carbonyl group more reactive toward nucleophiles favors the substitution process.

Steric and electronic factors are both important in determining reactivity. Sterically, we find within a series of similar acid derivatives that unhindered, accessible carbonyl groups react with nucleophiles more readily than do sterically hindered groups. The reactivity order is

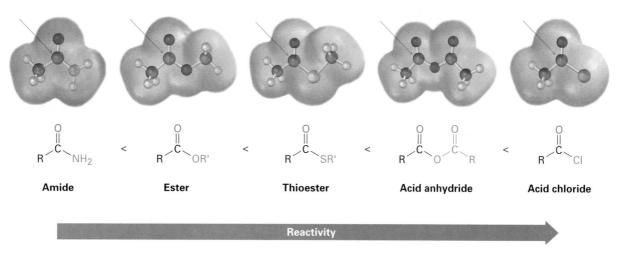

Electronically, we find that strongly polarized acyl compounds react more readily than less polar ones. Thus, acid chlorides are the most reactive because the electronegative chlorine atom withdraws electrons from the carbonyl carbon, whereas amides are the least reactive. Although subtle, electrostatic potential maps of various carboxylic acid derivatives indicate the differences by the relative blueness on the C=O carbons. Acyl phosphates are hard to place on this scale because they are not used in the laboratory, but in biological systems they appear to be somewhat more reactive than thioesters.

The way in which various substituents affect the polarization of a carbonyl group is similar to the way they affect the reactivity of an aromatic ring toward electrophilic substitution (Section 16.5). A chlorine substituent, for example, inductively *withdraws* electrons from an acyl group in the same way that it withdraws electrons from and thus deactivates an aromatic ring. Similarly, amino, methoxyl, and methylthio substituents *donate* electrons to acyl groups by resonance in the same way that they donate electrons to and thus activate aromatic rings.

As a consequence of these reactivity differences, it's usually possible to convert a more reactive acid derivative into a less reactive one. Acid chlorides, for instance, can be directly converted into anhydrides, thioesters, esters, and amides, but amides can't be directly converted into esters, thioesters, anhydrides, or acid chlorides. Remembering the reactivity order is therefore a way to keep track of a large number of reactions (Figure 21.2). Another consequence, as noted previously, is that only acyl phosphates, thioesters, esters, and amides are commonly found in nature. Acid halides and acid anhydrides react with water so rapidly that they can't exist for long in living organisms.

Figure 21.2 Interconversions of carboxylic acid derivatives. A more reactive acid derivative can be converted into a less reactive one, but not vice versa.

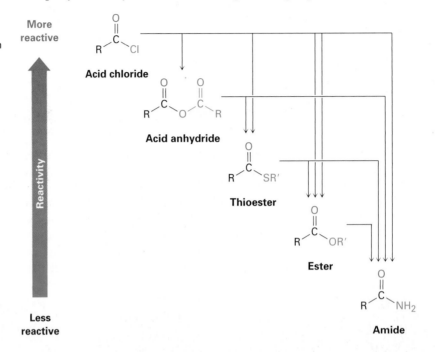

In studying the chemistry of carboxylic acid derivatives in the next few sections, we'll be concerned largely with the reactions of just a few nucleophiles and will see that the same kinds of reactions keep occurring (Figure 21.3).

▌ **Hydrolysis**   Reaction with water to yield a carboxylic acid

▌ **Alcoholysis**   Reaction with an alcohol to yield an ester

▌ **Aminolysis**   Reaction with ammonia or an amine to yield an amide

▌ **Reduction**   Reaction with a hydride reducing agent to yield an aldehyde or an alcohol

▌ **Grignard reaction** Reaction with an organometallic reagent to yield a ketone or an alcohol

Figure 21.3 Some general reactions of carboxylic acid derivatives.

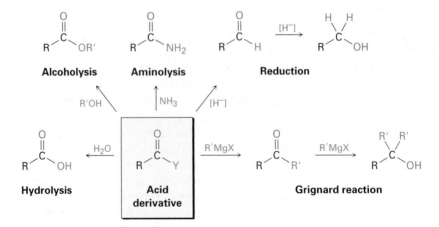

WORKED EXAMPLE 21.1   *Predicting the Product of a Nucleophilic Acyl Substitution Reaction*

Predict the product of the following nucleophilic acyl substitution reaction of benzoyl chloride with 2-propanol:

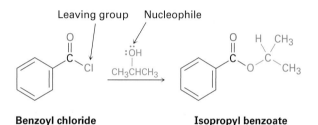

Benzoyl chloride

Strategy   A nucleophilic acyl substitution reaction involves the substitution of a nucleophile for a leaving group in a carboxylic acid derivative. Identify the leaving group (Cl^- in the case of an acid chloride) and the nucleophile (an alcohol in this case), and replace one by the other. The product is isopropyl benzoate.

Solution

Leaving group   Nucleophile

Benzoyl chloride   **Isopropyl benzoate**

Problem 21.4   Rank the compounds in each of the following sets in order of their expected reactivity toward nucleophilic acyl substitution:

(a) CH_3CCl, CH_3COCH_3, CH_3CNH_2
(each with an O double-bonded to the central C)

(b) CH_3COCH_3, $CH_3COCH_2CCl_3$, $CH_3COCH(CF_3)_2$
(each with an O double-bonded to the central C)

Problem 21.5   Predict the products of the following nucleophilic acyl substitution reactions:

(a) $H_3C-C(=O)-OCH_3$ $\xrightarrow[\text{H}_2\text{O}]{\text{NaOH}}$?

(b) $H_3C-C(=O)-Cl$ $\xrightarrow{\text{NH}_3}$?

(c) $H_3C-C(=O)-O-C(=O)-CH_3$ $\xrightarrow[\text{CH}_3\text{OH}]{\text{Na}^+ \text{ }^-\text{OCH}_3}$?

(d) $H_3C-C(=O)-SCH_3$ $\xrightarrow{\text{CH}_3\text{NH}_2}$?

Problem 21.6 | The following structure represents a tetrahedral alkoxide ion intermediate formed by addition of a nucleophile to a carboxylic acid derivative. Identify the nucleophile, the leaving group, the starting acid derivative, and the ultimate product.

21.3 | Nucleophilic Acyl Substitution Reactions of Carboxylic Acids

The direct nucleophilic acyl substitution of a carboxylic acid is difficult in the laboratory because −OH is a poor leaving group (Section 11.3). Thus, it's usually necessary to enhance the reactivity of the acid, either by using a strong acid catalyst to protonate the carboxyl and make it a better acceptor or by converting the −OH into a better leaving group. Under the right circumstances, however, acid chlorides, anhydrides, esters, and amides can all be prepared from carboxylic acids.

Conversion of Carboxylic Acids into Acid Chlorides

Carboxylic acids are converted into acid chlorides by treatment with thionyl chloride, $SOCl_2$.

2,4,6-Trimethylbenzoic acid → **2,4,6-Trimethylbenzoyl chloride (90%)** + HCl + SO_2

(reagents: $SOCl_2$ / $CHCl_3$)

The reaction occurs by a nucleophilic acyl substitution pathway in which the carboxylic acid is first converted into a chlorosulfite intermediate, thereby replacing the −OH of the acid with a much better leaving group. The chlorosulfite then reacts with a nucleophilic chloride ion. You might recall from Section 17.6 that an analogous chlorosulfite is involved in reaction of an alcohol with $SOCl_2$ to yield an alkyl chloride.

Carboxylic acid

An acyl chlorosulfite

Acid chloride

Conversion of Carboxylic Acids into Acid Anhydrides

Acid anhydrides can be derived from two molecules of carboxylic acid by strong heating to remove 1 equivalent of water. Because of the high temperatures needed, however, only acetic anhydride is commonly prepared this way.

Acetic acid   **Acetic anhydride**

Conversion of Carboxylic Acids into Esters

Perhaps the most useful reaction of carboxylic acids is their conversion into esters. There are many methods for accomplishing the transformation, including the S_N2 reaction of a carboxylate anion with a primary alkyl halide that we saw in Section 11.3.

Sodium butanoate   **Methyl butanoate (97%)**

Esters can also be synthesized by an acid-catalyzed nucleophilic acyl substitution reaction of a carboxylic acid with an alcohol, a process called the **Fischer esterification reaction**. Unfortunately, the need to use an excess of a liquid alcohol as solvent effectively limits the method to the synthesis of methyl, ethyl, propyl, and butyl esters.

Mandelic acid   **Ethyl mandelate (86%)**

Emil Fischer

Emil Fischer (1852–1919) was perhaps the finest organic chemist who has ever lived. Born in Euskirchen, Germany, he received his Ph.D. in 1874 at the University of Strasbourg with Adolf von Baeyer. He was professor of chemistry at the universities of Erlangen, Würzburg, and Berlin, where he carried out the research on sugars and purines that led to his receipt of the 1902 Nobel Prize in chemistry. During World War I, Fischer organized the German production of chemicals for the war effort, but the death of two sons in the war led to his depression and suicide.

The mechanism of the Fischer esterification reaction is shown in Figure 21.4. Carboxylic acids are not reactive enough to undergo nucleophilic addition directly, but their reactivity is greatly enhanced in the presence of a strong acid such as HCl or H_2SO_4. The mineral acid protonates the carbonyl-group oxygen atom, thereby giving the carboxylic acid a positive charge and rendering it much more reactive. Subsequent loss of water from the tetrahedral intermediate yields the ester product.

The net effect of Fischer esterification is substitution of an —OH group by —OR'. All steps are reversible, and the reaction can be driven in either direction by choice of reaction conditions. Ester formation is favored when a large excess of alcohol is used as solvent, but carboxylic acid formation is favored when a large excess of water is present.

Figure 21.4 MECHANISM:
Mechanism of Fischer esterification. The reaction is an acid-catalyzed, nucleophilic acyl substitution of a carboxylic acid.

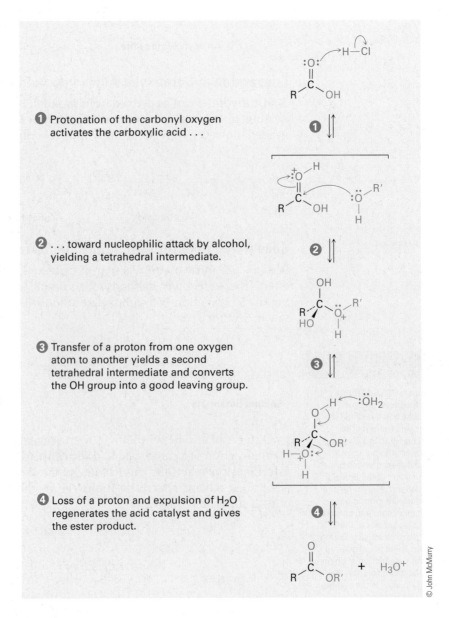

① Protonation of the carbonyl oxygen activates the carboxylic acid . . .

② . . . toward nucleophilic attack by alcohol, yielding a tetrahedral intermediate.

③ Transfer of a proton from one oxygen atom to another yields a second tetrahedral intermediate and converts the OH group into a good leaving group.

④ Loss of a proton and expulsion of H_2O regenerates the acid catalyst and gives the ester product.

© John McMurry

Evidence in support of the mechanism shown in Figure 21.4 comes from isotope-labeling experiments. When ^{18}O-labeled methanol reacts with benzoic acid, the methyl benzoate produced is found to be ^{18}O-labeled but the water produced is unlabeled. Thus, it is the C−OH bond of the carboxylic acid that is broken during the reaction rather than the CO−H bond and the RO−H bond of the alcohol that is broken rather than the R−OH bond.

Problem 21.7 | How might you prepare the following esters from the corresponding acids?

(a)

$$H_3C-\overset{\overset{\displaystyle O}{\|}}{C}-O-CH_2CH_2CH_2CH_3$$

(b)

$$CH_3CH_2CH_2-\overset{\overset{\displaystyle O}{\|}}{C}-O-CH_3$$

(c)

Problem 21.8 | If the following molecule is treated with acid catalyst, an intramolecular esterification reaction occurs. What is the structure of the product? (*Intramolecular* means within the same molecule.)

Conversion of Carboxylic Acids into Amides

Amides are difficult to prepare by direct reaction of carboxylic acids with amines because amines are bases that convert acidic carboxyl groups into their unreactive carboxylate anions. Thus, the −OH must be replaced by a better, nonacidic leaving group. In practice, amides are usually prepared by treating the carboxylic acid with dicyclohexylcarbodiimide (DCC) to activate it, followed by addition of the amine. The acid first adds to a C=N bond of DCC, and nucleophilic acyl substitution by amine then ensues, as shown in Figure 21.5. Alternatively, and depending on the reaction solvent, the reactive acyl intermediate might also react with a second equivalent of carboxylate ion to generate an acid anhydride that then reacts with the amine.

Figure 21.5 MECHANISM:
Mechanism of amide formation
by reaction of a carboxylic acid
and an amine with dicyclohexylcarbodiimide (DCC).

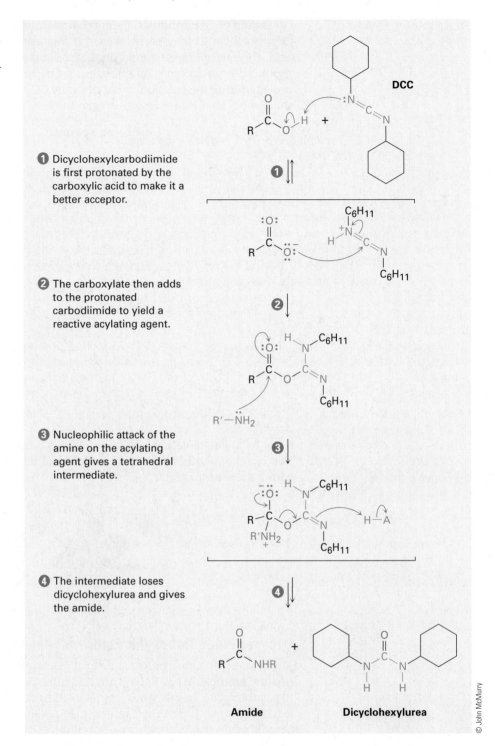

1 Dicyclohexylcarbodiimide
is first protonated by the
carboxylic acid to make it a
better acceptor.

2 The carboxylate then adds
to the protonated
carbodiimide to yield a
reactive acylating agent.

3 Nucleophilic attack of the
amine on the acylating
agent gives a tetrahedral
intermediate.

4 The intermediate loses
dicyclohexylurea and gives
the amide.

Amide   **Dicyclohexylurea**

© John McMurry

We'll see in Section 26.7 that this DCC-induced method of amide formation
is the key step in the laboratory synthesis of small proteins, or *peptides*. For
instance, when one amino acid with its NH$_2$ rendered unreactive and a second

amino acid with its $-CO_2H$ rendered unreactive are treated with DCC, a dipeptide is formed.

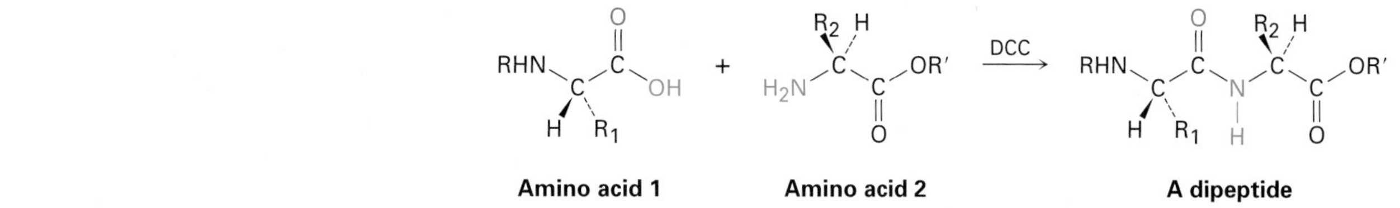

Amino acid 1   **Amino acid 2**   **A dipeptide**

Conversion of Carboxylic Acids into Alcohols

We said in Section 17.4 that carboxylic acids are reduced by $LiAlH_4$ to give primary alcohols, but we deferred a discussion of the reaction mechanism at that time. In fact, the reduction is a nucleophilic acyl substitution reaction in which $-H$ replaces $-OH$ to give an aldehyde, which is further reduced to a primary alcohol by nucleophilic addition. The aldehyde intermediate is much more reactive than the starting acid, so it reacts immediately and is not isolated.

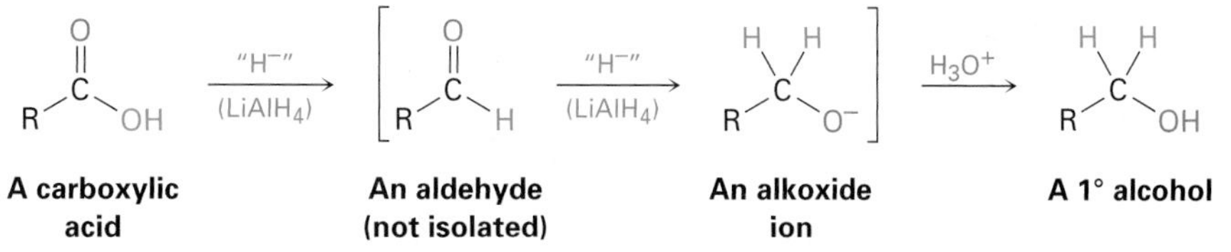

| A carboxylic acid | An aldehyde (not isolated) | An alkoxide ion | A 1° alcohol |

Because hydride ion is a base as well as a nucleophile, the actual nucleophilic acyl substitution step takes place on the carboxylate ion rather than on the free carboxylic acid and gives a high-energy *dianion* intermediate. In this intermediate, the two oxygens are undoubtedly complexed to a Lewis acidic aluminum species. Thus, the reaction is relatively difficult, and acid reductions require higher temperatures and extended reaction times.

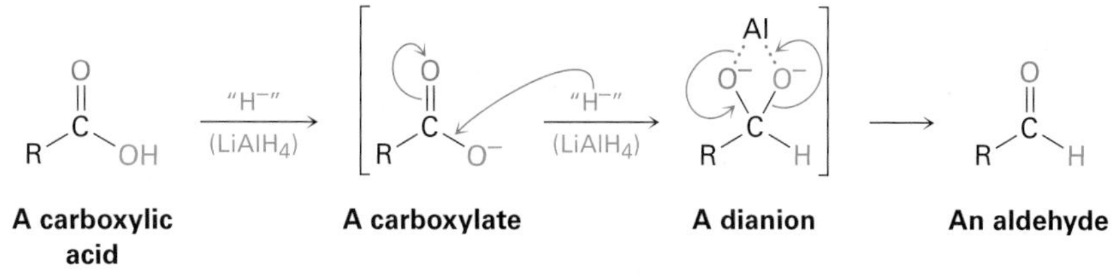

| A carboxylic acid | A carboxylate | A dianion | An aldehyde |

Alternatively, borane in tetrahydrofuran (BH_3/THF) is a useful reagent for reducing carboxylic acids to primary alcohols. Reaction of an acid with BH_3/THF occurs rapidly at room temperature, and the procedure is often preferred to reduction with $LiAlH_4$ because of its relative ease and safety. Borane reacts with carboxylic acids faster than with any other functional group, thereby allowing selective transformations such as that shown below on *p*-nitrophenylacetic acid. If the reduction of *p*-nitrophenylacetic acid were done with $LiAlH_4$, both nitro and carboxyl groups would be reduced.

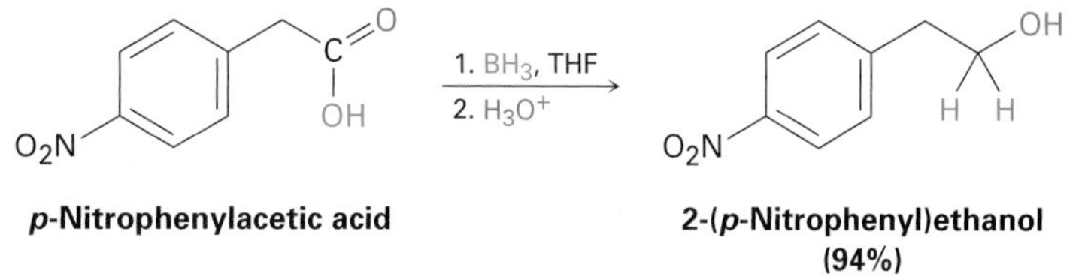

1. BH_3, THF
2. H_3O^+

***p*-Nitrophenylacetic acid**   **2-(*p*-Nitrophenyl)ethanol (94%)**

Biological Conversions of Carboxylic Acids

The direct conversion of a carboxylic acid to an acyl derivative by nucleophilic acyl substitution does not occur in biological chemistry. As in the laboratory, the acid must first be activated. This activation is often accomplished in living organisms by reaction of the acid with ATP to give an acyl adenosyl phosphate, or *acyl adenylate*. In the biosynthesis of fats, for example, a long-chain carboxylic acid reacts with ATP to give an acyl adenylate, which then reacts by subsequent nucleophilic acyl substitution of a thiol group in coenzyme A to give the corresponding acyl CoA (Figure 21.6).

Note that the first step in Figure 21.6—reaction of the carboxylate with ATP to give an acyl adenylate—is itself a nucleophilic acyl substitution on *phosphorus*. The carboxylate first adds to a P=O bond, giving a five-coordinate phosphorus intermediate that expels diphosphate ion as leaving group.

21.4 | Chemistry of Acid Halides

Preparation of Acid Halides

Acid chlorides are prepared from carboxylic acids by reaction with thionyl chloride ($SOCl_2$), as we saw in the previous section. Similar reaction of a carboxylic acid with phosphorus tribromide (PBr_3) yields the acid bromide.

Reactions of Acid Halides

Acid halides are among the most reactive of carboxylic acid derivatives and can be converted into many other kinds of compounds by nucleophilic acyl substitution mechanisms. The halogen can be replaced by $-OH$ to yield an acid, by $-OCOR$ to yield an anhydride, by $-OR$ to yield an ester, or by $-NH_2$ to yield an amide. In addition, the reduction of an acid halide yields a primary alcohol, and reaction with a Grignard reagent yields a tertiary alcohol. Although the reactions we'll be discussing in this section are illustrated only for acid chlorides, similar processes take place with other acid halides.

Figure 21.6 MECHANISM:
In fatty-acid biosynthesis, a carboxylic acid is activated by reaction with ATP to give an acyl adenylate, which undergoes nucleophilic acyl substitution with the −SH group on coenzyme A. (ATP = adenosine triphosphate; AMP = adenosine monophosphate.)

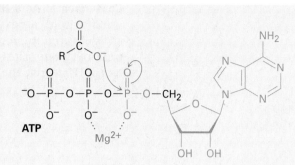

ATP

1 ATP is activated by coordination to magnesium ion, and nucleophilic addition of a fatty acid carboxylate to phosphorus then yields a pentacoordinate intermediate . . .

Pentacoordinate intermediate

2 . . . which expels diphosphate ion (PP$_i$) as leaving group and gives an acyl adenosyl phosphate in a process analogous to a nucleophilic acyl substitution reaction.

(PP$_i$)

Acetyl adenosyl phosphate (acyl adenylate)

3 The –SH group of coenzyme A adds to the acyl adenosyl phosphate, giving a tetrahedral alkoxide intermediate . . .

4 . . . which expels adenosine monophosphate (AMP) as leaving group and yields the fatty acyl CoA.

Fatty acyl CoA

AMP

© John McMurry

Conversion of Acid Halides into Acids: Hydrolysis Acid chlorides react with water to yield carboxylic acids. This hydrolysis reaction is a typical nucleophilic acyl substitution process and is initiated by attack of water on the acid chloride carbonyl group. The tetrahedral intermediate undergoes elimination of Cl^- and loss of H^+ to give the product carboxylic acid plus HCl.

An acid chloride

A carboxylic acid

Because HCl is generated during the hydrolysis, the reaction is often carried out in the presence of a base such as pyridine or NaOH to remove the HCl and prevent it from causing side reactions.

Conversion of Acid Halides into Anhydrides Nucleophilic acyl substitution reaction of an acid chloride with a carboxylate anion gives an acid anhydride. Both symmetrical and unsymmetrical acid anhydrides can be prepared in this way.

Sodium formate **Acetyl chloride** **Acetic formic anhydride (64%)**

Conversion of Acid Halides into Esters: Alcoholysis Acid chlorides react with alcohols to yield esters in a process analogous to their reaction with water to yield acids. In fact, this reaction is probably the most common method for preparing esters in the laboratory. As with hydrolysis, alcoholysis reactions are usually carried out in the presence of pyridine or NaOH to react with the HCl formed.

Benzoyl chloride **Cyclohexanol** **Cyclohexyl benzoate (97%)**

The reaction of an alcohol with an acid chloride is strongly affected by steric hindrance. Bulky groups on either partner slow down the reaction considerably, resulting in a reactivity order among alcohols of primary > secondary > tertiary. As a result, it's often possible to esterify an unhindered alcohol selectively in the presence of a more hindered one. This can be important in complex syntheses in which it's sometimes necessary to distinguish between similar functional groups. For example,

Primary alcohol
(less hindered
and more reactive)

Secondary alcohol
(more hindered
and less reactive)

Problem 21.9   How might you prepare the following esters using a nucleophilic acyl substitution reaction of an acid chloride?
(a) $CH_3CH_2CO_2CH_3$ (b) $CH_3CO_2CH_2CH_3$ (c) Ethyl benzoate

Problem 21.10   Which method would you choose if you wanted to prepare cyclohexyl benzoate— Fischer esterification or reaction of an acid chloride with an alcohol? Explain.

Conversion of Acid Halides into Amides: Aminolysis   Acid chlorides react rapidly with ammonia and amines to give amides. As with the acid chloride plus alcohol method for preparing esters, this reaction of acid chlorides with amines is the most commonly used laboratory method for preparing amides. Both monosubstituted and disubstituted amines can be used, but not trisubstituted amines (R_3N).

2-Methylpropanoyl
chloride

2-Methylpropanamide
(83%)

Benzoyl chloride

N,N-Dimethylbenzamide
(92%)

Because HCl is formed during the reaction, two equivalents of the amine must be used. One equivalent reacts with the acid chloride, and one equivalent reacts with the HCl by-product to form an ammonium chloride salt. If, however, the amine component is valuable, amide synthesis is often carried out using 1 equivalent of the amine plus 1 equivalent of an inexpensive base such as NaOH. For example, the sedative trimetozine is prepared commercially by reaction of 3,4,5-trimethoxybenzoyl chloride with the amine morpholine in the presence of one equivalent of NaOH.

| **3,4,5-Trimethoxy-**
benzoyl chloride | **Morpholine** | **Trimetozine**
(an amide) |

Problem 21.11   Write the mechanism of the reaction just shown between 3,4,5-trimethoxybenzoyl chloride and morpholine to form trimetozine. Use curved arrows to show the electron flow in each step.

Problem 21.12   How could you prepare the following amides using an acid chloride and an amine or ammonia?
(a) $CH_3CH_2CONHCH_3$ (b) *N,N*-Diethylbenzamide   (c) Propanamide

Conversion of Acid Chlorides into Alcohols: Reduction   Acid chlorides are reduced by $LiAlH_4$ to yield primary alcohols. The reaction is of little practical value, however, because the parent carboxylic acids are generally more readily available and can themselves be reduced by $LiAlH_4$ to yield alcohols. Reduction occurs via a typical nucleophilic acyl substitution mechanism in which a hydride ion ($H:^-$) adds to the carbonyl group, yielding a tetrahedral intermediate that expels Cl^-. The net effect is a substitution of $-Cl$ by $-H$ to yield an aldehyde, which is then immediately reduced by $LiAlH_4$ in a second step to yield the primary alcohol.

| **Benzoyl chloride** | **Benzyl alcohol (96%)** |

Reaction of Acid Chlorides with Organometallic Reagents   Grignard reagents react with acid chlorides to yield tertiary alcohols in which two of the substituents are the same.

| **Acid**
chloride | **3° Alcohol** |

The mechanism of this Grignard reaction is similar to that of $LiAlH_4$ reduction. The first equivalent of Grignard reagent adds to the acid chloride, loss of Cl^- from the tetrahedral intermediate yields a ketone, and a second equivalent of Grignard reagent immediately adds to the ketone to produce an alcohol.

Benzoyl chloride   **Acetophenone**   **2-Phenylpropan-2-ol**
(NOT isolated)   **(92%)**

The ketone intermediate can't usually be isolated because addition of the second equivalent of organomagnesium reagent occurs too rapidly. A ketone *can,* however, be isolated from the reaction of an acid chloride with a lithium diorganocopper (Gilman) reagent, $Li^+\ R_2Cu^-$. The reaction occurs by initial nucleophilic acyl substitution on the acid chloride by the diorganocopper anion to yield an acyl diorganocopper intermediate, followed by loss of R'Cu and formation of the ketone.

An acid   **An acyl**   **A ketone**
chloride   **diorganocopper**

The reaction is generally carried out at $-78\ °C$ in ether solution, and yields are often excellent. For example, manicone, a substance secreted by male ants to coordinate ant pairing and mating, has been synthesized by reaction of lithium diethylcopper with (E)-2,4-dimethyl-2-hexenoyl chloride.

2,4-Dimethylhex-2-enoyl   **4,6-Dimethyloct-4-en-3-one**
chloride   **(manicone, 92%)**

Note that the diorganocopper reaction occurs only with acid chlorides. Carboxylic acids, esters, acid anhydrides, and amides do not react with lithium diorganocopper reagents.

Problem 21.13   How could you prepare the following ketones by reaction of an acid chloride with a lithium diorganocopper reagent?

(a)   **(b)**

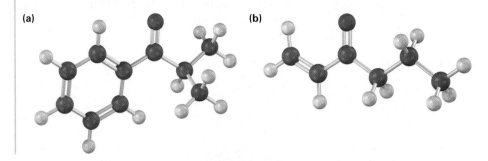

21.5 | Chemistry of Acid Anhydrides

Preparation of Acid Anhydrides

Acid anhydrides are typically prepared by nucleophilic acyl substitution reaction of an acid chloride with a carboxylate anion, as we saw in Section 21.4. Both symmetrical and unsymmetrical acid anhydrides can be prepared in this way.

Benzoyl chloride   **Sodium acetate**   **Acetic benzoic anhydride**

Reactions of Acid Anhydrides

The chemistry of acid anhydrides is similar to that of acid chlorides. Although anhydrides react more slowly than acid chlorides, the kinds of reactions the two groups undergo are the same. Thus, acid anhydrides react with water to form acids, with alcohols to form esters, with amines to form amides, and with LiAlH$_4$ to form primary alcohols. Only the ester and amide forming reactions are much used, however.

Alcoholysis   **Aminolysis**

R'OH   NH$_3$

H$_2$O   [H$^-$]   [H$^-$]

Hydrolysis   **Acid anhydride**   **Reduction**

Conversion of Acid Anhydrides into Esters   Acetic anhydride is often used to prepare acetate esters from alcohols. For example, aspirin (acetylsalicylic acid) is prepared commercially by the acetylation of *o*-hydroxybenzoic acid (salicylic acid) with acetic anhydride.

| Salicylic acid (*o*-hydroxybenzoic acid) | Acetic anhydride | Aspirin (an ester) |

Conversion of Acid Anhydrides into Amides   Acetic anhydride is also commonly used to prepare *N*-substituted acetamides from amines. For example, acetaminophen, a drug used in over-the-counter analgesics such as Tylenol, is prepared by reaction of *p*-hydroxyaniline with acetic anhydride. Note that the more nucleophilic $-NH_2$ group reacts rather than the less nucleophilic $-OH$ group.

| *p*-Hydroxyaniline | Acetic anhydride | Acetaminophen |

Notice in both of the previous reactions that only "half" of the anhydride molecule is used; the other half acts as the leaving group during the nucleophilic acyl substitution step and produces acetate ion as a by-product. Thus, anhydrides are inefficient to use, and acid chlorides are normally preferred for introducing acyl substituents other than acetyl groups.

Problem 21.14 | Write the mechanism of the reaction between *p*-hydroxyaniline and acetic anhydride to prepare acetaminophen.

Problem 21.15 | What product would you expect from reaction of 1 equivalent of methanol with a cyclic anhydride, such as phthalic anhydride (1,2-benzenedicarboxylic anhydride)? What is the fate of the second "half" of the anhydride?

Phthalic anhydride

21.6 | Chemistry of Esters

Esters are among the most widespread of all naturally occurring compounds. Many simple esters are pleasant-smelling liquids that are responsible for the fragrant odors of fruits and flowers. For example, methyl butanoate is found in pineapple oil, and isopentyl acetate is a constituent of banana oil. The ester linkage is also present in animal fats and in many biologically important molecules.

$$CH_3CH_2CH_2\overset{\overset{\displaystyle O}{\|}}{C}OCH_3$$

Methyl butanoate
(from pineapples)

$$CH_3\overset{\overset{\displaystyle O}{\|}}{C}OCH_2CH_2\overset{\overset{\displaystyle CH_3}{|}}{C}HCH_3$$

Isopentyl acetate
(from bananas)

$$\begin{array}{l} CH_2O\overset{\overset{\displaystyle O}{\|}}{C}R \\ CHO\overset{\overset{\displaystyle O}{\|}}{C}R \\ CH_2O\overset{\overset{\displaystyle O}{\|}}{C}R \end{array}$$

A fat
(R = C$_{11-17}$ chains)

The chemical industry uses esters for a variety of purposes. Ethyl acetate, for instance, is a commonly used solvent, and dialkyl phthalates are used as plasticizers to keep polymers from becoming brittle. You may be aware that there is current concern about possible toxicity of phthalates at high concentrations, although a recent assessment by the U.S. Food and Drug Administration found the risk to be minimal for most people, with the possible exception of male infants.

$$\text{OCH}_2\text{CH}_2\text{CH}_2\text{CH}_3$$
$$\text{OCH}_2\text{CH}_2\text{CH}_2\text{CH}_3$$

Dibutyl phthalate
(a plasticizer)

Preparation of Esters

CENGAGENOW Click *Organic Process* to **view an animation of the steps involved in Fischer esterification.**

Esters are usually prepared from carboxylic acids by the methods already discussed. Thus, carboxylic acids are converted directly into esters by S$_N$2 reaction of a carboxylate ion with a primary alkyl halide or by Fischer esterification of a carboxylic acid with an alcohol in the presence of a mineral acid catalyst. In addition, acid chlorides are converted into esters by treatment with an alcohol in the presence of base (Section 21.4).

$$R\overset{\overset{\displaystyle O}{\|}}{C}OH \xrightarrow{\text{SOCl}_2} R\overset{\overset{\displaystyle O}{\|}}{C}Cl$$

1. NaOH
2. R'X

R'OH
HCl

R'OH
Pyridine

$$R\overset{\overset{\displaystyle O}{\|}}{C}OR'$$

Method limited to primary alkyl halides

$$R\overset{\overset{\displaystyle O}{\|}}{C}OR'$$

Method limited to simple alcohols

$$R\overset{\overset{\displaystyle O}{\|}}{C}OR'$$

Method is very general

Reactions of Esters

Esters undergo the same kinds of reactions that we've seen for other carboxylic acid derivatives, but they are less reactive toward nucleophiles than either acid chlorides or anhydrides. All their reactions are equally applicable to both acyclic and cyclic esters, called **lactones**.

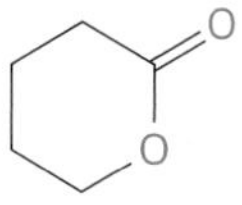

A lactone
(cyclic ester)

Conversion of Esters into Carboxylic Acids: Hydrolysis   An ester is hydrolyzed, either by aqueous base or by aqueous acid, to yield a carboxylic acid plus an alcohol.

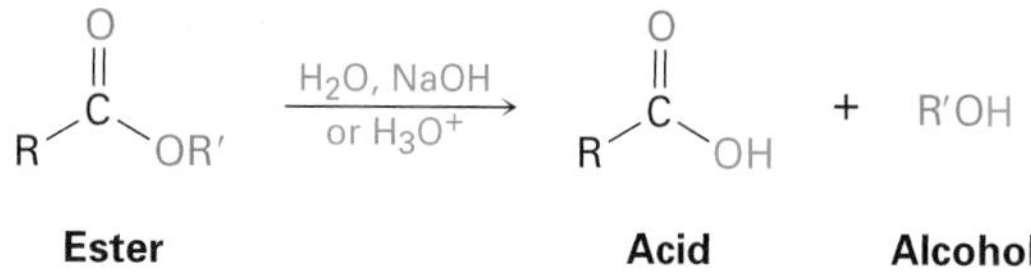

Ester   **Acid**   **Alcohol**

CENGAGENOW™ Click *Organic Process* to **view an animation of the steps involved in base-catalyzed ester hydrolysis.**

Ester hydrolysis in basic solution is called **saponification**, after the Latin word *sapo*, meaning "soap." As we'll see in Section 27.2, soap is in fact made by boiling animal fat with base to hydrolyze the ester linkages.

Ester hydrolysis occurs through a typical nucleophilic acyl substitution pathway in which hydroxide ion is the nucleophile that adds to the ester carbonyl group to give a tetrahedral intermediate. Loss of alkoxide ion then gives a carboxylic acid, which is deprotonated to give the carboxylate ion. Addition of aqueous HCl in a separate step after the saponification is complete then protonates the carboxylate ion and gives the carboxylic acid (Figure 21.17).

The mechanism shown in Figure 21.7 is supported by isotope-labeling studies. When ethyl propanoate labeled with ^{18}O in the ether-like oxygen is hydrolyzed in aqueous NaOH, the ^{18}O label shows up exclusively in the ethanol product. None of the label remains with the propanoic acid, indicating that saponification occurs by cleavage of the C—OR′ bond rather than the CO—R′ bond.

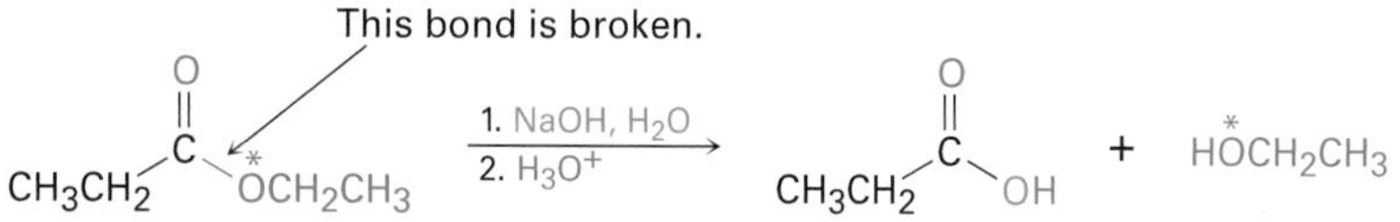

This bond is broken.

CH_3CH_2 ... OCH_2CH_3 1. NaOH, H_2O / 2. H_3O^+ CH_3CH_2 ... OH + $HOCH_2CH_3$

CENGAGENOW™ Click *Organic Process* to **view an animation of the steps involved in acid-catalyzed ester hydrolysis.**

Acid-catalyzed ester hydrolysis can occur by more than one mechanism, depending on the structure of the ester. The usual pathway, however, is just the reverse of a Fischer esterification reaction (Section 21.3). The ester is first activated toward nucleophilic attack by protonation of the carboxyl oxygen atom, and nucleophilic addition of water then occurs. Transfer of a proton and elimination of alcohol yields the carboxylic acid (Figure 21.8). Because this hydrolysis reaction is the reverse of a Fischer esterification reaction, Figure 21.8 is the reverse of Figure 21.4.

Ester hydrolysis is common in biological chemistry, particularly in the digestion of dietary fats and oils. We'll save a complete discussion of the mechanistic details of fat hydrolysis until Section 29.2 but will note for now that the reaction is catalyzed by various lipase enzymes and involves two sequential nucleophilic acyl substitution reactions. The first is a *transesterification* reaction in which an alcohol group on the lipase adds to an ester linkage in the fat molecule to give a tetrahedral intermediate that expels alcohol and forms an acyl

Figure 21.7 MECHANISM:
Mechanism of base-induced
ester hydrolysis (saponification).

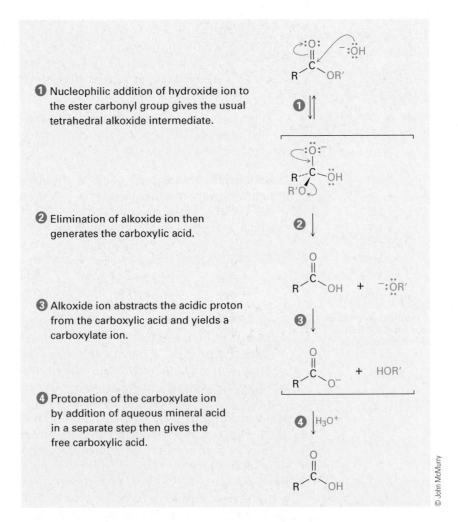

1 Nucleophilic addition of hydroxide ion to
the ester carbonyl group gives the usual
tetrahedral alkoxide intermediate.

2 Elimination of alkoxide ion then
generates the carboxylic acid.

3 Alkoxide ion abstracts the acidic proton
from the carboxylic acid and yields a
carboxylate ion.

4 Protonation of the carboxylate ion
by addition of aqueous mineral acid
in a separate step then gives the
free carboxylic acid.

© John McMurry

enzyme intermediate. The second is an addition of water to the acyl enzyme, followed by expulsion of the enzyme to give a hydrolyzed acid.

© John McMurry

Active Figure 21.8
MECHANISM: Mechanism of acid-catalyzed ester hydrolysis. The forward reaction is a hydrolysis; the back-reaction is a Fischer esterification and is thus the reverse of Figure 21.4. *Sign in at* **www.cengage.com/login** *to see a simulation based on this figure and to take a short quiz.*

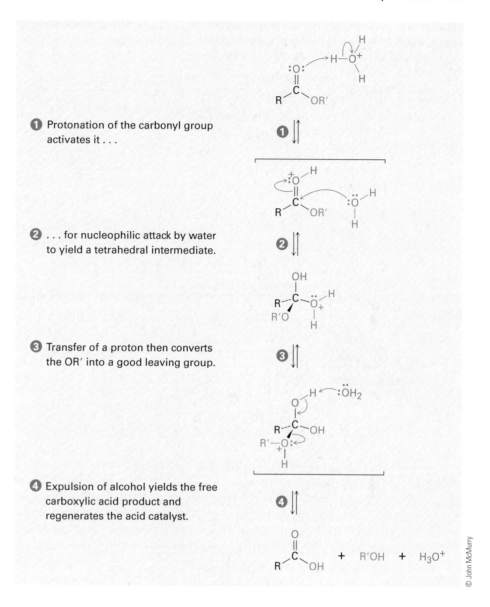

① Protonation of the carbonyl group activates it . . .

② . . . for nucleophilic attack by water to yield a tetrahedral intermediate.

③ Transfer of a proton then converts the OR′ into a good leaving group.

④ Expulsion of alcohol yields the free carboxylic acid product and regenerates the acid catalyst.

Problem 21.16 Why is the saponification of an ester irreversible? In other words, why doesn't treatment of a carboxylic acid with an alkoxide ion yield an ester?

Conversion of Esters into Amides: Aminolysis   Esters react with ammonia and amines to yield amides. The reaction is not often used, however, because it's usually easier to start with an acid chloride (Section 21.4).

Methyl benzoate   **Benzamide**

Conversion of Esters into Alcohols: Reduction Esters are easily reduced by treatment with $LiAlH_4$ to yield primary alcohols (Section 17.4).

$$CH_3CH_2CH=CHCOCH_2CH_3 \xrightarrow[\text{2. } H_3O^+]{\text{1. } LiAlH_4, \text{ ether}} CH_3CH_2CH=CHCH_2OH + CH_3CH_2OH$$

Ethyl pent-2-enoate   **Pent-2-en-1-ol (91%)**

$$\xrightarrow[\text{2. } H_3O^+]{\text{1. } LiAlH_4, \text{ ether}} HOCH_2CH_2CH_2CHCH_3$$

A lactone   **Pentane-1,4-diol (86%)**

The mechanism of ester (and lactone) reduction is similar to that of acid chloride reduction in that a hydride ion first adds to the carbonyl group, followed by elimination of alkoxide ion to yield an aldehyde. Further reduction of the aldehyde gives the primary alcohol.

$$\xrightarrow[\text{Ether}]{LiAlH_4} \left[\ \right] \longrightarrow + R'O^- \xrightarrow[\text{2. } H_3O^+]{\text{1. } LiAlH_4}$$

A primary alcohol

The aldehyde intermediate can be isolated if 1 equivalent of diisobutyl-aluminum hydride (DIBAH) is used as the reducing agent instead of $LiAlH_4$. The reaction has to be carried out at $-78\ °C$ to avoid further reduction to the alcohol. Such *partial* reductions of carboxylic acid derivatives to aldehydes also occur in numerous biological pathways, although the substrate is either a thioester or acyl phosphate rather than an ester.

$$CH_3(CH_2)_9CH_2\text{---}COCH_2CH_3 \xrightarrow[\text{2. } H_3O^+]{\text{1. DIBAH in toluene}} CH_3(CH_2)_9CH_2\text{---}CH + CH_3CH_2OH$$

Ethyl dodecanoate   **Dodecanal (88%)**

where DIBAH =

Problem 21.17 | What product would you expect from the reaction of butyrolactone with $LiAlH_4$? With DIBAH?

Butyrolactone

Problem 21.18   Show the products you would obtain by reduction of the following esters with LiAlH$_4$:

(a)

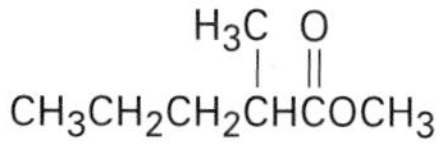

H$_3$C O
| ||
CH$_3$CH$_2$CH$_2$CHCOCH$_3$

(b)

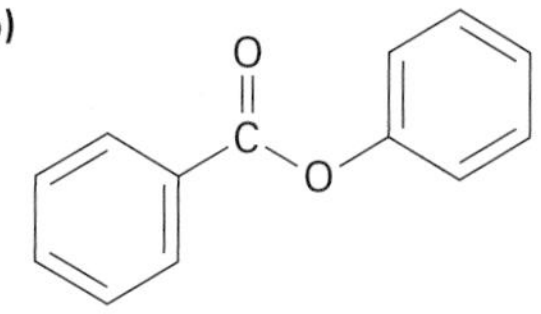

Conversion of Esters into Alcohols: Grignard Reaction   Esters and lactones react with 2 equivalents of a Grignard reagent to yield a tertiary alcohol in which two of the substituents are identical (Section 17.5). The reaction occurs by the usual nucleophilic substitution mechanism to give an intermediate ketone, which reacts further with the Grignard reagent to yield a tertiary alcohol.

1. 2 [PhMgBr]

2. H$_3$O$^+$

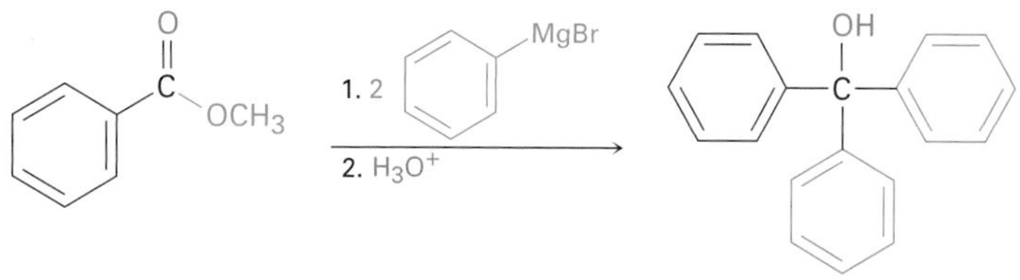

Methyl benzoate   **Triphenylmethanol (96%)**

Problem 21.19   What ester and what Grignard reagent might you start with to prepare the following alcohols?

(a)

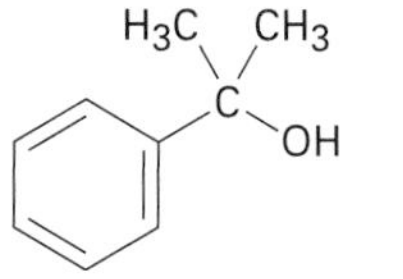

H$_3$C CH$_3$

(b)

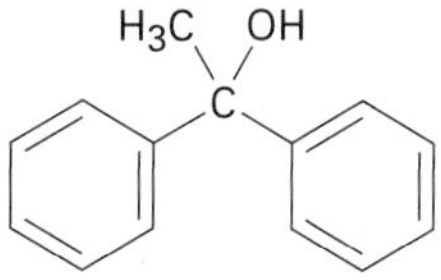

H$_3$C OH

(c)

OH
|
CH$_3$CH$_2$CH$_2$CH$_2$CCH$_2$CH$_3$
|
CH$_2$CH$_3$

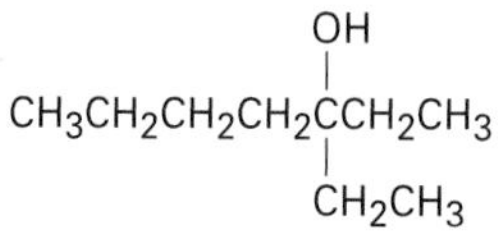

21.7 | Chemistry of Amides

Amides, like esters, are abundant in all living organisms—proteins, nucleic acids, and many pharmaceuticals have amide functional groups. The reason for this abundance of amides, of course, is that they are stable to the conditions found in living organisms. Amides are the least reactive of the common acid derivatives and undergo relatively few nucleophilic acyl substitution reactions.

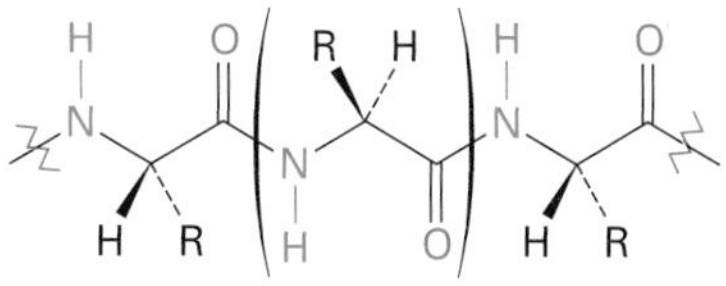

A protein segment

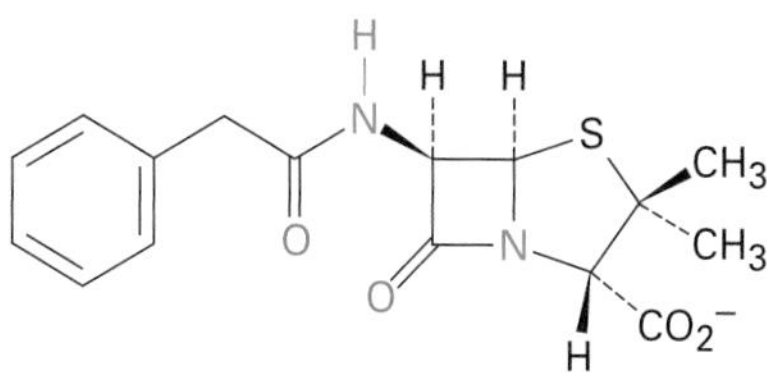

Benzylpenicillin (penicillin G)

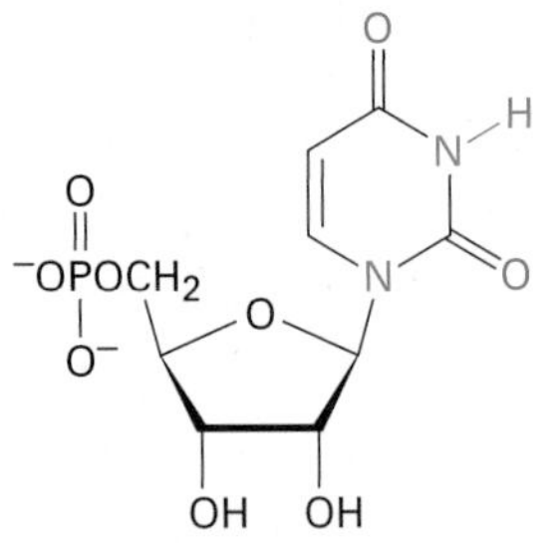

Uridine 5′-phosphate
(a ribonucleotide)

Preparation of Amides

Amides are usually prepared by reaction of an acid chloride with an amine (Section 21.4). Ammonia, monosubstituted amines, and disubstituted amines all undergo the reaction.

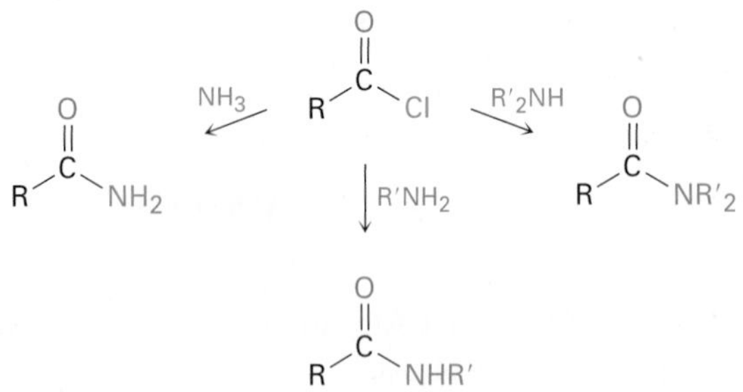

Reactions of Amides

Conversion of Amides into Carboxylic Acids: Hydrolysis Amides undergo hydrolysis to yield carboxylic acids plus ammonia or an amine on heating in either aqueous acid or aqueous base. The conditions required for amide hydrolysis are more severe than those required for the hydrolysis of acid chlorides or esters, but the mechanisms are similar. Acidic hydrolysis reaction occurs by nucleophilic addition of water to the protonated amide, followed by transfer of a proton from oxygen to nitrogen to make the nitrogen a better leaving group and subsequent elimination. The steps are reversible, with the equilibrium shifted toward product by protonation of NH_3 in the final step.

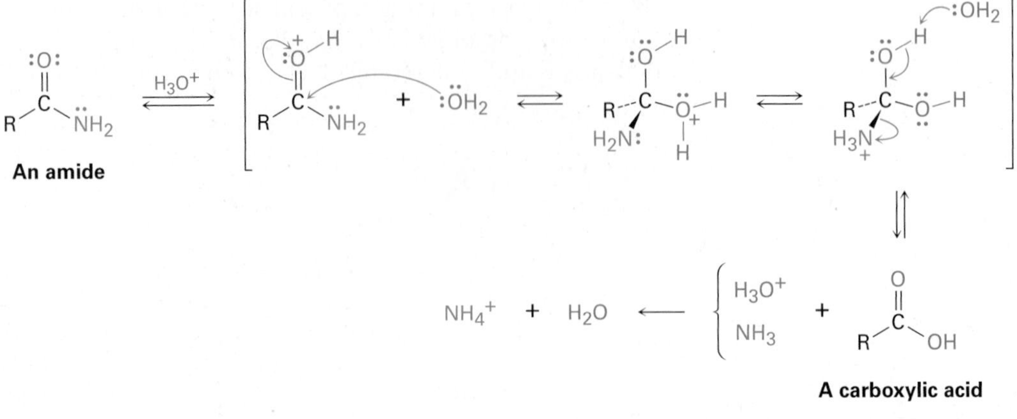

An amide

A carboxylic acid

Basic hydrolysis occurs by nucleophilic addition of OH$^-$ to the amide carbonyl group, followed by elimination of amide ion ($^-$NH$_2$) and subsequent deprotonation of the initially formed carboxylic acid by amide ion. The steps are reversible, with the equilibrium shifted toward product by the final deprotonation of the carboxylic acid. Basic hydrolysis is substantially more difficult than the analogous acid-catalyzed reaction because amide ion is a very poor leaving group, making the elimination step difficult.

An amide   **A carboxylate ion**

Amide hydrolysis is common in biological chemistry. Just as the hydrolysis of esters is the initial step in the digestion of dietary fats, the hydrolysis of amides is the initial step in the digestion of dietary proteins. The reaction is catalyzed by protease enzymes and occurs by a mechanism almost identical to that we just saw for fat hydrolysis. That is, an initial nucleophilic acyl substitution of an alcohol group in the enzyme on an amide linkage in the protein gives an acyl enzyme intermediate that then undergoes hydrolysis.

A protein   **Tetrahedral intermediate**   **An acyl enzyme**

Tetrahedral intermediate   **A cleaved protein fragment**

Conversion of Amides into Amines: Reduction   Like other carboxylic acid derivatives, amides can be reduced by LiAlH$_4$. The product of the reduction, however, is an *amine* rather than an alcohol. The net effect of an amide reduction reaction is thus the conversion of the amide carbonyl group into a methylene group (C=O → CH$_2$). This kind of reaction is specific for amides and does not occur with other carboxylic acid derivatives.

N-Methyldodecanamide   **Dodecylmethylamine (95%)**

Amide reduction occurs by nucleophilic addition of hydride ion to the amide carbonyl group, followed by expulsion of the *oxygen* atom as an aluminate anion leaving group to give an iminium ion intermediate. The intermediate iminium ion is then further reduced by LiAlH$_4$ to yield the amine.

Amide   **Iminium ion**

The reaction is effective with both acyclic and cyclic amides, or **lactams**, and is a good method for preparing cyclic amines.

A lactam   **A cyclic amine (80%)**

Problem 21.20 | How would you convert *N*-ethylbenzamide to each of the following products?
(a) Benzoic acid   (b) Benzyl alcohol   (c) $C_6H_5CH_2NHCH_2CH_3$

Problem 21.21 | How would you use the reaction of an amide with LiAlH$_4$ as the key step in going from bromocyclohexane to (*N*,*N*-dimethylaminomethyl)cyclohexane? Write all the steps in the reaction sequence.

(***N*,*N*-Dimethylaminomethyl)cyclohexane**

21.8 | Chemistry of Thioesters and Acyl Phosphates: Biological Carboxylic Acid Derivatives

As mentioned in the chapter introduction, the substrate for nucleophilic acyl substitution reactions in living organisms is generally either a thioester (RCOSR′) or an acyl phosphate (RCO$_2$PO$_3$$^{2-}$ or RCO$_2$PO$_3$R′$^-$). Neither is as reactive as an acid chloride or acid anhydride, yet both are stable enough to exist in living organisms while still reactive enough to undergo acyl substitution.

Acyl CoA's, such as acetyl CoA, are the most common thioesters in nature. Coenzyme A, abbreviated CoA, is a thiol formed by a phosphoric anhydride linkage (O=P−O−P=O) between phosphopantetheine and adenosine 3′,5′-bisphosphate. (The prefix "bis" means "two" and indicates that adenosine 3′,5′-bisphosphate has two phosphate groups, one on C3′ and one on C5′.) Reaction of coenzyme A with an acyl phosphate or acyl adenylate gives the acyl CoA (Figure 21.9). As we saw in Section 21.3 (Figure 21.6), formation of the acyl adenylate occurs by reaction of a carboxylic acid with ATP and is itself a nucleophilic acyl substitution reaction that takes place on phosphorus.

Figure 21.9 Formation of the thioester acetyl CoA by nucleophilic acyl substitution reaction of coenzyme A (CoA) with acetyl adenylate.

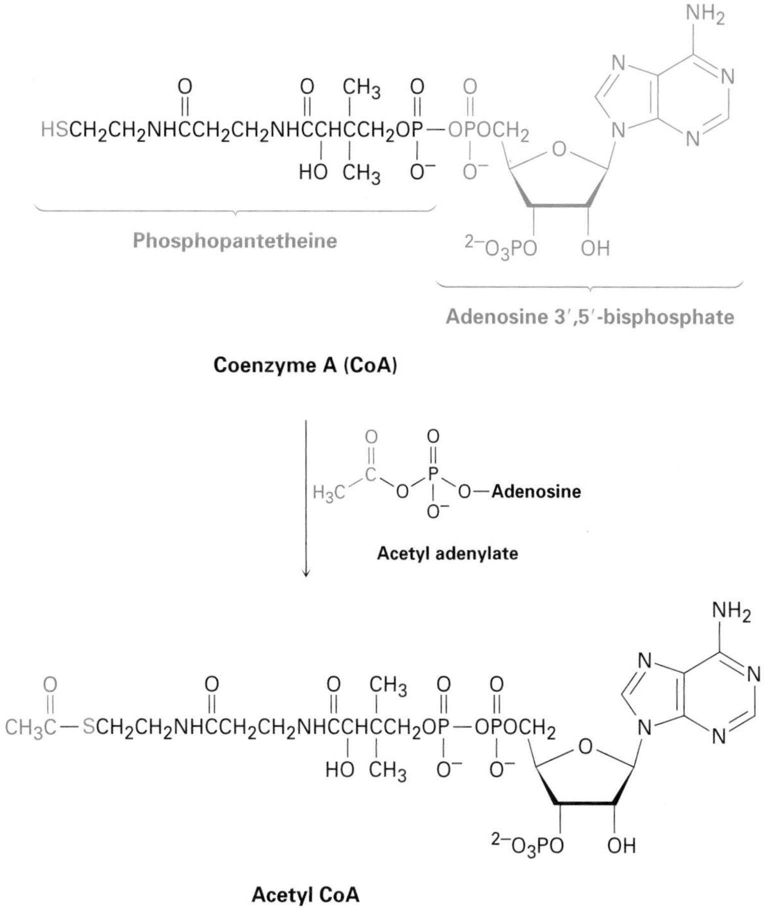

Coenzyme A (CoA)

Acetyl adenylate

Acetyl CoA

Once formed, an acyl CoA is a substrate for further nucleophilic acyl substitution reactions. For example, *N*-acetylglucosamine, a component of cartilage and other connective tissues, is synthesized by an aminolysis reaction between glucosamine and acetyl CoA.

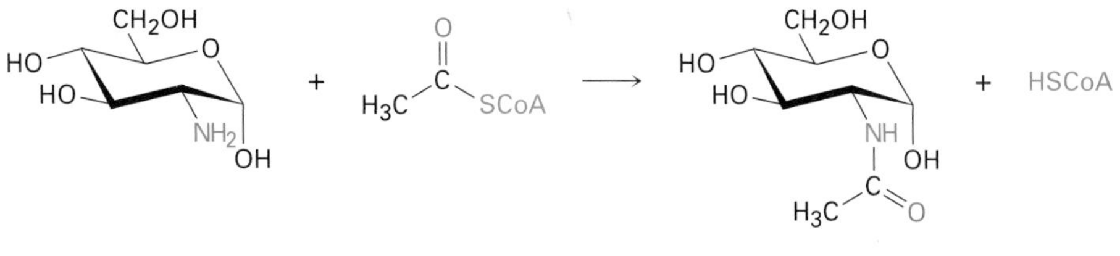

Glucosamine
(an amine)

N-**Acetylglucosamine**
(an amide)

Another example of a nucleophilic acyl substitution reaction, this one a substitution by hydride ion to effect partial reduction of a thioester to an aldehyde, occurs in the biosynthesis of mevaldehyde, an intermediate in terpenoid synthesis (Chapter 6 *Focus On*). In this reaction, (3*R*)-3-hydroxy-3-methyl-glutaryl CoA is reduced by hydride donation from NADPH.

(3S)-3-Hydroxy-3-methylglutaryl CoA

(R)-Mevaldehyde

Problem 21.22 | Write the mechanism of the reaction shown in Figure 21.9 between coenzyme A and acetyl adenylate to give acetyl CoA.

21.9 | Polyamides and Polyesters: Step-Growth Polymers

When an amine reacts with an acid chloride, an amide is formed. What would happen, though, if a *diamine* and a *diacid chloride* were allowed to react? Each partner could form *two* amide bonds, linking more and more molecules together until a giant polyamide resulted. In the same way, reaction of a diol with a diacid would lead to a polyester.

The alkene and diene polymers discussed in Sections 7.10 and 14.6 are called *chain-growth polymers* because they are produced by chain reactions. An initiator adds to a C=C bond to give a reactive intermediate, which adds to a second alkene molecule to produce a new intermediate, which adds to a third molecule, and so on. By contrast, polyamides and polyesters are called **step-growth polymers** because each bond in the polymer is formed independently of the others. A large number of different step-growth polymers have been made; some of the more important ones are shown in Table 21.2.

Table 21.2 | Some Common Step-Growth Polymers and Their Uses

Monomers	Structure	Polymer	Uses
Adipic acid + Hexamethylenediamine		Nylon 66	Fibers, clothing, tire cord
Dimethyl terephthalate + Ethylene glycol		Dacron, Mylar, Terylene	Fibers, clothing, films, tire cord
Caprolactam		Nylon 6, Perlon	Fibers, castings
Diphenyl carbonate + Bisphenol A		Lexan, polycarbonate	Equipment housing, molded articles
Toluene-2,6-diisocyanate + Poly(2-butene-1,4-diol)		Polyurethane, Spandex	Fibers, coatings, foams

Polyamides (Nylons)

The best known step-growth polymers are the polyamides, or *nylons,* first prepared by Wallace Carothers at the DuPont Company by heating a diamine with a diacid. For example, nylon 66 is prepared by reaction of adipic acid (hexanedioic acid) with hexamethylenediamine (1,6-hexanediamine) at 280 °C. The designation "66" tells the number of carbon atoms in the diamine (the first 6) and the diacid (the second 6).

Adipic acid   **Hexamethylenediamine**

Heat

Nylon 66

Nylons are used both in engineering applications and in making fibers. A combination of high impact strength and abrasion resistance makes nylon an excellent metal substitute for bearings and gears. As fiber, nylon is used in a variety of applications, from clothing to tire cord to ropes.

Polyesters

The most generally useful polyester is that made by reaction between dimethyl terephthalate (dimethyl 1,4-benzenedicarboxylate) and ethylene glycol (1,2-ethanediol). The product is used under the trade name Dacron to make clothing fiber and tire cord and under the name Mylar to make recording tape. The tensile strength of poly(ethylene terephthalate) film is nearly equal to that of steel.

Dimethyl terephthalate   **Ethylene glycol**   **A polyester
(Dacron, Mylar)**

Lexan, a polycarbonate prepared from diphenyl carbonate and bisphenol A, is another commercially valuable polyester. Lexan has an unusually high impact strength, making it valuable for use in telephones, bicycle safety helmets, and laptop computer cases.

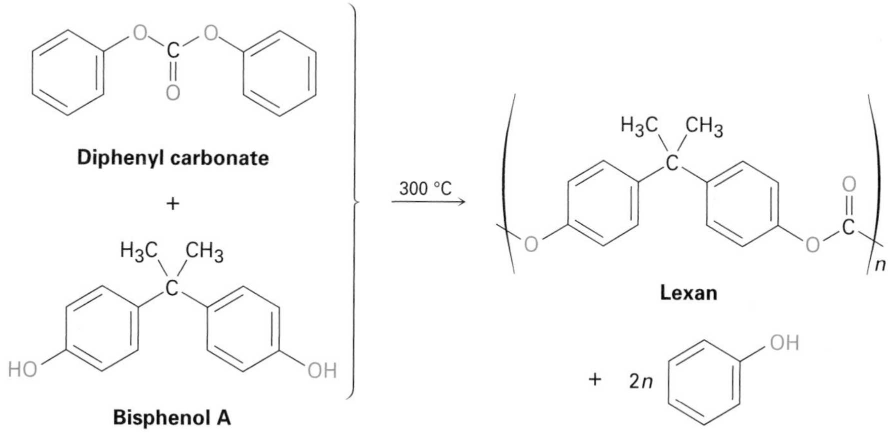

Diphenyl carbonate

+

Bisphenol A

$\xrightarrow{300\ °C}$

Lexan

+   2n

Sutures and Biodegradable Polymers

Because plastics are too often thrown away rather than recycled, much work has been carried out on developing *biodegradable* polymers, which can be broken down rapidly in landfills by soil microorganisms. Among the most common biodegradable polymers are poly(glycolic acid) (PGA), poly(lactic acid) (PLA), and polyhydroxybutyrate (PHB). All are polyesters and are therefore susceptible to hydrolysis of their ester links. Copolymers of PGA with PLA have found a particularly wide range of uses. A 90/10 copolymer of poly(glycolic acid) with poly(lactic acid) is used to make absorbable sutures, for instance. The sutures are entirely hydrolyzed and absorbed by the body within 90 days after surgery.

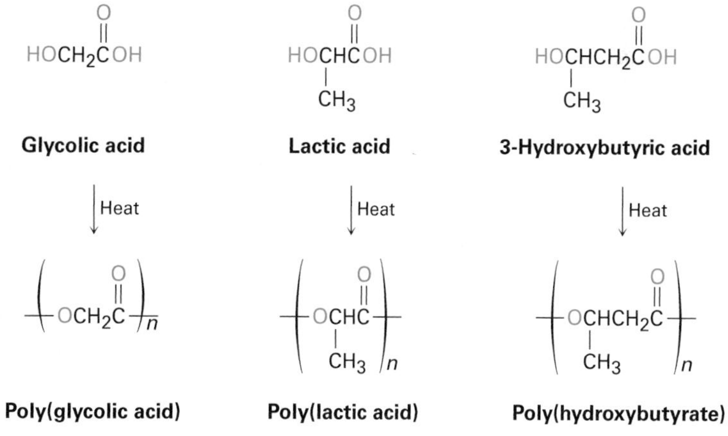

Glycolic acid   **Lactic acid**   **3-Hydroxybutyric acid**

Poly(glycolic acid)   **Poly(lactic acid)**   **Poly(hydroxybutyrate)**

In Europe, interest has centered particularly on polyhydroxybutyrate, which can be made into films for packaging as well as into molded items. The polymer degrades within 4 weeks in landfills, both by ester hydrolysis and by an E1cB elimination reaction of the oxygen atom β to the carbonyl group. The use of polyhydroxybutyrate is limited at present by its cost—about four times that of polypropylene.

Problem 21.23 Draw structures of the step-growth polymers you would expect to obtain from the following reactions:

(a) $BrCH_2CH_2CH_2Br$ + $HOCH_2CH_2CH_2OH$ $\xrightarrow{\text{Base}}$?

(b) $HOCH_2CH_2OH$ + $HO_2C(CH_2)_6CO_2H$ $\xrightarrow{\text{H}_2\text{SO}_4 \text{ catalyst}}$?

(c)

$H_2N(CH_2)_6NH_2$ + $\overset{\text{O}}{\overset{\|}{\text{C}}}lC(CH_2)_4\overset{\text{O}}{\overset{\|}{\text{C}}}Cl$ $\longrightarrow$?

Problem 21.24 Kevlar, a nylon polymer prepared by reaction of 1,4-benzenedicarboxylic acid (terephthalic acid) with 1,4-benzenediamine (*p*-phenylenediamine), is so strong that it's used to make bulletproof vests. Draw the structure of a segment of Kevlar.

Problem 21.25 Draw the structure of the polymer you would expect to obtain from reaction of dimethyl terephthalate with a triol such as glycerol. What structural feature would this new polymer have that was not present in Dacron? How do you think this new feature might affect the properties of the polymer?

21.10 | Spectroscopy of Carboxylic Acid Derivatives

Infrared Spectroscopy

All carbonyl-containing compounds have intense IR absorptions in the range 1650 to 1850 cm^{-1}. As shown in Table 21.3, the exact position of the absorption provides information about the specific kind of carbonyl group. For comparison, the IR absorptions of aldehydes, ketones, and carboxylic acids are included in the table, along with values for carboxylic acid derivatives.

Table 21.3 | **Infrared Absorptions of Some Carbonyl Compounds**

Carbonyl type	Example	Absorption (cm^{-1})
Saturated acid chloride	Acetyl chloride	1810
Aromatic acid chloride	Benzoyl chloride	1770
Saturated acid anhydride	Acetic anhydride	1820, 1760
Saturated ester	Ethyl acetate	1735
Aromatic ester	Ethyl benzoate	1720
Saturated amide	Acetamide	1690
Aromatic amide	Benzamide	1675
N-Substituted amide	*N*-Methylacetamide	1680
N,N-Disubstituted amide	*N,N*-Dimethylacetamide	1650
(Saturated aldehyde	Acetaldehyde	1730)
(Saturated ketone	Acetone	1715)
(Saturated carboxylic acid	Acetic acid	1710)

Acid chlorides are easily detected by their characteristic absorption near 1800 cm^{-1}. Acid anhydrides can be identified by the fact that they show two absorptions in the carbonyl region, one at 1820 cm^{-1} and another at 1760 cm^{-1}. Esters are detected by their absorption at 1735 cm^{-1}, a position somewhat higher than that for either aldehydes or ketones. Amides, by contrast, absorb near the low wavenumber end of the carbonyl region, with the degree of substitution on nitrogen affecting the exact position of the IR band.

Problem 21.26   What kinds of functional groups might compounds have if they show the following IR absorptions?
(a) Absorption at 1735 cm^{-1} (b) Absorption at 1810 cm^{-1}
(c) Absorptions at 2500–3300 cm^{-1} and 1710 cm^{-1} (d) Absorption at 1715 cm^{-1}

Problem 21.27   Propose structures for compounds that have the following formulas and IR absorptions:
(a) $C_6H_{12}O_2$, 1735 cm^{-1} (b) C_4H_9NO, 1650 cm^{-1}
(c) C_4H_5ClO, 1780 cm^{-1}

Nuclear Magnetic Resonance Spectroscopy

Hydrogens on the carbon next to a carbonyl group are slightly deshielded and absorb near $2\,\delta$ in the ^{1}H NMR spectrum. The exact nature of the carbonyl group can't be determined by ^{1}H NMR, however, because the α hydrogens of all acid derivatives absorb in the same range. Figure 21.10 shows the ^{1}H NMR spectrum of ethyl acetate.

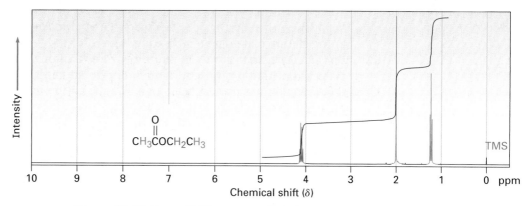

Figure 21.10 Proton NMR spectrum of ethyl acetate.

Although ^{13}C NMR is useful for determining the presence or absence of a carbonyl group in a molecule, the identity of the carbonyl group is difficult to determine. Aldehydes and ketones absorb near $200\,\delta$, while the carbonyl carbon atoms of various acid derivatives absorb in the range 160 to $180\,\delta$ (Table 21.4).

Table 21.4 | ¹³C NMR Absorptions in Some Carbonyl Compounds

Compound	Absorption (δ)	Compound	Absorption (δ)
Acetic acid	177.3	Acetic anhydride	166.9
Ethyl acetate	170.7	Acetone	205.6
Acetyl chloride	170.3	Acetaldehyde	201.0
Acetamide	172.6		

Focus On . . .

β-Lactam Antibiotics

Penicillium mold growing in a petri dish.

© Biophoto Associates/Photo Researchers, Inc.

The value of hard work and logical thinking shouldn't be underestimated, but pure luck also plays a role in most real scientific breakthroughs. What has been called "the supreme example [of luck] in all scientific history" occurred in the late summer of 1928, when the Scottish bacteriologist Alexander Fleming went on vacation, leaving in his lab a culture plate recently inoculated with the bacterium *Staphylococcus aureus*.

While Fleming was away, an extraordinary chain of events occurred. First, a 9-day cold spell lowered the laboratory temperature to a point where the *Staphylococcus* on the plate could not grow. During this time, spores from a colony of the mold *Penicillium notatum* being grown on the floor below wafted up into Fleming's lab and landed in the culture plate. The temperature then rose, and both *Staphylococcus* and *Penicillium* began to grow. On returning from vacation, Fleming discarded the plate into a tray of antiseptic, intending to sterilize it. Evidently, though, the plate did not sink deeply enough into the antiseptic, because when Fleming happened to glance at it a few days later, what he saw changed the course of human history. He noticed that the growing *Penicillium* mold appeared to dissolve the colonies of staphylococci.

Fleming realized that the *Penicillium* mold must be producing a chemical that killed the *Staphylococcus* bacteria, and he spent several years trying to isolate the substance. Finally, in 1939, the Australian pathologist Howard Florey and the German refugee Ernst Chain managed to isolate the active substance, called *penicillin*. The dramatic ability of penicillin to cure infections in mice was soon demonstrated, and successful tests in humans followed shortly thereafter. By 1943, penicillin was being produced on a large scale for military use in World War II, and by 1944 it was being used on civilians. Fleming, Florey, and Chain shared the 1945 Nobel Prize in medicine.

(continued)

Now called benzylpenicillin, or penicillin G, the substance first discovered by Fleming is but one member of a large class of so-called β-lactam antibiotics, compounds with a four-membered lactam (cyclic amide) ring. The four-membered lactam ring is fused to a five-membered, sulfur-containing ring, and the carbon atom next to the lactam carbonyl group is bonded to an acylamino substituent, RCONH$-$. This acylamino side chain can be varied in the laboratory to provide many hundreds of penicillin analogs with different biological activity profiles. Ampicillin, for instance, has an α-aminophenyl-acetamido substituent [PhCH(NH$_2$)CONH$-$].

Closely related to the penicillins are the *cephalosporins,* a group of β-lactam antibiotics that contain an unsaturated six-membered, sulfur-containing ring. Cephalexin, marketed under the trade name Keflex, is an example. Cephalosporins generally have much greater antibacterial activity than penicillins, particularly against resistant strains of bacteria.

The biological activity of penicillins and cephalosporins is due to the presence of the strained β-lactam ring, which reacts with and deactivates the transpeptidase enzyme needed to synthesize and repair bacterial cell walls. With the wall either incomplete or weakened, the bacterial cell ruptures and dies.

SUMMARY AND KEY WORDS

acid anhydride (RCO$_2$COR$'$), 785

acid halide (RCOX), 785

acyl phosphate (RCOPO$_3^{2-}$), 785

amide (RCONH$_2$), 785

Carboxylic acids can be transformed into a variety of **carboxylic acid derivatives** in which the carboxyl $-$OH group has been replaced by another substituent. **Acid halides**, **acid anhydrides**, **esters**, and **amides** are the most common such derivatives in the laboratory; **thioesters** and **acyl phosphates** are common in biological molecules.

The chemistry of carboxylic acid derivatives is dominated by the **nucleophilic acyl substitution reaction**. Mechanistically, these substitutions take place by addition of a nucleophile to the polar carbonyl group of the acid derivative to give a tetrahedral intermediate, followed by expulsion of a leaving group.

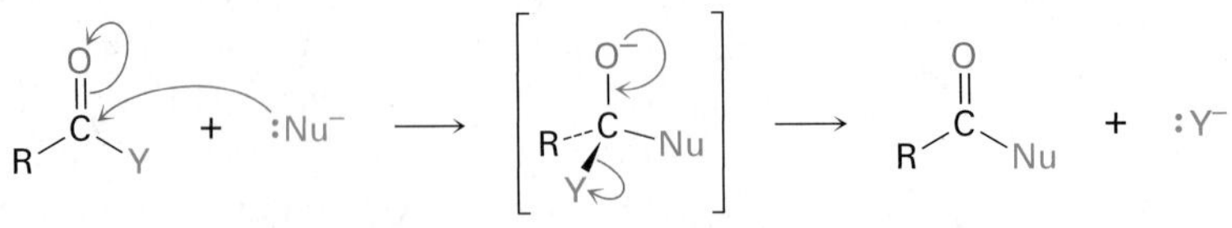

The reactivity of an acid derivative toward substitution depends both on the steric environment near the carbonyl group and on the electronic nature of the substituent, Y. The reactivity order is acid halide > acid anhydride > thioester > ester > amide.

The most common reactions of carboxylic acid derivatives are substitution by water *(hydrolysis)* to yield an acid, by an alcohol *(alcoholysis)* to yield an ester, by an amine *(aminolysis)* to yield an amide, by hydride ion to yield an alcohol *(reduction)*, and by an organometallic reagent to yield an alcohol *(Grignard reaction)*.

Step-growth polymers, such as polyamides and polyesters, are prepared by reactions between difunctional molecules. Polyamides (nylons) are formed by reaction between a diacid and a diamine; polyesters are formed from a diacid and a diol.

IR spectroscopy is a valuable tool for the structural analysis of acid derivatives. Acid chlorides, anhydrides, esters, and amides all show characteristic IR absorptions that can be used to identify these functional groups.

SUMMARY OF REACTIONS

1. Reactions of carboxylic acids (Section 21.3)
   (a) Conversion into acid chlorides

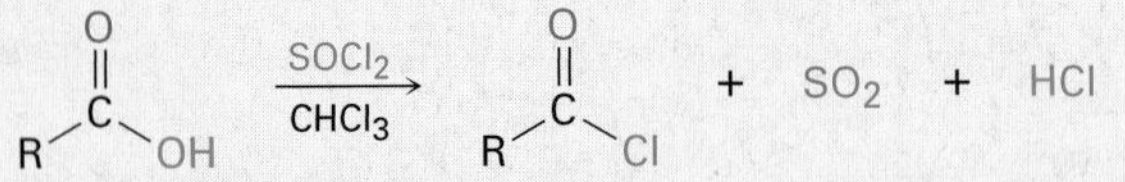

   (b) Conversion into esters

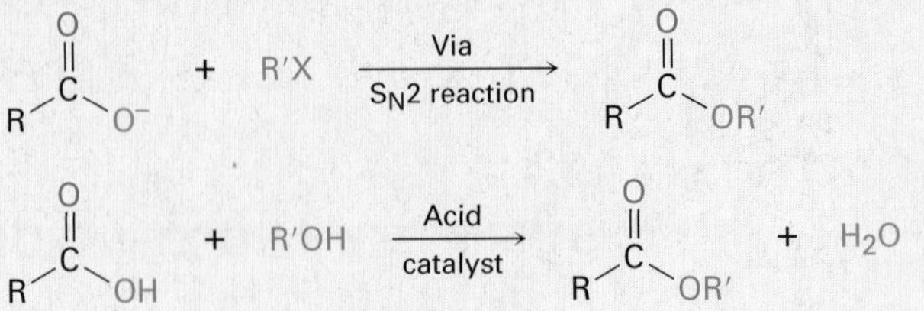

   (c) Conversion into amides

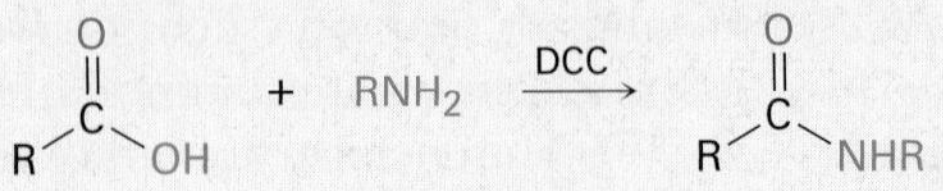

(d) Reduction to yield primary alcohols

2. Reactions of acid chlorides (Section 21.4)
  (a) Hydrolysis to yield acids

(b) Reaction with carboxylates to yield anhydrides

(c) Alcoholysis to yield esters

(d) Aminolysis to yield amides

(e) Reduction to yield primary alcohols

(f) Grignard reaction to yield tertiary alcohols

(e) Diorganocopper reaction to yield ketones

3. Reactions of acid anhydrides (Section 21.5)
   (a) Hydrolysis to yield acids

   (b) Alcoholysis to yield esters

   (c) Aminolysis to yield amides

4. Reactions of esters and lactones (Section 21.6)
   (a) Hydrolysis to yield acids

   (b) Reduction to yield primary alcohols

   (c) Partial reduction to yield aldehydes

   (d) Grignard reaction to yield tertiary alcohols

5. Reactions of amides (Section 21.7)
   (a) Hydrolysis to yield acids

(b) Reduction to yield amines

Organic KNOWLEDGE TOOLS

CENGAGENOW™ Sign in at **www.cengage.com/login** to assess your knowledge of this chapter's topics by taking a pre-test. The pre-test will link you to interactive organic chemistry resources based on your score in each concept area.

⊙WL Online homework for this chapter may be assigned in Organic OWL.

■ indicates problems assignable in Organic OWL.

VISUALIZING CHEMISTRY

(Problems 21.1–21.27 appear within the chapter.)

21.28 ■ Name the following compounds:

(a)

(b)

21.29 ■ How would you prepare the following compounds starting with an appropriate carboxylic acid and any other reagents needed? (Reddish brown = Br.)

(a)

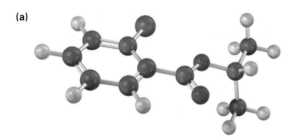

(b)

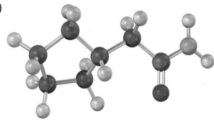

21.30 ■ The following structure represents a tetrahedral alkoxide-ion intermediate formed by addition of a nucleophile to a carboxylic acid derivative. Identify the nucleophile, the leaving group, the starting acid derivative, and the ultimate product (yellow-green = Cl):

21.31 Electrostatic potential maps of a typical amide (acetamide) and an acyl azide (acetyl azide) are shown. Which of the two do you think is more reactive in nucleophilic acyl substitution reactions? Explain.

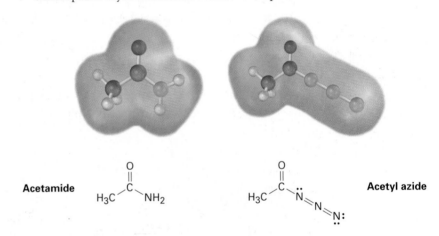

Acetamide

$$H_3C-\overset{\overset{\displaystyle O}{\|}}{C}-NH_2$$

Acetyl azide

$$H_3C-\overset{\overset{\displaystyle O}{\|}}{C}-\ddot{N}=N=\ddot{N}:$$

ADDITIONAL PROBLEMS

21.32 ■ Give IUPAC names for the following compounds:

(a)

(structure: 4-methylbenzamide — a benzene ring with CH_3 substituent and a $C(=O)NH_2$ group)

(b)

$$\begin{array}{c} CH_2CH_3 \quad O \\ | \qquad \| \\ CH_3CH_2CHCH=CHCCl \end{array}$$

(c)

$$\begin{array}{cc} O \qquad\qquad O \\ \| \qquad\qquad \| \\ CH_3OCCH_2CH_2COCH_3 \end{array}$$

(d)

$$\begin{array}{c} O \\ \| \\ \text{—}CH_2CH_2COCHCH_3 \\ \qquad\qquad | \\ \qquad\qquad CH_3 \end{array}$$

(e)

$$\begin{array}{c} O \\ \| \\ CH_3CHCH_2CNHCH_3 \\ | \\ Br \end{array}$$

(f)

$$\begin{array}{c} O \\ \| \\ \text{—}COCH_3 \end{array}$$

(g)

(structure: phenyl benzoate — benzene ring $C(=O)O$ phenyl)

(h)

$$\begin{array}{c} O \\ \| \\ \text{—}C\text{—}SCH(CH_3)_2 \end{array}$$

21.33 ■ Draw structures corresponding to the following names:
   (a) *p*-Bromophenylacetamide
   (b) *m*-Benzoylbenzamide
   (c) 2,2-Dimethylhexanamide
   (d) Cyclohexyl cyclohexanecarboxylate
   (e) Ethyl 2-cyclobutenecarboxylate
   (f) Succinic anhydride

21.34 Draw and name compounds that meet the following descriptions:
   (a) Three acid chlorides having the formula C_6H_9ClO
   (b) Three amides having the formula $C_7H_{11}NO$

21.35 ■ How might you prepare the following compounds from butanoic acid?
   (a) 1-Butanol   (b) Butanal   (c) 1-Bromobutane
   (d) Pentanenitrile   (e) 1-Butene   (f) N-Methylpentanamide
   (g) 2-Hexanone   (h) Butylbenzene   (i) Butanenitrile

21.36 ■ Predict the product(s) of the following reactions:

(a)

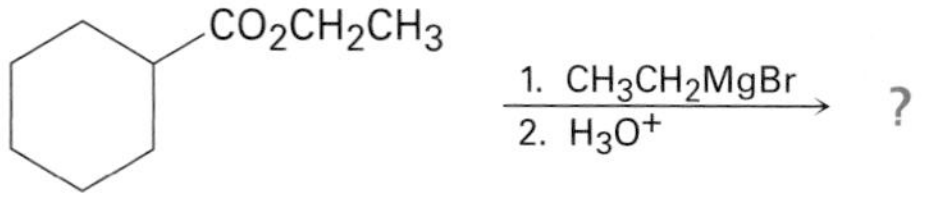

$CO_2CH_2CH_3$

$\xrightarrow[\text{2. }H_3O^+]{\text{1. }CH_3CH_2MgBr}}$?

(b) CH_3
       |
$CH_3CHCH_2CH_2CO_2CH_3$ $\xrightarrow[\text{2. }H_3O^+]{\text{1. DIBAH}}$?

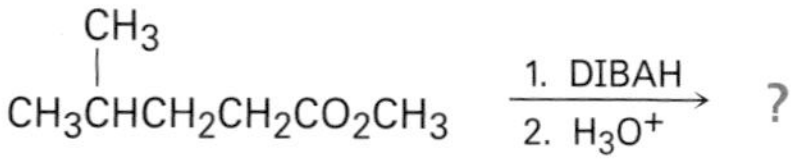

(c)

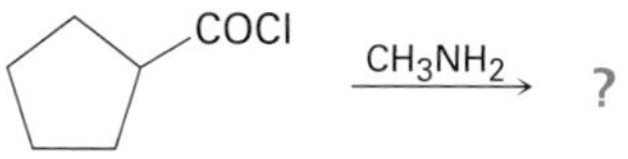

COCl

$\xrightarrow{CH_3NH_2}$?

(d) CO_2H
          ⁄ H

          ⁝CH_3
          H

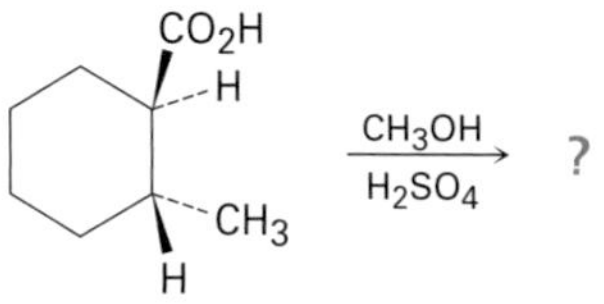

$\xrightarrow[\text{H}_2SO_4]{CH_3OH}$?

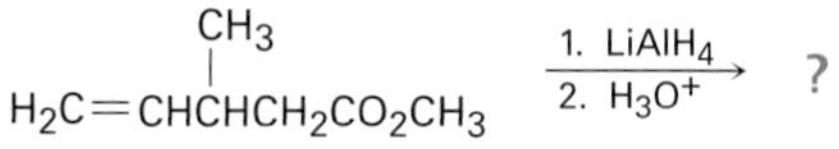

(e) CH_3
          |
$H_2C{=}CHCHCH_2CO_2CH_3$ $\xrightarrow[\text{2. }H_3O^+]{\text{1. LiAlH}_4}$?

(f)

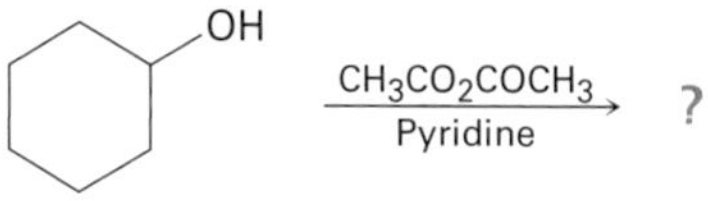

OH

$\xrightarrow[\text{Pyridine}]{CH_3CO_2COCH_3}$?

(g)

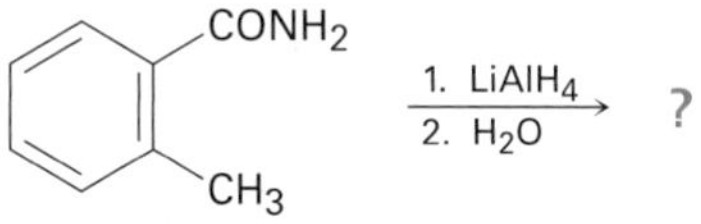

$CONH_2$

CH_3

$\xrightarrow[\text{2. }H_2O]{\text{1. LiAlH}_4}$?

(h)

CO_2H

Br

$\xrightarrow{SOCl_2}$?

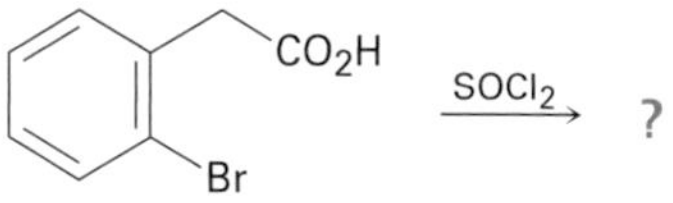

21.37 ■ Predict the product, if any, of reaction between propanoyl chloride and the following reagents:
   (a) Li(Ph)$_2$Cu in ether   (b) LiAlH$_4$, then H$_3$O$^+$
   (c) CH$_3$MgBr, then H$_3$O$^+$ (d) H$_3$O$^+$
   (e) Cyclohexanol   (f) Aniline
   (g) CH$_3$CO$_2$$^-$ $^+$Na

21.38 ■ Answer Problem 21.37 for reaction of the listed reagents with methyl propanoate.

21.39 ■ Answer Problem 21.37 for reaction of the listed reagents with propanamide.

21.40 What product would you expect to obtain from Grignard reaction of an excess of phenylmagnesium bromide with dimethyl carbonate, CH$_3$OCO$_2$CH$_3$?

21.41 Treatment of 5-aminopentanoic acid with DCC (dicyclohexylcarbodiimide) yields a lactam. Show the structure of the product and the mechanism of the reaction.

21.42 ■ Outline methods for the preparation of acetophenone (phenyl methyl ketone) starting from the following:
(a) Benzene   (b) Bromobenzene   (c) Methyl benzoate
(d) Benzonitrile   (e) Styrene

21.43 ■ The following reactivity order has been found for the basic hydrolysis of *p*-substituted methyl benzoates:

$$Y = NO_2 > Br > H > CH_3 > OCH_3$$

How can you explain this reactivity order? Where would you expect $Y = C\equiv N$, $Y = CHO$, and $Y = NH_2$ to be in the reactivity list?

21.44 ■ The following reactivity order has been found for the saponification of alkyl acetates by aqueous NaOH. Explain.

$$CH_3CO_2CH_3 > CH_3CO_2CH_2CH_3 > CH_3CO_2CH(CH_3)_2 > CH_3CO_2C(CH_3)_3$$

21.45 Explain the observation that attempted Fischer esterification of 2,4,6-trimethylbenzoic acid with methanol and HCl is unsuccessful. No ester is obtained, and the acid is recovered unchanged. What alternative method of esterification might be successful?

21.46 ■ Fats are biosynthesized from glycerol 3-phosphate and fatty-acyl CoA's by a reaction sequence that begins with the following step. Show the mechanism of the reaction.

21.47 When a carboxylic acid is dissolved in isotopically labeled water, the label rapidly becomes incorporated into *both* oxygen atoms of the carboxylic acid. Explain.

21.48 ■ When *ethyl* benzoate is heated in methanol containing a small amount of HCl, *methyl* benzoate is formed. Propose a mechanism for the reaction.

21.49 ■ *tert*-Butoxycarbonyl azide, a reagent used in protein synthesis, is prepared by treating *tert*-butoxycarbonyl chloride with sodium azide. Propose a mechanism for this reaction.

$$H_3C\underset{H_3C}{\overset{CH_3}{\underset{|}{C}}}O\overset{O}{\overset{||}{C}}Cl \; + \; NaN_3 \; \longrightarrow \; H_3C\underset{H_3C}{\overset{CH_3}{\underset{|}{C}}}O\overset{O}{\overset{||}{C}}N_3 \; + \; NaCl$$

21.50 We said in Section 21.6 that mechanistic studies on ester hydrolysis have been carried out using ethyl propanoate labeled with ^{18}O in the ether-like oxygen. Assume that ^{18}O-labeled acetic acid is your only source of isotopic oxygen, and then propose a synthesis of the labeled ethyl propanoate.

21.51 ■ Treatment of a carboxylic acid with trifluoroacetic anhydride leads to an unsymmetrical anhydride that rapidly reacts with alcohol to give an ester.

$$R\overset{O}{\overset{||}{C}}OH \;\xrightarrow{(CF_3CO)_2O}\; R\overset{O}{\overset{||}{C}}O\overset{O}{\overset{||}{C}}CF_3 \;\xrightarrow{R'OH}\; R\overset{O}{\overset{||}{C}}OR' \; + \; CF_3CO_2H$$

(a) Propose a mechanism for formation of the unsymmetrical anhydride.
(b) Why is the unsymmetrical anhydride unusually reactive?
(c) Why does the unsymmetrical anhydride react as indicated rather than giving a trifluoroacetate ester plus carboxylic acid?

21.52 Treatment of an α-amino acid with DCC yields a 2,5-diketopiperazine. Propose a mechanism.

An α-amino acid   **A 2,5-diketopiperazine**

21.53 ■ Succinic anhydride yields the cyclic imide succinimide when heated with ammonium chloride at 200 °C. Propose a mechanism for this reaction. Why do you suppose such a high reaction temperature is required?

21.54 Butacetin is an analgesic (pain-killing) agent that is synthesized commercially from *p*-fluoronitrobenzene. Propose a synthesis.

Butacetin

21.55 Phenyl 4-aminosalicylate is a drug used in the treatment of tuberculosis. Propose a synthesis of this compound starting from 4-nitrosalicylic acid.

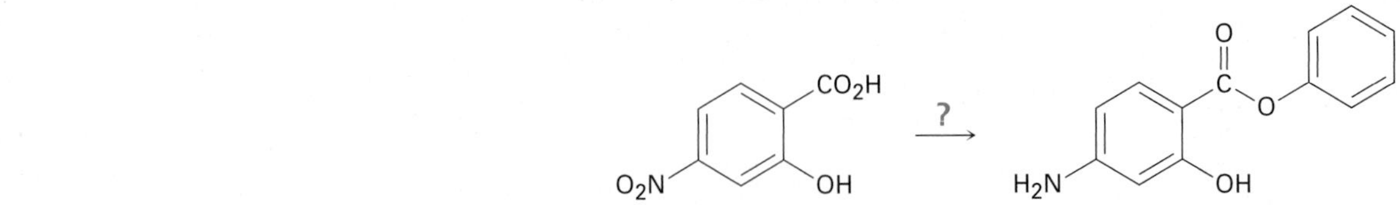

4-Nitrosalicylic acid   **Phenyl 4-aminosalicylate**

21.56 N,N-Diethyl-m-toluamide (DEET) is the active ingredient in many insect-repellent preparations. How might you synthesize this substance from m-bromotoluene?

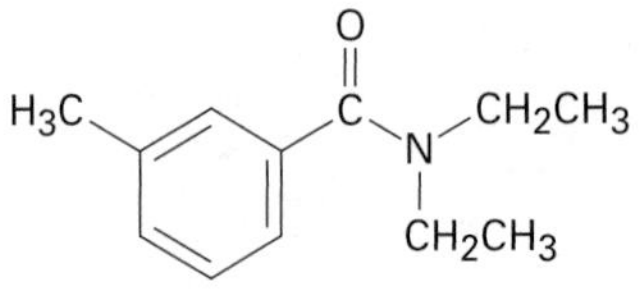

N,N-Diethyl-m-toluamide

21.57 Tranexamic acid, a drug useful against blood clotting, is prepared commercially from p-methylbenzonitrile. Formulate the steps likely to be used in the synthesis. (Don't worry about cis–trans isomers; heating to 300 °C interconverts the isomers.)

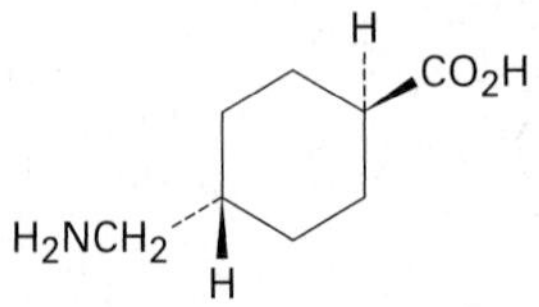

Tranexamic acid

21.58 One frequently used method for preparing methyl esters is by reaction of carboxylic acids with diazomethane, CH_2N_2.

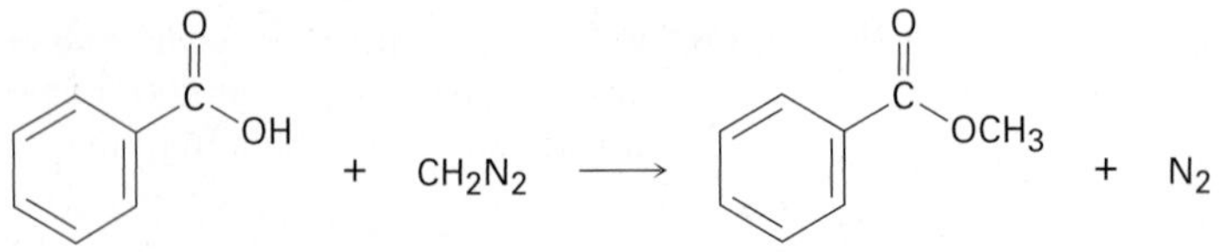

Benzoic acid   **Diazomethane**   **Methyl benzoate (100%)**

The reaction occurs in two steps: (1) protonation of diazomethane by the carboxylic acid to yield methyldiazonium ion, $CH_3N_2^+$, plus a carboxylate ion; and (2) reaction of the carboxylate ion with $CH_3N_2^+$.
(a) Draw two resonance structures of diazomethane, and account for step 1.
(b) What kind of reaction occurs in step 2?

21.59 ■ The hydrolysis of a biological thioester to the corresponding carboxylate is often more complex than the overall result might suggest. The conversion of succinyl CoA to succinate in the citric acid cycle, for instance, occurs by initial formation of an acyl phosphate, followed by reaction with guanosine diphosphate (GDP, a relative of ADP) to give succinate and guanosine triphosphate (GTP, a relative of ATP). Suggest mechanisms for both steps.

21.60 ■ One step in the *gluconeogenesis* pathway for the biosynthesis of glucose is the partial reduction of 3-phosphoglycerate to give glyceraldehyde 3-phosphate. The process occurs by phosphorylation with ATP to give 1,3-bisphosphoglycerate, reaction with a thiol group on the enzyme to give an enzyme-bound thioester, and reduction with NADH. Suggest mechanisms for all three reactions.

21.61 Penicillins and other β-lactam antibiotics (see the *Focus On* in this chapter) typically develop a resistance to bacteria due to bacterial synthesis of β-lactamase enzymes. Tazobactam, however, is able to inhibit the activity of the β-lactamase by trapping it, thereby preventing resistance from developing.

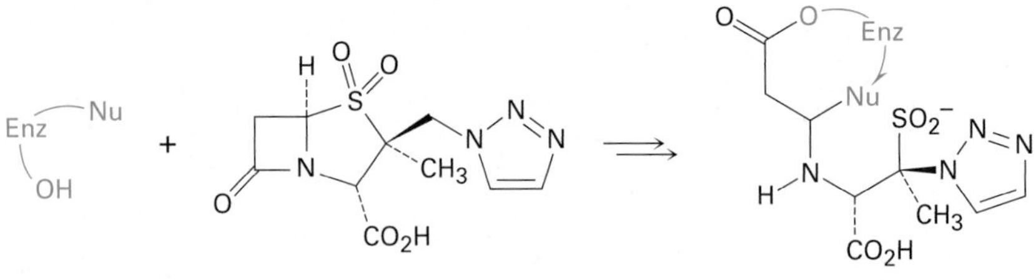

β-Lactamase   **Tazobactam**   **Trapped β-lactamase**

(a) The first step in trapping is reaction of a hydroxyl group on the β-lactamase to open the β-lactam ring of tazobactam. Show the mechanism.
(b) The second step is opening of the sulfur-containing ring in tazobactam to give an acyclic iminium ion intermediate. Show the mechanism.
(c) Cyclization of the iminium ion intermediate gives the trapped β-lactamase product. Show the mechanism.

21.62 ■ The following reaction, called the *benzilic acid rearrangement,* takes place by typical carbonyl-group reactions. Propose a mechanism (Ph = phenyl).

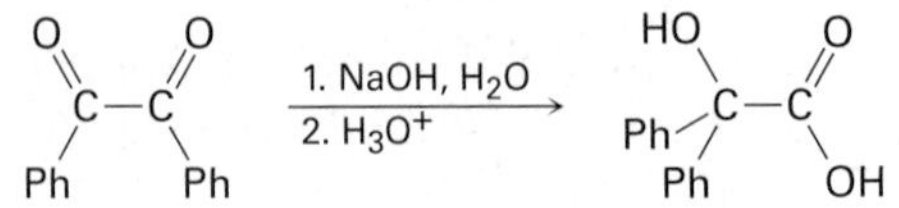

Benzil   **Benzylic acid**

21.63 ■ The step-growth polymer nylon 6 is prepared from caprolactam. The reaction involves initial reaction of caprolactam with water to give an intermediate open-chain amino acid, followed by heating to form the polymer. Propose mechanisms for both steps, and show the structure of nylon 6.

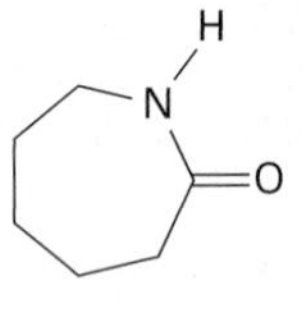

Caprolactam

21.64 *Qiana,* a polyamide fiber with a silky texture, has the following structure. What are the monomer units used in the synthesis of Qiana?

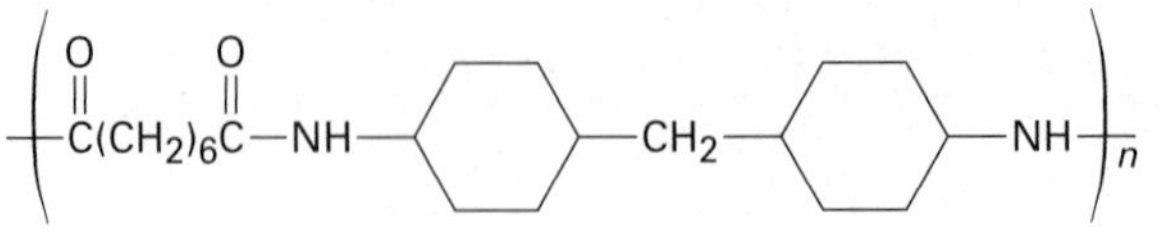

Qiana

21.65 What is the structure of the polymer produced by treatment of β-propiolactone with a small amount of hydroxide ion?

β-Propiolactone

21.66 Polyimides having the structure shown are used as coatings on glass and plastics to improve scratch resistance. How would you synthesize a polyimide? (See Problem 21.53.)

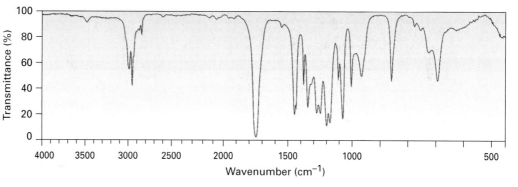

A polyimide

21.67 How would you distinguish spectroscopically between the following isomer pairs? Tell what differences you would expect to see.
(a) *N*-Methylpropanamide and *N,N*-dimethylacetamide
(b) 5-Hydroxypentanenitrile and cyclobutanecarboxamide
(c) 4-Chlorobutanoic acid and 3-methoxypropanoyl chloride
(d) Ethyl propanoate and propyl acetate

21.68 ■ Propose a structure for a compound, $C_4H_7ClO_2$, that has the following IR and ^1H NMR spectra:

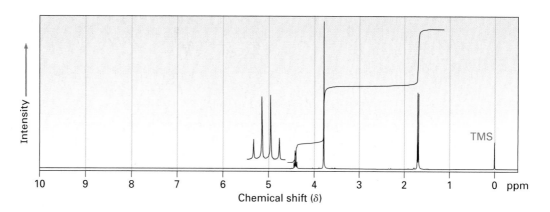

21.69 ■ Assign structures to compounds with the following ^{1}H NMR spectra:

(a) C_4H_7ClO
   IR: 1810 cm^{-1}

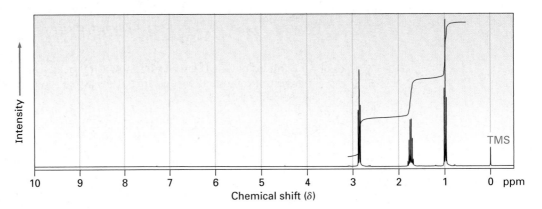

(b) $C_5H_7NO_2$
   IR: 2250, 1735 cm^{-1}

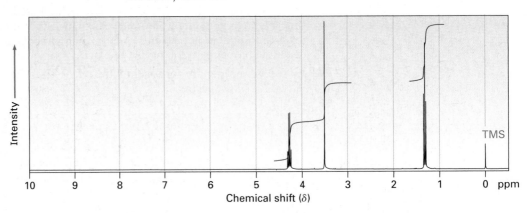

(c) $C_5H_{10}O_2$
   IR: 1735 cm^{-1}

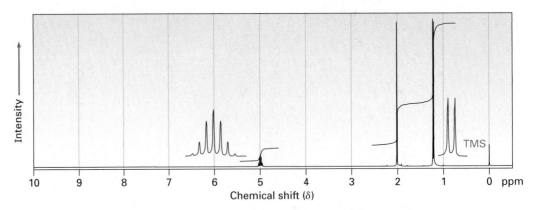

21.70 ■ Propose structures for compounds with the following ^1H NMR spectra:

(a) $C_5H_9ClO_2$
   IR: 1735 cm^{-1}

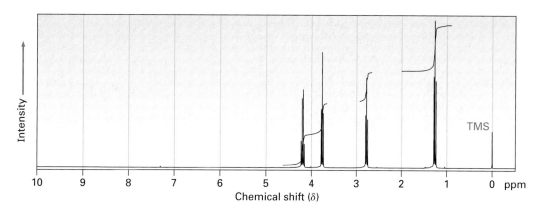

(b) $C_7H_{12}O_4$
   IR: 1735 cm^{-1}

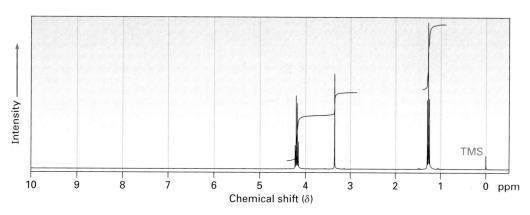

(c) $C_{11}H_{12}O_2$
   IR: 1710 cm^{-1}

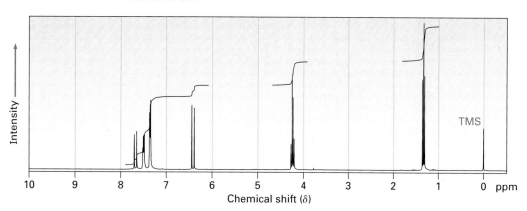

21.71 Epoxy adhesives are prepared in two steps. S_N2 reaction of the disodium salt of bisphenol A with epichlorohydrin forms a "prepolymer," which is then "cured" by treatment with a triamine such as $H_2NCH_2CH_2NHCH_2CH_2NH_2$.

Bisphenol A   **Epichlorohydrin**

"Prepolymer"

Draw structures to show how addition of the triamine results in strengthening the polymer. Amines are good nucleophiles and can open epoxide rings in the same way other bases can.

21.72 In the *iodoform reaction*, a triiodomethyl ketone reacts with aqueous NaOH to yield a carboxylate ion and iodoform (triiodomethane). Propose a mechanism for this reaction.

22

Carbonyl Alpha-Substitution Reactions

Organic KNOWLEDGE TOOLS

CENGAGENOW™ Throughout this chapter, sign in at **www.cengage.com/login** for online self-study and interactive tutorials based on your level of understanding.

⚙WL Online homework for this chapter may be assigned in Organic OWL.

We said in *A Preview of Carbonyl Compounds* that much of the chemistry of carbonyl compounds can be explained by just four fundamental reaction types: nucleophilic additions, nucleophilic acyl substitutions, α substitutions, and carbonyl condensations. Having studied the first two of these reactions in the past three chapters, let's now look in more detail at the third major carbonyl-group process—the **α-substitution reaction**.

Alpha-substitution reactions occur at the position *next to* the carbonyl group—the *α position*—and involve the substitution of an α hydrogen atom by an electrophile, E, through either an *enol* or *enolate ion* intermediate. Let's begin by learning more about these two species.

An enolate ion

A carbonyl compound

An enol

An alpha-substituted carbonyl compound

WHY THIS CHAPTER?

As with nucleophilic additions and nucleophilic acyl substitutions, many laboratory schemes, pharmaceutical syntheses, and biochemical pathways make frequent use of carbonyl α-substitution reactions. Their great value is that they constitute one of the few general methods for forming carbon–carbon bonds, thereby making it possible to build larger molecules from smaller precursors. We'll see how and why these reactions occur in this chapter.

22.1 | Keto–Enol Tautomerism

A carbonyl compound with a hydrogen atom on its α carbon rapidly equilibrates with its corresponding **enol** (Section 8.4). This rapid interconversion between two substances is a special kind of isomerism known as *keto–enol tautomerism*, from the Greek *tauto*, meaning "the same," and *meros*, meaning "part." The individual isomers are called **tautomers**.

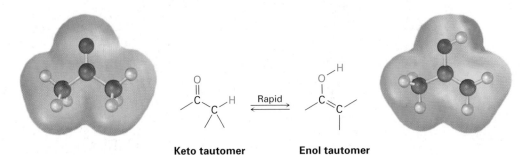

Keto tautomer   **Enol tautomer**

Note the difference between tautomers and resonance forms. Tautomers are constitutional isomers—different compounds with different structures—while resonance forms are different representations of a single structure. Tautomers have their *atoms* arranged differently, while resonance forms differ only in the position of their *electrons*. Note also that tautomers are *rapidly* interconvertible. Thus, keto and enol isomers are tautomers, but alkene isomers such as 1-butene and 2-butene are not, because they don't interconvert rapidly under normal circumstances.

But-1-ene   **But-2-ene**

Most carbonyl compounds exist almost exclusively in the keto form at equilibrium, and it's usually difficult to isolate the pure enol. For example, cyclohexanone contains only about 0.0001% of its enol tautomer at room temperature, and acetone contains only about 0.000 000 1% enol. The percentage of enol tautomer is even less for carboxylic acids, esters, and amides. Even though enols are difficult to isolate and are present only to a small extent at equilibrium, they are nevertheless responsible for much of the chemistry of carbonyl compounds because they are so reactive.

99.999 9%   0.000 1%   99.999 999 9%   0.000 000 1%

Cyclohexanone   **Acetone**

Keto–enol tautomerism of carbonyl compounds is catalyzed by both acids and bases. Acid catalysis occurs by protonation of the carbonyl oxygen atom to give an intermediate cation that loses H⁺ from its α carbon to yield a neutral enol (Figure 22.1). This proton loss from the cation intermediate is similar to what occurs during an E1 reaction when a carbocation loses H⁺ to form an alkene (Section 11.10).

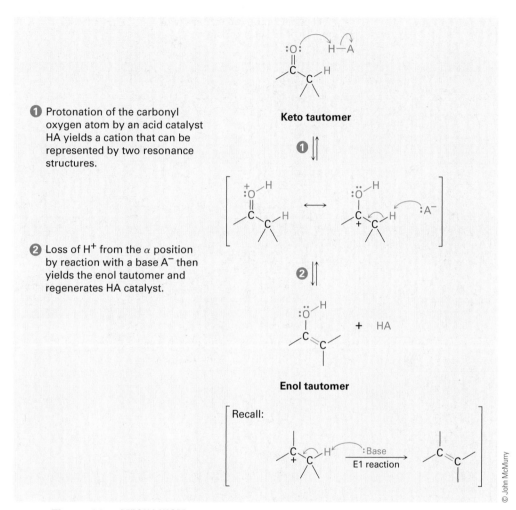

1 Protonation of the carbonyl oxygen atom by an acid catalyst HA yields a cation that can be represented by two resonance structures.

2 Loss of H⁺ from the α position by reaction with a base A⁻ then yields the enol tautomer and regenerates HA catalyst.

Figure 22.1 MECHANISM: Mechanism of acid-catalyzed enol formation. The protonated intermediate can lose H⁺, either from the oxygen atom to regenerate the keto tautomer or from the α carbon atom to yield an enol.

Base-catalyzed enol formation occurs because the carbonyl group makes the hydrogens on the α carbon weakly acidic. Thus, a carbonyl compound can donate one of its α hydrogens to the base, giving an **enolate ion** that is then protonated. Because the enolate ion is a resonance hybrid of two forms, it can be protonated either on the α carbon to regenerate the keto tautomer or on oxygen to give the enol tautomer (Figure 22.2).

Note that only the hydrogens on the α position of a carbonyl compound are acidic. Hydrogens at β, γ, δ, and so on, are not acidic and can't be removed by

Figure 22.2 MECHANISM:
Mechanism of base-catalyzed enol formation. The intermediate enolate ion, a resonance hybrid of two forms, can be protonated either on carbon to regenerate the starting keto tautomer or on oxygen to give an enol.

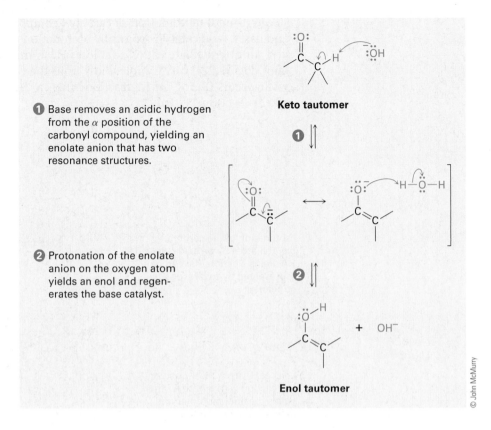

1 Base removes an acidic hydrogen from the α position of the carbonyl compound, yielding an enolate anion that has two resonance structures.

Keto tautomer

2 Protonation of the enolate anion on the oxygen atom yields an enol and regenerates the base catalyst.

Enol tautomer

© John McMurry

base. This unique behavior of α hydrogens is due to the fact that the resultant enolate ion is stabilized by a resonance form that places the charge on the electronegative oxygen.

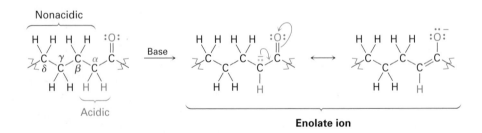

Problem 22.1 | Draw structures for the enol tautomers of the following compounds:
(a) Cyclopentanone   (b) Methyl thioacetate   (c) Ethyl acetate
(d) Propanal   (e) Acetic acid   (f) Phenylacetone

Problem 22.2 | How many acidic hydrogens does each of the molecules listed in Problem 22.1 have? Identify them.

Problem 22.3   Draw structures for all monoenol forms of the following molecule. Which would you expect to be most stable? Explain.

22.2 | Reactivity of Enols: The Mechanism of Alpha-Substitution Reactions

CENGAGENOW⁻ Click *Organic Process* to **view an animation showing substitution occurring alpha to a carbonyl**.

What kind of chemistry do enols have? Because their double bonds are electron-rich, enols behave as nucleophiles and react with electrophiles in much the same way that alkenes do. But because of resonance electron donation of a lone-pair of electrons on the neighboring oxygen, enols are more electron-rich and correspondingly more reactive than alkenes. Notice in the following electrostatic potential map of ethenol ($H_2C = CHOH$) how there is a substantial amount of electron density (yellow–red) on the α carbon.

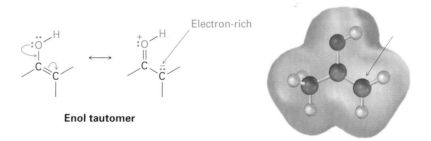

Enol tautomer

When an *alkene* reacts with an electrophile, such as HCl, initial addition of H^+ gives an intermediate cation and subsequent reaction with Cl^- yields an addition product (Section 6.7). When an *enol* reacts with an electrophile, however, only the initial addition step is the same. Instead of reacting with Cl^- to give an addition product, the intermediate cation loses the $-OH$ proton to give an α-substituted carbonyl compound. The general mechanism is shown in Figure 22.3.

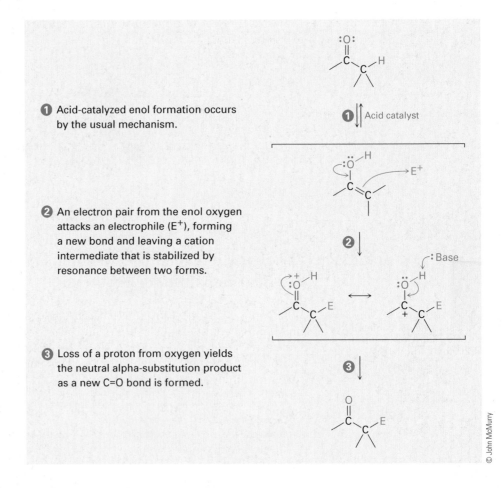

① Acid-catalyzed enol formation occurs by the usual mechanism.

② An electron pair from the enol oxygen attacks an electrophile (E$^+$), forming a new bond and leaving a cation intermediate that is stabilized by resonance between two forms.

③ Loss of a proton from oxygen yields the neutral alpha-substitution product as a new C=O bond is formed.

© John McMurry

22.3 | Alpha Halogenation of Aldehydes and Ketones

A particularly common α-substitution reaction in the laboratory is the halogenation of aldehydes and ketones at their α positions by reaction with Cl_2, Br_2, or I_2 in acidic solution. Bromine in acetic acid solvent is often used.

Acetophenone $\xrightarrow[\text{Acetic acid}]{Br_2}$ **α-Bromoacetophenone (72%)**

Remarkably, ketone halogenation also occurs in biological systems, particularly in marine alga, where dibromoacetaldehyde, bromoacetone, 1,1,1-tribromoacetone, and other related compounds have been found.

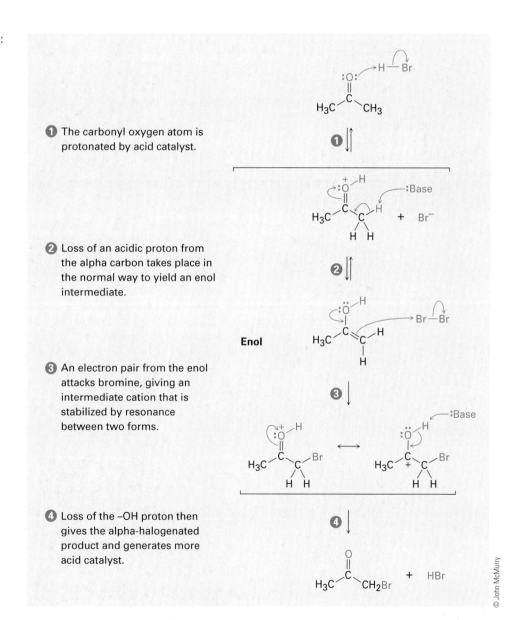

From the Hawaiian alga *Asparagopsis taxiformis*

The halogenation is a typical α-substitution reaction that proceeds by acid-catalyzed formation of an enol intermediate, as shown in Figure 22.4.

Figure 22.4 MECHANISM:
Mechanism of the acid-catalyzed bromination of acetone.

① The carbonyl oxygen atom is protonated by acid catalyst.

② Loss of an acidic proton from the alpha carbon takes place in the normal way to yield an enol intermediate.

③ An electron pair from the enol attacks bromine, giving an intermediate cation that is stabilized by resonance between two forms.

④ Loss of the –OH proton then gives the alpha-halogenated product and generates more acid catalyst.

© John McMurry

Evidence for the mechanism shown in Figure 22.4 includes the observation that acid-catalyzed halogenations show second-order kinetics and follow the rate law

$$\text{Reaction rate} = k\,[\text{Ketone}]\,[\text{H}^+]$$

In other words, the rate of halogenation depends only on the concentrations of ketone and acid and is independent of halogen concentration. Halogen is not involved in the rate-limiting step, so chlorination, bromination, and iodination of a given substrate all occur at the same rate.

Furthermore, if an aldehyde or ketone is treated with D_3O^+, the acidic α hydrogens are replaced by deuterium. For a given ketone, the rate of deuterium exchange is identical to the rate of halogenation, implying that a common intermediate is involved in both processes.

Enol

α-Bromo ketones are useful in the laboratory because they can be dehydrobrominated by base treatment to yield α,β-unsaturated ketones. For example, 2-methylcyclohexanone gives 2-bromo-2-methylcyclohexanone on halogenation, and the α-bromo ketone gives 2-methyl-2-cyclohexenone when heated in pyridine. The reaction takes place by an E2 elimination pathway (Section 11.8) and is a good method for introducing C=C bonds into molecules. Note that bromination of 2-methylcyclohexanone occurs primarily on the more highly substituted α position because the more highly substituted enol is favored over the less highly substituted one (Section 6.6).

| 2-Methylcyclo-hexanone | 2-Bromo-2-methyl-cyclohexanone | 2-Methylcyclo-hex-2-enone (63%) |

Problem 22.4 | Write the complete mechanism of the deuteration of acetone on treatment with D_3O^+.

Problem 22.5 | Show how you might prepare 1-penten-3-one from 3-pentanone.

22.4 | Alpha Bromination of Carboxylic Acids: The Hell–Volhard–Zelinskii Reaction

CENGAGENOW™ Click *Organic Interactive* to **use a web-based palette to predict products from a variety of halogenation reactions of carbonyls and carboxylic acids.**

The α bromination of carbonyl compounds by Br_2 in acetic acid is limited to aldehydes and ketones because acids, esters, and amides don't enolize to a sufficient extent. Carboxylic acids, however, can be α brominated by a mixture of Br_2 and PBr_3 in the **Hell–Volhard–Zelinskii (HVZ) reaction**.

$$CH_3CH_2CH_2CH_2CH_2CH_2\overset{\overset{\displaystyle O}{\|}}{C}OH \xrightarrow[\text{2. H}_2\text{O}]{\text{1. Br}_2\text{, PBr}_3} CH_3CH_2CH_2CH_2CH_2\overset{\overset{\displaystyle O}{\|}}{\underset{\underset{\displaystyle Br}{|}}{C}H}\overset{}{C}OH$$

Heptanoic acid　　　　　　　　　**2-Bromoheptanoic acid (90%)**

The Hell–Volhard–Zelinskii reaction is a bit more complex than it looks and actually involves α substitution of an *acid bromide enol* rather than a carboxylic acid enol. The process begins with reaction of the carboxylic acid with PBr_3 to form an acid bromide plus HBr (Section 21.4). The HBr then catalyzes enolization of the acid bromide, and the resultant enol reacts with Br_2 in an α-substitution reaction to give an α-bromo acid bromide. Addition of water hydrolyzes the acid bromide in a nucleophilic acyl substitution reaction and yields the α-bromo carboxylic acid product.

Carboxylic　　　　**Acid bromide**　　　　**Acid bromide**
acid　　　　　　　　　　　　　　　　　**enol**

Problem 22.6 | If methanol rather than water is added at the end of a Hell–Volhard–Zelinskii reaction, an ester rather than an acid is produced. Show how you could carry out the following transformation, and propose a mechanism for the ester-forming step.

$$CH_3CH_2\overset{\overset{\displaystyle CH_3}{|}}{C}H CH_2\overset{\overset{\displaystyle O}{\|}}{C}OH \xrightarrow{?} CH_3CH_2\overset{\overset{\displaystyle CH_3}{|}}{C}H\overset{\underset{\underset{\displaystyle Br}{|}}{}}{C}H\overset{\overset{\displaystyle O}{\|}}{C}OCH_3$$

22.5 | Acidity of Alpha Hydrogen Atoms: Enolate Ion Formation

As noted in Section 22.1, a hydrogen on the α position of a carbonyl compound is weakly acidic and can be removed by a strong base to yield an enolate ion. In comparing acetone ($pK_a = 19.3$) with ethane ($pK_a \approx 60$), for instance, the

CENGAGENOW™ Click *Organic Interactive* to **learn to draw the structures of carbonyl enolates and predict their reactivity.**

presence of a neighboring carbonyl group increases the acidity of the ketone over the alkane by a factor of 10^{40}.

Acetone
(pK_a = 19.3)

Ethane
(p$K_a \approx$ 60)

Abstraction of a proton from a carbonyl compound occurs when the α C—H bond is oriented roughly parallel to the p orbitals of the carbonyl group. The α carbon atom of the enolate ion is sp^2-hybridized and has a p orbital that overlaps the neighboring carbonyl p orbitals. Thus, the negative charge is shared by the electronegative oxygen atom, and the enolate ion is stabilized by resonance (Figure 22.5).

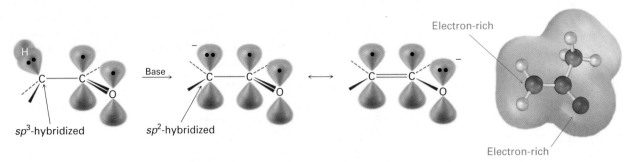

Figure 22.5 Mechanism of enolate ion formation by abstraction of an α proton from a carbonyl compound. The enolate ion is stabilized by resonance, and the negative charge (red) is shared by the oxygen and the α carbon atom, as indicated by the electrostatic potential map.

Carbonyl compounds are more acidic than alkanes for the same reason that carboxylic acids are more acidic than alcohols (Section 20.2). In both cases, the anions are stabilized by resonance. Enolate ions differ from carboxylate ions, however, in that their two resonance forms are not equivalent—the form with the negative charge on oxygen is lower in energy than the form with the charge on carbon. Nevertheless, the principle behind resonance stabilization is the same in both cases.

Acetone
(pK_a = 19.3)

Nonequivalent resonance forms

Acetic acid
(pK_a = 4.76)   **Equivalent resonance forms**

Because carbonyl compounds are only weakly acidic, a strong base is needed for enolate ion formation. If an alkoxide such as sodium ethoxide is used as base, deprotonation takes place only to the extent of about 0.1% because acetone is a weaker acid than ethanol (pK_a = 16). If, however, a more powerful base such as sodium hydride (NaH) or lithium diisopropylamide [LiN(i-C$_3$H$_7$)$_2$] is used, a carbonyl compound can be completely converted into its enolate ion. Lithium diisopropylamide (LDA), which is easily prepared by reaction of the strong base butyllithium with diisopropylamine, is widely used in the laboratory as a base for preparing enolate ions from carbonyl compounds.

Many types of carbonyl compounds, including aldehydes, ketones, esters, thioesters, acids, and amides, can be converted into enolate ions by reaction with LDA. Table 22.1 lists the approximate pK_a values of different types of carbonyl compounds and shows how these values compare to other acidic substances we've seen. Note that nitriles, too, are acidic and can be converted into enolate-like anions.

When a hydrogen atom is flanked by two carbonyl groups, its acidity is enhanced even more. Table 22.1 thus shows that compounds such as 1,3-diketones (β-diketones), 3-oxo esters (β-keto esters), and 1,3-diesters are more acidic than water. This enhanced acidity of β-dicarbonyl compounds is due to the stabilization of the resultant enolate ions by delocalization of the negative charge over both carbonyl groups. The enolate ion of 2,4-pentanedione, for instance, has three resonance forms. Similar resonance forms can be drawn for other doubly stabilized enolate ions.

Pentane-2,4-dione (pK_a = 9)

Base

Table 22.1 | Acidity Constants for Some Organic Compounds

Functional group	Example	pK_a
Carboxylic acid	CH₃COH	5
1,3-Diketone	CH₃CCH₂CCH₃	9
3-Keto ester	CH₃CCH₂COCH₃	11
1,3-Diester	CH₃OCCH₂COCH₃	13
Alcohol	CH₃OH	16
Acid chloride	CH₃CCl	16
Aldehyde	CH₃CH	17
Ketone	CH₃CCH₃	19
Thioester	CH₃CSCH₃	21
Ester	CH₃COCH₃	25
Nitrile	CH₃C≡N	25
N,N-Dialkylamide	CH₃CN(CH₃)₂	30
Dialkylamine	HN(i-C₃H₇)₂	40

| **WORKED EXAMPLE 22.1** | ***Identifying the Acidic Hydrogens in a Compound*** |

Identify the most acidic hydrogens in each of the following compounds, and rank the compounds in order of increasing acidity:

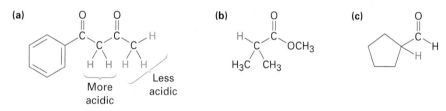

Strategy   Hydrogens on carbon next to a carbonyl group are acidic. In general, a β-dicarbonyl compound is most acidic, a ketone or aldehyde is next most acidic, and a carboxylic acid derivative is least acidic. Remember that alcohols, phenols, and carboxylic acids are also acidic because of their $-OH$ hydrogens.

Solution   The acidity order is (a) > (c) > (b). Acidic hydrogens are shown in red.

Problem 22.7   Identify the most acidic hydrogens in each of the following molecules:
(a) CH_3CH_2CHO (b) $(CH_3)_3CCOCH_3$ (c) CH_3CO_2H
(d) Benzamide   (e) $CH_3CH_2CH_2CN$ (f) $CH_3CON(CH_3)_2$

Problem 22.8   Draw a resonance structure of the acetonitrile anion, $^-{:}CH_2C{\equiv}N$, and account for the acidity of nitriles.

22.6 | Reactivity of Enolate Ions

Enolate ions are more useful than enols for two reasons. First, pure enols can't normally be isolated but are instead generated only as short-lived intermediates in low concentration. By contrast, stable solutions of pure enolate ions are easily prepared from most carbonyl compounds by reaction with a strong base. Second, enolate ions are more reactive than enols and undergo many reactions that enols don't. Whereas enols are neutral, enolate ions are negatively charged, making them much better nucleophiles. As a result, enolate ions are more common than enols in both laboratory and biological chemistry.

Because they are resonance hybrids of two nonequivalent forms, enolate ions can be looked at either as vinylic alkoxides ($C{=}C{-}O^-$) or as α-keto carbanions ($^-$C−C=O). Thus, enolate ions can react with electrophiles either on oxygen or on carbon. Reaction on oxygen yields an enol derivative, while reaction on carbon yields an α-substituted carbonyl compound (Figure 22.6). Both kinds of reactivity are known, but reaction on carbon is more common.

Active Figure 22.6 The electrostatic potential map of acetone enolate ion shows how the negative charge is delocalized over both the oxygen and the α carbon. As a result, two modes of reaction of an enolate ion with an electrophile E$^+$ are possible. Reaction on carbon to yield an α-substituted carbonyl product is more common. *Sign in at* **www.cengage.com/login** *to see a simulation based on this figure and to take a short quiz.*

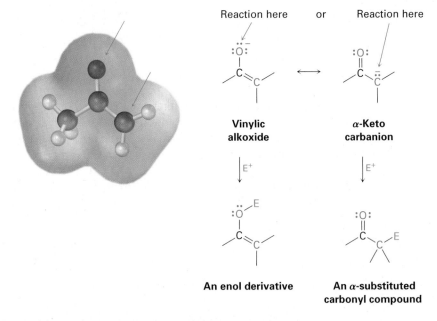

As an example of enolate-ion reactivity, aldehydes and ketones undergo base-promoted α halogenation. Even relatively weak bases such as hydroxide ion are effective for halogenation because it's not necessary to convert the ketone completely into its enolate ion. As soon as a small amount of enolate is generated, it reacts immediately with the halogen, removing it from the reaction and driving the equilibrium for further enolate ion formation.

Base-promoted halogenation of aldehydes and ketones is little used in practice because it's difficult to stop the reaction at the monosubstituted product. An α-halogenated ketone is generally more acidic than the starting, unsubstituted ketone because of the electron-withdrawing inductive effect of the halogen atom. Thus, the monohalogenated products are themselves rapidly turned into enolate ions and further halogenated.

If excess base and halogen are used, a methyl ketone is triply halogenated and then cleaved by base in the *haloform reaction*. The products are a carboxylic acid plus a so-called haloform (chloroform, $CHCl_3$; bromoform,

$CHBr_3$; or iodoform, CHI_3). Note that the second step of the reaction is a nucleophilic acyl substitution of $^-CX_3$ by ^-OH. That is, a halogen-stabilized *carbanion* acts as a leaving group.

A methyl ketone

where X = Cl, Br, I

Problem 22.9 | Why do you suppose ketone halogenations in acidic media referred to as being acid-*catalyzed*, whereas halogenations in basic media are base-*promoted*? In other words, why is a full equivalent of base required for halogenation?

22.7 | Alkylation of Enolate Ions

CENGAGENOW™ Click *Organic Interactive* to **use a web-based palette to predict products in halogenation and alkylation reactions of carbonyl enolates**.

Perhaps the single most important reaction of enolate ions is their *alkylation* by treatment with an alkyl halide or tosylate, thereby forming a new C−C bond and joining two smaller pieces into one larger molecule. Alkylation occurs when the nucleophilic enolate ion reacts with the electrophilic alkyl halide in an S_N2 reaction and displaces the leaving group by backside attack.

Enolate ion

Alkylation reactions are subject to the same constraints that affect all S_N2 reactions (Section 11.3). Thus, the leaving group X in the alkylating agent R−X can be chloride, bromide, iodide, or tosylate. The alkyl group R should be primary or methyl, and preferably should be allylic or benzylic. Secondary halides react poorly, and tertiary halides don't react at all because a competing E2 elimination of HX occurs instead. Vinylic and aryl halides are also unreactive because backside approach is sterically prevented.

$$R-X \begin{cases} -X: & \text{Tosylate} \; > \; -I \; > \; -Br \; > \; -Cl \\ R-: & \text{Allylic} \approx \text{Benzylic} \; > \; H_3C- \; > \; RCH_2- \end{cases}$$

The Malonic Ester Synthesis

One of the oldest and best known carbonyl alkylation reactions is the **malonic ester synthesis**, a method for preparing a carboxylic acid from an alkyl halide while lengthening the carbon chain by two atoms.

$$R-X \xrightarrow[\text{synthesis}]{\text{Malonic ester}} \underset{\text{H H}}{\overset{R}{\text{C}}}\text{CO}_2\text{H}$$

Diethyl propanedioate, commonly called diethyl malonate or *malonic ester,* is more acidic than monocarbonyl compounds ($pK_a = 13$) because its α hydrogens are flanked by two carbonyl groups. Thus, malonic ester is easily converted into its enolate ion by reaction with sodium ethoxide in ethanol. The enolate ion, in turn, is a good nucleophile that reacts rapidly with an alkyl halide to give an α-substituted malonic ester. Note in the following examples that the abbreviation "Et" is used for an ethyl group, $-\text{CH}_2\text{CH}_3$.

Diethyl propanedioate   **Sodio malonic ester**   **An alkylated**
(malonic ester)   **malonic ester**

The product of malonic ester alkylation has one acidic α hydrogen atom left, so the alkylation process can be repeated a second time to yield a dialkylated malonic ester.

An alkylated   **A dialkylated**
malonic ester   **malonic ester**

On heating with aqueous hydrochloric acid, the alkylated (or dialkylated) malonic ester undergoes hydrolysis of its two ester groups followed by *decarboxylation* (loss of CO_2) to yield a substituted monoacid.

An alkylated   **A carboxylic**
malonic ester   **acid**

Decarboxylation is not a general reaction of carboxylic acids. Rather, it is unique to compounds that have a *second* carbonyl group two atoms away from the $-CO_2H$. That is, only substituted malonic acids and β-keto acids undergo loss of CO_2 on heating. The decarboxylation reaction occurs by a cyclic mechanism and involves initial formation of an enol, thereby accounting for the need to have a second carbonyl group appropriately positioned.

A diacid   **An acid enol**   **A carboxylic acid**

A β-keto acid   **An enol**   **A ketone**

As noted previously, the overall effect of the malonic ester synthesis is to convert an alkyl halide into a carboxylic acid while lengthening the carbon chain by two atoms.

The malonic ester synthesis can also be used to prepare *cyclo*alkane-carboxylic acids. For example, when 1,4-dibromobutane is treated with diethyl malonate in the presence of 2 equivalents of sodium ethoxide base, the second alkylation step occurs *intramolecularly* to yield a cyclic product. Hydrolysis and decarboxylation then give cyclopentanecarboxylic acid. Three-, four-, five-, and six-membered rings can be prepared in this way, but yields decrease for larger ring sizes.

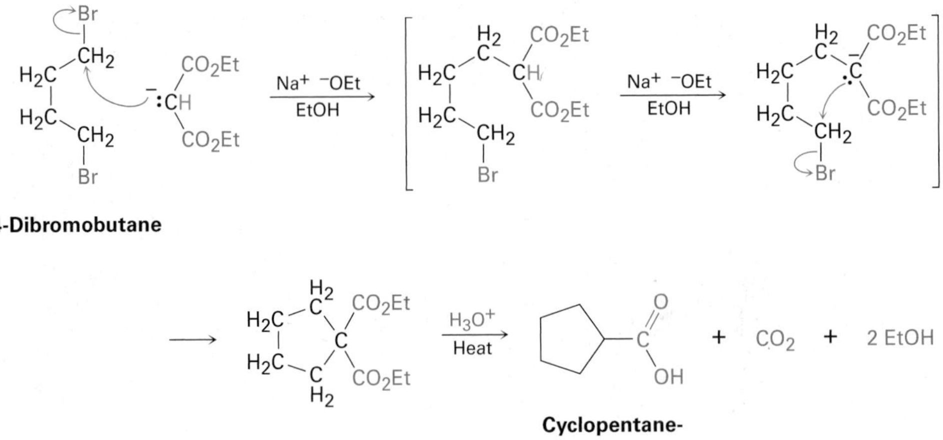

1,4-Dibromobutane

Cyclopentane-carboxylic acid

WORKED EXAMPLE 22.2	***Using the Malonic Ester Synthesis to Prepare a Carboxylic Acid***

How would you prepare heptanoic acid using a malonic ester synthesis?

Strategy   The malonic ester synthesis converts an alkyl halide into a carboxylic acid having two more carbons. Thus, a *seven*-carbon acid chain must be derived from the *five*-carbon alkyl halide 1-bromopentane.

Solution

Problem 22.10   How could you use a malonic ester synthesis to prepare the following compounds? Show all steps.

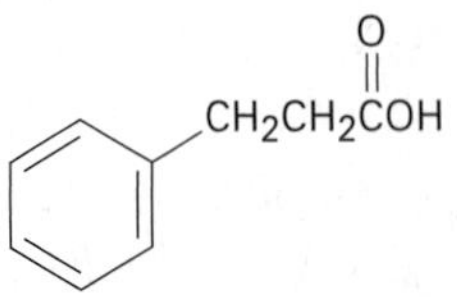

Problem 22.11   Monoalkylated and dialkylated acetic acids can be prepared by the malonic ester synthesis, but trialkylated acetic acids (R_3CCO_2H) can't be prepared. Explain.

Problem 22.12 | How could you use a malonic ester synthesis to prepare the following compound?

The Acetoacetic Ester Synthesis

Just as the malonic ester synthesis converts an alkyl halide into a carboxylic acid, the **acetoacetic ester synthesis** converts an alkyl halide into a methyl ketone having three more carbons.

$$R-X \xrightarrow[\text{synthesis}]{\text{Acetoacetic ester}} \underset{\underset{H \quad H}{\quad}}{R-\underset{}{C}-\overset{\overset{O}{\parallel}}{C}-CH_3}$$

Ethyl 3-oxobutanoate, commonly called ethyl acetoacetate or *acetoacetic ester,* is much like malonic ester in that its α hydrogens are flanked by two carbonyl groups. It is therefore readily converted into its enolate ion, which can be alkylated by reaction with an alkyl halide. A second alkylation can also be carried out if desired, since acetoacetic ester has two acidic α hydrogens.

EtO₂C—C(H)(H)—C(=O)—CH₃ $\xrightarrow[\text{EtOH}]{\text{Na}^+ \ ^-\text{OEt}}$ [Na⁺ EtO₂C—C⁻(H)—C(=O)—CH₃] $\xrightarrow{RX}$ EtO₂C—C(H)(R)—C(=O)—CH₃

Ethyl acetoacetate (acetoacetic ester)   **Sodio acetoacetic ester**   **A monoalkylated acetoacetic ester**

EtO₂C—C(H)(R)—C(=O)—CH₃ $\xrightarrow[\text{EtOH}]{\text{Na}^+ \ ^-\text{OEt}}$ [Na⁺ EtO₂C—C⁻(R)—C(=O)—CH₃] $\xrightarrow{R'X}$ EtO₂C—C(R)(R')—C(=O)—CH₃

A monoalkylated acetoacetic ester   **A dialkylated acetoacetic ester**

On heating with aqueous HCl, the alkylated (or dialkylated) acetoacetic ester is hydrolyzed to a β-keto acid, which then undergoes decarboxylation to yield a ketone product. The decarboxylation occurs in the same way as in the malonic ester synthesis and involves a ketone enol as the initial product.

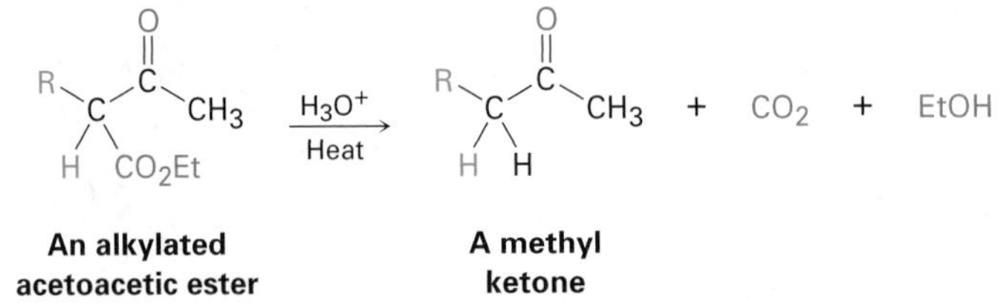

An alkylated acetoacetic ester

$\xrightarrow[\text{Heat}]{\text{H}_3\text{O}^+}$

A methyl ketone + CO_2 + EtOH

The three-step sequence of (1) enolate ion formation, (2) alkylation, and (3) hydrolysis/decarboxylation is applicable to all β-keto esters with acidic α hydrogens, not just to acetoacetic ester itself. For example, *cyclic* β-keto esters such as ethyl 2-oxocyclohexanecarboxylate can be alkylated and decarboxylated to give 2-substituted cyclohexanones.

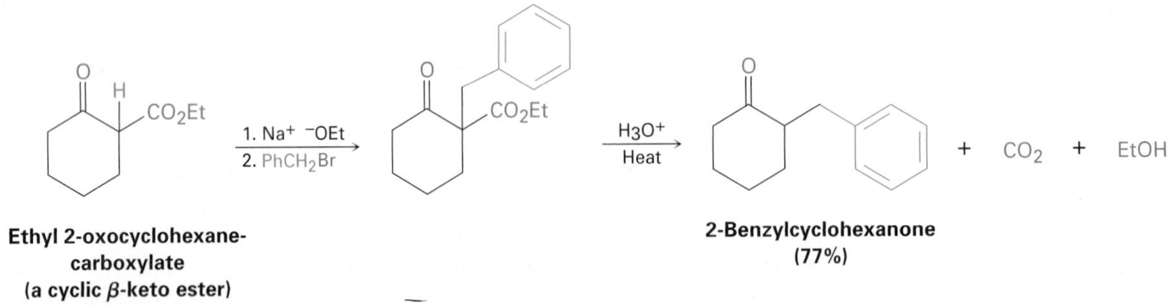

Ethyl 2-oxocyclohexane-carboxylate (a cyclic β-keto ester)

1. Na$^+$ $^-$OEt
2. PhCH$_2$Br

$\xrightarrow[\text{Heat}]{\text{H}_3\text{O}^+}$

2-Benzylcyclohexanone (77%) + CO_2 + EtOH

WORKED EXAMPLE 22.3 | ***Using the Acetoacetic Ester Synthesis to Prepare a Ketone***

How would you prepare 2-pentanone by an acetoacetic ester synthesis?

Strategy The acetoacetic ester synthesis yields a methyl ketone by adding three carbons to an alkyl halide.

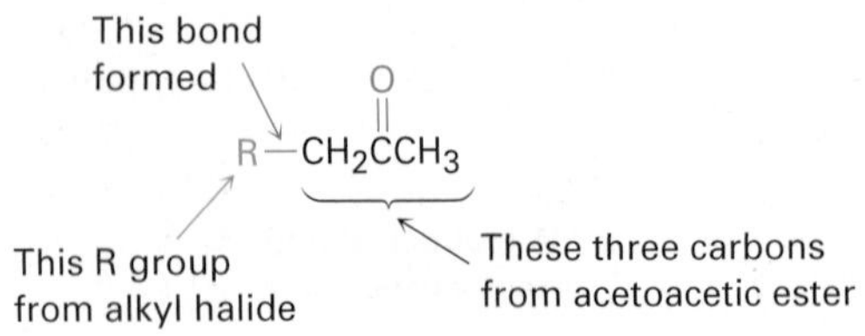

This bond formed

$$R - CH_2\overset{O}{\overset{\|}{C}}CH_3$$

This R group from alkyl halide

These three carbons from acetoacetic ester

Thus, the acetoacetic ester synthesis of 2-pentanone must involve reaction of bromoethane.

Solution

$$CH_3CH_2Br \ + \ EtOCCH_2CCH_3 \xrightarrow[\text{2. H}_3O^+,\text{ heat}]{\text{1. Na}^+ \ ^-\text{OEt}} CH_3CH_2CH_2CCH_3$$

Pentan-2-one

Problem 22.13

What alkyl halides would you use to prepare the following ketones by an acetoacetic ester synthesis?

(a)
$$\begin{array}{c} CH_3 \quad\quad O \\ | \quad\quad\quad\; \| \\ CH_3CHCH_2CH_2CCH_3 \end{array}$$

(b)

$$-CH_2CH_2CH_2CCH_3$$

Problem 22.14

Which of the following compounds *cannot* be prepared by an acetoacetic ester synthesis? Explain.
(a) Phenylacetone   (b) Acetophenone   (c) 3,3-Dimethyl-2-butanone

Problem 22.15

How would you prepare the following compound using an acetoacetic ester synthesis?

Direct Alkylation of Ketones, Esters, and Nitriles

Both the malonic ester synthesis and the acetoacetic ester synthesis are easy to carry out because they involve unusually acidic dicarbonyl compounds. As a result, relatively mild bases such as sodium ethoxide in ethanol as solvent can be used to prepare the necessary enolate ions. Alternatively, however, it's also possible in many cases to directly alkylate the α position of *mono*carbonyl compounds. A strong, sterically hindered base such as LDA is needed so that complete conversion to the enolate ion takes place rather than a nucleophilic addition, and a nonprotic solvent must be used.

Ketones, esters, and nitriles can all be alkylated using LDA or related dialkylamide bases in THF. Aldehydes, however, rarely give high yields of pure products because their enolate ions undergo carbonyl condensation reactions instead of alkylation. (We'll study this condensation reaction in the next chapter.) Some specific examples of alkylation reactions are shown.

Lactone

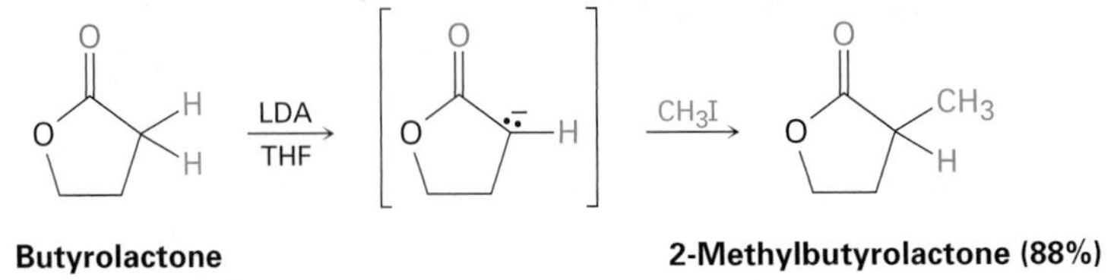

Butyrolactone

2-Methylbutyrolactone (88%)

Ester

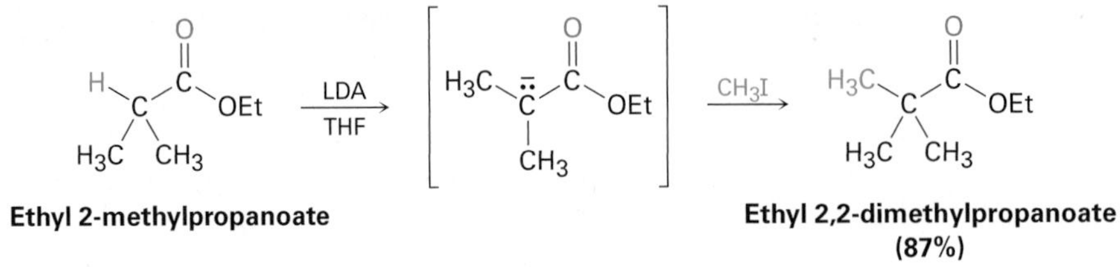

Ethyl 2-methylpropanoate

**Ethyl 2,2-dimethylpropanoate
(87%)**

Ketone

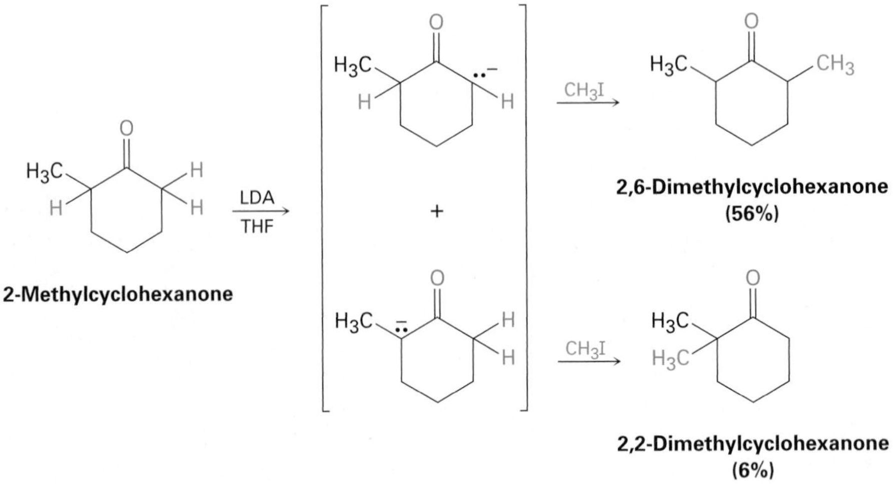

2-Methylcyclohexanone

**2,6-Dimethylcyclohexanone
(56%)**

**2,2-Dimethylcyclohexanone
(6%)**

Nitrile

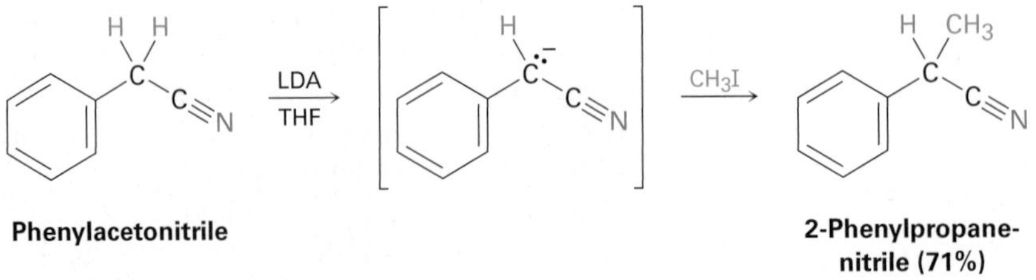

Phenylacetonitrile

**2-Phenylpropane-
nitrile (71%)**

Note in the ketone example that alkylation of 2-methylcyclohexanone leads to a mixture of products because both possible enolate ions are formed. In general, the major product in such cases occurs by alkylation at the less hindered, more accessible position. Thus, alkylation of 2-methylcyclohexanone occurs primarily at C6 (secondary) rather than C2 (tertiary).

WORKED EXAMPLE 22.4   *Using an Alkylation Reaction to Prepare a Substituted Ester*

How might you use an alkylation reaction to prepare ethyl 1-methylcyclohexane-carboxylate?

Ethyl 1-methylcyclohexanecarboxylate

Strategy   An alkylation reaction is used to introduce a methyl or primary alkyl group onto the α position of a ketone, ester, or nitrile by S_N2 reaction of an enolate ion with an alkyl halide. Thus, we need to look at the target molecule and identify any methyl or primary alkyl groups attached to an α carbon. In the present instance, the target has an α methyl group, which might be introduced by alkylation of an ester enolate ion with iodomethane.

Solution

Ethyl cyclohexane-   **Ethyl 1-methylcyclo-**
carboxylate   **hexanecarboxylate**

Problem 22.16   How might you prepare the following compounds using an alkylation reaction as the key step?

Biological Alkylations

Alkylations are rare but not unknown in biological chemistry. One example occurs during biosynthesis of the antibiotic indolmycin from indolyl-pyruvate when a base abstracts an acidic hydrogen from an α position and the resultant enolate ion carries out an S_N2 alkylation reaction on the methyl group of *S*-adenosylmethionine (SAM; Section 11.6). Although it's convenient to speak of "enolate ion" intermediates in biological pathways, it's unlikely that they exist for long in an aqueous cellular environment. Rather, proton removal and alkylation probably occur at essentially the same time (Figure 22.7).

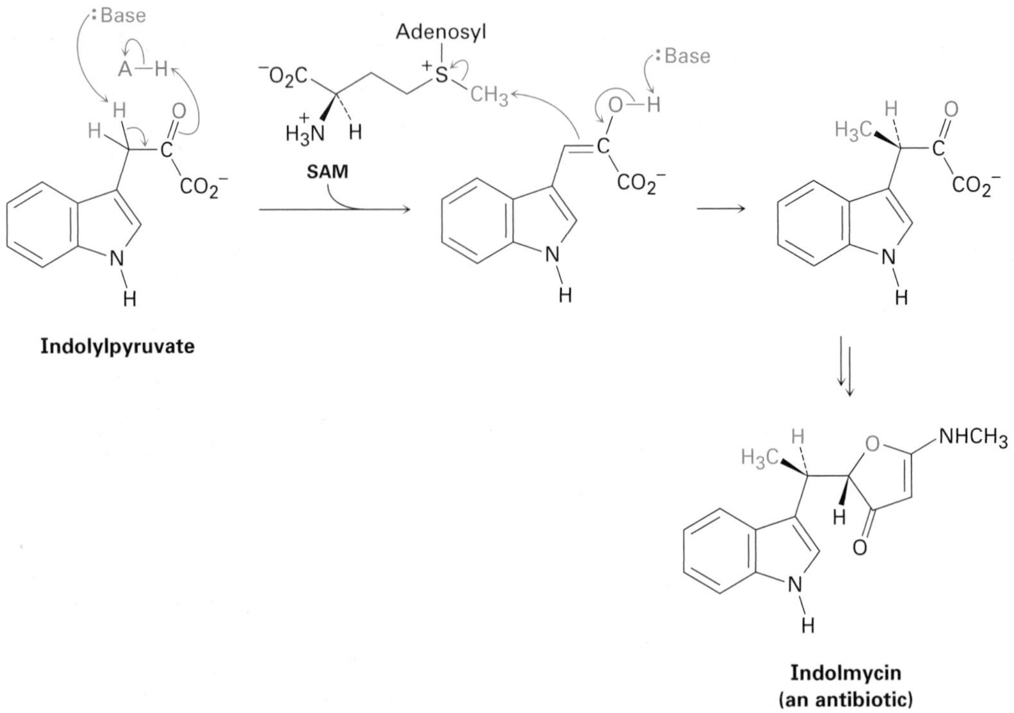

Figure 22.7 The biosynthesis of indolmycin from indolylpyruvate occurs through a pathway that includes an alkylation reaction of a short-lived enolate ion intermediate.

Focus On . . .

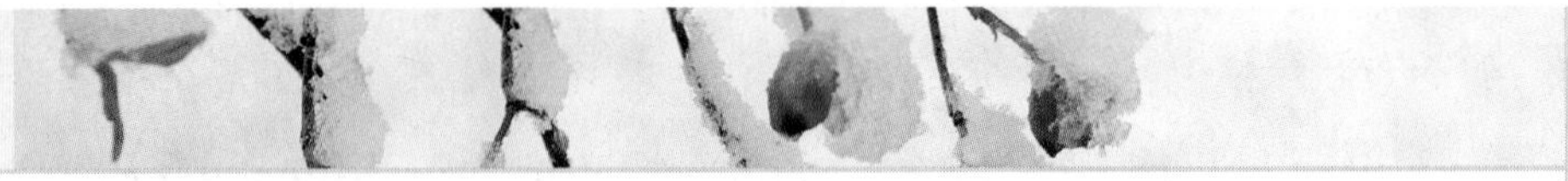

X-Ray Crystallography

Determining the three-dimensional shape of an object around you is easy—you just look at it, let your eyes focus the light rays reflected from the object, and let your brain assemble the data into a recognizable image. If the object is small, you use a microscope and let the microscope lens focus the visible light. Unfortunately, there is a limit to what you can see, even with the best optical microscope. Called the "diffraction limit," you can't see anything smaller than the wavelength of light you are using for the observation. Visible light has wavelengths of several hundred nanometers, but atoms in molecules have dimension on the order of 0.1 nm. Thus, to "see" a molecule—whether a small one in the laboratory or a large, complex enzyme with a molecular weight in the tens of thousands—you need wavelengths in the 0.1 nm range, which corresponds to X rays.

Let's say that we want to determine the structure and shape of an enzyme or other biological molecule. The technique used is called *X-ray crystallography*.

(continued)

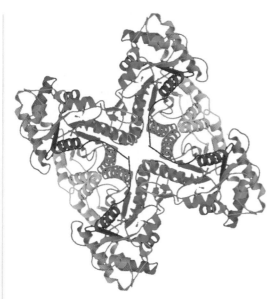

The structure of human muscle fructose-1,6-bisphosphate aldolase, as determined by X-ray crystallography and downloaded from the Protein Data Bank. (PDB ID: 1ALD; Gamblin, S. J., Davies, G. J., Grimes, J. M., Jackson, R. M., Littlechild, J. A., Watson, H. C. Activity and specificity of human aldolases. *J. Mol. Biol.* v219, pp. 573–576, 1991.)

First, the molecule is crystallized (which often turns out to be the most difficult and time-consuming part of the entire process) and a small crystal with a dimension of 0.4 to 0.5 mm on its longest axis is glued to the end of a glass fiber. The fiber and attached crystal are then mounted in an instrument called an *X-ray diffractometer*, consisting of a radiation source, a sample positioning and orienting device that can rotate the crystal in any direction, a detector, and a controlling computer.

Once mounted in the diffractometer, the crystal is irradiated with X rays, usually so-called Cu$K\alpha$ radiation with a wavelength of 0.154 nm. When the X rays strike the enzyme crystal, they interact with electrons in the molecule and are scattered into a diffraction pattern, which, when detected and visualized, appears as a series of intense spots against a null background.

Manipulation of the diffraction pattern to extract three-dimensional molecular data is a complex process, but the final result is that an electron-density map of the molecule is produced. Because electrons are largely localized around atoms, any two centers of electron density located within bonding distance of each other are assumed to represent bonded atoms, leading to a recognizable chemical structure. So important is this structural information for biochemistry that an online database of more than 40,000 biological substances has been created. Operated by Rutgers University and funded by the U.S. National Science Foundation, the Protein Data Bank (PDB) is a worldwide repository for processing and distributing three-dimensional structural data for biological macromolecules. We'll see how to access the PDB in the Chapter 26 *Focus On.*

SUMMARY AND KEY WORDS

The **α-substitution reaction** of a carbonyl compound through either an **enol** or **enolate ion** intermediate is one of the four fundamental reaction types in carbonyl-group chemistry.

Carbonyl compounds are in a rapid equilibrium with their enols, a process called keto–enol tautomerism. Although enol **tautomers** are normally present to only a small extent at equilibrium and can't usually be isolated in pure form, they nevertheless contain a highly nucleophilic double bond and react with electrophiles. For example, aldehydes and ketones are rapidly halogenated at the α position by reaction with Cl_2, Br_2, or I_2 in acetic acid solution. Alpha bromination of carboxylic acids can be similarly accomplished by the **Hell–Volhard–Zelinskii (HVZ) reaction**, in which an acid is treated with Br_2 and PBr_3. The α-halogenated products can then undergo base-induced E2 elimination to yield α,β-unsaturated carbonyl compounds.

Alpha hydrogen atoms of carbonyl compounds are weakly acidic and can be removed by strong bases, such as lithium diisopropylamide (LDA), to yield nucleophilic enolate ions. The most important reaction of enolate ions is their S_N2 alkylation with alkyl halides. The **malonic ester synthesis** converts an alkyl halide into a carboxylic acid with the addition of two carbon atoms. Similarly, the **acetoacetic ester synthesis** converts an alkyl halide into a methyl ketone. In addition, many carbonyl compounds, including ketones, esters, and nitriles, can be directly alkylated by treatment with LDA and an alkyl halide.

SUMMARY OF REACTIONS

1. Aldehyde/ketone halogenation (Section 22.3)

2. Hell–Volhard–Zelinskii bromination of acids (Section 22.4)

3. Dehydrobromination of α-bromo ketones (Section 22.3)

4. Haloform reaction (Section 22.6)

5. Alkylation of enolate ions (Section 22.7)
   (a) Malonic ester synthesis

   (b) Acetoacetic ester synthesis

   (c) Direct alkylation of ketones

   (d) Direct alkylation of esters

   (e) Direct alkylation of nitriles

EXERCISES

Organic KNOWLEDGE TOOLS

CENGAGENOW˙ Sign in at **www.cengage.com/login** to assess your knowledge of this chapter's topics by taking a pre-test. The pre-test will link you to interactive organic chemistry resources based on your score in each concept area.

⌂WL Online homework for this chapter may be assigned in Organic OWL.

■ indicates problems assignable in Organic OWL.

VISUALIZING CHEMISTRY

(Problems 22.1–22.16 appear within the chapter.)

22.17 ■ Show the steps in preparing each of the following substances, using either a malonic ester synthesis or an acetoacetic ester synthesis:

(a)   **(b)**

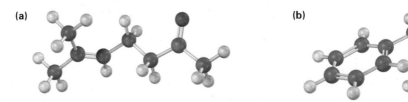

22.18 Unlike most β-diketones, the following β-diketone has no detectable enol content and is about as acidic as acetone. Explain.

22.19 ■ For a given α hydrogen atom to be acidic, the C—H bond must be parallel to the *p* orbitals of the C=O bond (that is, perpendicular to the plane of the adjacent carbonyl group). Identify the most acidic hydrogen atom in the conformation shown for the following structure. Is it axial or equatorial?

ADDITIONAL PROBLEMS

22.20 ■ Identify all the acidic hydrogens (pK_a < 25) in the following molecules:

(a)

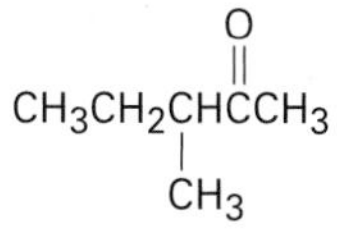

$$CH_3CH_2CHCCH_3$$
(with O double bond on the C, and CH_3 substituent below)

(b) (cyclopentane-1,3-dione, O=⬠=O)

(c)

$$HOCH_2CH_2CC\equiv CCH_3$$
(with O double bond)

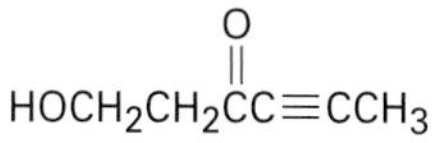

(d) 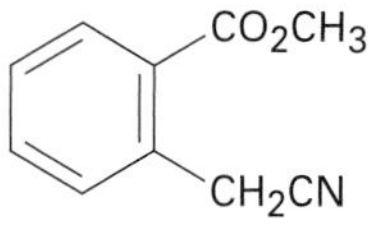 (benzene ring with CO_2CH_3 and CH_2CN substituents)

(e) 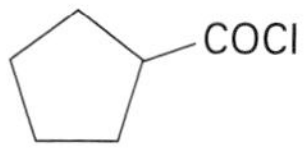 (cyclopentane with COCl substituent)

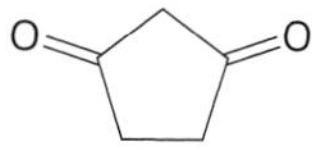

(f)

$$CH_3CH_2CC=CH_2$$
(with O double bond and CH_3 substituent below)

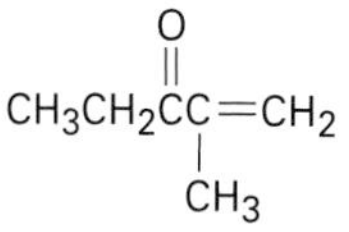

22.21 ■ Rank the following compounds in order of increasing acidity:

(a) $CH_3CH_2CO_2H$

(b) CH_3CH_2OH

(c) $(CH_3CH_2)_2NH$

(d) CH_3COCH_3

(e)

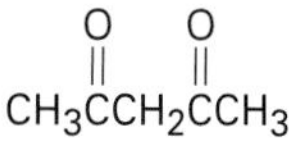

$$CH_3CCH_2CCH_3$$
(with two O double bonds)

(f) CCl_3CO_2H

22.22 ■ Write resonance structures for the following anions:

(a)

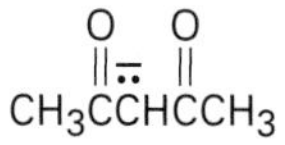

$$CH_3CCHCCH_3$$
(with two O double bonds)

(b)

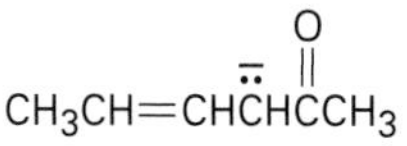

$$CH_3CH=CHCHCCH_3$$
(with O double bond)

(c)

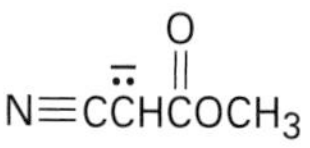

$$N\equiv CCHCOCH_3$$
(with O double bond)

(d)

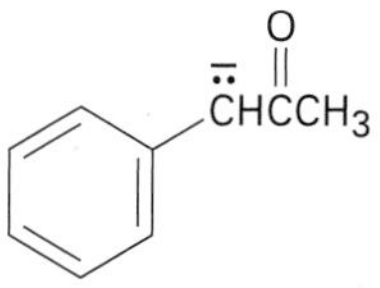

(benzene ring with) $CHCCH_3$ (with O double bond)

(e)

(indane-1,3-dione ring system with) $-C$ (with O double bond and OCH_3)

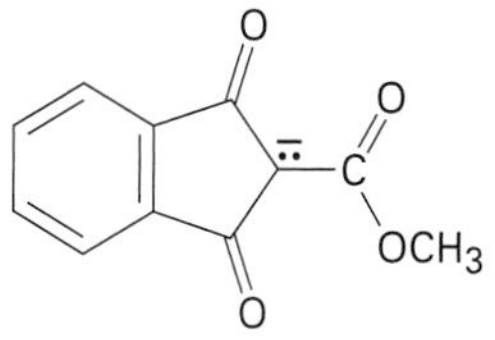

22.23 ■ Predict the product(s) of the following reactions:

(a) 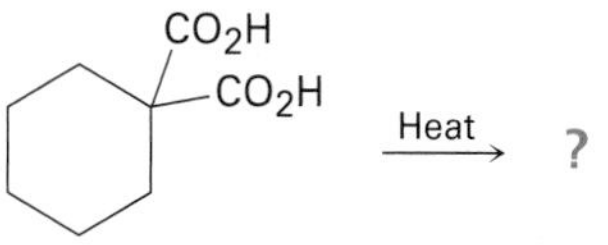 (cyclohexane ring with CO_2H and CO_2H substituents) $\xrightarrow{\text{Heat}}$?

(b) 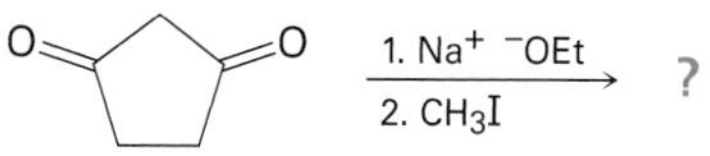 (cyclopentane-1,3-dione, O=⬠=O) $\xrightarrow[\text{2. CH}_3\text{I}]{\text{1. Na}^+ \; ^-\text{OEt}}$?

(c)

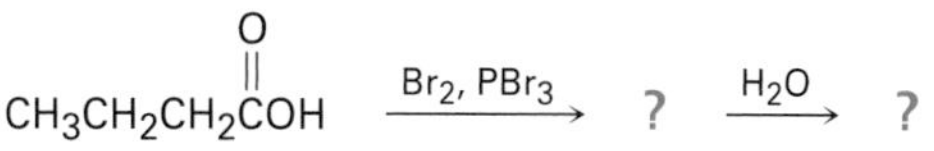

$$CH_3CH_2CH_2COH$$
(with O double bond) $\xrightarrow{\text{Br}_2, \text{ PBr}_3}$? $\xrightarrow{\text{H}_2\text{O}}$?

(d)

(benzene ring with) C (with O double bond) CH_3 $\xrightarrow[\text{I}_2]{\text{NaOH, H}_2\text{O}}$?

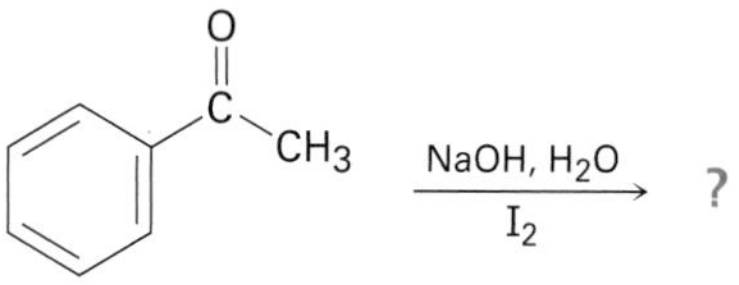

22.24 ■ Which, if any, of the following compounds can be prepared by a malonic ester synthesis? Show the alkyl halide you would use in each case.

(a) Ethyl pentanoate

(b) Ethyl 3-methylbutanoate

(c) Ethyl 2-methylbutanoate

(d) Ethyl 2,2-dimethylpropanoate

22.25 ■ Which, if any, of the following compounds can be prepared by an acetoacetic ester synthesis? Explain.

(a) Br.

(b)

(c) CH_3 $CH_3-C-CH_2CCH_3$ CH_3

22.26 ■ How would you prepare the following ketones using an acetoacetic ester synthesis?

(a) $CH_3CH_2CHCCH_3$ CH_2CH_3

(b) $CH_3CH_2CH_2CHCCH_3$ CH_3

22.27 ■ How would you prepare the following compounds using either an acetoacetic ester synthesis or a malonic ester synthesis?

(a) CH_3 CH_3CCO_2Et CO_2Et

(b)

(c)

(d) $H_2C=CHCH_2CH_2CCH_3$

22.28 ■ Which of the following substances would undergo the haloform reaction?

(a) CH_3COCH_3

(b) Acetophenone

(c) CH_3CH_2CHO

(d) CH_3CO_2H

(e) $CH_3C\equiv N$

22.29 ■ One way to determine the number of acidic hydrogens in a molecule is to treat the compound with NaOD in D_2O, isolate the product, and determine its molecular weight by mass spectrometry. For example, if cyclohexanone is treated with NaOD in D_2O, the product has MW = 102. Explain how this method works.

22.30 Base treatment of the following α,β-unsaturated carbonyl compound yields an anion by removal of H$^+$ from the γ carbon. Why are hydrogens on the γ carbon atom acidic?

22.31 Treatment of 1-phenyl-2-propenone with a strong base such as LDA does not yield an anion, even though it contains a hydrogen on the carbon atom next to the carbonyl group. Explain.

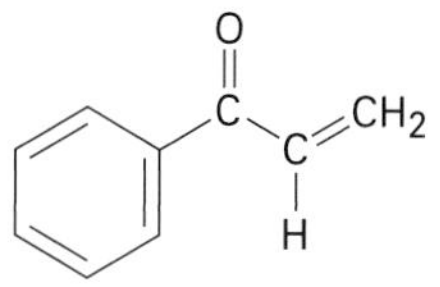

1-Phenylprop-2-enone

22.32 ■ When optically active (R)-2-methylcyclohexanone is treated with either aqueous base or acid, racemization occurs. Explain.

22.33 ■ Would you expect optically active (S)-3-methylcyclohexanone to be racemized on acid or base treatment in the same way as 2-methylcyclohexanone (Problem 22.32)? Explain.

22.34 When an optically active carboxylic acid such as (R)-2-phenylpropanoic acid is brominated under Hell–Volhard–Zelinskii conditions, is the product optically active or racemic? Explain.

22.35 Fill in the reagents **a–c** that are missing from the following scheme:

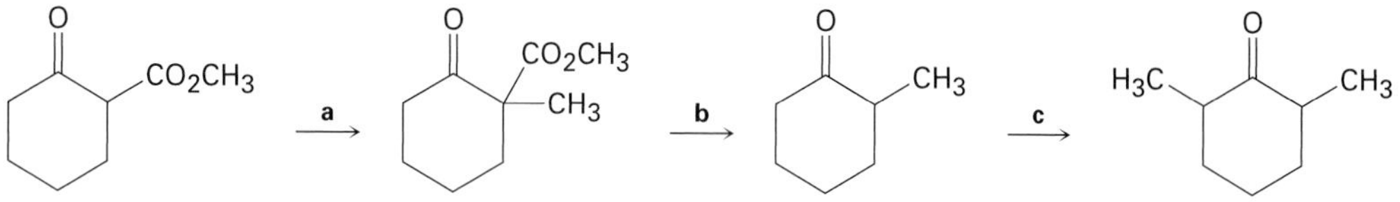

22.36 ■ Nonconjugated β,γ-unsaturated ketones, such as 3-cyclohexenone, are in an acid-catalyzed equilibrium with their conjugated α,β-unsaturated isomers. Propose a mechanism for this isomerization.

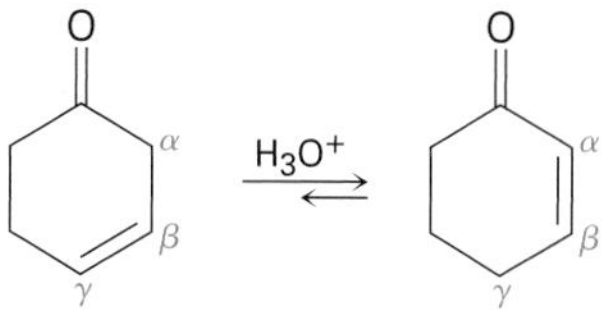

22.37 ■ The interconversion of unsaturated ketones described in Problem 22.36 is also catalyzed by base. Explain.

22.38 ■ An interesting consequence of the base-catalyzed isomerization of unsaturated ketones described in Problem 22.37 is that 2-substituted 2-cyclopentenones can be interconverted with 5-substituted 2-cyclopentenones. Propose a mechanism for this isomerization.

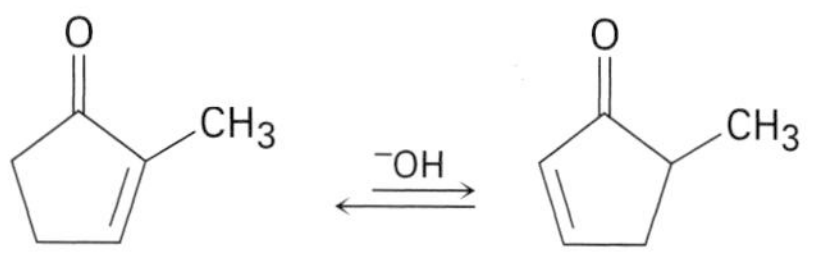

22.39 Although 2-substituted 2-cyclopentenones are in a base-catalyzed equilibrium with their 5-substituted 2-cyclopentenone isomers (Problem 22.38), the analogous isomerization is not observed for 2-substituted 2-cyclohexenones. Explain.

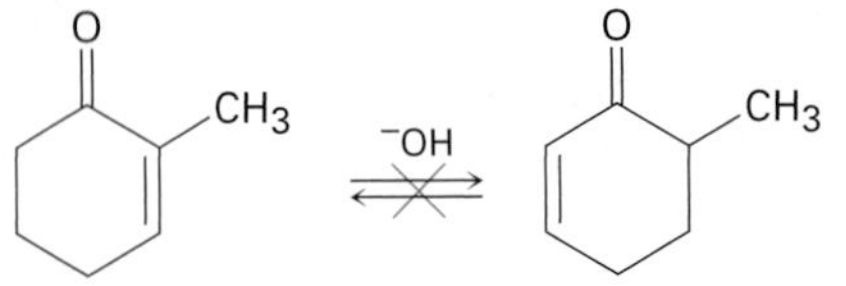

22.40 Using curved arrows, propose a mechanism for the following reaction, one of the steps in the metabolism of the amino acid alanine.

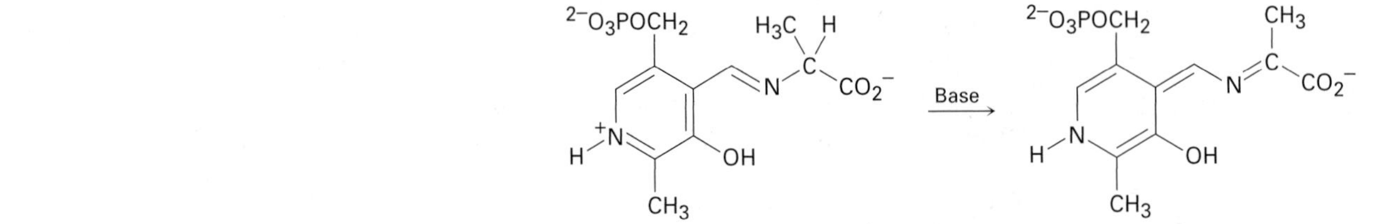

22.41 Using curved arrows, propose a mechanism for the following reaction, one of the steps in the biosynthesis of the amino acid tyrosine.

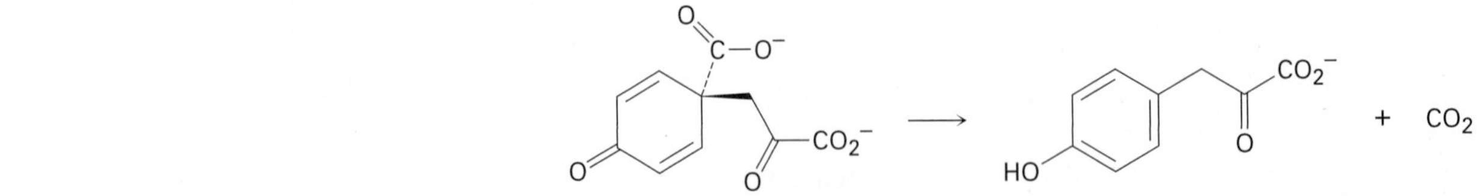

22.42 All attempts to isolate primary and secondary nitroso compounds result only in the formation of oximes. Tertiary nitroso compounds, however, are stable. Explain.

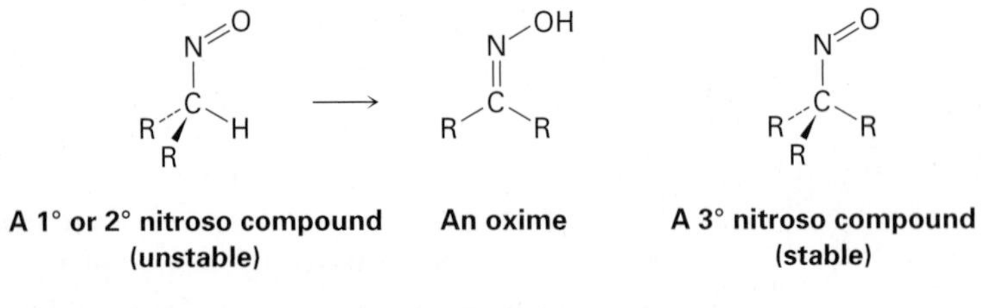

A 1° or 2° nitroso compound   **An oxime**   **A 3° nitroso compound**
(unstable)   **(stable)**

22.43 How might you convert geraniol into either ethyl geranylacetate or geranyl-acetone?

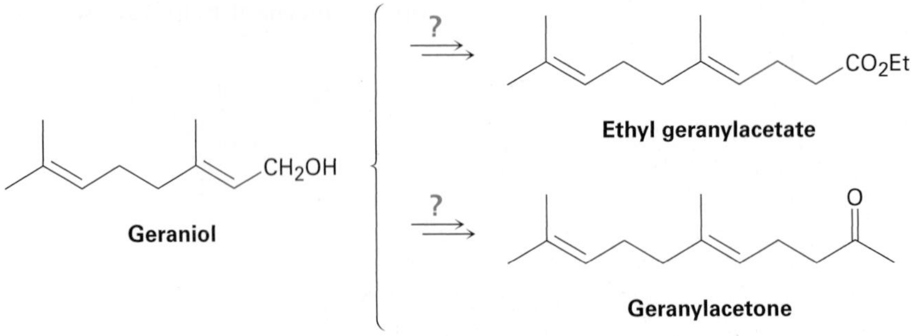

Ethyl geranylacetate

Geraniol

Geranylacetone

22.44 ■ How would you synthesize the following compounds from cyclo-hexanone? More than one step may be required.

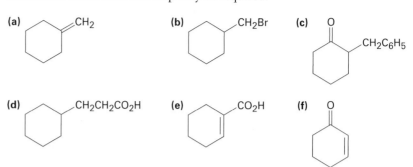

22.45 The two isomers *cis*- and *trans*-4-*tert*-butyl-2-methylcyclohexanone are interconverted by base treatment. Which isomer do you think is more stable, and why?

22.46 The following synthetic routes are incorrect. What is wrong with each?

(a)

$$CH_3CH_2CH_2CH_2\overset{\overset{\displaystyle O}{\|}}{C}OEt \xrightarrow[\text{2. Pyridine, heat}]{\text{1. Br}_2\text{, CH}_3\text{CO}_2\text{H}} CH_3CH_2CH=CHCOEt$$

(b)

$$\underset{\underset{\displaystyle CH_3CHCO_2Et}{|}}{\overset{\displaystyle CO_2Et}{}} \xrightarrow[\substack{\text{2. PhBr} \\ \text{3. H}_3\text{O}^+\text{, heat}}]{\text{1. Na}^+ \ ^-\text{OEt}} \text{(benzene ring)}\overset{\underset{\displaystyle CH_3}{|}}{CHCO_2H}$$

(c)

$$CH_3\overset{\overset{\displaystyle O}{\|}}{C}CH_2\overset{\overset{\displaystyle O}{\|}}{C}OEt \xrightarrow[\substack{\text{2. H}_2\text{C=CHCH}_2\text{Br} \\ \text{3. H}_3\text{O}^+\text{, heat}}]{\text{1. Na}^+ \ ^-\text{OEt}} H_2C=CHCH_2CH_2\overset{\overset{\displaystyle O}{\|}}{C}OH$$

22.47 Attempted Grignard reaction of cyclohexanone with *tert*-butylmagnesium bromide gives only about 1% yield of the expected addition product along with 99% unreacted cyclohexanone. If D_3O^+ is added to the reaction mixture after a suitable period, however, the "unreacted" cyclohexanone is found to have one deuterium atom incorporated into it. Explain.

22.48 One of the later steps in glucose biosynthesis is the isomerization of fructose 6-phosphate to glucose 6-phosphate. Propose a mechanism, using acid or base catalysis as needed.

Fructose 6-phosphate → **Glucose 6-phosphate**

22.49 The *Favorskii reaction* involves treatment of an α-bromo ketone with base to yield a ring-contracted product. For example, reaction of 2-bromocyclo-hexanone with aqueous NaOH yields cyclopentanecarboxylic acid. Propose a mechanism.

22.50 Treatment of a cyclic ketone with diazomethane is a method for accomplishing a *ring-expansion reaction*. For example, treatment of cyclohexanone with diazomethane yields cycloheptanone. Propose a mechanism.

22.51 Ketones react slowly with benzeneselenenyl chloride in the presence of HCl to yield α-phenylseleno ketones. Propose a mechanism for this acid-catalyzed α-substitution reaction.

22.52 Pentobarbital, marketed under the name Nembutal, is a barbiturate used in treating insomnia. It is synthesized in three steps from diethyl malonate. Show how you would synthesize the dialkylated intermediate, and then propose a mechanism for the reaction of that intermediate with urea to give pentobarbital.

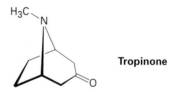

**Diethyl
malonate**

**Dialkylated
diethyl malonate**

**Pentobarbital
(Nembutal)**

22.53 As far back as the 16th century, South American Incas chewed the leaves of the coca bush, *Erythroxylon coca,* to combat fatigue. Chemical studies of *Erythroxylon coca* by Friedrich Wöhler in 1862 resulted in the discovery of *cocaine,* $C_{17}H_{21}NO_4$, as the active component. Basic hydrolysis of cocaine leads to methanol, benzoic acid, and another compound called *ecgonine,* $C_9H_{15}NO_3$. Oxidation of ecgonine with CrO_3 yields a keto acid that readily loses CO_2 on heating, giving tropinone.

Tropinone

(a) What is a likely structure for the keto acid?
(b) What is a likely structure for ecgonine, neglecting stereochemistry?
(c) What is a likely structure for cocaine, neglecting stereochemistry?

22.54 The final step in an attempted synthesis of laurene, a hydrocarbon isolated from the marine alga *Laurencia glandulifera,* involved the Wittig reaction shown. The product obtained, however, was not laurene but an isomer. Propose a mechanism to account for these unexpected results.

**Laurene
(NOT formed)**

22.55 The key step in a reported laboratory synthesis of sativene, a hydrocarbon isolated from the mold *Helminthosporium sativum,* involves the following base treatment of a keto tosylate. What kind of reaction is occurring? How would you complete the synthesis?

A keto tosylate   **Sativene**

22.56 Amino acids can be prepared by reaction of alkyl halides with diethyl acetamidomalonate, followed by heating the initial alkylation product with aqueous HCl. Show how you would prepare alanine, $CH_3CH(NH_2)CO_2H$, one of the twenty amino acids found in proteins, and propose a mechanism for acid-catalyzed conversion of the initial alkylation product to the amino acid.

Diethyl acetamidomalonate

22.57 Amino acids can also be prepared by a two-step sequence that involves Hell–Volhard–Zelinskii reaction of a carboxylic acid followed by treatment with ammonia. Show how you would prepare leucine, $(CH_3)_2CHCH_2CH(NH_2)CO_2H$, and identify the mechanism of the second step.

22.58 Heating carvone with aqueous sulfuric acid converts it into carvacrol. Propose a mechanism for the isomerization.

Carvone   **Carvacrol**

23

Carbonyl Condensation Reactions

We've now studied three of the four general kinds of carbonyl-group reactions and have seen two general kinds of behavior. In nucleophilic addition and nucleophilic acyl substitution reactions, a carbonyl compound behaves as an electrophile. In α-substitution reactions, however, a carbonyl compound behaves as a nucleophile when it is converted into its enol or enolate ion. In the carbonyl condensation reaction that we'll study in this chapter, the carbonyl compound behaves *both* as an electrophile and as a nucleophile.

Electrophilic carbonyl group reacts with nucleophiles.

Nucleophilic enolate ion reacts with electrophiles.

WHY THIS CHAPTER?

We'll see later in this chapter and again in Chapter 29 that carbonyl condensation reactions occur frequently in metabolic pathways. In fact, almost all classes of bio-molecules—carbohydrates, lipids, proteins, nucleic acids, and many others—are biosynthesized through pathways that involve carbonyl condensation reactions. As with the α-substitution reaction discussed in the previous chapter, the great value of carbonyl condensations is that they are one of the few general methods for forming carbon–carbon bonds, thereby making it possible to build larger molecules from smaller precursors. We'll see how and why these reactions occur in this chapter.

23.1 | Carbonyl Condensations: The Aldol Reaction

Carbonyl condensation reactions take place between two carbonyl partners and involve a *combination* of nucleophilic addition and α-substitution steps. One partner is converted into an enolate-ion nucleophile and adds to the electrophilic carbonyl group of the second partner. In so doing, the nucleophilic partner undergoes an α-substitution reaction and the electrophilic partner undergoes a nucleophilic addition. The general mechanism of the process is shown in Figure 23.1.

Active Figure 23.1
MECHANISM: The general mechanism of a carbonyl condensation reaction. One partner becomes a nucleophilic donor and adds to the second partner as an electrophilic acceptor. The product is a β-hydroxy carbonyl compound. *Sign in at* **www.cengage.com/login** *to see a simulation based on this figure and to take a short quiz.*

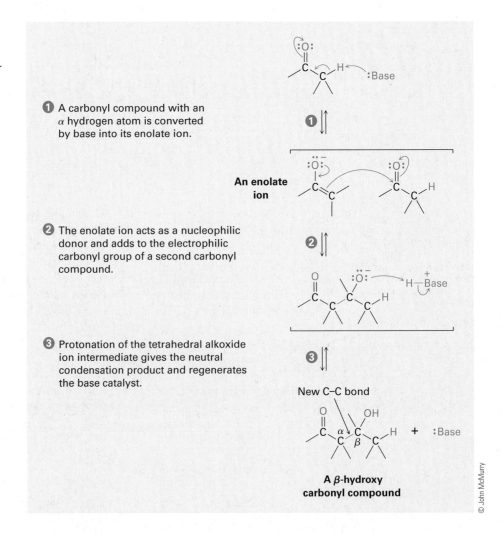

1 A carbonyl compound with an α hydrogen atom is converted by base into its enolate ion.

An enolate ion

2 The enolate ion acts as a nucleophilic donor and adds to the electrophilic carbonyl group of a second carbonyl compound.

3 Protonation of the tetrahedral alkoxide ion intermediate gives the neutral condensation product and regenerates the base catalyst.

New C–C bond

A β-hydroxy carbonyl compound

© John McMurry

CENGAGENOW Click *Organic Interactive* to **learn to draw the structures of products from aldol-type condensation reactions.**

Aldehydes and ketones with an α hydrogen atom undergo a base-catalyzed carbonyl condensation reaction called the **aldol reaction**. For example, treatment of acetaldehyde with a base such as sodium ethoxide or sodium hydroxide in a protic solvent leads to rapid and reversible formation of 3-hydroxybutanal, known commonly as *aldol* (*ald*ehyde + alco*hol*), hence the general name of the reaction.

Acetaldehyde   **3-Hydroxybutanal (aldol)**

The exact position of the aldol equilibrium depends both on reaction conditions and on substrate structure. The equilibrium generally favors condensation product in the case of aldehydes with no α substituent (RCH_2CHO) but favors reactant for disubstituted aldehydes (R_2CHCHO) and for most ketones. Steric factors are probably responsible for these trends, since increased substitution near the reaction site increases steric congestion in the aldol product.

Aldehydes

Phenylacetaldehyde
(10%)

(90%)

Ketones

Cyclohexanone
(78%)

(22%)

Aldol reactions, like all carbonyl condensations, occur by nucleophilic addition of the enolate ion of the donor molecule to the carbonyl group of the acceptor molecule. The resultant tetrahedral intermediate is then protonated to give an alcohol product (Figure 23.2). The reverse process occurs in exactly the opposite manner: base abstracts the −OH hydrogen from the aldol to yield a β-keto alkoxide ion, which cleaves to give one molecule of enolate ion and one molecule of neutral carbonyl compound.

WORKED EXAMPLE 23.1 | *Predicting the Product of an Aldol Reaction*

What is the structure of the aldol product from propanal?

Strategy An aldol reaction combines two molecules of reactant by forming a bond between the α carbon of one partner and the carbonyl carbon of the second partner. The product is a β-hydroxy aldehyde or ketone, meaning that the two oxygen atoms in the product have a 1,3 relationship.

Solution

Bond formed here

Figure 23.2 MECHANISM:
Mechanism of the aldol reaction, a typical carbonyl condensation.

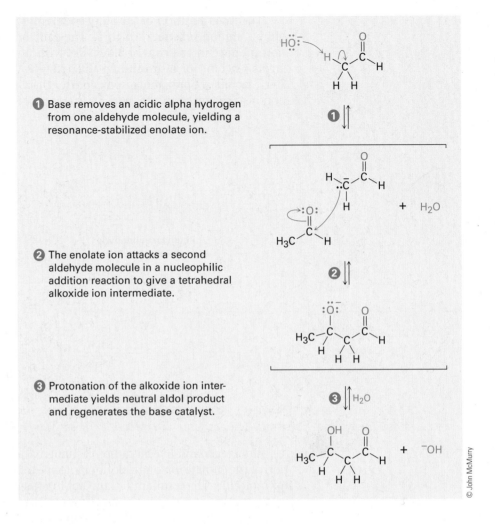

① Base removes an acidic alpha hydrogen from one aldehyde molecule, yielding a resonance-stabilized enolate ion.

② The enolate ion attacks a second aldehyde molecule in a nucleophilic addition reaction to give a tetrahedral alkoxide ion intermediate.

③ Protonation of the alkoxide ion intermediate yields neutral aldol product and regenerates the base catalyst.

© John McMurry

Problem 23.1 | Predict the aldol reaction product of the following compounds:

(a)

$$CH_3CH_2CH_2CH=O$$

(b)

(c)

Problem 23.2 | Using curved arrows to indicate the electron flow in each step, show how the base-catalyzed reverse aldol reaction of 4-hydroxy-4-methyl-2-pentanone takes place to yield 2 equivalents of acetone.

23.2 | Carbonyl Condensations versus Alpha Substitutions

Two of the four general carbonyl-group reactions—carbonyl condensations and α substitutions—take place under basic conditions and involve enolate-ion intermediates. Because the experimental conditions for the two reactions are similar, how can we predict which will occur in a given case? When we generate an enolate ion with the intention of carrying out an α alkylation, how can we be sure that a carbonyl condensation reaction won't occur instead?

There is no simple answer to this question, but the exact experimental conditions usually have much to do with the result. Alpha-substitution reactions require a full equivalent of strong base and are normally carried out so that the carbonyl compound is rapidly and completely converted into its enolate ion at a low temperature. An electrophile is then added rapidly to ensure that the reactive enolate ion is quenched quickly. In a ketone alkylation reaction, for instance, we might use 1 equivalent of lithium diisopropylamide (LDA) in tetrahydrofuran solution at −78 °C. Rapid and complete generation of the ketone enolate ion would occur, and no unreacted ketone would be left so that no condensation reaction could take place. We would then immediately add an alkyl halide to complete the alkylation reaction.

On the other hand, carbonyl condensation reactions require only a *catalytic* amount of a relatively weak base rather than a full equivalent so that a small amount of enolate ion is generated in the presence of unreacted carbonyl compound. Once a condensation has occurred, the basic catalyst is regenerated. To carry out an aldol reaction on propanal, for instance, we might dissolve the aldehyde in methanol, add 0.05 equivalent of sodium methoxide, and then warm the mixture to give the aldol product.

23.3 | Dehydration of Aldol Products: Synthesis of Enones

CENGAGENOW™ Click *Organic Process* to **view an animation showing the aldol condensation reaction.**

The β-hydroxy aldehydes or ketones formed in aldol reactions can be easily dehydrated to yield α,β-unsaturated products, or conjugated enones. In fact, it's this loss of water that gives the *condensation* reaction its name, because water condenses out of the reaction when the enone product forms.

A β-hydroxy ketone or aldehyde   **A conjugated enone**

Most alcohols are resistant to dehydration by base (Section 17.6) because hydroxide ion is a poor leaving group, but aldol products dehydrate easily because of the carbonyl group. Under *basic* conditions, an acidic α hydrogen is removed, yielding an enolate ion that expels the ⁻OH leaving group in an E1cB reaction (Section 11.10). Under *acidic* conditions, an enol is formed, the −OH group is protonated, and water is expelled in an E1 or E2 reaction.

Base-catalyzed

Enolate ion

Acid-catalyzed

Enol

The reaction conditions needed for aldol dehydration are often only a bit more vigorous (slightly higher temperature, for instance) than the conditions needed for the aldol formation itself. As a result, conjugated enones are usually obtained directly from aldol reactions without isolating the intermediate β-hydroxy carbonyl compounds.

Conjugated enones are more stable than nonconjugated enones for the same reason that conjugated dienes are more stable than nonconjugated dienes (Section 14.1). Interaction between the π electrons of the C=C bond and the π electrons of the C=O group leads to a molecular orbital description for a conjugated enone that shows an interaction of the π electrons over all four atomic centers (Figure 23.3).

The real value of aldol dehydration is that removal of water from the reaction mixture can be used to drive the aldol equilibrium toward product. Even though the initial aldol step itself may be unfavorable, as it usually is for ketones, the subsequent dehydration step nevertheless allows many aldol condensations to be

Figure 23.3 The π bonding molecular orbitals of a conjugated enone (propenal) and a conjugated diene (1,3-butadiene) are similar in shape and are spread over the entire π system.

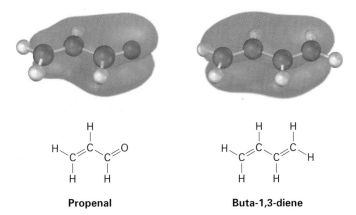

Propenal

Buta-1,3-diene

carried out in good yield. Cyclohexanone, for example, gives cyclohexylidenecyclohexanone in 92% yield even though the initial equilibrium is unfavorable.

Cyclohexanone

Cyclohexylidenecyclohexanone (92%)

WORKED EXAMPLE 23.2 · *Predicting the Product of an Aldol Reaction*

What is the structure of the enone obtained from aldol condensation of acetaldehyde?

Strategy In the aldol reaction, H_2O is eliminated and a double bond is formed by removing two hydrogens from the acidic α position of one partner and the carbonyl oxygen from the second partner. The product is thus an α,β-unsaturated aldehyde or ketone.

Solution

But-2-enal

Problem 23.3 What enone product would you expect from aldol condensation of each of the following compounds?

(a)

(b)

(c)

$$CH_3CHCH_2CH$$
  (with O double bond on terminal CH, and CH_3 substituent)

Problem 23.4 | Aldol condensation of 3-methylcyclohexanone leads to a mixture of two enone products, not counting double-bond isomers. Draw them.

23.4 | Using Aldol Reactions in Synthesis

The aldol reaction yields either a β-hydroxy aldehyde/ketone or an α,β-unsaturated aldehyde/ketone, depending on the experimental conditions. By learning how to think *backward,* it's possible to predict when the aldol reaction might be useful in synthesis. Whenever the target molecule contains either a β-hydroxy aldehyde/ketone or a conjugated enone functional group, it might come from an aldol reaction.

Aldol products   **Aldol reactants**

We can extend this kind of reasoning even further by imagining that subsequent transformations might be carried out on the aldol products. For example, a saturated ketone might be prepared by catalytic hydrogenation of the enone product. A good example can be found in the industrial preparation of 2-ethyl-1-hexanol, an alcohol used in the synthesis of plasticizers for polymers. Although 2-ethyl-1-hexanol bears little resemblance to an aldol product at first glance, it is in fact prepared commercially from butanal by an aldol reaction. Working backward, we can reason that 2-ethyl-1-hexanol might come from 2-ethylhexanal by a reduction. 2-Ethylhexanal, in turn, might be prepared by catalytic reduction of 2-ethyl-2-hexenal, which is the aldol condensation product of butanal. The reactions that follow show the sequence in reverse order.

Target: 2-Ethyl-1-hexanol   **2-Ethylhexanal**

Problem 23.5 | Which of the following compounds are aldol condensation products? What is the aldehyde or ketone precursor of each?
(a) 2-Hydroxy-2-methylpentanal   (b) 5-Ethyl-4-methyl-4-hepten-3-one

Problem 23.6 | 1-Butanol is prepared commercially by a route that begins with an aldol reaction. Show the steps that are likely to be involved.

Problem 23.7 | Show how you would synthesize the following compound using an aldol reaction:

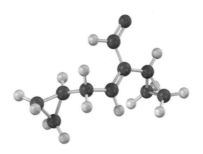

23.5 | Mixed Aldol Reactions

Until now, we've considered only *symmetrical* aldol reactions, in which the two carbonyl components have been the same. What would happen, though, if a *mixed* aldol reaction were carried out between two different carbonyl partners?

In general, a mixed aldol reaction between two similar aldehyde or ketone partners leads to a mixture of four possible products. For example, base treatment of a mixture of acetaldehyde and propanal gives a complex product mixture containing two "symmetrical" aldol products and two "mixed" aldol products. Clearly, such a reaction is of no practical value.

$$CH_3CHO \ + \ CH_3CH_2CHO \ \xrightarrow{\text{Base}}$$

OH
|
CH_3CHCH_2CHO + $CH_3CH_2CHCHCHO$
                                              |
                                              CH_3

OH

Symmetrical products

OH
|
$CH_3CHCHCHO$ + $CH_3CH_2CHCH_2CHO$
          |
          CH_3

OH

Mixed products

On the other hand, mixed aldol reactions *can* lead cleanly to a single product if either of two conditions is met:

▌ If one of the carbonyl partners contains no α hydrogens, and thus can't form an enolate ion to become a donor, but does contain an unhindered carbonyl group and so is a good acceptor of nucleophiles, then a mixed aldol reaction is likely to be successful. This is the case, for instance, when either benzaldehyde or formaldehyde is used as one of the carbonyl partners.

Neither benzaldehyde nor formaldehyde can form an enolate ion to add to another partner, yet both compounds have an unhindered carbonyl group.

The ketone 2-methylcyclohexanone, for instance, gives the mixed aldol product on reaction with benzaldehyde.

2-Methylcyclohexanone
(donor)

Benzaldehyde
(acceptor)

78%

▌ If one of the carbonyl partners is much more acidic than the other and so is transformed into its enolate ion in preference to the other, then a mixed aldol reaction is likely to be successful. Ethyl acetoacetate, for instance, is completely converted into its enolate ion in preference to enolate ion formation from monocarbonyl partners. Thus, aldol condensations of monoketones with ethyl acetoacetate occur preferentially to give the mixed product.

Cyclohexanone
(acceptor)

Ethyl acetoacetate
(donor)

80%

The situation can be summarized by saying that a mixed aldol reaction leads to a mixture of products unless one of the partners either has no α hydrogens but is a good electrophilic acceptor (such as benzaldehyde) or is an unusually acidic nucleophilic donor (such as ethyl acetoacetate).

Problem 23.8 | Which of the following compounds can probably be prepared by a mixed aldol reaction? Show the reactants you would use in each case.

(a)

$C_6H_5CH = CHCCH_3$

(b)

$C_6H_5C = CHCCH_3$
        |
        CH_3

(c)

23.6 | Intramolecular Aldol Reactions

The aldol reactions we've seen thus far have all been intermolecular, meaning that they have taken place between two different molecules. When certain *di*carbonyl compounds are treated with base, however, an *intra*molecular aldol reaction can occur, leading to the formation of a cyclic product. For example, base treatment of a 1,4-diketone such as 2,5-hexanedione yields a cyclopentenone product, and base treatment of a 1,5-diketone such as 2,6-heptanedione yields a cyclohexenone.

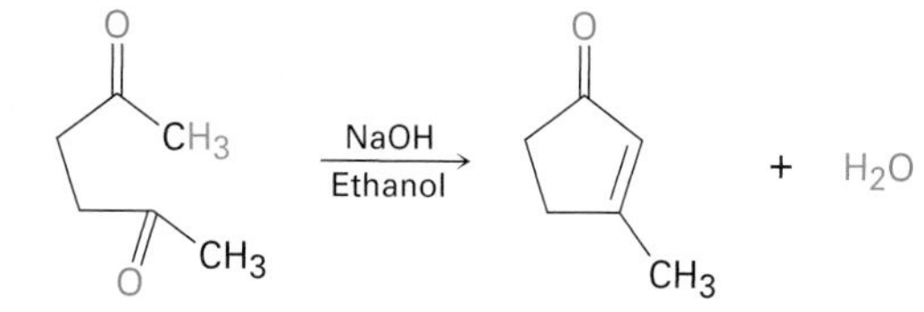

Hexane-2,5-dione
(a 1,4-diketone) → **3-Methylcyclopent-2-enone** + H_2O

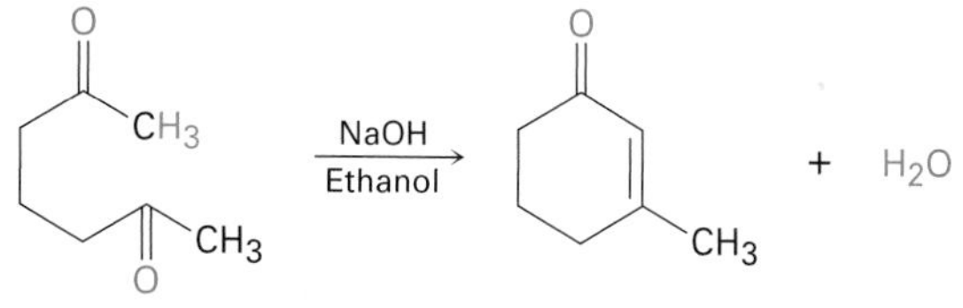

Heptane-2,6-dione
(a 1,5-diketone) → **3-Methylcyclohex-2-enone** + H_2O

The mechanism of intramolecular aldol reactions is similar to that of inter-molecular reactions. The only difference is that both the nucleophilic carbonyl anion donor and the electrophilic carbonyl acceptor are now in the same mol-ecule. One complication, however, is that intramolecular aldol reactions might lead to a mixture of products, depending on which enolate ion is formed. For example, 2,5-hexanedione might yield either the five-membered-ring product 3-methyl-2-cyclopentenone or the three-membered-ring product (2-methyl-cyclopropenyl)ethanone (Figure 23.4). In practice, though, only the cyclo-pentenone is formed.

Figure 23.4 Intramolecular aldol reaction of 2,5-hexanedione yields 3-methyl-2-cyclopentenone rather than the alternative cyclopropene.

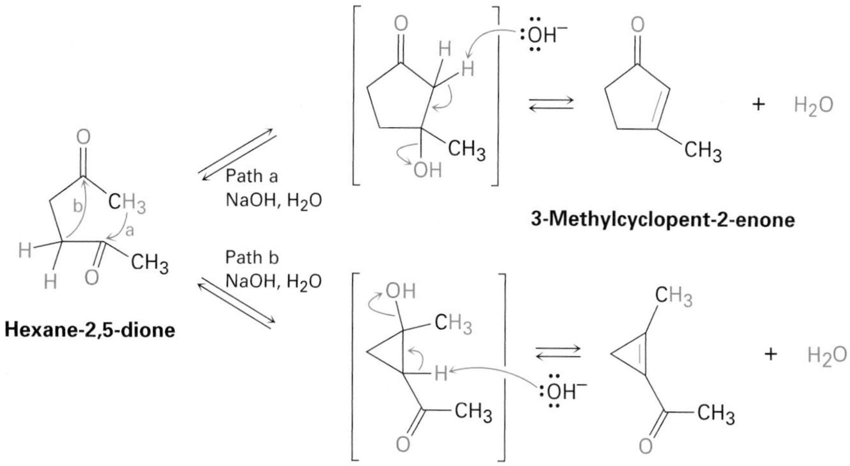

3-Methylcyclopent-2-enone

Hexane-2,5-dione

(2-Methylcyclopropenyl)ethanone
(NOT formed)

The selectivity observed in the intramolecular aldol reaction of 2,5-hexane-dione is due to the fact that all steps in the mechanism are reversible, so an equilibrium is reached. Thus, the relatively strain-free cyclopentenone product is considerably more stable than the highly strained cyclopropene alternative. For similar reasons, intramolecular aldol reactions of 1,5-diketones lead only to cyclohexenone products rather than to acylcyclobutenes.

Problem 23.9 Treatment of a 1,3-diketone such as 2,4-pentanedione with base does not give an aldol condensation product. Explain.

Problem 23.10 What product would you expect to obtain from base treatment of 1,6-cyclodecanedione?

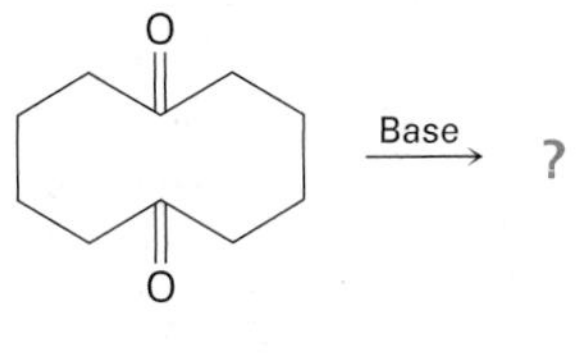

1,6-Cyclodecanedione

23.7 The Claisen Condensation Reaction

CENGAGENOW™ Click *Organic Process* to **view an animation showing the mechanism of the Claisen condensation reaction.**

Esters, like aldehydes and ketones, are weakly acidic. When an ester with an α hydrogen is treated with 1 equivalent of a base such as sodium ethoxide, a reversible carbonyl condensation reaction occurs to yield a β-keto ester. For example, ethyl acetate yields ethyl acetoacetate on base treatment. This reaction between two ester molecules is known as the **Claisen condensation reaction.** (We'll use ethyl esters, abbreviated "Et," for consistency, but other esters will also work.)

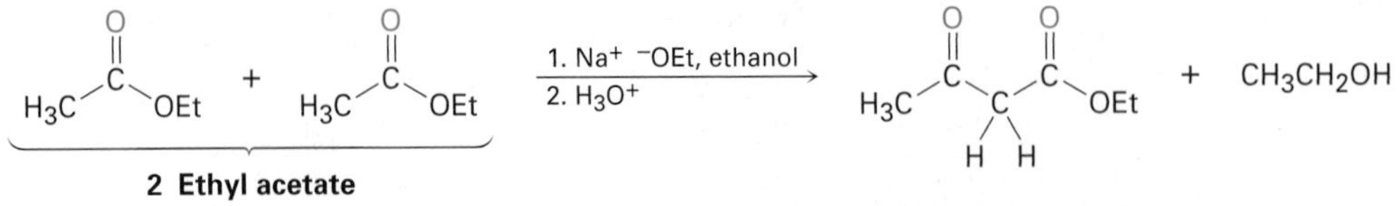

2 Ethyl acetate

Ethyl acetoacetate, a β-keto ester (75%)

The mechanism of the Claisen condensation is similar to that of the aldol condensation and involves the nucleophilic addition of an ester enolate ion to the carbonyl group of a second ester molecule. The only difference between the aldol condensation of an aldehyde or ketone and the Claisen condensation of an ester involves the fate of the initially formed tetrahedral intermediate. The tetrahedral intermediate in the aldol reaction is protonated to give an alcohol product—exactly the behavior previously seen for aldehydes and ketones (Section 19.4). The tetrahedral intermediate in the Claisen reaction, however, expels an alkoxide leaving group to yield an acyl substitution product—exactly the behavior previously seen for esters (Section 21.6). The mechanism of the Claisen condensation reaction is shown in Figure 23.5.

Active Figure 23.5
MECHANISM: Mechanism of the Claisen condensation reaction. *Sign in at* **www.cengage.com/login** *to see a simulation based on this figure and to take a short quiz.*

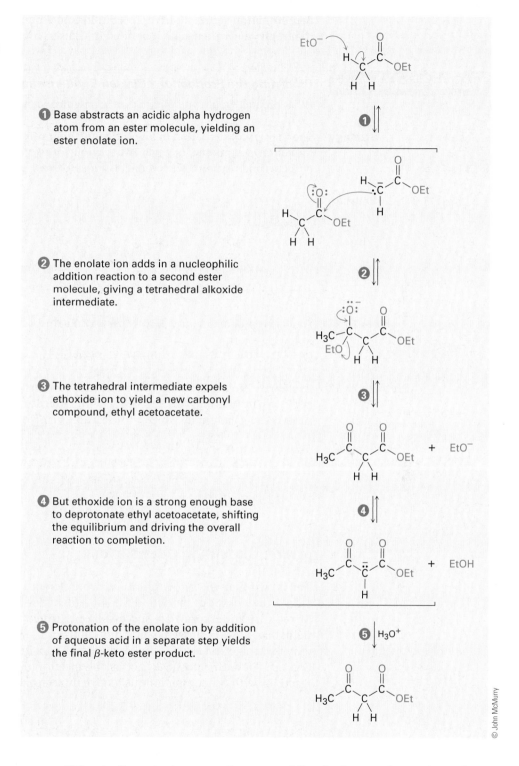

① Base abstracts an acidic alpha hydrogen atom from an ester molecule, yielding an ester enolate ion.

② The enolate ion adds in a nucleophilic addition reaction to a second ester molecule, giving a tetrahedral alkoxide intermediate.

③ The tetrahedral intermediate expels ethoxide ion to yield a new carbonyl compound, ethyl acetoacetate.

④ But ethoxide ion is a strong enough base to deprotonate ethyl acetoacetate, shifting the equilibrium and driving the overall reaction to completion.

⑤ Protonation of the enolate ion by addition of aqueous acid in a separate step yields the final β-keto ester product.

© John McMurry

If the starting ester has more than one acidic α hydrogen, the product β-keto ester has a highly acidic, doubly activated hydrogen atom that can be abstracted by base. This deprotonation of the product requires that a full equivalent of base rather than a catalytic amount be used in the reaction. Furthermore, the deprotonation serves to drive the equilibrium completely to the product side so that high yields are usually obtained in Claisen condensations.

WORKED EXAMPLE 23.3   ***Predicting the Product of a Claisen Condensation Reaction***

What product would you obtain from Claisen condensation of ethyl propanoate?

Strategy   The Claisen condensation of an ester results in loss of one molecule of alcohol and formation of a product in which an acyl group of one reactant bonds to the α carbon of the second reactant. The product is a β-keto ester.

Solution

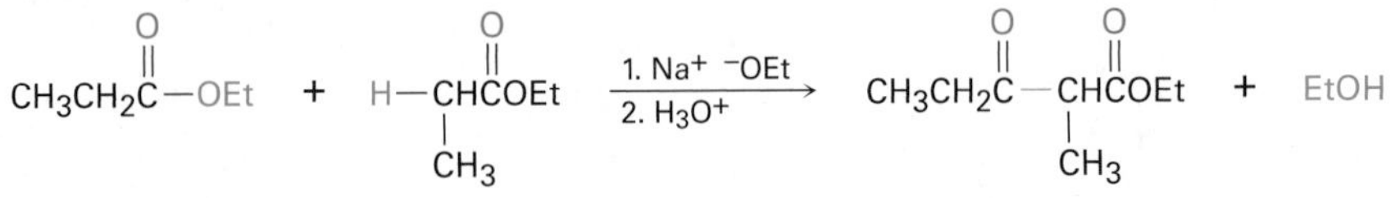

$$CH_3CH_2\overset{\overset{\displaystyle O}{\|}}{C}-OEt \quad + \quad H-\overset{\displaystyle |}{\underset{\displaystyle CH_3}{C}}H\overset{\overset{\displaystyle O}{\|}}{C}OEt \quad \xrightarrow[\text{2. H}_3\text{O}^+]{\text{1. Na}^+ \ ^-\text{OEt}} \quad CH_3CH_2\overset{\overset{\displaystyle O}{\|}}{C}-\overset{\displaystyle |}{\underset{\displaystyle CH_3}{C}}H\overset{\overset{\displaystyle O}{\|}}{C}OEt \quad + \quad EtOH$$

2 Ethyl propanoate   **Ethyl 2-methyl-3-oxopentanoate**

Problem 23.11   Show the products you would expect to obtain by Claisen condensation of the following esters:
(a) $(CH_3)_2CHCH_2CO_2Et$ (b) Ethyl phenylacetate   (c) Ethyl cyclohexylacetate

Problem 23.12   As shown in Figure 23.5, the Claisen reaction is reversible. That is, a β-keto ester can be cleaved by base into two fragments. Using curved arrows to indicate electron flow, show the mechanism by which this cleavage occurs.

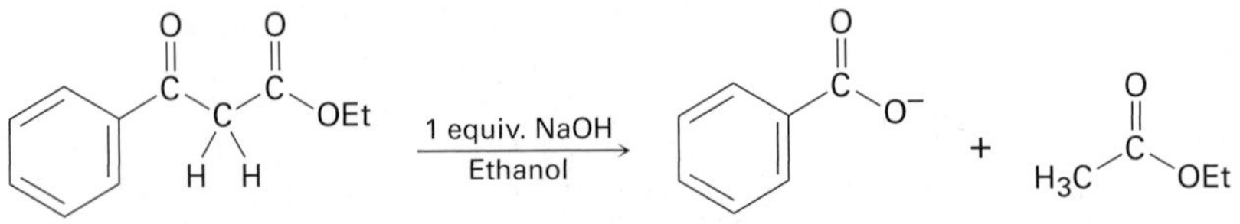

23.8 | Mixed Claisen Condensations

The mixed Claisen condensation of two different esters is similar to the mixed aldol condensation of two different aldehydes or ketones (Section 23.5). Mixed Claisen reactions are successful only when one of the two ester components has no α hydrogens and thus can't form an enolate ion. For example, ethyl benzoate and ethyl formate can't form enolate ions and thus can't serve as donors. They can, however, act as the electrophilic acceptor components in reactions with other ester anions to give mixed β-keto ester products.

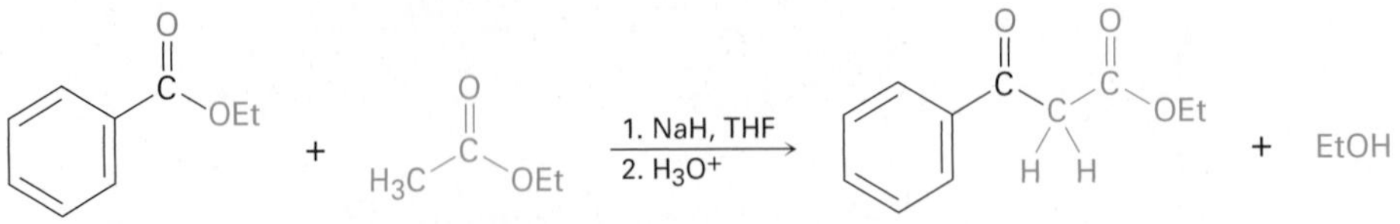

Ethyl benzoate   **Ethyl acetate**   **Ethyl benzoylacetate**
(acceptor)   **(donor)**

Mixed Claisen-like reactions can also be carried out between an ester and a ketone, resulting in the synthesis of a β-diketone. The reaction works best when the ester component has no α hydrogens and thus can't act as the nucleophilic donor. For example, ethyl formate gives high yields in mixed Claisen condensations with ketones.

2,2-Dimethylcyclohexanone (donor)   Ethyl formate (acceptor)   A β-keto aldehyde (91%)

WORKED EXAMPLE 23.4

Predicting the Product of a Mixed Claisen Condensation Reaction

Diethyl oxalate, $(CO_2Et)_2$, often gives high yields in mixed Claisen reactions. What product would you expect to obtain from a mixed Claisen reaction of ethyl acetate with diethyl oxalate?

Strategy

A mixed Claisen reaction is effective when only one of the two partners has an acidic α hydrogen atom. In the present case, ethyl acetate can be converted into its enolate ion, but diethyl oxalate cannot. Thus, ethyl acetate acts as the donor and diethyl oxalate as the acceptor.

Solution

Diethyl oxalate   Ethyl acetate

Problem 23.13 What product would you expect from the following mixed Claisen-like reaction?

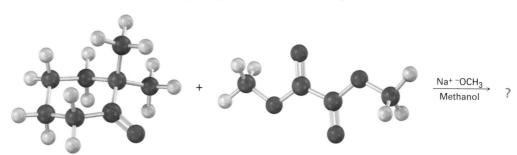

23.9 Intramolecular Claisen Condensations: The Dieckmann Cyclization

Intramolecular Claisen condensations can be carried out with diesters, just as intramolecular aldol condensations can be carried out with diketones (Section 23.6). Called the **Dieckmann cyclization**, the reaction works best on 1,6-diesters and 1,7-diesters. Intramolecular Claisen cyclization of a 1,6-diester gives a five-membered cyclic β-keto ester, and cyclization of a 1,7-diester gives a six-membered cyclic β-keto ester.

Diethyl hexanedioate
(a 1,6-diester)

Ethyl 2-oxocyclopentanecarboxylate
(82%)

Diethyl heptanedioate
(a 1,7-diester)

Ethyl 2-oxocyclohexanecarboxylate

The mechanism of the Dieckmann cyclization, shown in Figure 23.6, is the same as that of the Claisen condensation. One of the two ester groups is converted into an enolate ion, which then carries out a nucleophilic acyl substitution on the second ester group at the other end of the molecule. A cyclic β-keto ester product results.

The cyclic β-keto ester produced in a Dieckmann cyclization can be further alkylated and decarboxylated by a series of reactions analogous to those used in the acetoacetic ester synthesis (Section 22.7). For example, alkylation and subsequent decarboxylation of ethyl 2-oxocyclohexanecarboxylate yields a 2-alkylcyclohexanone. The overall sequence of (1) Dieckmann cyclization, (2) β-keto ester alkylation, and (3) decarboxylation is a powerful method for preparing 2-substituted cyclohexanones and cyclopentanones.

Ethyl 2-oxocyclo-hexanecarboxylate

2-Allylcyclohexanone
(83%)

Figure 23.6 MECHANISM:
Mechanism of the Dieckmann cyclization of a 1,7-diester to yield a cyclic β-keto ester product.

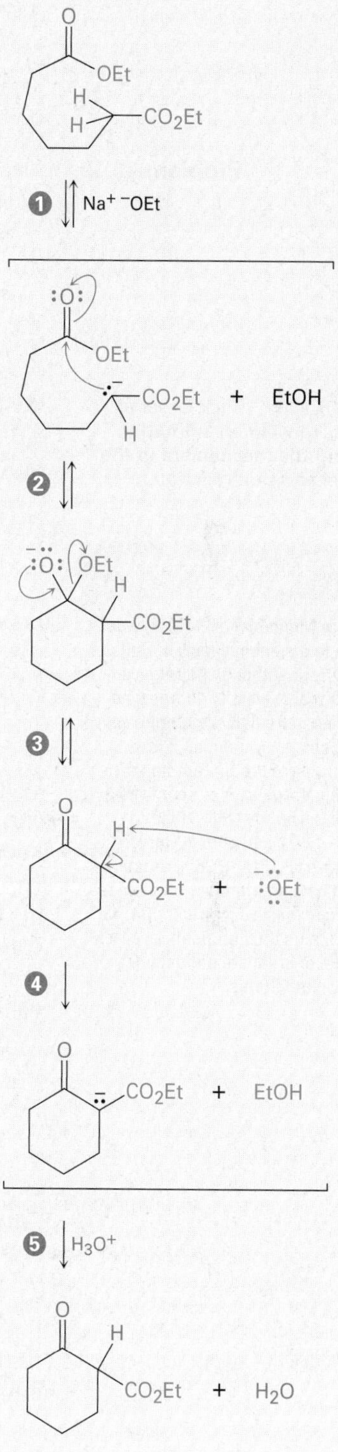

① Base abstracts an acidic α proton from the carbon atom next to one of the ester groups, yielding an enolate ion.

② Intramolecular nucleophilic addition of the ester enolate ion to the carbonyl group of the second ester at the other end of the chain then gives a cyclic tetrahedral intermediate.

③ Loss of alkoxide ion from the tetrahedral intermediate forms a cyclic β-keto ester.

④ Deprotonation of the acidic β-keto ester gives an enolate ion . . .

⑤ . . . which is protonated by addition of aqueous acid at the end of the reaction to generate the neutral β-keto ester product.

© John McMurry

Problem 23.14 | What product would you expect from the following reaction?

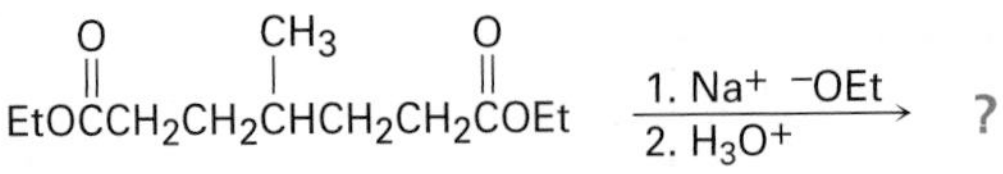

$$\underset{EtOCCH_2CH_2CHCH_2CH_2COEt}{\overset{\overset{O}{\parallel}\qquad\overset{CH_3}{|}\qquad\overset{O}{\parallel}}{}} \quad \xrightarrow[\text{2. H}_3\text{O}^+]{\text{1. Na}^+ \ ^-\text{OEt}} \quad ?$$

Problem 23.15 | Dieckmann cyclization of diethyl 3-methylheptanedioate gives a mixture of two β-keto ester products. What are their structures, and why is a mixture formed?

23.10 | Conjugate Carbonyl Additions: The Michael Reaction

We saw in Section 19.13 that certain nucleophiles, such as amines, react with α,β-unsaturated aldehydes and ketones to give the conjugate addition product, rather than the direct addition product.

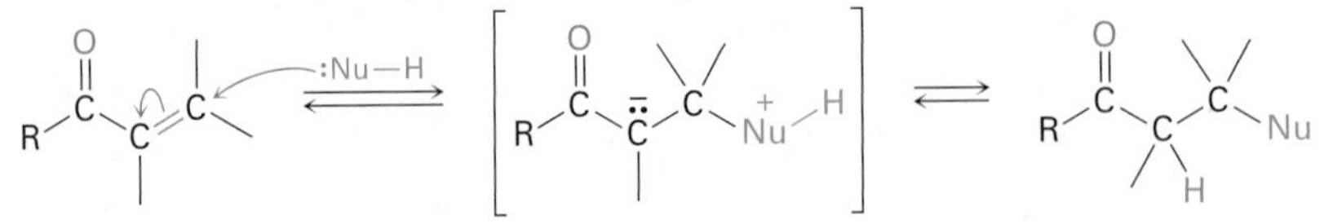

Conjugate addition product

Arthur Michael

Arthur Michael (1853–1942) was born to a wealthy family in Buffalo, New York. Although he received no formal university degrees, he studied in Heidelberg, Berlin, and the École de Médecine, Paris. Returning to the United States, he became professor of chemistry at Tufts University (1882–1889, 1894–1907), and then at Harvard University (1912–1936). Perhaps his most important contribution to science was his instrumental role in bringing the European research model of graduate education to the United States.

Exactly the same kind of conjugate addition can occur when a nucleophilic enolate ion reacts with an α,β-unsaturated carbonyl compound—a process known as the **Michael reaction**.

The best Michael reactions are those that take place when a particularly stable enolate ion such as that derived from a β-keto ester or other 1,3-dicarbonyl compound adds to an unhindered α,β-unsaturated ketone. For example, ethyl acetoacetate reacts with 3-buten-2-one in the presence of sodium ethoxide to yield the conjugate addition product.

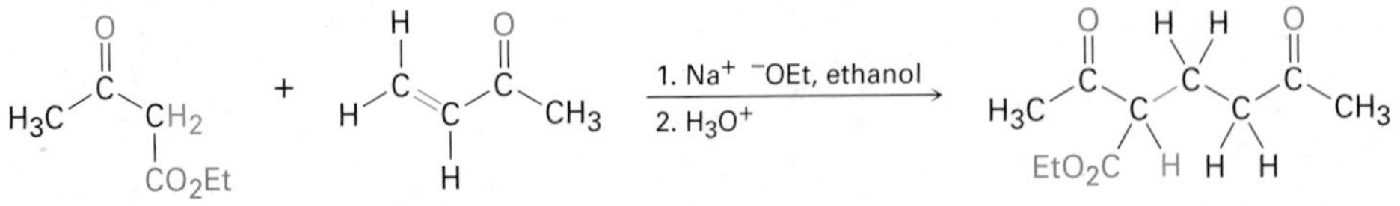

Ethyl acetoacetate   **3-Buten-2-one**

Michael reactions take place by addition of a nucleophilic enolate ion donor to the β carbon of an α,β-unsaturated carbonyl acceptor, according to the mechanism shown in Figure 23.7.

The Michael reaction occurs with a variety of α,β-unsaturated carbonyl compounds, not just conjugated ketones. Unsaturated aldehydes, esters, thioesters, nitriles, amides, and nitro compounds can all act as the electrophilic acceptor component in Michael reactions (Table 23.1). Similarly, a variety of different donors can be used, including β-diketones, β-keto esters, malonic esters, β-keto nitriles, and nitro compounds.

Active Figure 23.7
MECHANISM: Mechanism of the Michael reaction between a β-keto ester and an α,β-unsaturated ketone. *Sign in at* **www.cengage.com/login** *to see a simulation based on this figure and to take a short quiz.*

① The base catalyst removes an acidic alpha proton from the starting β-keto ester to generate a stabilized enolate ion nucleophile.

② The nucleophile adds to the α,β-unsaturated ketone electrophile in a Michael reaction to generate a new enolate as product.

③ The enolate product abstracts an acidic proton, either from solvent or from starting keto ester, to yield the final addition product.

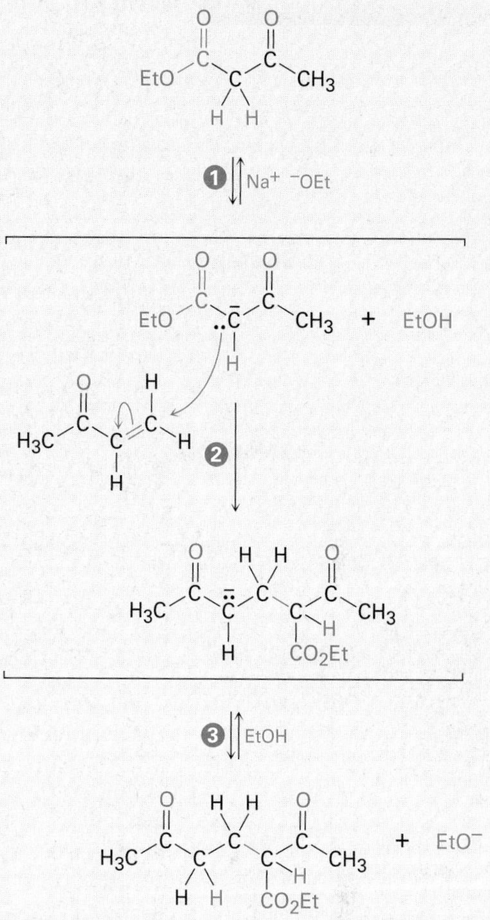

© John McMurry

CENGAGENOW Click *Organic Interactive* to **learn to predict products in Michael-style addition reactions**.

Table 23.1 Some Michael Acceptors and Michael Donors

Michael acceptors		Michael donors	
H₂C=CHCH (O)	Propenal	RCCH₂CR' (O O)	β-Diketone
H₂C=CHCCH₃ (O)	3-Buten-2-one	RCCH₂COEt (O O)	β-Keto ester
H₂C=CHCOEt (O)	Ethyl propenoate	EtOCCH₂COEt (O O)	Diethyl malonate
H₂C=CHCNH₂ (O)	Propenamide	RCCH₂C≡N (O)	β-Keto nitrile
H₂C=CHC≡N	Propenenitrile	RCH₂NO₂	Nitro compound
H₂C=CH (NO₂)	Nitroethylene		

| WORKED EXAMPLE 23.5 | ***Using the Michael Reaction*** |

How might you obtain the following compound using a Michael reaction?

Strategy A Michael reaction involves the conjugate addition of a stable enolate ion donor to an α,β-unsaturated carbonyl acceptor, yielding a 1,5-dicarbonyl product. Usually, the stable enolate ion is derived from a β-diketone, β-keto ester, malonic ester, or similar compound. The C–C bond made in the conjugate addition step is the one between the α carbon of the acidic donor and the β carbon of the unsaturated acceptor.

Solution

Problem 23.16 What product would you obtain from a base-catalyzed Michael reaction of 2,4-pentanedione with each of the following α,β-unsaturated acceptors?
(a) 2-Cyclohexenone (b) Propenenitrile (c) Ethyl 2-butenoate

Problem 23.17 What product would you obtain from a base-catalyzed Michael reaction of 3-buten-2-one with each of the following nucleophilic donors?

(a)
$$\underset{\text{EtOCCH}_2\text{COEt}}{\overset{O \quad \; O}{\underset{\parallel \quad \parallel}{}}}$$

(b)

Problem 23.18 How would you prepare the following compound using a Michael reaction?

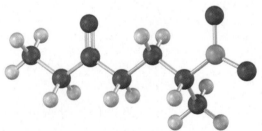

23.11 | Carbonyl Condensations with Enamines: The Stork Reaction

In addition to enolate ions, other kinds of carbon nucleophiles also add to α,β-unsaturated acceptors in Michael-like reactions. Among the most important such nucleophiles, particularly in biological chemistry, are *enamines*, which are readily prepared by reaction between a ketone and a secondary amine, as we saw in Section 19.8. For example:

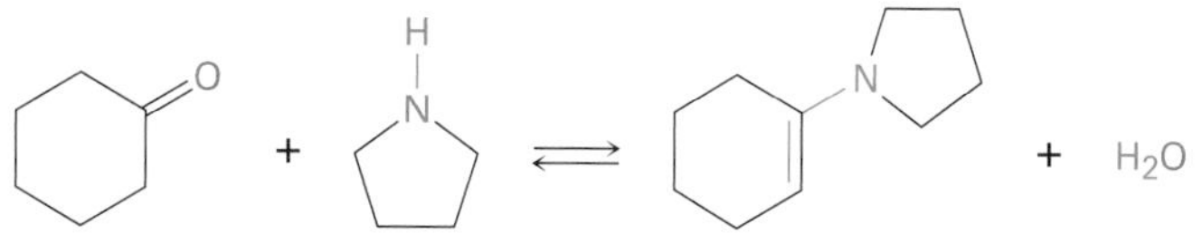

Cyclohexanone   **Pyrrolidine**   **1-Pyrrolidino-cyclohexene (87%)**

As the following resonance structures indicate, enamines are electronically similar to enolate ions. Overlap of the nitrogen lone-pair orbital with the double-bond p orbitals leads to an increase in electron density on the α carbon atom, making that carbon nucleophilic. An electrostatic potential map of N,N-dimethyl-aminoethylene shows this shift of electron density (red) toward the α position.

An enolate ion

An enamine

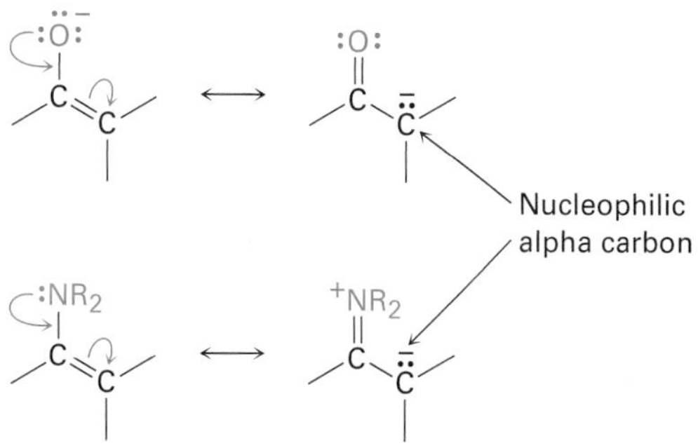

Nucleophilic alpha carbon

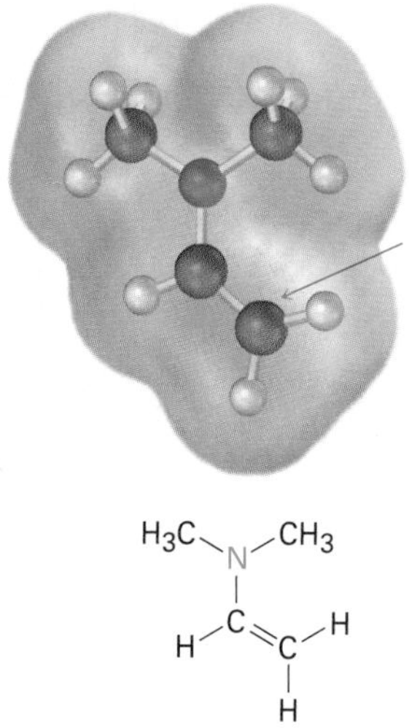

Gilbert Stork

Gilbert Stork (1921–) was born on New Year's eve in Brussels, Belgium. He received his secondary education in France, his undergraduate degree at the University of Florida, and his Ph.D. with Samuel McElvain at the University of Wisconsin in 1945. Following a period on the faculty at Harvard University, he has been professor of chemistry at Columbia University since 1953. A world leader in the development of organic synthesis, Stork has devised many useful new synthetic procedures and has accomplished the laboratory synthesis of many complex molecules.

Enamines behave in much the same way as enolate ions and enter into many of the same kinds of reactions. In the **Stork reaction**, for example, an enamine adds to an α,β-unsaturated carbonyl acceptor in a Michael-like process. The initial product is then hydrolyzed by aqueous acid (Section 19.8) to yield a 1,5-dicarbonyl compound. The overall reaction is thus a three-step sequence of (1) enamine formation from a ketone, (2) Michael addition to an α,β-unsaturated carbonyl compound, and (3) enamine hydrolysis back to a ketone.

The net effect of the Stork reaction is the Michael addition of a ketone to an α,β-unsaturated carbonyl compound. For example, cyclohexanone reacts with the cyclic amine pyrrolidine to yield an enamine; further reaction with an enone such as 3-buten-2-one yields a Michael adduct; and aqueous hydrolysis completes the sequence to provide a 1,5-diketone (Figure 23.8).

There are two advantages to the enamine–Michael reaction versus the enolate-ion–Michael that make enamines so useful in biological pathways. First, an enamine is neutral, easily prepared, and easily handled, while an enolate ion is charged, sometimes difficult to prepare, and must be handled with care.

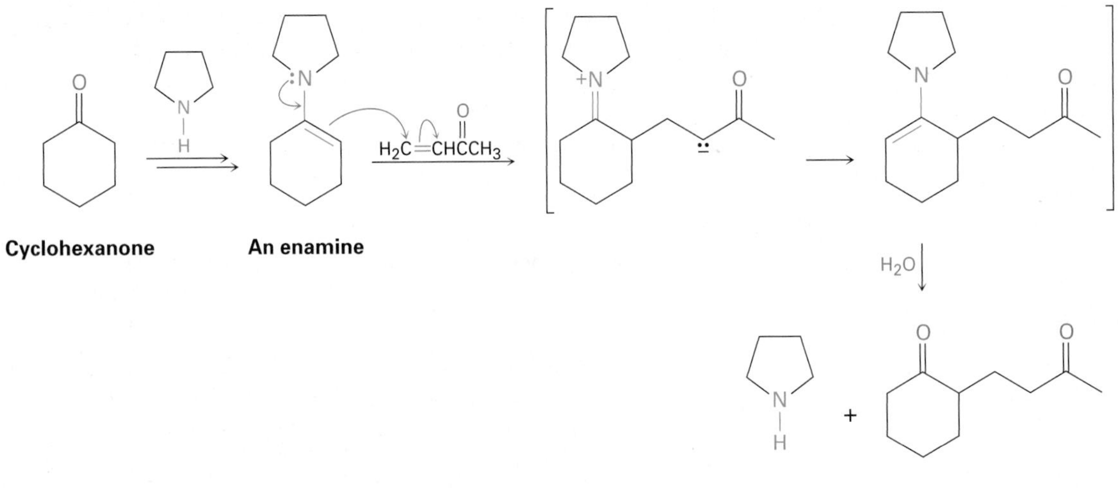

Cyclohexanone   **An enamine**

A 1,5-diketone (71%)

Figure 23.8 The Stork reaction between cyclohexanone and 3-buten-2-one. Cyclohexanone is first converted into an enamine, the enamine adds to the α,β-unsaturated ketone in a Michael reaction, and the conjugate addition product is hydrolyzed to yield a 1,5-diketone.

Second, an enamine from a *mono*ketone can be used in the Michael addition, whereas enolate ions only from β-*di*carbonyl compounds can be used.

WORKED EXAMPLE 23.6

Using the Stork Enamine Reaction

How might you use an enamine reaction to prepare the following compound?

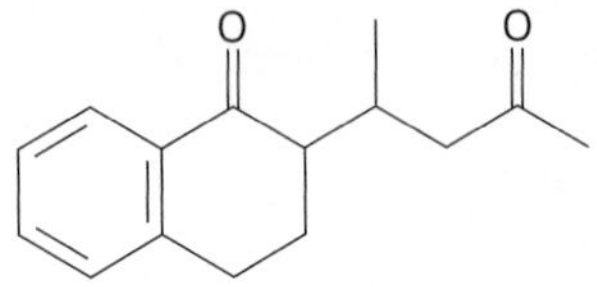

Strategy The overall result of an enamine reaction is the Michael addition of a ketone as donor to an α,β-unsaturated carbonyl compound as acceptor, yielding a 1,5-dicarbonyl product. The C–C bond made in the Michael addition step is the one between the α carbon of the ketone donor and the β carbon of the unsaturated acceptor.

Solution

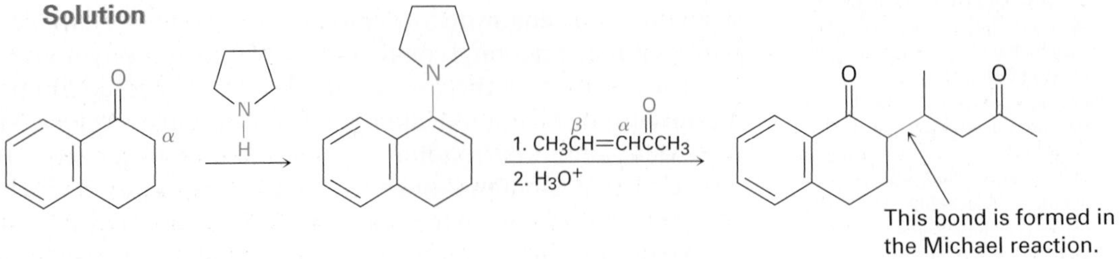

This bond is formed in the Michael reaction.

Problem 23.19 What products would result after hydrolysis from reaction of the enamine prepared from cyclopentanone and pyrrolidine with the following α,β-unsaturated acceptors?
(a) $CH_2{=}CHCO_2Et$ (b) $H_2C{=}CHCHO$ (c) $CH_3CH{=}CHCOCH_3$

Problem 23.20 Show how you might use an enamine reaction to prepare each of the following compounds:

(a)

CH_2CH_2CN

(b)

$CH_2CH_2CO_2CH_3$

23.12 | The Robinson Annulation Reaction

Carbonyl condensation reactions are perhaps the most versatile methods available for synthesizing complex molecules. By putting a few fundamental reactions together in the proper sequence, some remarkably useful transformations can be carried out. One such example is the **Robinson annulation reaction** for the synthesis of polycyclic molecules. The word *annulation* comes from the Latin *annulus,* meaning "ring," so an annulation reaction builds a new ring onto a molecule.

The Robinson annulation is a two-step process that combines a Michael reaction with an intramolecular aldol reaction. It takes place between a nucleophilic donor, such as a β-keto ester, an enamine, or a β-diketone, and an α,β-unsaturated ketone acceptor, such as 3-buten-2-one. The product is a substituted 2-cyclohexenone.

| 3-Buten-2-one | Ethyl acetoacetate | Michael product | Annulation product |

Sir Robert Robinson

Sir Robert Robinson (1886–1975) was born in Chesterfield, England, and received his D.Sc. from the University of Manchester with William Henry Perkin, Jr. After various academic appointments, he moved in 1930 to Oxford University, where he remained until his retirement in 1955. An accomplished mountain climber, Robinson was instrumental in developing the mechanistic descriptions of reactions that we use today. He received the 1947 Nobel Prize in chemistry.

The first step of the Robinson annulation is simply a Michael reaction. An enamine or an enolate ion from a β-keto ester or β-diketone effects a conjugate addition to an α,β-unsaturated ketone, yielding a 1,5-diketone. But as we saw in Section 23.6, 1,5-diketones undergo intramolecular aldol condensation to yield cyclohexenones when treated with base. Thus, the final product contains a six-membered ring, and an annulation has been accomplished. An example occurs during the commercial synthesis of the steroid hormone estrone (Figure 23.9).

In this example, the β-diketone 2-methyl-1,3-cyclopentanedione is used to generate the enolate ion required for Michael reaction and an aryl-substituted α,β-unsaturated ketone is used as the acceptor. Base-catalyzed Michael reaction between the two partners yields an intermediate triketone, which then cyclizes in an intramolecular aldol condensation to give a Robinson annulation product. Several further transformations are required to complete the synthesis of estrone.

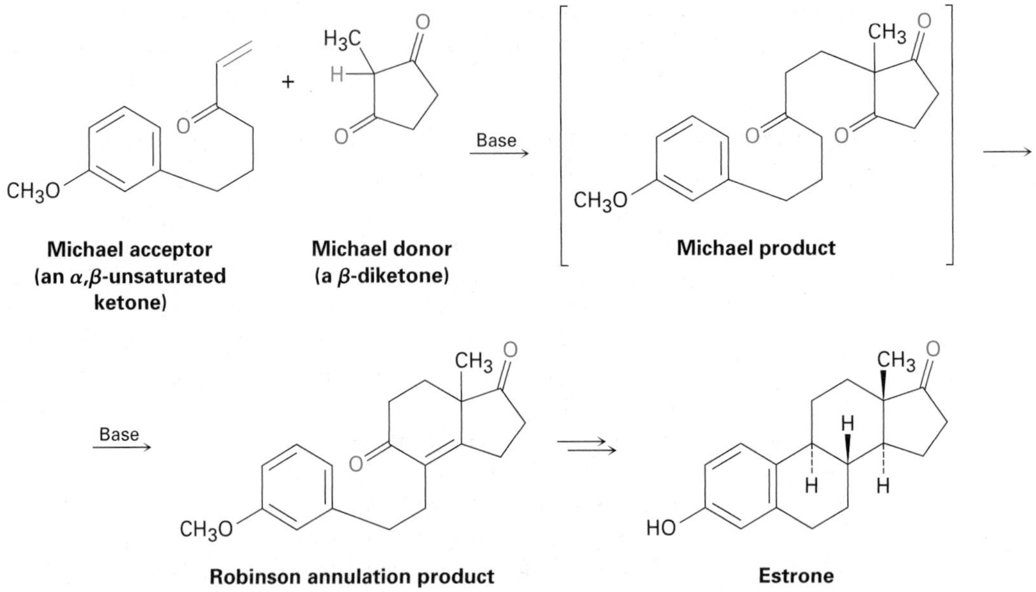

Figure 23.9 This Robinson annulation reaction is used in the commercial synthesis of the steroid hormone estrone. The nucleophilic donor is a β-diketone.

Problem 23.21 What product would you expect from a Robinson annulation reaction of 2-methyl-1,3-cyclopentanedione with 3-buten-2-one?

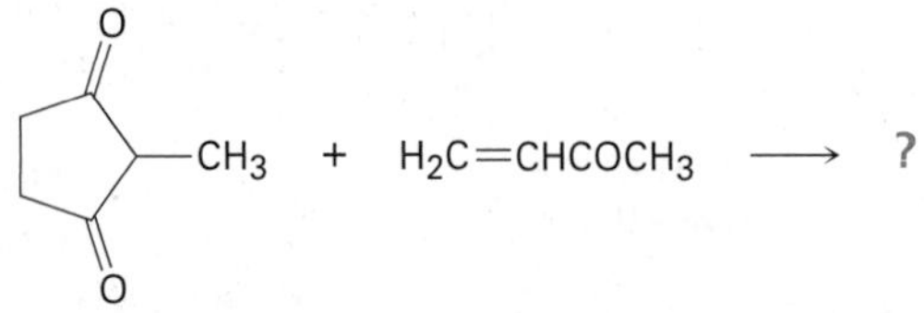

**2-Methylcyclo-
pentane-1,3-dione**   **But-3-en-2-one**

Problem 23.22 How would you prepare the following compound using a Robinson annulation reaction between a β-diketone and an α,β-unsaturated ketone? Draw the structures of both reactants and the intermediate Michael addition product.

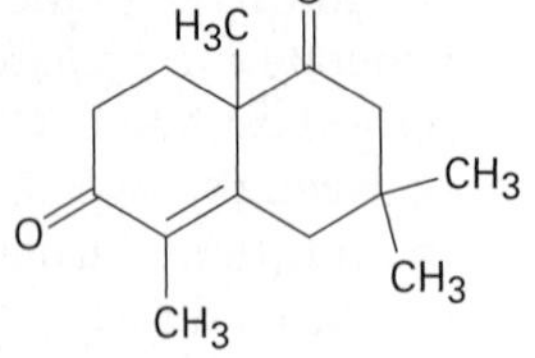

23.13 | Some Biological Carbonyl Condensation Reactions

Biological Aldol Reactions

Aldol reactions occur in many biological pathways, but are particularly important in carbohydrate metabolism, where enzymes called *aldolases* catalyze the addition of a ketone enolate ion to an aldehyde. Aldolases occur in all organisms and are of two types. Type I aldolases occur primarily in animals and higher plants; type II aldolases occur primarily in fungi and bacteria. Both types catalyze the same kind of reaction, but type I aldolases operate place through an enamine, while type II aldolases require a metal ion (usually Zn^{2+}) as Lewis acid and operate through an enolate ion.

An example of an aldolase-catalyzed reaction occurs in glucose biosynthesis when dihydroxyacetone phosphate reacts with glyceraldehyde 3-phosphate to give fructose 1,6-bisphosphate. In animals and higher plants, dihydroxyacetone phosphate is first converted into an enamine by reaction with the $-NH_2$ group on a lysine amino acid in the enzyme. The enamine then adds to glyceraldehyde 3-phosphate, and the iminium ion that results is hydrolyzed. In bacteria and fungi, the aldol reaction occurs directly, with the ketone carbonyl group of glyceraldehyde 3-phosphate complexed to a Zn^{2+} ion to make it a better acceptor (Figure 23.10, page 902).

Biological Claisen Condensations

Claisen condensations, like aldol reactions, also occur in a large number of biological pathways. In fatty-acid biosynthesis, for instance, an enolate ion generated by decarboxylation (Section 22.7) of malonyl ACP adds to the carbonyl group of another acyl group bonded through a thioester linkage to a synthase enzyme. The tetrahedral intermediate that results then expels the synthase, giving acetoacetyl ACP.

Malonyl ACP   **Enolate ion**   **Acetoacetyl ACP**

Acetyl synthase

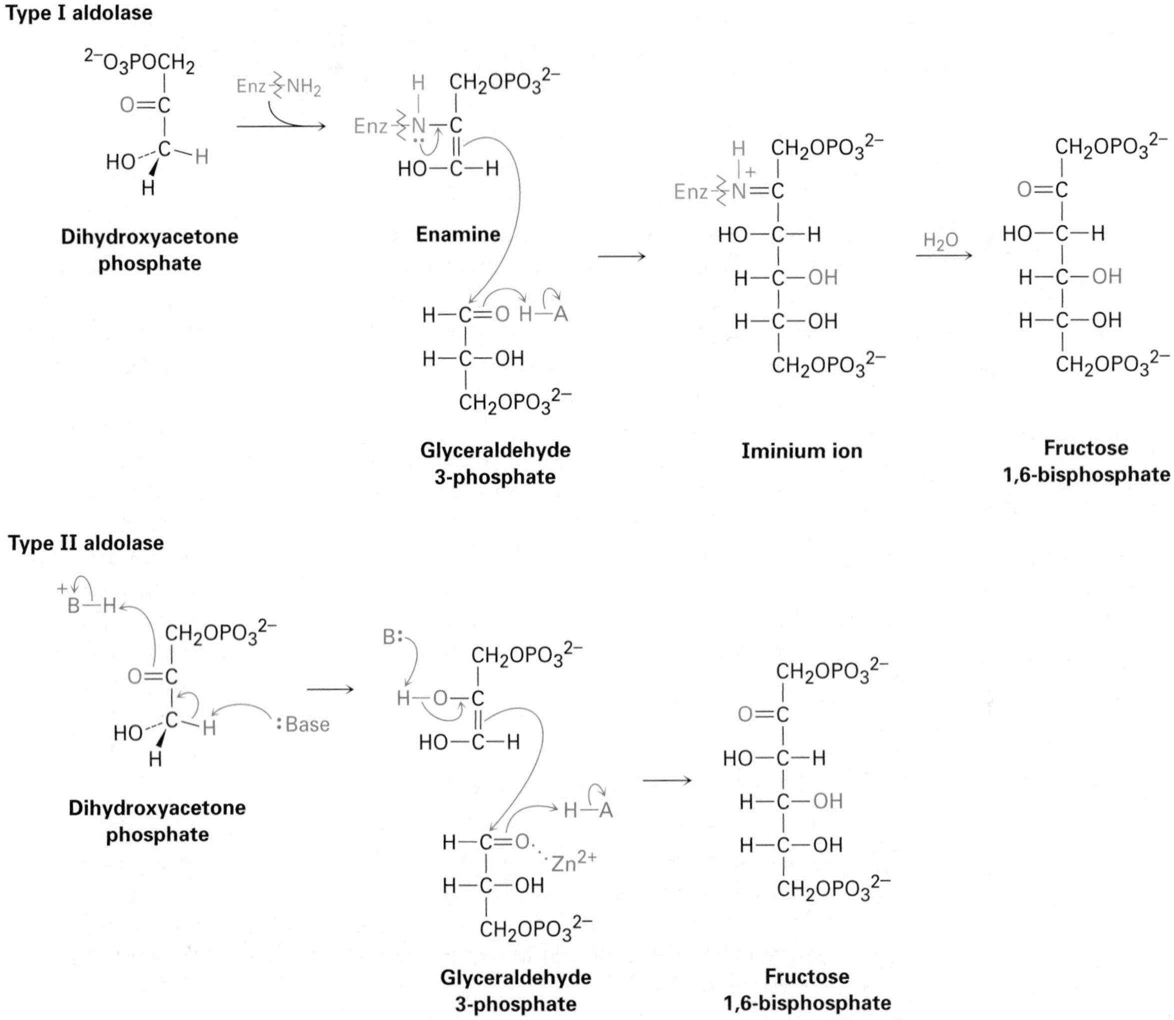

Type I aldolase

Type II aldolase

Figure 23.10 Mechanisms of type I and type II aldolase reactions in glucose biosynthesis.

Mixed Claisen condensations (Section 23.8) also occur frequently in living organisms, particularly in the pathway for fatty-acid biosynthesis that we'll discuss in Section 29.4. Butyryl synthase, for instance, reacts with malonyl ACP in a mixed Claisen condensation to give 3-ketohexanoyl ACP.

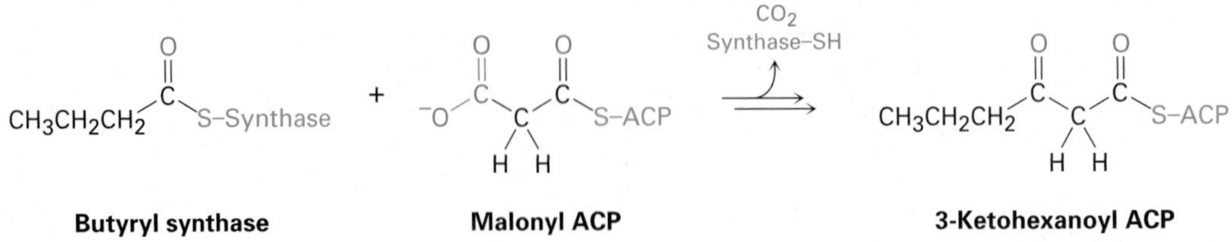

Focus On . . .

A Prologue to Metabolism

You are what you eat. Food molecules are metabolized by pathways that involve the four major carbonyl-group reactions.

Biochemistry *is* carbonyl chemistry. Almost all metabolic pathways used by living organisms involve one or more of the four fundamental carbonyl-group reactions we've seen in Chapters 19 through 23. The digestion and metabolic breakdown of all the major classes of food molecules—fats, carbohydrates, and proteins—take place by nucleophilic addition reactions, nucleophilic acyl substitutions, α substitutions, and carbonyl condensations. Similarly, hormones and other crucial biological molecules are built up from smaller precursors by these same carbonyl-group reactions.

Take *glycolysis,* for example, the metabolic pathway by which organisms convert glucose to pyruvate as the first step in extracting energy from carbohydrates.

Glucose →(Glycolysis)→ **Pyruvate**

Glycolysis is a ten-step process that begins with isomerization of glucose from its cyclic hemiacetal form to its open-chain aldehyde form—a reverse nucleophilic addition reaction. The aldehyde then undergoes tautomerization to yield an enol, which undergoes yet another tautomerization to give the ketone fructose.

Glucose (hemiacetal)   **Glucose (aldehyde)**   **Glucose (enol)**   **Fructose**

(continued)

Fructose, a β-hydroxy ketone, is then cleaved into two three-carbon molecules—one ketone and one aldehyde—by a reverse aldol reaction. Still further carbonyl-group reactions then occur until pyruvate results.

Fructose

These few examples are only an introduction; we'll look at several of the major metabolic pathways in much more detail in Chapter 29. The bottom line is that you haven't seen the end of carbonyl-group chemistry. A solid grasp of carbonyl-group reactions is crucial to an understanding of biochemistry.

SUMMARY AND KEY WORDS

A **carbonyl condensation reaction** takes place between two carbonyl partners and involves both nucleophilic addition and α-substitution steps. One carbonyl partner (the donor) is converted by base into a nucleophilic enolate ion, which adds to the electrophilic carbonyl group of the second partner (the acceptor). The donor molecule undergoes an α substitution, while the acceptor molecule undergoes a nucleophilic addition.

Nucleophilic donor **Electrophilic acceptor**

The **aldol reaction** is a carbonyl condensation that occurs between two aldehyde or ketone molecules. Aldol reactions are reversible, leading first to a β-hydroxy aldehyde or ketone and then to an α,β-unsaturated product. Mixed aldol condensations between two different aldehydes or ketones generally give a mixture of all four possible products. A mixed reaction can be successful, however, if one of the two partners is an unusually good donor (ethyl acetoacetate, for instance) or if it can act only as an acceptor (formaldehyde and benzaldehyde, for instance). Intramolecular aldol condensations of 1,4- and 1,5-diketones are also successful and provide a good way to make five- and six-membered rings.

The **Claisen reaction** is a carbonyl condensation that occurs between two ester molecules and gives a β-keto ester product. Mixed Claisen condensations between two different esters are successful only when one of the two partners has no acidic α hydrogens (ethyl benzoate and ethyl formate, for instance) and thus can function only as the acceptor partner. Intramolecular Claisen condensations, called **Dieckmann cyclization reactions**, provide excellent syntheses of five- and six-membered cyclic β-keto esters starting from 1,6- and 1,7-diesters.

The conjugate addition of a carbon nucleophile to an α,β-unsaturated acceptor is known as the **Michael reaction**. The best Michael reactions take place between unusually acidic donors (β-keto esters or β-diketones) and unhindered α,β-unsaturated acceptors. Enamines, prepared by reaction of a ketone with a disubstituted amine, are also good Michael donors.

Carbonyl condensation reactions are widely used in synthesis. One example of their versatility is the **Robinson annulation reaction**, which leads to the formation of an substituted cyclohexenone. Treatment of a β-diketone or β-keto ester with an α,β-unsaturated ketone leads first to a Michael addition, which is followed by intramolecular aldol cyclization. Condensation reactions are also used widely in nature for the biosynthesis of such molecules as fats and steroids.

SUMMARY OF REACTIONS

1. Aldol reaction (Section 23.1)

2. Mixed aldol reaction (Section 23.5)

3. Intramolecular aldol reaction (Section 23.6)

4. Dehydration of aldol products (Section 23.3)

5. Claisen condensation reaction (Section 23.7)

$$2\ \underset{\substack{\|\\O}}{RCH_2COR'} \quad \xrightleftharpoons{Na^+\ ^-OEt,\ ethanol} \quad RCH_2\underset{\substack{|\\R}}{\overset{\substack{O\\\|}}{C}}-\underset{}{\overset{\substack{O\\\|}}{CHCOR'}} \quad + \quad HOR'$$

6. Mixed Claisen condensation reaction (Section 23.8)

$$\underset{\substack{\|\\O}}{RCH_2COEt} \quad + \quad \underset{\substack{\|\\O}}{HCOEt} \quad \xrightleftharpoons{Na^+\ ^-OEt,\ ethanol} \quad HC-\underset{\substack{|\\R}}{\overset{\substack{O\\\|}}{CHCOEt}} \quad + \quad HOEt$$

7. Intramolecular Claisen condensation (Dieckmann cyclization; Section 23.9)

$$EtOC(CH_2)_4COEt \quad \xrightleftharpoons{Na^+\ ^-OEt,\ ethanol} \quad + \quad HOEt$$

$$EtOC(CH_2)_5COEt \quad \xrightleftharpoons{Na^+\ ^-OEt,\ ethanol} \quad + \quad HOEt$$

8. Michael reaction (Section 23.10)

9. Carbonyl condensations with enamines (Stork reaction; Section 23.11)

EXERCISES

Organic KNOWLEDGE TOOLS

CENGAGENOW™ Sign in at **www.cengage.com/login** to assess your knowledge of this chapter's topics by taking a pre-test. The pre-test will link you to interactive organic chemistry resources based on your score in each concept area.

⚙WL Online homework for this chapter may be assigned in Organic OWL.

■ indicates problems assignable in Organic OWL.

VISUALIZING CHEMISTRY

(Problems 23.1–23.22 appear within the chapter.)

23.23 ■ What ketones or aldehydes might the following enones have been prepared from by aldol reaction?

(a)   (b)

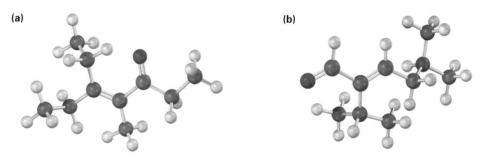

23.24 ■ The following structure represents an intermediate formed by addition of an ester enolate ion to a second ester molecule. Identify the reactant, the leaving group, and the product.

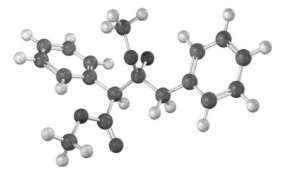

23.25 ■ The following molecule was formed by an intramolecular aldol reaction. What dicarbonyl precursor was used for its preparation?

■ Assignable in OWL

23.26 The following molecule was formed by a Robinson annulation reaction. What reactants were used?

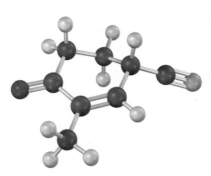

ADDITIONAL PROBLEMS

23.27 ■ Which of the following compounds would you expect to undergo aldol self-condensation? Show the product of each successful reaction.

   (a) Trimethylacetaldehyde   **(b)** Cyclobutanone

   (c) Benzophenone (diphenyl ketone)   **(d)** 3-Pentanone

   (e) Decanal   **(f)** 3-Phenyl-2-propenal

23.28 ■ How might you synthesize each of the following compounds using an aldol reaction? Show the structure of the starting aldehyde(s) or ketone(s) you would use in each case.

(a)

(b)

(c)

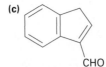

(d)

23.29 What product would you expect to obtain from aldol cyclization of hexanedial, $OHCCH_2CH_2CH_2CH_2CHO$?

23.30 Intramolecular aldol cyclization of 2,5-heptanedione with aqueous NaOH yields a mixture of two enone products in the approximate ratio 9:1. Write their structures, and show how each is formed.

23.31 The major product formed by intramolecular aldol cyclization of 2,5-heptanedione (Problem 23.30) has two singlet absorptions in the ^1H NMR spectrum at 1.65 δ and 1.90 δ, and has no absorptions in the range 3 to 10 δ. What is its structure?

23.32 Treatment of the minor product formed in the intramolecular aldol cyclization of 2,5-heptanedione (Problems 23.30 and 23.31) with aqueous NaOH converts it into the major product. Propose a mechanism to account for this base-catalyzed isomerization.

23.33 ■ The aldol reaction is catalyzed by acid as well as by base. What is the reactive nucleophile in the acid-catalyzed aldol reaction? Propose a mechanism.

23.34 How can you account for the fact that 2,2,6-trimethylcyclohexanone yields no detectable aldol product even though it has an acidic α hydrogen?

23.35 Cinnamaldehyde, the aromatic constituent of cinnamon oil, can be synthesized by a mixed aldol condensation. Show the starting materials you would use, and write the reaction.

Cinnamaldehyde

23.36 The bicyclic ketone shown below does not undergo aldol self-condensation even though it has two α hydrogen atoms. Explain.

23.37 ■ What condensation products would you expect to obtain by treatment of the following substances with sodium ethoxide in ethanol?
(a) Ethyl butanoate   (b) Cycloheptanone
(c) 3,7-Nonanedione   (d) 3-Phenylpropanal

23.38 In the mixed Claisen reaction of cyclopentanone with ethyl formate, a much higher yield of the desired product is obtained by first mixing the two carbonyl components and then adding base, rather than by first mixing base with cyclopentanone and then adding ethyl formate. Explain.

23.39 Give the structures of the possible Claisen condensation products from the following reactions. Tell which, if any, you would expect to predominate in each case.
(a) $CH_3CO_2Et + CH_3CH_2CO_2Et$ (b) $C_6H_5CO_2Et + C_6H_5CH_2CO_2Et$
(c) $EtOCO_2Et + Cyclohexanone$ (d) $C_6H_5CHO + CH_3CO_2Et$

23.40 Ethyl dimethylacetoacetate reacts instantly at room temperature when treated with ethoxide ion to yield two products, ethyl acetate and ethyl 2-methylpropanoate. Propose a mechanism for this cleavage reaction.

$$EtO_2C\diagdown\underset{\underset{\underset{H_3C\quad CH_3}{}}{|}}{C}\diagup CO_2Et \quad \xrightarrow[\text{Ethanol, 25 °C}]{Na^+ \ ^-OEt} \quad CH_3CO_2Et \quad + \quad CH_3\overset{\overset{CH_3}{|}}{CH}CO_2Et$$

23.41 In contrast to the rapid reaction shown in Problem 23.40, ethyl acetoacetate requires a temperature over 150 °C to undergo the same kind of cleavage reaction. How can you explain the difference in reactivity?

$$EtO_2C\diagdown\underset{\underset{\underset{H\quad H}{}}{|}}{C}\diagup CO_2Et \quad \xrightarrow[\text{Ethanol, 150 °C}]{Na^+ \ ^-OEt} \quad 2 \ CH_3CO_2Et$$

23.42 ■ How might the following compounds be prepared using Michael reactions? Show the nucleophilic donor and the electrophilic acceptor in each case.

(a)

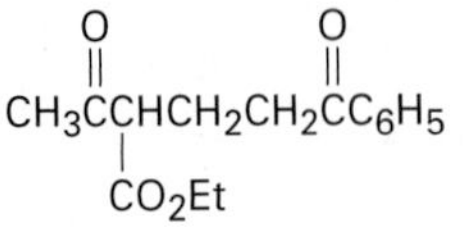

$$CH_3\overset{O}{\overset{||}{C}}CH CH_2CH_2\overset{O}{\overset{||}{C}}C_6H_5$$
$$|\\ CO_2Et$$

(b)

$$CH_3\overset{O}{\overset{||}{C}}CH_2CH_2CH_2\overset{O}{\overset{||}{C}}CH_3$$

(c)

$$EtO\overset{O}{\overset{||}{C}}CH CH_2CH_2C\equiv N$$
$$|\\ CO_2Et$$

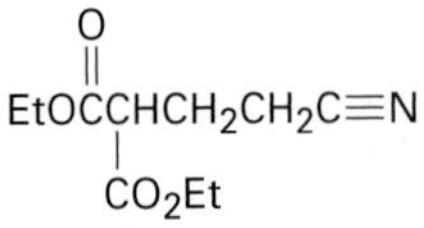

(d)

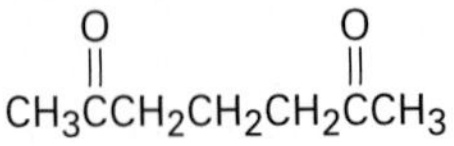

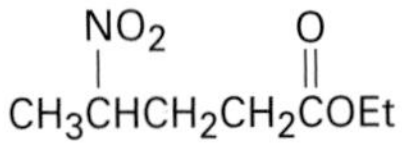

$$\overset{NO_2}{\underset{|}{CH_3CH}}CH_2CH_2\overset{O}{\overset{||}{C}}OEt$$

(e)

$$EtO\overset{O}{\overset{||}{C}}CH CH_2CH_2NO_2$$
$$|\\ CO_2Et$$

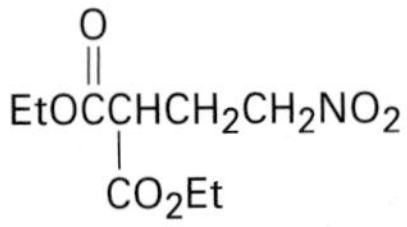

(f)

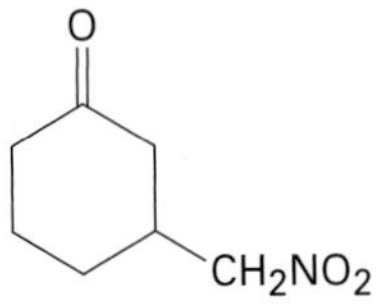

23.43 The so-called Wieland–Miescher ketone is a valuable starting material used in the synthesis of steroid hormones. How might you prepare it from 1,3-cyclo-hexanedione?

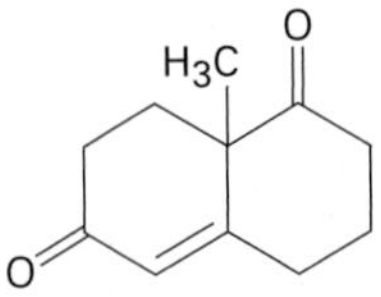

Wieland–Miescher ketone

23.44 The following reactions are unlikely to provide the indicated product in high yield. What is wrong with each?

(a)

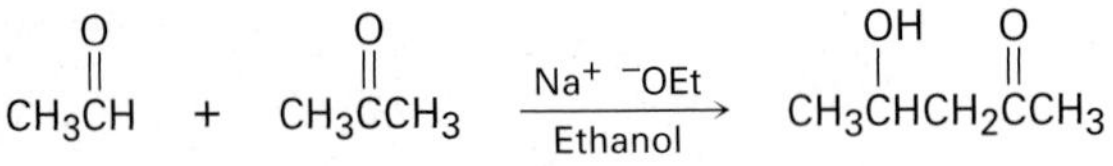

$$CH_3\overset{O}{\overset{||}{C}}H \quad + \quad CH_3\overset{O}{\overset{||}{C}}CH_3 \quad \xrightarrow[\text{Ethanol}]{Na^+ \ ^-OEt} \quad CH_3\overset{OH}{\underset{|}{C}}H CH_2\overset{O}{\overset{||}{C}}CH_3$$

(b)

$$+ \quad H_2C=CH\overset{O}{\overset{||}{C}}CH_3 \quad \xrightarrow[\text{Ethanol}]{Na^+ \ ^-OEt}$$

(c)

$$CH_3\overset{O}{\overset{||}{C}}CH_2CH_2CH_2\overset{O}{\overset{||}{C}}CH_3 \quad \xrightarrow[\text{Ethanol}]{Na^+ \ ^-OEt}$$

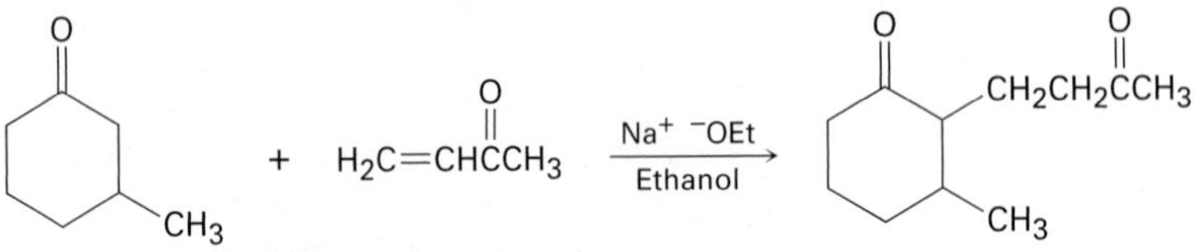

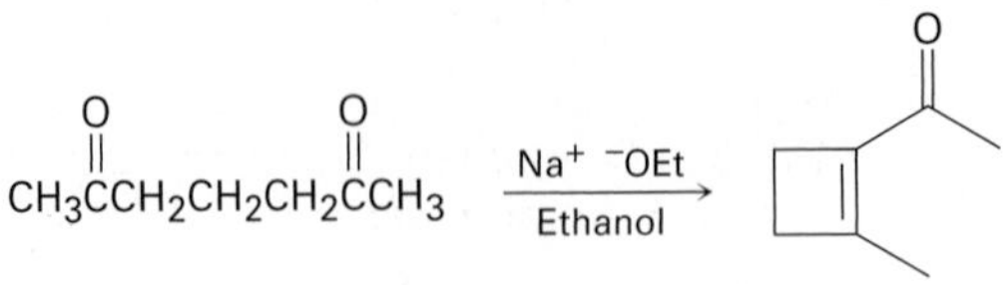

23.45 ■ Fill in the missing reagents **a–h** in the following scheme:

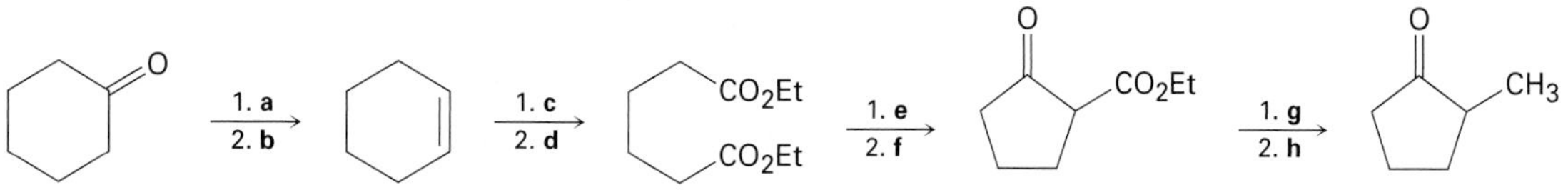

23.46 How would you prepare the following compounds from cyclohexanone?

(a)

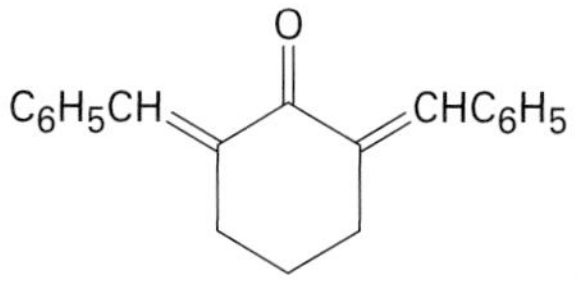

C_6H_5CH ═══ CHC_6H_5

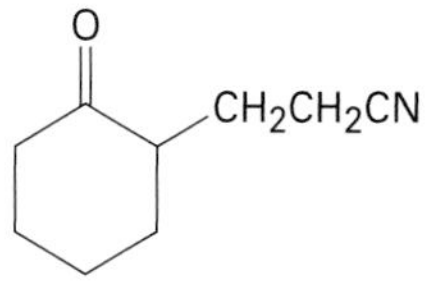

(b)

CH_2CH_2CN

(c)

CH_2CH ═ CH_2

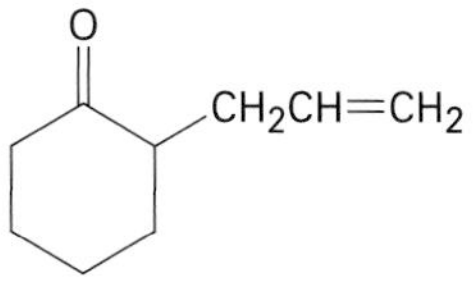

(d)

CO_2Et

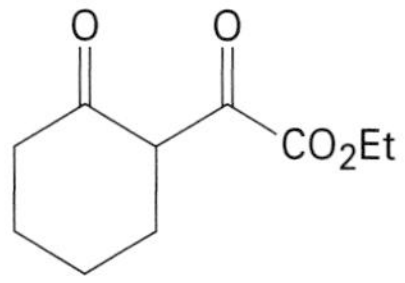

23.47 Leucine, one of the twenty amino acids found in proteins, is metabolized by a pathway that includes the following step. Propose a mechanism.

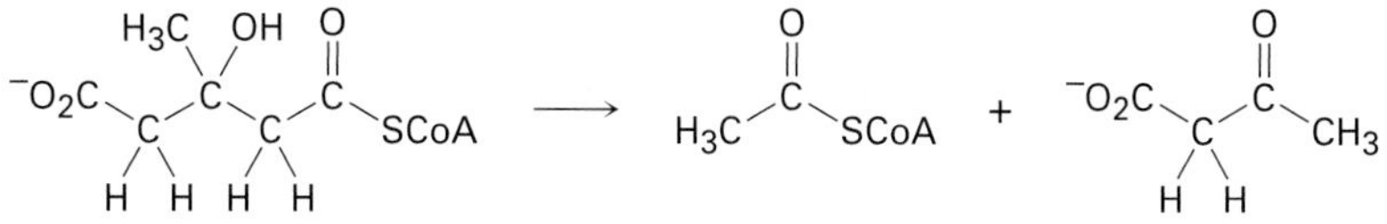

3-Hydroxy-3-methyl-glutaryl CoA ⟶ **Acetyl CoA** + **Acetoacetate**

23.48 Isoleucine, another of the twenty amino acids found in proteins, is metabolized by a pathway that includes the following step. Propose a mechanism.

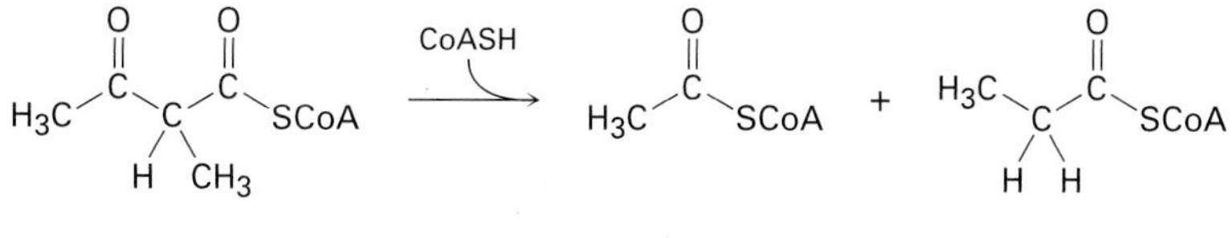

2-Methyl-3-keto-butyryl CoA → **Acetyl CoA** + **Propionyl CoA (propanoyl CoA)**

23.49 The first step in the citric acid cycle is reaction of oxaloacetate with acetyl CoA to give citrate. Propose a mechanism, using acid or base catalysis as needed.

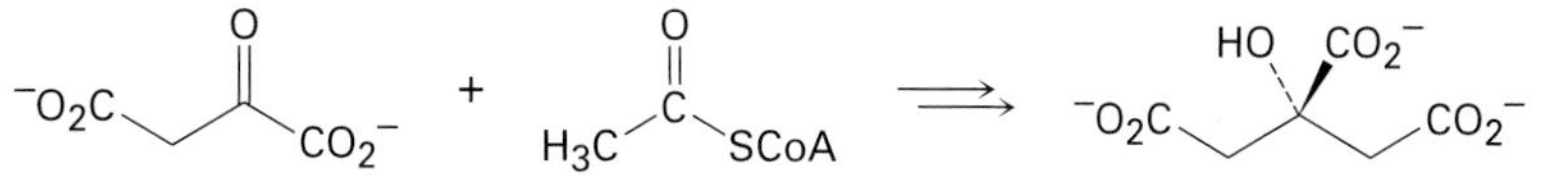

Oxaloacetate + **Acetyl CoA** ⟹ **Citrate**

23.50 The compound known as *Hagemann's ester* is prepared by treatment of a mixture of formaldehyde and ethyl acetoacetate with base, followed by acid-catalyzed decarboxylation.

$$CH_3COCH_2CO_2Et \quad + \quad CH_2O \quad \xrightarrow[\text{2. } H_3O^+]{\text{1. } Na^+ \ ^-OEt, \text{ ethanol}} \quad$$ $$\quad + \quad CO_2 \quad + \quad HOEt$$

Hagemann's ester

(a) The first step is an aldol-like condensation between ethyl acetoacetate and formaldehyde to yield an α,β-unsaturated product. Write the reaction, and show the structure of the product.

(b) The second step is a Michael reaction between ethyl acetoacetate and the unsaturated product of the first step. Show the structure of the product.

23.51 The third and fourth steps in the synthesis of Hagemann's ester from ethyl acetoacetate and formaldehyde (Problem 23.50) are an intramolecular aldol cyclization to yield a substituted cyclohexenone, and a decarboxylation reaction. Write both reactions, and show the products of each step.

23.52 When 2-methylcyclohexanone is converted into an enamine, only one product is formed despite the fact that the starting ketone is unsymmetrical. Build molecular models of the two possible products, and explain the fact that the sole product is the one with the double bond away from the methyl-substituted carbon.

NOT formed

23.53 The Stork enamine reaction and the intramolecular aldol reaction can be carried out in sequence to allow the synthesis of cyclohexenones. For example, reaction of the pyrrolidine enamine of cyclohexanone with 3-buten-2-one, followed by enamine hydrolysis and base treatment, yields the product indicated. Write each step, and show the mechanism of each.

1. $H_2C=CHCOCH_3$
2. H_3O^+
3. $NaOH, H_2O$

23.54 ■ How could you prepare the following cyclohexenones by combining a Stork enamine reaction with an intramolecular aldol condensation? (See Problem 23.53.)

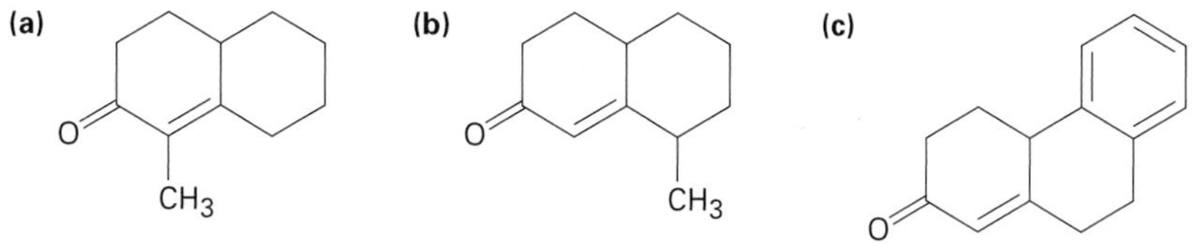

(a)   (b)   (c)

23.55 ■ The amino acid leucine is biosynthesized from α-ketoisovalerate by the following sequence of steps. Show the mechanism of each.

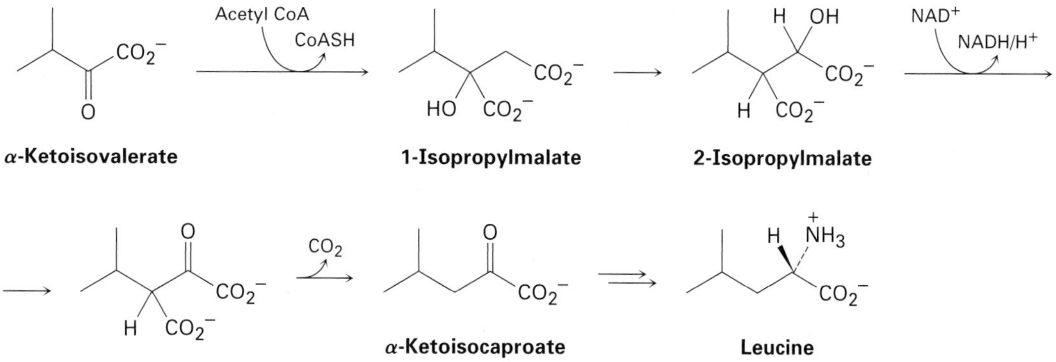

Acetyl CoA
CoASH

α-**Ketoisovalerate**   **1-Isopropylmalate**   **2-Isopropylmalate**

NAD$^+$
NADH/H$^+$

CO_2

α-**Ketoisocaproate**   **Leucine**

23.56 The *Knoevenagel reaction* is a carbonyl condensation reaction of an ester with an aldehyde or ketone to yield an α,β-unsaturated product. Show the mechanism of the Knoevenagel reaction of diethyl malonate with benzaldehyde.

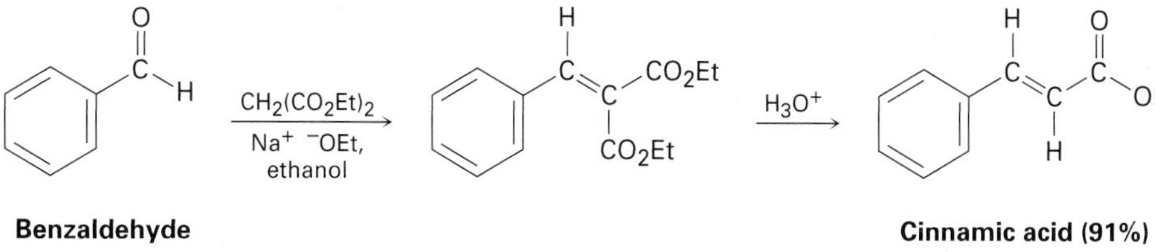

$CH_2(CO_2Et)_2$
Na^+ ^-OEt,
ethanol

H_3O^+

Benzaldehyde   **Cinnamic acid (91%)**

23.57 ■ The *Darzens reaction* involves a two-step, base-catalyzed condensation of ethyl chloroacetate with a ketone to yield an epoxy ester. The first step is a carbonyl condensation reaction, and the second step is an S_N2 reaction. Write both steps, and show their mechanisms.

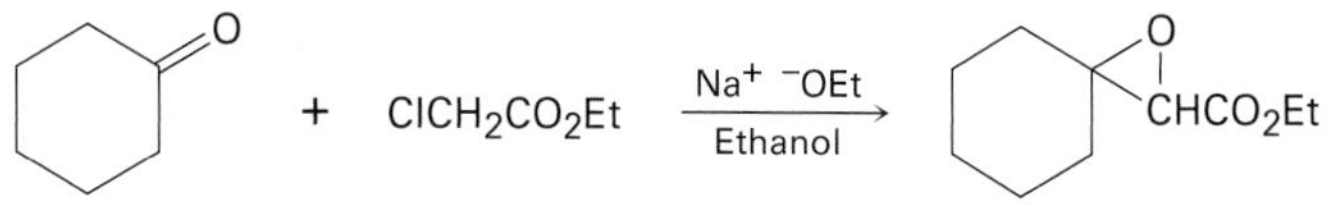

$+$ $ClCH_2CO_2Et$ $\xrightarrow[\text{Ethanol}]{Na^+ \ ^-OEt}$

■ Assignable in OWL

23.58 The following reaction involves a hydrolysis followed by an intramolecular nucleophilic acyl substitution reaction. Write both steps, and show their mechanisms.

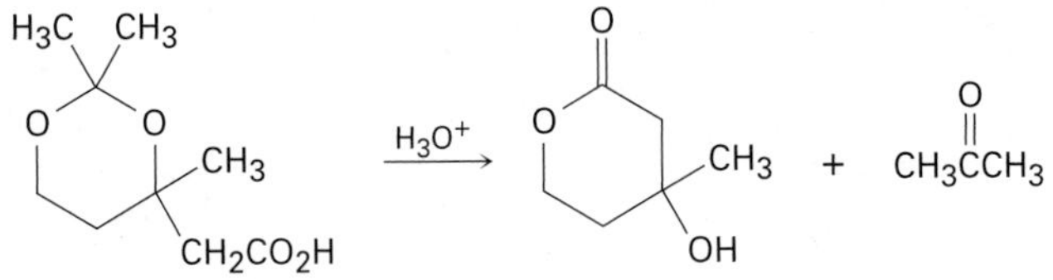

23.59 ■ The following reaction involves an intramolecular Michael reaction followed by an intramolecular aldol reaction. Write both steps, and show their mechanisms.

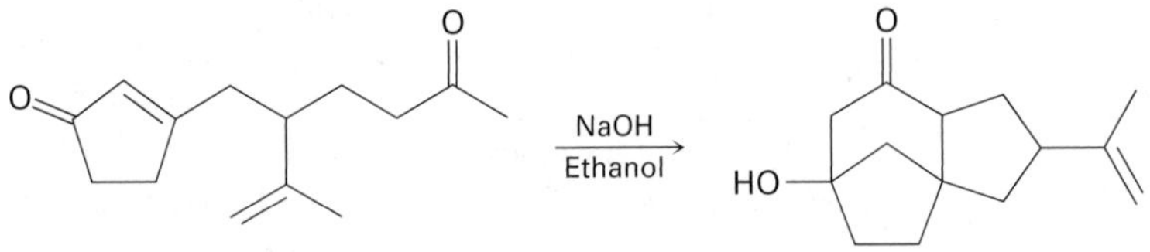

23.60 ■ The following reaction involves a conjugate addition reaction followed by an intramolecular Claisen condensation. Write both steps, and show their mechanisms.

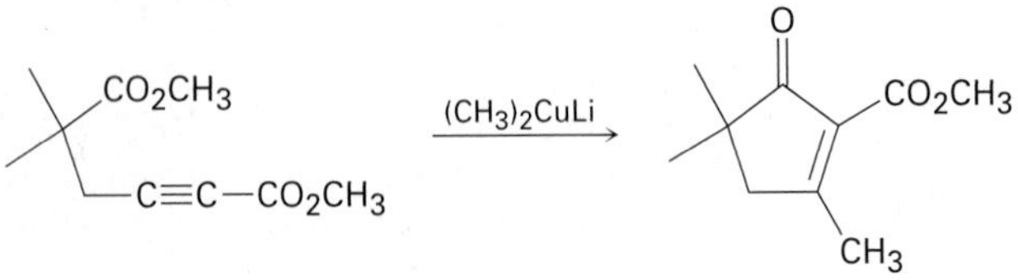

23.61 ■ The following reaction involves two successive intramolecular Michael reactions. Write both steps, and show their mechanisms.

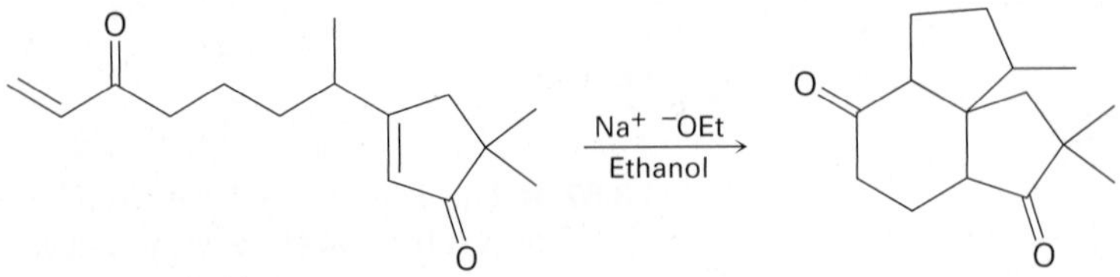

23.62 ■ The following reaction involves an intramolecular aldol reaction followed by a *retro* aldol-like reaction. Write both steps, and show their mechanisms.

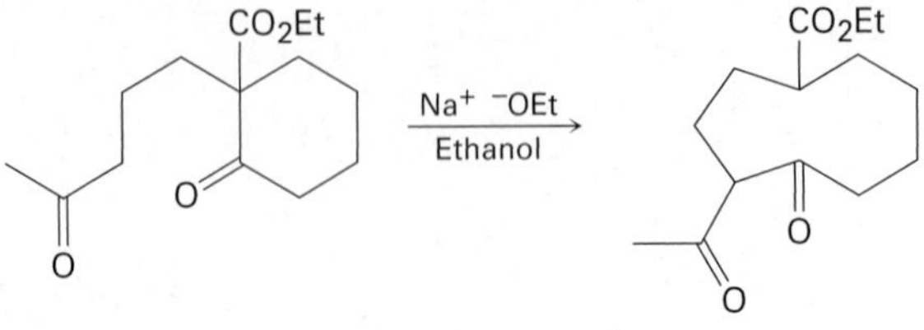

23.63 The *Mannich reaction* of a ketone, an amine, and an aldehyde is one of the few three-component reactions in organic chemistry. Cyclohexanone, for example, reacts with dimethylamine and acetaldehyde to yield an amino ketone. The reaction takes place in two steps, both of which are typical carbonyl-group reactions.

(a) The first step is reaction between the aldehyde and the amine to yield an intermediate iminium ion ($R_2C=NR_2^+$) plus water. Propose a mechanism, and show the structure of the intermediate iminium ion.

(b) The second step is reaction between the iminium ion intermediate and the ketone to yield the final product. Propose a mechanism.

23.64 Cocaine has been prepared by a sequence beginning with a Mannich reaction (Problem 23.63) between dimethyl acetonedicarboxylate, an amine, and a dialdehyde. Show the structures of the amine and dialdehyde.

24

Amines and Heterocycles

Amines are organic derivatives of ammonia in the same way that alcohols and ethers are organic derivatives of water. Like ammonia, amines contain a nitrogen atom with a lone pair of electrons, making amines both basic and nucleophilic. We'll soon see, in fact, that most of the chemistry of amines depends on the presence of this lone pair of electrons.

Amines occur widely in all living organisms. Trimethylamine, for instance, occurs in animal tissues and is partially responsible for the distinctive odor of fish, nicotine is found in tobacco, and cocaine is a stimulant found in the South American coca bush. In addition, amino acids are the building blocks from which all proteins are made, and cyclic amine bases are constituents of nucleic acids.

Trimethylamine   **Nicotine**   **Cocaine**

WHY THIS CHAPTER?

By the end of this chapter, we will have seen all the common functional groups. Of those groups, amines and carbonyl compounds are the most abundant and have the richest chemistry. In addition to the proteins and nucleic acids already mentioned, the majority of pharmaceutical agents contain amine functional groups, and many of the common coenzymes necessary for biological catalysis are amines.

24.1 | Naming Amines

Amines can be either alkyl-substituted (**alkylamines**) or aryl-substituted (**arylamines**). Although much of the chemistry of the two classes is similar, there are also substantial differences. Amines are classified as **primary (RNH_2)**,

secondary (R₂NH), or **tertiary (R₃N)**, depending on the number of organic substituents attached to nitrogen. Thus, methylamine (CH₃NH₂) is a primary amine, dimethylamine [(CH₃)₂NH] is a secondary amine, and trimethylamine [(CH₃)₃N] is a tertiary amine. Note that this usage of the terms *primary, secondary,* and *tertiary* is different from our previous usage. When we speak of a tertiary alcohol or alkyl halide, we refer to the degree of substitution at the alkyl carbon atom, but when we speak of a tertiary amine, we refer to the degree of substitution at the nitrogen atom.

tert-Butyl alcohol (a tertiary alcohol)   **Trimethylamine** (a tertiary amine)   **tert-Butylamine** (a primary amine)

Compounds containing a nitrogen atom with four attached groups also exist, but the nitrogen atom must carry a formal positive charge. Such compounds are called **quaternary ammonium salts**.

A quaternary ammonium salt

Primary amines are named in the IUPAC system in several ways. For simple amines, the suffix *-amine* is added to the name of the alkyl substituent. You might also recall from Chapter 15 that phenylamine, C₆H₅NH₂, has the common name *aniline*.

tert-Butylamine   **Cyclohexylamine**   **Aniline**

Alternatively, the suffix *-amine* can be used in place of the final *-e* in the name of the parent compound.

4,4-Dimethylcyclohexanamine   **1,4-Butanediamine**

Amines with more than one functional group are named by considering the $-NH_2$ as an *amino* substituent on the parent molecule.

$\underset{4\quad3\quad2\quad1}{CH_3CH_2\overset{\overset{\displaystyle NH_2}{|}}{C}HCO_2H}$

2,4-Diaminobenzoic acid structure with CO_2H, NH_2 at position 2, and NH_2 at position 4

$\underset{4\quad3\quad2\;1}{H_2NCH_2CH_2\overset{\overset{\displaystyle O}{\|}}{C}CH_3}$

2-Aminobutanoic acid   **2,4-Diaminobenzoic acid**   **4-Aminobutan-2-one**

Symmetrical secondary and tertiary amines are named by adding the prefix *di*- or *tri*- to the alkyl group.

Diphenylamine structure

$CH_3CH_2-\overset{\overset{\displaystyle H... }{}}{N}-CH_2CH_3$ with CH_2CH_3 below N

Diphenylamine   **Triethylamine**

Unsymmetrically substituted secondary and tertiary amines are named as *N*-substituted primary amines. The largest alkyl group is chosen as the parent name, and the other alkyl groups are *N*-substituents on the parent (*N* because they're attached to nitrogen).

$\underset{H_3C}{\overset{H_3C}{>}}N-CH_2CH_2CH_3$

N-Ethyl-N-methylcyclohexylamine structure

N,N-Dimethylpropylamine   **N-Ethyl-N-methylcyclohexylamine**

Heterocyclic amines—compounds in which the nitrogen atom occurs as part of a ring—are also common, and each different heterocyclic ring system has its own parent name. The heterocyclic nitrogen atom is always numbered as position 1.

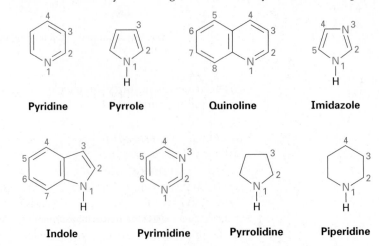

Pyridine   **Pyrrole**   **Quinoline**   **Imidazole**

Indole   **Pyrimidine**   **Pyrrolidine**   **Piperidine**

Problem 24.1 | Name the following compounds:

(a) CH₃NHCH₂CH₃

(b)

(c)

CH₂CH₃
|
N
  CH₃

(d)

N
|
CH₃

(e)

H
|
N

(f)

CH₃
|
H₂NCH₂CH₂CHNH₂

Problem 24.2 | Draw structures corresponding to the following IUPAC names:
(a) Triisopropylamine
(b) Triallylamine
(c) *N*-Methylaniline
(d) *N*-Ethyl-*N*-methylcyclopentylamine
(e) *N*-Isopropylcyclohexylamine
(f) *N*-Ethylpyrrole

Problem 24.3 | Draw structures for the following heterocyclic amines:
(a) 5-Methoxyindole
(b) 1,3-Dimethylpyrrole
(c) 4-(*N,N*-Dimethylamino)pyridine
(d) 5-Aminopyrimidine

24.2 | Properties of Amines

The bonding in alkylamines is similar to the bonding in ammonia. The nitrogen atom is *sp*³-hybridized, with the three substituents occupying three corners of a tetrahedron and the lone pair of electrons occupying the fourth corner. As you might expect, the C–N–C bond angles are close to the 109° tetrahedral value. For trimethylamine, the C–N–C bond angle is 108°, and the C–N bond length is 147 pm.

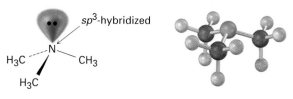

Trimethylamine

One consequence of tetrahedral geometry is that an amine with three different substituents on nitrogen is chiral, as we saw in Section 9.12. Unlike chiral carbon compounds, however, chiral amines can't usually be resolved because the two enantiomeric forms rapidly interconvert by a *pyramidal inversion*, much as an alkyl halide inverts in an S_N2 reaction. Pyramidal inversion occurs by a momentary rehybridization of the nitrogen atom to planar, *sp*² geometry, followed by rehybridization of the planar intermediate to tetrahedral, *sp*³ geometry

(Figure 24.1). The barrier to inversion is about 25 kJ/mol (6 kcal/mol), an amount only twice as large as the barrier to rotation about a C–C single bond.

Figure 24.1 Pyramidal inversion rapidly interconverts the two mirror-image (enantiomeric) forms of an amine.

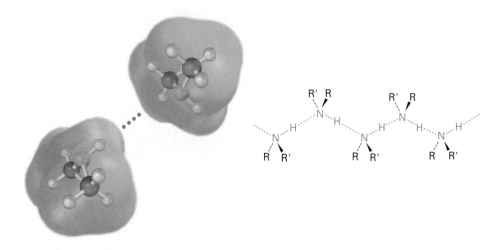

sp³-hybridized
(tetrahedral)

sp²-hybridized
(planar)

sp³-hybridized
(tetrahedral)

Alkylamines have a variety of applications in the chemical industry as starting materials for the preparation of insecticides and pharmaceuticals. Labetalol, for instance, a so-called β-blocker used for the treatment of high blood pressure, is prepared by S$_N$2 reaction of an epoxide with a primary amine. The substance marketed for drug use is a mixture of all four possible stereoisomers, but the biological activity derives primarily from the (*R,R*) isomer.

Labetalol

Like alcohols, amines with fewer than five carbon atoms are generally water-soluble. Also like alcohols, primary and secondary amines form hydrogen bonds and are highly associated. As a result, amines have higher boiling points than alkanes of similar molecular weight. Diethylamine (MW = 73 amu) boils at 56.3 °C, for instance, while pentane (MW = 72 amu) boils at 36.1 °C.

One other characteristic of amines is their odor. Low-molecular-weight amines such as trimethylamine have a distinctive fishlike aroma, while diamines such as 1,5-pentanediamine, commonly called cadaverine, have the appalling odors you might expect from their common names.

24.3 Basicity of Amines

The chemistry of amines is dominated by the lone pair of electrons on nitrogen, which makes amines both basic and nucleophilic. They react with acids to form acid–base salts, and they react with electrophiles in many of the polar reactions seen in past chapters. Note in the following electrostatic potential map of trimethylamine how the negative (red) region corresponds to the lone-pair of electrons on nitrogen.

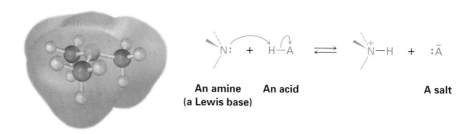

An amine   **An acid**   **A salt**
(a Lewis base)

Amines are much stronger bases than alcohols and ethers, their oxygen-containing analogs. When an amine is dissolved in water, an equilibrium is established in which water acts as an acid and transfers a proton to the amine. Just as the acid strength of a carboxylic acid can be measured by defining an acidity constant K_a (Section 2.8), the base strength of an amine can be measured by defining an analogous *basicity constant* K_b. The larger the value of K_b and the smaller the value of pK_b, the more favorable the proton-transfer equilibrium and the stronger the base.

For the reaction

$$RNH_2 + H_2O \rightleftharpoons RNH_3^+ + OH^-$$

$$K_b = \frac{[RNH_3^+]\,[OH^-]}{[RNH_2]}$$

$$pK_b = -\log K_b$$

In practice, K_b values are not often used. Instead, the most convenient way to measure the *basicity* of an amine (RNH_2) is to look at the *acidity* of the corresponding ammonium ion (RNH_3^+).

For the reaction

$$RNH_3^+ + H_2O \rightleftharpoons RNH_2 + H_3O^-$$

$$K_a = \frac{[RNH_2]\,[H_3O^+]}{[RNH_3^+]}$$

so

$$K_a \cdot K_b = \frac{[RNH_2]\ [H_3O^+]}{[RNH_3^+]} \cdot \frac{[RNH_3^+]\ [OH^-]}{[RNH_2]}$$

$$= [H_3O^+]\ [OH^-] = K_w = 1.00 \times 10^{-14}$$

Thus

$$K_a = \frac{K_w}{K_b} \quad \text{and} \quad K_b = \frac{K_w}{K_a}$$

and

$$pK_a + pK_b = 14$$

These equations say that the K_b of an amine multiplied by the K_a of the corresponding ammonium ion is equal to K_w, the ion-product constant for water (1.00×10^{-14}). Thus, if we know K_a for an ammonium ion, we also know K_b for the corresponding amine base because $K_b = K_w/K_a$. The more acidic the ammonium ion, the less tightly the proton is held and the weaker the corresponding base. That is, a weaker base has an ammonium ion with a smaller pK_a, and a stronger base has an ammonium ion with a larger pK_a.

Weaker base   Smaller pK_a for ammonium ion

Stronger base   Larger pK_a for ammonium ion

Table 24.1 lists pK_a values of some ammonium ions and indicates that there is a substantial range of amine basicities. Most simple alkylamines are similar in their base strength, with pK_a's for their ammonium ions in the narrow range 10 to 11. *Arylamines*, however, are considerably less basic than alkylamines, as are the heterocyclic amines pyridine and pyrrole.

In contrast with amines, *amides* ($RCONH_2$) are nonbasic. Amides don't undergo substantial protonation by aqueous acids, and they are poor nucleophiles. The main reason for this difference in basicity between amines and amides is that an amide is stabilized by delocalization of the nitrogen lone-pair electrons through orbital overlap with the carbonyl group. In resonance terms, amides are more stable and less reactive than amines because they are hybrids of two resonance forms. This amide resonance stabilization is lost when the nitrogen atom is protonated, so protonation is disfavored. Electrostatic potential maps show clearly the decreased electron density on the amide nitrogen.

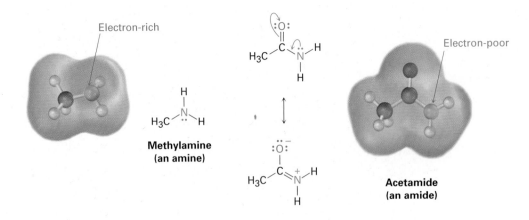

Electron-rich

Methylamine
(an amine)

Electron-poor

Acetamide
(an amide)

Table 24.1 | **Basicity of Some Common Amines**

Name	Structure	pK_a of ammonium ion
Ammonia	NH_3	9.26
Primary alkylamine		
Methylamine	CH_3NH_2	10.64
Ethylamine	$CH_3CH_2NH_2$	10.75
Secondary alkylamine		
Diethylamine	$(CH_3CH_2)_2NH$	10.98
Pyrrolidine		11.27
Tertiary alkylamine		
Triethylamine	$(CH_3CH_2)_3N$	10.76
Arylamine		
Aniline		4.63
Heterocyclic amine		
Pyridine		5.25
Pyrimidine		1.3
Pyrrole		0.4
Imidazole		6.95

It's often possible to take advantage of their basicity to purify amines. For example, if a mixture of a basic amine and a neutral compound such as a ketone or alcohol is dissolved in an organic solvent and aqueous acid is added, the basic amine dissolves in the water layer as its protonated salt, while the neutral compound remains in the organic solvent layer. Separation of the water layer and neutralization of the ammonium ion by addition of NaOH then provides the pure amine (Figure 24.2).

In addition to their behavior as bases, primary and secondary amines can also act as very weak acids because an N–H proton can be removed by a sufficiently strong base. We've seen, for example, how diisopropylamine (p$K_a \approx 40$) reacts with butyllithium to yield lithium diisopropylamide (LDA; Section 22.5). Dialkylamine anions like LDA are extremely powerful bases that are often used

Figure 24.2 Separation and purification of an amine component from a mixture.

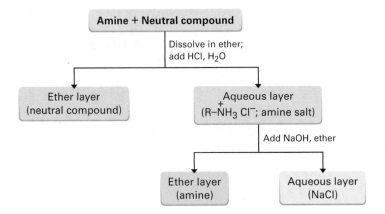

in laboratory organic chemistry for the generation of enolate ions from carbonyl compounds (Section 22.7).

C_4H_9Li + Diisopropylamine $\xrightarrow[\text{solvent}]{\text{THF}}$ Lithium diisopropylamide (LDA) + C_4H_{10}

Butyllithium

Diisopropylamine

Lithium diisopropylamide (LDA)

Problem 24.4 | Which compound in each of the following pairs is more basic?
(a) $CH_3CH_2NH_2$ or $CH_3CH_2CONH_2$ **(b)** NaOH or CH_3NH_2
(c) CH_3NHCH_3 or pyridine

Problem 24.5 | The benzylammonium ion ($C_6H_5CH_2NH_3^+$) has $pK_a = 9.33$, and the propyl-ammonium ion has $pK_a = 10.71$. Which is the stronger base, benzylamine or propylamine? What are the pK_b's of benzylamine and propylamine?

24.4 | Basicity of Substituted Arylamines

As noted previously, arylamines are generally less basic than alkylamines. Anilinium ion has $pK_a = 4.63$, for instance, whereas methylammonium ion has $pK_a = 10.64$. Arylamines are less basic than alkylamines because the nitrogen lone-pair electrons are delocalized by interaction with the aromatic ring π electron system and are less available for bonding to H^+. In resonance terms, arylamines are stabilized relative to alkylamines because of their five resonance forms.

Much of the resonance stabilization is lost on protonation, however, so the energy difference between protonated and nonprotonated forms is higher for arylamines than it is for alkylamines. As a result, arylamines are less basic. Figure 24.3 illustrates the difference.

Figure 24.3 Arylamines have a larger positive $\Delta G°$ for protonation and are therefore less basic than alkylamines, primarily because of resonance stabilization of the ground state. Electrostatic potential maps show that lone-pair electron density is delocalized in the amine but the charge is localized in the corresponding ammonium ion.

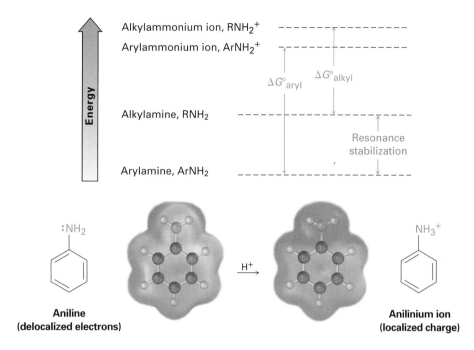

Substituted arylamines can be either more basic or less basic than aniline, depending on the substituent. Electron-donating substituents, such as $-CH_3$, $-NH_2$, and $-OCH_3$, which increase the reactivity of an aromatic ring toward electrophilic substitution (Section 16.4), also increase the basicity of the corresponding arylamine. Electron-withdrawing substituents, such as $-Cl$, $-NO_2$, and $-CN$, which decrease ring reactivity toward electrophilic substitution, also decrease arylamine basicity. Table 24.2 considers only p-substituted anilines, but similar trends are observed for ortho and meta derivatives.

Problem 24.6 Without looking at Table 24.2, rank the following compounds in order of ascending basicity.
(a) p-Nitroaniline, p-aminobenzaldehyde, p-bromoaniline
(b) p-Chloroaniline, p-aminoacetophenone, p-methylaniline
(c) p-(Trifluoromethyl)aniline, p-methylaniline, p-(fluoromethyl)aniline

24.5 Biological Amines and the Henderson–Hasselbalch Equation

We saw in Section 20.3 that the extent of dissociation of a carboxylic acid HA in an aqueous solution buffered to a given pH can be calculated with the Henderson–Hasselbalch equation. Furthermore, we concluded that at the physiological

Table 24.2 | **Base Strength of Some *p*-Substituted Anilines**

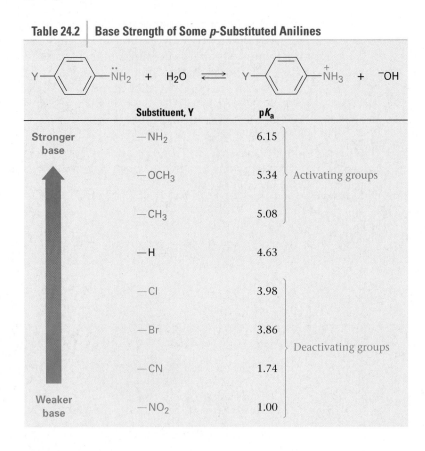

Substituent, Y	pK_a	
Stronger base		
—NH$_2$	6.15	
—OCH$_3$	5.34	Activating groups
—CH$_3$	5.08	
—H	4.63	
—Cl	3.98	
—Br	3.86	
		Deactivating groups
—CN	1.74	
Weaker base —NO$_2$	1.00	

pH of 7.3 inside living cells, carboxylic acids are almost entirely dissociated into their carboxylate anions, RCO_2^-.

Henderson–Hasselbalch equation: $\mathrm{pH} = \mathrm{p}K_a + \log\dfrac{[A^-]}{[HA]}$

$$\log\frac{[A^-]}{[HA]} = \mathrm{pH} - \mathrm{p}K_a$$

What about amine bases? In what form do they exist at the physiological pH inside cells—as the amine ($A^- = RNH_2$), or as the ammonium ion ($HA = RNH_3^+$)? Let's take a 0.0010 M solution of methylamine at pH = 7.3, for example. According to Table 24.1, the pK_a of methylammonium ion is 10.64, so from the Henderson–Hasselbalch equation, we have

$$\log\frac{[RNH_2]}{[RNH_3^+]} = \mathrm{pH} - \mathrm{p}K_a = 7.3 - 10.64 = -3.34$$

$$\frac{[RNH_2]}{[RNH_3^+]} = \mathrm{antilog}(-3.34) = 4.6 \times 10^{-4}$$

$$[RNH_2] = (4.6 \times 10^{-4})[RNH_3^+]$$

In addition, we know that

$$[RNH_2] + [RNH_3^+] = 0.0010 \text{ M}$$

Solving the two simultaneous equations gives $[RNH_3^+] = 0.0010$ M and $[RNH_2] = 5 \times 10^{-7}$ M. In other words, at a physiological pH of 7.3, essentially 100% of the methylamine in a 0.0010 M solution exists in its protonated form as methylammonium ion. The same is true of other amine bases, so we write cellular amines in their protonated form and amino acids in their ammonium carboxylate form to reflect their structures at physiological pH.

The amino group is protonated at pH = 7.3.

The carboxylic acid group is dissociated at pH = 7.3.

Alanine
(an amino acid)

Problem 24.7 | Calculate the percentages of neutral and protonated forms present in a solution of 0.0010 M pyrimidine at pH = 7.3. The pK_a of pyrimidinium ion is 1.3.

24.6 | Synthesis of Amines

Reduction of Nitriles, Amides, and Nitro Compounds

CENGAGENOW" Click *Organic Interactive* to **use a web-based palette to predict products from a variety of reactions that yield amines**.

We've already seen in Sections 20.7 and 21.7 how amines can be prepared by reduction of nitriles and amides with $LiAlH_4$. The two-step sequence of S_N2 displacement with CN^- followed by reduction thus converts an alkyl halide into a primary alkylamine having one more carbon atom. Amide reduction converts carboxylic acids and their derivatives into amines with the same number of carbon atoms.

Arylamines are usually prepared by nitration of an aromatic starting material, followed by reduction of the nitro group (Section 16.2). The reduction step can be carried out in many different ways, depending on the circumstances. Catalytic hydrogenation over platinum works well but is often incompatible with the presence elsewhere in the molecule of other reducible groups, such as C=C bonds or carbonyl groups. Iron, zinc, tin, and tin(II) chloride ($SnCl_2$) are also effective when used in acidic aqueous solution. Tin(II) chloride is particularly mild and is often used when other reducible functional groups are present.

p-tert-Butylnitrobenzene

p-tert-Butylaniline (100%)

_m_-Nitrobenzaldehyde

_m_-Aminobenzaldehyde (90%)

Problem 24.8 | Propose structures for either a nitrile or an amide that might be a precursor of each of the following amines:
(a) $CH_3CH_2CH_2NH_2$ (b) $(CH_3CH_2CH_2)_2NH$
(c) Benzylamine, $C_6H_5CH_2NH_2$ (d) _N_-Ethylaniline

S_N2 Reactions of Alkyl Halides

Ammonia and other amines are good nucleophiles in S_N2 reactions. As a result, the simplest method of alkylamine synthesis is by S_N2 alkylation of ammonia or an alkylamine with an alkyl halide. If ammonia is used, a primary amine results; if a primary amine is used, a secondary amine results; and so on. Even tertiary amines react rapidly with alkyl halides to yield quaternary ammonium salts, $R_4N^+ X^-$.

Unfortunately, these reactions don't stop cleanly after a single alkylation has occurred. Because ammonia and primary amines have similar reactivity, the initially formed monoalkylated substance often undergoes further reaction to yield a mixture of products. Even secondary and tertiary amines undergo further alkylation, although to a lesser extent. For example, treatment of 1-bromooctane with a twofold excess of ammonia leads to a mixture containing only 45% of octyl-amine. A nearly equal amount of dioctylamine is produced by double alkylation, along with smaller amounts of trioctylamine and tetraoctylammonium bromide.

$$CH_3(CH_2)_6CH_2Br \ + \ :NH_3 \longrightarrow CH_3(CH_2)_6CH_2\ddot{N}H_2 \ + \ [CH_3(CH_2)_6CH_2]_2\ddot{N}H$$

1-Bromooctane   **Octylamine (45%)**   **Dioctylamine (43%)**

$$+ \ [CH_3(CH_2)_6CH_2]_3N\text{:} \ + \ [CH_3(CH_2)_6CH_2]_4\overset{+}{N} \ \overset{-}{Br}$$

Trace   **Trace**

A better method for preparing primary amines is to use the *azide synthesis,* in which azide ion, N_3^-, is used for S_N2 reaction with a primary or secondary alkyl halide to give an alkyl azide, RN_3. Because alkyl azides are not nucleophilic, overalkylation can't occur. Subsequent reduction of the alkyl azide, either by catalytic hydrogenation over a palladium catalyst or by reaction with $LiAlH_4$, then leads to the desired primary amine. Although the method works well, low-molecular-weight alkyl azides are explosive and must be handled carefully.

1-Bromo-2-phenylethane   **2-Phenylethyl azide**   **2-Phenylethylamine (89%)**

Another alternative for preparing a primary amine from an alkyl halide is the **Gabriel amine synthesis,** which uses a *phthalimide* alkylation. An **imide** (−CONHCO−) is similar to a β-keto ester in that the acidic N−H hydrogen is flanked by two carbonyl groups. Thus, imides are deprotonated by such bases as KOH, and the resultant anions are readily alkylated in a reaction similar to the acetoacetic ester synthesis (Section 22.7). Basic hydrolysis of the N-alkylated imide then yields a primary amine product. The imide hydrolysis step is analogous to the hydrolysis of an amide (Section 21.7).

Phthalimide

Problem 24.9 | Write the mechanism of the last step in the Gabriel amine synthesis, the base-promoted hydrolysis of a phthalimide to yield an amine plus phthalate ion.

Problem 24.10 | Show two methods for the synthesis of dopamine, a neurotransmitter involved in regulation of the central nervous system. Use any alkyl halide needed.

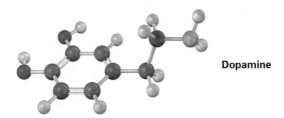

Dopamine

Reductive Amination of Aldehydes and Ketones

Amines can be synthesized in a single step by treatment of an aldehyde or ketone with ammonia or an amine in the presence of a reducing agent, a process called **reductive amination**. For example, amphetamine, a central nervous system stimulant, is prepared commercially by reductive amination of phenyl-2-propanone with ammonia, using hydrogen gas over a nickel catalyst as the reducing agent.

$$\text{Phenylpropan-2-one} \xrightarrow[\substack{\text{H}_2/\text{Ni} \\ (\text{or NaBH}_4)}]{\text{NH}_3} \text{Amphetamine} + \text{H}_2\text{O}$$

Phenylpropan-2-one   **Amphetamine**

Reductive amination takes place by the pathway shown in Figure 24.4. An imine intermediate is first formed by a nucleophilic addition reaction (Section 19.8), and the C=N bond of the imine is then reduced.

Ammonia, primary amines, and secondary amines can all be used in the reductive amination reaction, yielding primary, secondary, and tertiary amines, respectively.

Primary amine   **Secondary amine**   **Tertiary amine**

Active Figure 24.4
MECHANISM: Mechanism of reductive amination of a ketone to yield an amine. Details of the imine-forming step were shown in Figure 19.8 on page 711. *Sign in at* **www.cengage.com/login** *to see a simulation based on this figure and to take a short quiz.*

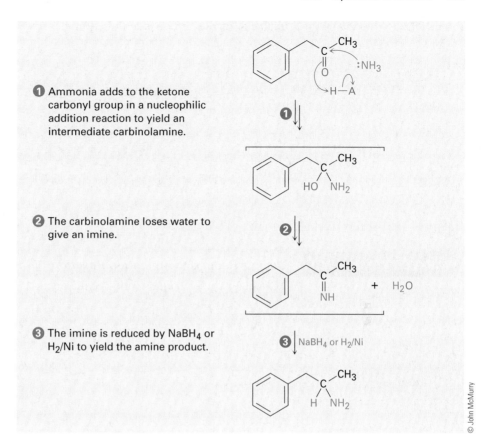

1 Ammonia adds to the ketone carbonyl group in a nucleophilic addition reaction to yield an intermediate carbinolamine.

2 The carbinolamine loses water to give an imine.

3 The imine is reduced by NaBH₄ or H₂/Ni to yield the amine product.

© John McMurry

Many different reducing agents are effective, but the most common choice in the laboratory is sodium cyanoborohydride, $NaBH_3CN$. Sodium cyanoborohydride is similar in reactivity to sodium borohydride ($NaBH_4$) but is more stable in weak acid solution.

Cyclohexanone

N,N-Dimethylcyclohexylamine (85%)

Reductive aminations also occur in various biological pathways. In the biosynthesis of the amino acid proline, for instance, glutamate 5-semialdehyde undergoes internal imine formation to give 1-pyrrolinium 5-carboxylate, which is then reduced by nucleophilic addition of hydride ion to the C=N bond.

Reduced nicotinamide adenine dinucleotide, NADH, acts as the biological reducing agent.

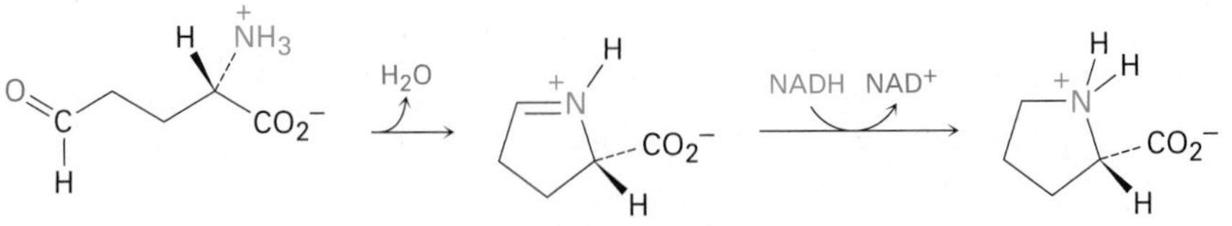

H
NH_3^+
CO_2^-
H_2O
NADH NAD+

Glutamate 5-semialdehyde | 1-Pyrrolinium 5-carboxylate | Proline

WORKED EXAMPLE 24.1

Using a Reductive Amination Reaction

How might you prepare *N*-methyl-2-phenylethylamine using a reductive amination reaction?

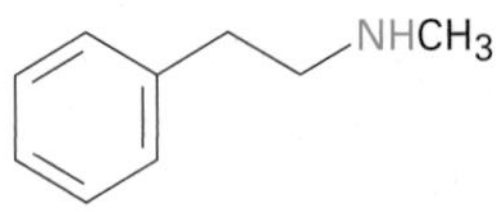

$NHCH_3$

***N*-Methyl-2-phenylethylamine**

Strategy Look at the target molecule, and identify the groups attached to nitrogen. One of the groups must be derived from the aldehyde or ketone component, and the other must be derived from the amine component. In the case of *N*-methyl-2-phenylethylamine, there are two combinations that can lead to the product: phenylacetaldehyde plus methylamine or formaldehyde plus 2-phenylethylamine. In general, it's usually better to choose the combination with the simpler amine component—methylamine in this case—and to use an excess of that amine as reactant.

Solution

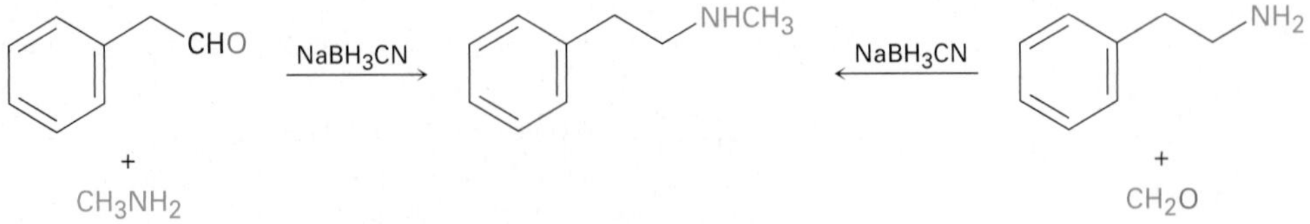

CHO
+
CH_3NH_2
$NaBH_3CN$
$NHCH_3$
$NaBH_3CN$
NH_2
+
CH_2O

Problem 24.11 How might the following amines be prepared using reductive amination reactions? Show all precursors if more than one is possible.

(a)

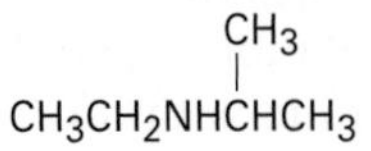

CH_3
$CH_3CH_2NHCHCH_3$

(b)

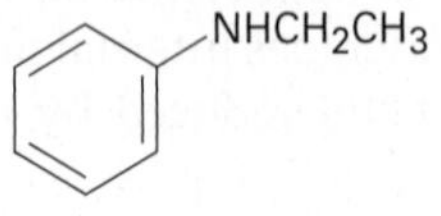

$NHCH_2CH_3$

(c)

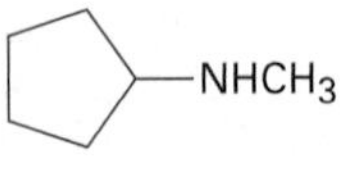

$NHCH_3$

Problem 24.12 How could you prepare the following amine using a reductive amination reaction?

Hofmann and Curtius Rearrangements

Carboxylic acid derivatives can be converted into primary amines with loss of one carbon atom by both the **Hofmann rearrangement** and the **Curtius rearrangement**. Although the Hofmann rearrangement involves a primary amide and the Curtius rearrangement involves an acyl azide, both proceed through similar mechanisms.

Hofmann
rearrangement

$$R-C(=O)-NH_2 \xrightarrow[H_2O]{NaOH, Br_2} R-NH_2 + CO_2$$

An amide

Curtius
rearrangement

$$R-C(=O)-N=\overset{+}{N}=\overset{-}{N} \xrightarrow[Heat]{H_2O} R-NH_2 + CO_2 + N_2$$

An acyl azide

Hofmann rearrangement occurs when a primary amide, $RCONH_2$, is treated with Br_2 and base (Figure 24.5). The overall mechanism is lengthy, but most of the individual steps have been encountered before. Thus, the bromination of an amide in steps 1 and 2 is analogous to the base-promoted bromination of a ketone enolate ion (Section 22.6), and the rearrangement of the bromoamide anion in step 4 is analogous to a carbocation rearrangement (Section 6.11). Nucleophilic addition of water to the isocyanate carbonyl group in step 5 is a typical carbonyl-group process (Section 19.4), as is the final decarboxylation step (Section 22.7).

Despite its mechanistic complexity, the Hofmann rearrangement often gives high yields of both arylamines and alkylamines. For example, the appetite-suppressant drug phentermine is prepared commercially by Hofmann rearrangement of a primary amide. Commonly known by the name *Fen-Phen*, the combination of phentermine with another appetite-suppressant, fenflur-amine, is suspected of causing heart damage.

$$\xrightarrow[H_2O]{NaOH, Cl_2}$$

2,2-Dimethyl-3-phenyl-propanamide   **Phetermine** $+$ CO_2

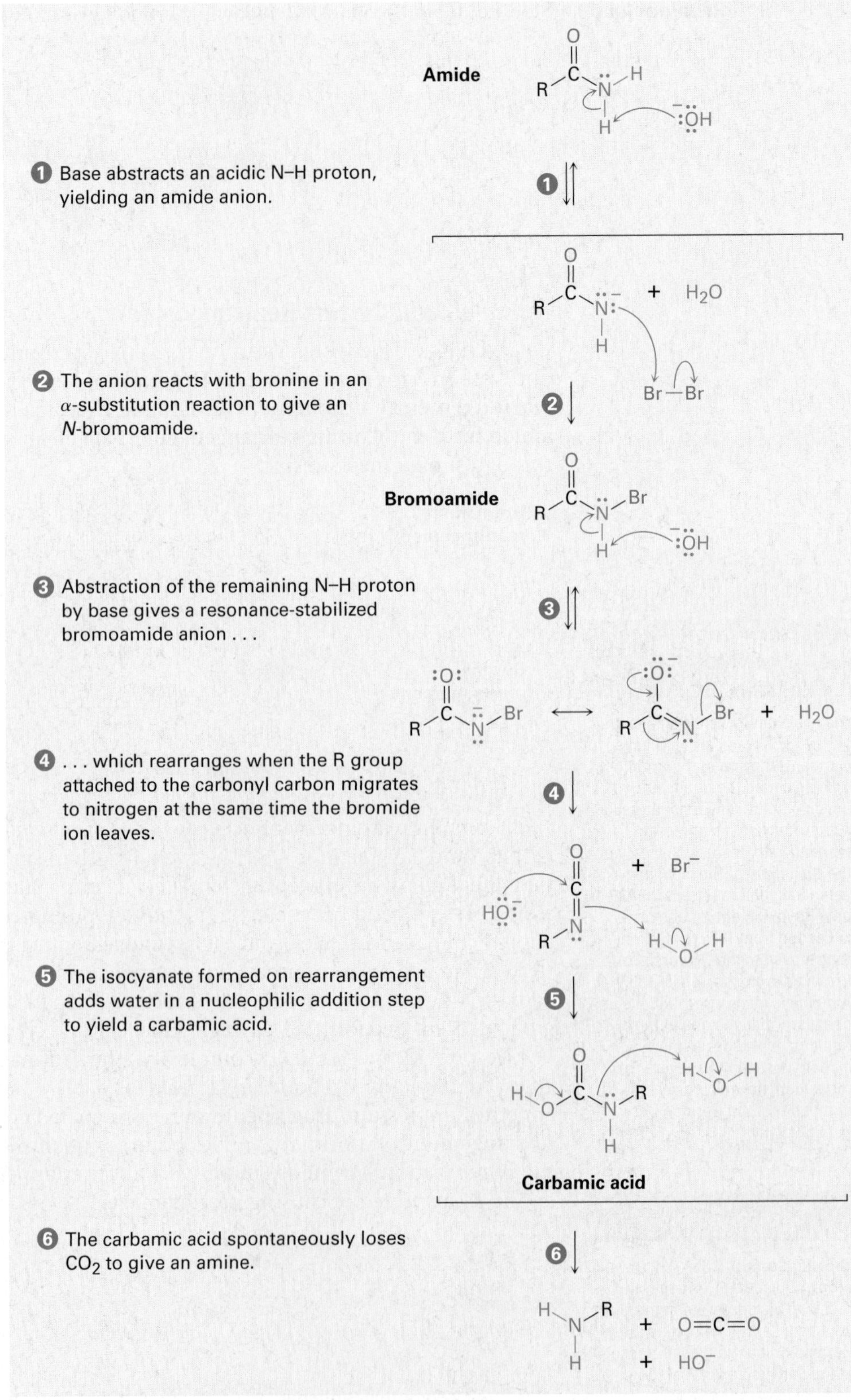

Amide

① Base abstracts an acidic N–H proton, yielding an amide anion.

② The anion reacts with bronine in an α-substitution reaction to give an N-bromoamide.

Bromoamide

③ Abstraction of the remaining N–H proton by base gives a resonance-stabilized bromoamide anion . . .

④ . . . which rearranges when the R group attached to the carbonyl carbon migrates to nitrogen at the same time the bromide ion leaves.

⑤ The isocyanate formed on rearrangement adds water in a nucleophilic addition step to yield a carbamic acid.

Carbamic acid

⑥ The carbamic acid spontaneously loses CO_2 to give an amine.

© John McMurry

Figure 24.5 MECHANISM: Mechanism of the Hofmann rearrangement of an amide to an amine. Each step is analogous to a reaction studied previously.

The Curtius rearrangement, like the Hofmann rearrangement, involves migration of an $-R$ group from the $C=O$ carbon atom to the neighboring nitrogen with simultaneous loss of a leaving group. The reaction takes place on heating an acyl azide that is itself prepared by nucleophilic acyl substitution of an acid chloride.

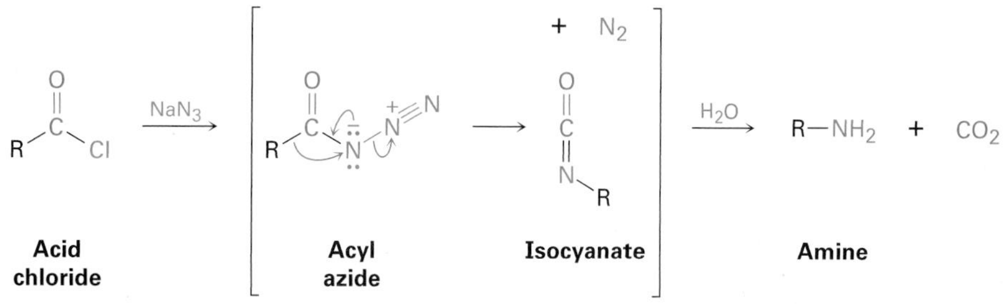

Like the Hofmann rearrangement, the Curtius rearrangement is often used commercially. For example, the antidepressant drug tranylcypromine is made by Curtius rearrangement of 2-phenylcyclopropanecarbonyl chloride.

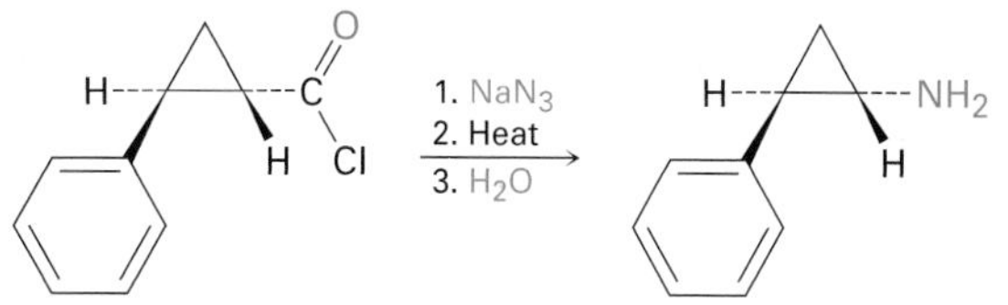

trans-2-Phenylcyclo-
propanecarbonyl chloride

Tranylcypromine

WORKED EXAMPLE 24.2

Using the Hofmann and Curtius Reactions

How would you prepare *o*-methylbenzylamine from a carboxylic acid, using both Hofmann and Curtius rearrangements?

Strategy   Both Hofmann and Curtius rearrangements convert a carboxylic acid derivative—either an amide (Hofmann) or an acid chloride (Curtius)—into a primary amine with loss of one carbon, $RCOY \rightarrow RNH_2$. Both reactions begin with the same carboxylic acid, which can be identified by replacing the $-NH_2$ group of the amine product by a $-CO_2H$ group. In the present instance, *o*-methylphenylacetic acid is needed.

Solution

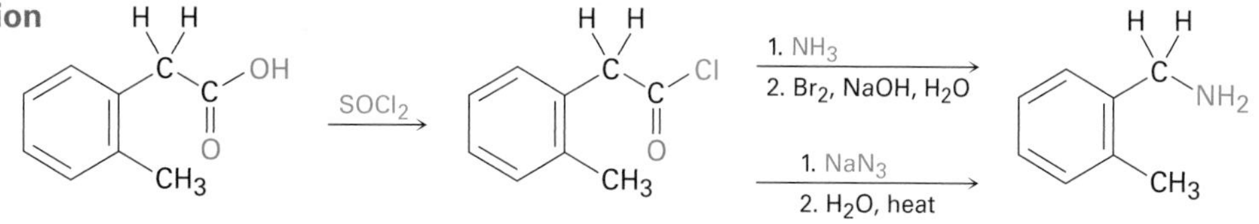

o-Methylphenyl-
acetic acid

o-Methylbenzylamine

Problem 24.13 How would you prepare the following amines, using both Hofmann and Curtius rearrangements on a carboxylic acid derivative?

(a)

$$CH_3CCH_2CH_2NH_2$$

with CH_3 groups above and below

(b)

An aromatic ring with NH_2 and H_3C substituents.

24.7 | Reactions of Amines

Alkylation and Acylation

We've already studied the two most general reactions of amines—alkylation and acylation. As we saw earlier in this chapter, primary, secondary, and tertiary amines can be alkylated by reaction with a primary alkyl halide. Alkylations of primary and secondary amines are difficult to control and often give mixtures of products, but tertiary amines are cleanly alkylated to give quaternary ammonium salts. Primary and secondary (but not tertiary) amines can also be acylated by nucleophilic acyl substitution reaction with an acid chloride or an acid anhydride to yield an amide (Sections 21.4 and 21.5). Note that overacylation of the nitrogen does not occur because the amide product is much less nucleophilic and less reactive than the starting amine.

Hofmann Elimination

Like alcohols, amines can be converted into alkenes by an elimination reaction. Because an amide ion, NH_2^-, is such a poor leaving group, however, it must first be converted into a better leaving group. In the **Hofmann elimination reaction**, an amine is methylated by reaction with excess iodomethane to produce a quaternary ammonium salt, which then undergoes elimination to give an alkene on heating with a base, typically silver oxide, Ag_2O. For example, 1-methylpentylamine is converted into 1-hexene in 60% yield.

$$CH_3CH_2CH_2CH_2CHCH_3$$

1-Methylpentylamine

$\xrightarrow[\text{CH}_3\text{I}]{\text{Excess}}$

$$CH_3CH_2CH_2CH_2CHCH_3$$ with $^+N(CH_3)_3\ I^-$

(1-Methylpentyl)trimethyl-ammonium iodide

$\xrightarrow[\text{H}_2\text{O, heat}]{\text{Ag}_2\text{O}}$

$$CH_3CH_2CH_2CH_2CH{=}CH_2$$ + $N(CH_3)_3$

1-Hexene (60%)

Silver oxide acts by exchanging hydroxide ion for iodide ion in the quaternary salt, thus providing the base necessary to cause elimination. The actual elimination step is an E2 reaction (Section 11.8) in which hydroxide ion removes a proton at the same time that the positively charged nitrogen atom leaves.

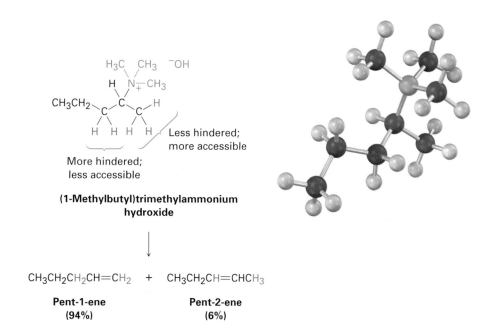

Quaternary
ammonium salt   **Alkene**

An interesting feature of the Hofmann elimination is that it gives products different from those of most other E2 reactions. Whereas the *more* highly substituted alkene product generally predominates in the E2 reaction of an alkyl halide (Zaitsev's rule; Section 11.7), the *less* highly substituted alkene predominates in the Hofmann elimination of a quaternary ammonium salt. The reason for this selectivity is probably steric. Because of the large size of the trialkylamine leaving group, the base must abstract a hydrogen from the most sterically accessible, least hindered position.

H₃C CH₃ ⁻OH [see figure]

Less hindered; more accessible

More hindered; less accessible

(1-Methylbutyl)trimethylammonium hydroxide

$CH_3CH_2CH_2CH{=}CH_2$ + $CH_3CH_2CH{=}CHCH_3$

Pent-1-ene   **Pent-2-ene**
(94%)   **(6%)**

The Hofmann elimination reaction is not often used today in the laboratory, but analogous biological eliminations occur frequently, although usually with protonated ammonium ions rather than quaternary ammonium salts. In the biosynthesis of nucleic acids, for instance, a substance called adenylosuccinate undergoes an elimination of a positively charged nitrogen to give fumarate plus adenosine monophosphate.

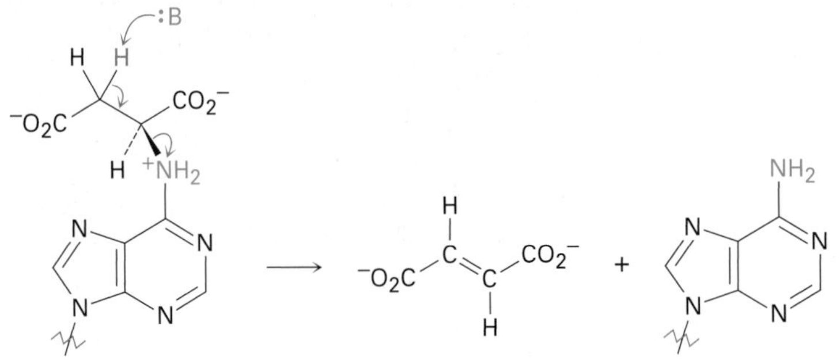

| Adenylosuccinate | Fumarate | Adenosine monophosphate |

WORKED EXAMPLE 24.3   ***Predicting the Product of a Hofmann Elimination***

What product would you expect from Hofmann elimination of the following amine?

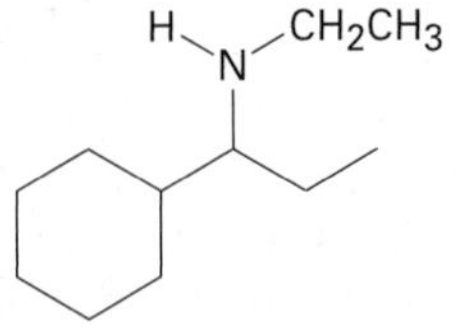

Strategy   The Hofmann elimination is an E2 reaction that converts an amine into an alkene and occurs with non-Zaitsev regiochemistry to form the least highly substituted double bond. To predict the product, look at the reactant and identify the positions from which elimination might occur (the positions two carbons removed from nitrogen). Then carry out an elimination using the most accessible hydrogen. In the present instance, there are three possible positions from which elimination might occur— one primary, one secondary, and one tertiary. The primary position is the most accessible and leads to the least highly substituted alkene, ethylene.

Solution

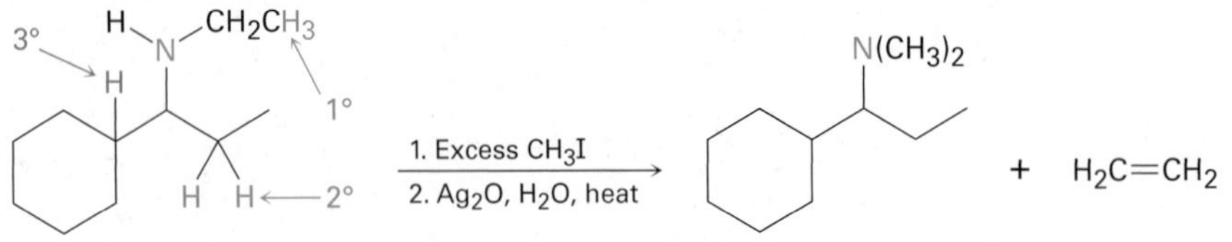

Problem 24.14   What products would you expect from Hofmann elimination of the following amines? If more than one product is formed, indicate which is major.

(a) NH_2
$CH_3CH_2CH_2CHCH_2CH_2CH_2CH_3$

(b)

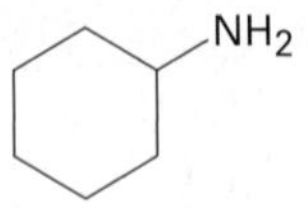

(c) NH_2
$CH_3CH_2CH_2CHCH_2CH_2CH_3$

(d)

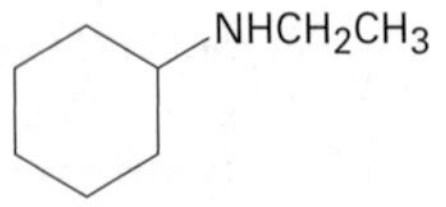

Problem 24.15 What product would you expect from Hofmann elimination of a heterocyclic amine such as piperidine? Write all the steps.

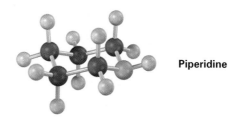

Piperidine

24.8 Reactions of Arylamines

Electrophilic Aromatic Substitution

An amino group is strongly activating and ortho- and para-directing in electrophilic aromatic substitution reactions (Section 16.4). This high reactivity can be a drawback at times because it's often difficult to prevent polysubstitution. For instance, reaction of aniline with Br_2 takes place rapidly and yields the 2,4,6-tribrominated product. The amino group is so strongly activating that it's not possible to stop at the monobromo stage.

$$\text{Aniline} \xrightarrow[\text{H}_2\text{O}]{3 \text{ Br}_2} \text{2,4,6-Tribromoaniline (100\%)}$$

Another drawback to the use of amino-substituted benzenes in electrophilic aromatic substitution reactions is that Friedel–Crafts reactions are not successful (Section 16.3). The amino group forms an acid–base complex with the $AlCl_3$ catalyst, which prevents further reaction from occurring. Both drawbacks can be overcome, however, by carrying out electrophilic aromatic substitution reactions on the corresponding *amide* rather than on the free amine.

As we saw in Section 21.5, treatment of an amine with acetic anhydride yields the corresponding acetyl amide, or acetamide. Although still activating and ortho-, para-directing, amido substituents (−NHCOR) are less strongly activating and less basic than amino groups because their nitrogen lone-pair electrons are delocalized by the neighboring carbonyl group. As a result, bromination of an *N*-arylamide occurs cleanly to give a monobromo product, and hydrolysis with aqueous base then gives the free amine. For example, *p*-toluidine (4-methylaniline) can be acetylated, brominated, and hydrolyzed to yield 2-bromo-4-methylaniline. None of the 2,6-dibrominated product is obtained.

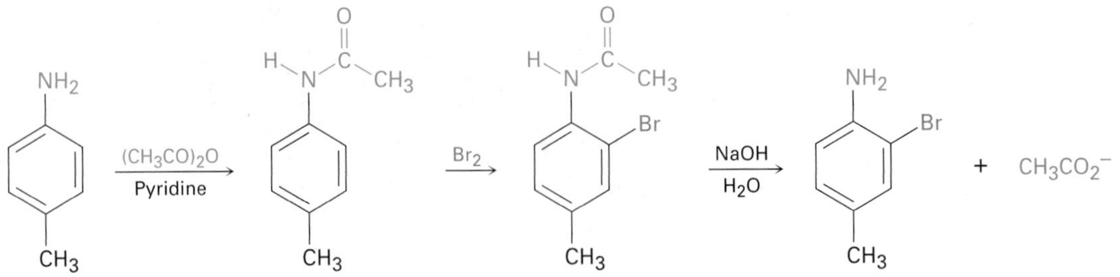

p-Toluidine

**2-Bromo-4-methyl-
aniline (79%)**

Friedel–Crafts alkylations and acylations of N-arylamides also proceed normally. For example, benzoylation of acetanilide (N-acetylaniline) under Friedel–Crafts conditions gives 4-aminobenzophenone in 80% yield after hydrolysis.

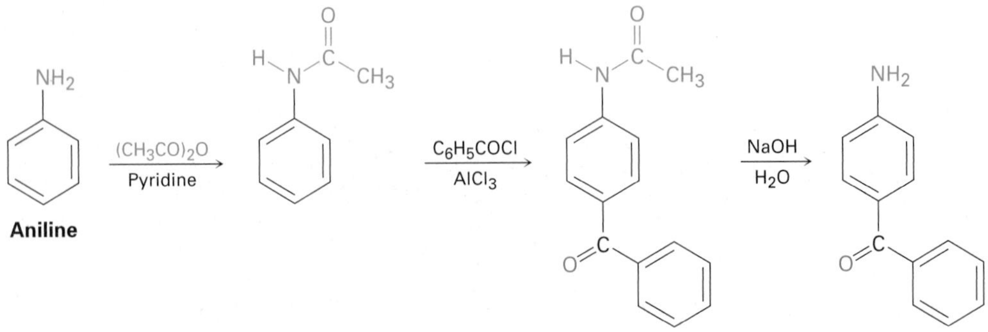

Aniline

**4-Aminobenzophenone
(80%)**

Modulating the reactivity of an amino-substituted benzene by forming an amide is a useful trick that allows many kinds of electrophilic aromatic substitutions to be carried out that would otherwise be impossible. A good example is the preparation of the sulfa drugs. Sulfa drugs, such as sulfanilamide, were among the first pharmaceutical agents to be used clinically against bacterial infection. Although they have largely been replaced by safer and more powerful antibiotics, sulfa drugs are credited with saving the lives of thousands of wounded during World War II, and they are still prescribed for infections of the urinary tract. They are prepared by chlorosulfonation of acetanilide, followed by reaction of p-(N-acetylamino)benzenesulfonyl chloride with ammonia or some other amine to give a sulfonamide. Hydrolysis of the amide then yields the sulfa drug. Note that this amide hydrolysis can be carried out in the presence of the sulfonamide group because sulfonamides hydrolyze very slowly.

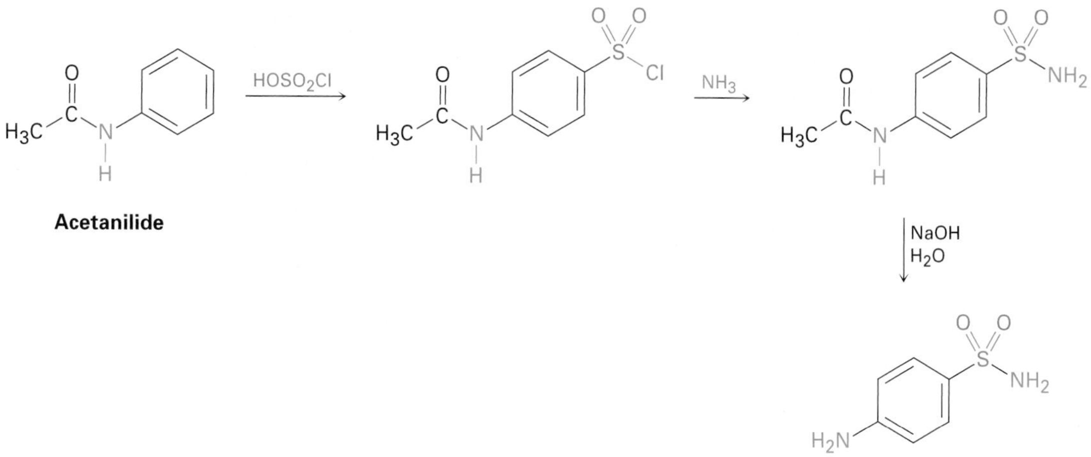

Acetanilide

Sulfanilamide
(a sulfa drug)

Problem 24.16 Propose a synthesis of the drug sulfathiazole from benzene and any necessary amine.

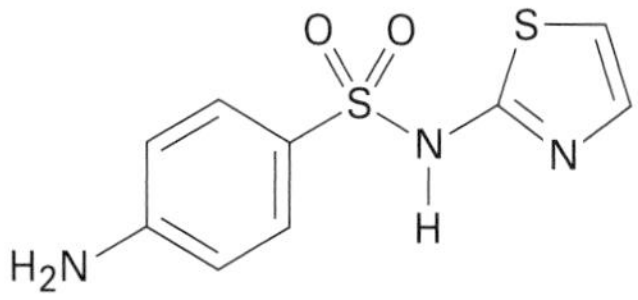

Sulfathiazole

Problem 24.17 Propose syntheses of the following compounds from benzene:
(a) *N,N*-Dimethylaniline   (b) *p*-Chloroaniline
(c) *m*-Chloroaniline   (d) 2,4-Dimethylaniline

Diazonium Salts: The Sandmeyer Reaction

Primary arylamines react with nitrous acid, HNO_2, to yield stable **arenediazonium salts**, $Ar\overset{+}{-}N\equiv N\,X^-$, a process called a **diazotization reaction**. *Alkyl*amines also react with nitrous acid, but the alkanediazonium products of these reactions are so reactive they can't be isolated. Instead, they lose nitrogen instantly to yield carbocations. The analogous loss of N_2 from an arenediazonium ion to yield an aryl cation is disfavored by the instability of the cation.

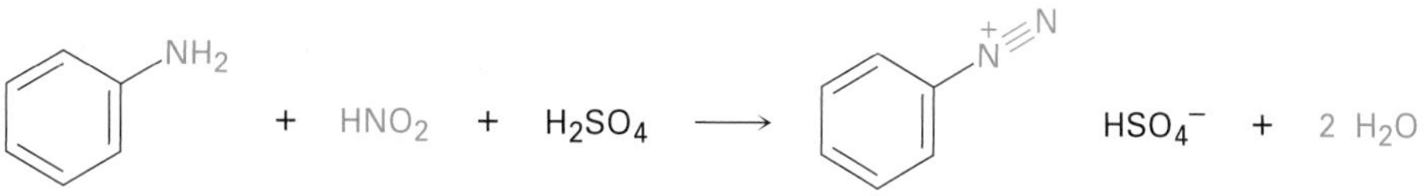

Arenediazonium salts are extremely useful because the diazonio group (N_2) can be replaced by a nucleophile in a substitution reaction.

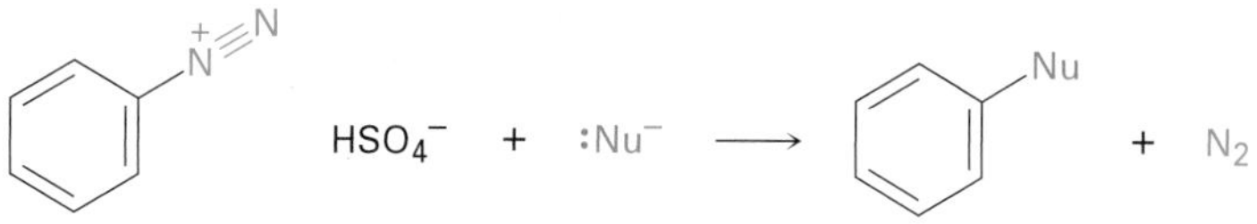

Many different nucleophiles—halide, hydride, cyanide, and hydroxide among others—react with arenediazonium salts, yielding many different kinds of substituted benzenes. The overall sequence of (1) nitration, (2) reduction, (3) diazotization, and (4) nucleophilic substitution is perhaps the single most versatile method of aromatic substitution.

Aryl chlorides and bromides are prepared by reaction of an arenediazonium salt with the corresponding copper(I) halide, CuX, a process called the **Sandmeyer reaction**. Aryl iodides can be prepared by direct reaction with NaI without using a copper(I) salt. Yields generally fall between 60 and 80%.

p-Methylaniline → **p-Bromotoluene (73%)**

Aniline → **Iodobenzene (67%)**

Similar treatment of an arenediazonium salt with CuCN yields the nitrile, ArCN, which can then be further converted into other functional groups such as carboxyl. For example, Sandmeyer reaction of *o*-methylbenzenediazonium bisulfate with CuCN yields *o*-methylbenzonitrile, which can be hydrolyzed to give *o*-methylbenzoic acid. This product can't be prepared from *o*-xylene by the usual side-chain oxidation route because both methyl groups would be oxidized.

o-Methylaniline → **o-Methylbenzene-diazonium bisulfate** → **o-Methylbenzonitrile** → **o-Methylbenzoic acid**

The diazonio group can also be replaced by −OH to yield a phenol and by −H to yield an arene. A phenol is prepared by reaction of the arenediazonium salt with copper(I) oxide in an aqueous solution of copper(II) nitrate, a reaction that is especially useful because few other general methods exist for introducing an −OH group onto an aromatic ring.

p-Methylaniline (p-Toluidine) → **p-Cresol (93%)**

Reduction of a diazonium salt to give an arene occurs on treatment with hypophosphorous acid, H_3PO_2. This reaction is used primarily when there is a need for temporarily introducing an amino substituent onto a ring to take advantage of its directing effect. Suppose, for instance, that you needed to make 3,5-dibromotoluene. The product can't be made by direct bromination of toluene because reaction would occur at positions 2 and 4. Starting with *p*-methylaniline (*p*-toluidine), however, dibromination occurs ortho to the strongly directing amino substituent, and diazotization followed by treatment with H_3PO_2 yields the desired product.

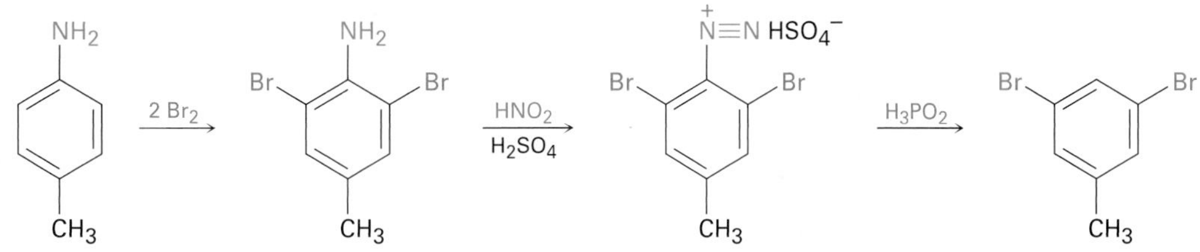

p-Methylaniline   **3,5-Dibromotoluene**

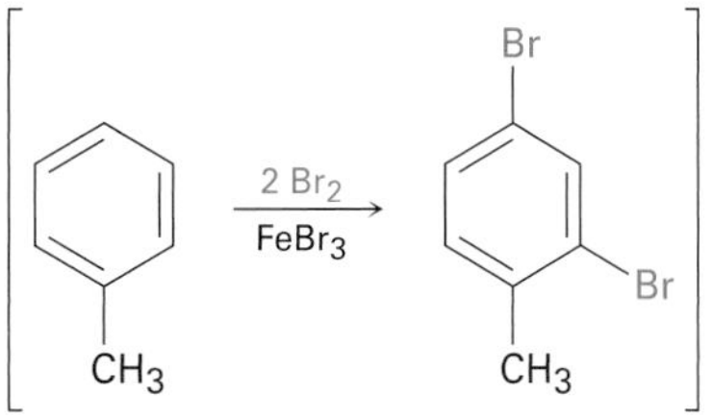

Toluene   **2,4-Dibromotoluene**

Mechanistically, these diazonio replacement reactions occur through radical rather than polar pathways. In the presence of a copper(I) compound, for instance, it's thought that the arenediazonium ion is first converted to an aryl radical plus copper(II), followed by subsequent reaction to give product plus regenerated copper(I) catalyst.

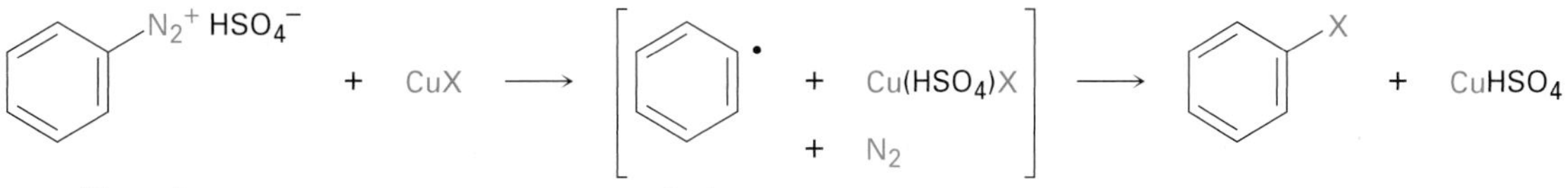

Diazonium   **Aryl**
compound   **radical**

WORKED EXAMPLE 24.4   *Using Diazonium Replacement Reactions*

How would you prepare *m*-hydroxyacetophenone from benzene, using a diazonium replacement reaction in your scheme?

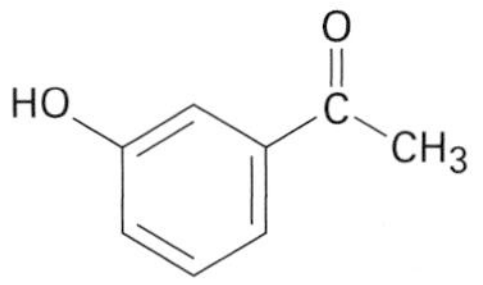

m-Hydroxyacetophenone

Strategy As always, organic syntheses are planned by working backward from the final product, one step at a time. First, identify the functional groups in the product and recall how those groups can be synthesized. *m*-Hydroxyacetophenone has an −OH group and a −COCH$_3$ group in a meta relationship on a benzene ring. A hydroxyl group is generally introduced onto an aromatic ring by a four-step sequence of nitration, reduction, diazotization, and diazonio replacement. An acetyl group is introduced by a Friedel–Crafts acylation reaction.

Next, ask yourself what an immediate precursor of the target might be. Since an acetyl group is a meta director while a hydroxyl group is an ortho and para director, acetophenone might be a precursor of *m*-hydroxyacetophenone. Benzene, in turn, is a precursor of acetophenone.

Solution

Benzene   Acetophenone   *m*-Hydroxyacetophenone

1. HNO$_3$, H$_2$SO$_4$
2. SnCl$_2$, H$_3$O$^+$
3. HNO$_2$, H$_2$SO$_4$
4. Cu$_2$O, Cu(NO$_3$)$_2$, H$_2$O

CH$_3$COCl / AlCl$_3$

Problem 24.18 | How would you prepare the following compounds from benzene, using a diazonium replacement reaction in your scheme?
(a) *p*-Bromobenzoic acid   (b) *m*-Bromobenzoic acid
(c) *m*-Bromochlorobenzene   (d) *p*-Methylbenzoic acid
(e) 1,2,4-Tribromobenzene

Diazonium Coupling Reactions

Arenediazonium salts undergo a coupling reaction with activated aromatic rings such as phenols and arylamines to yield brightly colored **azo compounds**, Ar−N=N−Ar′.

An azo compound

where   Y = −OH or −NR$_2$

Diazonium coupling reactions are typical electrophilic aromatic substitutions in which the positively charged diazonium ion is the electrophile that reacts with the electron-rich ring of a phenol or arylamine. Reaction usually occurs at the para position, although ortho reaction can take place if the para position is blocked.

**Benzenediazonium
bisulfate**　　　　　**Phenol**

**_p_-Hydroxyazobenzene
(orange crystals, mp 152 °C)**

Azo-coupled products are widely used as dyes for textiles because their extended conjugated π electron system causes them to absorb in the visible region of the electromagnetic spectrum (Section 14.9). _p_-(Dimethylamino)azobenzene, for instance, is a bright yellow compound that was at one time used as a coloring agent in margarine.

**Benzenediazonium
bisulfate**　　　　**_N,N_-Dimethylaniline**　　　　**_p_-(Dimethylamino)azobenzene
(yellow crystals, mp 127 °C)**

Problem 24.19 | Propose a synthesis of _p_-(dimethylamino)azobenzene from benzene as your only organic starting material.

24.9 | Heterocycles

A **heterocycle** is a cyclic compound that contains atoms of two or more elements in its ring, usually carbon along with nitrogen, oxygen, or sulfur. Heterocyclic amines are particularly common, and many have important biological properties. Pyridoxal phosphate, a coenzyme; sildenafil (Viagra), a well-known pharmaceutical; and heme, the oxygen carrier in blood, are examples.

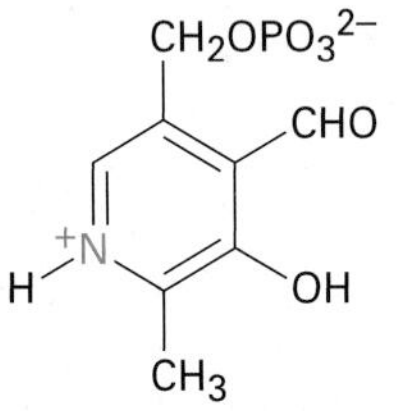

Pyridoxal phosphate
(a coenzyme)

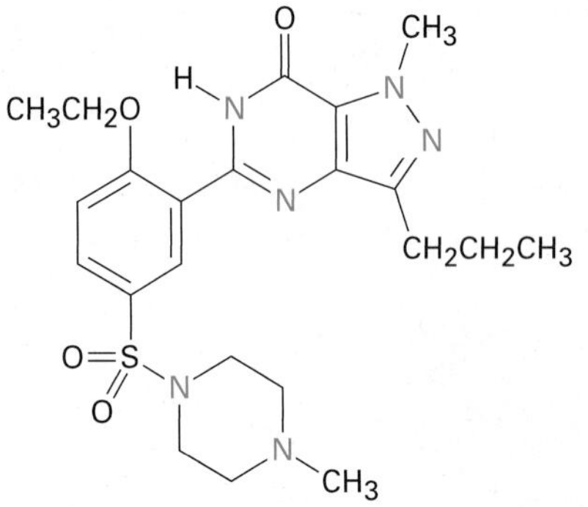

Sildenafil
(Viagra)

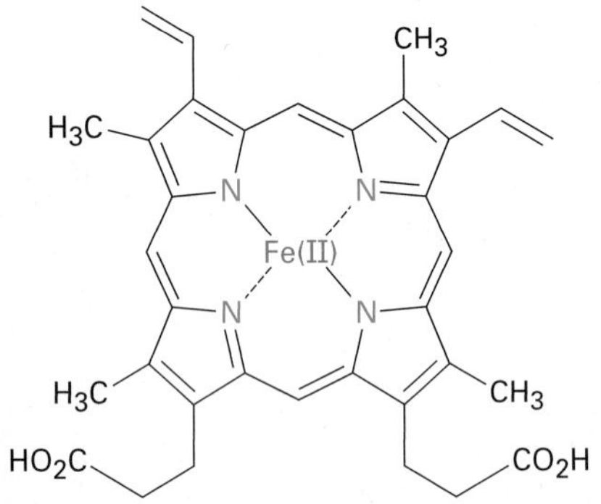

Heme

Most heterocycles have the same chemistry as their open-chain counterparts. Lactones and acyclic esters behave similarly, lactams and acyclic amides behave similarly, and cyclic and acyclic ethers behave similarly. In certain cases, however, particularly when the ring is unsaturated, heterocycles have unique and interesting properties.

Pyrrole and Imidazole

Pyrrole, the simplest five-membered unsaturated heterocyclic amine, is obtained commercially by treatment of furan with ammonia over an alumina catalyst at 400 °C. Furan, the oxygen-containing analog of pyrrole, is obtained by acid-catalyzed dehydration of the five-carbon sugars found in oat hulls and corncobs.

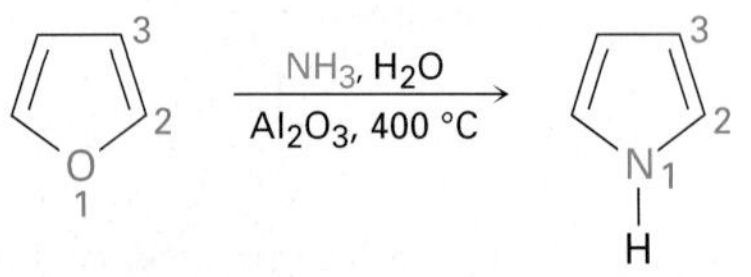

Furan   **Pyrrole**

Although pyrrole appears to be both an amine and a conjugated diene, its chemical properties are not consistent with either of these structural features. Unlike most other amines, pyrrole is not basic—the pK_a of the pyrrolinium ion is 0.4; unlike most other conjugated dienes, pyrrole undergoes electrophilic substitution reactions rather than additions. The reason for both these properties, as noted previously in Section 15.5, is that pyrrole has six π electrons and is aromatic. Each of the four carbons contributes one

π electron, and the sp^2-hybridized nitrogen contributes two more from its lone pair.

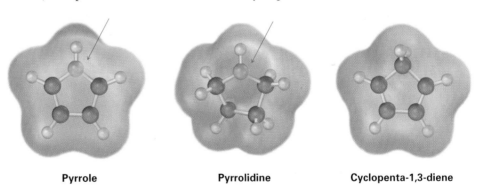

Pyrrole

Lone pair in *p* orbital

sp^2-hybridized

Six π electrons

Because the nitrogen lone pair is a part of the aromatic sextet, protonation on nitrogen would destroy the aromaticity of the ring. The nitrogen atom in pyrrole is therefore less electron-rich, less basic, and less nucleophilic than the nitrogen in an aliphatic amine. By the same token, the *carbon* atoms of pyrrole are *more* electron-rich and more nucleophilic than typical double-bond carbons. The pyrrole ring is therefore reactive toward electrophiles in the same way that enamines are (Section 23.11). Electrostatic potential maps show how the pyrrole nitrogen is electron-poor (less red) compared with the nitrogen in its saturated counterpart pyrrolidine, while the pyrrole carbon atoms are electron-rich (more red) compared with the carbons in 1,3-cyclopentadiene.

Pyrrole   **Pyrrolidine**   **Cyclopenta-1,3-diene**

The chemistry of pyrrole is similar to that of activated benzene rings. In general, however, the heterocycles are more reactive toward electrophiles than benzene rings are, and low temperatures are often necessary to control the reactions. Halogenation, nitration, sulfonation, and Friedel–Crafts acylation can all be accomplished. For example:

$$\xrightarrow[\text{0 °C}]{\text{Br}_2}$$

+   HBr

Pyrrole   **2-Bromopyrrole**
(92%)

Electrophilic substitutions normally occur at C2, the position next to the nitrogen, because reaction at this position leads to a more stable intermediate cation having three resonance forms, whereas reaction at C3 gives a less stable cation with only two resonance forms (Figure 24.6).

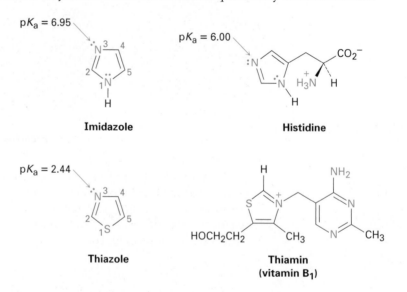

2-Nitropyrrole

3-Nitropyrrole
(NOT formed)

Figure 24.6 Electrophilic nitration of pyrrole. The intermediate produced by reaction at C2 is more stable than that produced by reaction at C3.

Other common five-membered heterocyclic amines include imidazole and thiazole. Imidazole, a constituent of the amino acid histidine, has two nitrogens, only one of which is basic. Thiazole, the five-membered ring system on which the structure of thiamin (vitamin B_1) is based, also contains a basic nitrogen that is alkylated in thiamin to form a quaternary ammonium ion.

$pK_a = 6.95$

Imidazole

$pK_a = 6.00$

Histidine

$pK_a = 2.44$

Thiazole

Thiamin
(vitamin B_1)

Problem 24.20 Draw an orbital picture of thiazole. Assume that both the nitrogen and sulfur atoms are sp^2-hybridized, and show the orbitals that the lone pairs occupy.

Problem 24.21 What is the percent protonation of the imidazole nitrogen atom in histidine at a physiological pH of 7.3? (See Section 24.5.)

Pyridine and Pyrimidine

Pyridine is the nitrogen-containing heterocyclic analog of benzene. Like benzene, pyridine is a flat, aromatic molecule, with bond angles of 120° and C–C bond lengths of 139 pm, intermediate between typical single and double bonds. The five carbon atoms and the sp^2-hybridized nitrogen atom each contribute one π electron to the aromatic sextet, and the lone-pair electrons occupy an sp^2 orbital in the plane of the ring (Section 15.5).

As shown in Table 24.1, pyridine ($pK_a = 5.25$) is a stronger base than pyrrole but a weaker base than alkylamines. The diminished basicity of pyridine compared with an alkylamine is due to the fact that the lone-pair electrons on the pyridine nitrogen are in an sp^2 orbital, while those on an alkylamine nitrogen are in an sp^3 orbital. Because s orbitals have their maximum electron density at the nucleus but p orbitals have a node at the nucleus, electrons in an orbital with more s character are held more closely to the positively charged nucleus and are less available for bonding. As a result, the sp^2-hybridized nitrogen atom (33% s character) in pyridine is less basic than the sp^3-hybridized nitrogen in an alkylamine (25% s character).

Pyridine

Unlike benzene, pyridine undergoes electrophilic aromatic substitution reactions with great difficulty. Halogenation can be carried out under drastic conditions, but nitration occurs in very low yield, and Friedel–Crafts reactions are not successful. Reactions usually give the 3-substituted product.

Pyridine   **3-Bromopyridine**
                          (30%)

The low reactivity of pyridine toward electrophilic aromatic substitution is caused by a combination of factors. One is that acid–base complexation between the basic ring nitrogen atom and the incoming electrophile places a positive charge on the ring, thereby deactivating it. Equally important is that the electron density of the ring is decreased by the electron-withdrawing inductive effect of the electronegative nitrogen atom. Thus, pyridine has a substantial dipole moment ($\mu = 2.26$ D), with the ring carbons acting as the positive end of the dipole. Reaction of an electrophile with the positively polarized carbon atoms is therefore difficult.

$$\mu = 2.26 \text{ D}$$

In addition to pyridine, the six-membered diamine pyrimidine is also found commonly in biological molecules, particularly as a constituent of nucleic acids. With a pK_a of 1.3, pyrimidine is substantially less basic than pyridine because of the inductive effect of the second nitrogen.

Pyrimidine
$pK_a = 1.3$

Problem 24.22 | Electrophilic aromatic substitution reactions of pyridine normally occur at C3. Draw the carbocation intermediates resulting from reaction of an electrophile at C1, C2, and C3, and explain the observed result.

Polycyclic Heterocycles

As we saw in Section 15.7, quinoline, isoquinoline, indole, and purine are common polycyclic heterocycles. The first three contain both a benzene ring and a heterocyclic aromatic ring, while purine contains two heterocyclic rings joined together. All four ring systems occur commonly in nature, and many compounds with these rings have pronounced physiological activity. The quinoline alkaloid quinine, for instance, is widely used as an antimalarial drug, tryptophan is a common amino acid, and the purine adenine is a constituent of nucleic acids.

Quinoline **Isoquinoline** **Indole** **Purine**

Quinine
(antimalarial)

Tryptophan
(amino acid)

Adenine
(DNA constituent)

The chemistry of these polycyclic heterocycles is just what you might expect from a knowledge of the simpler heterocycles pyridine and pyrrole. Quinoline and isoquinoline both have basic, pyridine-like nitrogen atoms, and both undergo electrophilic substitutions, although less easily than benzene. Reaction occurs on the benzene ring rather than on the pyridine ring, and a mixture of substitution products is obtained.

Quinoline → Br₂ / H₂SO₄ → **5-Bromoquinoline** + **8-Bromoquinoline** + HBr

A 51 : 49 ratio

Isoquinoline → HNO₃ / H₂SO₄, 0 °C → **5-Nitroisoquinoline** + **8-Nitroisoquinoline** + H₂O

A 90 : 10 ratio

Indole has a nonbasic, pyrrole-like nitrogen and undergoes electrophilic substitution more easily than benzene. Substitution occurs at C3 of the electron-rich pyrrole ring, rather than on the benzene ring.

Indole → Br₂ / Dioxane, 0 °C → **3-Bromoindole** + HBr

Purine has three basic, pyridine-like nitrogens with lone-pair electrons in sp^2 orbitals in the plane of the ring. The remaining purine nitrogen is nonbasic and pyrrole-like, with its lone-pair electrons as part of the aromatic π electron system.

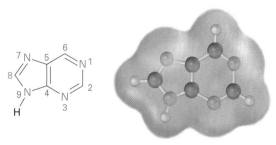

Purine

Problem 24.23 Which nitrogen atom in the hallucinogenic indole alkaloid *N,N*-dimethyltryptamine is more basic? Explain.

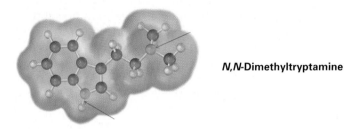

***N,N*-Dimethyltryptamine**

Problem 24.24 Indole reacts with electrophiles at C3 rather than at C2. Draw resonance forms of the intermediate cations resulting from reaction at C2 and C3, and explain the observed results.

24.10 | Spectroscopy of Amines

Infrared Spectroscopy

Primary and secondary amines can be identified by a characteristic N−H stretching absorption in the 3300 to 3500 cm^{-1} range of the IR spectrum. Alcohols also absorb in this range (Section 17.11), but amine absorption bands are generally sharper and less intense than hydroxyl bands. Primary amines show a pair of bands at about 3350 and 3450 cm^{-1}, and secondary amines show a single band at 3350 cm^{-1}. Tertiary amines have no absorption in this region because they have no N−H bonds. An IR spectrum of cyclohexylamine is shown in Figure 24.7.

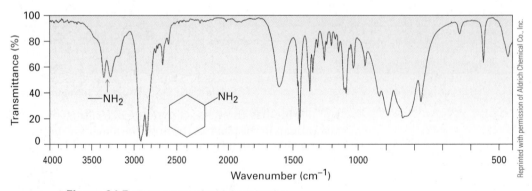

Figure 24.7 IR spectrum of cyclohexylamine.

In addition to looking for a characteristic N−H absorption, there is also a simple trick for telling whether a compound is an amine. Addition of a small amount of HCl produces a broad and strong ammonium band in the 2200 to 3000 cm^{-1} range if the sample contains an amino group. Figure 24.8 gives an example.

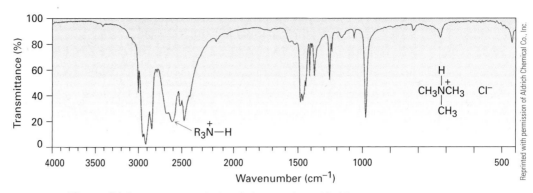

Figure 24.8 IR spectrum of trimethylammonium chloride.

Nuclear Magnetic Resonance Spectroscopy

Amines are difficult to identify solely by ^{1}H NMR spectroscopy because N–H hydrogens tend to appear as broad signals without clear-cut coupling to neighboring C–H hydrogens. As with O–H absorptions (Section 17.11), amine N–H absorptions can appear over a wide range and are best identified by adding a small amount of D_2O to the sample tube. Exchange of N–D for N–H occurs, and the N–H signal disappears from the NMR spectrum.

$$\backslash \text{N–H} \xrightleftharpoons{\text{D}_2\text{O}} \backslash \text{N–D} + \text{HDO}$$

Hydrogens on the carbon next to nitrogen are deshielded because of the electron-withdrawing effect of the nitrogen, and they therefore absorb at lower field than alkane hydrogens. *N*-Methyl groups are particularly distinctive because they absorb as a sharp three-proton singlet at 2.2 to 2.6 δ. This *N*-methyl resonance at 2.42 δ is easily seen in the ^{1}H NMR spectrum of *N*-methylcyclohexylamine (Figure 24.9).

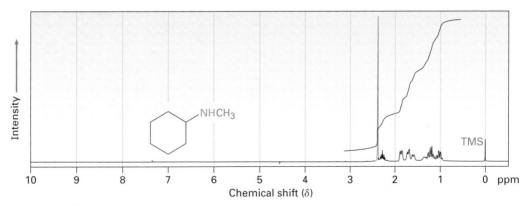

Figure 24.9 Proton NMR spectrum of *N*-methylcyclohexylamine.

Carbons next to amine nitrogens are slightly deshielded in the ^{13}C NMR spectrum and absorb about 20 ppm downfield from where they would absorb in an alkane of similar structure. In *N*-methylcyclohexylamine, for example, the ring carbon to which nitrogen is attached absorbs at a position 24 ppm lower than that of any other ring carbon.

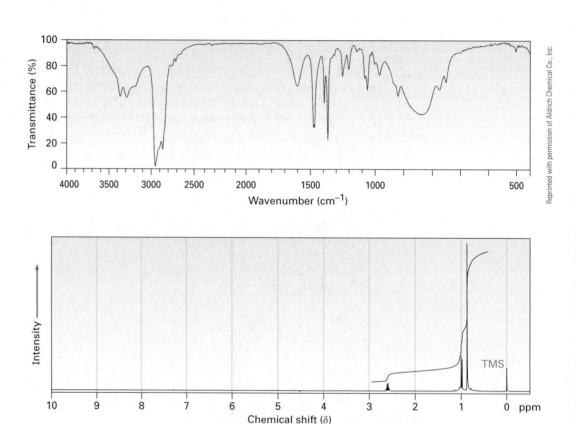

Problem 24.25 Compound **A**, $C_6H_{12}O$, has an IR absorption at $1715 \ cm^{-1}$ and gives compound **B**, $C_6H_{15}N$, when treated with ammonia and $NaBH_3CN$. The IR and ^1H NMR spectra of **B** are shown. What are the structures of **A** and **B**?

Mass Spectrometry

The *nitrogen rule* of mass spectrometry says that a compound with an odd number of nitrogen atoms has an odd-numbered molecular weight. Thus, the presence of nitrogen in a molecule is detected simply by observing its mass spectrum. An odd-numbered molecular ion usually means that the unknown compound has one or three nitrogen atoms, and an even-numbered molecular ion usually means that a compound has either zero or two nitrogen atoms. The logic behind the rule derives from the fact that nitrogen is trivalent, thus requiring an odd number of hydrogen atoms. For example, morphine has the formula $C_{17}H_{19}NO_3$ and a molecular weight of 285 amu.

Alkylamines undergo a characteristic α cleavage in the mass spectrometer, similar to the cleavage observed for alcohols (Section 17.11). A C–C bond nearest the nitrogen atom is broken, yielding an alkyl radical and a resonance-stabilized, nitrogen-containing cation.

As an example, the mass spectrum of N-ethylpropylamine shown in Figure 24.10 has peaks at $m/z = 58$ and $m/z = 72$, corresponding to the two possible modes of α cleavage.

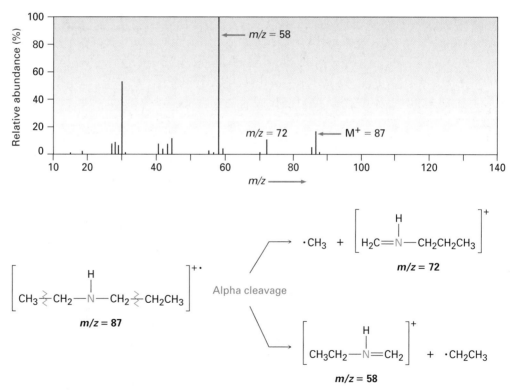

Figure 24.10 Mass spectrum of N-ethylpropylamine. The two possible modes of α cleavage lead to the observed fragment ions at $m/z = 58$ and $m/z = 72$.

Focus On . . .

Green Chemistry II: Ionic Liquids

Liquids made of ions? Usually when we think of ionic compounds, we think of high-melting solids: sodium chloride, magnesium sulfate, lithium carbonate, and so forth. But yes, there also ionic compounds that are liquid at room temperature, and they are gaining importance as reaction solvents, particularly for use in green chemistry processes (see the Chapter 11 *Focus On*).

Ionic liquids have been known for nearly a century; the first to be discovered was ethylammonium nitrate, $CH_3CH_2NH_3^+\ NO_3^-$, with a melting point of 12 °C. More generally, however, the ionic liquids in use today are salts in which the cation is unsymmetrical and in which one or both of the ions are bulky so that the charges are dispersed over a large volume. Both factors minimize the crystal lattice energy and disfavor formation of the solid. Typical cations are quaternary ammonium ions from heterocyclic amines, either 1,3-dialkylimidazolium ions, *N*-alkylpyridinium ions, or ring-substituted *N*-alkylpyridinium ions.

$$\left[\begin{array}{l} R\ =\ -CH_3, -CH_2CH_3, -CH_2CH_2CH_2CH_3, \\ \quad -CH_2CH_2CH_2CH_2CH_2CH_2CH_2CH_3 \end{array} \right]$$

1,3-Dialkylimidazolium ions

$$\left[\begin{array}{l} R\ =\ -CH_2CH_3, -CH_2CH_2CH_2CH_3, \\ \quad -CH_2CH_2CH_2CH_2CH_2CH_3 \end{array} \right]$$

***N*-Alkylpyridinium ions**

Anions are just as varied as the cations, and more than 250 different ionic liquids with different anion/cation combinations are commercially available. Hexafluorophosphate, tetrafluoroborate, alkyl sulfates, trifluoromethanesulfonates (triflates), and halides are some anion possibilities.

Hexafluorophosphate Tetrafluoroborate Methyl sulfate Trifluoromethanesulfonate Halide Cl^-, Br^-, I^-

(continued)

Yes, these liquids really do consist of ionic rather than molecular substances.

Ionic liquids have several important features that make them attractive for use as solvents, particularly in green chemistry:

▌ They dissolve both polar and nonpolar organic compounds, giving high solute concentrations and thereby minimizing the amount of solvent needed.

▌ They can be optimized for specific reactions by varying cation and anion structures.

▌ They are nonflammable.

▌ They are thermally stable.

▌ They have negligible vapor pressures and do not evaporate.

▌ They are generally recoverable and can be reused many times.

As an example of their use in organic chemistry, the analgesic drug Pravadoline has been synthesized in two steps using 1-butyl-3-methylimidazolium hexafluorophosphate, abbreviated [bmim][PF$_6$], as the solvent for both steps. The first step is a base-induced S$_N$2 reaction of 2-methylindole with a primary alkyl halide, and the second is a Friedel–Crafts acylation. Both steps take place in 95% yield, and the ionic solvent is recovered simply by washing the reaction mixture, first with toluene and then with water. We'll be hearing a lot more about ionic solvents in coming years.

Pravadoline

SUMMARY AND KEY WORDS

Amines are organic derivatives of ammonia. They are named in the IUPAC system either by adding the suffix -*amine* to the name of the alkyl substituent or by considering the amino group as a substituent on a more complex parent molecule.

The chemistry of amines is dominated by the lone-pair electrons on nitrogen, which makes amines both basic and nucleophilic. The base strength of **arylamines** is generally lower than that of **alkylamines** because the nitrogen lone-pair electrons are delocalized by interaction with the aromatic π system. Electron-withdrawing substituents on the aromatic ring further weaken the basicity of a substituted aniline, while electron-donating substituents increase basicity. Alkylamines are sufficiently basic that they exist almost entirely in their protonated form at the physiological pH of 7.3 inside cells.

Heterocyclic amines are compounds that contain one or more nitrogen atoms as part of a ring. Saturated heterocyclic amines usually have the same chemistry as their open-chain analogs, but unsaturated heterocycles such as pyrrole, imidazole, pyridine, and pyrimidine are aromatic. All four are unusually stable, and all undergo aromatic substitution on reaction with electrophiles. Pyrrole is nonbasic because its nitrogen lone-pair electrons are part of the aromatic π system. Fused-ring heterocycles such as quinoline, isoquinoline, indole, and purine are also commonly found in biological molecules.

Arylamines are prepared by nitration of an aromatic ring followed by reduction. Alkylamines are prepared by S_N2 reaction of ammonia or an amine with an alkyl halide. This method often gives poor yields, however, and an alternative such as the **Gabriel amine synthesis** is preferred. Amines can also be prepared by a number of reductive methods, including $LiAlH_4$ reduction of amides, nitriles, and azides. Also important is the **reductive amination** reaction in which a ketone or an aldehyde is treated with an amine in the presence of a reducing agent such as $NaBH_3CN$. In addition, amines result from the **Hofmann** and **Curtius rearrangements** of carboxylic acid derivatives. Both methods involve migration of the —R group bonded to the carbonyl carbon and yield a product that has one less carbon atom than the starting material.

Many of the reactions of amines are familiar from past chapters. Thus, amines react with alkyl halides in S_N2 reactions and with acid chlorides in nucleophilic acyl substitution reactions. Amines also undergo E2 elimination to yield alkenes if they are first quaternized by treatment with iodomethane and then heated with silver oxide, a process called the **Hofmann elimination**.

Arylamines are converted by diazotization with nitrous acid into **arenediazonium salts**, $ArN_2^+ X^-$. The diazonio group can then be replaced by many other substituents in the **Sandmeyer reaction** to give a wide variety of substituted aromatic compounds. Aryl chlorides, bromides, iodides, and nitriles can be prepared from arenediazonium salts, as can arenes and phenols. In addition to their reactivity toward substitution reactions, diazonium salts undergo coupling with phenols and arylamines to give brightly colored azo dyes.

SUMMARY OF REACTIONS

1. Synthesis of amines (Section 24.6)

   (a) Reduction of nitriles

$$RCH_2X \xrightarrow{NaCN} RCH_2C\equiv N \xrightarrow[\text{2. }H_2O]{\text{1. LiAlH}_4,\text{ ether}} RCH_2-\overset{\overset{H}{|}}{\underset{\underset{H}{|}}{C}}-NH_2$$

   (b) Reduction of amides

$$R-\overset{\overset{O}{\|}}{C}-NH_2 \xrightarrow[\text{2. }H_2O]{\text{1. LiAlH}_4,\text{ ether}} R-\overset{\overset{H}{|}}{\underset{\underset{H}{|}}{C}}-NH_2$$

   (c) Reduction of nitrobenzenes

$$\text{C}_6\text{H}_5\text{NO}_2 \xrightarrow[\substack{\text{or Fe, }H_3O^+ \\ \text{or SnCl}_2,\, H_3O^+}]{H_2,\text{ Pt}} \text{C}_6\text{H}_5\text{NH}_2$$

   (d) S_N2 Alkylation of alkyl halides

Ammonia	$\overset{..}{N}H_3$	+ R—X	⟶	$R\overset{+}{N}H_3$ X^-	$\xrightarrow{NaOH}$ RNH_2	Primary
Primary	$R\overset{..}{N}H_2$	+ R—X	⟶	$R_2\overset{+}{N}H_2$ X^-	$\xrightarrow{NaOH}$ R_2NH	Secondary
Secondary	$R_2\overset{..}{N}H$	+ R—X	⟶	$R_3\overset{+}{N}H$ X^-	$\xrightarrow{NaOH}$ R_3N	Tertiary
Tertiary	$R_3\overset{..}{N}$	+ R—X	⟶	$R_4\overset{+}{N}$ X^-		Quaternary ammonium

   (e) Gabriel amine synthesis

$$\text{phthalimide}\,N-H \xrightarrow[\text{2. R—X}]{\text{1. KOH}} \text{phthalimide}\,N-R \xrightarrow[H_2O]{NaOH} R-NH_2$$

   (f) Reduction of azides

$$RCH_2-X \xrightarrow[\text{ethanol}]{Na^+\ ^-N_3} RCH_2-N{=}\overset{+}{N}{=}\overset{-}{N} \xrightarrow[\text{2. }H_2O]{\text{1. LiAlH}_4,\text{ ether}} R-NH_2$$

   (g) Reductive amination of aldehydes/ketones

$$R-\overset{\overset{O}{\|}}{C}-R' \xrightarrow[NaBH_3CN]{NH_3} R-\overset{\overset{H}{|}}{\underset{\underset{R'}{|}}{C}}-NH_2$$

(h) Hofmann rearrangement of amides

(i) Curtius rearrangement of acyl azides

2. Reactions of amines
   (a) Alkylation with alkyl halides; see reaction 1(d)
   (b) Hofmann elimination (Section 24.7)

(c) Diazotization (Section 24.8)

3. Reactions of arenediazonium salts (Section 24.8)
   (a) Nucleophilic substitutions

(b) Diazonium coupling

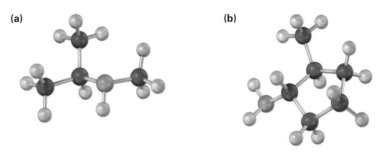

Organic **KNOWLEDGE TOOLS**

CENGAGENOW˜ Sign in at **www.cengage.com/login** to assess your knowledge of this chapter's topics by taking a pre-test. The pre-test will link you to interactive organic chemistry resources based on your score in each concept area.

☉WL Online homework for this chapter may be assigned in Organic OWL.

■ indicates problems assignable in Organic OWL.

VISUALIZING CHEMISTRY

(Problems 24.1–24.25 appear within the chapter.)

24.26 ■ Name the following amines, and identify each as primary, secondary, or tertiary:

(a)

(b)

(c)

24.27 ■ The following compound contains three nitrogen atoms. Rank them in order of increasing basicity.

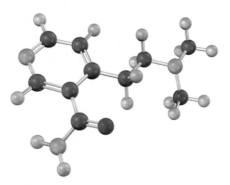

24.28 Name the following amine, including *R,S* stereochemistry, and draw the product of its reaction with excess iodomethane followed by heating with Ag_2O (Hofmann elimination). Is the stereochemistry of the alkene product *Z* or *E*? Explain.

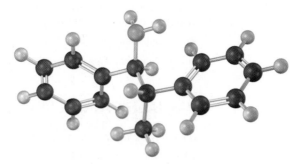

24.29 Which nitrogen atom in the following compound is more basic? Explain.

ADDITIONAL PROBLEMS

24.30 Classify each of the amine nitrogen atoms in the following substances as primary, secondary, or tertiary:

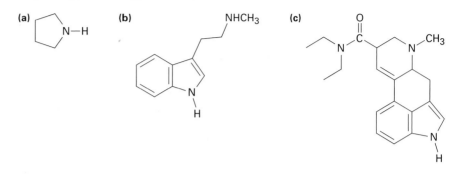

(a)   **(b)**   **(c)**

Lysergic acid diethylamide

24.31 ■ Draw structures corresponding to the following IUPAC names:
  (a) *N,N*-Dimethylaniline   **(b)** (Cyclohexylmethyl)amine
  (c) *N*-Methylcyclohexylamine   **(d)** (2-Methylcyclohexyl)amine
  (e) 3-(*N,N*-Dimethylamino)propanoic acid

24.32 ■ Name the following compounds:

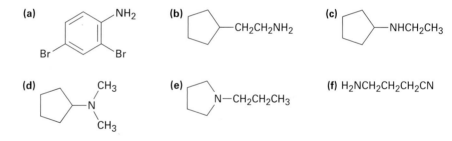

(a)   **(b)**   **(c)**

(d)   **(e)**   **(f)** $H_2NCH_2CH_2CH_2CN$

24.33 ■ Give the structures of the major organic products you would expect from reaction of *m*-toluidine (*m*-methylaniline) with the following reagents:
  (a) Br_2 (1 equivalent)   **(b)** CH_3I (excess)
  (c) CH_3COCl in pyridine   **(d)** The product of (c), then HSO_3Cl

24.34 ■ Show the products from reaction of *p*-bromoaniline with the following reagents:
  (a) CH_3I (excess)   **(b)** HCl   **(c)** HNO_2, H_2SO_4
  (d) CH_3COCl **(e)** CH_3MgBr **(f)** CH_3CH_2Cl, $AlCl_3$
  (g) Product of (c) with CuCl, HCl
  (h) Product of (d) with CH_3CH_2Cl, $AlCl_3$

24.35 ■ How would you prepare the following substances from 1-butanol?
  (a) Butylamine   **(b)** Dibutylamine   **(c)** Propylamine
  (d) Pentylamine   **(e)** *N,N*-Dimethylbutylamine **(f)** Propene

24.36 ■ How would you prepare the following substances from pentanoic acid?
  (a) Pentanamide   **(b)** Butylamine   **(c)** Pentylamine
  (d) 2-Bromopentanoic acid **(e)** Hexanenitrile   **(f)** Hexylamine

24.37 ■ How would you prepare aniline from the following starting materials?
(a) Benzene (b) Benzamide (c) Toluene

24.38 ■ How would you convert aniline into each of the products listed in Problem 24.37?

24.39 How would you prepare benzylamine, $C_6H_5CH_2NH_2$, from benzene? More than one step is needed.

24.40 ■ How might you prepare pentylamine from the following starting materials?
(a) Pentanamide (b) Pentanenitrile (c) 1-Butene
(d) Hexanamide (e) 1-Butanol (f) 5-Decene
(g) Pentanoic acid

24.41 ■ What are the major products you would expect from Hofmann elimination of the following amines?

(a)
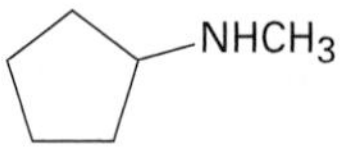
NHCH3

(b)
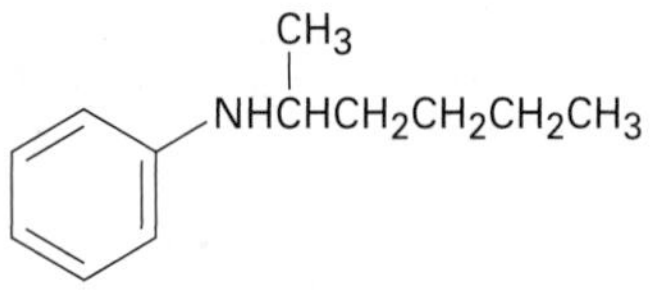
CH3
NHCHCH2CH2CH2CH3

(c)
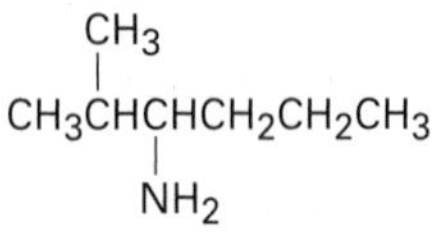
CH3
CH3CHCHCH2CH2CH3
NH2

24.42 ■ Predict the product(s) of the following reactions. If more than one product is formed, tell which is major.

(a)
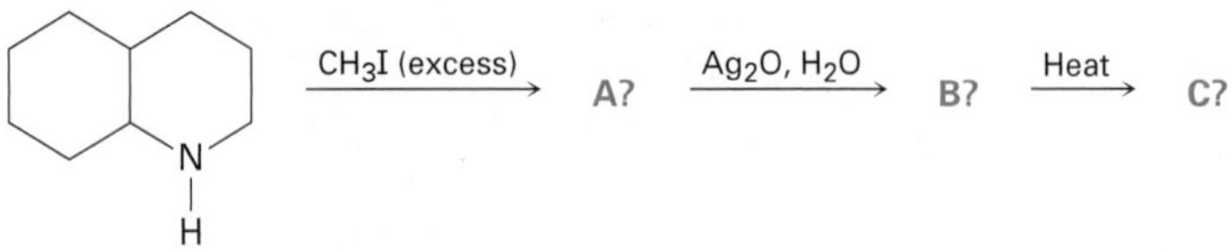
N
H
CH3I (excess)
A?
Ag2O, H2O
B?
Heat
C?

(b)
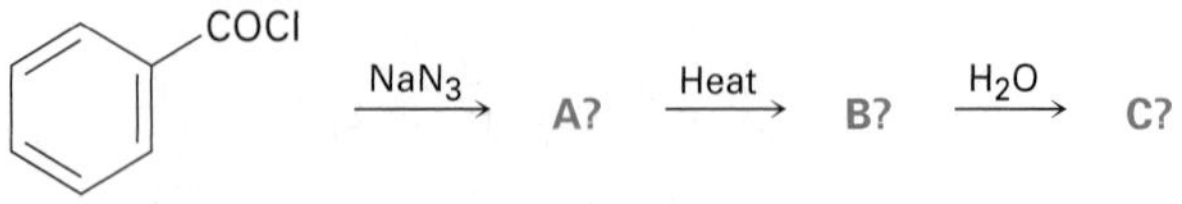
COCl
NaN3
A?
Heat
B?
H2O
C?

(c)
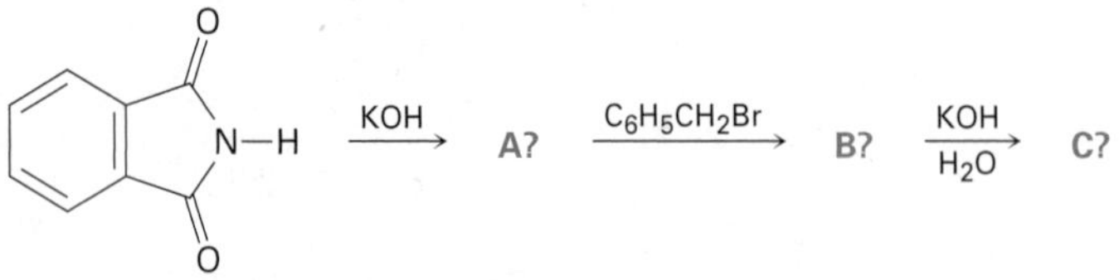
O
N—H
O
KOH
A?
C6H5CH2Br
B?
KOH
H2O
C?

(d) $BrCH_2CH_2CH_2CH_2Br \; + \; \text{1 equiv } CH_3NH_2 \xrightarrow[H_2O]{NaOH} \; ?$

24.43 Fill in the missing reagents **a–e** in the following scheme:

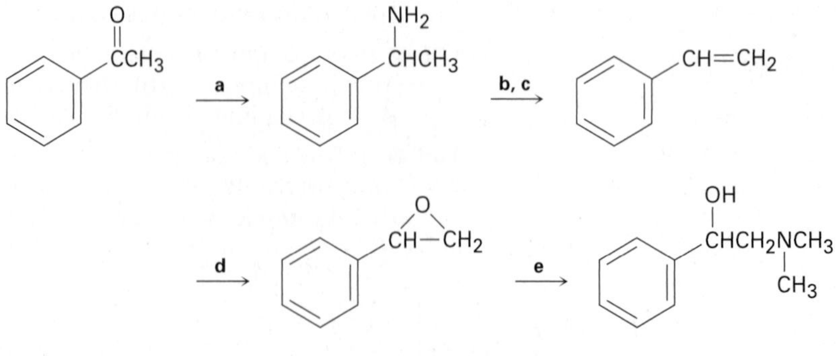
O
CCH3
a
NH2
CHCH3
b, c
CH=CH2
d
O
CH—CH2
e
OH
CHCH2NCH3
CH3

24.44 Although pyrrole is a much weaker base than most other amines, it is a much stronger acid ($pK_a \approx 15$ for the pyrrole versus 35 for diethylamine). The N–H proton is readily abstracted by base to yield the pyrrole anion, $C_4H_4N^-$. Explain.

24.45 Histamine, whose release in the body triggers nasal secretions and constricted airways, has three nitrogen atoms. List them in order of increasing basicity, and explain your ordering.

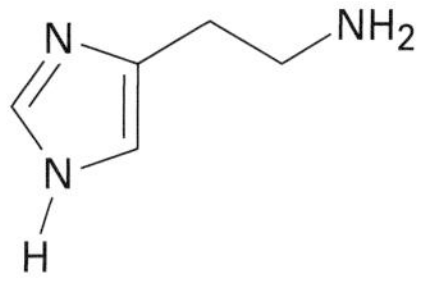

Histamine

24.46 Oxazole is a five-membered aromatic heterocycle. Would you expect oxazole to be more basic or less basic than pyrrole? Explain.

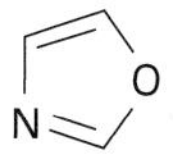

Oxazole

24.47 Protonation of an amide using strong acid occurs on oxygen rather than on nitrogen. Suggest a reason for this behavior, taking resonance into account.

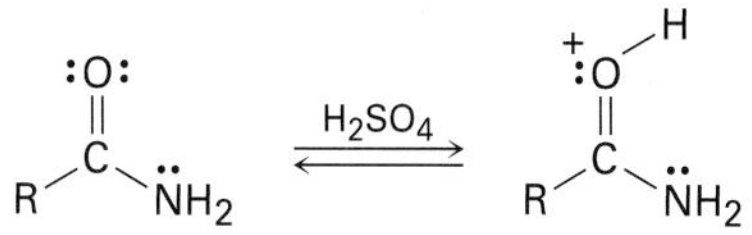

24.48 Substituted pyrroles are often prepared by treatment of a 1,4-diketone with ammonia. Propose a mechanism.

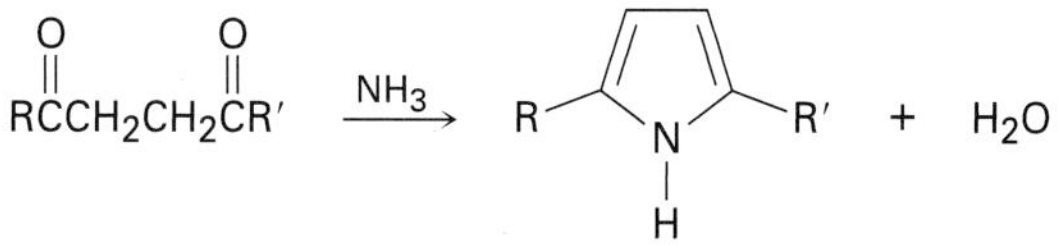

24.49 3,5-Dimethylisoxazole is prepared by reaction of 2,4-pentanedione with hydroxylamine. Propose a mechanism.

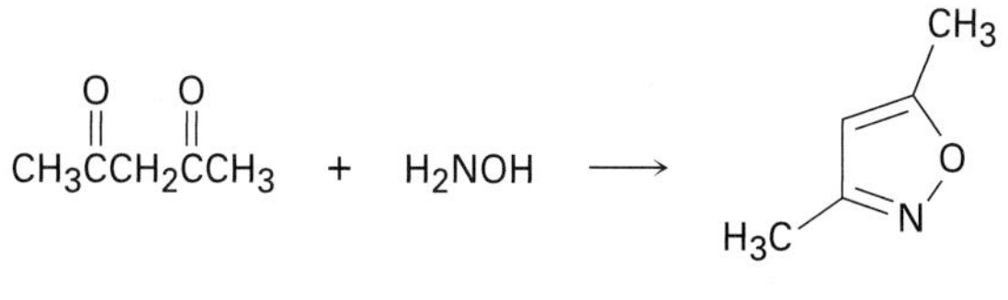

3,5-Dimethylisoxazole

24.50 Account for the fact that *p*-nitroaniline ($pK_a = 1.0$) is less basic than *m*-nitroaniline ($pK_a = 2.5$) by a factor of 30. Draw resonance structures to support your argument. (The pK_a values refer to the corresponding ammonium ions.)

24.51 Fill in the missing reagents **a–d** in the following synthesis of racemic methamphetamine from benzene.

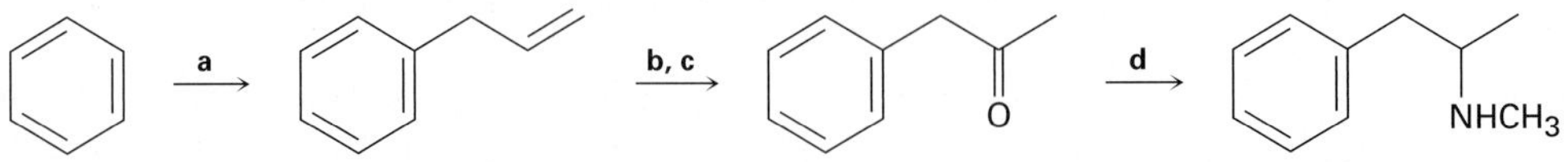

(R,S)-Methamphetamine

24.52 How might a reductive amination be used to synthesize ephedrine, an amino alcohol that is widely used for the treatment of bronchial asthma?

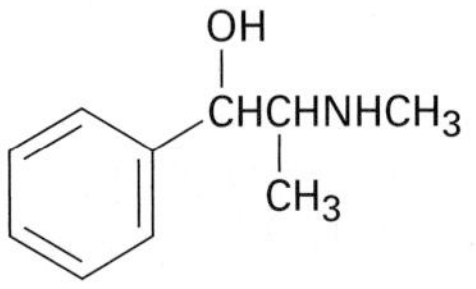

Ephedrine

24.53 One problem with reductive amination as a method of amine synthesis is that by-products are sometimes obtained. For example, reductive amination of benzaldehyde with methylamine leads to a mixture of *N*-methylbenzylamine and *N*-methyldibenzylamine. How do you suppose the tertiary amine by-product is formed? Propose a mechanism.

24.54 Chlorophyll, heme, vitamin B$_{12}$, and a host of other substances are biosynthesized from porphobilinogen (PBG), which is itself formed from condensation of two molecules of 5-aminolevulinate. The two 5-aminolevulinates are bound to lysine (Lys) amino acids in the enzyme, one in the enamine form and one in the imine form, and their condensation is thought to occur by the following steps. Using curved arrows, show the mechanism of each step.

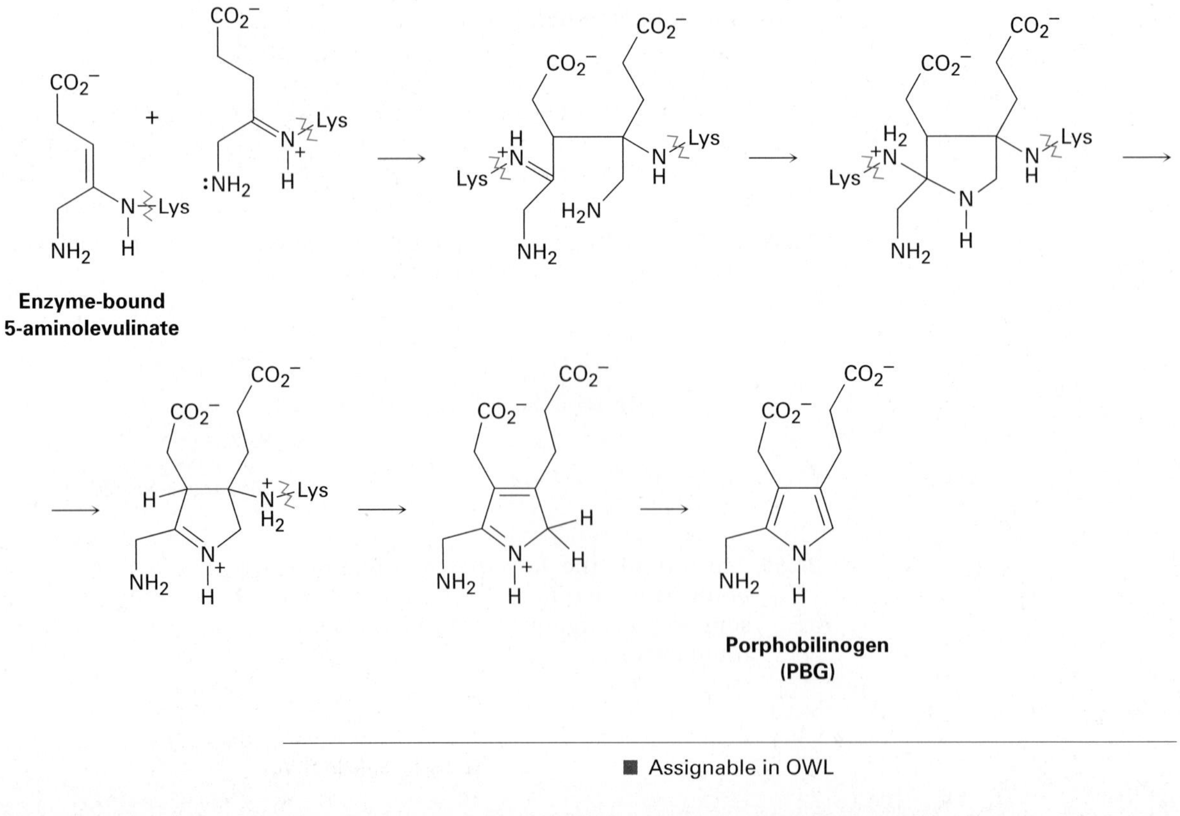

Enzyme-bound 5-aminolevulinate

Porphobilinogen (PBG)

■ Assignable in OWL

24.55 Choline, a component of the phospholipids in cell membranes, can be prepared by S_N2 reaction of trimethylamine with ethylene oxide. Show the structure of choline, and propose a mechanism for the reaction.

$$(CH_3)_3N \quad + \quad \underset{H_2C-CH_2}{\overset{O}{\triangle}} \quad \longrightarrow \quad \textbf{Choline}$$

24.56 Cyclopentamine is an amphetamine-like central nervous system stimulant. Propose a synthesis of cyclopentamine from materials of five carbons or less.

Cyclopentamine

24.57 Tetracaine is a substance used medicinally as a spinal anesthetic during lumbar punctures (spinal taps).

Tetracaine

(a) How would you prepare tetracaine from the corresponding aniline derivative, ArNH₂?
(b) How would you prepare tetracaine from *p*-nitrobenzoic acid?
(c) How would you prepare tetracaine from benzene?

24.58 Atropine, C₁₇H₂₃NO₃, is a poisonous alkaloid isolated from the leaves and roots of *Atropa belladonna,* the deadly nightshade. In small doses, atropine acts as a muscle relaxant; 0.5 ng (nanogram, 10^{-9} g) is sufficient to cause pupil dilation. On basic hydrolysis, atropine yields tropic acid, C₆H₅CH(CH₂OH)CO₂H, and tropine, C₈H₁₅NO. Tropine is an optically inactive alcohol that yields tropidene on dehydration with H₂SO₄. Propose a structure for atropine.

Tropidene

24.59 Tropidene (Problem 24.58) can be converted by a series of steps into tropilidene (1,3,5-cycloheptatriene). How would you accomplish this conversion?

24.60 Propose a structure for the product with formula C₉H₁₇N that results when 2-(2-cyanoethyl)cyclohexanone is reduced catalytically.

24.61 Coniine, $C_8H_{17}N$, is the toxic principle of the poison hemlock drunk by Socrates. When subjected to Hofmann elimination, coniine yields 5-(*N,N*-dimethylamino)-1-octene. If coniine is a secondary amine, what is its structure?

24.62 How would you synthesize coniine (Problem 24.61) from acrylonitrile ($H_2C\!=\!CHCN$) and ethyl 3-oxohexanoate ($CH_3CH_2CH_2COCH_2CO_2Et$)? (Hint: See Problem 24.60.)

24.63 Tyramine is an alkaloid found, among other places, in mistletoe and ripe cheese. How would you synthesize tyramine from benzene? From toluene?

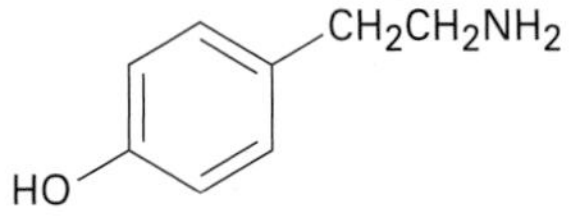

Tyramine

24.64 How would you prepare the following compounds from toluene? A diazonio replacement reaction is needed in some instances.

(a) (b) 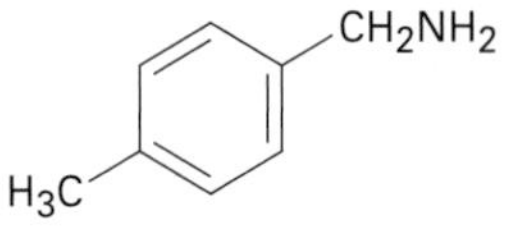 (c)

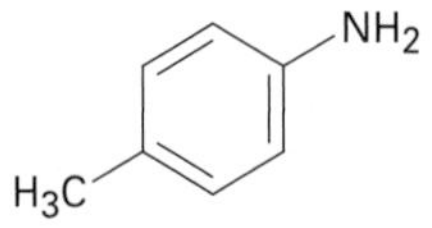

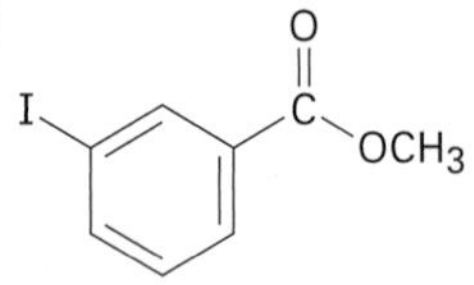

24.65 Reaction of anthranilic acid (*o*-aminobenzoic acid) with HNO_2 and H_2SO_4 yields a diazonium salt that can be treated with base to yield a neutral diazonium carboxylate.
(a) What is the structure of the neutral diazonium carboxylate?
(b) Heating the diazonium carboxylate results in the formation of CO_2, N_2, and an intermediate that reacts with 1,3-cyclopentadiene to yield the following product:

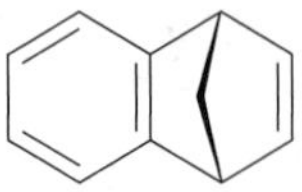

What is the structure of the intermediate, and what kind of reaction does it undergo with cyclopentadiene?

24.66 Cyclooctatetraene was first synthesized in 1911 by a route that involved the following transformation:

How might you use the Hofmann elimination to accomplish this reaction? How would you finish the synthesis by converting cyclooctatriene into cyclo-octatetraene?

24.67 When an α-hydroxy amide is treated with Br$_2$ in aqueous NaOH under Hofmann rearrangement conditions, loss of CO$_2$ occurs and a chain-shortened aldehyde is formed. Propose a mechanism.

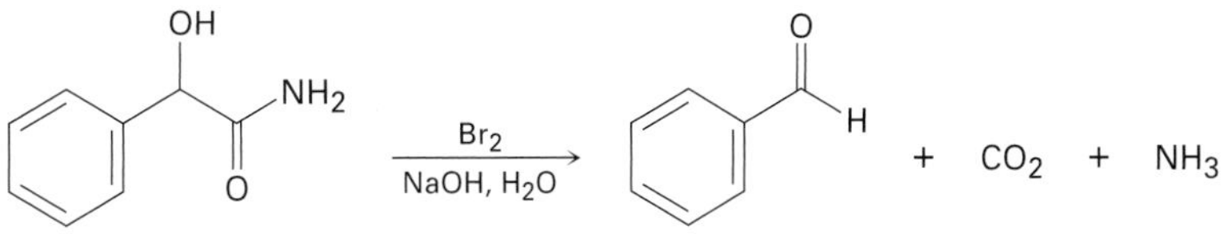

24.68 The following transformation involves a conjugate nucleophilic addition reaction (Section 19.13) followed by an intramolecular nucleophilic acyl substitution reaction (Section 21.2). Show the mechanism.

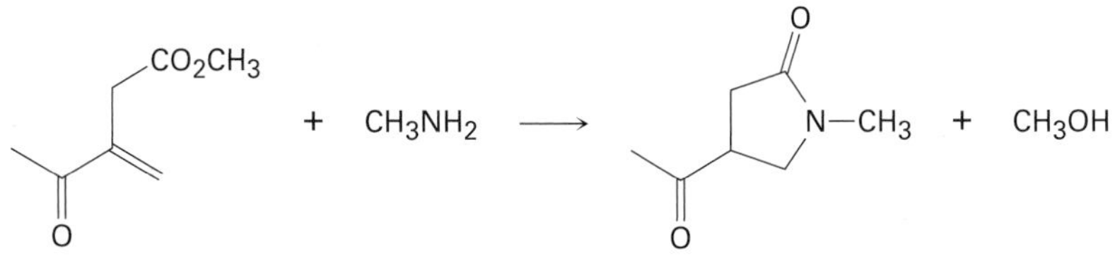

24.69 Propose a mechanism for the following reaction:

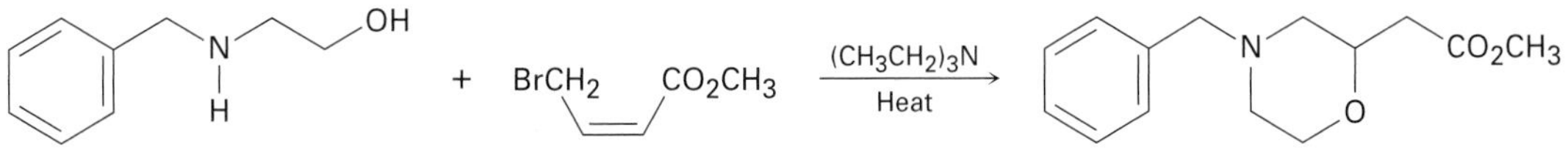

24.70 One step in the biosynthesis of morphine is the reaction of dopamine with p-hydroxyphenylacetaldehyde to give (S)-norcoclaurine. Assuming that the reaction is acid-catalyzed, propose a mechanism.

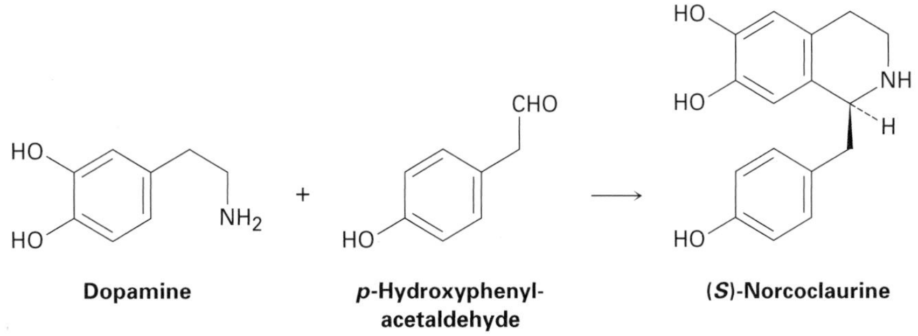

| **Dopamine** | **p-Hydroxyphenyl-** | **(S)-Norcoclaurine** |
| | **acetaldehyde** | |

24.71 The antitumor antibiotic mitomycin C functions by forming cross-links in DNA chains.

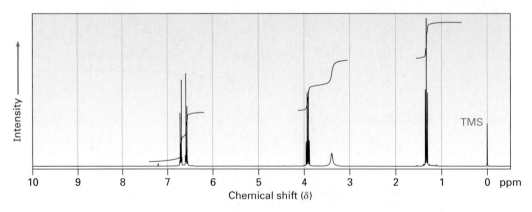

Mitomycin C   **Enamine**

(a) The first step is loss of methoxide and formation of an iminium ion intermediate that is deprotonated to give an enamine. Show the mechanism.

(b) The second step is reaction of the enamine with DNA to open the three-membered, nitrogen-containing (aziridine) ring. Show the mechanism.

(c) The third step is loss of carbamate ($NH_2CO_2^-$) and formation of an unsaturated iminium ion, followed by a conjugate addition of another part of the DNA chain. Show the mechanism.

24.72 Phenacetin, a substance formerly used in over-the-counter headache remedies, has the formula $C_{10}H_{13}NO_2$. Phenacetin is neutral and does not dissolve in either acid or base. When warmed with aqueous NaOH, phenacetin yields an amine, $C_8H_{11}NO$, whose ^1H NMR spectrum is shown. When heated with HI, the amine is cleaved to an aminophenol, C_6H_7NO. What is the structure of phenacetin, and what are the structures of the amine and the aminophenol?

24.73 Propose structures for amines with the following ^1H NMR spectra:
(a) C_3H_9NO

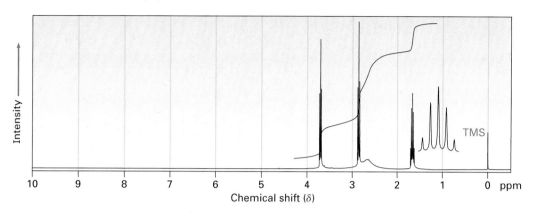

(b) $C_4H_{11}NO_2$

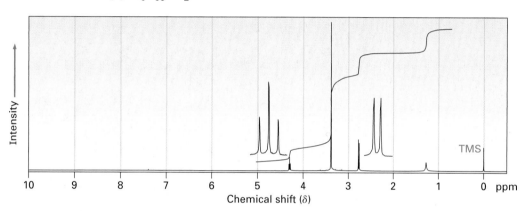

24.74 Propose structures for compounds that show the following ^1H NMR spectra.
(a) $C_9H_{13}N$

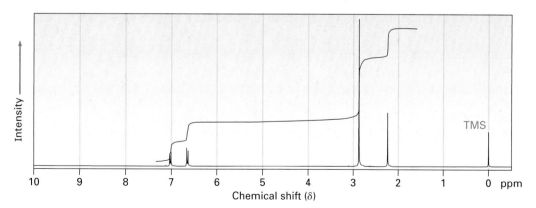

(b) $C_{15}H_{17}N$

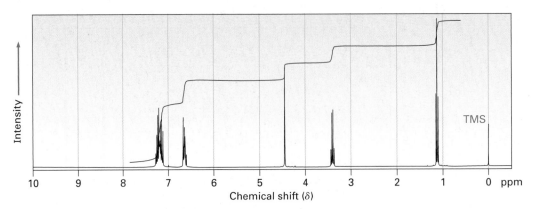

24.75 α-Amino acids can be prepared by the *Strecker synthesis,* a two-step process in which an aldehyde is treated with ammonium cyanide followed by hydrolysis of the amino nitrile intermediate with aqueous acid. Propose a mechanism for the reaction.

An α-amino acid

24.76 One of the reactions used in determining the sequence of nucleotides in a strand of DNA is reaction with hydrazine. Propose a mechanism for the following reaction, which occurs by an initial conjugate addition followed by internal amide formation.

25

Biomolecules: Carbohydrates

Organic KNOWLEDGE TOOLS

CENGAGENOW™ Throughout this chapter, sign in at www.cengage.com/login for online self-study and interactive tutorials based on your level of understanding.

☉WL Online homework for this chapter may be assigned in Organic OWL.

Carbohydrates occur in every living organism. The sugar and starch in food and the cellulose in wood, paper, and cotton are nearly pure carbohydrates. Modified carbohydrates form part of the coating around living cells, other carbohydrates are part of the nucleic acids that carry our genetic information, and still others are used as medicines.

The word **carbohydrate** derives historically from the fact that glucose, the first simple carbohydrate to be obtained pure, has the molecular formula $C_6H_{12}O_6$ and was originally thought to be a "hydrate of carbon, $C_6(H_2O)_6$." This view was soon abandoned, but the name persisted. Today, the term *carbohydrate* is used to refer loosely to the broad class of polyhydroxylated aldehydes and ketones commonly called *sugars*. Glucose, also known as *dextrose* in medical work, is the most familiar example.

Glucose (dextrose),
a pentahydroxyhexanal

Carbohydrates are synthesized by green plants during photosynthesis, a complex process in which sunlight provides the energy to convert CO_2 and H_2O into glucose plus oxygen. Many molecules of glucose are then chemically linked for storage by the plant in the form of either cellulose or starch. It has been estimated that more than 50% of the dry weight of the earth's biomass—all plants and animals—consists of glucose polymers. When eaten and metabolized, carbohydrates then provide animals with a source of readily available energy.

Sean Duggan

Thus, carbohydrates act as the chemical intermediaries by which solar energy is stored and used to support life.

$$6\,CO_2 \;+\; 6\,H_2O \xrightarrow{\text{Sunlight}} 6\,O_2 \;+\; C_6H_{12}O_6 \longrightarrow \text{Cellulose, starch}$$

Glucose

Because humans and most other mammals lack the enzymes needed for digestion of cellulose, they require starch as their dietary source of carbohydrates. Grazing animals such as cows, however, have microorganisms in their first stomach that are able to digest cellulose. The energy stored in cellulose is thus moved up the biological food chain when these ruminant animals eat grass and are then used for food.

WHY THIS CHAPTER?

Carbohydrates are the first major class of biomolecules we'll discuss. We'll see in this chapter what the structures and primary biological functions of carbohydrates are, and then in Chapter 29, we'll return to the subject to see how carbohydrates are biosynthesized and degraded in organisms.

25.1 | Classification of Carbohydrates

Carbohydrates are generally classed as either *simple* or *complex*. **Simple sugars**, or **monosaccharides**, are carbohydrates like glucose and fructose that can't be converted into smaller sugars by hydrolysis. **Complex carbohydrates** are made of two or more simple sugars linked together by acetal bonds (Section 19.10). Sucrose (table sugar), for example, is a *disaccharide* made up of one glucose linked to one fructose. Similarly, cellulose is a *polysaccharide* made up of several thousand glucose units linked together. Enzyme-catalyzed hydrolysis of a polysaccharide breaks it down into its constituent monosaccharides.

Sucrose
(a disaccharide)

Cellulose
(a polysaccharide)

Monosaccharides are further classified as either **aldoses** or **ketoses**. The -*ose* suffix designates a carbohydrate, and the *aldo-* and *keto-* prefixes identify the kind of carbonyl group present in the molecule, whether aldehyde or ketone. The number of carbon atoms in the monosaccharide is indicated by the appropriate numerical prefix *tri-, tetr-, pent-, hex-*, and so forth, in the name. Putting it all together, glucose is an *aldohexose,* a six-carbon aldehydo sugar; fructose is a *ketohexose,* a six-carbon keto sugar; ribose is an *aldopentose,* a five-carbon aldehydo sugar; and sedoheptulose is a *ketoheptose,* a seven-carbon keto sugar. Most of the common simple sugars are either pentoses or hexoses.

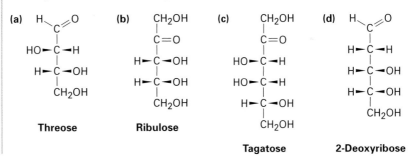

Glucose	**Fructose**	**Ribose**	**Sedoheptulose**
(an aldohexose)	**(a ketohexose)**	**(an aldopentose)**	**(a ketoheptose)**

Problem 25.1 Classify each of the following monosaccharides:

(a)

Threose

(b)

Ribulose

(c)

Tagatose

(d)

2-Deoxyribose

25.2 | Depicting Carbohydrate Stereochemistry: Fischer Projections

CENGAGENOW™ Click *Organic Interactive* to **learn to draw and interpret Fischer projections of simple monosaccharides.**

Because carbohydrates usually have numerous chirality centers, it was recognized long ago that a quick method for representing carbohydrate stereochemistry is needed. In 1891, Emil Fischer suggested a method based on the projection of a tetrahedral carbon atom onto a flat surface. These **Fischer projections** were soon adopted and are now a standard means of representing stereochemistry at chirality centers, particularly in carbohydrate chemistry.

A tetrahedral carbon atom is represented in a Fischer projection by two crossed lines. The horizontal lines represent bonds coming out of the page, and the vertical lines represent bonds going into the page.

For example, (R)-glyceraldehyde, the simplest monosaccharide, can be drawn as in Figure 25.1.

Figure 25.1 A Fischer projection of (R)-glyceraldehyde.

(R)-Glyceraldehyde
(Fischer projection)

Because a given chiral molecule can be drawn in many different ways, it's often necessary to compare two projections to see if they represent the same or different enantiomers. To test for identity, Fischer projections can be moved around on the paper, but only two kinds of motions are allowed; moving a Fischer projection in any other way inverts its meaning.

▌ A Fischer projection can be rotated on the page by 180°, but *not by 90° or 270°*. Only a 180° rotation maintains the Fischer convention by keeping the same substituent groups going into and coming out of the plane. In the following Fischer projection of (R)-glyceraldehyde, for example, the −H and −OH groups come out of the plane both before and after a 180° rotation.

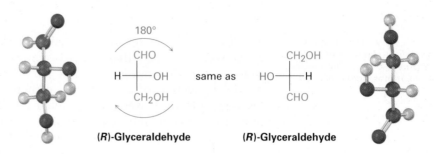

(R)-Glyceraldehyde   (R)-Glyceraldehyde

A 90° rotation breaks the Fischer convention by exchanging the groups that go into the plane and those that come out. In the following Fischer projections of (*R*)-glyceraldehyde, the −H and −OH groups come out of the plane before rotation but go into the plane after a 90° rotation. As a result, the rotated projection represents (*S*)-glyceraldehyde.

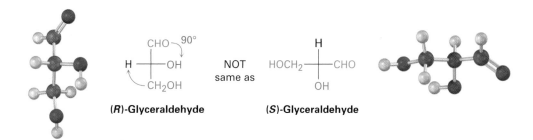

(*R*)-Glyceraldehyde   **(*S*)-Glyceraldehyde**

▎ A Fischer projection can have one group held steady while the other three rotate in either a clockwise or a counterclockwise direction. The effect is simply to rotate around a single bond, which does not change the stereochemistry.

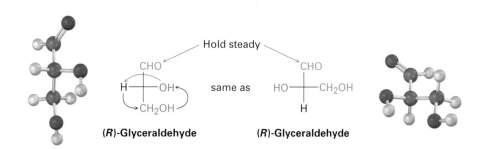

(*R*)-Glyceraldehyde   **(*R*)-Glyceraldehyde**

R,S stereochemical designations (Section 9.5) can be assigned to the chirality center in a Fischer projection by following three steps, as shown in Worked Example 25.1.

Step 1   Assign priorities to the four substituents in the usual way.

Step 2   Place the group of lowest priority, usually H, at the top of the Fischer projection by using one of the allowed motions. This means that the lowest-priority group is oriented back, away from the viewer, as required for assigning configuration.

Step 3   Determine the direction of rotation 1 → 2 → 3 of the remaining three groups, and assign *R* or *S* configuration.

Carbohydrates with more than one chirality center are shown in Fischer projections by stacking the centers on top of one another. By convention, the carbonyl carbon is always placed either at or near the top. Glucose, for example, has four chirality centers stacked on top of one another in a Fischer projection. Such representations don't, however, give an accurate picture of the true conformation of a molecule, which actually is curled around on itself like a bracelet.

Glucose
(carbonyl group at top)

WORKED EXAMPLE 25.1   *Assigning R or S Configuration to a Fischer Projection*

Assign *R* or *S* configuration to the following Fischer projection of alanine:

Strategy   Follow the steps in the text. (1) Assign priorities to the four substituents on the chiral carbon. (2) Manipulate the Fischer projection to place the group of lowest priority at the top by carrying out one of the allowed motions. (3) Determine the direction 1 → 2 → 3 of the remaining three groups.

Solution   The priorities of the groups are (1) $-NH_2$, (2) $-CO_2H$, (3) $-CH_3$, and (4) $-H$. To bring the group of lowest priority ($-H$) to the top, we might want to hold the $-CH_3$ group steady while rotating the other three groups counterclockwise.

Going from first- to second- to third-highest priority requires a counterclockwise turn, corresponding to *S* stereochemistry.

S configuration

Problem 25.2 | Convert the following Fischer projections into tetrahedral representations, and assign *R* or *S* stereochemistry to each:

(a)

$$CO_2H$$
$$H_2N \text{—} H$$
$$CH_3$$

(b)

$$CHO$$
$$H \text{—} OH$$
$$CH_3$$

(c)

$$CH_3$$
$$H \text{—} CHO$$
$$CH_2CH_3$$

Problem 25.3 | Which of the following Fischer projections of glyceraldehyde represent the same enantiomer?

$$CHO$$
$$HO \text{—} H$$
$$CH_2OH$$

A

$$OH$$
$$HOCH_2 \text{—} H$$
$$CHO$$

B

$$H$$
$$HO \text{—} CH_2OH$$
$$CHO$$

C

$$CH_2OH$$
$$H \text{—} CHO$$
$$OH$$

D

Problem 25.4 | Redraw the following molecule as a Fischer projection, and assign *R* or *S* configuration to the chirality center (yellow-green = Cl):

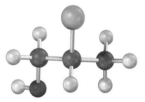

Problem 25.5 | Redraw the following aldotetrose as a Fischer projection, and assign *R* or *S* configuration to each chirality center:

25.3 | D,L Sugars

Glyceraldehyde, the simplest aldose, has only one chirality center and thus has two enantiomeric (mirror-image) forms. Only the dextrorotatory enantiomer occurs naturally, however. That is, a sample of naturally occurring glyceraldehyde placed in a polarimeter rotates plane-polarized light in a clockwise direction, denoted (+). Since (+)-glyceraldehyde has been found to have an *R* configuration at C2, it can be represented in a Fischer projection as shown in Figure 25.1. For historical reasons dating back long before the adoption of the *R,S* system, (*R*)-(+)-glyceraldehyde is also referred to as D-glyceraldehyde (D for dextrorotatory). The other enantiomer, (*S*)-(−)-glyceraldehyde, is known as L-glyceraldehyde (L for levorotatory).

Because of the way monosaccharides are biosynthesized in nature, glucose, fructose, and most (although not all) other naturally occurring monosaccharides have the same *R* stereochemical configuration as D-glyceraldehyde at the chirality center farthest from the carbonyl group. In Fischer projections, therefore, most naturally occurring sugars have the hydroxyl group at the bottom chirality center pointing to the right (Figure 25.2). All such compounds are referred to as **D sugars**.

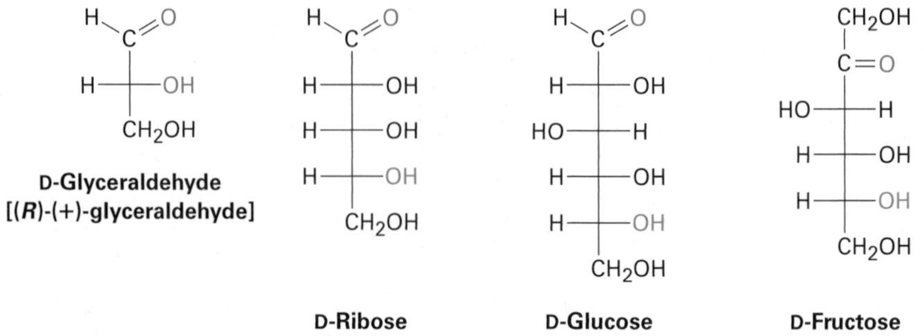

Figure 25.2 Some naturally occurring D sugars. The −OH group at the chirality center farthest from the carbonyl group has the same configuration as (*R*)-(+)-glyceraldehyde and points toward the right in Fischer projections.

In contrast with D sugars, **L sugars** have an *S* configuration at the lowest chirality center, with the bottom −OH group pointing to the *left* in Fischer projections. Thus, an L sugar is the mirror image (enantiomer) of the corresponding D sugar and has the opposite configuration from the D sugar at all chirality centers. Note that the D and L notations have no relation to the direction in which a given sugar rotates plane-polarized light; a D sugar can be either dextrorotatory or levorotatory. The prefix D indicates only that the −OH group at the lowest chirality center has *R* stereochemistry and points to the right when the molecule is drawn in a Fischer projection. Note also that the D,L system of carbohydrate nomenclature describes the configuration at only one chirality center and says nothing about the configuration of other chirality centers that may be present.

Mirror

L-Glyceraldehyde
[(S)-(–)-glyceraldehyde]

L-Glucose
(not naturally occurring)

D-Glucose

Problem 25.6 | Assign *R* or *S* configuration to each chirality center in the following monosaccharides, and tell whether each is a D sugar or an L sugar:

(a)
```
        CHO
  HO ──┼── H
  HO ──┼── H
       CH2OH
```

(b)
```
        CHO
   H ──┼── OH
  HO ──┼── H
   H ──┼── OH
       CH2OH
```

(c)
```
        CH2OH
        C=O
  HO ──┼── H
   H ──┼── OH
       CH2OH
```

Problem 25.7 | (+)-Arabinose, an aldopentose that is widely distributed in plants, is systematically named (2R,3S,4S)-2,3,4,5-tetrahydroxypentanal. Draw a Fischer projection of (+)-arabinose, and identify it as a D sugar or an L sugar.

25.4 | Configurations of the Aldoses

Louis F. Fieser

Louis F. Fieser (1899–1977) was born in Columbus, Ohio, and received his Ph.D. at Harvard University in 1924 with James B. Conant. He was professor of chemistry at Bryn Mawr College and then at Harvard University from 1930 to 1968. While at Bryn Mawr, he met his future wife, Mary, then a student. In collaboration, the two Fiesers wrote numerous chemistry texts and monographs. Among his scientific contributions, Fieser was known for his work in steroid chemistry and in carrying out the first synthesis of vitamin K. He was also the inventor of jellied gasoline, or napalm, which was developed at Harvard during World War II.

Aldotetroses are four-carbon sugars with two chirality centers and an aldehyde carbonyl group. Thus, there are $2^2 = 4$ possible stereoisomeric aldotetroses, or two D,L pairs of enantiomers named *erythrose* and *threose*.

Aldopentoses have three chirality centers and a total of $2^3 = 8$ possible stereoisomers, or four D,L pairs of enantiomers. These four pairs are called *ribose, arabinose, xylose,* and *lyxose.* All except lyxose occur widely. D-Ribose is an important constituent of RNA (ribonucleic acid), L-arabinose is found in many plants, and D-xylose is found in wood.

Aldohexoses have four chirality centers and a total of $2^4 = 16$ possible stereoisomers, or eight D,L pairs of enantiomers. The names of the eight are *allose, altrose, glucose, mannose, gulose, idose, galactose,* and *talose.* Only D-glucose, from starch and cellulose, and D-galactose, from gums and fruit pectins, are found widely in nature. D-Mannose and D-talose also occur naturally but in lesser abundance.

Fischer projections of the four-, five-, and six-carbon D aldoses are shown in Figure 25.3. Starting with D-glyceraldehyde, we can imagine constructing the two D aldotetroses by inserting a new chirality center just below the aldehyde carbon. Each of the two D aldotetroses then leads to two D aldopentoses (four total), and

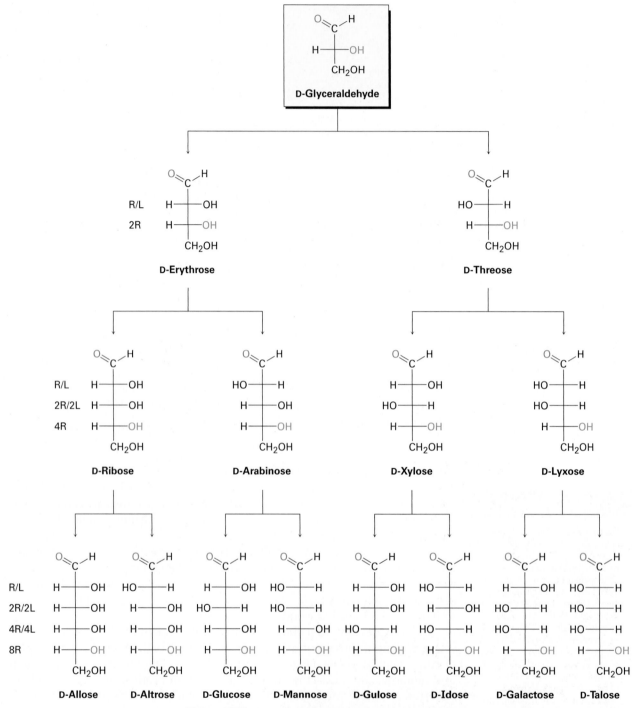

Figure 25.3 Configurations of D aldoses. The structures are arranged from left to right so that the —OH groups on C2 alternate right/left (R/L) in going across a series. Similarly, the —OH groups at C3 alternate two right/two left (2R/2L), the —OH groups at C4 alternate 4R/4L, and the —OH groups at C5 are to the right in all eight (8R). Each D aldose has a corresponding L enantiomer, which is not shown.

each of the four D aldopentoses leads to two D aldohexoses (eight total). In addition, each of the D aldoses in Figure 25.3 has an L enantiomer, which is not shown.

Louis Fieser of Harvard University suggested the following procedure for remembering the names and structures of the eight D aldohexoses:

Step 1 Set up eight Fischer projections with the −CHO group on top and the −CH₂OH group at the bottom.

Step 2 At C5, place all eight −OH groups to the right (D series).

Step 3 At C4, alternate four −OH groups to the right and four to the left.

Step 4 At C3, alternate two −OH groups to the right, two to the left.

Step 5 At C2, alternate −OH groups right, left, right, left.

Step 6 Name the eight isomers using the mnemonic "**All al**truists **gl**adly **ma**ke **gu**m in **gal**lon **ta**nks."

The structures of the four D aldopentoses can be generated in a similar way and named by the mnemonic suggested by a Cornell University undergraduate: "**ri**bs **a**re **ex**tra **l**ean."

WORKED EXAMPLE 25.2   ***Drawing a Fischer Projection***

Draw a Fischer projection of L-fructose.

Strategy   Because L-fructose is the enantiomer of D-fructose, simply look at the structure of D-fructose and reverse the configuration at each chirality center.

Solution

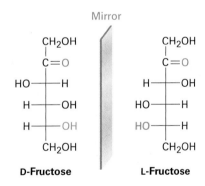

Problem 25.8   Only the D sugars are shown in Figure 25.3. Draw Fischer projections for the following L sugars:
(a) L-Xylose   **(b)** L-Galactose   **(c)** L-Allose

Problem 25.9   How many aldoheptoses are there? How many are D sugars, and how many are L sugars?

Problem 25.10 The following model is that of an aldopentose. Draw a Fischer projection of the sugar, name it, and identify it as a D sugar or an L sugar.

25.5 | Cyclic Structures of Monosaccharides: Anomers

We said in Section 19.10 that aldehydes and ketones undergo a rapid and reversible nucleophilic addition reaction with alcohols to form hemiacetals.

$$
\underset{\textbf{An aldehyde}}{\overset{\displaystyle O}{\underset{R}{\overset{\parallel}{C}}}\diagdown_H} \;+\; R'OH \;\;\underset{\text{catalyst}}{\overset{H^+}{\rightleftharpoons}}\;\; \underset{\textbf{A hemiacetal}}{\overset{\displaystyle OH}{\underset{H}{\overset{\mid}{\underset{R}{C}}}}\diagdown_{OR'}}
$$

If the carbonyl and the hydroxyl group are in the same molecule, an intramolecular nucleophilic addition can take place, leading to the formation of a cyclic hemiacetal. Five- and six-membered cyclic hemiacetals are relatively strain-free and particularly stable, and many carbohydrates therefore exist in an equilibrium between open-chain and cyclic forms. Glucose, for instance, exists in aqueous solution primarily in the six-membered, **pyranose** form resulting from intramolecular nucleophilic addition of the −OH group at C5 to the C1 carbonyl group (Figure 25.4). The name *pyranose* is derived from *pyran,* the name of the unsaturated six-membered cyclic ether.

Like cyclohexane rings (Section 4.6), pyranose rings have a chairlike geometry with axial and equatorial substituents. By convention, the rings are usually drawn by placing the hemiacetal oxygen atom at the right rear, as shown in Figure 25.4. Note that an −OH group on the *right* in a Fischer projection is on the *bottom* face of the pyranose ring, and an −OH group on the *left* in a Fischer projection is on the *top* face of the ring. For D sugars, the terminal −CH₂OH group is on the top of the ring, whereas for L sugars, the −CH₂OH group is on the bottom.

When an open-chain monosaccharide cyclizes to a pyranose form, a new chirality center is generated at the former carbonyl carbon and two diastereomers, called **anomers**, are produced. The hemiacetal carbon atom is referred to as the **anomeric center.** For example, glucose cyclizes reversibly in aqueous solution to a 37:63 mixture of two anomers (Figure 25.4). The compound with its newly generated −OH group at C1 *cis* to the −OH at the lowest chirality center in a Fischer projection is called the **α anomer**; its full name is α-D-glucopyranose. The compound with its newly generated −OH group *trans* to the −OH at the lowest chirality center in a Fischer projection is called the **β anomer**; its full name is β-D-glucopyranose. Note that in β-D-glucopyranose, all the

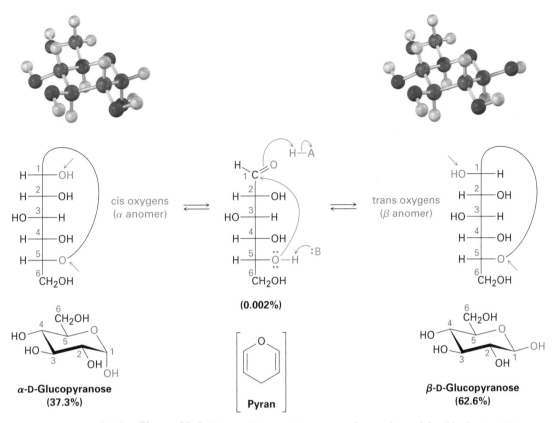

Active Figure 25.4 Glucose in its cyclic pyranose forms. As explained in the text, two anomers are formed by cyclization of glucose. The molecule whose newly formed −OH group at C1 is cis to the oxygen atom on the lowest chirality center (C5) in a Fischer projection is the α anomer. The molecule whose newly formed −OH group is trans to the oxygen atom on the lowest chirality center in a Fischer projection is the β anomer. *Sign in at* **www.cengage.com/login** *to see a simulation based on this figure and to take a short quiz.*

substituents on the ring are equatorial. Thus, β-D-glucopyranose is the least sterically crowded and most stable of the eight D aldohexoses.

Some monosaccharides also exist in a five-membered cyclic hemiacetal form called a **furanose** form. D-Fructose, for instance, exists in water solution as 70% β-pyranose, 2% α-pyranose, 0.7% open-chain, 23% β-furanose, and 5% α-furanose. The pyranose form results from addition of the −OH at C6 to the carbonyl group, while the furanose form results from addition of the −OH at C5 to the carbonyl group (Figure 25.5).

Both anomers of D-glucopyranose can be crystallized and purified. Pure α-D-glucopyranose has a melting point of 146 °C and a specific rotation, $[\alpha]_D$, of +112.2; pure β-D-glucopyranose has a melting point of 148 to 155 °C and a specific rotation of +18.7. When a sample of either pure anomer is dissolved in water, however, the optical rotation slowly changes and ultimately reaches a constant value of +52.6. That is, the specific rotation of the α-anomer solution decreases from +112.2 to +52.6, and the specific rotation of the β-anomer solution increases from +18.7 to +52.6. Called **mutarotation**, this change in optical rotation is due to the slow conversion of the pure anomers into a 37:63 equilibrium mixture.

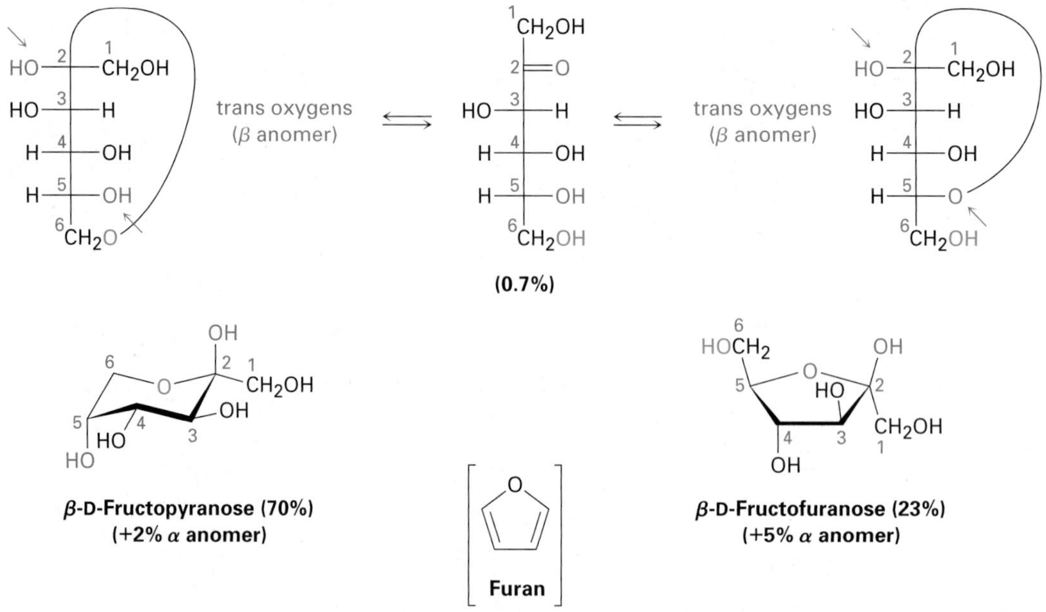

Figure 25.5 Pyranose and furanose forms of fructose in aqueous solution. The two pyranose anomers result from addition of the C6 —OH group to the C2 carbonyl; the two furanose anomers result from addition of the C5 —OH group to the C2 carbonyl.

Mutarotation occurs by a reversible ring-opening of each anomer to the open-chain aldehyde, followed by reclosure. Although equilibration is slow at neutral pH, it is catalyzed by both acid and base.

| WORKED EXAMPLE 25.3 | *Drawing the Chair Conformation of an Aldohexose* |

D-Mannose differs from D-glucose in its stereochemistry at C2. Draw D-mannose in its chairlike pyranose form.

Strategy First draw a Fischer projection of D-mannose. Then lay it on its side, and curl it around so that the —CHO group (C1) is on the right front and the —CH₂OH group (C6) is toward the left rear. Now, connect the —OH at C5 to the C1 carbonyl group to form the pyranose ring. In drawing the chair form, raise the leftmost carbon (C4) up and drop the rightmost carbon (C1) down.

Solution

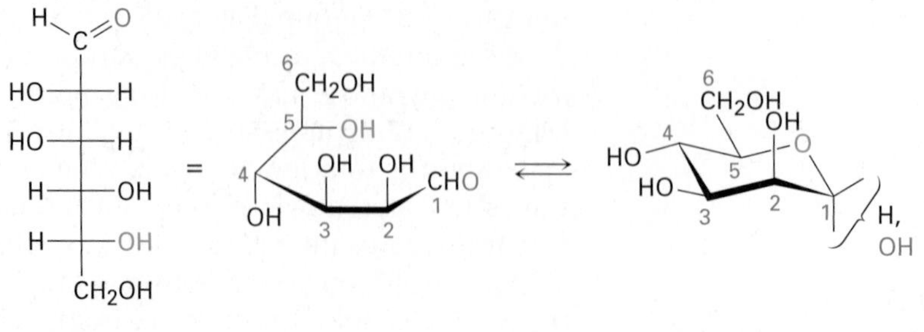

D-Mannose   (Pyranose form)

WORKED EXAMPLE 25.4   *Drawing the Chair Conformation of a Pyranose*

Draw β-L-glucopyranose in its more stable chair conformation.

Strategy   It's probably easiest to begin by drawing the chair conformation of β-D-gluco-pyranose. Then draw its mirror image by changing the stereochemistry at every position on the ring, and carry out a ring-flip to give the more stable chair conformation. Note that the −CH₂OH group is on the bottom face of the ring in the L enantiomer.

Solution

β-D-Glucopyranose   β-L-Glucopyranose

Problem 25.11   Ribose exists largely in a furanose form, produced by addition of the C4 −OH group to the C1 aldehyde. Draw D-ribose in its furanose form.

Problem 25.12   Figure 25.5 shows only the β-pyranose and β-furanose anomers of D-fructose. Draw the α-pyranose and α-furanose anomers.

Problem 25.13   Draw β-D-galactopyranose and β-D-mannopyranose in their more stable chair conformations. Label each ring substituent as either axial or equatorial. Which would you expect to be more stable, galactose or mannose?

Problem 25.14   Draw β-L-galactopyranose in its more stable chair conformation, and label the substituents as either axial or equatorial.

Problem 25.15   Identify the following monosaccharide, write its full name, and draw its open-chain form in Fischer projection.

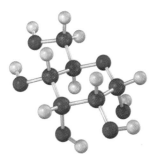

25.6 | Reactions of Monosaccharides

CENGAGENOW™ Click *Organic Interactive* to **predict products from reactions of simple monosaccharides**.

Because monosaccharides contain only two kinds of functional groups, hydroxyls and carbonyls, most of the chemistry of monosaccharides is the familiar chemistry of these two groups. Alcohols can be converted to esters and ethers and can be oxidized; carbonyl compounds can react with nucleophiles and can be reduced.

Ester and Ether Formation

Monosaccharides behave as simple alcohols in much of their chemistry. For example, carbohydrate −OH groups can be converted into esters and ethers, which are often easier to work with than the free sugars. Because of their many hydroxyl groups, monosaccharides are usually soluble in water but insoluble in organic solvents such as ether. They are also difficult to purify and have a tendency to form syrups rather than crystals when water is removed. Ester and ether derivatives, however, are soluble in organic solvents and are easily purified and crystallized.

Esterification is normally carried out by treating the carbohydrate with an acid chloride or acid anhydride in the presence of a base (Sections 21.4 and 21.5). All the −OH groups react, including the anomeric one. For example, β-D-glucopyranose is converted into its pentaacetate by treatment with acetic anhydride in pyridine solution.

β-D-Glucopyranose

Penta-O-acetyl-β-D-glucopyranose (91%)

Carbohydrates are converted into ethers by treatment with an alkyl halide in the presence of base—the Williamson ether synthesis (Section 18.2). Standard Williamson conditions using a strong base tend to degrade sensitive sugar molecules, but silver oxide works well as a mild base and gives high yields of ethers. For example, α-D-glucopyranose is converted into its pentamethyl ether in 85% yield on reaction with iodomethane and Ag_2O.

α-D-Glucopyranose

α-D-Glucopyranose pentamethyl ether (85%)

Problem 25.16 Draw the products you would obtain by reaction of β-D-ribofuranose with:
(a) CH_3I, Ag_2O (b) $(CH_3CO)_2O$, pyridine

β-D-Ribofuranose

Glycoside Formation

We saw in Section 19.10 that treatment of a hemiacetal with an alcohol and an acid catalyst yields an acetal.

A hemiacetal   **An acetal**

In the same way, treatment of a monosaccharide hemiacetal with an alcohol and an acid catalyst yields an acetal called a **glycoside**, in which the anomeric −OH has been replaced by an −OR group. For example, reaction of β-D-glucopyranose with methanol gives a mixture of α and β methyl D-glucopyranosides. (Note that a *gly*coside is the functional group name for any sugar, whereas a *glu*coside is a glycoside formed specifically from glucose.)

β-D-Glucopyranose   **Methyl α-D-glucopyranoside**   **Methyl β-D-glucopyranoside**
(a cyclic hemiacetal)   **(66%)**   **(33%)**

Glycosides are named by first citing the alkyl group and then replacing the -*ose* ending of the sugar with -*oside*. Like all acetals, glycosides are stable to neutral water. They aren't in equilibrium with an open-chain form, and they don't show mutarotation. They can, however, be converted back to the free monosaccharide by hydrolysis with aqueous acid (Section 19.10).

Glycosides are abundant in nature, and many biologically important molecules contain glycosidic linkages. For example, digitoxin, the active component of the digitalis preparations used for treatment of heart disease, is a glycoside consisting of a steroid alcohol linked to a trisaccharide. Note also that the three sugars are linked to one another by glycoside bonds.

Steroid

Trisaccharide

Digitoxigenin, a glycoside

The laboratory synthesis of glycosides can be difficult because of the numerous −OH groups on the sugar molecule. One method that is particularly suitable for preparing glucose β-glycosides involves treatment of glucose pentaacetate with HBr, followed by addition of the appropriate alcohol in the presence of silver oxide. Called the *Koenigs–Knorr reaction,* the sequence involves formation of a pyranosyl bromide, followed by nucleophilic substitution. For example, methylarbutin, a glycoside found in pears, has been prepared by reaction of tetraacetyl-α-D-glucopyranosyl bromide with *p*-methoxyphenol.

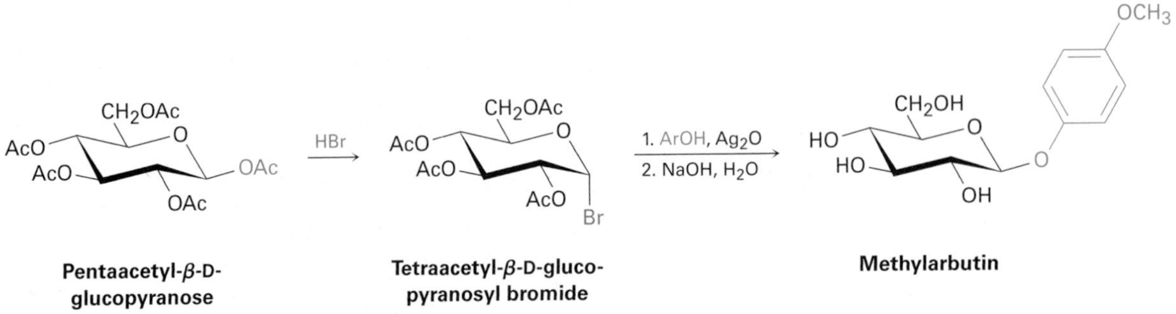

Pentaacetyl-β-D-glucopyranose   **Tetraacetyl-β-D-glucopyranosyl bromide**   **Methylarbutin**

Although the Koenigs–Knorr reaction appears to involve a simple backside S_N2 displacement of bromide ion by alkoxide ion, the situation is actually more complex. Both α and β anomers of tetraacetyl-D-glucopyranosyl bromide give the same β-glycoside product, implying that they react by a common pathway.

The results can be understood by assuming that tetraacetyl-D-glucopyranosyl bromide (either α or β anomer) undergoes a spontaneous S_N1-like loss of Br^-, followed by internal reaction with the ester group at C2 to form an oxonium ion. Since the acetate at C2 is on the bottom of the glucose ring, the C−O bond also forms from the bottom. Backside S_N2 displacement of the oxonium ion then occurs with the usual inversion of configuration, yielding a β-glycoside and regenerating the acetate at C2 (Figure 25.6).

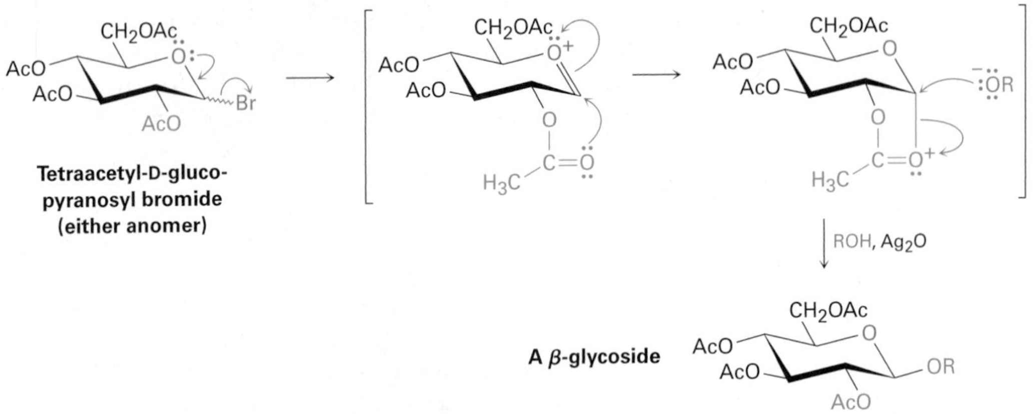

Tetraacetyl-D-glucopyranosyl bromide (either anomer)

A β-glycoside

Figure 25.6 Mechanism of the Koenigs–Knorr reaction, showing the neighboring-group effect of a nearby acetate.

The participation shown by the nearby acetate group in the Koenigs–Knorr reaction is referred to as a *neighboring-group effect* and is a common occurrence in organic chemistry. Neighboring-group effects are usually noticeable only because they affect the rate or stereochemistry of a reaction; the nearby group itself does not undergo any evident change during the reaction.

Biological Ester Formation: Phosphorylation

In living organisms, carbohydrates occur not only in the free form but also linked through their anomeric center to other molecules such as lipids *(glycolipids)* or proteins *(glycoproteins)*. Collectively called *glycoconjugates,* these sugar-linked molecules are components of cell walls and are crucial to the mechanism by which different cell types recognize one another.

Glycoconjugate formation occurs by reaction of the lipid or protein with a glycosyl nucleoside diphosphate, itself formed by initial phosphorylation of a monosaccharide with ATP to give a glycosyl phosphate. The glycosyl phosphate then reacts with a second nucleoside triphosphate, usually uridine triphosphate (UTP), to give a glycosyl uridine diphosphate. The purpose of the phosphorylation is to activate the anomeric $-OH$ group of the sugar and make it a better leaving group in a nucleophilic substitution reaction by a protein or lipid (Figure 25.7).

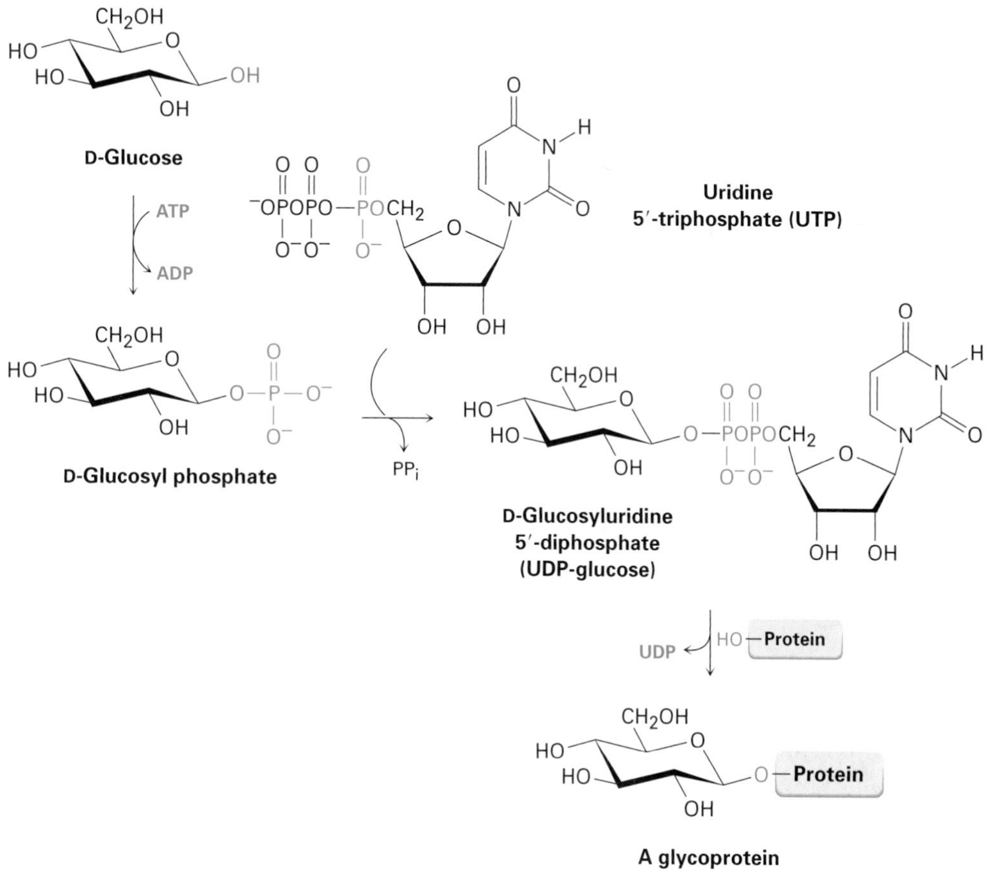

Figure 25.7 Glycoprotein formation occurs by initial phosphorylation of the starting carbohydrate to a glycosyl phosphate, followed by reaction with UTP to form a glycosyl uridine 5′-diphosphate. Nucleophilic substitution by an $-OH$ (or $-NH_2$) group on a protein then gives the glycoprotein.

Reduction of Monosaccharides

Treatment of an aldose or ketose with $NaBH_4$ reduces it to a polyalcohol called an **alditol**. The reduction occurs by reaction of the open-chain form present in the aldehyde/ketone $\rightleftarrows$ hemiacetal equilibrium. Although only a small amount of the open-chain form is present at any given time, that small amount is reduced, more is produced by opening of the pyranose form, that additional amount is reduced, and so on, until the entire sample has undergone reaction.

β-D-Glucopyranose   **D-Glucose**   **D-Glucitol (D-sorbitol), an alditol**

D-Glucitol, the alditol produced by reduction of D-glucose, is itself a naturally occurring substance present in many fruits and berries. It is used under its alternative name, D-sorbitol, as a sweetener and sugar substitute in foods.

Problem 25.17 | Reduction of D-glucose leads to an optically active alditol (D-glucitol), whereas reduction of D-galactose leads to an optically inactive alditol. Explain.

Problem 25.18 | Reduction of L-gulose with $NaBH_4$ leads to the same alditol (D-glucitol) as reduction of D-glucose. Explain.

Oxidation of Monosaccharides

Like other aldehydes, an aldose is easily oxidized to yield the corresponding carboxylic acid, called an **aldonic acid**. Many specialized reagents whose names you may have run across will oxidizes aldoses, including *Tollens' reagent* (Ag^+ in aqueous NH_3), *Fehling's reagent* (Cu^{2+} in aqueous sodium tartrate), and *Benedict's reagent* (Cu^{2+} in aqueous sodium citrate). All three reactions serve as simple chemical tests for what are called **reducing sugars**—*reducing* because the sugar reduces the metal oxidizing reagent.

If Tollens' reagent is used, metallic silver is produced as a shiny mirror on the walls of the reaction flask or test tube. In fact, the reaction is used commercially for manufacturing specialty mirrors. If Fehling's or Benedict's reagent is used, a reddish precipitate of Cu_2O signals a positive result. Some simple diabetes self-test kits sold in drugstores still use the Benedict test, although more modern methods have largely replaced the chemical test.

All aldoses are reducing sugars because they contain an aldehyde group, but some ketoses are reducing sugars as well. Fructose reduces Tollens' reagent, for example, even though it contains no aldehyde group. Reduction occurs because fructose is readily isomerized to an aldose in basic solution by a series of keto–enol tautomeric shifts (Figure 25.8). Glycosides, however, are nonreducing because the acetal group is not hydrolyzed to an aldehyde under basic conditions.

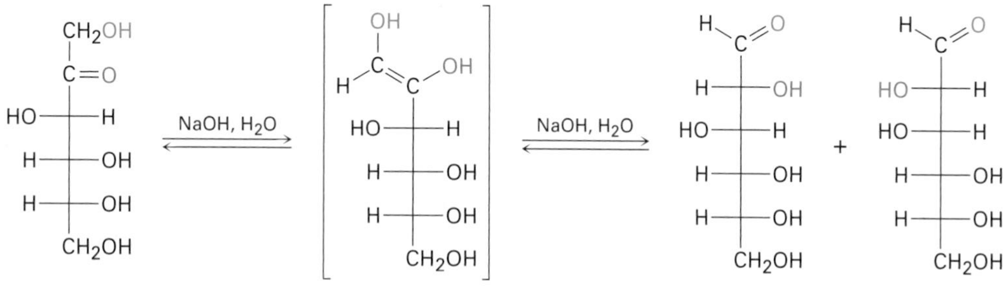

D-Fructose An enediol D-Glucose D-Mannose

Figure 25.8 Fructose, a ketose, is a reducing sugar because it undergoes two base-catalyzed keto–enol tautomerizations that result in conversion to an aldose.

Although the Tollens reaction is a useful test for reducing sugars, it doesn't give good yields of aldonic acid products because the alkaline conditions cause decomposition of the carbohydrate. For preparative purposes, a buffered solution of aqueous Br_2 is a better oxidant. The reaction is specific for aldoses; ketoses are not oxidized by aqueous Br_2.

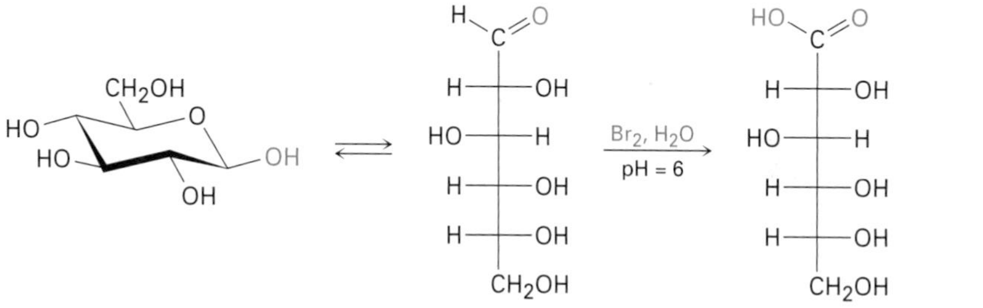

D-Glucose

D-Gluconic acid (an aldonic acid)

If a more powerful oxidizing agent such as warm dilute HNO_3 is used, an aldose is oxidized to a dicarboxylic acid, called an **aldaric acid**. Both the –CHO group at C1 and the terminal $-CH_2OH$ group are oxidized in this reaction.

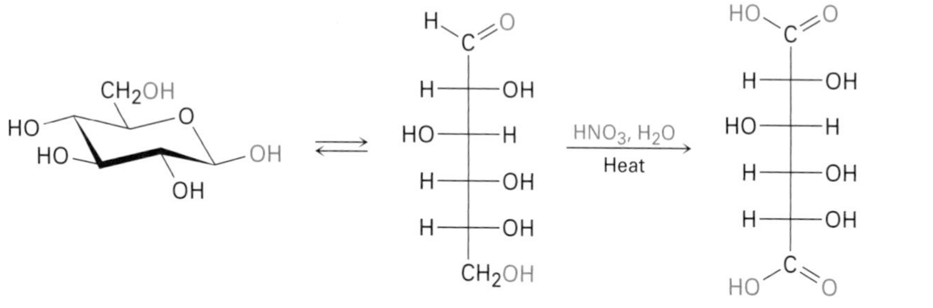

D-Glucose

D-Glucaric acid (an aldaric acid)

Finally, if only the $-CH_2OH$ end of the aldose is oxidized without affecting the $-CHO$ group, the product is a monocarboxylic acid called a **uronic acid**. The reaction must be done enzymatically; no satisfactory chemical reagent is known that can accomplish this selective oxidation in the laboratory.

D-Glucose → Enzyme → D-Glucuronic acid (a uronic acid)

$$\begin{array}{c} CHO \\ H\text{——}OH \\ HO\text{——}H \\ H\text{——}OH \\ H\text{——}OH \\ CO_2H \end{array}$$

Problem 25.19 | D-Glucose yields an optically active aldaric acid on treatment with HNO_3, but D-allose yields an optically inactive aldaric acid. Explain.

Problem 25.20 | Which of the other six D aldohexoses yield optically active aldaric acids on oxidation, and which yield optically inactive (meso) aldaric acids? (See Problem 25.19.)

Chain Lengthening: The Kiliani–Fischer Synthesis

Heinrich Kiliani

Heinrich Kiliani (1855–1945) was born in Würzburg, Germany, and received a Ph.D. at the University of Munich with Emil Erlenmeyer. He was professor of chemistry at the University of Freiburg, where he worked on the chemistry of the heart stimulant drug digitoxin.

Much early activity in carbohydrate chemistry was devoted to unraveling the stereochemical relationships among monosaccharides. One of the most important methods used was the *Kiliani–Fischer synthesis*, which results in the lengthening of an aldose chain by one carbon atom. The C1 aldehyde group of the starting sugar becomes C2 of the chain-lengthened sugar, and a new C1 carbon is added. For example, an aldo*pent*ose is converted by the Kiliani–Fischer synthesis into two aldo*hex*oses.

Discovery of the chain-lengthening sequence was initiated by the observation of Heinrich Kiliani in 1886 that aldoses react with HCN to form cyanohydrins (Section 19.6). Emil Fischer immediately realized the importance of Kiliani's discovery and devised a method for converting the cyanohydrin nitrile group into an aldehyde.

Fischer's original method for conversion of the nitrile into an aldehyde involved hydrolysis to a carboxylic acid, ring closure to a cyclic ester (lactone), and subsequent reduction. A modern improvement is to reduce the nitrile over a palladium catalyst, yielding an imine intermediate that is hydrolyzed to an aldehyde. Note that the cyanohydrin is formed as a mixture of stereoisomers at the new chirality center, so two new aldoses, differing only in their stereochemistry at C2, result from Kiliani–Fischer synthesis. Chain extension of D-arabinose, for example, yields a mixture of D-glucose and D-mannose.

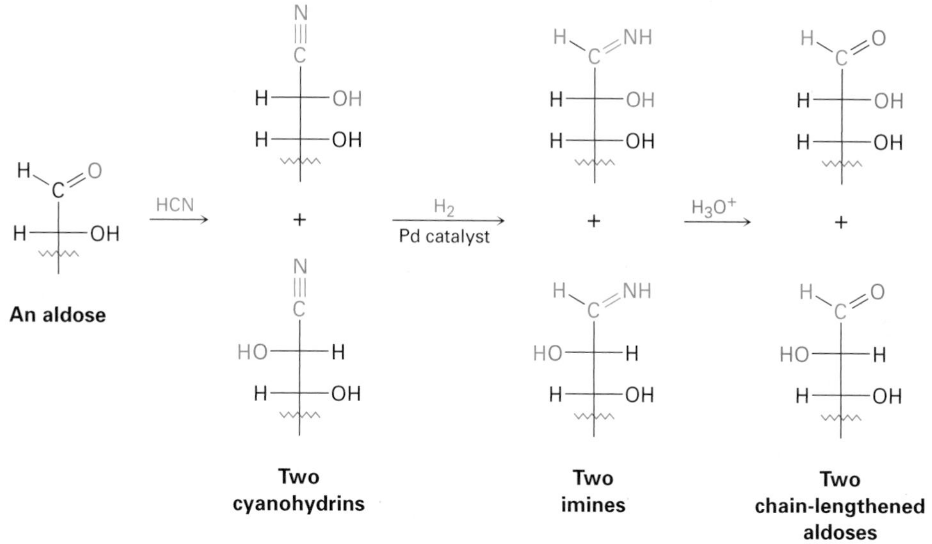

An aldose

Two cyanohydrins

Two imines

Two chain-lengthened aldoses

Problem 25.21 | What product(s) would you expect from Kiliani–Fischer reaction of D-ribose?

Problem 25.22 | What aldopentose would give a mixture of L-gulose and L-idose on Kiliani–Fischer chain extension?

Chain Shortening: The Wohl Degradation

Just as the Kiliani–Fischer synthesis lengthens an aldose chain by one carbon, the *Wohl degradation* shortens an aldose chain by one carbon. The Wohl degradation is almost the exact opposite of the Kiliani–Fischer sequence. That is, the aldose aldehyde carbonyl group is first converted into a nitrile, and the resulting cyanohydrin loses HCN under basic conditions—the reverse of a nucleophilic addition reaction.

Conversion of the aldehyde into a nitrile is accomplished by treatment of an aldose with hydroxylamine to give an *oxime* (Section 19.8), followed by dehydration of the oxime with acetic anhydride. The Wohl degradation does not give particularly high yields of chain-shortened aldoses, but the reaction is general for all aldopentoses and aldohexoses. For example, D-galactose is converted by Wohl degradation into D-lyxose.

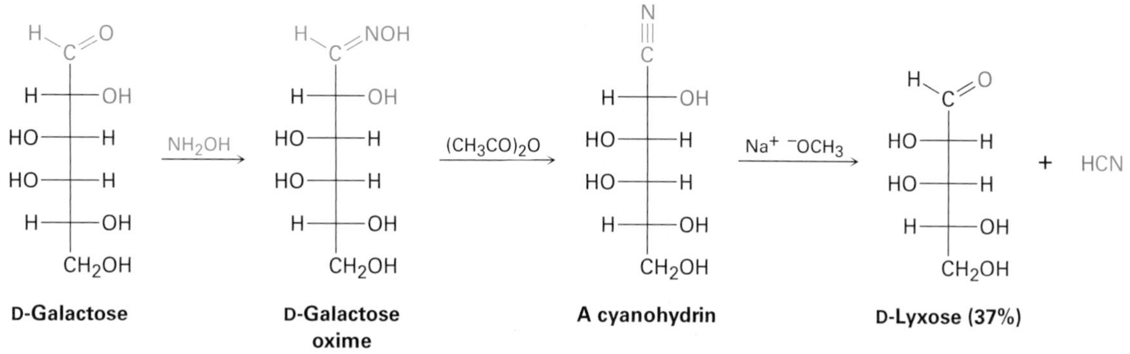

D-Galactose   **D-Galactose oxime**   **A cyanohydrin**   **D-Lyxose (37%)**   + HCN

Problem 25.23 | Two of the four D aldopentoses yield D-threose on Wohl degradation. What are their structures?

25.7 | The Eight Essential Monosaccharides

Our bodies need to obtain eight monosaccharides for proper functioning. Although all can be biosynthesized from simpler precursors if necessary, it's more energetically efficient to obtain them from the diet. The eight are L-fucose (6-deoxy-L-galactose), D-galactose, D-glucose, D-mannose, N-acetyl-D-glucosamine, N-acetyl-D-galactosamine, D-xylose, and N-acetyl-D-neuraminic acid (Figure 25.9). All are used for the synthesis of the glycoconjugate components of cell walls.

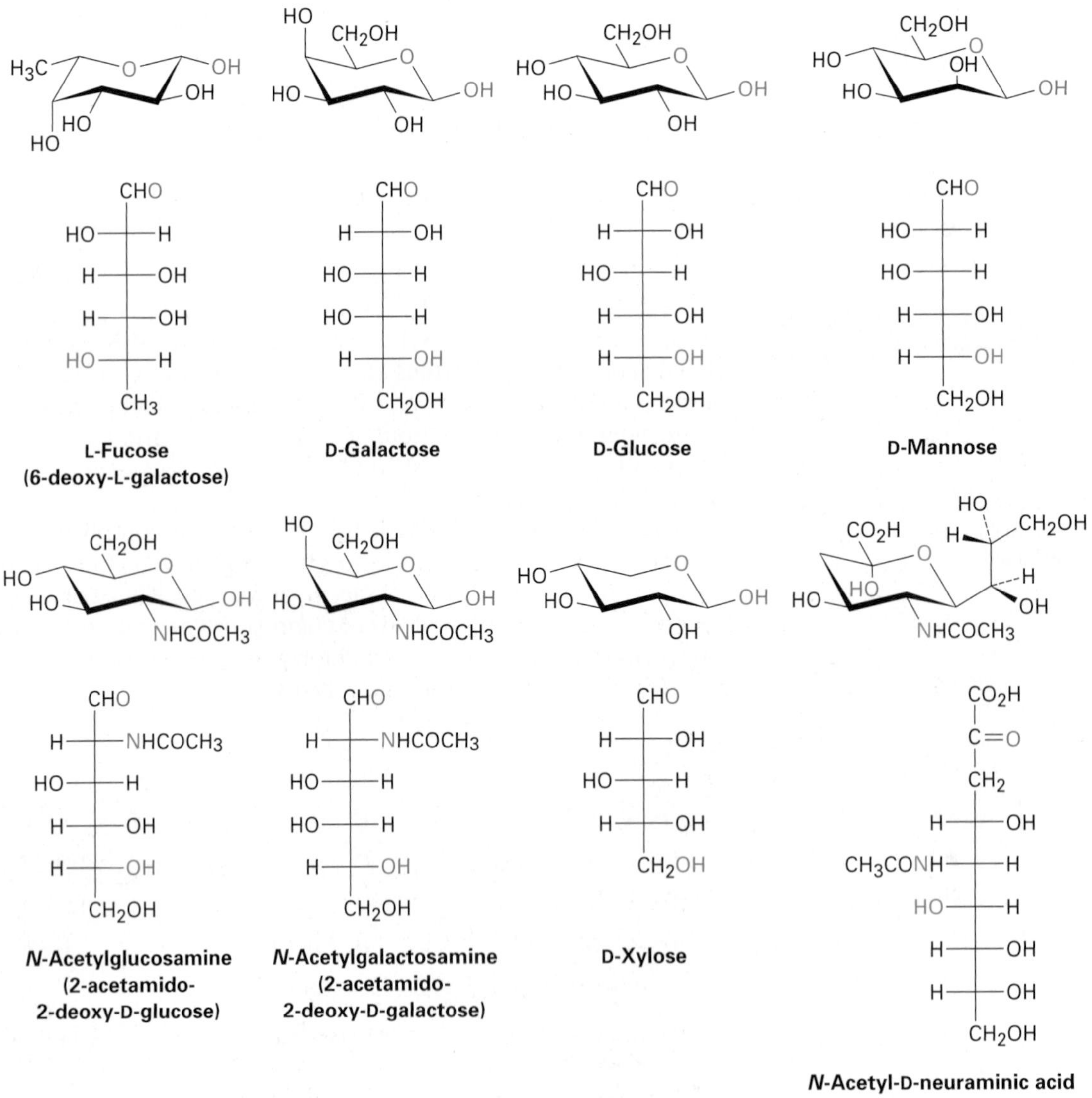

Figure 25.9 Structures of the eight monosaccharides essential to humans.

Of the eight essential monosaccharides, galactose, glucose, and mannose are simple aldohexoses, while xylose is an aldopentose. Fucose is a **deoxy sugar**, meaning that it has an oxygen atom "missing." That is, an −OH group (the one at C6) is replaced by an −H. N-Acetylglucosamine and N-acetylgalactosamine are amide derivatives of **amino sugars** in which an −OH (the one at C2) is replaced by an −NH$_2$ group. N-Acetylneuraminic acid is the parent compound of the *sialic acids,* a group of more than 30 compounds with different modifications, including various oxidations, acetylations, sulfations, and methylations. Note that neuraminic acid has nine carbons and is an aldol reaction product of N-acetylmannosamine with pyruvate (CH$_3$COCO$_2$$^-$).

All the essential monosaccharides arise from glucose, by the conversions summarized in Figure 25.10. We'll not look specifically at these conversions, but might note that end-of-chapter Problems 25.55 through 25.57 lead you through several of the biosynthetic pathways.

Figure 25.10 An overview of biosynthetic pathways for the eight essential monosaccharides.

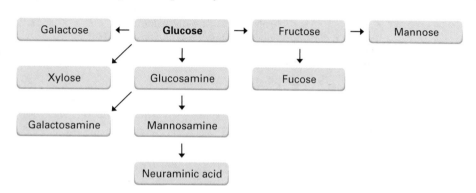

Problem 25.24   Show how neuraminic acid can arise by an aldol reaction of N-acetylmannosamine with pyruvate (CH$_3$COCO$_2$$^-$).

25.8 Disaccharides

We saw in Section 25.6 that reaction of a monosaccharide with an alcohol yields a glycoside in which the anomeric −OH group is replaced by an −OR substituent. If the alcohol is itself a sugar, the glycosidic product is a **disaccharide**.

Cellobiose and Maltose

Disaccharides contain a glycosidic acetal bond between the anomeric carbon of one sugar and an −OH group at any position on the other sugar. A glycosidic bond between C1 of the first sugar and the −OH at C4 of the second sugar is particularly common. Such a bond is called a *1→4 link.*

The glycosidic bond to an anomeric carbon can be either α or β. Maltose, the disaccharide obtained by enzyme-catalyzed hydrolysis of starch, consists of two α-D-glucopyranose units joined by a 1→4-α-glycoside bond. Cellobiose, the disaccharide obtained by partial hydrolysis of cellulose, consists of two β-D-glucopyranose units joined by a 1→4-β-glycoside bond.

Maltose, a 1→ 4-α-glycoside
[4-*O*-(α-D-glucopyranosyl)-α-D-glucopyranose]

Cellobiose, a 1→ 4-β-glycoside
[4-*O*-(β-D-glucopyranosyl)-β-D-glucopyranose]

Maltose and cellobiose are both reducing sugars because the anomeric carbons on the right-hand glucopyranose units have hemiacetal groups and are in equilibrium with aldehyde forms. For a similar reason, both maltose and cellobiose exhibit mutarotation of α and β anomers of the glucopyranose unit on the right.

Maltose or cellobiose   **Maltose or cellobiose**   **Maltose or cellobiose**
(β anomers)   **(aldehydes)**   **(α anomers)**

Despite the similarities of their structures, cellobiose and maltose have dramatically different biological properties. Cellobiose can't be digested by humans and can't be fermented by yeast. Maltose, however, is digested without difficulty and is fermented readily.

Problem 25.25 | Show the product you would obtain from the reaction of cellobiose with the following reagents:
(a) $NaBH_4$ (b) Br_2, H_2O (c) CH_3COCl, pyridine

Lactose

Lactose is a disaccharide that occurs naturally in both human and cow's milk. It is widely used in baking and in commercial milk formulas for infants. Like cellobiose and maltose, lactose is a reducing sugar. It exhibits mutarotation and is a 1→4-β-linked glycoside. Unlike cellobiose and maltose, however, lactose contains two different monosaccharides—D-glucose and D-galactose—joined by a β-glycosidic bond between C1 of galactose and C4 of glucose.

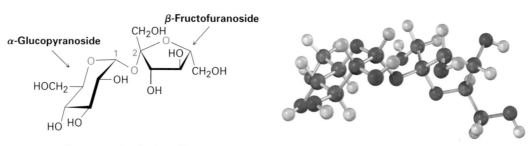

Lactose, a 1→4-β-glycoside
[4-*O*-(β-D-galactopyranosyl)-β-D-glucopyranose]

Sucrose

Sucrose, or ordinary table sugar, is among the most abundant pure organic chemicals in the world and is the one most widely known to nonchemists. Whether from sugar cane (20% by weight) or sugar beets (15% by weight) and whether raw or refined, all table sugar is sucrose.

Sucrose is a disaccharide that yields 1 equivalent of glucose and 1 equivalent of fructose on hydrolysis. This 1:1 mixture of glucose and fructose is often referred to as *invert sugar* because the sign of optical rotation inverts, or changes, during the hydrolysis from sucrose ($[\alpha]_D = +66.5$) to a glucose/fructose mixture ($[\alpha]_D = -22.0$). Insects such as honeybees have enzymes called *invertases* that catalyze the hydrolysis of sucrose to a glucose + fructose mixture. Honey, in fact, is primarily a mixture of glucose, fructose, and sucrose.

Unlike most other disaccharides, sucrose is not a reducing sugar and does not undergo mutarotation. These observations imply that sucrose is not a hemiacetal and suggest that glucose and fructose must both be glycosides. This can happen only if the two sugars are joined by a glycoside link between the anomeric carbons of both sugars—C1 of glucose and C2 of fructose.

Sucrose, a 1→2-glycoside
[2-*O*-(α-D-glucopyranosyl)-β-D-fructofuranoside]

25.9 | Polysaccharides and Their Synthesis

Polysaccharides are complex carbohydrates in which tens, hundreds, or even thousands of simple sugars are linked together through glycoside bonds. Because they have only the one free anomeric −OH group at the end of a very long chain, polysaccharides aren't reducing sugars and don't show noticeable mutarotation. Cellulose and starch are the two most widely occurring polysaccharides.

Cellulose

Cellulose consists of several thousand D-glucose units linked by 1→4-β-glyco-side bonds like those in cellobiose. Different cellulose molecules then interact to form a large aggregate structure held together by hydrogen bonds.

Cellulose, a 1→ 4-O-(β-D-glucopyranoside) polymer

Nature uses cellulose primarily as a structural material to impart strength and rigidity to plants. Leaves, grasses, and cotton, for instance, are primarily cellulose. Cellulose also serves as raw material for the manufacture of cellulose acetate, known commercially as acetate rayon, and cellulose nitrate, known as guncotton. Guncotton is the major ingredient in smokeless powder, the explosive propellant used in artillery shells and in ammunition for firearms.

Starch and Glycogen

Potatoes, corn, and cereal grains contain large amounts of *starch*, a polymer of glucose in which the monosaccharide units are linked by 1→4-α-glycoside bonds like those in maltose. Starch can be separated into two fractions: amylose, which is insoluble in cold water, and amylopectin, which *is* soluble in cold water. Amylose accounts for about 20% by weight of starch and consists of several hundred glucose molecules linked together by 1→4-α-glycoside bonds.

Amylose, a 1→ 4-O-(α-D-glucopyranoside) polymer

Amylopectin accounts for the remaining 80% of starch and is more complex in structure than amylose. Unlike cellulose and amylose, which are linear polymers, amylopectin contains 1→6-α-glycoside branches approximately every 25 glucose units.

**Amylopectin: α-(1 → 4) links
with α-(1 → 6) branches**

Starch is digested in the mouth and stomach by α-glycosidase enzymes, which catalyze the hydrolysis of glycoside bonds and release individual molecules of glucose. Like most enzymes, α-glycosidases are highly selective in their action. They hydrolyze only the α-glycoside links in starch and leave the β-glycoside links in cellulose untouched. Thus, humans can digest potatoes and grains but not grass and leaves.

Glycogen is a polysaccharide that serves the same energy storage function in animals that starch serves in plants. Dietary carbohydrates not needed for immediate energy are converted by the body to glycogen for long-term storage. Like the amylopectin found in starch, glycogen contains a complex branching structure with both 1→4 and 1→6 links (Figure 25.11). Glycogen molecules are larger than those of amylopectin—up to 100,000 glucose units—and contain even more branches.

Figure 25.11 A representation of the structure of glycogen. The hexagons represent glucose units linked by 1→4 and 1→6 glycoside bonds.

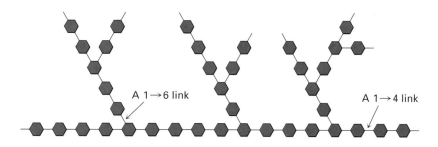

Polysaccharide Synthesis

With numerous −OH groups of similar reactivity, polysaccharides are so structurally complex that their laboratory synthesis has been a particularly difficult problem. Several methods have recently been devised, however, that have greatly simplified the problem. Among these new approaches is the *glycal assembly method,* developed by Samuel Danishefsky at Columbia University.

Easily prepared from the appropriate monosaccharide, a *glycal* is an unsaturated sugar with a C1–C2 double bond. To ready it for use in polysaccharide synthesis, the primary −OH group of the glycal is first protected at its primary −OH group by formation of a silyl ether (Section 17.8) and at its two adjacent secondary −OH groups by formation of a cyclic carbonate ester. Then, the protected glycal is epoxidized.

A glycal   **A protected glycal**   **An epoxide**

Treatment of the protected glycal epoxide in the presence of $ZnCl_2$ with a *second* glycal having a free −OH group causes acid-catalyzed opening of the epoxide ring by backside attack (Section 18.6) and yields a disaccharide. The disaccharide is itself a glycal, so it can be epoxidized and coupled again to yield a trisaccharide, and so on. Using the appropriate sugars at each step, a great variety of polysaccharides can be prepared. After the appropriate sugars are linked, the silyl ethers and cyclic carbonate protecting groups are removed by hydrolysis.

A disaccharide glycal

25.10 | Some Other Important Carbohydrates

In addition to the common carbohydrates mentioned in previous sections, there are a variety of important carbohydrate-derived materials. Their structural resemblance to sugars is clear, but they aren't simple aldoses or ketoses.

Deoxy sugars, as we saw in Section 25.7, have an oxygen atom "missing." That is, an −OH group is replaced by an −H. The most common deoxy sugar is 2-deoxyribose, a monosaccharide found in DNA (deoxyribonucleic acid). Note that 2-deoxyribose exists in water solution as a complex equilibrium mixture of both furanose and pyranose forms.

α-D-2-Deoxyribopyranose (40%)   **(0.7%)**   **α-D-2-Deoxyribofuranose (13%)**
(+ 35% β anomer)   **(+ 12% β anomer)**

Amino sugars, such as D-glucosamine, have an −OH group replaced by an −NH$_2$. The *N*-acetyl amide derived from D-glucosamine is the monosaccharide unit from which *chitin,* the hard crust that protects insects and shellfish, is made. Still other amino sugars are found in antibiotics such as streptomycin and gentamicin.

β-D-Glucosamine

**Gentamicin
(an antibiotic)**

25.11 Cell-Surface Carbohydrates and Carbohydrate Vaccines

It was once thought that carbohydrates were useful in nature only as structural materials and energy sources. Although carbohydrates do indeed serve these purposes, they have many other important biochemical functions as well. As noted in Section 25.6, for instance, glycoconjugates are centrally involved in cell–cell recognition, the critical process by which one type of cell distinguishes another. Small polysaccharide chains, covalently bound by glycosidic links to −OH or −NH$_2$ groups on proteins, act as biochemical markers on cell surfaces, as illustrated by the human blood-group antigens.

It has been known for more than a century that human blood can be classified into four blood-group types (A, B, AB, and O) and that blood from a donor of one type can't be transfused into a recipient with another type unless the two types are compatible (Table 25.1). Should an incompatible mix be made, the red blood cells clump together, or *agglutinate.*

The agglutination of incompatible red blood cells, which indicates that the body's immune system has recognized the presence of foreign cells in the body and has formed antibodies against them, results from the presence of polysaccharide markers on the surface of the cells. Types A, B, and O red blood cells

Table 25.1 | **Human Blood-Group Compatibilities**

	Acceptor blood type			
Donor blood type	A	B	AB	0
A	O	X	O	X
B	X	O	O	X
AB	X	X	O	X
O	O	O	O	O

each have their own unique markers, or *antigenic determinants;* type AB cells have both type A and type B markers. The structures of all three blood-group determinants are shown in Figure 25.12. Note that the monosaccharide constituents of each marker are among the eight essential sugars shown previously in Figure 25.9.

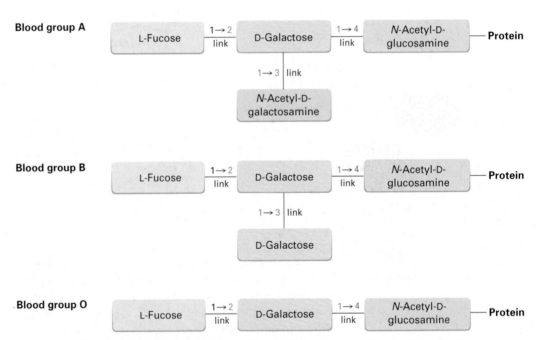

Figure 25.12 Structures of the A, B, and O blood-group antigenic determinants.

Elucidation of the role of carbohydrates in cell recognition is a vigorous area of current research that offers hope of breakthroughs in the understanding of a wide range of diseases from bacterial infections to cancer. Particularly exciting is the possibility of developing useful anticancer vaccines to help mobilize the body's immune system against tumor cells. Recent advances along these lines have included a laboratory synthesis of the so-called globo H hexasaccharide, found on the surface of human breast, prostate, colon, and pancreatic cancer cells. Preliminary studies have shown that patients treated with the synthetic globo H hexasaccharide linked to a carrier protein develop antibodies that recognize and kill tumor cells. Clinical trials against breast cancer are in progress.

Globo H hexasaccharide

Sweetness

The real thing comes from cane fields like this one.

Say the word *sugar* and most people immediately think of sweet-tasting candies, desserts, and such. In fact, most simple carbohydrates *do* taste sweet, but the degree of sweetness varies greatly from one sugar to another. With sucrose (table sugar) as a reference point, fructose is nearly twice as sweet, but lactose is only about one-sixth as sweet. Comparisons are difficult, though, because perceived sweetness varies depending on the concentration of the solution being tasted. Nevertheless, the ordering in Table 25.2 is generally accepted.

Table 25.2	Sweetness of Some Sugars and Sugar Substitutes	
Name	**Type**	**Sweetness**
Lactose	Disaccharide	0.16
Glucose	Monosaccharide	0.75
Sucrose	Disaccharide	1.00
Fructose	Monosaccharide	1.75
Aspartame	Synthetic	180
Acesulfame-K	Synthetic	200
Saccharin	Synthetic	350
Sucralose	Semisynthetic	600
Alitame	Semisynthetic	2000

(continued)

The desire of many people to cut their caloric intake has led to the development of synthetic sweeteners such as saccharin, aspartame, acesulfame, and sucralose. All are far sweeter than natural sugars, so the choice of one or another depends on personal taste, government regulations, and (for baked goods) heat stability. Saccharin, the oldest synthetic sweetener has been used for more than a century, although it has a somewhat metallic aftertaste. Doubts about its safety and potential carcinogenicity were raised in the early 1970s, but it has now been cleared of suspicion. Acesulfame potassium, one of the most recently approved sweeteners, is proving to be extremely popular in soft drinks because it has little aftertaste. Sucralose, another recently approved sweetener, is particularly useful in baked goods because of its stability at high temperatures. Alitame, not yet approved for sale in the United States but likely to be so soon, is claimed to be 2000 times as sweet as sucrose! Of the five synthetic sweeteners listed in Table 25.2, only sucralose has clear structural resemblance to a carbohydrate, but it differs dramatically in containing three chlorine atoms.

Saccharin   **Aspartame**   **Acesulfame potassium**

Sucralose   **Alitame**

SUMMARY AND KEY WORDS

Carbohydrates are polyhydroxy aldehydes and ketones. They are classified according to the number of carbon atoms and the kind of carbonyl group they contain. Glucose, for example, is an aldohexose, a six-carbon aldehydo sugar. **Monosaccharides** are further classified as either **D sugars** or **L sugars**, depending on the stereochemistry of the chirality center farthest from the carbonyl group. Carbohydrate stereochemistry is frequently depicted using **Fischer projections**, which represent a chirality center as the intersection of two crossed lines.

Monosaccharides normally exist as cyclic hemiacetals rather than as open-chain aldehydes or ketones. The hemiacetal linkage results from reaction of the carbonyl group with an −OH group three or four carbon atoms away. A five-membered cyclic hemiacetal is called a **furanose**, and a six-membered cyclic hemiacetal is called a **pyranose**. Cyclization leads to the formation of a new chirality center and production of two diastereomeric hemiacetals, called α and β **anomers**.

Much of the chemistry of monosaccharides is the familiar chemistry of alcohols and aldehydes/ketones. Thus, the hydroxyl groups of carbohydrates form esters and ethers. The carbonyl group of a monosaccharide can be reduced with $NaBH_4$ to form an **alditol**, oxidized with aqueous Br_2 to form an **aldonic acid**, oxidized with HNO_3 to form an **aldaric acid**, oxidized enzymatically to form a **uronic acid**, or treated with an alcohol in the presence of acid to form a **glycoside**. Monosaccharides can also be chain-lengthened by the multistep **Kiliani–Fischer synthesis** and can be chain-shortened by the **Wohl degradation**.

Disaccharides are complex carbohydrates in which simple sugars are linked by a glycoside bond between the **anomeric center** of one unit and a hydroxyl of the second unit. The sugars can be the same, as in maltose and cellobiose, or different, as in lactose and sucrose. The glycosidic bond can be either α (maltose) or β (cellobiose, lactose) and can involve any hydroxyl of the second sugar. A 1→4 link is most common (cellobiose, maltose), but others such as 1→2 (sucrose) are also known. **Polysaccharides**, such as cellulose, starch, and glycogen, are used in nature as structural materials, as a means of long-term energy storage, and as cell-surface markers.

SUMMARY OF REACTIONS

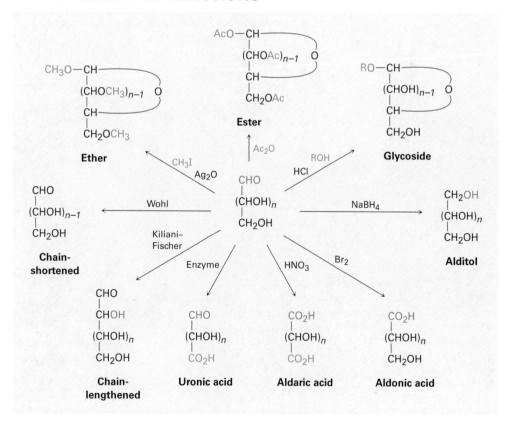

EXERCISES

Organic KNOWLEDGE TOOLS

CENGAGENOW˙ Sign in at **www.cengage.com/login** to assess your knowledge of this chapter's topics by taking a pre-test. The pre-test will link you to interactive organic chemistry resources based on your score in each concept area.

⬛WL Online homework for this chapter may be assigned in Organic OWL.

⬛ indicates problems assignable in Organic OWL.

VISUALIZING CHEMISTRY

(Problems 25.1–25.25 appear within the chapter.)

25.26 ⬛ Identify the following aldoses, and tell whether each is a D or L sugar:

(a) (b)

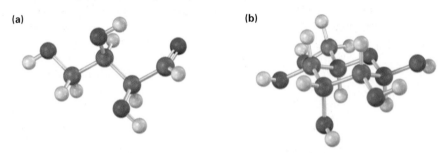

25.27 Draw Fischer projections of the following molecules, placing the carbonyl group at the top in the usual way. Identify each as a D or L sugar.

(a) (b)

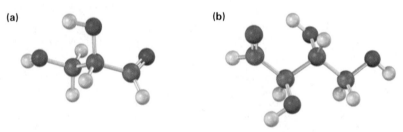

25.28 ⬛ The following structure is that of an L aldohexose in its pyranose form. Identify it, and tell whether it is an α or β anomer.

⬛ Assignable in OWL

25.29 ■ The following model is that of an aldohexose:

(a) Draw Fischer projections of the sugar, its enantiomer, and a diastereomer.
(b) Is this a D sugar or an L sugar? Explain.
(c) Draw the β anomer of the sugar in its furanose form.

ADDITIONAL PROBLEMS

25.30 ■ Classify each of the following sugars. (For example, glucose is an aldohexose.)

(a)
```
CH2OH
|
C=O
|
CH2OH
```

(b)
```
    CH2OH
H——OH
    C=O
H——OH
    CH2OH
```

(c)
```
     CHO
 H———OH
HO———H
 H———OH
HO———H
 H———OH
     CH2OH
```

25.31 Write open-chain structures for the following:
(a) A ketotetrose   (b) A ketopentose
(c) A deoxyaldohexose   (d) A five-carbon amino sugar

25.32 Does ascorbic acid (vitamin C) have a D or L configuration?

```
            OH
            |
HO     C====C
   C          C=O
H———O
HO———H
    CH2OH
```
Ascorbic acid

25.33 Draw the three-dimensional furanose form of ascorbic acid (Problem 25.32), and assign R or S stereochemistry to each chirality center.

25.34 ■ Assign R or S configuration to each chirality center in the following molecules:

(a)
```
      H
H3C———Br
Br———H
     CH3
```

(b)
```
      [benzene ring]
H3C———OH
H3C———H
     OH
```

(c)
```
      NH2
H———CO2H
H———OH
H———H
   [benzene ring]
```

25.35 Draw Fischer projections of the following molecules:
  (a) The *S* enantiomer of 2-bromobutane
  (b) The *R* enantiomer of alanine, $CH_3CH(NH_2)COOH$
  (c) The *R* enantiomer of 2-hydroxypropanoic acid
  (d) The *S* enantiomer of 3-methylhexane

25.36 Draw Fischer projections for the two D aldoheptoses whose stereochemistry at C3, C4, C5, and C6 is the same as that of D-glucose at C2, C3, C4, and C5.

25.37 ■ The following cyclic structure is that of allose. Is this a furanose or pyranose form? Is it an α or β anomer? Is it a D or L sugar?

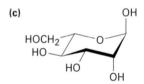

25.38 ■ What is the complete name of the following sugar?

25.39 ■ Write the following sugars in their open-chain forms:

(a)   (b)   (c)

25.40 ■ Draw D-ribulose in its five-membered cyclic β-hemiacetal form.

$$CH_2OH$$
$$C=O$$
H——OH   **Ribulose**
H——OH
$$CH_2OH$$

25.41 Look up the structure of D-talose in Figure 25.3, and draw the β anomer in its pyranose form. Identify the ring substituents as axial or equatorial.

25.42 ■ Draw structures for the products you would expect to obtain from reaction of β-D-talopyranose with each of the following reagents:
  (a) $NaBH_4$ in H_2O (b) Warm dilute HNO_3 (c) Br_2, H_2O
  (d) CH_3CH_2OH, HCl   (e) CH_3I, Ag_2O (f) $(CH_3CO)_2O$, pyridine

25.43 What is the stereochemical relationship of D-ribose to L-xylose? What generalizations can you make about the following properties of the two sugars?
  (a) Melting point   (b) Solubility in water
  (c) Specific rotation   (d) Density

■ Assignable in OWL

25.44 All aldoses exhibit mutarotation. For example, α-D-galactopyranose has $[\alpha]_D = +150.7$, and β-D-galactopyranose has $[\alpha]_D = +52.8$. If either anomer is dissolved in water and allowed to reach equilibrium, the specific rotation of the solution is +80.2. What are the percentages of each anomer at equilibrium? Draw the pyranose forms of both anomers.

25.45 How many D-2-ketohexoses are possible? Draw them.

25.46 One of the D-2-ketohexoses is called *sorbose*. On treatment with $NaBH_4$, sorbose yields a mixture of gulitol and iditol. What is the structure of sorbose?

25.47 Another D-2-ketohexose, *psicose,* yields a mixture of allitol and altritol when reduced with $NaBH_4$. What is the structure of psicose?

25.48 L-Gulose can be prepared from D-glucose by a route that begins with oxidation to D-glucaric acid, which cyclizes to form two six-membered-ring lactones. Separating the lactones and reducing them with sodium amalgam gives D-glucose and L-gulose. What are the structures of the two lactones, and which one is reduced to L-gulose?

25.49 ■ What other D aldohexose gives the same alditol as D-talose?

25.50 Which of the eight D aldohexoses give the same aldaric acids as their L enantiomers?

25.51 Which of the other three D aldopentoses gives the same aldaric acid as D-lyxose?

25.52 Draw the structure of L-galactose, and then answer the following questions:
(a) Which other aldohexose gives the same aldaric acid as L-galactose on oxidation with warm HNO_3?
(b) Is this other aldohexose a D sugar or an L sugar?
(c) Draw this other aldohexose in its most stable pyranose conformation.

25.53 Galactose, one of the eight essential monosaccharides (Section 25.7), is biosynthesized from UDP-glucose by galactose 4-epimerase, where UDP = uridylyl diphosphate (a ribonucleotide diphosphate; Section 28.1). The enzyme requires NAD^+ for activity (Section 17.7), but it is not a stoichiometric reactant, and NADH is not a final reaction product. Propose a mechanism.

UDP-Glucose   **UDP-Galactose**

25.54 Mannose, one of the eight essential monosaccharides (Section 25.7), is biosynthesized as its 6-phosphate derivative from fructose 6-phosphate. No enzyme cofactor is required. Propose a mechanism.

Fructose   **Mannose**
6-phosphate   **6-phosphate**

25.55 Glucosamine, one of the eight essential monosaccharides (Section 25.7), is biosynthesized as its 6-phosphate derivative from fructose 6-phosphate by reaction with ammonia. Propose a mechanism.

Fructose 6-phosphate

Glucosamine 6-phosphate

25.56 ■ Gentiobiose, a rare disaccharide found in saffron and gentian, is a reducing sugar and forms only D-glucose on hydrolysis with aqueous acid. Reaction of gentiobiose with iodomethane and Ag₂O yields an octamethyl derivative, which can be hydrolyzed with aqueous acid to give 1 equivalent of 2,3,4,6-tetra-O-methyl-D-glucopyranose and 1 equivalent of 2,3,4-tri-O-methyl-D-glucopyranose. If gentiobiose contains a β-glycoside link, what is its structure?

25.57 Amygdalin, or laetrile, is a cyanogenic glycoside isolated in 1830 from almond and apricot seeds. Acidic hydrolysis of amygdalin liberates HCN, along with benzaldehyde and 2 equivalents of D-glucose. If amygdalin is a β-glycoside of benzaldehyde cyanohydrin with gentiobiose (Problem 21.56), what is its structure?

25.58 Trehalose is a nonreducing disaccharide that is hydrolyzed by aqueous acid to yield 2 equivalents of D-glucose. Methylation followed by hydrolysis yields 2 equivalents of 2,3,4,6-tetra-O-methylglucose. How many structures are possible for trehalose?

25.59 Trehalose (Problem 25.58) is cleaved by enzymes that hydrolyze α-glycosides but not by enzymes that hydrolyze β-glycosides. What is the structure and systematic name of trehalose?

25.60 Isotrehalose and neotrehalose are chemically similar to trehalose (Problems 25.58 and 25.59) except that neotrehalose is hydrolyzed only by β-glycosidase enzymes, whereas isotrehalose is hydrolyzed by both α- and β-glycosidase enzymes. What are the structures of isotrehalose and neotrehalose?

25.61 D-Glucose reacts with acetone in the presence of acid to yield the nonreducing 1,2:5,6-diisopropylidene-D-glucofuranose. Propose a mechanism.

1,2:5,6-Diisopropylidene-D-glucofuranose

25.62 D-Mannose reacts with acetone to give a diisopropylidene derivative (Problem 25.61) that is still reducing toward Tollens' reagent. Propose a likely structure for this derivative.

25.63 ■ Glucose and mannose can be interconverted (in low yield) by treatment with dilute aqueous NaOH. Propose a mechanism.

25.64 Propose a mechanism to account for the fact that D-gluconic acid and D-mannonic acid are interconverted when either is heated in pyridine solvent.

25.65 The *cyclitols* are a group of carbocyclic sugar derivatives having the general formulation 1,2,3,4,5,6-cyclohexanehexol. How many stereoisomeric cyclitols are possible? Draw them in their chair forms.

25.66 Compound **A** is a D aldopentose that can be oxidized to an optically inactive aldaric acid **B**. On Kiliani–Fischer chain extension, **A** is converted into **C** and **D**; **C** can be oxidized to an optically active aldaric acid **E**, but **D** is oxidized to an optically inactive aldaric acid **F**. What are the structures of **A**–**F**?

25.67 Simple sugars undergo reaction with phenylhydrazine, $PhNHNH_2$, to yield crystalline derivatives called *osazones*. The reaction is a bit complex, however, as shown by the fact that glucose and fructose yield the same osazone.

D-Glucose $+$ NH_3 $+$ $PhNH_2$
$+$ $2 H_2O$

D-Fructose

(a) Draw the structure of a third sugar that yields the same osazone as glucose and fructose.

(b) Using glucose as the example, the first step in osazone formation is reaction of the sugar with phenylhydrazine to yield an imine called a *phenylhydrazone*. Draw the structure of the product.

(c) The second and third steps in osazone formation are tautomerization of the phenylhydrazone to give an enol, followed by elimination of aniline to give a keto imine. Draw the structures of both the enol tautomer and the keto imine.

(d) The final step is reaction of the keto imine with 2 equivalents of phenylhydrazine to yield the osazone plus ammonia. Propose a mechanism for this step.

25.68 When heated to 100 °C, D-idose undergoes a reversible loss of water and exists primarily as 1,6-anhydro-D-idopyranose.

D-Idose   **1,6-Anhydro-D-idopyranose**

(a) Draw D-idose in its pyranose form, showing the more stable chair conformation of the ring.
(b) Which is more stable, α-D-idopyranose or β-D-idopyranose? Explain.
(c) Draw 1,6-anhydro-D-idopyranose in its most stable conformation.
(d) When heated to 100 °C under the same conditions as those used for D-idose, D-glucose does not lose water and does not exist in a 1,6-anhydro form. Explain.

25.69 Acetyl coenzyme A (acetyl CoA) is the key intermediate in food metabolism. What sugar is present in acetyl CoA?

Acetyl coenzyme A

25.70 ■ One of the steps in the biological pathway for carbohydrate metabolism is the conversion of fructose 1,6-bisphosphate into dihydroxyacetone phosphate and glyceraldehyde 3-phosphate. Propose a mechanism for the transformation.

Fructose 1,6-bisphosphate   **Dihydroxyacetone phosphate**   **Glyceraldehyde 3-phosphate**

25.71 L-Fucose, one of the eight essential monosaccharides (Section 25.7), is bio-synthesized from GDP-D-mannose by the following three-step reaction sequence, where GDP = guanosine diphosphate (a ribonucleoside diphosphate; Section 28.1):

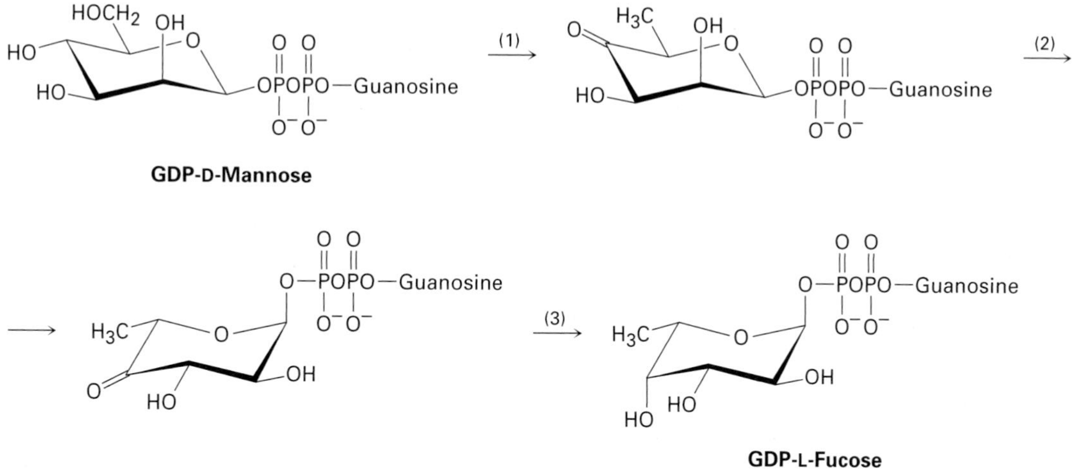

GDP-D-Mannose

GDP-L-Fucose

(a) Step 1 involves an oxidation to a ketone, a dehydration to an enone, and a conjugate reduction. The step requires $NADP^+$, but no NADPH is formed as a final reaction product. Propose a mechanism.

(b) Step 2 accomplishes two epimerizations and utilizes acidic and basic sites in the enzyme but does not require a coenzyme. Propose a mechanism.

(c) Step 3 requires NADPH as coenzyme. Show the mechanism.

26

Biomolecules: Amino Acids, Peptides, and Proteins

Organic KNOWLEDGE TOOLS

CENGAGENOW™ Throughout this chapter, sign in at **www.cengage.com/login** for online self-study and interactive tutorials based on your level of understanding.

ⓌWL Online homework for this chapter may be assigned in Organic OWL.

Proteins occur in every living organism, are of many different types, and have many different biological functions. The keratin of skin and fingernails, the fibroin of silk and spider webs, and the estimated 50,000 to 70,000 enzymes that catalyze the biological reactions in our bodies are all proteins. Regardless of their function, all proteins are made up of many *amino acids* linked together in a long chain.

Amino acids, as their name implies, are difunctional. They contain both a basic amino group and an acidic carboxyl group.

Alanine, an amino acid

Their value as building blocks to make proteins stems from the fact that amino acids can join together into long chains by forming amide bonds between the $-NH_2$ of one amino acid and the $-CO_2H$ of another. For classification purposes, chains with fewer than 50 amino acids are often called **peptides**, while the term **protein** is used for larger chains.

Amide bonds

Many

Sean Duggan

WHY THIS CHAPTER?

Continuing our look at the four main classes of biomolecules, we'll focus in this chapter on amino acids, the fundamental building blocks from which the 100,000 or so proteins in our bodies are made. We'll then see how amino acids are incorporated into proteins and the structures of those proteins. Any understanding of biological chemistry would be impossible without this study.

26.1 | Structures of Amino Acids

We saw in Sections 20.3 and 24.5 that a carboxyl group is deprotonated and exists as the carboxylate anion at a physiological pH of 7.3, while an amino group is protonated and exists as the ammonium cation. Thus, amino acids exist in aqueous solution primarily in the form of a dipolar ion, or **zwitterion** (German *zwitter*, meaning "hybrid").

(uncharged)   **(zwitterion)**

Alanine

Amino acid zwitterions are internal salts and therefore have many of the physical properties associated with salts. They have large dipole moments, are soluble in water but insoluble in hydrocarbons, and are crystalline substances with relatively high melting points. In addition, amino acids are *amphiprotic*; they can react either as acids or as bases, depending on the circumstances. In aqueous acid solution, an amino acid zwitterion is a base that *accepts* a proton to yield a cation; in aqueous base solution, the zwitterion is an acid that *loses* a proton to form an anion. Note that it is the carboxylate, $-CO_2^-$, that acts as the basic site and accepts a proton in acid solution, and it is the ammonium cation, $-NH_3^+$, that acts as the acidic site and donates a proton in base solution.

The structures, abbreviations (both three- and one-letter), and pK_a values of the 20 amino acids commonly found in proteins are shown in Table 26.1. All are

Table 26.1 | **The 20 Common Amino Acids in Proteins**

Name	Abbreviations		MW	Stucture	pK_a α-CO_2H	pK_a α-NH_3^+	pK_a side chain	pI
Neutral Amino Acids								
Alanine	Ala	A	89		2.34	9.69	—	6.01
Asparagine	Asn	N	132		2.02	8.80	—	5.41
Cysteine	Cys	C	121		1.96	10.28	8.18	5.07
Glutamine	Gln	Q	146		2.17	9.13	—	5.65
Glycine	Gly	G	75		2.34	9.60	—	5.97
Isoleucine	Ile	I	131		2.36	9.60	—	6.02
Leucine	Leu	L	131		2.36	9.60	—	5.98
Methionine	Met	M	149		2.28	9.21	—	5.74
Phenylalanine	Phe	F	165		1.83	9.13	—	5.48
Proline	Pro	P	115		1.99	10.60	—	6.30

(continued)

Table 26.1 | **The 20 Common Amino Acids in Proteins** *(continued)*

Name	Abbreviations		MW	Stucture	pK_a α-CO_2H	pK_a α-NH_3^+	pK_a side chain	pI
Neutral Amino Acids *continued*								
Serine	Ser	S	105		2.21	9.15	—	5.68
Threonine	Thr	T	119		2.09	9.10	—	5.60
Tryptophan	Trp	W	204		2.83	9.39	—	5.89
Tyrosine	Tyr	Y	181		2.20	9.11	10.07	5.66
Valine	Val	V	117		2.32	9.62	—	5.96
Acidic Amino Acids								
Aspartic acid	Asp	D	133		1.88	9.60	3.65	2.77
Glutamic acid	Glu	E	147		2.19	9.67	4.25	3.22
Basic Amino Acids								
Arginine	Arg	R	174		2.17	9.04	12.48	10.76
Histidine	His	H	155		1.82	9.17	6.00	7.59
Lysine	Lys	K	146		2.18	8.95	10.53	9.74

α-amino acids, meaning that the amino group in each is a substituent on the α carbon atom—the one next to the carbonyl group. Nineteen of the twenty amino acids are primary amines, RNH_2, and differ only in the nature of the substituent attached to the α carbon, called the **side chain**. Proline is a secondary amine and the only amino acid whose nitrogen and α carbon atoms are part of a ring.

A primary α-amino acid   **Proline, a secondary α-amino acid**

In addition to the twenty amino acids commonly found in proteins, two others—selenocysteine and pyrrolysine—are found in some organisms, and more than 700 nonprotein amino acids are also found in nature. γ-Aminobutyric acid (GABA), for instance, is found in the brain and acts as a neurotransmitter; homocysteine is found in blood and is linked to coronary heart disease; and thyroxine is found in the thyroid gland, where it acts as a hormone.

Selenocysteine   **Pyrrolysine**

γ-Aminobutyric acid   **Homocysteine**   **Thyroxine**

Except for glycine, $H_2NCH_2CO_2H$, the α carbons of amino acids are chirality centers. Two enantiomers of each are therefore possible, but nature uses only one to build proteins. In Fischer projections, naturally occurring amino acids are represented by placing the $-CO_2^-$ group at the top and the side chain down, as if drawing a carbohydrate (Section 25.2) and then placing the $-NH_3^+$ group on the left. Because of their stereochemical similarity to

L sugars (Section 25.3), the naturally occurring α-amino acids are often referred to as L amino acids.

L-Alanine   **L-Serine**   **L-Cysteine**   **L-Glyceraldehyde**
(S)-Alanine   **(S)-Serine**   **(R)-Cysteine**

The 20 common amino acids can be further classified as neutral, acidic, or basic, depending on the structure of their side chains. Fifteen of the twenty have neutral side chains, two (aspartic acid and glutamic acid) have an extra carboxylic acid function in their side chains, and three (lysine, arginine, and histidine) have basic amino groups in their side chains. Note that both cysteine (a thiol) and tyrosine (a phenol), although usually classified as neutral amino acids, nevertheless have weakly acidic side chains that can be deprotonated in strongly basic solution.

At the physiological pH of 7.3 within cells, the side-chain carboxyl groups of aspartic acid and glutamic acid are deprotonated and the basic side-chain nitrogens of lysine and arginine are protonated. Histidine, however, which contains a heterocyclic imidazole ring in its side chain, is not quite basic enough to be protonated at pH 7.3. Note that only the pyridine-like, doubly bonded nitrogen in histidine is basic. The pyrrole-like singly bonded nitrogen is nonbasic because its lone pair of electrons is part of the 6 π electron aromatic imidazole ring (Section 24.9).

Histidine

Humans are able to synthesize only 11 of the 20 amino acids in proteins, called *nonessential amino acids*. The other 9, called *essential amino acids,* are biosynthesized only in plants and microorganisms and must be obtained in our diet. The division between essential and nonessential amino acids is not clearcut, however: tyrosine, for instance, is sometimes considered nonessential because humans can produce it from phenylalanine, but phenylalanine itself is essential and must be obtained in the diet. Arginine can be synthesized by humans, but much of the arginine we need also comes from our diet.

Problem 26.1 | How many of the α-amino acids shown in Table 26.1 contain aromatic rings? How many contain sulfur? How many contain alcohols? How many contain hydrocarbon side chains?

Problem 26.2 | Of the 19 L amino acids, 18 have the *S* configuration at the α carbon. Cysteine is the only L amino acid that has an *R* configuration. Explain.

Problem 26.3 | The amino acid threonine, (2*S*,3*R*)-2-amino-3-hydroxybutanoic acid, has two chirality centers.
(a) Draw a Fischer projection of threonine.
(b) Draw a Fischer projection of a threonine diastereomer, and label its chirality centers as *R* or *S*.

26.2 | Amino Acids, the Henderson–Hasselbalch Equation, and Isoelectric Points

CENGAGENOW™ Click *Organic Interactive* to **learn to estimate isoelectric points for simple amino acids and peptides.**

According to the Henderson–Hasselbalch equation (Sections 20.3 and 24.5), if we know both the pH of a solution and the pK_a of an acid HA, we can calculate the ratio of $[A^-]$ to $[HA]$ in the solution. Furthermore, when $pH = pK_a$, the two forms A^- and HA are present in equal amounts because $\log 1 = 0$.

$$pH = pK_a + \log\frac{[A^-]}{[HA]} \quad \text{or} \quad \log\frac{[A^-]}{[HA]} = pH - pK_a$$

To apply the Henderson–Hasselbalch equation to an amino acid, let's find out what species are present in a 1.00 M solution of alanine at pH = 9.00. According to Table 26.1, protonated alanine $[^+H_3NCH(CH_3)CO_2H]$ has $pK_{a1} = 2.34$, and neutral zwitterionic alanine $[^+H_3NCH(CH_3)CO_2^-]$ has $pK_{a2} = 9.69$:

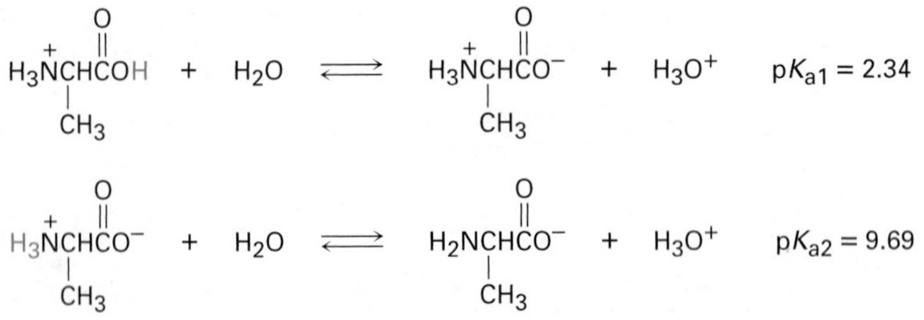

Since the pH of the solution is much closer to pK_{a2} than to pK_{a1}, we need to use pK_{a2} for the calculation. From the Henderson–Hasselbalch equation, we have:

$$\log\frac{[A^-]}{[HA]} = pH - pK_a = 9.00 - 9.69 = -0.69$$

so

$$\frac{[A^-]}{[HA]} = \text{antilog}(-0.69) = 0.20 \quad \text{and} \quad [A^-] = 0.20\,[HA]$$

In addition, we know that

$$[A^-] + [HA] = 1.00 \text{ M}$$

Solving the two simultaneous equations gives [HA] = 0.83 and [A$^-$] = 0.17. In other words, at pH = 9.00, 83% of alanine molecules in a 1.00 M solution are neutral (zwitterionic) and 17% are deprotonated. Similar calculations can be done at any other pH and the results plotted to give the *titration curve* shown in Figure 26.1.

Each leg of the titration curve is calculated separately. The first leg, from pH 1 to 6, corresponds to the dissociation of protonated alanine, H_2A^+. The second leg, from pH 6 to 11, corresponds to the dissociation of zwitterionic alanine, HA. It's as if we started with H_2A^+ at low pH and then titrated with NaOH. When 0.5 equivalent of NaOH is added, the deprotonation of H_2A^+ is 50% done; when 1.0 equivalent of NaOH is added, the deprotonation of H_2A^+ is complete and HA predominates; when 1.5 equivalent of NaOH is added, the deprotonation of HA is 50% done; and when 2.0 equivalents of NaOH is added, the deprotonation of HA is complete.

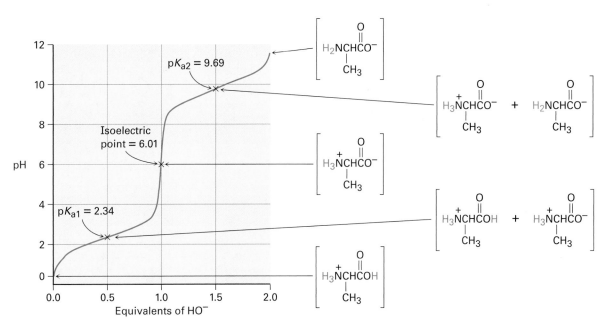

Figure 26.1 A titration curve for alanine, plotted using the Henderson–Hasselbalch equation. Each of the two legs is plotted separately. At pH < 1, alanine is entirely protonated; at pH = 2.34, alanine is a 50:50 mix of protonated and neutral forms; at pH 6.01, alanine is entirely neutral; at pH = 9.69, alanine is a 50:50 mix of neutral and deprotonated forms; at pH > 11.5, alanine is entirely deprotonated.

Look carefully at the titration curve in Figure 26.1. In acid solution, the amino acid is protonated and exists primarily as a cation. In basic solution, the amino acid is deprotonated and exists primarily as an anion. In between the two is an intermediate pH at which the amino acid is exactly balanced between anionic and cationic forms and exists primarily as the neutral, dipolar zwitterion. This pH is called the amino acid's **isoelectric point (pI)** and has a value of 6.01 for alanine.

The isoelectric point of an amino acid depends on its structure, with values for the 20 common amino acids given in Table 26.1. The 15 neutral amino acids have isoelectric points near neutrality, in the pH range 5.0 to 6.5. The two acidic amino acids have isoelectric points at lower pH so that deprotonation of the side-chain $-CO_2H$ does not occur at their pI, and the three basic amino acids have isoelectric points at higher pH so that protonation of the side-chain amino group does not occur at their pI.

More specifically, the pI of any amino acid is the average of the two acid-dissociation constants that involve the neutral zwitterion. For the 13 amino acids with a neutral side chain, pI is the average of pK_{a1} and pK_{a2}. For the four amino acids with either a strongly or weakly acidic side chain, pI is the average of the two *lowest* pK_a values. For the three amino acids with a basic side chain, pI is the average of the two *highest* pK_a values.

Just as individual amino acids have isoelectric points, proteins have an overall pI because of the acidic or basic amino acids they may contain. The enzyme lysozyme, for instance, has a preponderance of basic amino acids and thus has a high isoelectric point (pI = 11.0). Pepsin, however, has a preponderance of acidic amino acids and a low isoelectric point (pI ~ 1.0). Not surprisingly, the solubilities and properties of proteins with different pI's are strongly affected by the pH of the medium. Solubility is usually lowest at the isoelectric point, where the protein has no net charge, and is higher both above and below the pI, where the protein is charged.

We can take advantage of the differences in isoelectric points to separate a mixture of proteins into its pure constituents. Using a technique known as *electrophoresis,* a mixture of proteins is placed near the center of a strip of paper or gel. The paper or gel is moistened with an aqueous buffer of a given pH, and electrodes are connected to the ends of the strip. When an electric potential is applied, those proteins with negative charges (those that are deprotonated because the pH of the buffer is above their isoelectric point) migrate slowly toward the positive electrode. At the same time, those amino acids with positive charges (those that are protonated because the pH of the buffer is below their isoelectric point) migrate toward the negative electrode.

Different proteins migrate at different rates, depending on their isoelectric points and on the pH of the aqueous buffer, thereby separating the mixture into its pure components. Figure 26.2 illustrates this separation for a mixture containing basic, neutral, and acidic components.

Figure 26.2 Separation of a protein mixture by electrophoresis. At pH = 6.00, a neutral protein does not migrate, a basic protein is protonated and migrates toward the negative electrode, and an acidic protein is deprotonated and migrates toward the positive electrode.

Strip buffered to pH = 6.00

Basic	**Neutral**	**Acidic**
pI = 7.50	pI = 6.00	pI = 4.50

$-$ ←   → $+$

Problem 26.4   Hemoglobin has pI = 6.8. Does hemoglobin have a net negative charge or net positive charge at pH = 5.3? At pH = 7.3?

26.3 | Synthesis of Amino Acids

α-Amino acids can be synthesized in the laboratory using some of the reactions discussed in previous chapters. One of the oldest methods of α-amino acid synthesis begins with α bromination of a carboxylic acid by treatment with Br_2 and PBr_3 (the Hell–Volhard–Zelinskii reaction; Section 22.4). S_N2 substitution of the α-bromo acid with ammonia then yields an α-amino acid.

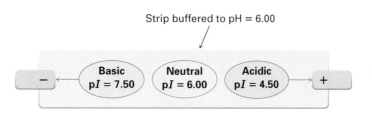

4-Methylpentanoic acid   1. Br_2, PBr_3 2. H_2O → **2-Bromo-4-methylpentanoic acid** $\xrightarrow[\text{(excess)}]{NH_3}$ **(R,S)-Leucine (45%)**

Problem 26.5   Show how you could prepare the following α-amino acids from the appropriate carboxylic acids:
(a) Phenylalanine   (b) Valine

The Amidomalonate Synthesis

A more general method for preparation of α-amino acids is the *amidomalonate synthesis,* a straightforward extension of the malonic ester synthesis (Section 22.7). The reaction begins with conversion of diethyl acetamidomalonate into an enolate ion by treatment with base, followed by S_N2 alkylation with a primary alkyl halide. Hydrolysis of both the amide protecting group and the esters occurs when the alkylated product is warmed with aqueous acid, and decarboxylation then takes place to yield an α-amino acid. For example, aspartic acid can be prepared from ethyl bromoacetate, $BrCH_2CO_2Et$:

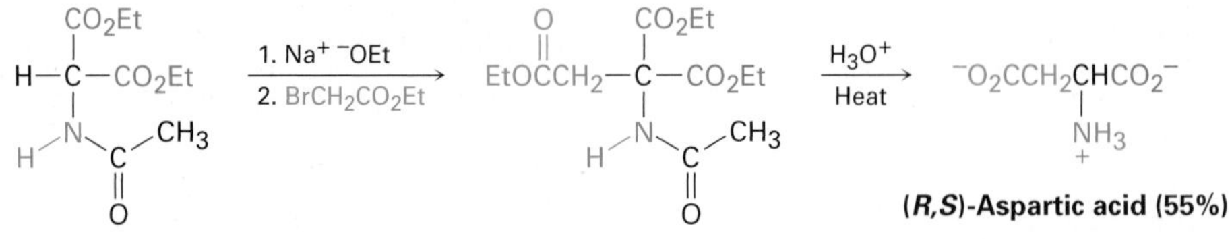

Diethyl acetamidomalonate

(*R,S*)-Aspartic acid (55%)

Problem 26.6 | What alkyl halides would you use to prepare the following α-amino acids by the amidomalonate method?
(a) Leucine   (b) Histidine   (c) Tryptophan   (d) Methionine

Reductive Amination of α-Keto Acids

Yet a third method for the synthesis of α-amino acids is by reductive amination of an α-keto acid with ammonia and a reducing agent. Alanine, for instance, is prepared by treatment of pyruvic acid with ammonia in the presence of $NaBH_4$. As described in Section 24.6, the reaction proceeds through formation of an intermediate imine that is then reduced.

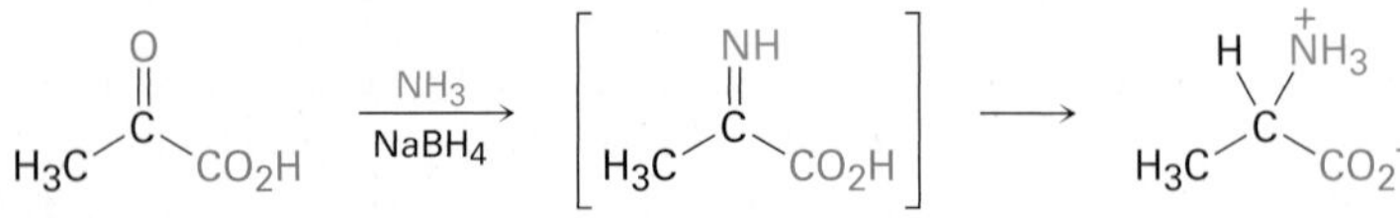

Pyruvic acid

Imine intermediate

(*R,S*)-Alanine

Enantioselective Synthesis

The synthesis of an α-amino acid from an achiral precursor by any of the methods described in the previous section yields a racemic mixture, with equal amounts of *S* and *R* enantiomers. To use an amino acid in the laboratory synthesis of a naturally occurring protein, however, the pure *S* enantiomer must be obtained.

Two methods are used in practice to obtain enantiomerically pure amino acids. One way is to resolve the racemic mixture into its pure enantiomers (Section 9.8). A more direct approach, however, is to use an *enantioselective synthesis* to prepare only the desired *S* enantiomer directly. As discussed in the Chapter 19 *Focus On,* the idea behind enantioselective synthesis is to find a chiral reaction catalyst that will temporarily hold a substrate molecule in an unsymmetrical environment. While in that chiral environment, the substrate may be more

William S. Knowles

William S. Knowles (1917–) was born in Taunton, Massachusetts, and received his Ph.D. from Columbia University in 1942. Following his graduate studies, he began work at the Monsanto Company in St. Louis, Missouri, where he remained until his retirement in 1986. He received the 2001 Nobel Prize in chemistry for his work on enantioselective synthesis, one of the few nonacademic scientists to be thus honored.

open to reaction on one side than on another, leading to an excess of one enantiomeric product over another.

William Knowles at the Monsanto Company discovered some years ago that α-amino acids can be prepared enantioselectively by hydrogenation of a Z enamido acid with a chiral hydrogenation catalyst. (S)-Phenylalanine, for instance, is prepared in 98.7% purity contaminated by only 1.3% of the (R) enantiomer when a chiral rhodium catalyst is used. For this discovery, Knowles shared the 2001 Nobel Prize in chemistry.

A (Z) enamido acid   **(S)-Phenylalanine**

The most effective catalysts for enantioselective amino acid synthesis are coordination complexes of rhodium(I) with 1,5-cyclooctadiene (COD) and a chiral diphosphine such as (R,R)-1,2-bis(o-anisylphenylphosphino)ethane, the so-called DiPAMP ligand. The complex owes its chirality to the presence of the trisubstituted phosphorus atoms (Section 9.12).

[Rh(R,R-DiPAMP)(COD)]⁺ BF₄⁻

Problem 26.7 | Show how you could prepare the following amino acid enantioselectively:

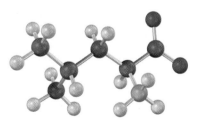

26.4 | Peptides and Proteins

Proteins and peptides are amino acid polymers in which the individual amino acids, called **residues**, are linked together by amide bonds, or *peptide bonds*. An amino group from one residue forms an amide bond with the carboxyl of a second residue, the amino group of the second forms an amide bond with the carboxyl of a third, and so on. For example, alanylserine is the dipeptide that results when an amide bond is formed between the alanine carboxyl and the serine amino group.

Alanine (Ala)

+

Serine (Ser)

⟶

Alanylserine (Ala-Ser)

Note that two dipeptides can result from reaction between alanine and serine, depending on which carboxyl group reacts with which amino group. If the alanine amino group reacts with the serine carboxyl, serylalanine results.

Serine (Ser)

+

Alanine (Ala)

⟶

Serylalanine (Ser-Ala)

The long, repetitive sequence of $-N-CH-CO-$ atoms that make up a continuous chain is called the protein's **backbone**. By convention, peptides are written with the **N-terminal amino acid** (the one with the free $-NH_3^+$ group) on the left and the **C-terminal amino acid** (the one with the free $-CO_2^-$ group) on the right. The name of the peptide is indicated by using the abbreviations listed in Table 26.1 for each amino acid. Thus, alanylserine is abbreviated Ala-Ser or A-S, and serylalanine is abbreviated Ser-Ala or S-A. Needless to say, the one-letter abbreviations are more convenient than the older three-letter abbreviations.

The amide bond that links different amino acids together in peptides is no different from any other amide bond (Section 24.3). Amide nitrogens are nonbasic because their unshared electron pair is delocalized by interaction with the carbonyl group. This overlap of the nitrogen p orbital with the p orbitals of the carbonyl group imparts a certain amount of double-bond character to the

C−N bond and restricts rotation around it. The amide bond is therefore planar, and the N−H is oriented 180° to the C=O.

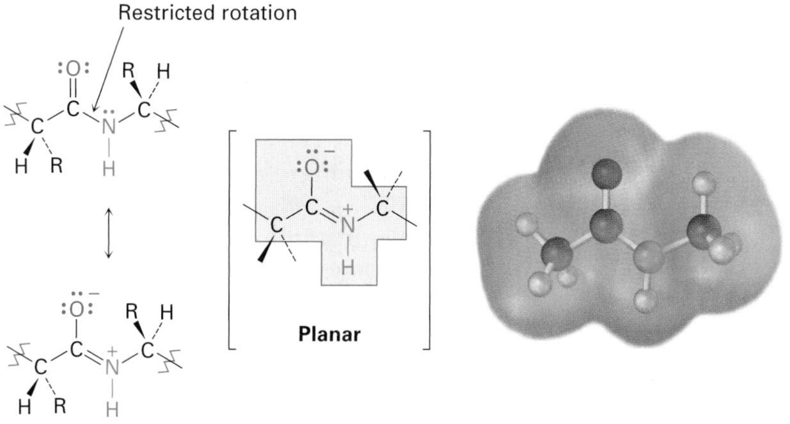

Restricted rotation

Planar

A second kind of covalent bonding in peptides occurs when a disulfide linkage, RS−SR, is formed between two cysteine residues. As we saw in Section 18.8, a disulfide is formed by mild oxidation of a thiol, RSH, and is cleaved by mild reduction.

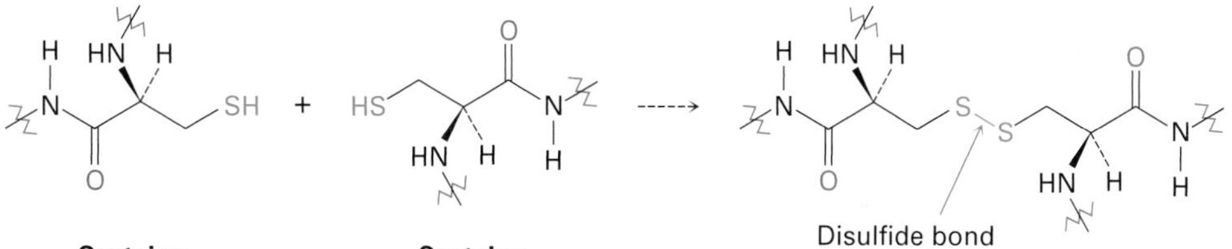

Cysteine   Cysteine

Disulfide bond

A disulfide bond between cysteine residues in different peptide chains links the otherwise separate chains together, while a disulfide bond between cysteine residues in the same chain forms a loop. Such is the case, for instance, with vasopressin, an antidiuretic hormone found in the pituitary gland. Note that the C-terminal end of vasopressin occurs as the primary amide, −CONH$_2$, rather than as the free acid.

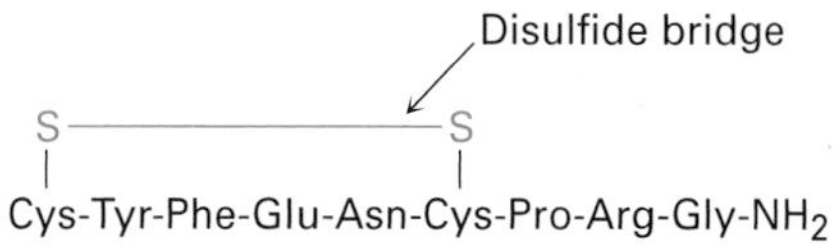

Disulfide bridge

S————————S
| |
Cys-Tyr-Phe-Glu-Asn-Cys-Pro-Arg-Gly-NH$_2$

Vasopressin

Problem 26.8 Six isomeric tripeptides contain valine, tyrosine, and glycine. Name them using both three- and one-letter abbreviations.

Problem 26.9 Draw the structure of M-P-V-G, and indicate the amide bonds.

26.5 | Amino Acid Analysis of Peptides

To determine the structure of a protein or peptide, we need to answer three questions: What amino acids are present? How much of each is present? In what sequence do the amino acids occur in the peptide chain? The answers to the first two questions are provided by an automated instrument called an *amino acid analyzer*.

An amino acid analyzer is an automated instrument based on analytical techniques worked out in the 1950s by William Stein and Stanford Moore at the Rockefeller Institute, now the Rockefeller University, in New York. In preparation for analysis, the peptide is broken into its constituent amino acids by reducing all disulfide bonds, capping the −SH groups of cysteine residues by S_N2 reaction with iodoacetic acid, and hydrolyzing the amide bonds by heating with aqueous 6 M HCl at 110 °C for 24 hours. The resultant amino acid mixture is then analyzed, either by high-pressure liquid chromatography (HPLC) as described in the Chapter 12 *Focus On,* or by a related technique called ion-exchange chromatography.

In the ion-exchange technique, separated amino acids exiting *(eluting)* from the end of the chromatography column mix with a solution of *ninhydrin* and undergo a rapid reaction that produces an intense purple color. The color is detected by a spectrometer, and a plot of elution time versus spectrometer absorbance is obtained.

Ninhydrin   α-Amino acid   (purple color)

Because the amount of time required for a given amino acid to elute from a standard column is reproducible, the identities of the amino acids in a peptide can be determined. The amount of each amino acid in the sample is determined by measuring the intensity of the purple color resulting from its reaction with ninhydrin. Figure 26.3 shows the results of amino acid analysis of a standard equimolar mixture of 17 α-amino acids. Typically, amino acid analysis requires about 100 picomoles (2–3 μg) of sample for a protein containing about 200 residues.

Problem 26.10 | Show the structure of the product you would expect to obtain by S_N2 reaction of a cysteine residue with iodoacetic acid.

Problem 26.11 | Show the structures of the products obtained on reaction of valine with ninhydrin.

Figure 26.3 Amino acid analysis of an equimolar mixture of 17 amino acids.

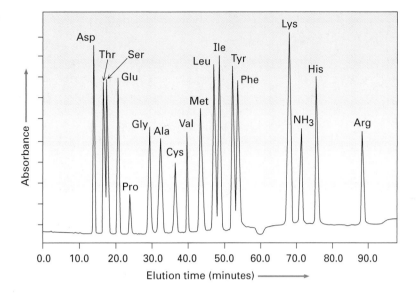

26.6 | Peptide Sequencing: The Edman Degradation

CENGAGENOW™ Click *Organic Interactive* to **predict products from degradation and modification reactions of simple peptides**.

With the identities and amounts of amino acids known, the peptide is *sequenced* to find out in what order the amino acids are linked together. Much peptide sequencing is now done by mass spectrometry, using either electrospray ionization (ESI) or matrix-assisted laser desorption ionization (MALDI) linked to a time-of-flight (TOF) mass analyzer, as described in Section 12.4. Also in common use is a chemical method of peptide sequencing called the *Edman degradation*.

The general idea of peptide sequencing by Edman degradation is to cleave one amino acid at a time from an end of the peptide chain. That terminal amino acid is then separated and identified, and the cleavage reactions are repeated on the chain-shortened peptide until the entire peptide sequence is known. Automated protein sequencers are available that allow as many as 50 repetitive sequencing cycles to be carried out before a buildup of unwanted by products interferes with the results. So efficient are these instruments that sequence information can be obtained from as little as 1 to 5 picomoles of sample—less than 0.1 μg.

Edman degradation involves treatment of a peptide with phenyl isothiocyanate (PITC), $C_6H_5-N=C=S$, followed by treatment with trifluoroacetic acid, as shown in Figure 26.4. The first step attaches the PITC to the $-NH_2$ group of the N-terminal amino acid, and the second step splits the N-terminal residue from the peptide chain, yielding an anilinothiazolinone (ATZ) derivative plus the chain-shortened peptide. Further acid-catalyzed rearrangement of the ATZ derivative with aqueous acid converts it into a phenylthiohydantoin (PTH), which is identified chromatographically by comparison of its elution time with the known elution times of PTH derivatives of the 20 common amino acids. The chain-shortened peptide is then automatically resubmitted to another round of Edman degradation.

Complete sequencing of large proteins by Edman degradation is impractical because of the buildup of unwanted by-products. To get around the problem, a large peptide chain is first cleaved by partial hydrolysis into a number of smaller fragments, the sequence of each fragment is determined, and the individual fragments are fitted together by matching the overlapping ends. In this way, protein chains with more than 400 amino acids have been sequenced.

Pehr Victor Edman

Pehr Victor Edman (1916–1977) was born in Stockholm, Sweden, and received an M.D. in 1946 at the Karolinska Institute. After a year in the United States at the Rockefeller Institute, he returned to Sweden as professor at the University of Lund. In 1957, he moved to St. Vincent's School of Medical Research in Melbourne, Australia, where he developed and automated the method of peptide sequencing that bears his name. A reclusive man, he never received the prizes or recognition merited by the importance of his work.

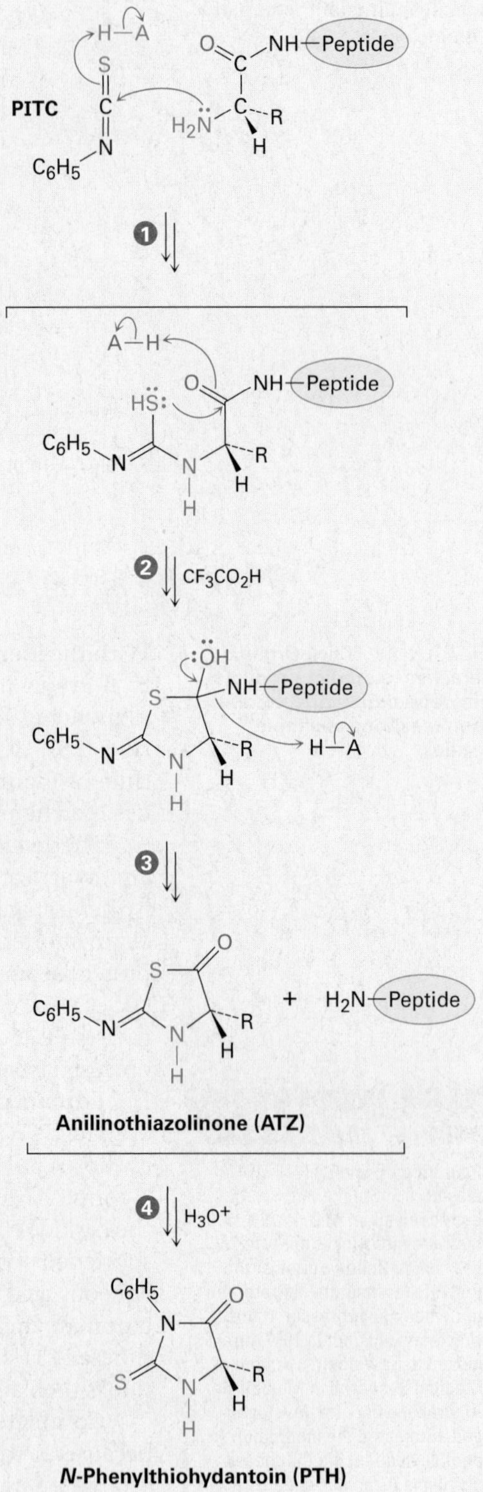

1 Nucleophilic addition of the peptide terminal amino group to phenyl isothiocyanate (PITC) gives an *N*-phenylthiourea derivative.

2 Acid-catalyzed cyclization of the phenylthiourea yields a tetrahedral intermediate . . .

3 . . . which expels the chain-shortened peptide and forms an anilino-thiazolinone (ATZ) derivative.

Anilinothiazolinone (ATZ)

4 The ATZ rearranges in the presence of aqueous acid to an isomeric *N*-phenylthiohydantoin (PTH) as the final product.

***N*-Phenylthiohydantoin (PTH)**

© John McMurry

Figure 26.4 MECHANISM: Mechanism of the Edman degradation for N-terminal analysis of peptides.

Partial hydrolysis of a peptide can be carried out either chemically with aqueous acid or enzymatically. Acidic hydrolysis is unselective and leads to a more or less random mixture of small fragments, but enzymatic hydrolysis is quite specific. The enzyme trypsin, for instance, catalyzes hydrolysis of peptides only at the carboxyl side of the basic amino acids arginine and lysine; chymotrypsin cleaves only at the carboxyl side of the aryl-substituted amino acids phenylalanine, tyrosine, and tryptophan.

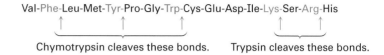

Val-Phe-Leu-Met-Tyr-Pro-Gly-Trp-Cys-Glu-Asp-Ile-Lys-Ser-Arg-His

Chymotrypsin cleaves these bonds.   Trypsin cleaves these bonds.

Problem 26.12   The octapeptide angiotensin II has the sequence Asp-Arg-Val-Tyr-Ile-His-Pro-Phe. What fragments would result if angiotensin II were cleaved with trypsin? With chymotrypsin?

Problem 26.13   What is the N-terminal residue on a peptide that gives the following PTH derivative on Edman degradation?

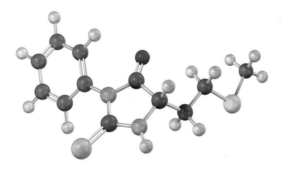

Problem 26.14   Draw the structure of the PTH derivative that would be formed on Edman degradation of angiotensin II (Problem 26.12).

Problem 26.15   Give the amino acid sequence of hexapeptides that produce the following sets of fragments on partial acid hydrolysis:
(a) Arg, Gly, Ile, Leu, Pro, Val gives Pro-Leu-Gly, Arg-Pro, Gly-Ile-Val
(b) N, L, M, W, V$_2$ gives V-L, V-M-W, W-N-V

26.7 | Peptide Synthesis

With its structure known, the synthesis of a peptide can then be undertaken—perhaps to obtain a larger amount for biological evaluation. A simple amide might be formed by treating an amine and a carboxylic acid with dicyclohexylcarbodiimide (DCC; Section 21.7), but peptide synthesis is a more difficult problem because many different amide bonds must be formed in a specific order rather than at random.

The solution to the specificity problem is to *protect* those functional groups we want to render unreactive while leaving exposed only those functional groups we want to react. For example, if we wanted to couple alanine with leucine to synthesize Ala-Leu, we could protect the −NH$_2$ group of alanine and the $-CO_2H$ group of leucine to render them unreactive, then form the desired amide bond, and then remove the protecting groups.

Many different amino- and carboxyl-protecting groups have been devised, but only a few are widely used. Carboxyl groups are often protected simply by converting them into methyl or benzyl esters. Both groups are easily introduced by standard methods of ester formation (Section 21.6) and are easily removed by mild hydrolysis with aqueous NaOH. Benzyl esters can also be cleaved by catalytic *hydrogenolysis* of the weak benzylic C–O bond (RCO_2—$CH_2Ph + H_2 \rightarrow RCO_2H + PhCH_3$).

Amino groups are often protected as their *tert*-butoxycarbonyl amide, or Boc, derivatives. The Boc protecting group is introduced by reaction of the amino acid with di-*tert*-butyl dicarbonate in a nucleophilic acyl substitution reaction and is removed by brief treatment with a strong organic acid such as trifluoroacetic acid, CF_3CO_2H.

Thus, five steps are needed to synthesize a dipeptide such as Ala-Leu:

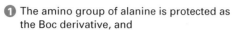

① The amino group of alanine is protected as the Boc derivative, and

② the carboxyl group of leucine is protected as the methyl ester.

③ The two protected amino acids are coupled using DCC.

④ The Boc protecting group is removed by acid treatment.

⑤ The methyl ester is removed by basic hydrolysis.

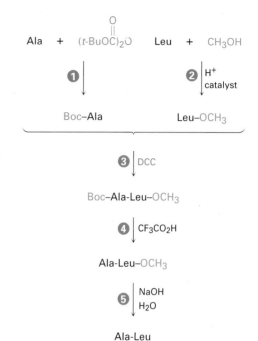

These steps can be repeated to add one amino acid at a time to the growing chain or to link two peptide chains together. Many remarkable achievements in peptide synthesis have been reported, including a complete synthesis of human insulin. Insulin is composed of two chains totaling 51 amino acids linked by two disulfide bridges. Its structure was determined by Frederick Sanger, who received the 1958 Nobel Prize in chemistry for his work.

Frederick Sanger

Frederick Sanger (1918–) was born in Gloucestershire, England, and received his Ph.D. at the University of Cambridge in 1943. After 10 years on the faculty at Cambridge, he joined the Medical Research Council in 1951, where he has remained. In 1958, he was awarded the Nobel Prize in chemistry for his determination of the structure of insulin, and in 1980 he became only the fourth person ever to win a second Nobel Prize. This second prize was awarded for his development of a method for determining the sequence of nucleotides in DNA.

A chain (21 units)
Gly
Ile
Val
Glu   S——————————S
Gln-Cys-Cys-Thr-Ser-Ile-Cys-Ser-Leu-Tyr-Gln-Leu-Glu-Asn-Tyr-Cys-Asn
                 S   S
                 S   S
His-Leu-Cys-Gly-Ser-His-Leu-Val-Glu-Ala-Leu-Tyr-Leu-Val-Cys   S
Glu   Gly
B chain (30 units) Asn   Glu
Val   Arg
Phe   Thr-Lys-Pro-Thr-Tyr-Phe-Phe-Gly

Insulin

Problem 26.16 | Show the mechanism for formation of a Boc derivative by reaction of an amino acid with di-*tert*-butyl dicarbonate.

Problem 26.17 | Write all five steps required for the synthesis of Leu-Ala from alanine and leucine.

26.8 | Automated Peptide Synthesis: The Merrifield Solid-Phase Method

Robert Bruce Merrifield

Robert Bruce Merrifield (1921–2006) was born in Fort Worth, Texas, and received his Ph.D. at the University of California, Los Angeles, in 1949. He then joined the faculty at the Rockefeller Institute, where he remained until his death. In 1984, he was awarded the Nobel Prize in chemistry for his development of methods for the automated synthesis of peptides.

The synthesis of large peptide chains by sequential addition of one amino acid at a time is long and arduous, but an immense simplification is possible using the *solid-phase* method introduced by R. Bruce Merrifield at the Rockefeller University. In the Merrifield method, peptide synthesis is carried out with the growing amino acid chain covalently bonded to small beads of a polymer resin rather than in solution. In the standard Merrifield procedure, polystyrene resin is used, prepared so that 1 of every 100 or so benzene rings contained a chloromethyl ($-CH_2Cl$) group, and a Boc-protected C-terminal amino acid is then bonded to the resin through an ester bond formed by S_N2 reaction.

Chloromethylated polystyrene resin

Resin-bound amino acid

With the first amino bonded to the resin, a repeating series of four steps is then carried out to build a peptide.

① A Boc-protected amino acid is covalently linked to the polystyrene polymer by formation of an ester bond (S_N2 reaction).

2 The polymer-bonded amino acid is washed free of excess reagent and then treated with trifluoroacetic acid to remove the Boc group.

2 | 1. Wash
2. CF$_3$CO$_2$H

$$H_2NCHCOCH_2-\boxed{Polymer}$$

R

3 A second Boc-protected amino acid is coupled to the first by reaction with DCC. Excess reagents are removed by washing them from the insoluble polymer.

3 | 1. DCC, Boc—NHCHCOH
2. Wash | R'

$$Boc-NHCHC-NHCHCOCH_2-\boxed{Polymer}$$

R'   R

4 The cycle of deprotection, coupling, and washing is repeated as many times as desired to add amino acid units to the growing chain.

4 | Repeat cycle many times

$$Boc-NHCHC\,\text{(}NHCHC\text{)}_{\overline{n}}NHCHCOCH_2-\boxed{Polymer}$$

R''   R'   R

5 After the desired peptide has been made, treatment with anhydrous HF removes the final Boc group and cleaves the ester bond to the polymer, yielding the free peptide.

5 | HF

$$H_2NCHC\,\text{(}NHCHC\text{)}_{\overline{n}}NHCHCOH + HOCH_2-\boxed{Polymer}$$

R''   R'   R

The details of the solid-phase technique have been improved substantially over the years, but the fundamental idea remains the same. The most commonly used resins at present are either the Wang resin or the PAM (phenylacetamidomethyl) resin, and the most commonly used N-protecting group is the fluorenylmethyloxycarbonyl, or Fmoc group, rather than Boc.

Wang resin

PAM resin

Fmoc-protected amino acid

Robotic peptide synthesizers are now used to automatically repeat the coupling, washing, and deprotection steps with different amino acids. Each step occurs in high yield, and mechanical losses are minimized because the peptide intermediates are never removed from the insoluble polymer until the final step. Using this procedure, up to 25 to 30 mg of a peptide with 20 amino acids can be routinely prepared.

26.9 | Protein Structure

CENGAGENOW Click *Organic Interactive* to **use interactive animations to view aspects of protein structure.**

Proteins are usually classified as either *fibrous* or *globular,* according to their three-dimensional shape. **Fibrous proteins**, such as the collagen in tendons and connective tissue and the myosin in muscle tissue, consist of polypeptide chains arranged side by side in long filaments. Because these proteins are tough and insoluble in water, they are used in nature for structural materials. **Globular proteins**, by contrast, are usually coiled into compact, roughly spherical shapes. These proteins are generally soluble in water and are mobile within cells. Most of the 3000 or so enzymes that have been characterized to date are globular proteins.

Proteins are so large that the word *structure* takes on a broader meaning than it does with simpler organic compounds. In fact, chemists speak of four different levels of structure when describing proteins.

▌ The **primary structure** of a protein is simply the amino acid sequence.

▌ The **secondary structure** of a protein describes how *segments* of the peptide backbone orient into a regular pattern.

▌ The **tertiary structure** describes how the *entire* protein molecule coils into an overall three-dimensional shape.

▌ The **quaternary structure** describes how different protein molecules come together to yield large aggregate structures.

Primary structure is determined, as we've seen, by sequencing the protein. Secondary, tertiary, and quaternary structures are determined by X-ray crystallography (Chapter 22 *Focus On*) because it's not yet possible to predict computationally how a given protein sequence will fold.

The most common secondary structures are the α helix and the β-pleated sheet. An **α helix** is a right-handed coil of the protein backbone, much like the coil of a telephone cord (Figure 26.5a). Each coil of the helix contains 3.6 amino acid residues, with a distance between coils of 540 pm, or 5.4 Å. The structure is stabilized by hydrogen bonds between amide N−H groups and C=O groups four residues away, with an N−H····O distance of 2.8 Å. The α helix is an extremely common secondary structure, and almost all globular proteins contain many helical segments. Myoglobin, a small globular protein containing 153 amino acid residues in a single chain, is an example (Figure 26.5b).

A **β-pleated sheet** differs from an α helix in that the peptide chain is extended rather than coiled and the hydrogen bonds occur between residues in adjacent chains (Figure 26.6a). The neighboring chains can run either in the same direction (parallel) or in opposite directions (antiparallel), although the antiparallel arrangement is more common and energetically somewhat more favorable. Concanavalin A, for instance, consists of two identical chains of 237 residues, each with extensive regions of antiparallel β sheets (Figure 26.6b).

(a)

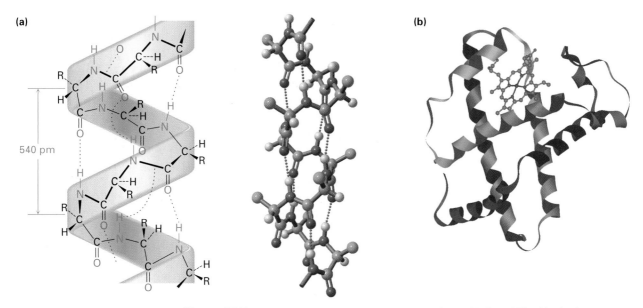

(b)

Figure 26.5 (a) The α-helical secondary structure of proteins is stabilized by hydrogen bonds between the N—H group of one residue and the C=O group four residues away. **(b)** The structure of myoglobin, a globular protein with extensive helical regions that are shown as coiled ribbons in this representation.

(a)

Chain 1

Chain 2

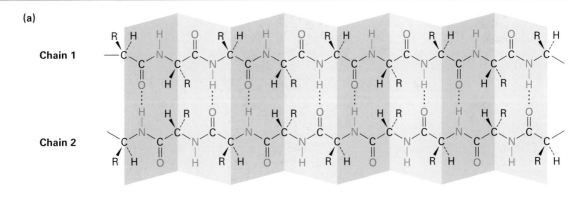

(b)

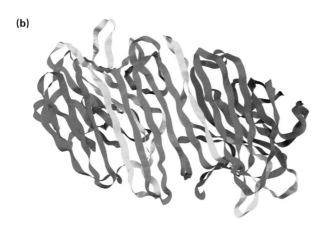

Figure 26.6 (a) The β-pleated sheet secondary structure of proteins is stabilized by hydrogen bonds between parallel or antiparallel chains. **(b)** The structure of concanavalin A, a protein with extensive regions of antiparallel β sheets, shown as flat ribbons.

What about tertiary structure? Why does any protein adopt the shape it does? The forces that determine the tertiary structure of a protein are the same forces that act on all molecules, regardless of size, to provide maximum stability. Particularly important are the hydrophilic (water-loving; Section 2.13) interactions of the polar side chains on acidic or basic amino acids. Those acidic or basic amino acids with charged side chains tend to congregate on the exterior of the protein, where they can be solvated by water. Those amino acids with neutral, nonpolar side chains tend to congregate on the hydrocarbon-like interior of a protein molecule, away from the aqueous medium.

Also important for stabilizing a protein's tertiary structure are the formation of disulfide bridges between cysteine residues, the formation of hydrogen bonds between nearby amino acid residues, and the presence of ionic attractions, called *salt bridges,* between positively and negatively charged sites on various amino acid side chains within the protein.

Because the tertiary structure of a globular protein is delicately held together by weak intramolecular attractions, a modest change in temperature or pH is often enough to disrupt that structure and cause the protein to become **denatured**. Denaturation occurs under such mild conditions that the primary structure remains intact but the tertiary structure unfolds from a specific globular shape to a randomly looped chain (Figure 26.7).

Figure 26.7 A representation of protein denaturation. A globular protein loses its specific three-dimensional shape and becomes randomly looped.

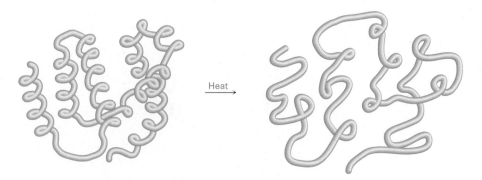

Heat

Denaturation is accompanied by changes in both physical and biological properties. Solubility is drastically decreased, as occurs when egg white is cooked and the albumins unfold and coagulate. Most enzymes also lose all catalytic activity when denatured, since a precisely defined tertiary structure is required for their action. Although most denaturation is irreversible, some cases are known where spontaneous *renaturation* of an unfolded protein to its stable tertiary structure occurs. Renaturation is accompanied by a full recovery of biological activity.

26.10 | Enzymes and Coenzymes

An **enzyme**—usually a large protein—is a substance that acts as a catalyst for a biological reaction. Like all catalysts, an enzyme doesn't affect the equilibrium constant of a reaction and can't bring about a chemical change that is otherwise unfavorable. An enzyme acts only to lower the activation energy for a reaction, thereby making the reaction take place more rapidly. Sometimes, in fact, the rate acceleration brought about by enzymes is extraordinary. Millionfold rate increases are common, and the glycosidase enzymes that hydrolyze polysaccharides increase the reaction rate by a factor of more than 10^{17}, changing the time required for the reaction from millions of years to milliseconds.

Unlike many of the catalysts that chemists use in the laboratory, enzymes are usually specific in their action. Often, in fact, an enzyme will catalyze only a single reaction of a single compound, called the enzyme's *substrate*. For example, the enzyme amylase, found in the human digestive tract, catalyzes only the hydrolysis of starch to yield glucose; cellulose and other polysaccharides are untouched by amylase.

Different enzymes have different specificities. Some, such as amylase, are specific for a single substrate, but others operate on a range of substrates. Papain, for instance, a globular protein of 212 amino acids isolated from papaya fruit, catalyzes the hydrolysis of many kinds of peptide bonds. In fact, it's this ability to hydrolyze peptide bonds that makes papain useful as a meat tenderizer and a cleaner for contact lenses.

Enzymes function through a pathway that involves initial formation of an enzyme–substrate complex $E \cdot S$, a multistep chemical conversion of the enzyme-bound substrate into enzyme-bound product $E \cdot P$, and final release of product from the complex.

$$E + S \;\rightleftharpoons\; E \cdot S \;\rightleftharpoons\; E \cdot P \;\rightleftharpoons\; E + P$$

The overall rate constant for conversion of the $E \cdot S$ complex to products $E + P$ is called the **turnover number** because it represents the number of substrate molecules the enzyme turns over into product per unit time. A value of about 10^3 per second is typical.

The rate acceleration achieved by enzymes is due to several factors. Particularly important is the ability of the enzyme to stabilize and thus lower the energy of the transition state(s). That is, it's not the ability of the enzyme to bind the *substrate* that matters but rather its ability to bind and thereby stabilize the *transition state*. Often, in fact, the enzyme binds the transition structure as much as 10^{12} times more tightly than it binds the substrate or products. As a result, the transition state is substantially lowered in energy. An energy diagram for an enzyme-catalyzed process might look like that in Figure 26.8.

Enzymes are classified into six categories depending on the kind of reaction they catalyze, as shown in Table 26.2. *Oxidoreductases* catalyze oxidations and reductions; *transferases* catalyze the transfer of a group from one substrate to another; *hydrolases* catalyze hydrolysis reactions of esters, amides, and related substrates; *lyases* catalyze the elimination or addition of a small molecule such as H_2O from or to a substrate; *isomerases* catalyze isomerizations; and *ligases* catalyze the bonding together of two molecules, often coupled with the hydrolysis

Figure 26.8 Energy diagrams for uncatalyzed (red) and enzyme-catalyzed (blue) processes. The enzyme makes available an alternative, lower-energy pathway. Rate enhancement is due to the ability of the enzyme to bind to the transition state for product formation, thereby lowering its energy.

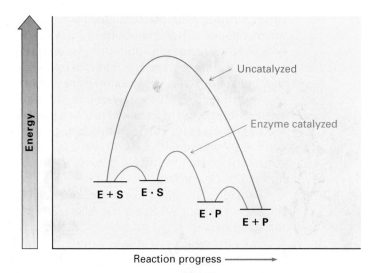

of ATP. The systematic name of an enzyme has two parts, ending with *-ase*. The first part identifies the enzyme's substrate, and the second part identifies its class. For example, hexose kinase is a transferase that catalyzes the transfer of a phosphate group from ATP to a hexose sugar.

Table 26.2 | Classification of Enzymes

Class	Some subclasses	Function
Oxidoreductases	Dehydrogenases	Introduction of double bond
	Oxidases	Oxidation
	Reductases	Reduction
Transferases	Kinases	Transfer of phosphate group
	Transaminases	Transfer of amino group
Hydrolases	Lipases	Hydrolysis of ester
	Nucleases	Hydrolysis of phosphate
	Proteases	Hydrolysis of amide
Lyases	Decarboxylases	Loss of CO_2
	Dehydrases	Loss of H_2O
Isomerases	Epimerases	Isomerization of chirality center
Ligases	Carboxylases	Addition of CO_2
	Synthetases	Formation of new bond

In addition to their protein part, most enzymes also contain a small nonprotein part called a *cofactor*. A **cofactor** can be either an inorganic ion, such as Zn^{2+}, or a small organic molecule, called a **coenzyme**. A coenzyme is not a catalyst but is a reactant that undergoes chemical change during the reaction and requires an additional step to return to its initial state. Many, although not all, coenzymes are derived from *vitamins*—substances that an organism requires for growth but is unable to synthesize and must receive in its diet. Coenzyme A from pantothenate (vitamin B_3), NAD^+ from niacin, FAD from riboflavin (vitamin B_2), tetrahydrofolate from folic acid, pyridoxal phosphate from pyridoxine (vitamin B_6), and thiamin diphosphate from thiamin (vitamin B_1) are examples (Table 26.3 on pages 1044–1045). We'll discuss the chemistry and mechanisms of coenzyme reactions at appropriate points later in the text.

Problem 26.18 To what classes do the following enzymes belong?
(a) Pyruvate decarboxylase   (b) Chymotrypsin   (c) Alcohol dehydrogenase

26.11 | How Do Enzymes Work? Citrate Synthase

Enzymes work by bringing reactant molecules together, holding them in the orientation necessary for reaction, and providing any necessary acidic or basic sites to catalyze specific steps. As an example, let's look at citrate synthase, an enzyme that catalyzes the aldol-like addition of acetyl CoA to oxaloacetate to give citrate. The reaction is the first step in the *citric acid cycle,* in which acetyl groups produced by degradation of food molecules are metabolized to yield CO_2 and H_2O. We'll look at the details of the citric acid cycle in Section 29.7.

Oxaloacetate   **Acetyl CoA**   **Citrate**

Citrate synthase is a globular protein of 433 amino acids with a deep cleft lined by an array of functional groups that can bind to oxaloacetate. On binding oxaloacetate, the original cleft closes and another opens up to bind acetyl CoA. This second cleft is also lined by appropriate functional groups, including a histidine at position 274 and an aspartic acid at position 375. The two reactants are now held by the enzyme in close proximity and with a suitable orientation for reaction. Figure 26.9 on page 1046 shows the structure of citrate synthase as determined by X-ray crystallography, along with a close-up of the active site.

As shown in Figure 26.10 on page 1047, the first step in the aldol reaction is generation of the enol of acetyl CoA. The side-chain carboxyl of an aspartate residue acts as base to abstract an acidic α proton, while at the same time the side-chain imidazole ring of a histidine donates H^+ to the carbonyl oxygen. The enol thus produced then does a nucleophilic addition to the ketone carbonyl group of oxaloacetate. The first histidine acts as a base to remove the $-OH$ hydrogen from the enol, while a second histidine residue simultaneously donates a proton to the oxaloacetate carbonyl group, giving citryl CoA. Water then hydrolyzes the thiol ester group in citryl CoA in a nucleophilic acyl substitution reaction, releasing citrate and coenzyme A as the final products.

Table 26.3 | **Structures of Some Common Coenzymes**

Adenosine triphosphate—ATP (phosphorylation)

Coenzyme A (acyl transfer)

Nicotinamide adenine dinucleotide—NAD⁺ (oxidation/reduction)
(NADP⁺)

Flavin adenine dinucleotide—FAD (oxidation/reduction)

(continued)

Table 26.3 | **Structures of Some Common Coenzymes** *(continued)*

Tetrahydrofolate (transfer of C$_1$ units)

S-Adenosylmethionine (methyl transfer)

Lipoic acid (acyl transfer)

Pyridoxal phosphate (amino acid metabolism)

Thiamin diphosphate (decarboxylation)

Biotin (carboxylation)

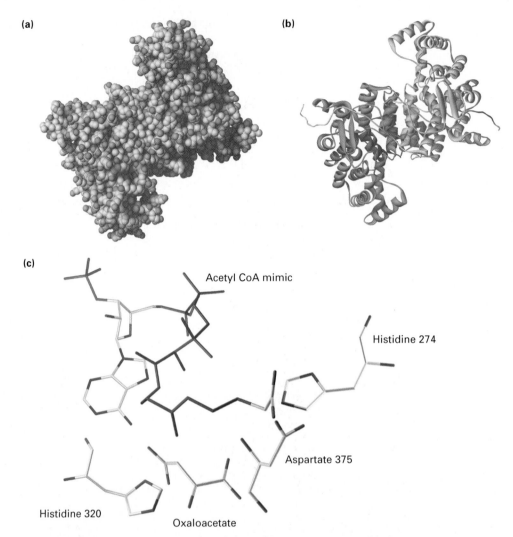

(a)

(b)

(c)

Acetyl CoA mimic

Histidine 274

Aspartate 375

Histidine 320

Oxaloacetate

Figure 26.9 X-ray crystal structure of citrate synthase. Part **(a)** is a space-filling model and part **(b)** is a ribbon model, which emphasizes the α-helical segments of the protein chain and indicates that the enzyme is dimeric; that is, it consists of two identical chains held together by hydrogen bonds and other intermolecular attractions. Part **(c)** is a close-up of the active site in which oxaloacetate and an unreactive acetyl CoA mimic are bound.

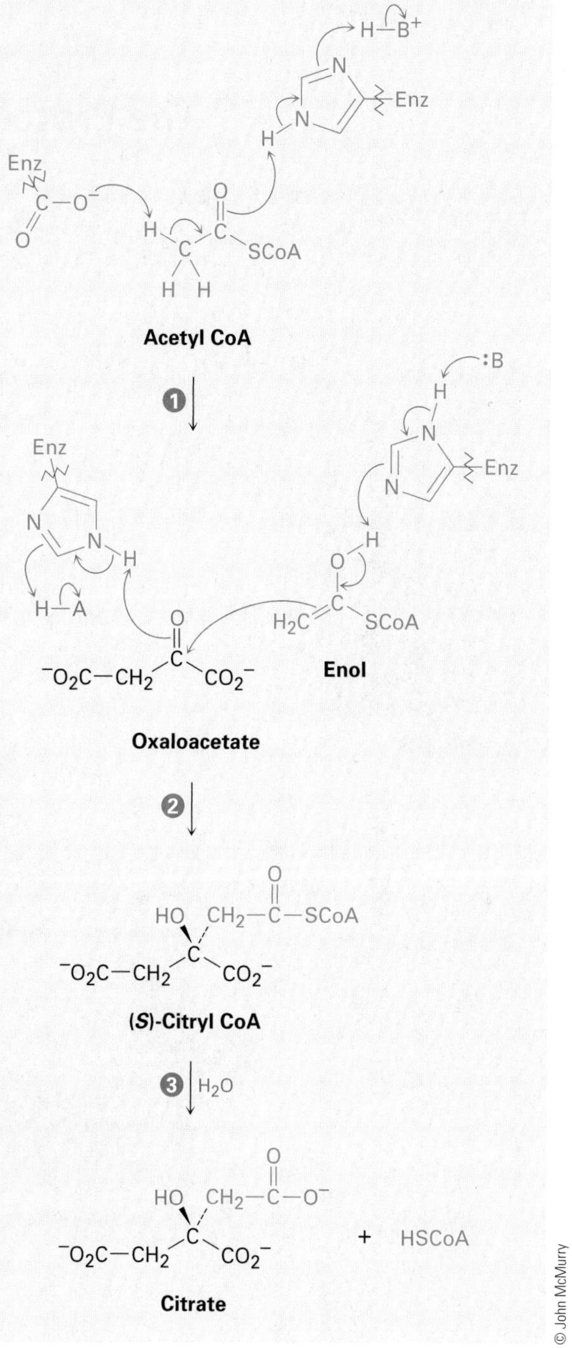

1 The side-chain carboxylate group of an aspartic acid acts as a base and removes an acidic α proton from acetyl CoA, while the N–H group on the side chain of a histidine acts as an acid and donates a proton to the carbonyl oxygen, giving an enol.

2 A histidine deprotonates the acetyl-CoA enol, which adds to the ketone carbonyl group of oxaloacetate in an aldol-like reaction. Simultaneously, an acid N–H proton of another histidine protonates the carbonyl oxygen, producing (S)-citryl CoA.

3 The thioester group of citryl CoA is hydrolyzed by a typical nucleophilic acyl substitution reaction to produce citrate plus coenzyme A.

Figure 26.10 MECHANISM: Mechanism of the addition of acetyl CoA to oxaloacetate to give (S)-citryl CoA, catalyzed by citrate synthase.

© John McMurry

Focus On . . .

The Protein Data Bank

Enzymes are so large, so structurally complex, and so numerous that the use of computer databases and molecular visualization programs has become an essential tool for studying biological chemistry. Of the various databases available online, the Kyoto Encyclopedia of Genes and Genomes (KEGG) database (http://www.genome.ad.jp/kegg), maintained by the Kanehisa Laboratory of Kyoto University Bioinformatics Center, is useful for obtaining information on biosynthetic pathways of the sort we'll be describing in the next few chapters. For obtaining information on a specific enzyme, the BRENDA database (http://www.brenda.uni-koeln.de), maintained by the Institute of Biochemistry at the University of Cologne, Germany, is particularly valuable.

Perhaps the most useful of all biological databases is the Protein Data Bank (PDB), operated by the Research Collaboratory for Structural Bioinformatics (RCSB). The PDB is a worldwide repository of X-ray and NMR structural data for biological macromolecules. In early 2007, data for more than 40,000 structures were available, and more than 6000 new structures were being added yearly. To access the Protein Data Bank, go to http://www.rcsb.org/pdb/ and a home page like that shown in Figure 26.11 will appear. As with much that is available online, however, the PDB site is changing rapidly, so you may not see quite the same thing.

Figure 26.11 The Protein Data Bank home page.

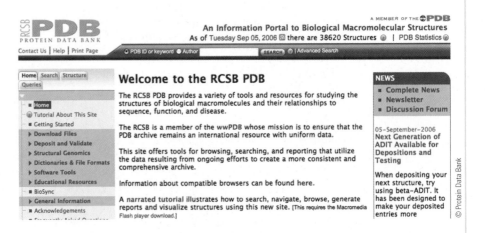

To learn how to use the PDB, begin by running the short tutorial listed near the top of the blue sidebar on the left of the screen. After that introduction, start exploring. Let's say you want to view citrate synthase, the enzyme shown previously in Figure 26.9 that catalyzes the addition of acetyl CoA to oxaloacetate to give citrate. Type "citrate synthase" into the small

(continued)

search window on the top line, click on "Search," and a list of 30 or so structures will appear. Scroll down near the end of the list until you find the entry with a PDB code of 5CTS and the title "Proposed Mechanism for the Condensation Reaction of Citrate Synthase: 1.9 Å Structure of the Ternary Complex with Oxaloacetate and Carboxymethyl Coenzyme A." Alternatively, if you know the code of the enzyme you want, you can enter it directly into the search window. Click on the PDB code of entry 5CTS, and a new page containing information about the enzyme will open.

If you choose, you can download the structure file to your computer and open it with any of numerous molecular graphics programs to see an image like that in Figure 26.12. The biologically active molecule is a dimer of two identical subunits consisting primarily of α-helical regions displayed as coiled ribbons. For now, just click on "Display Molecule," followed by "Image Gallery," to see some of the tools for visualizing and further exploring the enzyme.

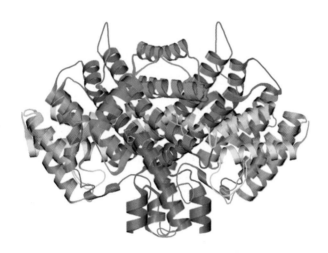

Figure 26.12 An image of citrate synthase, downloaded from the Protein Data Bank.

SUMMARY AND KEY WORDS

Proteins are large biomolecules made up of **α-amino acid residues** linked together by amide, or *peptide,* bonds. Chains with fewer than 50 amino acids are often called **peptides**, while the term **protein** is reserved for larger chains. Twenty amino acids are commonly found in proteins; all are α-amino acids, and all except glycine have stereochemistry similar to that of L sugars. In neutral solution, amino acids exist as dipolar **zwitterions**.

Amino acids can be synthesized in racemic form by several methods, including ammonolysis of an α-bromo acid, alkylation of diethyl acetamidomalonate, and reductive amination of an α-keto acid. Alternatively, an enantioselective synthesis of amino acids can be carried out using a chiral hydrogenation catalyst.

To determine the structure of a peptide or protein, the identity and amount of each amino acid present is first found by amino acid analysis. The peptide is hydrolyzed to its constituent α-amino acids, which are separated and identified. Next, the peptide is sequenced. **Edman degradation** by treatment with phenyl isothiocyanate (PITC) cleaves one residue from the N terminus of the peptide and forms an easily identifiable phenylthiohydantoin (PTH) derivative of the **N-terminal amino acid**. A series of sequential Edman degradations allows the sequencing of a peptide chain up to 50 residues in length.

Peptide synthesis requires the use of selective protecting groups. An N-protected amino acid with a free carboxyl group is coupled to an O-protected amino acid with a free amino group in the presence of dicyclohexylcarbodiimide (DCC). Amide formation occurs, the protecting groups are removed, and the sequence is repeated. Amines are usually protected as their *tert*-butoxycarbonyl (Boc) derivatives, and acids are protected as esters. This synthetic sequence is often carried out by the Merrifield solid-phase method, in which the peptide is esterified to an insoluble polymeric support.

Proteins have four levels of structure. **Primary structure** describes a protein's amino acid sequence; **secondary structure** describes how segments of the protein chain orient into regular patterns—either α-**helix** or β-**pleated sheet**; **tertiary structure** describes how the entire protein molecule coils into an overall three-dimensional shape; and **quaternary structure** describes how individual protein molecules aggregate into larger structures.

Proteins are classified as either globular or fibrous. **Fibrous proteins** such as α-keratin are tough, rigid, and water-insoluble; **globular proteins** such as myoglobin are water-soluble and roughly spherical in shape. Many globular proteins are **enzymes**—substances that act as catalysts for biological reactions. Enzymes are grouped into six classes according to the kind of reaction they catalyze. They function by bringing reactant molecules together, holding them in the orientation necessary for reaction, and providing any necessary acidic or basic sites to catalyze specific steps.

SUMMARY OF REACTIONS

1. Amino acid synthesis (Section 26.3)
  (a) From α-bromo acids

  (b) Diethyl acetamidomalonate synthesis

(c) Reductive amination of an α-keto acid

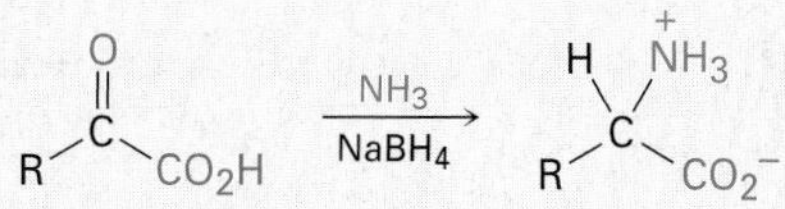

(d) Enantioselective synthesis

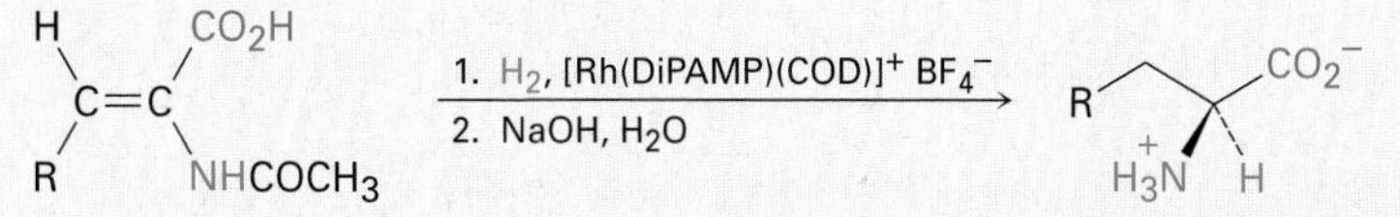

A (Z) enamido acid   **An (S)-amino acid**

2. Peptide sequencing by Edman degradation (Section 26.6)

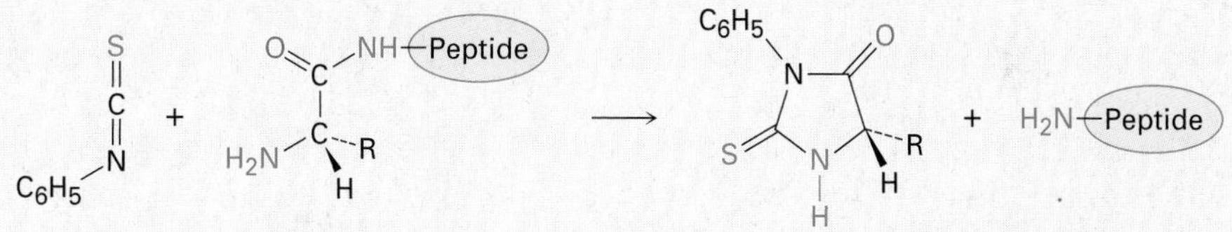

3. Peptide synthesis (Section 26.7)
   (a) Amine protection

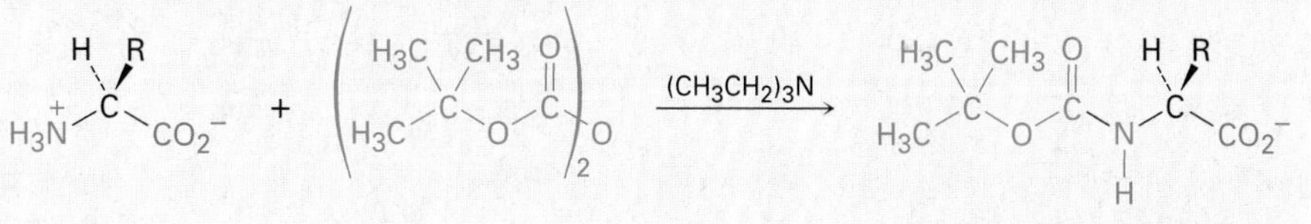

Boc-protected amino acid

   (b) Carboxyl protection

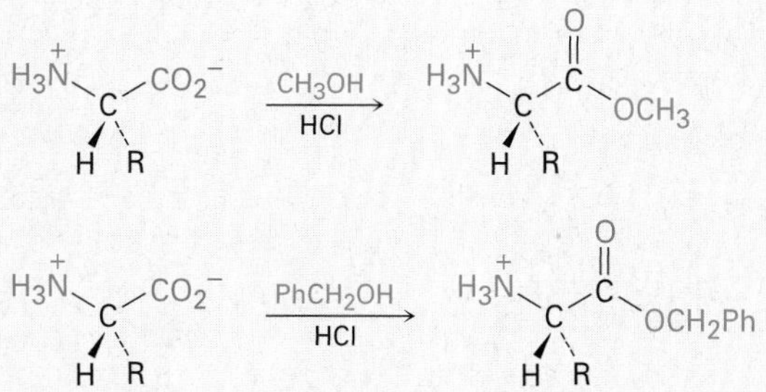

EXERCISES

Organic KNOWLEDGE TOOLS

CENGAGENOW˜ Sign in at **www.cengage.com/login** to assess your knowledge of this chapter's topics by taking a pre-test. The pre-test will link you to interactive organic chemistry resources based on your score in each concept area.

⚘WL Online homework for this chapter may be assigned in Organic OWL.

■ indicates problems assignable in Organic OWL.

VISUALIZING CHEMISTRY

(Problems 26.1–26.18 appear within the chapter.)

26.19 ■ Identify the following amino acids:

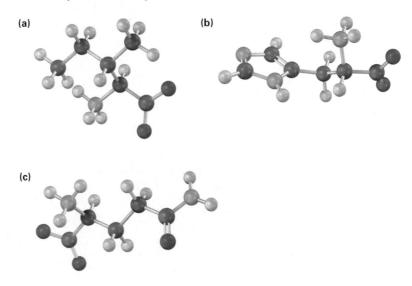

26.20 ■ Give the sequence of the following tetrapeptide (yellow = S):

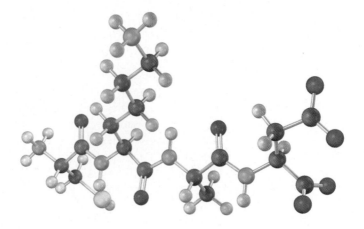

26.21 Isoleucine and threonine (Problem 26.3) are the only two amino acids with two chirality centers. Assign *R* or *S* configuration to the methyl-bearing carbon atom of isoleucine.

26.22 ■ Is the following structure a D amino acid or an L amino acid? Identify it.

26.23 ■ Give the sequence of the following tetrapeptide:

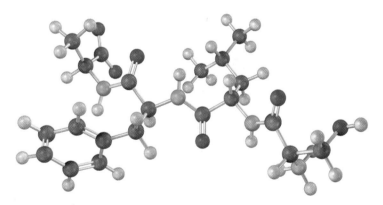

ADDITIONAL PROBLEMS

26.24 Except for cysteine, only *S* amino acids occur in proteins. Several *R* amino acids are also found in nature, however. (*R*)-Serine is found in earthworms, and (*R*)-alanine is found in insect larvae. Draw Fischer projections of (*R*)-serine and (*R*)-alanine. Are these D or L amino acids?

26.25 Cysteine is the only amino acid that has L stereochemistry but an *R* configuration. Make up a structure for another L amino acid of your own creation that also has an *R* configuration.

26.26 Draw a Fischer projection of (*S*)-proline.

26.27 ■ Show the structures of the following amino acids in their zwitterionic forms:
(a) Trp   (b) Ile   (c) Cys   (d) His

26.28 ■ Proline has pK_{a1} = 1.99 and pK_{a2} = 10.60. Use the Henderson–Hasselbalch equation to calculate the ratio of protonated and neutral forms at pH = 2.50. Calculate the ratio of neutral and deprotonated forms at pH = 9.70.

26.29 Using both three- and one-letter codes for amino acids, write the structures of all possible peptides containing the following amino acids:
(a) Val, Ser, Leu   (b) Ser, Leu$_2$, Pro

26.30 ■ Predict the product of the reaction of valine with the following reagents:
(a) CH$_3$CH$_2$OH, acid   (b) Di-*tert*-butyl dicarbonate
(c) KOH, H$_2$O   (d) CH$_3$COCl, pyridine; then H$_2$O

26.31 ■ Show how you could use the acetamidomalonate method to prepare the following amino acids:
(a) Leucine   (b) Tryptophan

26.32 Show how you could prepare the following amino acids using a reductive amination:
(a) Methionine   (b) Isoleucine

26.33 Show how you could prepare the following amino acids enantioselectively:
(a) Pro   (b) Val

26.34 Serine can be synthesized by a simple variation of the amidomalonate method using formaldehyde rather than an alkyl halide. How might this be done?

26.35 ■ Write full structures for the following peptides:
(a) C-H-E-M   (b) P-E-P-T-I-D-E

26.36 ■ Propose two structures for a tripeptide that gives Leu, Ala, and Phe on hydrolysis but does not react with phenyl isothiocyanate.

26.37 Show the steps involved in a synthesis of Phe-Ala-Val using the Merrifield procedure.

26.38 ■ Draw the structure of the PTH derivative product you would obtain by Edman degradation of the following peptides:
(a) I-L-P-F   (b) D-T-S-G-A

26.39 Look at the side chains of the 20 amino acids in Table 26.1, and then think about what is *not* present. None of the 20 contain either an aldehyde or a ketone carbonyl group, for instance. Is this just one of nature's oversights, or is there a likely chemical reason? What complications might an aldehyde or ketone carbonyl group cause?

26.40 The α-helical parts of myoglobin and other proteins stop whenever a proline residue is encountered in the chain. Why is proline never present in a protein α-helix?

26.41 ■ Which amide bonds in the following polypeptide are cleaved by trypsin? By chymotrypsin?

Phe-Leu-Met-Lys-Tyr-Asp-Gly-Gly-Arg-Val-Ile-Pro-Tyr

26.42 What kinds of reactions do the following classes of enzymes catalyze?
(a) Hydrolases   (b) Lyases   (c) Transferases

26.43 ■ Which of the following amino acids are more likely to be found on the outside of a globular protein, and which on the inside? Explain.
(a) Valine   (b) Aspartic acid   (c) Phenylalanine   (d) Lysine

26.44 The chloromethylated polystyrene resin used for Merrifield solid-phase peptide synthesis is prepared by treatment of polystyrene with chloromethyl methyl ether and a Lewis acid catalyst. Propose a mechanism for the reaction.

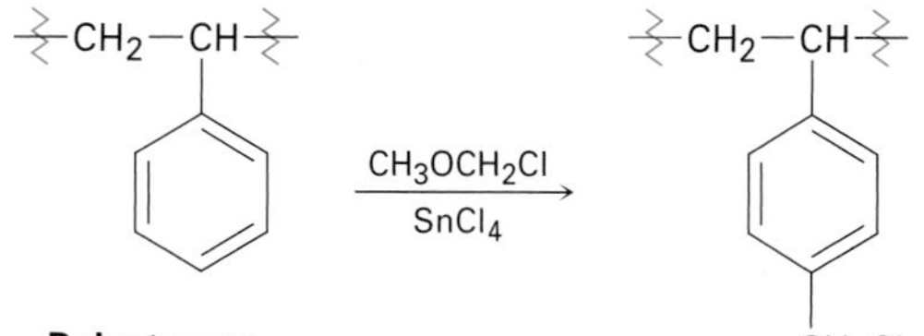

Polystyrene

$$\xrightarrow[\text{SnCl}_4]{\text{CH}_3\text{OCH}_2\text{Cl}}$$

CH₂Cl

26.45 ■ An Fmoc protecting group can be removed from an amino acid by treatment with the amine base piperidine. Propose a mechanism.

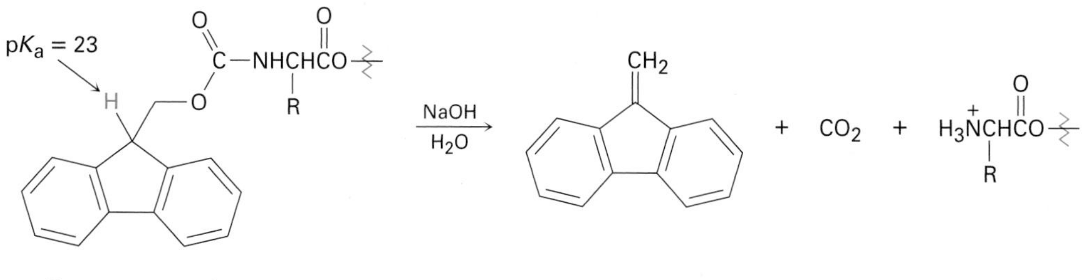

$pK_a = 23$

Fmoc-protected amino acid

$$\xrightarrow[\text{H}_2\text{O}]{\text{NaOH}}$$

$+ \ CO_2 \ + \ H_3\overset{+}{N}CHCO$

26.46 Leuprolide is a synthetic nonapeptide used to treat both endometriosis in women and prostate cancer in men.

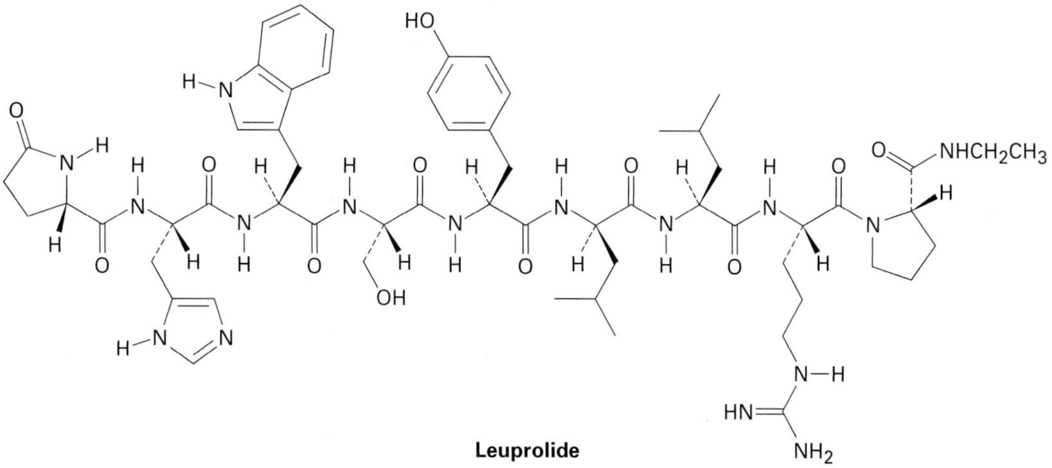

Leuprolide

(a) Both C-terminal and N-terminal amino acids in leuprolide have been structurally modified. Identify the modifications.
(b) One of the nine amino acids in leuprolide has D stereochemistry rather than the usual L. Which one?
(c) Write the structure of leuprolide using both one- and three-letter abbreviations.
(d) What charge would you expect leuprolide to have at neutral pH?

26.47 ■ Proteins can be cleaved specifically at the amide bond on the carboxyl side of methionine residues by reaction with cyanogen bromide, $BrC \equiv N$.

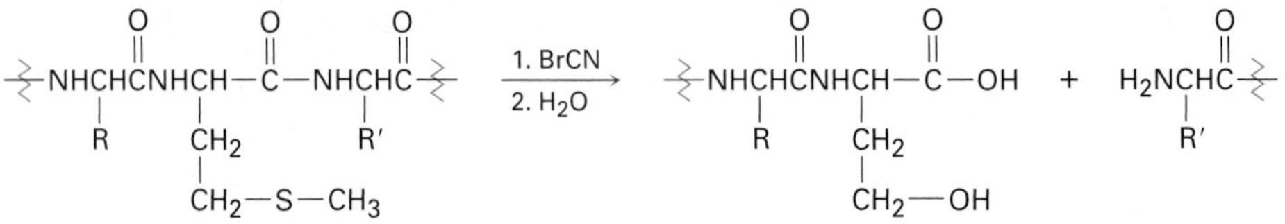

The reaction occurs in several steps:

(a) The first step is a nucleophilic substitution reaction of the sulfur on the methionine side chain with BrCN to give a cyanosulfonium ion, $[R_2SCN]^+$. Show the structure of the product, and propose a mechanism for the reaction.

(b) The second step is an internal S_N2 reaction, with the carbonyl oxygen of the methionine residue displacing the positively charged sulfur leaving group and forming a five-membered ring product. Show the structure of the product and the mechanism of its formation.

(c) The third step is a hydrolysis reaction to split the peptide chain. The carboxyl group of the former methionine residue is now part of a lactone (cyclic ester) ring. Show the structure of the lactone product and the mechanism of its formation.

(d) The final step is a hydrolysis of the lactone to give the product shown. Show the mechanism of the reaction.

26.48 A clever new method of peptide synthesis involves formation of an amide bond by reaction of an α-keto acid with an N-alkylhydroxylamine:

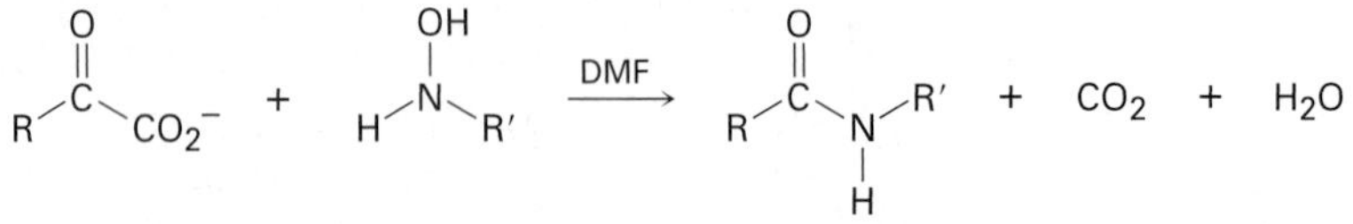

An α-keto acid   A hydroxylamine   An amide

The reaction is thought to occur by nucleophilic addition of the N-alkyl-hydroxylamine to the keto acid as if forming an oxime (Section 19.8), followed by decarboxylation and elimination of water. Show the mechanism.

26.49 Arginine, the most basic of the 20 common amino acids, contains a *guanidino* functional group in its side chain. Explain, using resonance structures to show how the protonated guanidino group is stabilized.

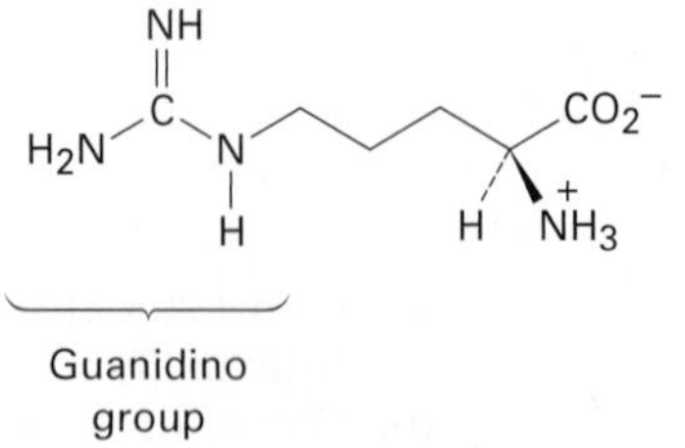

Arginine

Guanidino
group

26.50 Cytochrome *c* is an enzyme found in the cells of all aerobic organisms. Elemental analysis of cytochrome *c* shows that it contains 0.43% iron. What is the minimum molecular weight of this enzyme?

26.51 Evidence for restricted rotation around amide CO—N bonds comes from NMR studies. At room temperature, the ^{1}H NMR spectrum of *N,N*-dimethylformamide shows three peaks: 2.9 δ (singlet, 3 H), 3.0 δ (singlet, 3 H), 8.0 δ (singlet, 1 H). As the temperature is raised, however, the two singlets at 2.9 δ and 3.0 δ slowly merge. At 180 °C, the ^{1}H NMR spectrum shows only two peaks: 2.95 δ (singlet, 6 H) and 8.0 δ (singlet, 1 H). Explain this temperature-dependent behavior.

N,N-Dimethylformamide

26.52 ■ Propose a structure for an octapeptide that shows the composition Asp, Gly$_2$, Leu, Phe, Pro$_2$, Val on amino acid analysis. Edman analysis shows a glycine N-terminal group, and leucine is the C-terminal group. Acidic hydrolysis gives the following fragments:

Val-Pro-Leu, Gly, Gly-Asp-Phe-Pro, Phe-Pro-Val

26.53 The reaction of ninhydrin with an α-amino acid occurs in several steps.
(a) The first step is formation of an imine by reaction of the amino acid with ninhydrin. Show its structure and the mechanism of its formation.
(b) The second step is a decarboxylation. Show the structure of the product and the mechanism of the decarboxylation reaction.
(c) The third step is hydrolysis of an imine to yield an amine and an aldehyde. Show the structures of both products and the mechanism of the hydrolysis reaction.
(d) The final step is formation of the purple anion. Show the mechanism of the reaction.

Ninhydrin

26.54 Draw resonance forms for the purple anion obtained by reaction of ninhydrin with an α-amino acid (Problem 26.53).

26.55 Look up the structure of human insulin (Section 26.7), and indicate where in each chain the molecule is cleaved by trypsin and chymotrypsin.

26.56 ■ What is the structure of a nonapeptide that gives the following fragments when cleaved?

Trypsin cleavage: Val-Val-Pro-Tyr-Leu-Arg, Ser-Ile-Arg

Chymotrypsin cleavage: Leu-Arg, Ser-Ile-Arg-Val-Val-Pro-Tyr

26.57 Oxytocin, a nonapeptide hormone secreted by the pituitary gland, functions by stimulating uterine contraction and lactation during childbirth. Its sequence was determined from the following evidence:
1. Oxytocin is a cyclic compound containing a disulfide bridge between two cysteine residues.
2. When the disulfide bridge is reduced, oxytocin has the constitution Asn, Cys$_2$, Gln, Gly, Ile, Leu, Pro, Tyr.
3. Partial hydrolysis of reduced oxytocin yields seven fragments: Asp-Cys, Ile-Glu, Cys-Tyr, Leu-Gly, Tyr-Ile-Glu, Glu-Asp-Cys, Cys-Pro-Leu.
4. Gly is the C-terminal group.
5. Both Glu and Asp are present as their side-chain amides (Gln and Asn) rather than as free side-chain acids.

What is the amino acid sequence of reduced oxytocin? What is the structure of oxytocin itself?

26.58 *Aspartame,* a nonnutritive sweetener marketed under the trade name Nutra-Sweet (among others), is the methyl ester of a simple dipeptide, Asp-Phe-OCH$_3$.
(a) Draw the structure of aspartame.
(b) The isoelectric point of aspartame is 5.9. Draw the principal structure present in aqueous solution at this pH.
(c) Draw the principal form of aspartame present at physiological pH = 7.3.

26.59 Refer to Figure 26.2 and propose a mechanism for the final step in the Edman degradation—the acid-catalyzed rearrangement of the ATZ derivative to the PTH derivative.

26.60 Amino acids are metabolized by a transamination reaction in which the −NH$_2$ group of the amino acid changes places with the keto group of an α-keto acid. The products are a new amino acid and a new α-keto acid. Show the product from transamination of isoleucine.

26.61 ■ The first step in the biological degradation of histidine is formation of a 4-methylideneimidazol-5-one (MIO) by cyclization of a segment of the peptide chain in the histidine ammonia lyase enzyme. Propose a mechanism.

4-Methylideneimidazol-5-one (MIO)

26.62 The first step in the biological degradation of lysine is reductive amination with α-ketoglutarate to give saccharopine. Nicotinamide adenine dinucleotide phosphate (NADPH), a relative of NADH, is the reducing agent. Show the mechanism.

27

Biomolecules: Lipids

Lipids are naturally occurring organic molecules that have limited solubility in water and can be isolated from organisms by extraction with nonpolar organic solvents. Fats, oils, waxes, many vitamins and hormones, and most nonprotein cell-membrane components are examples. Note that this definition differs from the sort used for carbohydrates and proteins in that lipids are defined by a physical property (solubility) rather than by structure. Of the many kinds of lipids, we'll be concerned in this chapter only with a few: triacylglycerols, eicosanoids, terpenoids, and steroids.

Lipids are classified into two broad types: those like fats and waxes, which contain ester linkages and can be hydrolyzed, and those like cholesterol and other steroids, which don't have ester linkages and can't be hydrolyzed.

Animal fat—a triester
(R, R′, R″ = C$_{11}$–C$_{19}$ chains)

Cholesterol

WHY THIS CHAPTER?

We've now covered two of the four major classes of biomolecules—proteins and carbohydrates—and have two remaining. We'll cover lipids, the largest and most diverse class of biomolecules, in this chapter, looking both at their structure and function and at their metabolism.

Sean Duggan

1060

27.1 | Waxes, Fats, and Oils

Waxes are mixtures of esters of long-chain carboxylic acids with long-chain alcohols. The carboxylic acid usually has an even number of carbons from 16 through 36, while the alcohol has an even number of carbons from 24 through 36. One of the major components of beeswax, for instance, is triacontyl hexadecanoate, the ester of the C_{30} alcohol triacontanol and the C_{16} acid hexadecanoic acid. The waxy protective coatings on most fruits, berries, leaves, and animal furs have similar structures.

$$CH_3(CH_2)_{14}\overset{\overset{\displaystyle O}{\|}}{C}O(CH_2)_{29}CH_3$$

Triacontyl hexadecanoate (from beeswax)

Animal **fats** and vegetable **oils** are the most widely occurring lipids. Although they appear different—animal fats like butter and lard are solids, whereas vegetable oils like corn and peanut oil are liquid—their structures are closely related. Chemically, fats and oils are *triglycerides,* or **triacylglycerols**—triesters of glycerol with three long-chain carboxylic acids called **fatty acids**. Animals use fats for long-term energy storage because they are much less highly oxidized than carbohydrates and provide about six times as much energy as an equal weight of stored, hydrated glycogen.

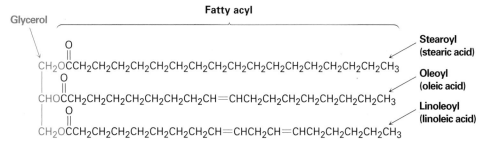

A triacylglycerol

Hydrolysis of a fat or oil with aqueous NaOH yields glycerol and three fatty acids. The fatty acids are generally unbranched and contain an even number of carbon atoms between 12 and 20. If double bonds are present, they have largely, although not entirely, *Z*, or cis, geometry. The three fatty acids of a specific triacylglycerol molecule need not be the same, and the fat or oil from a given source is likely to be a complex mixture of many different triacylglycerols. Table 27.1 lists some of the commonly occurring fatty acids, and Table 27.2 lists the approximate composition of fats and oils from different sources.

More than 100 different fatty acids are known, and about 40 occur widely. Palmitic acid (C_{16}) and stearic acid (C_{18}) are the most abundant saturated fatty acids; oleic and linoleic acids (both C_{18}) are the most abundant unsaturated ones. Oleic acid is *monounsaturated* since it has only one double bond, whereas linoleic, linolenic, and arachidonic acids are **polyunsaturated fatty acids** because they have more than one double bond. Linoleic and linolenic

Table 27.1 | **Structures of Some Common Fatty Acids**

Name	No. of carbons	Melting point (°C)	Structure
Saturated			
Lauric	12	43.2	$CH_3(CH_2)_{10}CO_2H$
Myristic	14	53.9	$CH_3(CH_2)_{12}CO_2H$
Palmitic	16	63.1	$CH_3(CH_2)_{14}CO_2H$
Stearic	18	68.8	$CH_3(CH_2)_{16}CO_2H$
Arachidic	20	76.5	$CH_3(CH_2)_{18}CO_2H$
Unsaturated			
Palmitoleic	16	−0.1	$(Z)\text{-}CH_3(CH_2)_5CH{=}CH(CH_2)_7CO_2H$
Oleic	18	13.4	$(Z)\text{-}CH_3(CH_2)_7CH{=}CH(CH_2)_7CO_2H$
Linoleic	18	−12	$(Z,Z)\text{-}CH_3(CH_2)_4(CH{=}CHCH_2)_2(CH_2)_6CO_2H$
Linolenic	18	−11	$(\text{all } Z)\text{-}CH_3CH_2(CH{=}CHCH_2)_3(CH_2)_6CO_2H$
Arachidonic	20	−49.5	$(\text{all } Z)\text{-}CH_3(CH_2)_4(CH{=}CHCH_2)_4CH_2CH_2CO_2H$

Table 27.2 | **Approximate Composition of Some Fats and Oils**

	Saturated fatty acids (%)				Unsaturated fatty acids (%)	
Source	C_{12} lauric	C_{14} myristic	C_{16} palmitic	C_{18} stearic	C_{18} oleic	C_{18} linoleic
Animal fat						
Lard	—	1	25	15	50	6
Butter	2	10	25	10	25	5
Human fat	1	3	25	8	46	10
Whale blubber	—	8	12	3	35	10
Vegetable oil						
Coconut	50	18	8	2	6	1
Corn	—	1	10	4	35	45
Olive	—	1	5	5	80	7
Peanut	—	—	7	5	60	20

acids occur in cream and are essential in the human diet; infants grow poorly and develop skin lesions if fed a diet of nonfat milk for prolonged periods.

$$CH_3CH_2CH_2CH_2CH_2CH_2CH_2CH_2CH_2CH_2CH_2CH_2CH_2CH_2CH_2CH_2CH_2\overset{\displaystyle O}{\overset{\|}{C}}OH$$

Stearic acid

CH₃CH₂CH=CHCH₂CH=CHCH₂CH=CHCH₂CH₂CH₂CH₂CH₂CH₂CH₂COH

$$CH_3CH_2CH{=}CHCH_2CH{=}CHCH_2CH{=}CHCH_2CH_2CH_2CH_2CH_2CH_2CH_2\overset{\displaystyle O}{\overset{\|}{C}}OH$$

Linolenic acid, a polyunsaturated fatty acid

The data in Table 27.1 show that unsaturated fatty acids generally have lower melting points than their saturated counterparts, a trend that is also true for triacylglycerols. Since vegetable oils generally have a higher proportion of unsaturated to saturated fatty acids than animal fats (Table 27.2), they have lower melting points. The difference is a consequence of structure. Saturated fats have a uniform shape that allows them to pack together efficiently in a crystal lattice. In unsaturated vegetable oils, however, the C=C bonds introduce bends and kinks into the hydrocarbon chains, making crystal formation more difficult. The more double bonds there are, the harder it is for the molecules to crystallize and the lower the melting point of the oil.

The C=C bonds in vegetable oils can be reduced by catalytic hydrogenation, typically carried out at high temperature using a nickel catalyst, to produce saturated solid or semisolid fats. Margarine and shortening are produced by hydrogenating soybean, peanut, or cottonseed oil until the proper consistency is obtained. Unfortunately, the hydrogenation reaction is accompanied by some cis–trans isomerization of the double bonds that remain, producing fats with about 10% to 15% trans unsaturated fatty acids. Dietary intake of trans fatty acids increases cholesterol levels in the blood, thereby increasing the risk of heart problems. The conversion of linoleic acid into elaidic acid is an example.

Linoleic acid

Elaidic acid

Problem 27.1 Carnauba wax, used in floor and furniture polishes, contains an ester of a C_{32} straight-chain alcohol with a C_{20} straight-chain carboxylic acid. Draw its structure.

Problem 27.2 Draw structures of glyceryl tripalmitate and glyceryl trioleate. Which would you expect to have a higher melting point?

27.2 | Soap

Soap has been known since at least 600 BC, when the Phoenicians prepared a curdy material by boiling goat fat with extracts of wood ash. The cleansing properties of soap weren't generally recognized, however, and the use of soap did not become widespread until the 18th century. Chemically, soap is a mixture of the sodium or potassium salts of the long-chain fatty acids produced by hydrolysis *(saponification)* of animal fat with alkali. Wood ash was used as a source of alkali until the early 1800s, when the development of the LeBlanc process for making Na_2CO_3 by heating sodium sulfate with limestone became available.

A fat
(R = C_{11}–C_{19} aliphatic chains)

Crude soap curds contain glycerol and excess alkali as well as soap but can be purified by boiling with water and adding NaCl or KCl to precipitate the pure carboxylate salts. The smooth soap that precipitates is dried, perfumed, and pressed into bars for household use. Dyes are added to make colored soaps, antiseptics are added for medicated soaps, pumice is added for scouring soaps, and air is blown in for soaps that float. Regardless of these extra treatments and regardless of price, though, all soaps are basically the same.

Soaps act as cleansers because the two ends of a soap molecule are so different. The carboxylate end of the long-chain molecule is ionic and therefore hydrophilic (Section 2.13), or attracted to water. The long hydrocarbon portion of the molecule, however, is nonpolar and hydrophobic, avoiding water and therefore more soluble in oils. The net effect of these two opposing tendencies is that soaps are attracted to both oils and water and are therefore useful as cleansers.

When soaps are dispersed in water, the long hydrocarbon tails cluster together on the inside of a tangled, hydrophobic ball, while the ionic heads on the surface of the cluster stick out into the water layer. These spherical clusters, called **micelles**, are shown schematically in Figure 27.1. Grease and oil droplets are solubilized in water when they are coated by the nonpolar tails of soap molecules in the center of micelles. Once solubilized, the grease and dirt can be rinsed away.

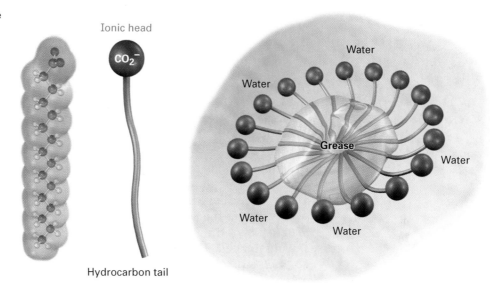

Figure 27.1 A soap micelle solubilizing a grease particle in water. An electrostatic potential map of a fatty acid carboxylate shows how the negative charge is located in the head group.

As useful as they are, soaps also have some drawbacks. In hard water, which contains metal ions, soluble sodium carboxylates are converted into insoluble magnesium and calcium salts, leaving the familiar ring of scum around bathtubs and the gray tinge on white clothes. Chemists have circumvented these problems by synthesizing a class of synthetic detergents based on salts of long-chain alkylbenzenesulfonic acids. The principle of synthetic detergents is the same as that of soaps: the alkylbenzene end of the molecule is attracted to grease, while the anionic sulfonate end is attracted to water. Unlike soaps, though, sulfonate detergents don't form insoluble metal salts in hard water and don't leave an unpleasant scum.

**A synthetic detergent
(R = a mixture of C$_{12}$ chains)**

Problem 27.3 Draw the structure of magnesium oleate, a component of bathtub scum.

Problem 27.4 Write the saponification reaction of glyceryl dioleate monopalmitate with aqueous NaOH.

27.3 | Phospholipids

CENGAGENOW™ Click *Organic Interactive* to **learn to identify common phospholipids by their charge and type.**

Just as waxes, fats, and oils are esters of carboxylic acids, **phospholipids** are diesters of phosphoric acid, H_3PO_4.

A phosphoric acid monoester   **A phosphoric acid diester**   **A phosphoric acid triester**   **A carboxylic acid ester**

Phospholipids are of two general kinds: *glycerophospholipids* and *sphingomyelins*. Glycerophospholipids are based on phosphatidic acid, which contains a glycerol backbone linked by ester bonds to two fatty acids and one phosphoric acid. Although the fatty-acid residues can be any of the C_{12}–C_{20} units typically present in fats, the acyl group at C1 is usually saturated and the one at C2 is usually unsaturated. The phosphate group at C3 is also bonded to an amino alcohol such as choline $[HOCH_2CH_2N(CH_3)_3]^+$, ethanolamine $(HOCH_2CH_2NH_2)$, or serine $[HOCH_2CH(NH_2)CO_2H]$. The compounds are chiral and have an L, or R, configuration at C2.

Phosphatidic acid   **Phosphatidylcholine**   **Phosphatidylethanolamine**   **Phosphatidylserine**

Sphingomyelins are the second major group of phospholipids. These compounds have sphingosine or a related dihydroxyamine as their backbone and are particularly abundant in brain and nerve tissue, where they are a major constituent of the coating around nerve fibers.

CH$_2$(CH$_2$)$_{15-23}$CH$_3$

O=C

Sphingosine

A sphingomyelin

Phospholipids are found widely in both plant and animal tissues and make up approximately 50% to 60% of cell membranes. Because they are like soaps in having a long, nonpolar hydrocarbon tail bound to a polar ionic head, phospholipids in the cell membrane organize into a **lipid bilayer** about 5.0 nm (50 Å) thick. As shown in Figure 27.2, the nonpolar tails aggregate in the center of the bilayer in much the same way that soap tails aggregate in the center of a micelle. This bilayer serves as an effective barrier to the passage of water, ions, and other components into and out of cells.

Figure 27.2 Aggregation of glycerophospholipids into the lipid bilayer that composes cell membranes.

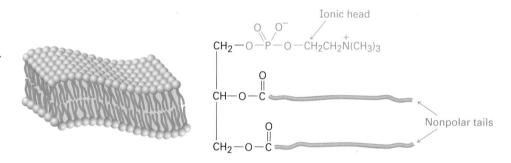

27.4 Prostaglandins and Other Eicosanoids

The **prostaglandins** are a group of C$_{20}$ lipids that contain a five-membered ring with two long side chains. First isolated in the 1930s by Ulf von Euler at the Karolinska Institute in Sweden, much of the structural and chemical work on the prostaglandins was carried out by Sune Bergström and Bengt Samuelsson. The name *prostaglandin* derives from the fact that the compounds were first isolated from sheep prostate glands, but they have subsequently been shown to be present in small amounts in all body tissues and fluids.

The several dozen known prostaglandins have an extraordinarily wide range of biological effects. Among their many properties, they can lower blood pressure, affect blood-platelet aggregation during clotting, lower gastric secretions, control inflammation, affect kidney function, affect reproductive systems, and stimulate uterine contractions during childbirth.

Prostaglandins, together with related compounds called thromboxanes and leukotrienes, make up a class of compounds called **eicosanoids** because they are derived biologically from 5,8,11,14-eicosatetraenoic acid, or arachidonic acid (Figure 27.3). Prostaglandins (PG) have a cyclopentane ring with two long side chains; thromboxanes (TX) have a six-membered, oxygen-containing ring; and leukotrienes (LT) are acyclic.

Figure 27.3 Structures of some representative eicosanoids. All are derived biologically from arachidonic acid.

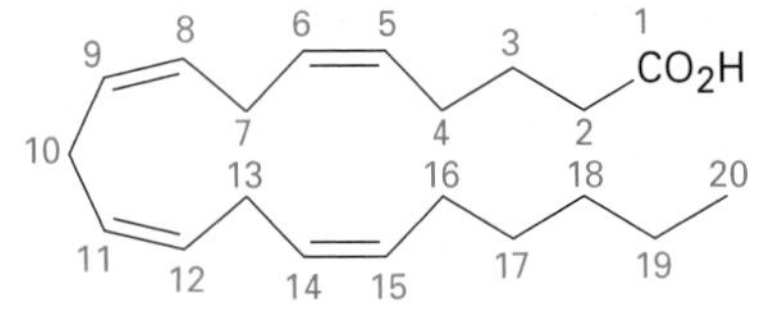

Arachidonic acid

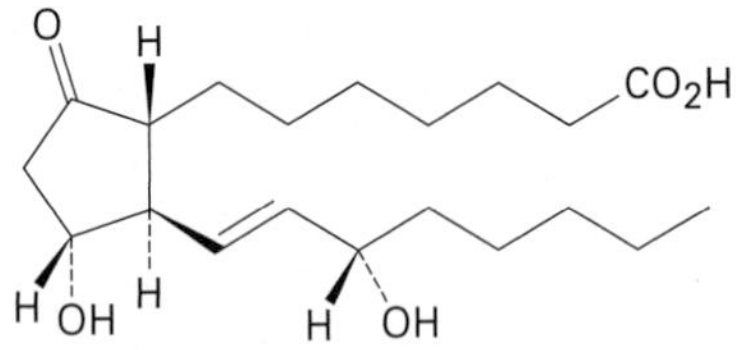

Prostaglandin E$_1$ (PGE$_1$)

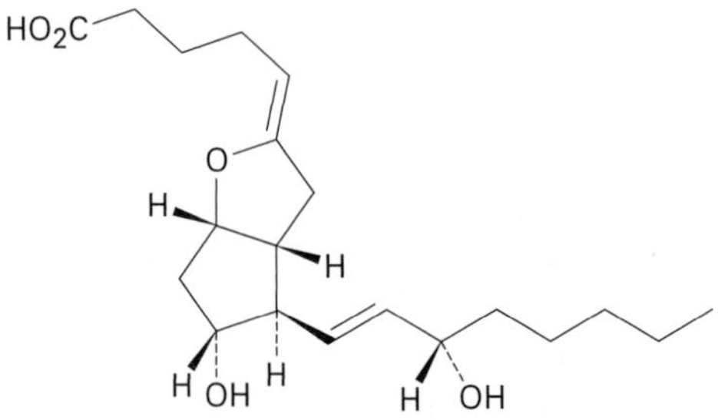

Prostaglandin I$_2$ (PGI$_2$) (prostacyclin)

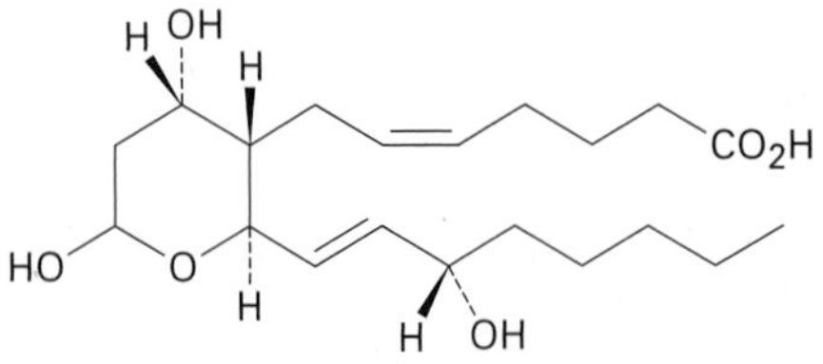

Thromboxane B$_2$ (TXB$_2$)

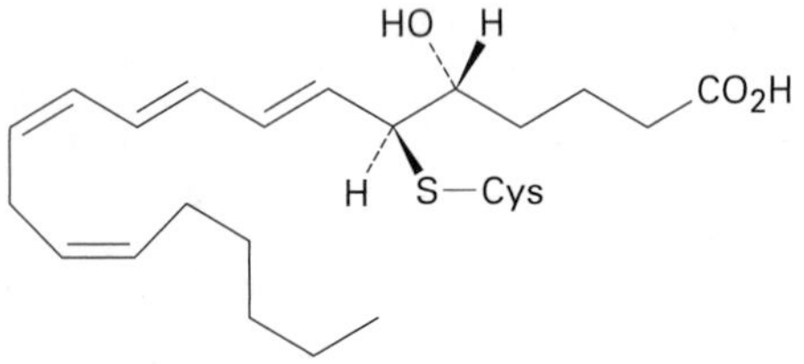

Leukotriene E$_4$ (LTE$_4$)

Ulf Svante von Euler	Sune K. Bergström	Bengt Samuelsson
Ulf Svante von Euler (1905–1983) was born in Stockholm, Sweden, to a distinguished academic family. His father, Hans von Euler-Chelpin, received the 1929 Nobel Prize in chemistry; his godfather, Svante Arrhenius, received the 1903 Nobel Prize in chemistry; and his mother had a Ph.D. in botany. Von Euler received an M.D. from the Karolinska Institute in 1930, and then remained there his entire career (1930–1971). He received the 1970 Nobel Prize in medicine for his work on the chemical transmission of nerve impulses.	**Sune K. Bergström** (1916–2004) was born in Stockholm, Sweden, and received an M.D. from the Karolinska Institute in 1944. He was professor at the University of Lund (1947–1958) before moving back to the Karolinska Institute in 1958. He shared the 1982 Nobel Prize in medicine for his work on identifying and studying the prostaglandins.	**Bengt Samuelsson** (1934–) was born in Halmstad, Sweden, and received both Ph.D. (1960) and M.D. (1961) degrees from the Karolinska Institute, where he worked with Sune Bergström. He remained at the Karolinska Institute as professor and shared the 1982 Nobel Prize in medicine with Bergström and John R. Vane.

Eicosanoids are named based on their ring system (PG, TX, or LT), substitution pattern, and number of double bonds. The various substitution patterns on the ring are indicated by letter as in Figure 27.4, and the number of double bonds is indicated by a subscript. Thus, PGE_1 is a prostaglandin with the "E" substitution pattern and one double bond. The numbering of the atoms in the various eicosanoids is the same as in arachidonic acid, starting with the $-CO_2H$ carbon as C1, continuing around the ring, and ending with the $-CH_3$ carbon at the other end of the chain as C20.

Figure 27.4 The nomenclature system for eicosanoids.

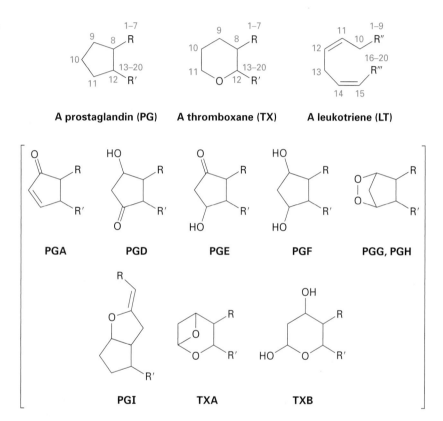

Eicosanoid biosynthesis begins with the conversion of arachidonic acid to PGH_2, catalyzed by the multifunctional PGH synthase (PGHS), also called cyclooxygenase (COX). There are two distinct enzymes, PGHS-1 and PGHS-2 (or COX-1 and COX-2), both of which accomplish the same reaction but appear to function independently. COX-1 carries out the normal physiological production of prostaglandins, and COX-2 produces additional prostaglandin in response to arthritis or other inflammatory conditions. Vioxx, Celebrex, Bextra, and several other drugs selectively inhibit the COX-2 enzyme but also appear to cause potentially serious heart problems in weakened patients. (See the Chapter 15 *Focus On*.)

PGHS accomplishes two transformations, an initial reaction of arachidonic acid with O_2 to yield PGG_2 and a subsequent reduction of the hydroperoxide group ($-OOH$) to the alcohol PGH_2. The sequence of steps involved in these transformations was shown in Figure 7.9, page 244.

Further processing of PGH$_2$ then leads to other eicosanoids. PGE$_2$, for instance, arises by an isomerization of PGH$_2$ catalyzed by PGE synthase (PGES). The coenzyme glutathione is needed for enzyme activity, although it is not chemically changed during the isomerization and its role is not fully understood. One possibility is that the glutathione thiolate anion breaks the O−O bond in PGH$_2$ by an S$_N$2-like attack on one of the oxygen atoms, giving a thioperoxy intermediate (R−S−O−R′) that eliminates glutathione to give the ketone (Figure 27.5).

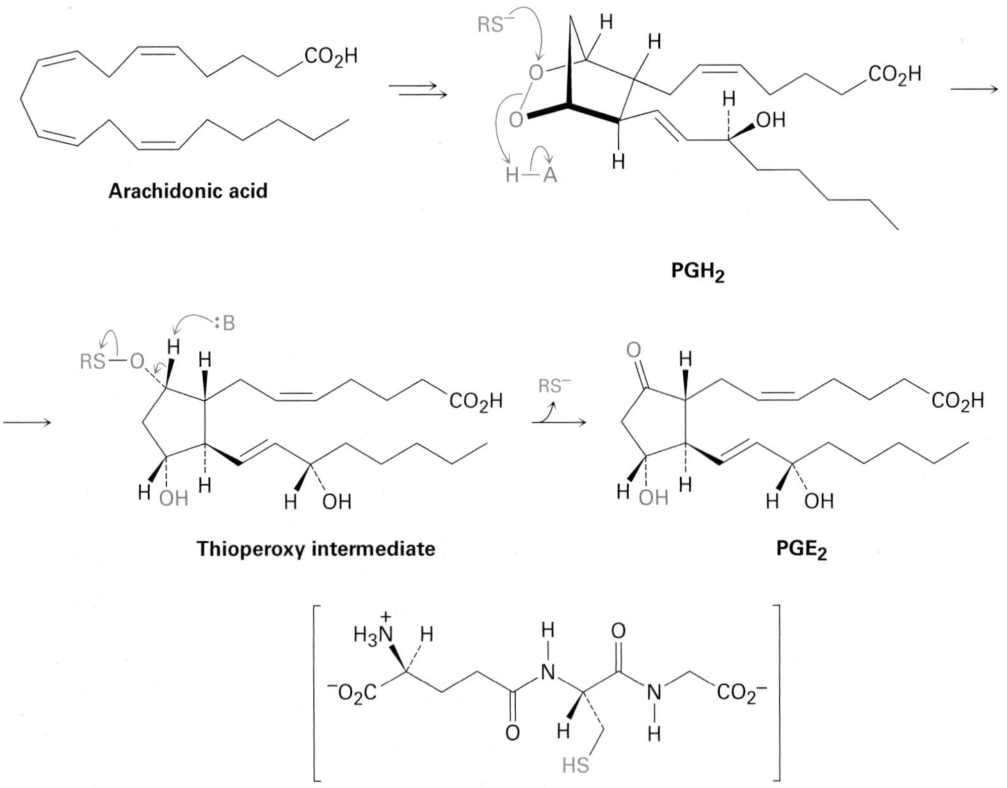

Arachidonic acid

PGH$_2$

Thioperoxy intermediate

PGE$_2$

Glutathione

Figure 27.5 Mechanism of the conversion of PGH$_2$ into PGE$_2$.

Problem 27.5 Assign *R* or *S* configuration to each chirality center in prostaglandin E$_2$ (Figure 27.5), the most abundant and biologically potent of mammalian prostaglandins.

27.5 | Terpenoids

In the Chapter 6 *Focus On,* "Terpenes: Naturally Occurring Alkenes," we looked briefly at **terpenoids,** a vast and diverse group of lipids found in all living organisms. Despite their apparent structural differences, all terpenoids are related. All contain a multiple of five carbons and are derived biosynthetically from the five-carbon precursor isopentenyl diphosphate (Figure 27.6). Note that formally, a *terpenoid* contains oxygen, while a *terpene* is a hydrocarbon. For simplicity, we'll use the term *terpenoid* to refer to both.

Figure 27.6 Structures of some representative terpenoids.

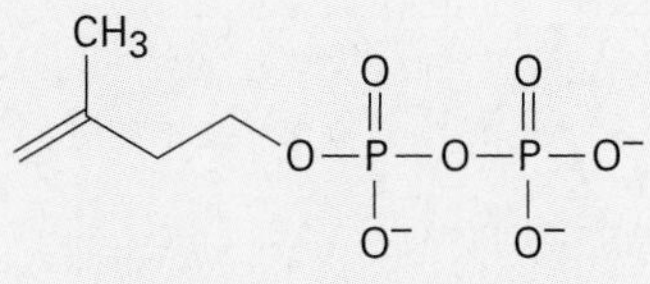

Isopentenyl diphosphate

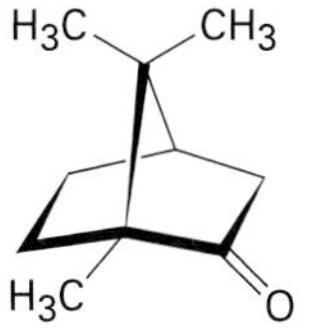

Camphor
(a monoterpenoid—C_{10})

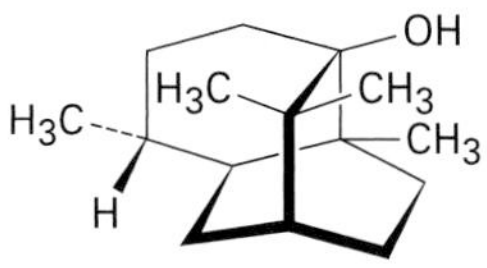

Patchouli alcohol
(a sesquiterpenoid—C_{15})

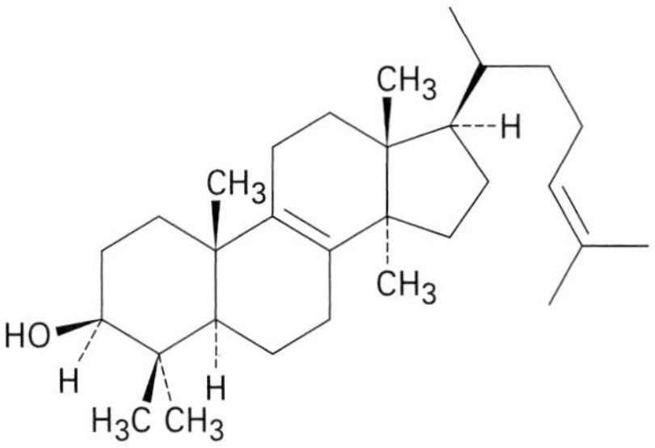

Lanosterol
(a triterpenoid—C_{30})

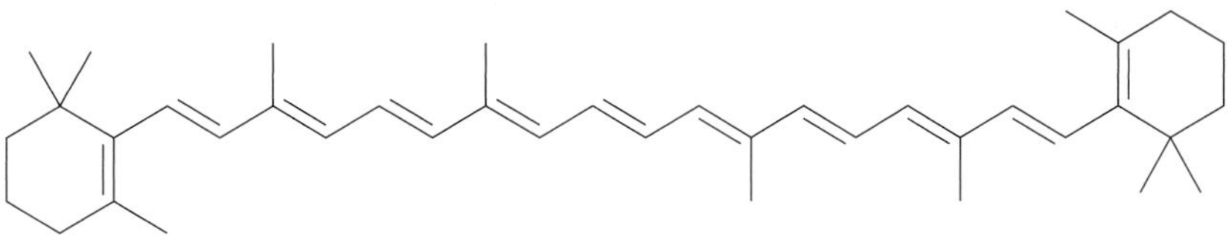

β-Carotene
(a tetraterpenoid—C_{40})

Terpenoids are classified according to the number of five-carbon multiples they contain. *Monoterpenoids* contain 10 carbons and are derived from two isopentenyl diphosphates, *sesquiterpenoids* contain 15 carbons and are derived from three isopentenyl diphosphates, *diterpenoids* contain 20 carbons and are derived from four isopentenyl diphosphates, and so on, up to triterpenoids (C_{30}) and tetraterpenoids (C_{40}). Monoterpenoids and sesquiterpenoids are found primarily in plants, bacteria, and fungi, but the higher terpenoids occur in both plants and animals. The triterpenoid lanosterol, for example, is the precursor from which steroid hormones are made, and the tetraterpenoid β-carotene is a dietary source of vitamin A (Figure 27.6).

The terpenoid precursor isopentenyl diphosphate, formerly called isopentenyl pyrophosphate and abbreviated IPP, is biosynthesized by two different pathways depending on the organism and the structure of the final product. In animals and higher plants, sesquiterpenoids and triterpenoids arise primarily from the *mevalonate* pathway, whereas monoterpenoids, diterpenoids, and tetraterpenoids are biosynthesized by the *1-deoxyxylulose 5-phosphate (DXP)* pathway. In bacteria, both pathways are used. We'll look only at the mevalonate pathway, which is more common and better understood at present.

(R)-Mevalonate

Isopentenyl diphosphate (IPP) $\longrightarrow$ **Terpenoids**

1-Deoxy-D-xylulose 5-phosphate

The Mevalonate Pathway to Isopentenyl Diphosphate

As summarized in Figure 27.7, the mevalonate pathway begins with the conversion of acetate to acetyl CoA, followed by Claisen condensation to yield acetoacetyl CoA. A second carbonyl condensation reaction with a third molecule of acetyl CoA, this one an aldol-like process, then yields the six-carbon compound 3-hydroxy-3-methylglutaryl CoA, which is reduced to give mevalonate. Phosphorylation, followed by loss of CO_2 and phosphate ion, completes the process.

Step 1 of Figure 27.7: Claisen Condensation The first step in mevalonate biosynthesis is a Claisen condensation (Section 23.7) to yield acetoacetyl CoA, a reaction catalyzed by acetoacetyl-CoA acetyltransferase. An acetyl group is first bound to the enzyme by a nucleophilic acyl substitution reaction with a cysteine −SH group. Formation of an enolate ion from a second molecule of acetyl CoA, followed by Claisen condensation, then yields the product.

Acetyl CoA

Acetoacetyl CoA

Step 2 of Figure 27.7: Aldol Condensation Acetoacetyl CoA next undergoes an aldol-like addition (Section 23.1) of an acetyl CoA enolate ion in a reaction catalyzed by 3-hydroxy-3-methylglutaryl-CoA synthase. The reaction again occurs

Figure 27.7 MECHANISM:
The mevalonate pathway for the biosynthesis of isopentenyl diphosphate from three molecules of acetyl CoA. Individual steps are explained in the text.

1 Claisen condensation of two molecules of acetyl CoA gives acetoacetyl CoA.

2 Aldol-like condensation of acetoacetyl CoA with a third molecule of acetyl CoA, followed by hydrolysis, gives (3S)-3-hydroxy-3-methylglutaryl CoA.

3 Reduction of the thioester group by 2 equivalents of NADPH gives (R)-mevalonate, a dihydroxy acid.

4 Phosphorylation of the tertiary hydroxyl and diphosphorylation of the primary hydroxyl, followed by decarboxylation and simultaneous expulsion of phosphate, gives isopentenyl diphosphate, the precursor of terpenoids.

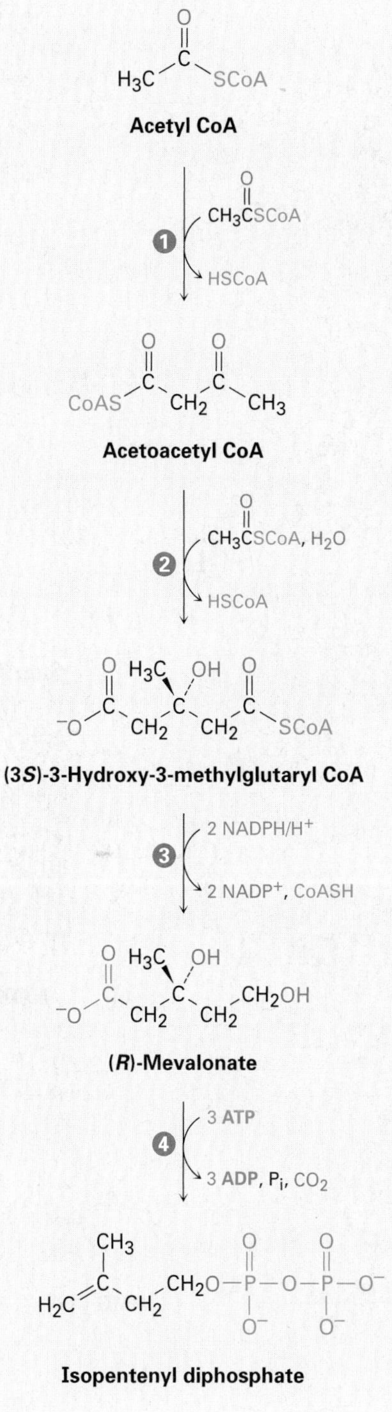

Acetyl CoA

Acetoacetyl CoA

(3S)-3-Hydroxy-3-methylglutaryl CoA

(R)-Mevalonate

Isopentenyl diphosphate

© John McMurry

by initial formation of a thioester bond between the substrate and a cysteine —SH group in the enzyme, followed by enolate-ion addition and subsequent hydrolysis to give (3S)-3-hydroxy-3-methylglutaryl CoA (HMG-CoA).

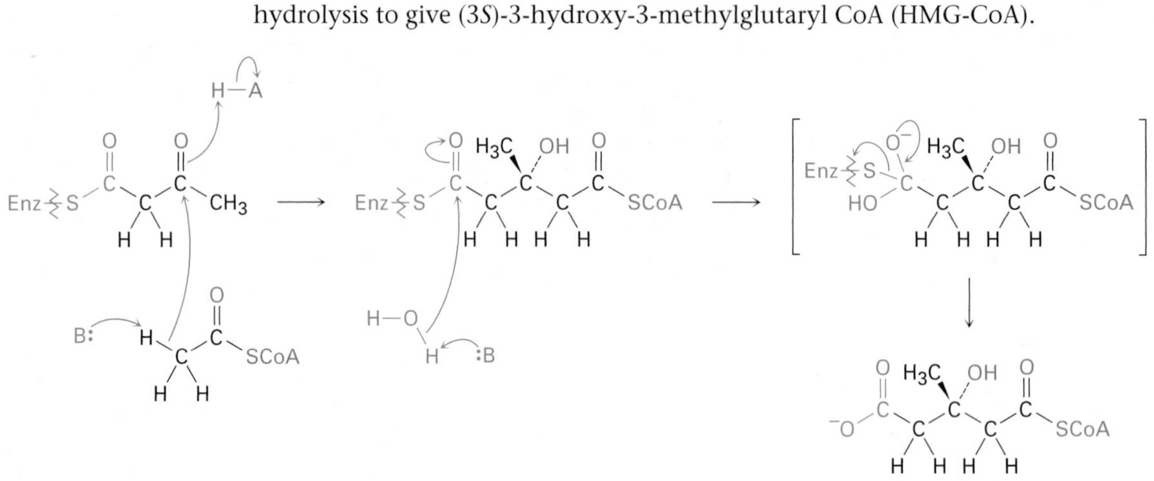

(3S)-3-Hydroxy-3-methylglutaryl CoA (HMG-CoA)

Step 3 of Figure 27.7: Reduction   Reduction of HMG-CoA to give (R)-mevalonate is catalyzed by 3-hydroxy-3-methylglutaryl-CoA reductase and requires two equivalents of reduced nicotinamide adenine dinucleotide phosphate (NADPH), a close relative of NADH (Section 19.12). The reaction occurs in several steps and proceeds through an aldehyde intermediate. The first step is a nucleophilic acyl substitution reaction involving hydride transfer from NADPH to the thioester carbonyl group of HMG-CoA. Following expulsion of HSCoA as leaving group, the aldehyde intermediate undergoes a second hydride addition to give mevalonate.

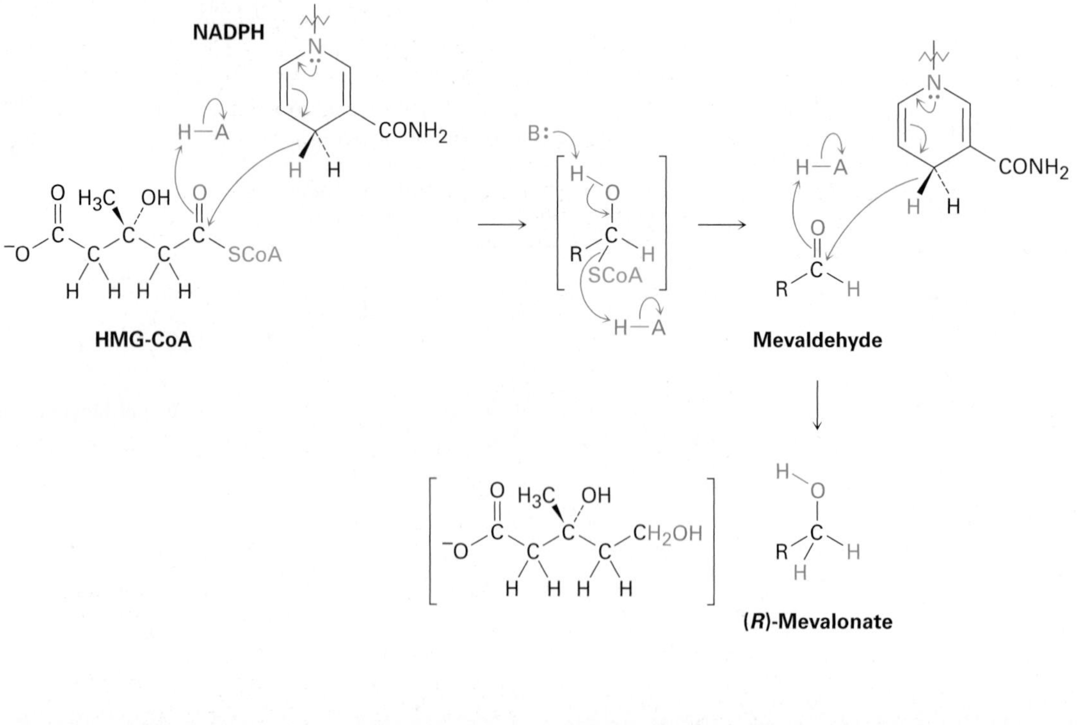

HMG-CoA

Mevaldehyde

(R)-Mevalonate

Step 4 of Figure 27.7: Phosphorylation and Decarboxylation Three additional reactions are needed to convert mevalonate to isopentenyl diphosphate. The first two are straightforward phosphorylations that occur by nucleophilic substitution reactions on the terminal phosphorus of ATP. Mevalonate is first converted to mevalonate 5-phosphate (phosphomevalonate) by reaction with ATP in a process catalyzed by mevalonate kinase. Mevalonate 5-phosphate then reacts with a second ATP to give mevalonate 5-diphosphate (diphosphomevalonate). The third reaction results in phosphorylation of the tertiary hydroxyl group, followed by decarboxylation and loss of phosphate ion.

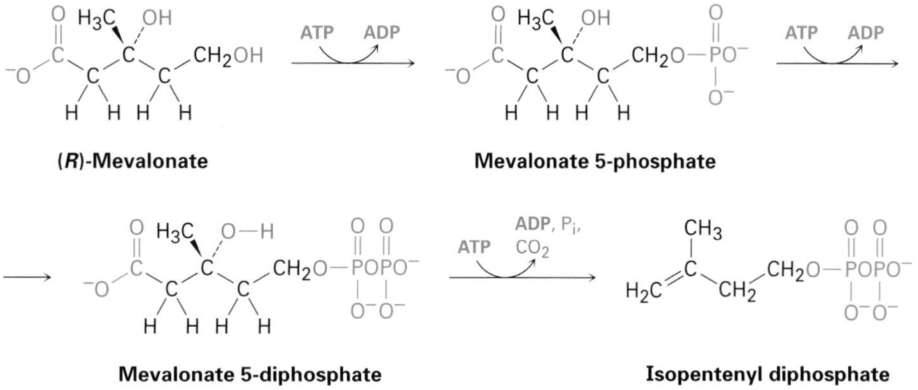

(*R*)-Mevalonate

Mevalonate 5-phosphate

Mevalonate 5-diphosphate

Isopentenyl diphosphate

The final decarboxylation of mevalonate 5-diphosphate appears unusual because decarboxylations of acids do not typically occur except in β-keto acids and malonic acids, in which the carboxylate group is two atoms away from an additional carbonyl group (Section 22.7). The function of this second carbonyl group is to act as an electron acceptor and stabilize the charge resulting from loss of CO_2. In fact, though, the decarboxylation of a β-keto acid and the decarboxylation of mevalonate 5-diphosphate are closely related.

Catalyzed by mevalonate-5-diphosphate decarboxylase, the substrate is first phosphorylated on the free —OH group by reaction with ATP to give a tertiary phosphate, which undergoes spontaneous dissociation to give a tertiary carbocation. The positive charge then acts as an electron acceptor to facilitate decarboxylation in exactly the same way a β carbonyl group does, giving isopentenyl diphosphate. (In the following structures, the diphosphate group is abbreviated OPP.)

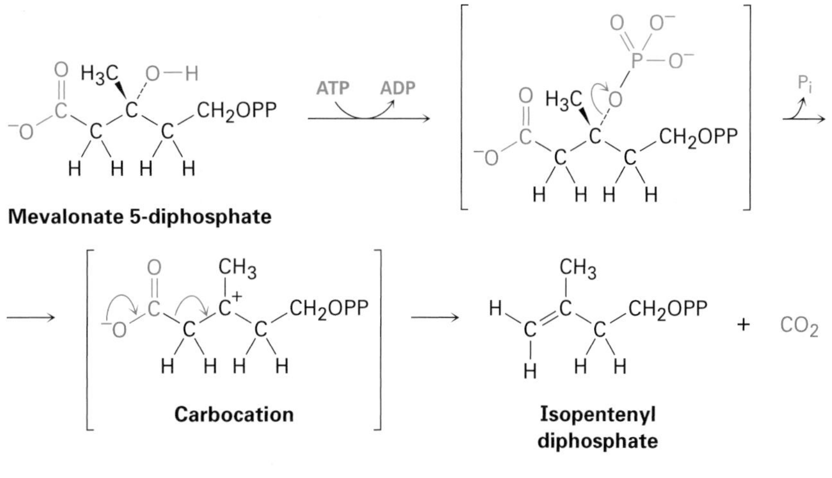

Mevalonate 5-diphosphate

Carbocation

Isopentenyl
diphosphate

+ CO_2

Problem 27.6 Studies of the conversion of mevalonate 5-phosphate to isopentenyl diphosphate have shown the following result. Which hydrogen, *pro-R* or *pro-S,* ends up cis to the methyl group, and which ends up trans?

Mevalonate 5-diphosphate   **Isopentenyl diphosphate**

Conversion of Isopentenyl Diphosphate to Terpenoids

The conversion of isopentenyl diphosphate (IPP) to terpenoids begins with its isomerization to dimethylallyl diphosphate, abbreviated DMAPP and formerly called dimethylallyl pyrophosphate. These two C_5 building blocks then combine to give the C_{10} unit geranyl diphosphate (GPP). The corresponding alcohol, geraniol, is itself a fragrant terpenoid that occurs in rose oil.

Further combination of GPP with another IPP gives the C_{15} unit farnesyl diphosphate (FPP), and so on, up to C_{25}. Terpenoids with more than 25 carbons—that is, triterpenoids (C_{30}) and tetraterpenoids (C_{40})—are synthesized by dimerization of C_{15} and C_{20} units, respectively (Figure 27.8). Triterpenoids and

Figure 27.8 An overview of terpenoid biosynthesis from isopentenyl diphosphate.

Isopentenyl diphosphate (IPP)   **Dimethylallyl diphosphate (DMAPP)**

PP_i

Geranyl diphosphate (GPP)   ⟹ **Monoterpenes (C_{10})**

IPP

PP_i

Farnesyl diphosphate (FPP)   ⟹ **Sesquiterpenes (C_{15})**

Dimerization

Squalene   ⟹ **Triterpenes (C_{30})**

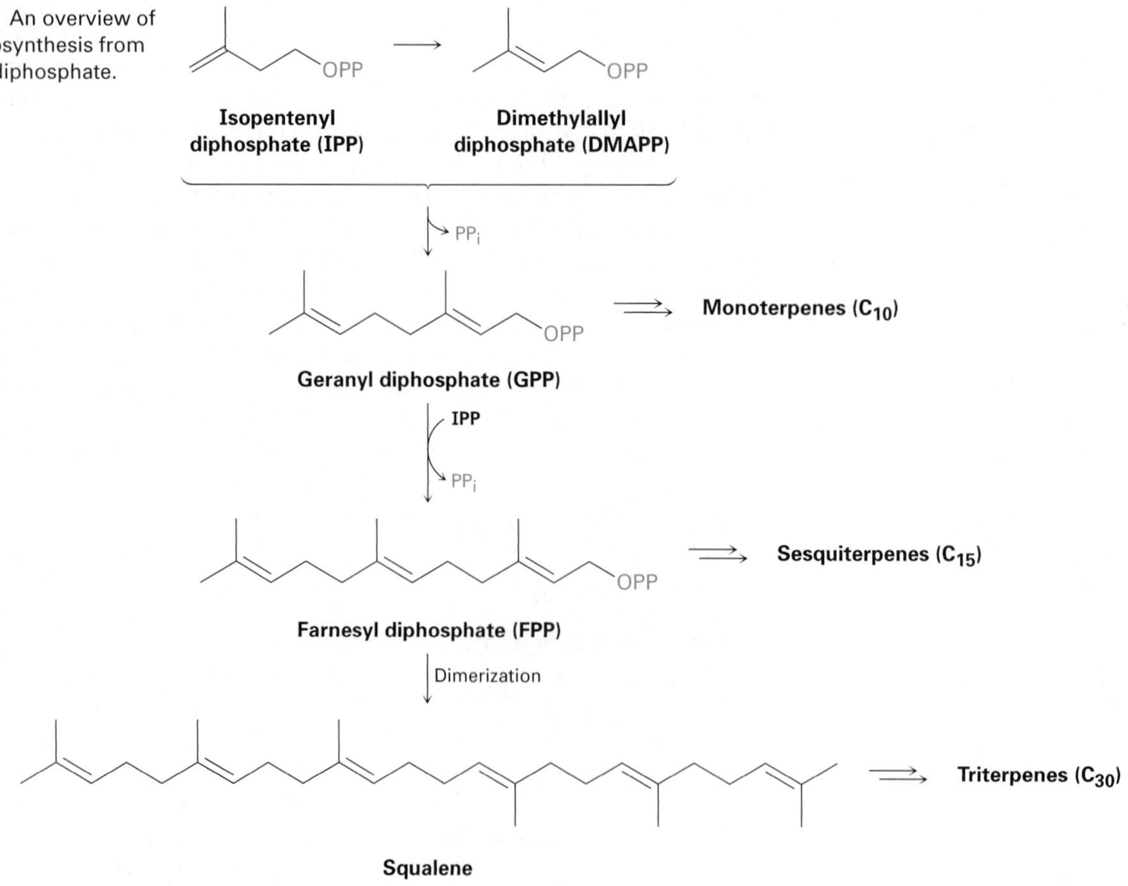

steroids, in particular, arise from reductive dimerization of farnesyl diphosphate to give squalene.

The isomerization of isopentenyl diphosphate to dimethylallyl diphosphate is catalyzed by IPP isomerase and occurs through a carbocation pathway. Protonation of the IPP double bond by a hydrogen-bonded cysteine residue in the enzyme gives a tertiary carbocation intermediate, which is deprotonated by a glutamate residue as base to yield DMAPP. X-ray structural studies on the enzyme show that it holds the substrate in an unusually deep, well-protected pocket to shield the highly reactive carbocation from reaction with solvent or other external substances.

Both the initial coupling of DMAPP with IPP to give geranyl diphosphate and the subsequent coupling of GPP with a second molecule of IPP to give farnesyl diphosphate are catalyzed by farnesyl diphosphate synthase. The process requires Mg^{2+} ion, and the key step is a nucleophilic substitution reaction in which the double bond of IPP behaves as a nucleophile in displacing diphosphate ion leaving group (PP_i). The exact mechanism of the nucleophilic substitution step—whether S_N1 or S_N2—is difficult to establish conclusively. Available evidence suggests, however, that the substrate develops considerable cationic character and that spontaneous dissociation of the allylic diphosphate ion in an S_N1-like pathway probably occurs (Figure 27.9).

The further conversion of geranyl diphosphate into monoterpenoids typically involves carbocation intermediates and multistep reaction pathways that are catalyzed by terpene cyclases. Monoterpene cyclases function by first isomerizing geranyl diphosphate to its allylic isomer linalyl diphosphate (LPP), a process that occurs by spontaneous S_N1-like dissociation to an allylic carbocation, followed by recombination. The effect of this isomerization is to convert the C2–C3 double bond of GPP into a single bond, thereby making cyclization possible and allowing *E/Z* isomerization of the double bond. Further dissociation and cyclization by electrophilic addition of the cationic carbon to the terminal double bond then gives a cyclic cation, which might either rearrange, undergo a hydride shift, be captured by a nucleophile, or be deprotonated to give any of the several hundred known monoterpenoids. As just one example, limonene, a monoterpene found in many citrus oils, arises by the biosynthetic pathway shown in Figure 27.10.

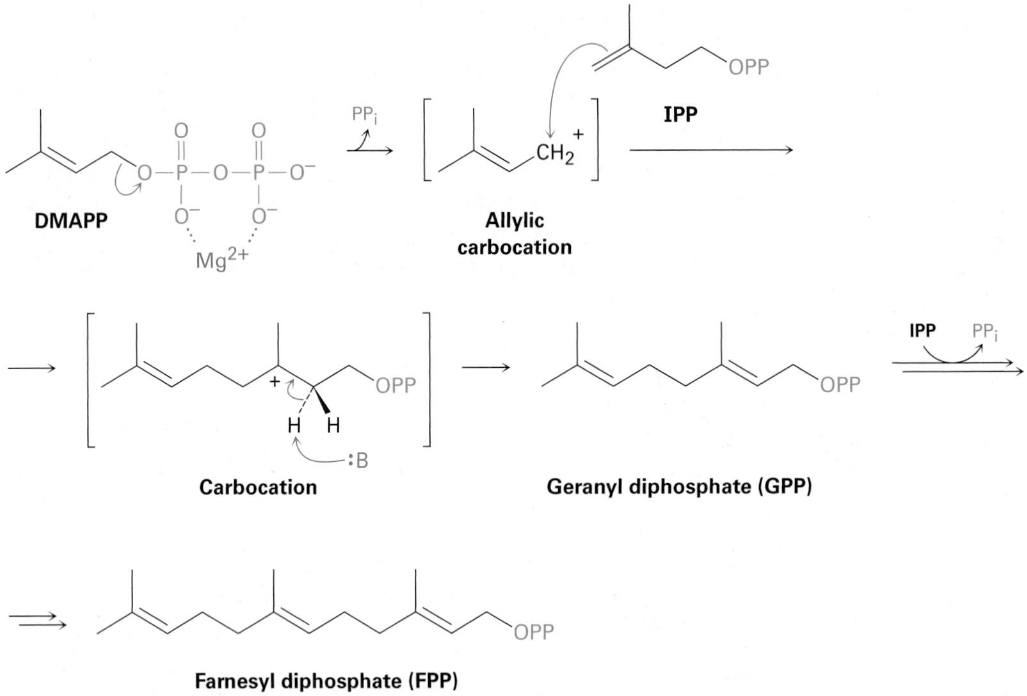

Figure 27.9 Mechanism of the coupling reaction of dimethylallyl diphosphate (DMAPP) and isopentenyl diphosphate (IPP), to give geranyl diphosphate (GPP).

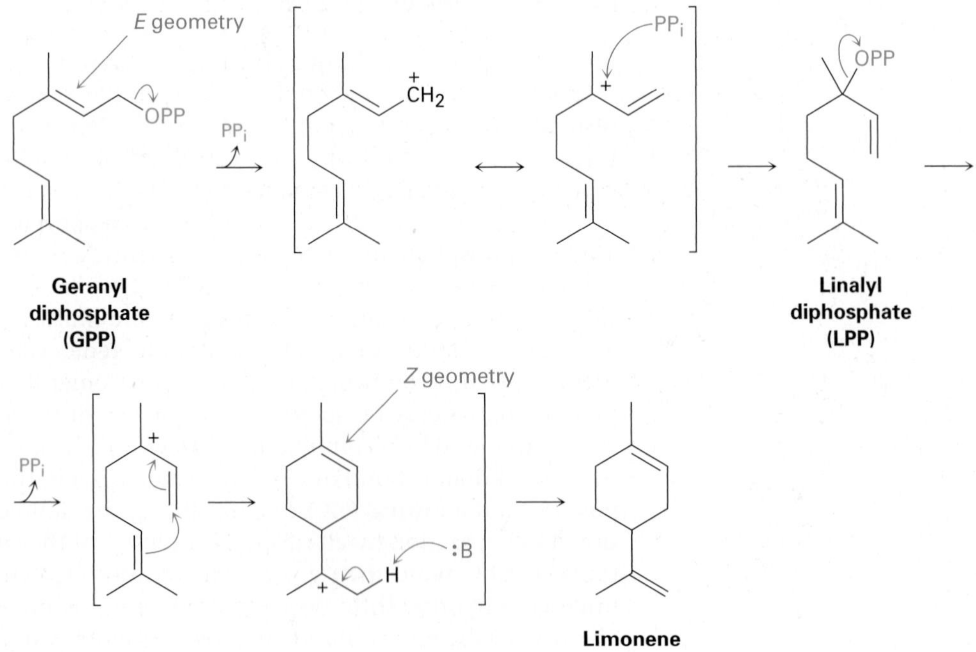

Figure 27.10 Mechanism of the formation of the monoterpene limonene from geranyl diphosphate.

WORKED EXAMPLE 27.1 ***Proposing a Terpenoid Biosynthesis Pathway***

Propose a mechanistic pathway for the biosynthesis of α-terpineol from geranyl diphosphate.

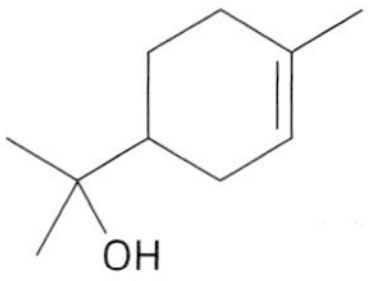

α-Terpineol

Strategy α-Terpineol, a monoterpenoid, must be derived biologically from geranyl diphosphate through its isomer linalyl diphosphate. Draw the precursor in a conformation that approximates the structure of the target molecule, and then carry out a cationic cyclization, using the appropriate double bond to displace the diphosphate leaving group. Since the target is an alcohol, the carbocation resulting from cyclization must react with water.

Solution

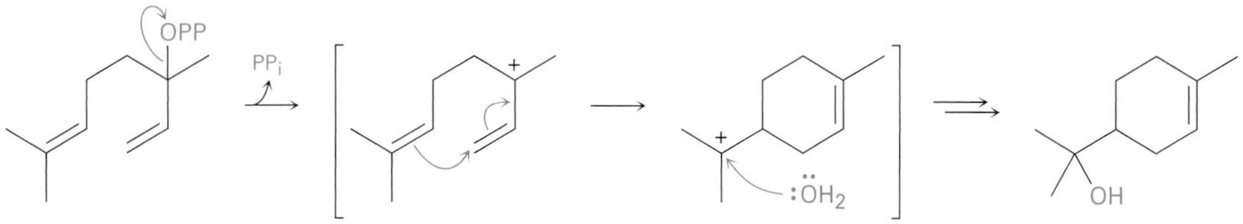

Linalyl diphosphate **α-Terpineol**

Problem 27.7 Propose mechanistic pathways for the biosynthetic formation of the following terpenes:

(a)

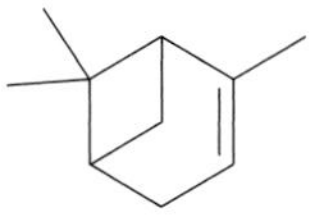

α-Pinene

(b)

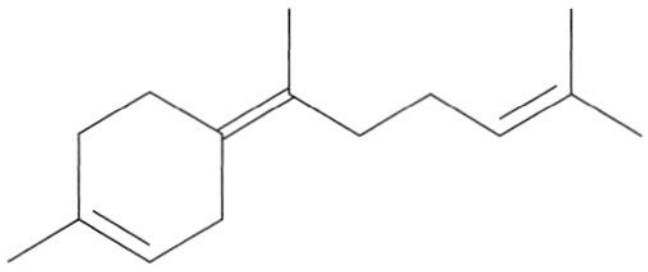

γ-Bisabolene

27.6 Steroids

CENGAGENOW™ Click *Organic Interactive* to **use a web-based palette to assign *R,S* designations to chiral centers in steroids**.

In addition to fats, phospholipids, eicosanoids, and terpenoids, the lipid extracts of plants and animals also contain **steroids**, molecules that are derived from the triterpene lanosterol (Figure 27.6) and whose structures are based on a tetracyclic ring system. The four rings are designated A, B, C, and D, beginning at the lower left, and the carbon atoms are numbered beginning in the A ring. The three six-membered rings (A, B, and C) adopt chair conformations but are prevented by their rigid geometry from undergoing the usual cyclohexane ring-flips (Section 4.6).

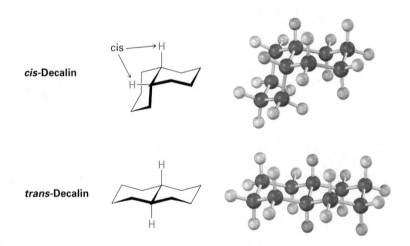

A steroid
(R = various side chains)

Two cyclohexane rings can be joined in either a cis or a trans manner. With cis fusion to give *cis*-decalin, both groups at the ring-junction positions (the *angular* groups) are on the same side of the two rings. With trans fusion to give *trans*-decalin, the groups at the ring junctions are on opposite sides.

cis-Decalin

trans-Decalin

As shown in Figure 27.11, steroids can have either a cis or a trans fusion of the A and B rings, but the other ring fusions (B–C and C–D) are usually trans. An A–B trans steroid has the C19 angular methyl group up, denoted β, and the hydrogen atom at C5 down, denoted α, on opposite sides of the molecule. An A–B cis steroid, by contrast, has both the C19 angular methyl group and the C5 hydrogen atom on the same side (β) of the molecule. Both kinds of steroids are relatively long, flat molecules that have their two methyl groups (C18 and C19) protruding axially above the ring system. The A–B trans steroids are the more common, although A–B cis steroids are found in liver bile.

Figure 27.11 Steroid conformations. The three six-membered rings have chair conformations but are unable to undergo ring-flips. The A and B rings can be either cis-fused or trans-fused.

An A–B trans steroid

An A–B cis steroid

Substituent groups on the steroid ring system can be either axial or equatorial. As with simple cyclohexanes (Section 4.7), equatorial substitution is generally more favorable than axial substitution for steric reasons. The hydroxyl group at C3 of cholesterol, for example, has the more stable equatorial orientation. Unlike what happens with simple cyclohexanes, however, steroids are rigid molecules whose geometry prevents cyclohexane ring-flips.

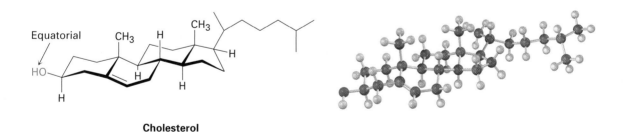

Cholesterol

Problem 27.8 Draw the following molecules in chair conformations, and tell whether the ring substituents are axial or equatorial:

Problem 27.9 | Lithocholic acid is an A–B cis steroid found in human bile. Draw lithocholic acid showing chair conformations as in Figure 27.11, and tell whether the hydroxyl group at C3 is axial or equatorial.

Lithocholic acid

Steroid Hormones

In humans, most steroids function as **hormones**, chemical messengers that are secreted by endocrine glands and carried through the bloodstream to target tissues. There are two main classes of steroid hormones: the *sex hormones,* which control maturation, tissue growth, and reproduction, and the *adrenocortical hormones,* which regulate a variety of metabolic processes.

Sex Hormones   Testosterone and androsterone are the two most important male sex hormones, or *androgens*. Androgens are responsible for the development of male secondary sex characteristics during puberty and for promoting tissue and muscle growth. Both are synthesized in the testes from cholesterol. Androstenedione is another minor hormone that has received particular attention because of its use by prominent athletes.

Testosterone

Androsterone

Androstenedione

(Androgens)

Estrone and estradiol are the two most important female sex hormones, or *estrogens*. Synthesized in the ovaries from testosterone, estrogenic hormones are responsible for the development of female secondary sex characteristics and for regulation of the menstrual cycle. Note that both have a benzene-like aromatic A ring. In addition, another kind of sex hormone called a *progestin* is essential for preparing the uterus for implantation of a fertilized ovum during pregnancy. Progesterone is the most important progestin.

Estrone

Estradiol

**Progesterone
(a progestin)**

(Estrogens)

Adrenocortical Hormones Adrenocortical steroids are secreted by the adrenal glands, small organs located near the upper end of each kidney. There are two types of adrenocortical steroids, called *mineralocorticoids* and *glucocorticoids*. Mineralocorticoids, such as aldosterone, control tissue swelling by regulating cellular salt balance between Na^+ and K^+. Glucocorticoids, such as hydrocortisone, are involved in the regulation of glucose metabolism and in the control of inflammation. Glucocorticoid ointments are widely used to bring down the swelling from exposure to poison oak or poison ivy.

**Aldosterone
(a mineralocorticoid)**

**Hydrocortisone
(a glucocorticoid)**

Synthetic Steroids In addition to the many hundreds of steroids isolated from plants and animals, thousands more have been synthesized in pharmaceutical laboratories in a search for new drugs. Among the best-known synthetic steroids are the oral contraceptives and anabolic agents. Most birth-control pills are a mixture of two compounds, a synthetic estrogen, such as ethynylestradiol, and a synthetic progestin, such as norethindrone. Anabolic steroids, such as methandrostenolone (Dianabol), are synthetic androgens that mimic the tissue-building effects of natural testosterone.

**Ethynylestradiol
(a synthetic estrogen)**

**Norethindrone
(a synthetic progestin)**

**Methandrostenolone
(Dianabol)**

27.7 | Biosynthesis of Steroids

Steroids are heavily modified triterpenoids that are biosynthesized in living organisms from farnesyl diphosphate (C_{15}) by a reductive dimerization to the acyclic hydrocarbon squalene (C_{30}), which is converted into lanosterol (Figure 27.12). Further rearrangements and degradations then take place to yield various steroids. The conversion of squalene to lanosterol is among the most intensively studied of all biosynthetic transformations, with notable contributions by Konrad Bloch and J. W. Cornforth, who received Nobel Prizes for their work. Starting from an achiral, open-chain polyene, the entire process requires only two enzymes and results in the formation of six carbon–carbon bonds, four rings, and seven chirality centers.

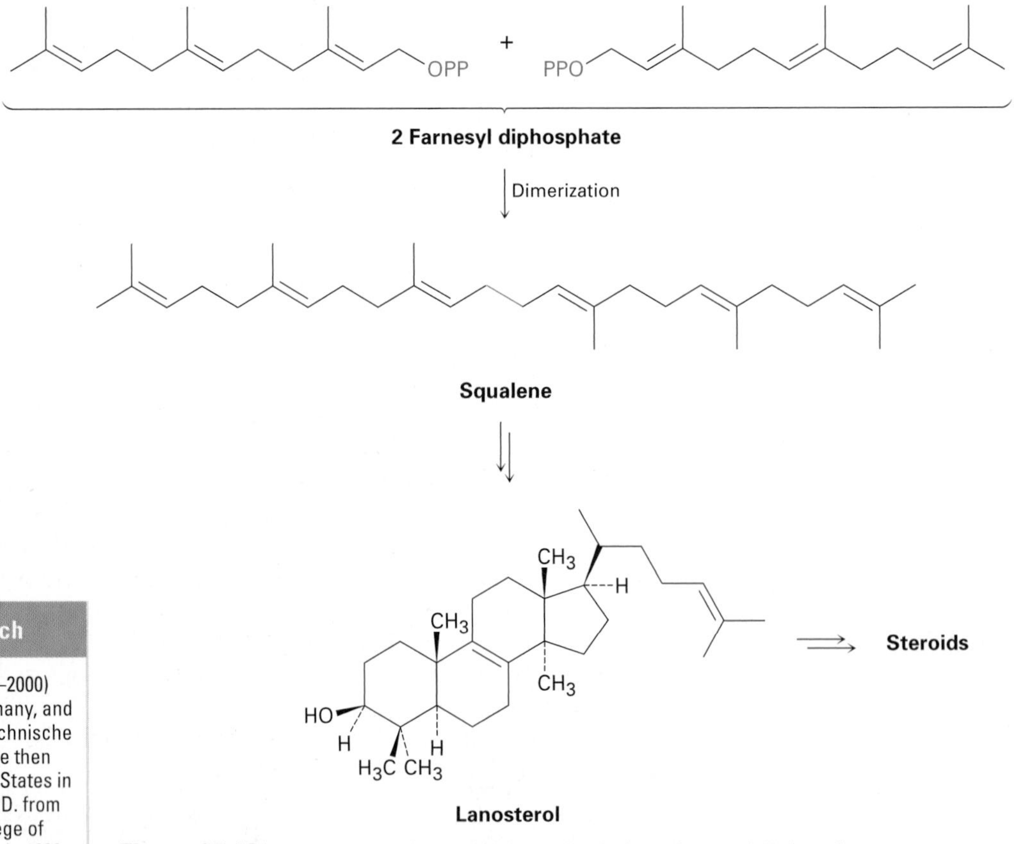

2 Farnesyl diphosphate

Dimerization

Squalene

Steroids

Lanosterol

Figure 27.12 An overview of steroid biosynthesis from farnesyl diphosphate.

Lanosterol biosynthesis begins with the selective conversion of squalene to its epoxide, (3S)-2,3-oxidosqualene, catalyzed by squalene epoxidase. Molecular O_2 provides the source of the epoxide oxygen atom, and NADPH is required, along with a flavin coenzyme. The proposed mechanism involves reaction of $FADH_2$ with O_2 to produce a flavin hydroperoxide intermediate (ROOH), which transfers an oxygen to squalene in a pathway initiated by nucleophilic attack of the squalene double bond on the terminal hydroperoxide oxygen (Figure 27.13). The flavin alcohol formed as a by-product loses H_2O to give FAD, which is reduced back to $FADH_2$ by NADPH. As noted in Section 7.8, such an epoxidation mechanism is closely analogous to that by which peroxyacids (RCO_3H) react with alkenes to give epoxides in the laboratory.

Squalene   (**3S**)-2,3-Oxidosqualene

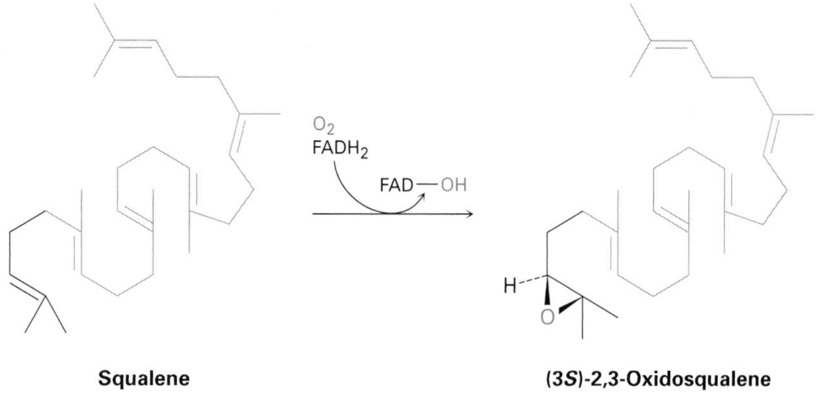

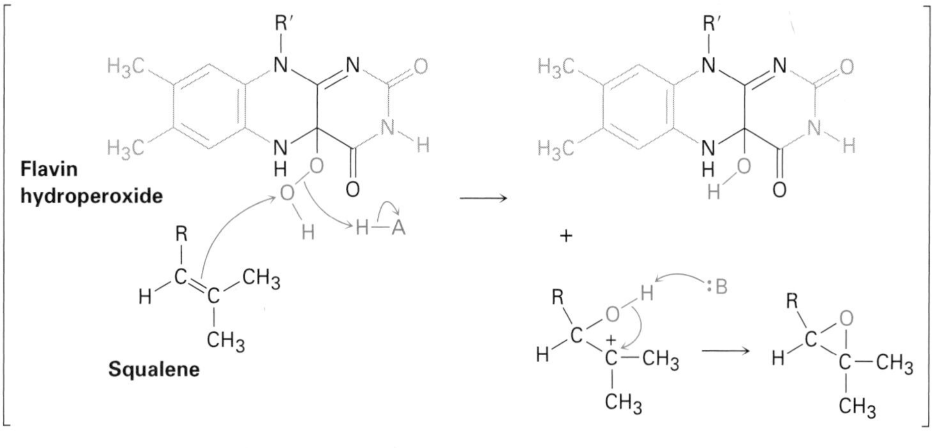

Figure 27.13 Proposed mechanism of the oxidation of squalene by flavin hydroperoxide.

The second part of lanosterol biosynthesis is catalyzed by oxidosqualene: lanosterol cyclase and occurs as shown in Figure 27.14. Squalene is folded by the enzyme into a conformation that aligns the various double bonds for undergoing a cascade of successive intramolecular electrophilic additions, followed by a series of hydride and methyl migrations. Except for the initial epoxide protonation/cyclization, the process is probably stepwise and appears to involve discrete carbocation intermediates that are stabilized by electrostatic interactions with electron-rich aromatic amino acids in the enzyme.

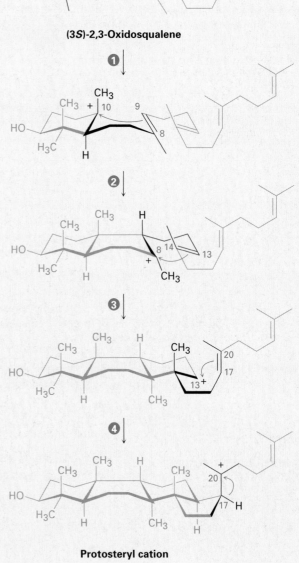

1 Protonation on oxygen opens the epoxide ring and gives a tertiary carbocation at C4. Intramolecular electrophilic addition of C4 to the 5,10 double bond then yields a tertiary monocyclic carbocation at C10.

2 The C10 carbocation adds to the 8,9 double bond, giving a C8 tertiary bicyclic carbocation.

3 Further intramolecular addition of the C8 carbocation to the 13,14 double bond occurs with non-Markovnikov regio-chemistry and gives a tricyclic *secondary* carbocation at C13.

4 The fourth and final cyclization occurs by addition of the C13 cation to the 17,20 double bond, giving the protosteryl cation with 17β stereochemistry.

Figure 27.14 MECHANISM: Mechanism of the conversion of 2,3-oxidosqualene to lanosterol. Four cationic cyclizations are followed by four rearrangements and a final loss of H$^+$ from C9. The steroid numbering system is used for referring to specific positions in the intermediates (Section 27.6). Individual steps are explained in the text.

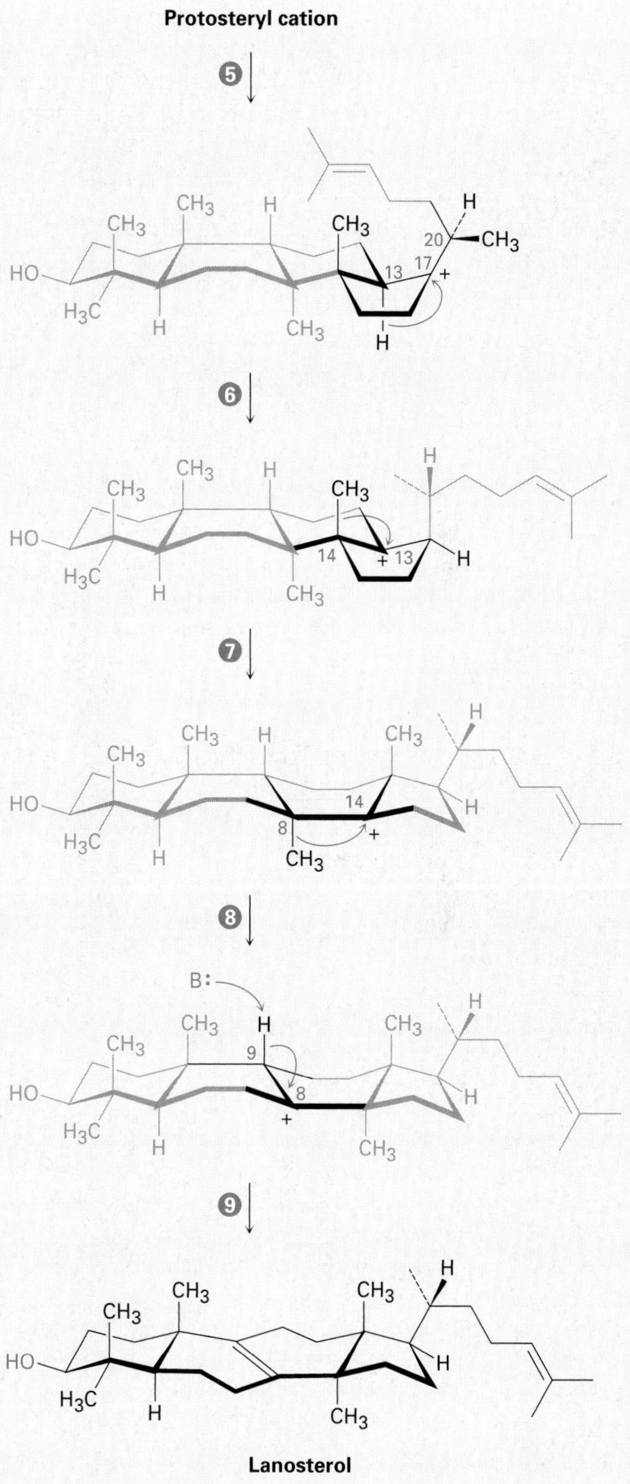

Protosteryl cation

⑤ Hydride migration from C17 to C20 occurs, establishing *R* stereochemistry at C20.

⑥ A second hydride migration takes place, from C13 to C17, establishing the final 17β stereochemistry of the side chain.

⑦ Methyl migration from C14 to C13 occurs.

⑧ A second methyl migration occurs, from C8 to C14.

⑨ Loss of a proton from C9 forms an 8,9 double bond and gives lanosterol.

Lanosterol

© John McMurry

Figure 27.14 *(continued)*

Steps 1–2 of Figure 27.14: Epoxide Opening and Initial Cyclizations Cyclization is initiated in step 1 by protonation of the epoxide ring by an aspartic acid residue in the enzyme. Nucleophilic opening of the protonated epoxide by the nearby 5,10 double bond (steroid numbering; Section 27.6) then yields a tertiary carbocation at C10. Further addition of C10 to the 8,9 double bond in step 2 next gives a bicyclic tertiary cation at C8.

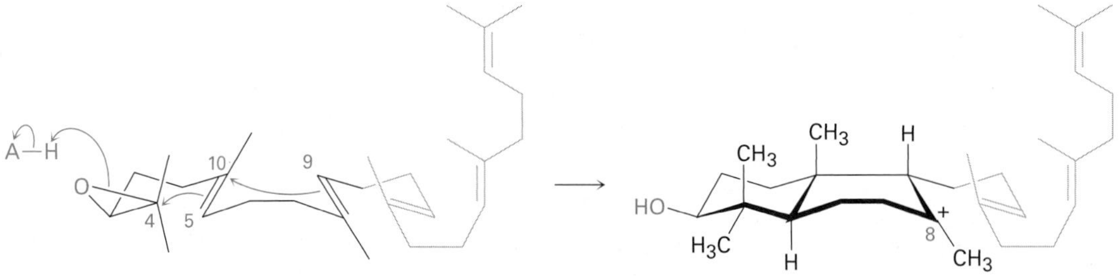

(3*S*)-2,3-Oxidosqualene

Step 3 of Figure 27.14: Third Cyclization The third cationic cyclization is somewhat unusual because it occurs with non-Markovnikov regiochemistry and gives a secondary cation at C13 rather than the alternative tertiary cation at C14. There is growing evidence, however, that the tertiary carbocation may in fact be formed initially and that the secondary cation arises by subsequent rearrangement. The secondary cation is probably stabilized in the enzyme pocket by the proximity of an electron-rich aromatic ring.

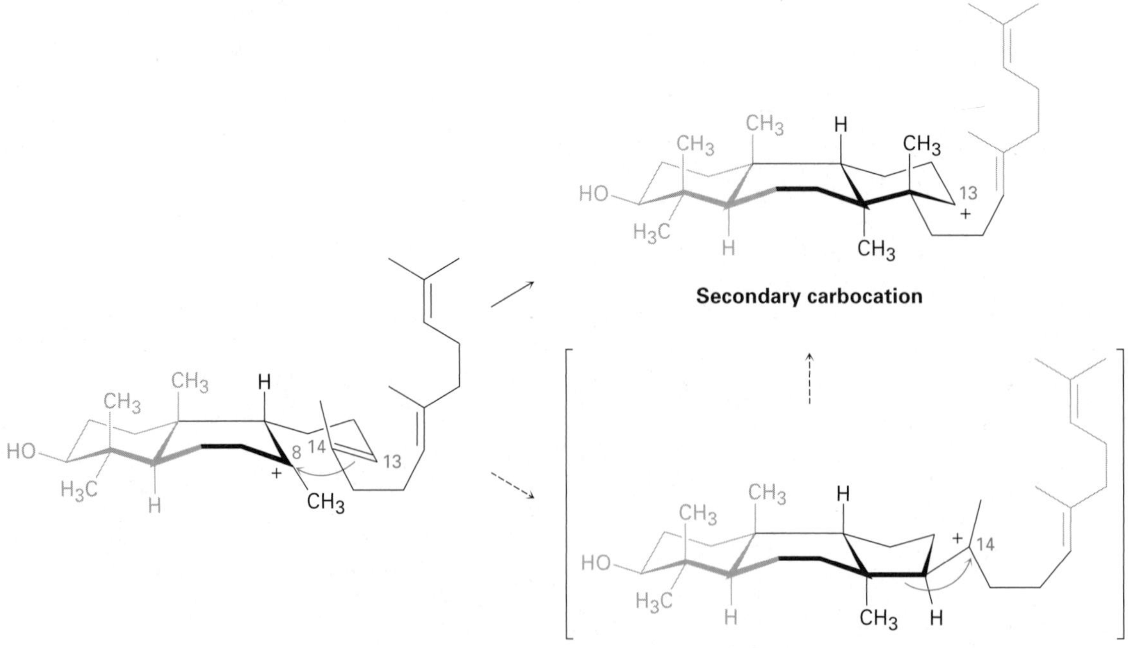

Secondary carbocation

Tertiary carbocation

Step 4 of Figure 27.14: Final Cyclization The fourth and last cyclization occurs in step 4 by addition of the cationic center at C13 to the 17,20 double bond, giving what is known as the *protosteryl* cation. The side-chain alkyl group at

C17 has β (up) stereochemistry, although this stereochemistry is lost in step 5 and then reset in step 6.

Protosteryl cation

Steps 5–9 of Figure 27.14: Carbocation Rearrangements   Once the tetracyclic carbon skeleton of lanosterol has been formed, a series of carbocation rearrangements occur (Section 6.11). The first rearrangement, hydride migration from C17 to C20, occurs in step 5 and results in establishment of R stereochemistry at C20 in the side chain. A second hydride migration then occurs from C13 to C17 on the α (bottom) face of the ring in step 6 and reestablishes the 17β orientation of the side chain. Finally, two methyl group migrations, the first from C14 to C13 on the top (β) face and the second from C8 to C14 on the bottom (α) face, place the positive charge at C8. A basic histidine residue in the enzyme then removes the neighboring β proton from C9 to give lanosterol.

Protosteryl cation   **Lanosterol**

From lanosterol, the pathway for steroid biosynthesis continues on to yield cholesterol. Cholesterol then becomes a branch point, serving as the common precursor from which all other steroids are derived.

Lanosterol   **Cholesterol**

Problem 27.10 | Compare the structures of lanosterol and cholesterol, and catalog the changes needed for the transformation.

Focus On . . .

Saturated Fats, Cholesterol, and Heart Disease

It's hard to resist, but a high intake of saturated animal fat doesn't do much for your cholesterol level.

We hear a lot these days about the relationships between saturated fats, cholesterol, and heart disease. What are the facts? It's well established that a diet rich in saturated animal fats often leads to an increase in blood serum cholesterol, particularly in sedentary, overweight people. Conversely, a diet lower in saturated fats and higher in polyunsaturated fats leads to a lower serum cholesterol level. Studies have shown that a serum cholesterol level greater than 240 mg/dL (a desirable value is <200 mg/dL) is correlated with an increased incidence of coronary artery disease, in which cholesterol deposits build up on the inner walls of coronary arteries, blocking the flow of blood to the heart muscles.

A better indication of a person's risk of heart disease comes from a measurement of blood lipoprotein levels. *Lipoproteins* are complex molecules with both lipid and protein parts that transport lipids through the body. They can be divided into three types according to density, as shown in Table 27.3. Very-low-density lipoproteins (VLDLs) act primarily as carriers of triglycerides from the intestines to peripheral tissues, whereas low-density lipoproteins (LDLs) and high-density lipoproteins (HDLs) act as carriers of cholesterol to and from the liver. Evidence suggests that LDLs transport cholesterol as its fatty-acid ester *to* peripheral tissues, whereas HDLs remove cholesterol as its stearate ester *from* dying cells. If LDLs deliver more cholesterol than is needed, and if insufficient HDLs are present to remove it, the excess is deposited in arteries. Thus, a *low* level of *low*-density lipoproteins is good because it means that less cholesterol is being transported, and a *high* level of *high*-density lipoproteins is good because it means that more cholesterol is being removed. In addition, HDL contains an enzyme that has antioxidant properties, offering further protection against heart disease.

As a rule of thumb, a person's risk drops about 25% for each increase of 5 mg/dL in HDL concentration. Normal values are about 45 mg/dL for men and 55 mg/dL for women, perhaps explaining why premenopausal women appear to be somewhat less susceptible than men to heart disease.

Not surprisingly, the most important factor in gaining high HDL levels is a generally healthful lifestyle. Obesity, smoking, and lack of exercise lead to low HDL levels, whereas regular exercise and a sensible diet lead to high HDL levels. Distance runners and other endurance athletes have HDL levels nearly 50% higher than the general population. Failing that—not everyone wants to run 50 miles per week—diet is also important. Diets high in cold-water fish

(continued)

Table 27.3 | Serum Lipoproteins

Name	Density (g/mL)	% Lipid	% Protein	Optimal (mg/dL)	Poor (mg/dL)
VLDL	0.940–1.006	90	10	—	—
LDL	1.006–1.063	75	25	<100	>130
HDL	1.063–1.210	60	40	>60	<40

like salmon and whitefish, raise HDL and lower blood cholesterol because these fish contain almost entirely polyunsaturated fat. Animal fat from red meat and cooking fats should be minimized because saturated fats and monounsaturated trans fats raise blood cholesterol.

SUMMARY AND KEY WORDS

eicosanoid, 1067

fat, 1061

fatty acid, 1061

hormone, 1082

lipid, 1060

lipid bilayer, 1067

micelle, 1064

oil, 1061

phospholipid, 1066

polyunsaturated fatty acid, 1061

prostaglandin, 1067

steroid, 1079

terpenoid, 1070

triacylglycerol, 1061

wax, 1061

Lipids are the naturally occurring materials isolated from plants and animals by extraction with nonpolar organic solvents. Animal **fats** and vegetable **oils** are the most widely occurring lipids. Both are **triacylglycerols**—triesters of glycerol with long-chain **fatty acids**. Animal fats are usually saturated, whereas vegetable oils usually have unsaturated fatty acid residues.

Phospholipids are important constituents of cell membranes and are of two kinds. *Glycerophospholipids,* such as phosphatidylcholine and phosphatidylethanolamine, are closely related to fats in that they have a glycerol backbone esterified to two fatty acids (one saturated and one unsaturated) and to one phosphate ester. *Sphingomyelins* have the amino alcohol sphingosine for their backbone.

Eicosanoids and **terpenoids** are still other classes of lipids. Eicosanoids, of which prostaglandins are the most abundant kind, are derived biosynthetically from arachidonic acid, are found in all body tissues, and have a wide range of physiological activity. Terpenoids are often isolated from the essential oils of plants, have an immense diversity of structure, and are produced biosynthetically from the five-carbon precursor isopentenyl diphosphate (IPP). Isopentenyl diphosphate is itself biosynthesized from 3 equivalents of acetate in the mevalonate pathway.

Steroids are plant and animal lipids with a characteristic tetracyclic carbon skeleton. Like the eicosanoids, steroids occur widely in body tissues and have a large variety of physiological activities. Steroids are closely related to terpenoids and arise biosynthetically from the triterpene lanosterol. Lanosterol, in turn, arises from cationic cyclization of the acyclic hydrocarbon squalene.

EXERCISES

Organic KNOWLEDGE TOOLS

CENGAGENOW™ Sign in at **www.cengage.com/login** to assess your knowledge of this chapter's topics by taking a pre-test. The pre-test will link you to interactive organic chemistry resources based on your score in each concept area.

⚙WL Online homework for this chapter may be assigned in Organic OWL.

■ indicates problems assignable in Organic OWL.

VISUALIZING CHEMISTRY

(Problems 27.1–27.10 appear within the chapter.)

27.11 ■ The following model is that of cholic acid, a constituent of human bile. Locate the three hydroxyl groups, and identify each as axial or equatorial. Is cholic acid an A–B trans steroid or an A–B cis steroid?

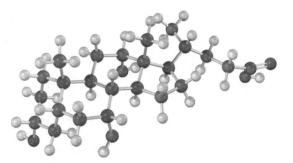

27.12 Propose a biosynthetic pathway for the sesquiterpene helminthogermacrene from farnesyl diphosphate.

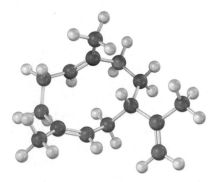

27.13 ■ Identify the following fatty acid, and tell whether it is more likely to be found in peanut oil or in red meat:

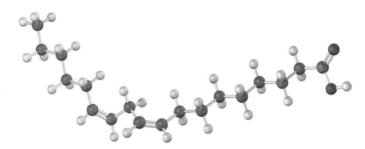

ADDITIONAL PROBLEMS

27.14 Fats can be either optically active or optically inactive, depending on their structure. Draw the structure of an optically active fat that yields 2 equivalents of stearic acid and 1 equivalent of oleic acid on hydrolysis. Draw the structure of an optically inactive fat that yields the same products.

27.15 Spermaceti, a fragrant substance from sperm whales, was much used in cosmetics until it was banned in 1976 to protect the whales from extinction. Chemically, spermaceti is cetyl palmitate, the ester of cetyl alcohol (n-$C_{16}H_{33}OH$) with palmitic acid. Draw its structure.

27.16 The *plasmalogens* are a group of lipids found in nerve and muscle cells. How do plasmalogens differ from fats?

$$CH_2OCH{=}CHR$$
$$|$$
$$\underset{O}{CHOCR'} \qquad \textbf{A plasmalogen}$$
$$|$$
$$CH_2OCR''$$

27.17 What products would you obtain from hydrolysis of a plasmalogen (Problem 27.16) with aqueous NaOH? With H_3O^+?

27.18 *Cardiolipins* are a group of lipids found in heart muscles. What products would be formed if all ester bonds, including phosphates, were saponified by treatment with aqueous NaOH?

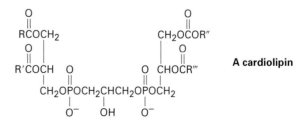

A cardiolipin

27.19 ■ Stearolic acid, $C_{18}H_{32}O_2$, yields stearic acid on catalytic hydrogenation and undergoes oxidative cleavage with ozone to yield nonanoic acid and nonanedioic acid. What is the structure of stearolic acid?

27.20 ■ How would you synthesize stearolic acid (Problem 27.19) from 1-decyne and 1-chloro-7-iodoheptane?

27.21 ■ Show the products you would expect to obtain from reaction of glyceryl trioleate with the following reagents:

(a) Excess Br_2 in CH_2Cl_2 (b) H_2/Pd
(c) $NaOH/H_2O$ (d) O_3, then Zn/CH_3CO_2H
(e) $LiAlH_4$, then H_3O^+ (f) CH_3MgBr, then H_3O^+

27.22 ■ How would you convert oleic acid into the following substances?

(a) Methyl oleate (b) Methyl stearate
(c) Nonanal (d) Nonanedioic acid
(e) 9-Octadecynoic acid (stearolic acid) (f) 2-Bromostearic acid
(g) 18-Pentatriacontanone, $CH_3(CH_2)_{16}CO(CH_2)_{16}CH_3$

27.23 Cold-water fish like salmon are rich in *omega-3* fatty acids, which have a double bond three carbons in from the noncarboxyl end of the chain and have been shown to lower blood cholesterol levels. Draw the structure of 5,8,11,14,17-eicosapentaenoic acid, a common example. (Eicosane = $C_{20}H_{42}$.)

27.24 Without proposing an entire biosynthetic pathway, draw the appropriate precursor, either geranyl diphosphate or farnesyl diphosphate, in a conformation that shows a likeness to each of the following terpenoids:

Guaiol   Sabinene   Cedrene

27.25 ■ Indicate by asterisks the chirality centers present in each of the terpenoids shown in Problem 27.24. What is the maximum possible number of stereoisomers for each?

27.26 ■ Assume that the three terpenoids in Problem 27.24 are derived biosynthetically from isopentenyl diphosphate and dimethylallyl diphosphate, each of which was isotopically labeled at the diphosphate-bearing carbon atom (C1). At what positions would the terpenoids be isotopically labeled?

27.27 ■ Assume that acetyl CoA containing a ^{14}C isotopic label in the carboxyl carbon atom is used as starting material for the biosynthesis of mevalonate, as shown in Figure 27.7. At what positions in mevalonate would the isotopic label appear?

27.28 ■ Assume that acetyl CoA containing a ^{14}C isotopic label in the carboxyl carbon atom is used as starting material and that the mevalonate pathway is followed. Identify the positions in α-cadinol where the label would appear.

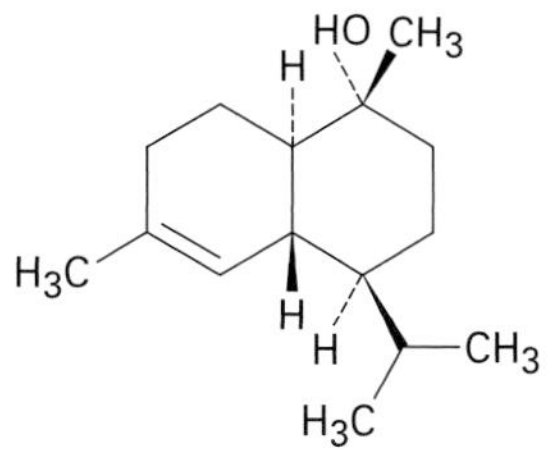

α-**Cadinol**

27.29 Assume that acetyl CoA containing a ^{14}C isotopic label in the carboxyl carbon atom is used as starting material and that the mevalonate pathway is followed. Identify the positions in squalene where the label would appear.

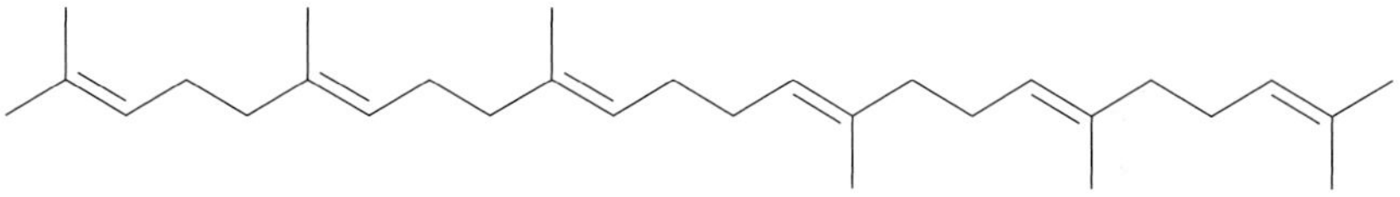

Squalene

27.30 Assume that acetyl CoA containing a ^{14}C isotopic label in the carboxyl carbon atom is used as starting material and that the mevalonate pathway is followed. Identify the positions in lanosterol where the label would appear.

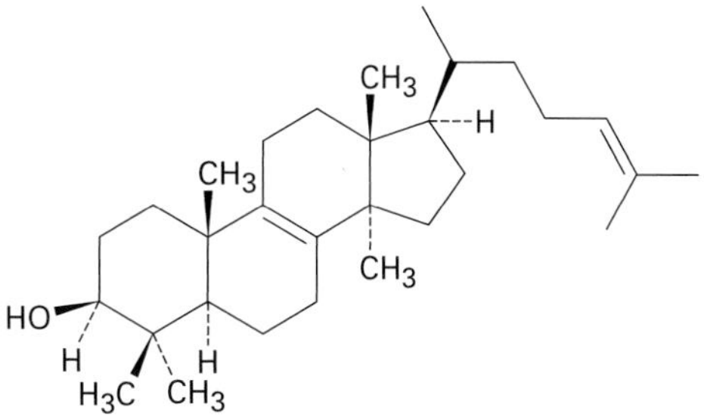

Lanosterol

27.31 Propose a mechanistic pathway for the biosynthesis of caryophyllene, a substance found in clove oil.

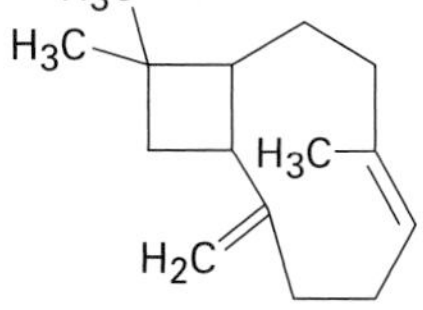

Caryophyllene

27.32 Flexibilene, a compound isolated from marine coral, is the only known terpenoid to contain a 15-membered ring. What is the structure of the acyclic biosynthetic precursor of flexibilene? Show the mechanistic pathway for the biosynthesis.

Flexibilene

27.33 ■ Suggest a mechanism by which ψ-ionone is transformed into β-ionone on treatment with acid.

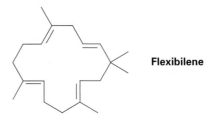

H_3O^+

ψ-Ionone   **β-Ionone**

27.34 ■ Draw the most stable chair conformation of dihydrocarvone.

Dihydrocarvone

27.35 ■ Draw the most stable chair conformation of menthol, and label each substituent as axial or equatorial.

Menthol (from peppermint oil)

27.36 ■ As a general rule, equatorial alcohols are esterified more readily than axial alcohols. What product would you expect to obtain from reaction of the following two compounds with 1 equivalent of acetic anhydride?

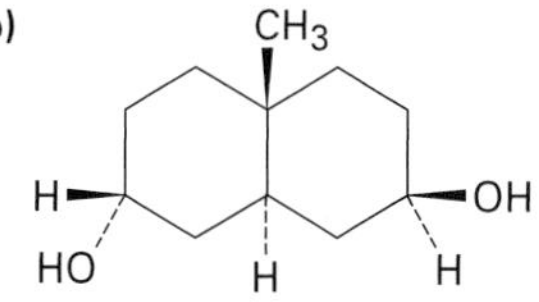

27.37 ■ Propose a mechanistic pathway for the biosynthesis of isoborneol. A carbocation rearrangement is needed at one point in the scheme.

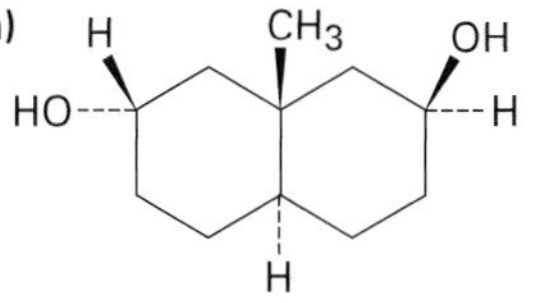

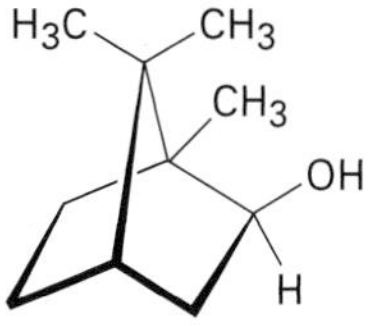

Isoborneol

27.38 ■ Isoborneol (Problem 27.37) is converted into camphene on treatment with dilute sulfuric acid. Propose a mechanism for the reaction, which involves a carbocation rearrangement.

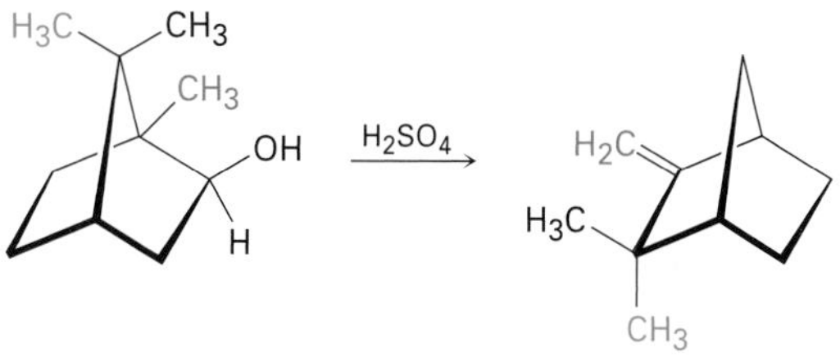

Isoborneol   **Camphene**

27.39 Digitoxigenin is a heart stimulant obtained from the purple foxglove *Digitalis purpurea* and used in the treatment of heart disease. Draw the three-dimensional conformation of digitoxigenin, and identify the two −OH groups as axial or equatorial.

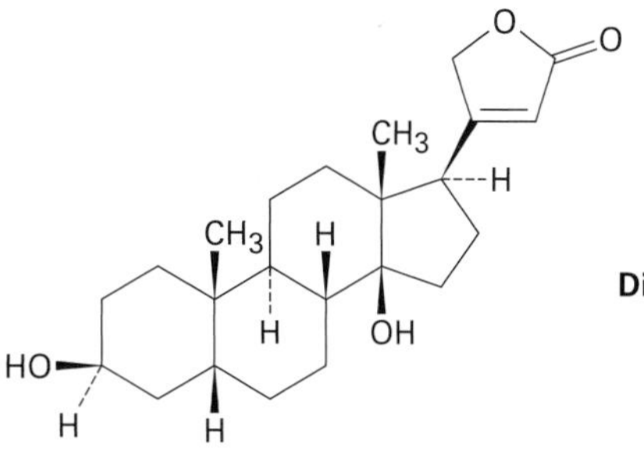

Digitoxigenin

27.40 What product would you obtain by reduction of digitoxigenin (Problem 27.39) with $LiAlH_4$? By oxidation with pyridinium chlorochromate?

27.41 ■ Vaccenic acid, $C_{18}H_{34}O_2$, is a rare fatty acid that gives heptanal and 11-oxoundecanoic acid [$OHC(CH_2)_9CO_2H$] on ozonolysis followed by zinc treatment. When allowed to react with $CH_2I_2/Zn(Cu)$, vaccenic acid is converted into lactobacillic acid. What are the structures of vaccenic and lactobacillic acids?

27.42 Eleostearic acid, $C_{18}H_{30}O_2$, is a rare fatty acid found in the tung oil used for finishing furniture. On ozonolysis followed by treatment with zinc, eleostearic acid furnishes one part pentanal, two parts glyoxal ($OHC—CHO$), and one part 9-oxononanoic acid [$OHC(CH_2)_7CO_2H$]. What is the structure of eleostearic acid?

27.43 Diterpenoids are derived biosynthetically from geranylgeranyl diphosphate (GGPP), which is itself biosynthesized by reaction of farnesyl diphosphate with isopentenyl diphosphate. Show the structure of GGPP, and propose a mechanism for its biosynthesis from FPP and IPP.

27.44 Diethylstilbestrol (DES) has estrogenic activity even though it is structurally unrelated to steroids. Once used as an additive in animal feed, DES has been implicated as a causative agent in several types of cancer. Show how DES can be drawn so that it is sterically similar to estradiol.

Diethylstilbestrol

Estradiol

27.45 Propose a synthesis of diethylstilbestrol (Problem 27.44) from phenol and any other organic compound required.

27.46 ■ What products would you expect from reaction of estradiol (Problem 27.44) with the following reagents?
(a) NaH, then CH_3I (b) CH_3COCl, pyridine
(c) Br_2, $FeBr_3$ (d) Pyridinium chlorochromate in CH_2Cl_2

27.47 Cembrene, $C_{20}H_{32}$, is a diterpene hydrocarbon isolated from pine resin. Cembrene has a UV absorption at 245 nm, but dihydrocembrene ($C_{20}H_{34}$), the product of hydrogenation with 1 equivalent H_2, has no UV absorption. On exhaustive hydrogenation, 4 equivalents H_2 react, and octahydrocembrene, $C_{20}H_{40}$, is produced. On ozonolysis of cembrene, followed by treatment of the ozonide with zinc, four carbonyl-containing products are obtained:

Propose a structure for cembrene that is consistent with its formation from geranylgeranyl diphosphate.

27.48 α-Fenchone is a pleasant-smelling terpenoid isolated from oil of lavender. Propose a pathway for the formation of α-fenchone from geranyl diphosphate. A carbocation rearrangement is required.

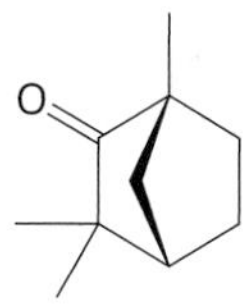

α-**Fenchone**

27.49 Fatty acids are synthesized by a multistep route that starts with acetate. The first step is a reaction between protein-bound acetyl and malonyl units to give a protein-bound 3-ketobutyryl unit. Show the mechanism, and tell what kind of reaction is occurring.

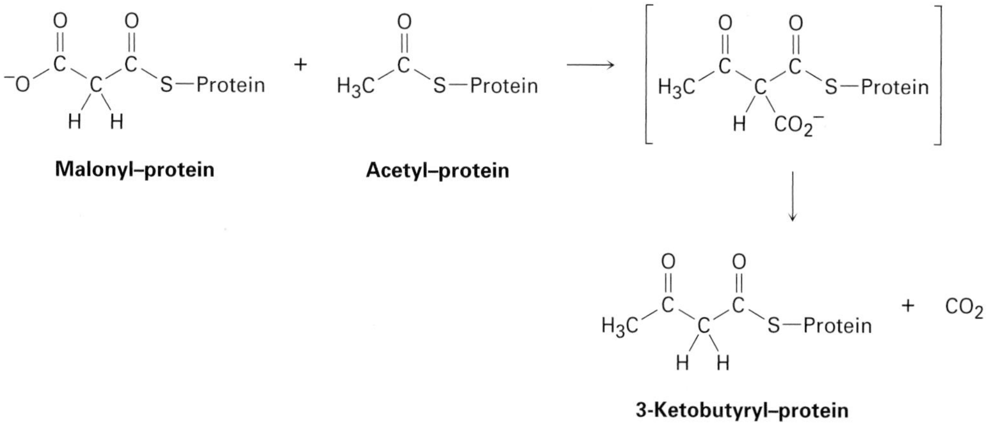

Malonyl–protein   **Acetyl–protein**

3-Ketobutyryl–protein

27.50 Propose a mechanism for the biosynthesis of the sesquiterpene trichodiene from farnesyl diphosphate. The process involves cyclization to give an intermediate secondary carbocation, followed by several carbocation rearrangements.

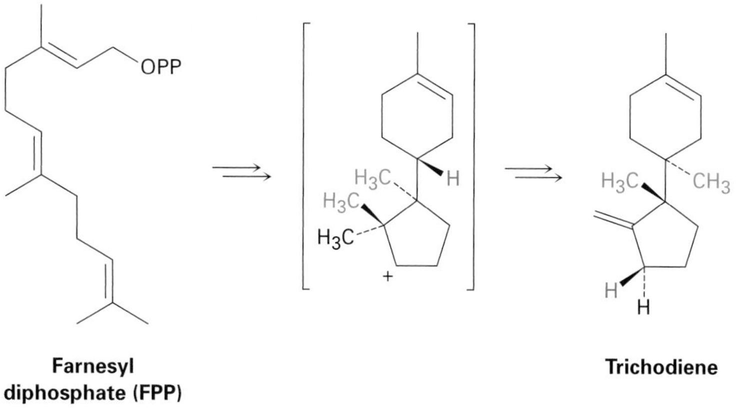

Farnesyl diphosphate (FPP)   **Trichodiene**

28

Biomolecules: Nucleic Acids

Organic KNOWLEDGE TOOLS

CENGAGENOW™ Throughout this chapter, sign in at **www.cengage.com/login** for online self-study and interactive tutorials based on your level of understanding.

ⓦWL Online homework for this chapter may be assigned in Organic OWL.

The nucleic acids, **deoxyribonucleic acid (DNA)** and **ribonucleic acid (RNA)**, are the chemical carriers of a cell's genetic information. Coded in a cell's DNA is the information that determines the nature of the cell, controls the cell's growth and division, and directs biosynthesis of the enzymes and other proteins required for cellular functions.

In addition to the nucleic acids themselves, nucleic acid derivatives such as ATP are involved as phosphorylating agents in many biochemical pathways, and several important coenzymes, including NAD^+, FAD, and coenzyme A, have nucleic acid components.

WHY THIS CHAPTER?

Nucleic acids are the last of the four major classes of biomolecules we'll consider. So much has been written and spoken about DNA in the media that the basics of DNA replication and transcription are probably known to you. Thus, we'll move fairly quickly through the fundamentals and then focus more closely on the chemical details of DNA sequencing and synthesis.

28.1 | Nucleotides and Nucleic Acids

Just as proteins are biopolymers made of amino acids, nucleic acids are biopolymers made of **nucleotides** joined together to form a long chain. Each nucleotide is composed of a **nucleoside** bonded to a phosphate group, and each nucleoside is composed of an aldopentose sugar linked through its anomeric carbon to the nitrogen atom of a heterocyclic purine or pyrimidine base.

The sugar component in RNA is ribose, and the sugar in DNA is 2′-deoxyribose. (The prefix *2′-deoxy* indicates that oxygen is missing from the 2′ position of ribose.) DNA contains four different amine bases, two substituted purines (adenine and guanine) and two substituted pyrimidines (cytosine and thymine). Adenine, guanine, and cytosine also occur in RNA, but thymine is replaced in RNA by a closely related pyrimidine base called uracil.

The structures of the four deoxyribonucleotides and the four ribonucleotides are shown in Figure 28.1. Note that in naming and numbering nucleotides, positions on the sugars are given a prime superscript to distinguish them from positions on the amine base. Position 3 would be on the base, for instance, while position 3′ would be on the sugar. Although similar chemically, DNA and RNA differ dramatically in size. Molecules of DNA are enormous, with molecular weights up to several billion. Molecules of RNA, by contrast, are much smaller, containing as few as 60 nucleotides and having molecular weights as low as 22,000.

CENGAGENOW™ Click *Organic Interactive* to **learn to recognize classes of nucleic acids and their base-pair partners**.

Figure 28.1 Structures of the four deoxyribonucleotides and the four ribonucleotides.

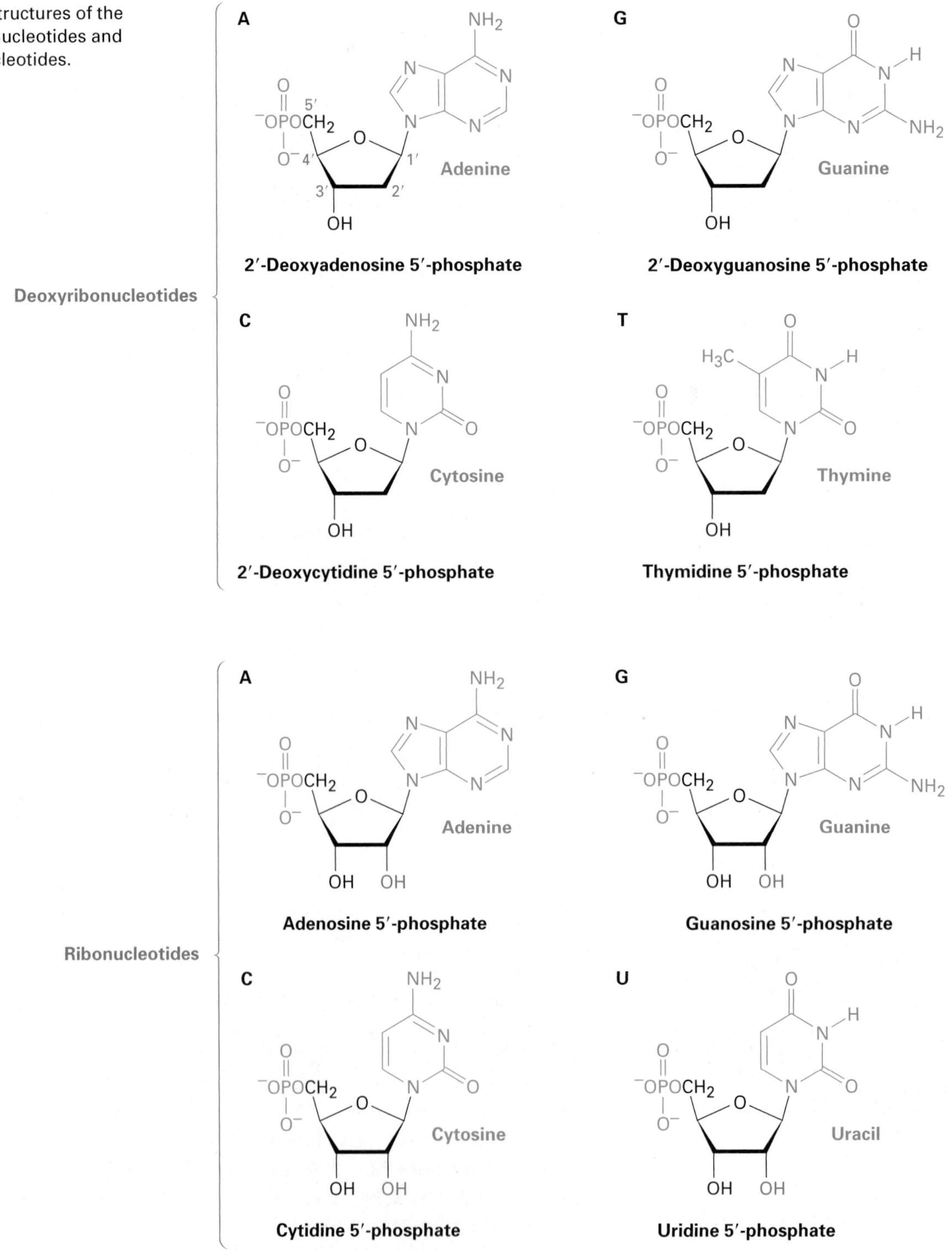
Deoxyribonucleotides
A
Adenine
2′-Deoxyadenosine 5′-phosphate
G
Guanine
2′-Deoxyguanosine 5′-phosphate
C
Cytosine
2′-Deoxycytidine 5′-phosphate
T
Thymine
Thymidine 5′-phosphate
Ribonucleotides
A
Adenine
Adenosine 5′-phosphate
G
Guanine
Guanosine 5′-phosphate
C
Cytosine
Cytidine 5′-phosphate
U
Uracil
Uridine 5′-phosphate

Nucleotides are linked together in DNA and RNA by phosphodiester bonds [RO—(PO_2^-)—OR'] between phosphate, the 5' hydroxyl group on one nucleoside, and the 3'-hydroxyl group on another nucleoside. One end of the nucleic acid polymer has a free hydroxyl at C3' (the **3' end**), and the other end has a phosphate at C5' (the **5' end**). The sequence of nucleotides in a chain is described by starting at the 5' end and identifying the bases in order of occurrence, using the abbreviations G, C, A, T (or U for RNA). Thus, a typical DNA sequence might be written as TAGGCT.

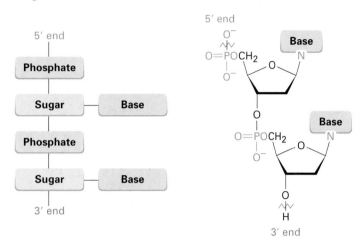

Problem 28.1 Draw the full structure of the DNA dinucleotide AG.

Problem 28.2 Draw the full structure of the RNA dinucleotide UA.

28.2 | Base Pairing in DNA: The Watson–Crick Model

Samples of DNA isolated from different tissues of the same species have the same proportions of heterocyclic bases, but samples from different species often have greatly different proportions of bases. Human DNA, for example, contains about 30% each of adenine and thymine and about 20% each of guanine and cytosine. The bacterium *Clostridium perfringens,* however, contains about 37% each of adenine and thymine and only 13% each of guanine and cytosine. Note that in both examples the bases occur in pairs. Adenine and thymine are present in equal amounts, as are cytosine and guanine. Why?

In 1953, James Watson and Francis Crick made their classic proposal for the secondary structure of DNA. According to the Watson–Crick model, DNA under physiological conditions consists of two polynucleotide strands, running in opposite directions and coiled around each other in a **double helix** like the handrails on a spiral staircase. The two strands are complementary rather than identical and are held together by hydrogen bonds between specific pairs of

CENGAGENOW˙ Click *Organic Interactive* to **use interactive animations to view aspects of DNA structure**.

bases, A with T and C with G. That is, whenever an A base occurs in one strand, a T base occurs opposite it in the other strand; when a C base occurs in one, a G occurs in the other (Figure 28.2). This complementary base-pairing thus explains why A and T are always found in equal amounts, as are G and C.

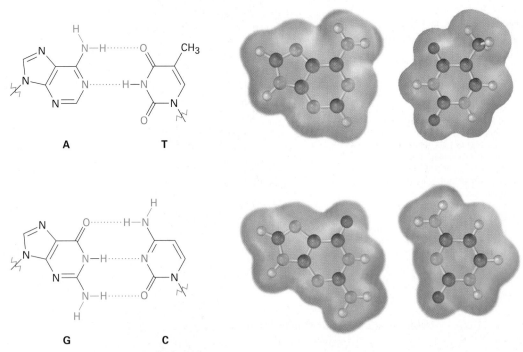

Active Figure 28.2 Hydrogen-bonding between base pairs in the DNA double helix. Electrostatic potential maps show that the faces of the bases are relatively neutral (green), while the edges have positive (blue) and negative (red) regions. Pairing G with C and A with T brings together oppositely charged regions. *Sign in at* **www.cengage.com/login** *to see a simulation based on this figure and to take a short quiz.*

A full turn of the DNA double helix is shown in Figure 28.3. The helix is 20 Å wide, there are 10 base pairs per turn, and each turn is 34 Å in length. Notice in Figure 28.3 that the two strands of the double helix coil in such a way that two kinds of "grooves" result, a *major groove* 12 Å wide and a *minor groove* 6 Å wide. The major groove is slightly deeper than the minor groove, and both are lined by hydrogen bond donors and acceptors. As a result, a variety of flat, polycyclic aromatic molecules are able to slip sideways, or *intercalate*, between the stacked bases. Many cancer-causing and cancer-preventing agents function by interacting with DNA in this way.

An organism's genetic information is stored as a sequence of deoxyribonucleotides strung together in the DNA chain. For the information to be preserved and passed on to future generations, a mechanism must exist for copying DNA. For the information to be used, a mechanism must exist for decoding the DNA message and implementing the instructions it contains.

What Crick called the "central dogma of molecular genetics" says that the function of DNA is to store information and pass it on to RNA. The function of

Active Figure 28.3 A turn of the DNA double helix in both space-filling and wire-frame formats. The sugar–phosphate backbone runs along the outside of the helix, and the amine bases hydrogen bond to one another on the inside. Both major and minor grooves are visible. *Sign in at* **www.cengage.com/login** *to see a simulation based on this figure and to take a short quiz.*

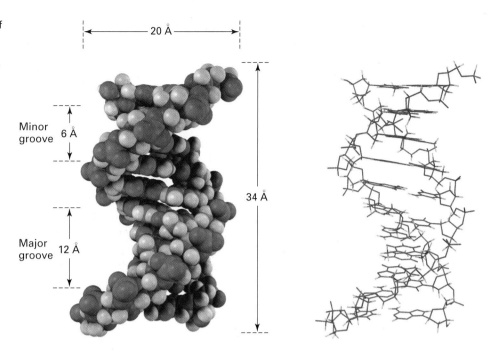

RNA, in turn, is to read, decode, and use the information received from DNA to make proteins. Thus, three fundamental processes take place.

▌ Replication—the process by which identical copies of DNA are made so that information can be preserved and handed down to offspring

▌ Transcription—the process by which the genetic messages are read and carried out of the cell nucleus to ribosomes, where protein synthesis occurs

▌ Translation—the process by which the genetic messages are decoded and used to synthesize proteins

| **WORKED EXAMPLE 28.1** | ***Predicting the Complementary Base Sequence in Double-Stranded DNA*** |

What sequence of bases on one strand of DNA is complementary to the sequence TATGCAT on another strand?

Strategy   Remember that A and G form complementary pairs with T and C, respectively, and then go through the sequence replacing A by T, G by C, T by A, and C by G. Remember also that the 5′ end is on the left and the 3′ end is on the right in the original strand.

Solution   Original: (5′) TATGCAT (3′)
Complement: (3′) ATACGTA (5′) or (5′) ATGCATA (3′)

Problem 28.3 | What sequence of bases on one strand of DNA is complementary to the following sequence on another strand?

(5′) GGCTAATCCGT (3′)

28.3 | Replication of DNA

DNA **replication** is an enzyme-catalyzed process that begins with a partial untwisting of the double helix and breaking of the hydrogen bonds between strands, brought about by enzymes called *helicases*. As the strands separate and bases are exposed, new nucleotides line up on each strand in a complementary manner, A to T and G to C, and two new strands begin to grow. Each new strand is complementary to its old template strand, and two identical DNA double helices are produced (Figure 28.4). Because each of the new DNA molecules contains one old strand and one new strand, the process is described as *semiconservative replication*.

Figure 28.4 A representation of semiconservative DNA replication. The original double-stranded DNA partially unwinds, bases are exposed, nucleotides line up on each strand in a complementary manner, and two new strands begin to grow. Both strands are synthesized in the same 5′ → 3′ direction, one continuously and one in fragments.

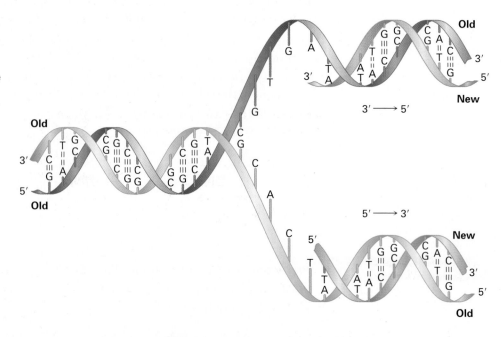

Addition of nucleotides to the growing chain takes place in the 5′ → 3′ direction and is catalyzed by DNA polymerase. The key step is the addition of a nucleoside 5′-triphosphate to the free 3′-hydroxyl group of the growing chain with loss of a diphosphate leaving group.

Template strand

Because both new DNA strands are synthesized in the 5′ → 3′ direction, they can't be made in exactly the same way. One new strand must have its 3′ end nearer a point of unraveling (the *replication fork*), while the other new strand has its 5′ end nearer the replication fork. What happens is that the complement of the original 5′ → 3′ strand is synthesized continuously in a single piece to give a newly synthesized copy called the *leading strand*, while the complement of the original 3′ → 5′ strand is synthesized discontinuously in small pieces called *Okazaki fragments* that are subsequently linked by DNA ligases to form the *lagging strand*.

The magnitude of the replication process is staggering. The nucleus of every human cell contains 46 chromosomes (23 pairs), each of which consists of one very large DNA molecule. Each chromosome, in turn, is made up of hundreds of DNA segments called *genes,* and the sum of all genes in a human cell (the human *genome*) is estimated to be 2.9 billion base pairs. Despite the size of these enormous molecules, their base sequence is faithfully copied during replication. The copying process takes only minutes, and an error occurs only about once each 10 to 100 billion bases.

28.4 | Transcription of DNA

As noted previously, RNA is structurally similar to DNA but contains ribose rather than deoxyribose and uracil rather than thymine. There are three major kinds of RNA, each of which serves a specific function. All three are much smaller molecules than DNA, and all remain single-stranded rather than double-stranded.

▮ **Messenger RNA (mRNA)** carries genetic messages from DNA to ribosomes, small granular particles in the cytoplasm of a cell where protein synthesis takes place.

▌**Ribosomal RNA (rRNA)** complexed with protein provides the physical makeup of the ribosomes.

▌**Transfer RNA (tRNA)** transports amino acids to the ribosomes, where they are joined together to make proteins.

The conversion of the information in DNA into proteins begins in the nucleus of cells with the synthesis of mRNA by **transcription** of DNA. In bacteria, the process begins when RNA polymerase recognizes and binds to a *promoter sequence* on DNA, typically consisting of around 40 base pairs located upstream (5′) of the transcription start site. Within the promoter are two hexameric *consensus sequences,* one located 10 base pairs upstream of the start and the second located 35 base pairs upstream.

Following formation of the polymerase–promoter complex, several turns of the DNA double helix untwist, forming a "bubble" and exposing 14 or so base pairs of the two strands. Appropriate ribonucleotides then line up by hydrogen-bonding to their complementary bases on DNA, bond formation occurs in the 5′ → 3′ direction, the RNA polymerase moves along the DNA chain, and the growing RNA molecule unwinds from DNA (Figure 28.5). At any one time, about 12 base pairs of the growing RNA remain hydrogen-bonded to the DNA template.

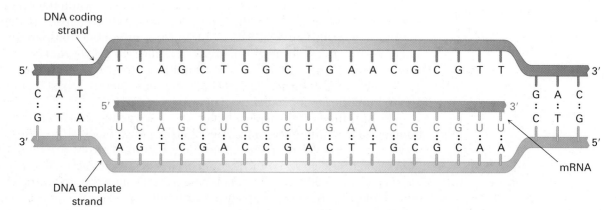

Figure 28.5 Biosynthesis of RNA using a DNA segment as a template.

Unlike what happens in DNA replication, where both strands are copied, only one of the two DNA strands is transcribed into mRNA. The strand that contains the gene is often called the **coding strand**, or *primer strand,* and the strand that gets transcribed is called the **template strand**. Because the template strand and the coding strand are complementary, and because the template strand and the transcribed RNA are also complementary, *the RNA molecule produced during transcription is a copy of the DNA coding strand.* The only difference is that the RNA molecule has a U everywhere the DNA coding strand has a T.

Another part of the picture in vertebrates and flowering plants is that genes are often not continuous segments of the DNA chain. Instead, a gene will begin in one small section of DNA called an *exon,* then be interrupted by a noncoding section called an *intron,* and then take up again farther down the chain in another exon. The final mRNA molecule results only after the noncoded sections are cut out and the remaining pieces are joined together by spliceosomes. The gene for triose phosphate isomerase in maize, for instance, contains nine exons accounting for approximately 80% of the DNA base pairs and eight introns accounting for only 20% of the base pairs.

Problem 28.4 | Show how uracil can form strong hydrogen bonds to adenine.

Problem 28.5 | What RNA base sequence is complementary to the following DNA base sequence?

(5′) GATTACCGTA (3′)

Problem 28.6 | From what DNA base sequence was the following RNA sequence transcribed?

(5′) UUCGCAGAGU (3′)

28.5 | Translation of RNA: Protein Biosynthesis

The primary cellular function of mRNA is to direct biosynthesis of the thousands of diverse peptides and proteins required by an organism—perhaps 100,000 in a human. The mechanics of protein biosynthesis take place on ribosomes, small granular particles in the cytoplasm of a cell that consist of about 60% ribosomal RNA and 40% protein.

The specific ribonucleotide sequence in mRNA forms a message that determines the order in which amino acid residues are to be joined. Each "word," or **codon**, along the mRNA chain consists of a sequence of three ribonucleotides that is specific for a given amino acid. For example, the series UUC on mRNA is a codon directing incorporation of the amino acid phenylalanine into the growing protein. Of the $4^3 = 64$ possible triplets of the four bases in RNA, 61 code for specific amino acids and 3 code for chain termination. Table 28.1 shows the meaning of each codon.

The message embedded in mRNA is read by transfer RNA (tRNA) in a process called **translation**. There are 61 different tRNAs, one for each of the 61 codons that specifies an amino acid. A typical tRNA is single-stranded and has roughly the shape of a cloverleaf, as shown in Figure 28.6 on page 1111. It consists of about 70 to 100 ribonucleotides and is bonded to a specific amino acid by an ester linkage through the 3′ hydroxyl on ribose at the 3′ end of the tRNA. Each tRNA also contains on its middle leaf a segment called an **anticodon**, a sequence of three ribonucleotides complementary to the codon sequence. For example, the codon sequence UUC present on mRNA is read by a phenylalanine-bearing tRNA having the complementary anticodon base sequence GAA. [Remember that nucleotide sequences are written in the 5′ → 3′ direction, so the sequence in an anticodon must be reversed. That is, the complement to (5′)-UUC-(3′) is (3′)-AAG-(5′), which is written as (5′)-GAA-(3′).]

As each successive codon on mRNA is read, different tRNAs bring the correct amino acids into position for enzyme-mediated transfer to the growing

Table 28.1 | **Codon Assignments of Base Triplets**

First base (5′ end)	Second base	Third base (3′ end)			
		U	**C**	**A**	**G**
U	U	Phe	Phe	Leu	Leu
	C	Ser	Ser	Ser	Ser
	A	Tyr	Tyr	Stop	Stop
	G	Cys	Cys	Stop	Trp
C	U	Leu	Leu	Leu	Leu
	C	Pro	Pro	Pro	Pro
	A	His	His	Gln	Gln
	G	Arg	Arg	Arg	Arg
A	U	Ile	Ile	Ile	Met
	C	Thr	Thr	Thr	Thr
	A	Asn	Asn	Lys	Lys
	G	Ser	Ser	Arg	Arg
G	U	Val	Val	Val	Val
	C	Ala	Ala	Ala	Ala
	A	Asp	Asp	Glu	Glu
	G	Gly	Gly	Gly	Gly

peptide. When synthesis of the proper protein is completed, a "stop" codon signals the end and the protein is released from the ribosome. The process is illustrated in Figure 28.7.

WORKED EXAMPLE 28.2

Predicting the Amino Acid Sequence Transcribed from DNA

What amino acid sequence is coded by the following segment of a DNA coding strand?

(5′) CTA-ACT-AGC-GGG-TCG-CCG (3′)

Strategy The mRNA produced during translation is a copy of the DNA coding strand, with each T replaced by U. Thus, the mRNA has the sequence

(5′) CUA-ACU-AGC-GGG-UCG-CCG (3′)

Each set of three bases forms a codon, whose meaning can be found in Table 28.1.

Solution Leu-Thr-Ser-Gly-Ser-Pro.

Figure 28.6 Structure of a tRNA molecule. The tRNA is a roughly cloverleaf-shaped molecule containing an anticodon triplet on one "leaf" and an amino acid unit attached covalently at its 3' end. The example shown is a yeast tRNA that codes for phenylalanine. The nucleotides not specifically identified are chemically modified analogs of the four common nucleotides.

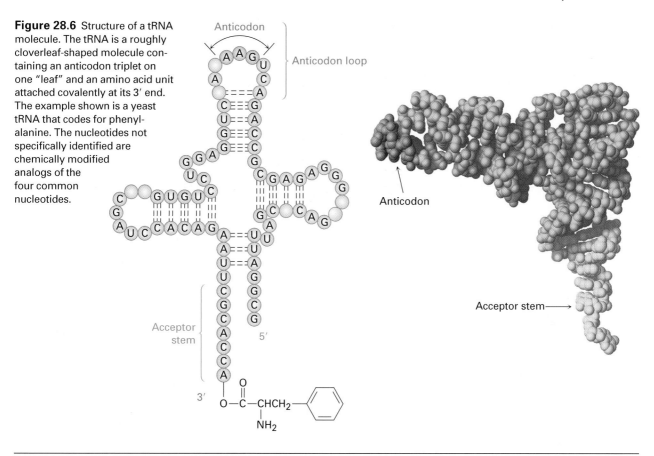

Figure 28.7 A representation of protein biosynthesis. The codon base sequences on mRNA are read by tRNAs containing complementary anticodon base sequences. Transfer RNAs assemble the proper amino acids into position for incorporation into the growing peptide.

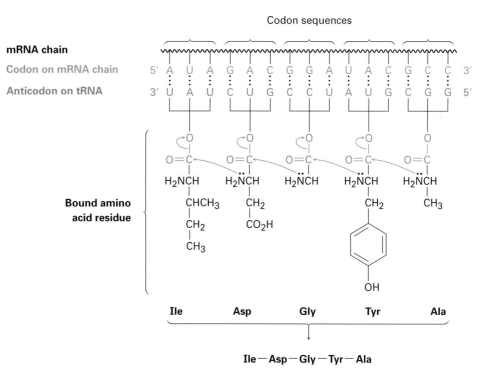

Problem 28.7 | List codon sequences for the following amino acids:
(a) Ala   (b) Phe   (c) Leu   (d) Tyr

Problem 28.8 | List anticodon sequences on the tRNAs carrying the amino acids shown in Problem 28.7.

Problem 28.9 | What amino acid sequence is coded by the following mRNA base sequence?

CUU-AUG-GCU-UGG-CCC-UAA

Problem 28.10 | What is the base sequence in the original DNA strand on which the mRNA sequence in Problem 28.9 was made?

28.6 | DNA Sequencing

One of the greatest scientific revolutions in history is now under way in molecular biology, as scientists are learning how to manipulate and harness the genetic machinery of organisms. None of the extraordinary advances of the past two decades would have been possible, however, were it not for the discovery in 1977 of methods for sequencing immense DNA chains.

The first step in DNA sequencing is to cleave the enormous chain at known points to produce smaller, more manageable pieces, a task accomplished by the use of *restriction endonucleases*. Each different restriction enzyme, of which more than 3500 are known and approximately 200 are commercially available, cleaves a DNA molecule at a point in the chain where a specific base sequence occurs. For example, the restriction enzyme *Alu*I cleaves between G and C in the four-base sequence AG-CT. Note that the sequence is a *palindrome,* meaning that the *sequence* (5′)-AGCT-(3′) is the same as its *complement* (3′)-TCGA-(5′) when both are read in the same 5′ → 3′ direction. The same is true for other restriction endonucleases.

If the original DNA molecule is cut with another restriction enzyme having a different specificity for cleavage, still other segments are produced whose sequences partially overlap those produced by the first enzyme. Sequencing of all the segments, followed by identification of the overlapping regions, allows complete DNA sequencing.

Two methods of DNA sequencing are available. The *Maxam–Gilbert method* uses chemical techniques, while the **Sanger dideoxy method** uses enzymatic reactions. The Sanger method is the more commonly used of the two and was the method responsible for sequencing the entire human genome of 2.9 billion base pairs. In commercial sequencing instruments, the dideoxy method begins with a mixture of the following:

▮ The restriction fragment to be sequenced

▮ A small piece of DNA called a *primer,* whose sequence is complementary to that on the 3′ end of the restriction fragment

▮ The four 2′-deoxyribonucleoside triphosphates (dNTPs)

▌ Very small amounts of the four 2′,3′-dideoxyribonucleoside triphosphates (ddNTPs), each of which is labeled with a fluorescent dye of a different color (A 2′,3′-*dideoxy*ribonucleoside triphosphate is one in which both 2′ and 3′ −OH groups are missing from ribose.)

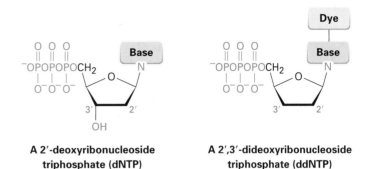

**A 2′-deoxyribonucleoside
triphosphate (dNTP)**

**A 2′,3′-dideoxyribonucleoside
triphosphate (ddNTP)**

DNA polymerase is added to the mixture, and a strand of DNA complementary to the restriction fragment begins to grow from the end of the primer. Most of the time, only normal deoxyribonucleotides are incorporated into the growing chain because of their much higher concentration in the mixture, but every so often, a dideoxyribonucleotide is incorporated. When that happens, DNA synthesis stops because the chain end no longer has a 3′-hydroxyl group for adding further nucleotides.

When reaction is complete, the product consists of a mixture of DNA fragments of all possible lengths, each terminated by one of the four dye-labeled dideoxyribonucleotides. This product mixture is then separated according to the size of the pieces by gel electrophoresis (Section 26.2), and the identity of the terminal dideoxyribonucleotide in each piece—and thus the sequence of the restriction fragment—is identified simply by noting the color with which the attached dye fluoresces. Figure 28.8 shows a typical result.

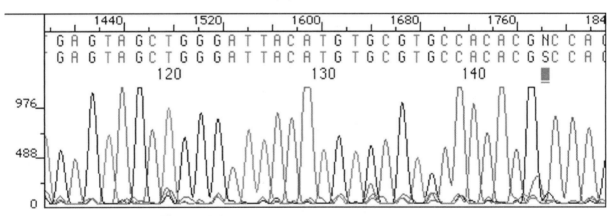

Figure 28.8 The sequence of a restriction fragment determined by the Sanger dideoxy method can be read simply by noting the colors of the dye attached to each of the various terminal nucleotides.

So efficient is the automated dideoxy method that sequences up to 1100 nucleotides in length, with a throughput of up to 19,000 bases per hour, can be sequenced with 98% accuracy. After a decade of work, preliminary sequence information for the entire human genome of 2.9 billion base pairs was announced early in 2001. Remarkably, our genome appears to contain only about 30,000 genes, less than one-third the previously predicted number and only twice the number found in the common roundworm.

28.7 | DNA Synthesis

The ongoing revolution in molecular biology has brought with it an increased demand for the efficient chemical synthesis of short DNA segments, called *oligonucleotides,* or simply *oligos.* The problems of DNA synthesis are similar to those of protein synthesis (Section 26.7) but are more difficult because of the complexity of the nucleotide monomers. Each nucleotide has multiple reactive sites that must be selectively protected and deprotected at the proper times, and coupling of the four nucleotides must be carried out in the proper sequence. Automated DNA synthesizers are available, however, that allow the fast and reliable synthesis of DNA segments up to 200 nucleotides in length.

DNA synthesizers operate on a principle similar to that of the Merrifield solid-phase peptide synthesizer (Section 26.8). In essence, a protected nucleotide is covalently bonded to a solid support, and one nucleotide at a time is added to the growing chain by the use of a coupling reagent. After the final nucleotide has been added, all the protecting groups are removed and the synthetic DNA is cleaved from the solid support. Five steps are needed:

Step 1   The first step in DNA synthesis is to attach a protected deoxynucleoside to a silica (SiO_2) support by an ester linkage to the 3' −OH group of the deoxynucleoside. Both the 5' −OH group on the sugar and free −NH_2 groups on the heterocyclic bases must be protected. Adenine and cytosine bases are protected by benzoyl groups, guanine is protected by an isobutyryl group, and thymine requires no protection. The deoxyribose 5' −OH is protected as its *p*-dimethoxytrityl (DMT) ether.

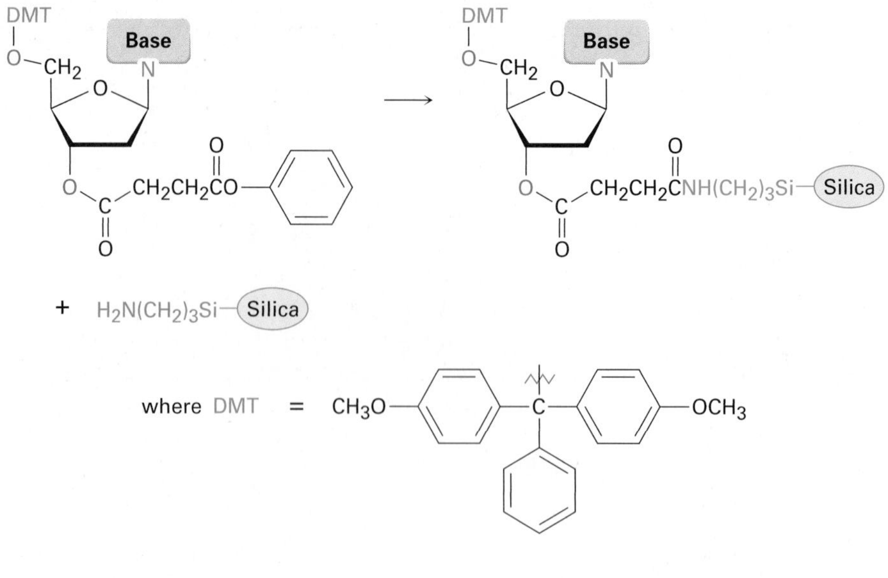

N-protected adenine **N-protected guanine** **N-protected cytosine** **Thymine**

Step 2 The second step is removal of the DMT protecting group by treatment with dichloroacetic acid in CH_2Cl_2. The reaction occurs by an S_N1 mechanism and proceeds rapidly because of the stability of the tertiary, benzylic dimethoxytrityl cation.

Step 3 The third step is the coupling of the polymer-bonded deoxynucleoside with a protected deoxynucleoside containing a *phosphoramidite* group at its 3′ position. [A phosphoramidite has the structure $R_2NP(OR)_2$.] The coupling reaction takes place in the polar aprotic solvent acetonitrile; requires catalysis by the heterocyclic amine tetrazole; and yields a *phosphite*, $P(OR)_3$, as product. Note that one of the phosphorus oxygen atoms is protected by a β-cyanoethyl group, $-OCH_2CH_2C\equiv N$. The coupling step takes place in better than 99% yield.

Step 4 With the coupling accomplished, the phosphite product is oxidized to a phosphate by treatment with iodine in aqueous tetrahydrofuran in the presence of 2,6-dimethylpyridine. The cycle (1) deprotection, (2) coupling, and (3) oxidation is then repeated until an oligonucleotide chain of the desired sequence has been built.

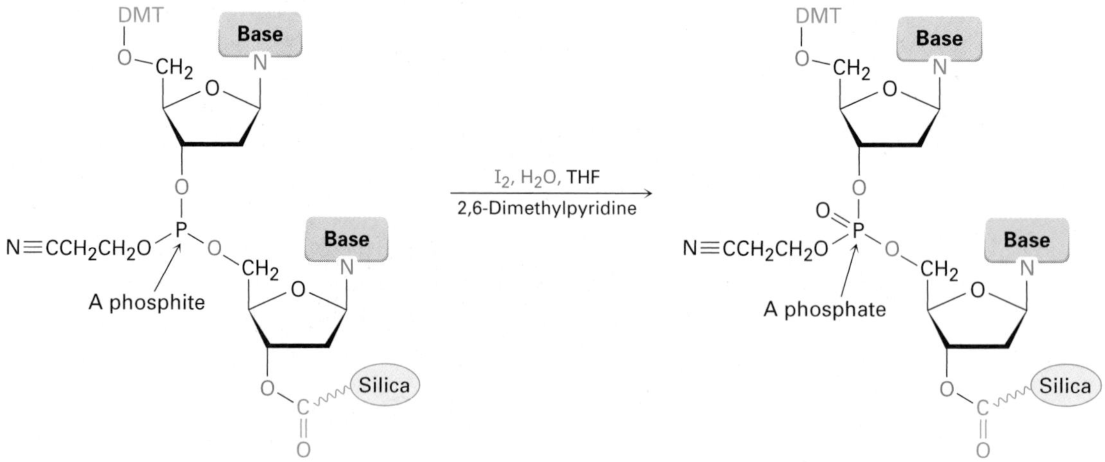

I_2, H_2O, THF
2,6-Dimethylpyridine

A phosphite

A phosphate

Step 5 The final step is removal of all protecting groups and cleavage of the ester bond holding the DNA to the silica. All these reactions are done at the same time by treatment with aqueous NH_3. Purification by electrophoresis then yields the synthetic DNA.

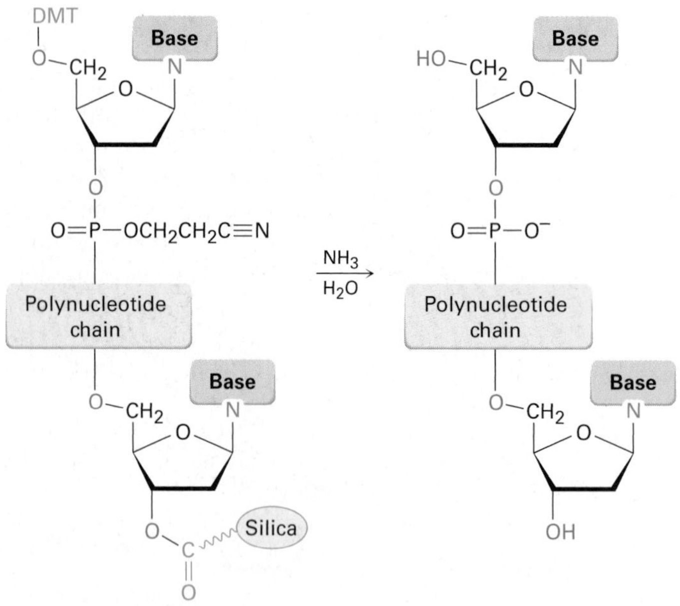

$O{=}P{-}OCH_2CH_2C{\equiv}N$

$\dfrac{NH_3}{H_2O}$

$O{=}P{-}O^-$

Problem 28.11 *p*-Dimethoxytrityl (DMT) ethers are easily cleaved by mild acid treatment. Show the mechanism of the cleavage reaction.

Problem 28.12   Propose a mechanism to account for cleavage of the β-cyanoethyl protecting group from the phosphate groups on treatment with aqueous ammonia. (Acrylonitrile, $H_2C{=}CHCN$, is a by-product.) What kind of reaction is occurring?

28.8 | The Polymerase Chain Reaction

Kary Banks Mullis

Kary Banks Mullis (1944–) was born in rural Lenoir, North Carolina; did undergraduate work at Georgia Tech.; and received his Ph.D. at the University of California, Berkeley, in 1973. From 1979 to 1986 he worked at Cetus Corp., where his work on developing PCR was carried out. Since 1988, he has followed his own drummer as self-employed consultant and writer. He received the 1993 Nobel Prize in chemistry.

It often happens that only tiny amounts of a gene sequence can be obtained directly from an organism's DNA, so methods for obtaining larger amounts are sometimes needed to carry out the sequencing. The invention of the **polymerase chain reaction (PCR)** by Kary Mullis in 1986 has been described as being to genes what Gutenberg's invention of the printing press was to the written word. Just as the printing press produces multiple copies of a book, PCR produces multiple copies of a given DNA sequence. Starting from less than 1 *picogram* of DNA with a chain length of 10,000 nucleotides (1 pg = 10^{-12} g; about 100,000 molecules), PCR makes it possible to obtain several micrograms (1 μg = 10^{-6} g; about 10^{11} molecules) in just a few hours.

The key to the polymerase chain reaction is *Taq* DNA polymerase, a heat-stable enzyme isolated from the thermophilic bacterium *Thermus aquaticus* found in a hot spring in Yellowstone National Park. *Taq* polymerase is able to take a single strand of DNA that has a short, primer segment of complementary chain at one end and then finish constructing the entire complementary strand. The overall process takes three steps, as shown schematically in Figure 28.9. (More recently, improved heat-stable DNA polymerase enzymes have become available, including Vent polymerase and *Pfu* polymerase, both isolated from bacteria growing near geothermal vents in the ocean floor. The error rate of both enzymes is substantially less than that of *Taq*.)

Step 1   The double-stranded DNA to be amplified is heated in the presence of *Taq* polymerase, Mg^{2+} ion, the four deoxynucleotide triphosphate monomers (dNTPs), and a large excess of two short oligonucleotide primers of about 20 bases each. Each primer is complementary to the sequence at the end of one of the target DNA segments. At a temperature of 95 °C, double-stranded DNA denatures, spontaneously breaking apart into two single strands.

Step 2   The temperature is lowered to between 37 and 50 °C, allowing the primers, because of their relatively high concentration, to anneal by hydrogen-bonding to their complementary sequence at the end of each target strand.

Step 3   The temperature is then raised to 72 °C, and *Taq* polymerase catalyzes the addition of further nucleotides to the two primed DNA strands. When replication of each strand is finished, *two* copies of the original DNA now exist. Repeating the denature–anneal–synthesize cycle a second time yields four DNA copies, repeating a third time yields eight copies, and so on, in an exponential series.

PCR has been automated, and 30 or so cycles can be carried out in an hour, resulting in a theoretical amplification factor of 2^{30} (~10^9). In practice, however, the efficiency of each cycle is less than 100%, and an experimental amplification of about 10^6 to 10^8 is routinely achieved for 30 cycles.

Figure 28.9 The polymerase chain reaction. Details are explained in the text.

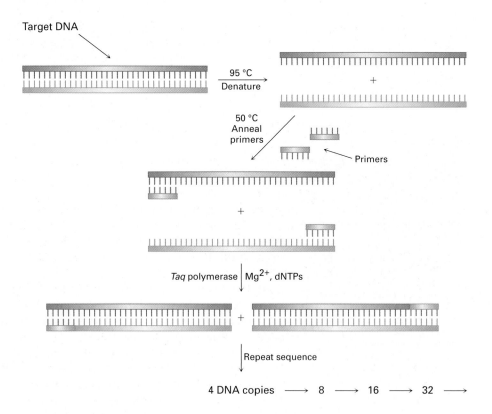

DNA Fingerprinting

The invention of DNA sequencing has affected society in many ways, few more dramatic than those stemming from the development of *DNA finger-printing*. DNA fingerprinting arose from the discovery in 1984 that human genes contain short, repeating sequences of noncoding DNA, called *short tandem repeat* (STR) loci. Furthermore, the STR loci are slightly different for every individual, except identical twins. By sequencing these loci, a pattern unique to each person can be obtained.

Perhaps the most common and well-publicized use of DNA fingerprinting is that carried out by crime laboratories to link suspects to biological evidence—blood, hair follicles, skin, or semen—found at a crime scene. Thousands of court cases have now been decided based on DNA evidence.

For use in criminal cases, forensic laboratories in the United States have agreed on 13 core STR loci that are most accurate for identification of an individual. Based on these 13 loci, a Combined DNA Index System (CODIS) has

(continued)

Historians have wondered for many years whether Thomas Jefferson fathered a child by Sally Hemings. DNA fingerprinting evidence obtained in 1998 is inconclusive but strongly suggestive.

been established to serve as a registry of convicted offenders. When a DNA sample is obtained from a crime scene, the sample is subjected to cleavage with restriction endonucleases to cut out fragments containing the STR loci, the fragments are amplified using the polymerase chain reaction, and the sequences of the fragments are determined.

If the profile of sequences from a known individual and the profile from DNA obtained at a crime scene match, the probability is approximately 82 billion to 1 that the DNA is from the same individual. In paternity cases, where the DNA of father and offspring are related but not fully identical, the identity of the father can be established with a probability of 100,000 to 1. Even after several generations have passed, paternity can still be implied by DNA analysis of the Y chromosome of direct male-line descendants. The most well-known such case is that of Thomas Jefferson, who may have fathered a child by his slave Sally Hemings. Although Jefferson himself has no male-line descendants, DNA analysis of the male-line descendants of Jefferson's paternal uncle contained the same Y chromosome as a male-line descendant of Eston Hemings, youngest son of Sally Hemings. Thus, a mixing of the two genomes is clear, although the male individual responsible for that mixing can't be conclusively identified.

Among its many other applications, DNA fingerprinting is widely used for the diagnosis of genetic disorders, both prenatally and in newborns. Cystic fibrosis, hemophilia, Huntington's disease, Tay–Sachs disease, sickle cell anemia, and thalassemia are among the many diseases that can be detected, enabling early treatment of an affected child. Furthermore, by studying the DNA fingerprints of relatives with a history of a particular disorder, it's possible to identify DNA patterns associated with the disease and perhaps obtain clues for eventual cure. In addition, the U.S. Department of Defense now requires blood and saliva samples from all military personnel. The samples are stored, and DNA is extracted should the need for identification of a casualty arise.

SUMMARY AND KEY WORDS

anticodon, 1109

coding strand, 1108

codon, 1109

deoxyribonucleic acid (DNA), 1100

double helix, 1103

3′ end, 1103

5′ end, 1103

messenger RNA (mRNA), 1107

nucleoside, 1100

nucleotide, 1100

The nucleic acids **DNA (deoxyribonucleic acid)** and **RNA (ribonucleic acid)** are biological polymers that act as chemical carriers of an organism's genetic information. Enzyme-catalyzed hydrolysis of nucleic acids yields **nucleotides**, the monomer units from which RNA and DNA are constructed. Further enzyme-catalyzed hydrolysis of the nucleotides yields **nucleosides** plus phosphate. Nucleosides, in turn, consist of a purine or pyrimidine base linked to C1 of an aldopentose sugar—ribose in RNA and 2-deoxyribose in DNA. The nucleotides are joined by phosphate links between the 5′ phosphate of one nucleotide and the 3′ hydroxyl on the sugar of another nucleotide.

Molecules of DNA consist of two complementary polynucleotide strands held together by hydrogen bonds between heterocyclic bases on the different strands and coiled into a **double helix**. Adenine and thymine form hydrogen bonds to each other, as do cytosine and guanine.

Three processes take place in deciphering the genetic information of DNA:

▌ **Replication** of DNA is the process by which identical DNA copies are made. The DNA double helix unwinds, complementary deoxyribonucleotides line up in order, and two new DNA molecules are produced.

▌ **Transcription** is the process by which RNA is produced to carry genetic information from the nucleus to the ribosomes. A short segment of the DNA double helix unwinds, and complementary ribonucleotides line up to produce **messenger RNA (mRNA)**.

▌ **Translation** is the process by which mRNA directs protein synthesis. Each mRNA is divided into **codons**, ribonucleotide triplets that are recognized by small amino acid–carrying molecules of **transfer RNA (tRNA)**, which deliver the appropriate amino acids needed for protein synthesis.

Sequencing of DNA is carried out by the **Sanger dideoxy method**, and small DNA segments can be synthesized in the laboratory by automated instruments. Small amounts of DNA can be amplified by factors of 10^6 using the **polymerase chain reaction (PCR)**.

EXERCISES

VISUALIZING CHEMISTRY

(Problems 28.1–28.12 appear within the chapter.)

28.13 ■ Identify the following bases, and tell whether each is found in DNA, RNA, or both:

(a)   (b)   (c)

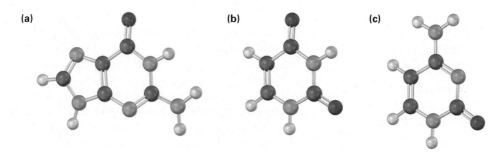

28.14 ■ Identify the following nucleotide, and tell how it is used:

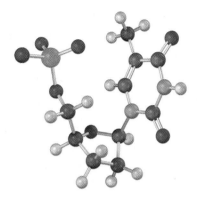

28.15 Amine bases in nucleic acids can react with alkylating agents in typical S_N2 reactions. Look at the following electrostatic potential maps, and tell which is the better nucleophile, guanine or adenine. The reactive positions in each are indicated.

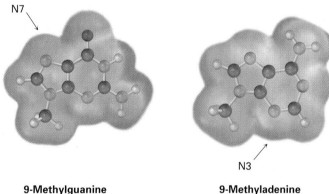

9-Methylguanine   **9-Methyladenine**

ADDITIONAL PROBLEMS

28.16 Human brain natriuretic peptide (BNP) is a small peptide of 32 amino acids used in the treatment of congestive heat failure. How many nitrogen bases are present in the DNA that codes for BNP?

28.17 Human and horse insulin both have two polypeptide chains, with one chain containing 21 amino acids and the other containing 30 amino acids. They differ in primary structure at two places. At position 9 in one chain, human insulin has Ser and horse insulin has Gly; at position 30 in the other chain, human insulin has Thr and horse insulin has Ala. How must the DNA for the two insulins differ?

28.18 ■ The DNA of sea urchins contains about 32% A. What percentages of the other three bases would you expect in sea urchin DNA? Explain.

28.19 The codon UAA stops protein synthesis. Why does the sequence UAA in the following stretch of mRNA not cause any problems?

-GCA-UUC-GAG-GUA-ACG-CCC-

28.20 ■ Which of the following base sequences would most likely be recognized by a restriction endonuclease? Explain.
(a) GAATTC **(b)** GATTACA **(c)** CTCGAG

28.21 ■ For what amino acids do the following ribonucleotide triplets code?
(a) AAU **(b)** GAG **(c)** UCC **(d)** CAU

28.22 ■ From what DNA sequences were each of the mRNA codons in Problem 28.21 transcribed?

28.23 ■ What anticodon sequences of tRNAs are coded for by the codons in Problem 28.21?

28.24 Draw the complete structure of the ribonucleotide codon UAC. For what amino acid does this sequence code?

28.25 Draw the complete structure of the deoxyribonucleotide sequence from which the mRNA codon in Problem 28.24 was transcribed.

28.26 Give an mRNA sequence that will code for synthesis of metenkephalin.

Tyr-Gly-Gly-Phe-Met

28.27 Give an mRNA sequence that will code for the synthesis of angiotensin II.

Asp-Arg-Val-Tyr-Ile-His-Pro-Phe

28.28 ■ What amino acid sequence is coded for by the following DNA coding strand?

(5′) CTT-CGA-CCA-GAC-AGC-TTT (3′)

28.29 ■ What amino acid sequence is coded for by the following mRNA base sequence?

(5′) CUA-GAC-CGU-UCC-AAG-UGA (3′)

28.30 If the DNA coding sequence -CAA-CCG-GAT- were miscopied during replication and became -CGA-CCG-GAT-, what effect would there be on the sequence of the protein produced?

28.31 ■ Show the steps involved in a laboratory synthesis of the DNA fragment with the sequence CTAG.

28.32 ■ The final step in DNA synthesis is deprotection by treatment with aqueous ammonia. Show the mechanisms by which deprotection occurs at the points indicated in the following structure:

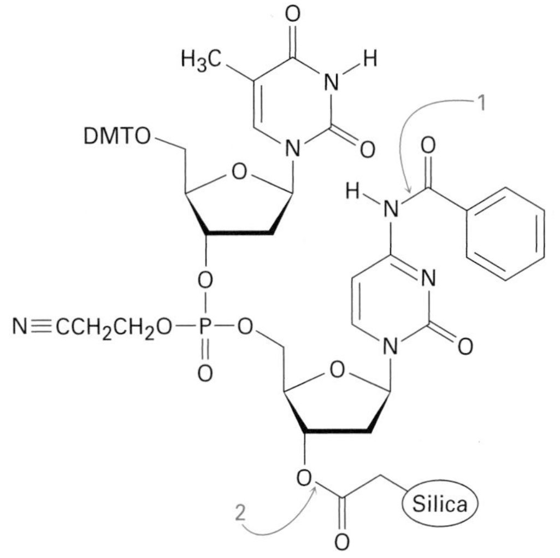

28.33 ■ Draw the structure of cyclic adenosine monophosphate (cAMP), a messenger involved in the regulation of glucose production in the body. Cyclic AMP has a phosphate ring connecting the 3′ and 5′ hydroxyl groups on adenosine.

28.34 The final step in the metabolic degradation of uracil is the oxidation of malonic semialdehyde to give malonyl CoA. Propose a mechanism.

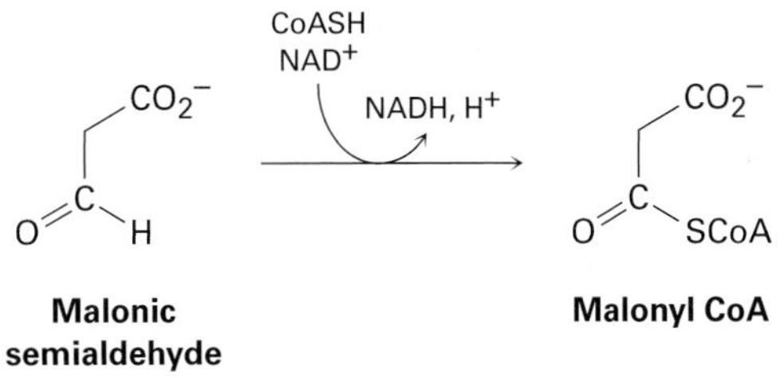

Malonic semialdehyde   **Malonyl CoA**

28.35 One of the steps in the biosynthesis of a nucleotide called inosine monophosphate is the formation of aminoimidazole ribonucleotide from formylglycinamidine ribonucleotide. Propose a mechanism.

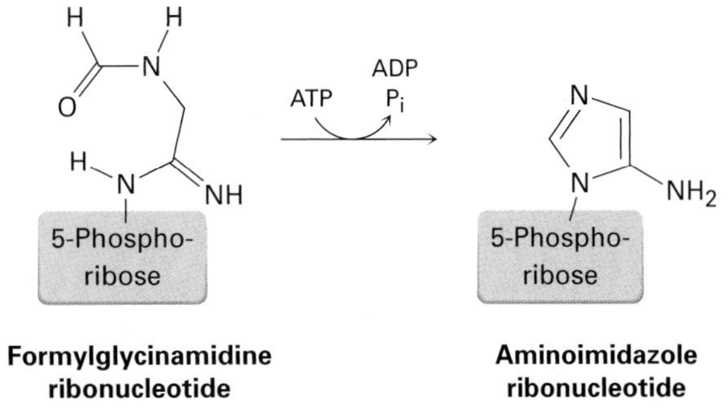

Formylglycinamidine ribonucleotide   **Aminoimidazole ribonucleotide**

28.36 One of the steps in the metabolic degradation of guanine is hydrolysis to give xanthine. Propose a mechanism.

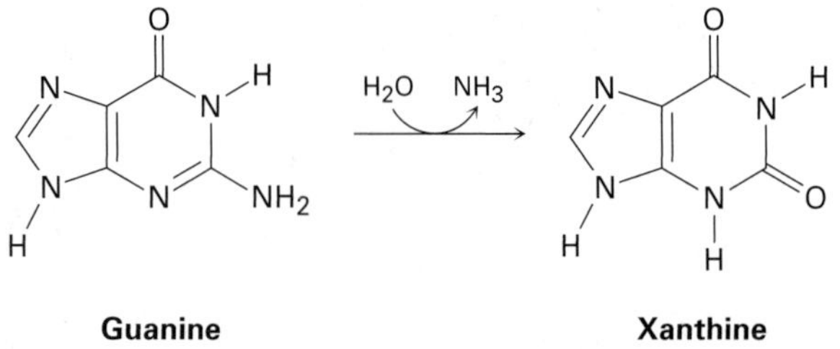

Guanine   Xanthine

28.37 One of the steps in the biosynthesis of uridine monophosphate is the reaction of aspartate with carbamoyl phosphate to give carbamoyl aspartate followed by cyclization to form dihydroorotate. Propose mechanisms for both steps.

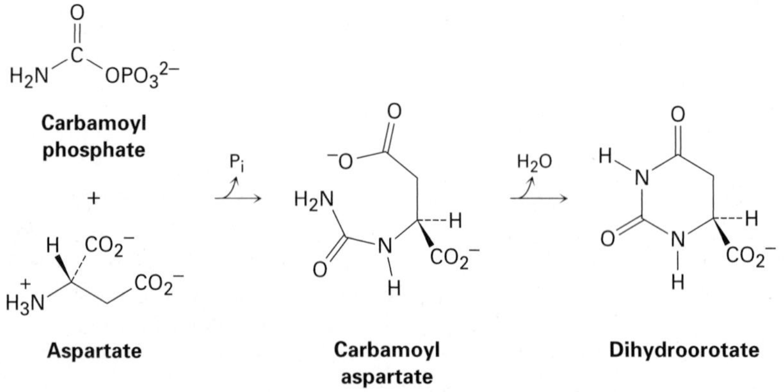

Carbamoyl
phosphate

Aspartate

Carbamoyl
aspartate

Dihydroorotate

29

The Organic Chemistry of Metabolic Pathways

Anyone who wants to understand or contribute to the revolution now taking place in the biological sciences must first understand life processes at the molecular level. This understanding, in turn, must be based on a detailed knowledge of the chemical reactions and paths used by living organisms. Just knowing *what* occurs is not enough; it's also necessary to understand *how* and *why* organisms use the chemistry they do.

Biochemical reactions are not mysterious. It's true that many of the biological reactions occurring in even the simplest living organism are more complex than those carried out in any laboratory, yet they follow the same rules of reactivity as laboratory reactions and they take place by the same mechanisms. In past chapters, we've seen many biological reactions used as examples, but it's now time to focus specifically on biological reactions, with particular attention to some typical metabolic pathways that organisms use to synthesize and degrade biomolecules.

A word of warning: biological molecules are often larger and more complex than the substances we've been dealing with thus far. As always, though, keep your focus on the functional groups in those parts of the molecules where changes occur. The reactions themselves are the same sorts of additions, eliminations, substitutions, carbonyl condensations, and so forth, that we've been dealing with all along. By the end of this chapter, a fundamental conclusion should be clear: the chemistry of living organisms *is* organic chemistry.

WHY THIS CHAPTER?

In this chapter, we'll look at some of the pathways by which organisms carry out their chemistry, focusing primarily on how they metabolize fats and carbohydrates. The treatment will be far from complete, but it should give you an idea of the kinds of processes that occur.

Sean Duggan

1125

29.1 | An Overview of Metabolism and Biochemical Energy

The many reactions that go on in the cells of living organisms are collectively called **metabolism**. The pathways that break down larger molecules into smaller ones are called **catabolism**, and the pathways that synthesize larger biomolecules from smaller ones are known as **anabolism**. Catabolic reaction pathways are usually exergonic and release energy, while anabolic pathways are often endergonic and absorb energy. Catabolism can be divided into the four stages shown in Figure 29.1.

Stage 1 Bulk food is digested in the stomach and small intestine to give small molecules.

Stage 2 Fatty acids, monosaccharides, and amino acids are degraded in cells to yield acetyl CoA.

Stage 3 Acetyl CoA is oxidized in the citric acid cycle to give CO_2.

Stage 4 The energy released in the citric acid cycle is used by the electron-transport chain to oxidatively phosphorylate ADP and produce ATP.

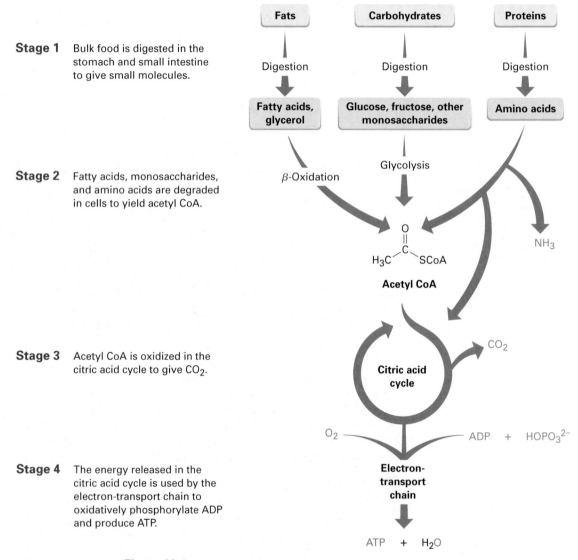

Figure 29.1 An overview of catabolic pathways for the degradation of food and the production of biochemical energy. The ultimate products of food catabolism are CO_2 and H_2O, with the energy released in the citric acid cycle used to drive the endergonic synthesis of adenosine triphosphate (ATP) from adenosine diphosphate (ADP) plus phosphate ion, $HOPO_3^{2-}$.

In *digestion*, the first catabolic stage, food is broken down in the mouth, stomach, and small intestine by hydrolysis of ester, glycoside (acetal), and peptide (amide) bonds to yield primarily fatty acids plus glycerol, simple sugars, and amino acids. These smaller molecules are further degraded in the cytoplasm of cells in the second stage of catabolism to yield acetyl groups attached by a thioester bond to the large carrier molecule coenzyme A. The resultant compound, acetyl coenzyme A (acetyl CoA), is a key substance both in the metabolism of food molecules and in numerous other biological pathways. As noted in Section 21.8, the acetyl group in acetyl CoA is linked to the sulfur atom of phosphopantetheine, which is itself linked to adenosine 3′,5′-bisphosphate.

Acetyl CoA—a thioester

Acetyl groups are oxidized inside cellular mitochondria in the third stage of catabolism, the *citric acid cycle*, to yield CO_2. (We'll see the details of the process in Section 29.7.) Like many oxidations, this stage releases a large amount of energy, which is used in the fourth stage, the *electron-transport chain*, to accomplish the endergonic phosphorylation of ADP with hydrogen phosphate ion ($HOPO_3^{2-}$, abbreviated P_i) to give ATP. The final result of food catabolism, ATP has been called the "energy currency" of the cell. Catabolic reactions "pay off" in ATP by synthesizing it from ADP plus phosphate ion, and anabolic reactions "spend" ATP by transferring a phosphate group to another molecule, thereby regenerating ADP. Energy production and use in living organisms thus revolves around the ATP $\rightleftarrows$ ADP interconversion.

Adenosine diphosphate (ADP)   **Adenosine triphosphate (ATP)**

ADP and ATP are both phosphoric acid anhydrides, which contain $-\overset{O}{\overset{\|}{P}}-O-\overset{O}{\overset{\|}{P}}-$ linkages analogous to the $-\overset{O}{\overset{\|}{C}}-O-\overset{O}{\overset{\|}{C}}-$ linkage in carboxylic acid anhydrides. Just as carboxylic anhydrides react with alcohols by breaking a

C—O bond and forming a carboxylic ester (Section 21.5), phosphoric anhydrides react with alcohols by breaking a P—O bond and forming a phosphate ester, $ROPO_3^{2-}$. Note that phosphorylation reactions with ATP generally require the presence of a divalent metal cation in the enzyme, usually Mg^{2+}, to form a Lewis acid/base complex with the phosphate oxygen atoms and neutralize some negative charge.

A phosphate ester **ADP**

How does the body use ATP? Recall from Section 5.7 that the free-energy change ΔG must be negative and energy must be released for a reaction to occur spontaneously. If ΔG is positive, the reaction is unfavorable and the process can't occur spontaneously.

What typically happens for an energetically unfavorable reaction to occur is that it is "coupled" to an energetically favorable reaction so that the overall free-energy change for the two reactions together is favorable. To understand what it means for reactions to be coupled, imagine that reaction 1 does not occur to any reasonable extent because it has a small equilibrium constant and is energetically unfavorable; that is, the reaction has $\Delta G > 0$.

$$(1)\ \mathbf{A}\ +\ m\ \rightleftarrows\ \mathbf{B}\ +\ n \qquad \Delta G\ >\ 0$$

where **A** and **B** are the biochemically "interesting" substances undergoing transformation, while m and n are enzyme cofactors, H_2O, or various other substances.

Imagine also that product n can react with substance o to yield p and q in a second, strongly favorable reaction that has a large equilibrium constant and $\Delta G \ll 0$.

$$(2)\ n\ +\ o\ \rightleftarrows\ p\ +\ q \qquad \Delta G\ <\ 0$$

Considering the two reactions together, they share, or are coupled through, the common intermediate n, which is a product in the first reaction and a reactant in the second. When even a tiny amount of n is formed in reaction 1, it undergoes essentially complete conversion in reaction 2, thereby removing it from the first equilibrium and forcing reaction 1 to continually replenish n until the reactant **A** is gone. That is, the two reactions added together have a favorable $\Delta G < 0$, and we say that the favorable reaction 2 "drives" the unfavorable reaction 1. Because the two reactions are coupled through n, the transformation of **A** to **B** becomes possible.

(1)	**A** + m ⇌ **B** + $n̸$	$\Delta G > 0$
(2)	$n̸$ + o ⇌ p + q	$\Delta G \ll 0$
Net:	**A** + m + o ⇌ **B** + p + q	$\Delta G < 0$

As an example of two reactions that are coupled, look at the phosphorylation reaction of glucose to yield glucose 6-phosphate plus water, an important step in the breakdown of dietary carbohydrates. The reaction of glucose with $HOPO_3{}^{2-}$ does not occur spontaneously because it is energetically unfavorable, with $\Delta G^{\circ\prime} = +13.8$ kJ/mol. (The standard free-energy change for a biological reaction is denoted $\Delta G^{\circ\prime}$ and refers to a process in which reactants and products have a concentration of 1.0 M in a solution with pH = 7.)

$$HOCH_2CHCHCHCHCH \xrightarrow{\ HOPO_3{}^{2-}\ } {}^-OPOCH_2CHCHCHCHCH + H_2O \qquad \Delta G^{\circ\prime} = +13.8 \text{ kJ}$$

Glucose   **Glucose 6-phosphate**

With ATP, however, glucose undergoes an energetically favorable reaction to yield glucose 6-phosphate plus ADP. The overall effect is as if $HOPO_3{}^{2-}$ reacted with glucose and ATP then reacted with the water by-product, making the coupled process favorable by about 16.7 kJ/mol (4.0 kcal/mol). That is, ATP drives the phosphorylation reaction of glucose.

Glucose + $HOPO_3{}^{2-}$ ⟶ Glucose 6-phosphate + H_2O	$\Delta G^{\circ\prime} = +13.8$ kJ/mol
ATP + H_2O ⟶ ADP + $HOPO_3{}^{2-}$ + H^+	$\Delta G^{\circ\prime} = -30.5$ kJ/mol
Net: Glucose + ATP ⟶ Glucose 6-phosphate + ADP + H^+	$\Delta G^{\circ\prime} = -16.7$ kJ/mol

It's this ability to drive otherwise unfavorable phosphorylation reactions that makes ATP so useful. The resultant phosphates are much more reactive as leaving groups in nucleophilic substitutions and eliminations than the corresponding alcohols they're derived from and are therefore more likely to be chemically useful.

Problem 29.1 | One of the steps in fat metabolism is the reaction of glycerol (1,2,3-propanetriol) with ATP to yield glycerol 1-phosphate. Write the reaction, and draw the structure of glycerol 1-phosphate.

29.2 | Catabolism of Triacylglycerols: The Fate of Glycerol

The metabolic breakdown of triacylglycerols begins with their hydrolysis to yield glycerol plus fatty acids. The reaction is catalyzed by a lipase, whose mechanism of action is shown in Figure 29.2. The active site of the enzyme contains a catalytic triad of aspartic acid, histidine, and serine residues, which act cooperatively to provide the necessary acid and base catalysis for the individual steps. Hydrolysis is accomplished by two sequential nucleophilic acyl substitution reactions, one that covalently binds an acyl group to the side chain —OH of a serine residue on the enzyme and a second that frees the fatty acid from the enzyme.

Steps 1–2 of Figure 29.2: Acyl Enzyme Formation   The first nucleophilic acyl substitution step—reaction of the triacylglycerol with the active-site serine to give an acyl enzyme—begins with deprotonation of the serine alcohol by histidine to form the more strongly nucleophilic alkoxide ion. This proton transfer is facilitated by a nearby side-chain carboxylate anion of aspartic acid, which makes the histidine more basic and stabilizes the resultant histidine cation by electrostatic interactions. The deprotonated serine adds to a carbonyl group of a triacylglycerol to give a tetrahedral intermediate.

The tetrahedral intermediate expels a diacylglycerol as the leaving group and produces an acyl enzyme. The step is catalyzed by a proton transfer from histidine to make the leaving group a neutral alcohol.

Steps 3–4 of Figure 29.2: Hydrolysis   The second nucleophilic acyl substitution step hydrolyzes the acyl enzyme and gives the free fatty acid by a mechanism analogous to that of the first two steps. Water is deprotonated by histidine to give hydroxide ion, which adds to the enzyme-bound acyl group. The tetrahedral

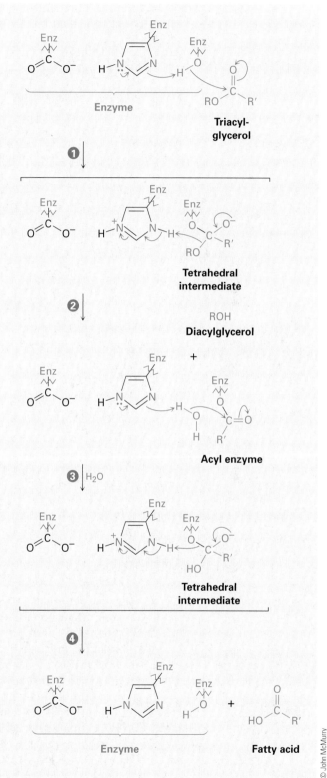

1 The enzyme active site contains an aspartic acid, a histidine, and a serine. First, histidine acts as a base to deprotonate the –OH group of serine, with the negatively charged carboxylate of aspartic acid stabilizing the nearby histidine cation that results. Serine then adds to the carbonyl group of the triacylglycerol, yielding a tetrahedral intermediate.

2 This intermediate expels a diacylglycerol as leaving group in a nucleophilic acyl substitution reaction, giving an acyl enzyme. The diacylglycerol is protonated by the histidine cation.

3 Histidine deprotonates a water molecule, which adds to the acyl group. A tetrahedral intermediate is again formed, and the histidine cation is again stabilized by the nearby carboxylate.

4 The tetrahedral intermediate expels the serine as leaving group in a second nucleophilic acyl substitution reaction, yielding a free fatty acid. The serine accepts a proton from histidine, and the enzyme has now returned to its starting structure.

Figure 29.2 MECHANISM: Mechanism of action of lipase. The active site of the enzyme contains a catalytic triad of aspartic acid, histidine, and serine, which react cooperatively to carry out two nucleophilic acyl substitution reactions. Individual steps are explained in the text.

intermediate then expels the neutral serine residue as the leaving group, freeing the fatty acid and returning the enzyme to its active form.

Acyl enzyme

Tetrahedral intermediate

Enzyme   **Fatty acid**

The fatty acids released on triacylglycerol hydrolysis are transported to mitochondria and degraded to acetyl CoA, while the glycerol is carried to the liver for further metabolism. In the liver, glycerol is first phosphorylated by reaction with ATP. Oxidation by NAD^+ then yields dihydroxyacetone phosphate (DHAP), which enters the carbohydrate metabolic pathway. We'll discuss this carbohydrate pathway in more detail in Section 29.5.

Glycerol   ***sn*-Glycerol 3-phosphate**   **Dihydroxyacetone phosphate (DHAP)**

You might note that C2 of glycerol is a prochiral center (Section 9.13) with two identical "arms." As is typical for enzyme-catalyzed reactions, the phosphorylation of glycerol is selective. Only the *pro-R* arm undergoes reaction, although this can't be predicted in advance. Note also that the phosphorylation product is named *sn*-glycerol 3-phosphate, where the *sn*- prefix means "stereospecific numbering." In this convention, the molecule is drawn in Fischer projection with the −OH group at C2 pointing to the left and the glycerol carbon atoms numbered beginning at the top.

Oxidation of *sn*-glycerol 3-phosphate to give dihydroxyacetone phosphate is catalyzed by *sn*-glycerol-3-phosphate dehydrogenase, with NAD^+ as cofactor. The reaction is stereospecific, occurring exclusively on the *Re* face of the nicotinamide ring and adding a hydrogen with *pro-R* stereochemistry. All alcohol dehydrogenases are stereospecific, although their specificity differs depending on the enzyme.

29.3 | Catabolism of Triacylglycerols: *β*-Oxidation

The fatty acids that result from triacylglycerol hydrolysis are catabolized by a repetitive four-step sequence of enzyme-catalyzed reactions called the **β-oxidation pathway**, shown in Figure 29.3 on page 1134. Each passage along the pathway results in the cleavage of an acetyl group from the end of the fatty-acid chain, until the entire molecule is ultimately degraded. As each acetyl group is produced, it enters the citric acid cycle and is further degraded, as we'll see in Section 29.7.

Step 1 of Figure 29.3: Introduction of a Double Bond   The *β*-oxidation pathway begins when a fatty acid forms a thioester with coenzyme A to give a fatty acyl CoA. Two hydrogen atoms are then removed from C2 and C3 of the fatty acyl CoA by one of a family of acyl-CoA dehydrogenases to yield an *α,β*-unsaturated acyl CoA. This kind of oxidation—the introduction of a conjugated double bond into a carbonyl compound—occurs frequently in biochemical pathways and usually involves the coenzyme flavin adenine dinucleotide (FAD). Reduced FADH$_2$ is the by-product.

Figure 29.3 MECHANISM:
The four steps of the β-oxidation pathway, resulting in the cleavage of an acetyl group from the end of the fatty-acid chain. The key chain-shortening step is a retro-Claisen reaction of a β-keto thioester. Individual steps are explained in the text.

① A conjugated double bond is introduced by removal of hydrogens from C2 and C3 by the coenzyme flavin adenine dinucleotide (FAD).

② Conjugate nucleophilic addition of water to the double bond gives a β-hydroxyacyl CoA.

③ The alcohol is oxidized by NAD$^+$ to give a β-keto thioester.

④ Nucleophilic addition of coenzyme A to the keto group occurs, followed by a retro-Claisen condensation reaction. The products are acetyl CoA and a chain-shortened fatty acyl CoA.

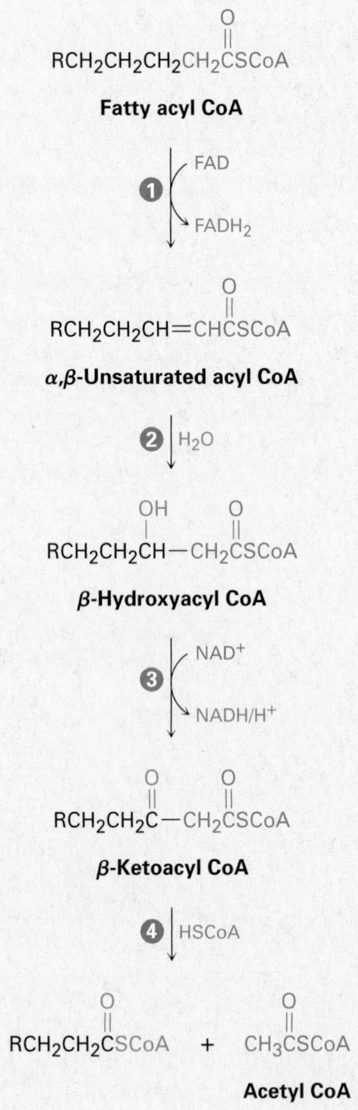

$$\underset{\text{Fatty acyl CoA}}{RCH_2CH_2CH_2CH_2\overset{\overset{\displaystyle O}{\|}}{C}SCoA}$$

① $\quad$ FAD $\quad\searrow$ FADH$_2$

$$\underset{\alpha,\beta\text{-Unsaturated acyl CoA}}{RCH_2CH_2CH=CH\overset{\overset{\displaystyle O}{\|}}{C}SCoA}$$

② $\quad$ H$_2$O

$$\underset{\beta\text{-Hydroxyacyl CoA}}{RCH_2CH_2\overset{\overset{\displaystyle OH}{|}}{C}H-CH_2\overset{\overset{\displaystyle O}{\|}}{C}SCoA}$$

③ $\quad$ NAD$^+$ $\quad\searrow$ NADH/H$^+$

$$\underset{\beta\text{-Ketoacyl CoA}}{RCH_2CH_2\overset{\overset{\displaystyle O}{\|}}{C}-CH_2\overset{\overset{\displaystyle O}{\|}}{C}SCoA}$$

④ $\quad$ HSCoA

$$RCH_2CH_2\overset{\overset{\displaystyle O}{\|}}{C}SCoA \quad + \quad \underset{\text{Acetyl CoA}}{CH_3\overset{\overset{\displaystyle O}{\|}}{C}SCoA}$$

© John McMurry

The mechanisms of FAD-catalyzed reactions are often difficult to establish because flavin coenzymes can operate by both two-electron (polar) and one-electron (radical) pathways. As a result, extensive studies of the family of acyl-CoA dehydrogenases have not yet provided a clear picture of how these enzymes function. What is known is that: (1) The first step is abstraction of the *pro-R* hydrogen from the acidic α position of the acyl CoA to give a thioester enolate ion. Hydrogen-bonding between the acyl carbonyl group and the ribitol hydroxyls of FAD increases the acidity of the acyl group. (2) The *pro-R* hydrogen at the β position is transferred to FAD. (3) The α,β-unsaturated acyl CoA that results has a trans double bond.

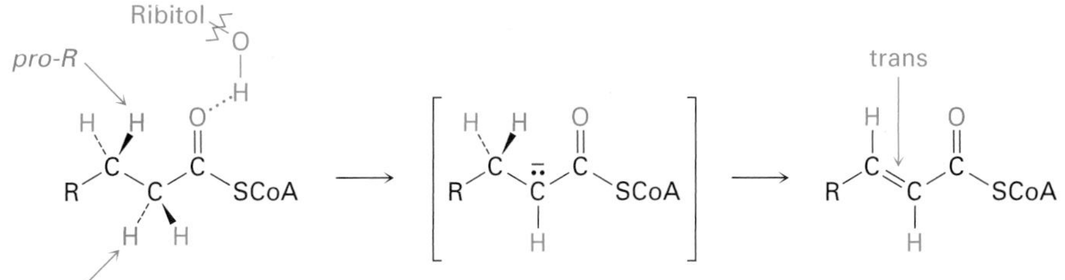

Thioester enolate

One suggested mechanism is that the reaction may take place by a conjugate hydride-transfer mechanism, analogous to what occurs during alcohol oxidations with NAD⁺. Electrons on the enolate ion might expel a β hydride ion, which could add to the doubly bonded N5 nitrogen on FAD. Protonation of the intermediate at N1 would give the product.

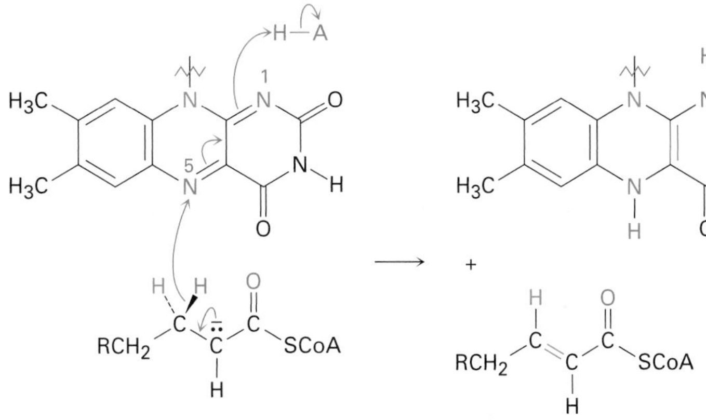

Step 2 of Figure 29.3: Conjugate Addition of Water The α,β-unsaturated acyl CoA produced in step 1 reacts with water by a conjugate addition pathway (Section 19.13) to yield a β-hydroxyacyl CoA in a process catalyzed by enoyl CoA hydratase. Water as nucleophile adds to the β carbon of the double bond, yielding an enolate ion intermediate that is protonated on the α position.

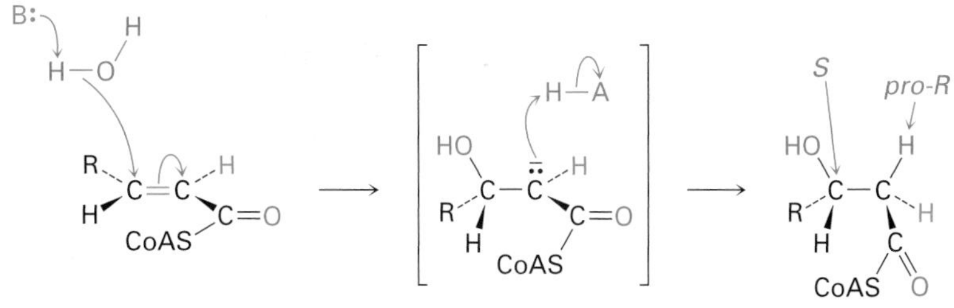

(3S)-Hydroxyacyl CoA

Step 3 of Figure 29.3: Alcohol Oxidation   The β-hydroxyacyl CoA from step 2 is oxidized to a β-ketoacyl CoA in a reaction catalyzed by one of a family of L-3-hydroxyacyl-CoA dehydrogenases, which differ in substrate specificity according to the chain length of the acyl group. As in the oxidation of *sn*-glycerol 3-phosphate to dihydroxyacetone phosphate mentioned at the end of Section 29.2, this alcohol oxidation requires NAD$^+$ as a coenzyme and yields reduced NADH/H$^+$ as by-product. Deprotonation of the hydroxyl group is carried out by a histidine residue at the active site.

β-Hydroxyacyl CoA   **β-Ketoacyl CoA**

Step 4 of Figure 29.3: Chain Cleavage   Acetyl CoA is split off from the chain in the final step of β-oxidation, leaving an acyl CoA that is two carbon atoms shorter than the original. The reaction is catalyzed by β-ketoacyl-CoA thiolase and is mechanistically the reverse of a Claisen condensation reaction (Section 23.7). In the *forward* direction, a Claisen condensation joins two esters together to form a β-keto ester product. In the *reverse* direction, a retro-Claisen reaction splits a β-keto ester (or β-keto thioester) apart to form two esters (or two thioesters).

The retro-Claisen reaction occurs by initial nucleophilic addition of a cysteine –SH group on the enzyme to the keto group of the β-ketoacyl CoA to yield an alkoxide ion intermediate. Cleavage of the C2–C3 bond then follows, with expulsion of an acetyl CoA enolate ion. Protonation of the enolate ion gives acetyl CoA, and the enzyme-bound acyl group undergoes nucleophilic acyl substitution by reaction with a molecule of coenzyme A. The chain-shortened acyl CoA that results then enters another round of the β-oxidation pathway for further degradation.

β-Ketoacyl CoA

Acetyl CoA

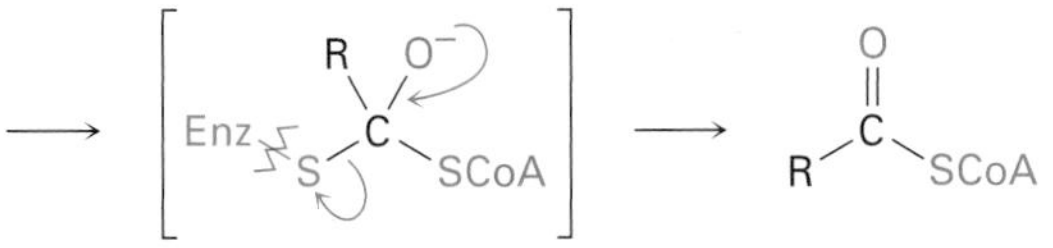

**Chain-shortened
acyl CoA**

Look at the catabolism of myristic acid shown in Figure 29.4 to see the over-all results of the β-oxidation pathway. The first passage converts the 14-carbon myristoyl CoA into the 12-carbon lauroyl CoA plus acetyl CoA, the second pas-sage converts lauroyl CoA into the 10-carbon caproyl CoA plus acetyl CoA, the third passage converts caproyl CoA into the 8-carbon capryloyl CoA, and so on. Note that the final passage produces *two* molecules of acetyl CoA because the precursor has four carbons.

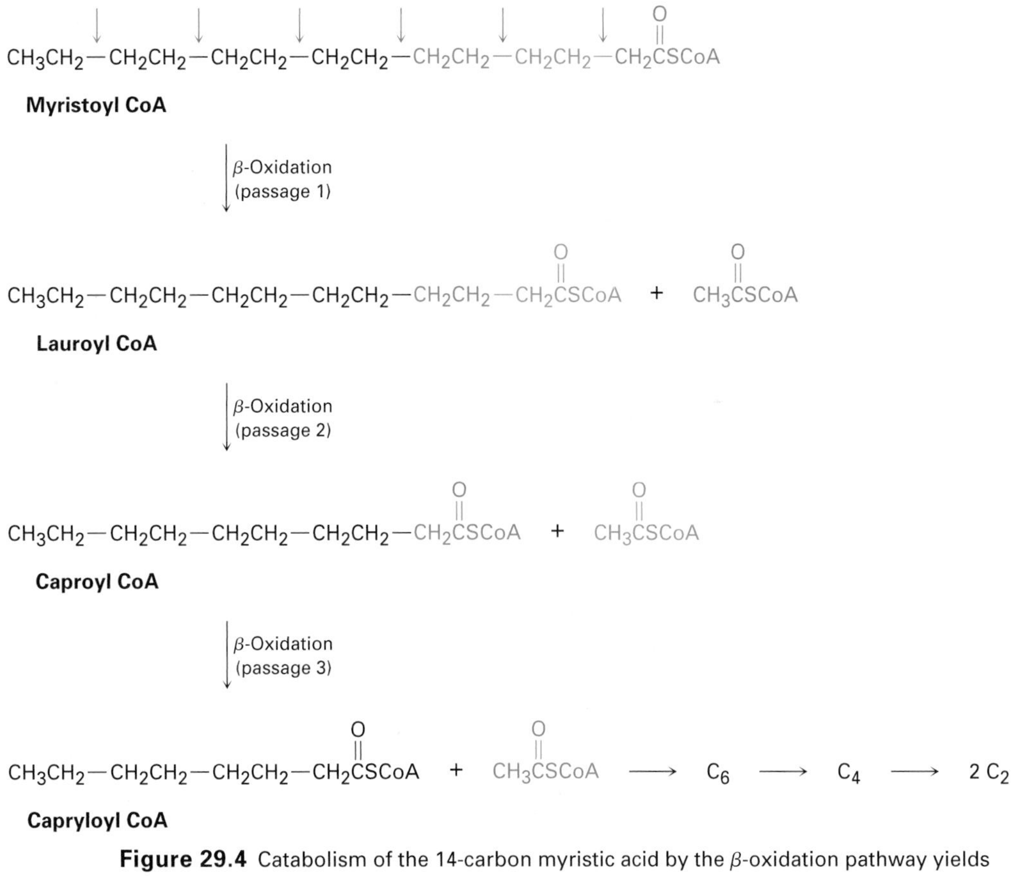

Figure 29.4 Catabolism of the 14-carbon myristic acid by the β-oxidation pathway yields seven molecules of acetyl CoA after six passages.

Most fatty acids have an even number of carbon atoms, so none are left over after β-oxidation. Those fatty acids with an odd number of carbon atoms yield the three-carbon propionyl CoA in the final β-oxidation. Propionyl CoA is then converted to succinate by a multistep radical pathway, and succinate enters the citric acid cycle (Section 29.7). Note that the three-carbon propionyl group should properly be called *propanoyl,* but biochemists generally use the non-systematic name.

Problem 29.2 | Write the equations for the remaining passages of the β-oxidation pathway following those shown in Figure 29.4.

Problem 29.3 | How many molecules of acetyl CoA are produced by catabolism of the following fatty acids, and how many passages of the β-oxidation pathway are needed?
(a) Palmitic acid, $CH_3(CH_2)_{14}CO_2H$
(b) Arachidic acid, $CH_3(CH_2)_{18}CO_2H$

29.4 | Biosynthesis of Fatty Acids

One of the most striking features of the common fatty acids is that they have an even number of carbon atoms (Table 27.1, p. 1062). This even number results because all fatty acids are derived biosynthetically from acetyl CoA by sequential addition of two-carbon units to a growing chain. The acetyl CoA, in turn, arises primarily from the metabolic breakdown of carbohydrates in the glycolysis pathway that we'll see in Section 29.5. Thus, dietary carbohydrates consumed in excess of immediate energy needs are turned into fats for storage.

As a rule, the anabolic pathway by which a substance is made is not the reverse of the catabolic pathway by which the same substance is degraded. The two paths *must* differ in some respects for both to be energetically favorable. Thus, the β-oxidation pathway for converting fatty acids *into* acetyl CoA and the biosynthesis of fatty acids *from* acetyl CoA are related but are not exact opposites. Differences include the identity of the acyl-group carrier, the stereochemistry of the β-hydroxyacyl reaction intermediate, and the identity of the redox coenzyme. FAD is used to introduce a double bond in β-oxidation, while NADPH is used to reduce the double bond in fatty-acid biosynthesis.

In bacteria, each step in fatty-acid synthesis is catalyzed by separate enzymes. In vertebrates, however, fatty-acid synthesis is catalyzed by a large, multienzyme complex called a *synthase* that contains two identical subunits of 2505 amino acids each and catalyzes all steps in the pathway. An overview of fatty-acid biosynthesis is shown in Figure 29.5.

Steps 1–2 of Figure 29.5: Acyl Transfers   The starting material for fatty-acid synthesis is the thioester acetyl CoA, the ultimate product of carbohydrate breakdown, as we'll see in Section 29.6. The synthetic pathway begins with several *priming reactions,* which transport acetyl CoA and convert it into more reactive species. The first priming reaction is a nucleophilic acyl substitution reaction that converts acetyl CoA into acetyl ACP (acyl carrier protein). The reaction is catalyzed by ACP transacylase.

Notice that the mechanism of the nucleophilic acyl substitution step can be given in an abbreviated form that saves space by not explicitly showing the tetrahedral reaction intermediate. Instead, electron movement is shown as a heart-shaped path around the carbonyl oxygen to imply the full mechanism.

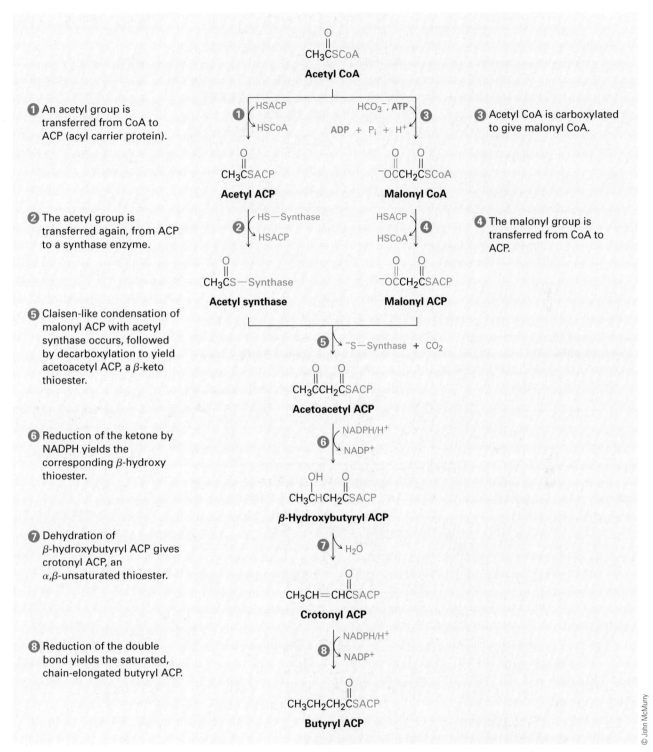

1 An acetyl group is transferred from CoA to ACP (acyl carrier protein).

3 Acetyl CoA is carboxylated to give malonyl CoA.

2 The acetyl group is transferred again, from ACP to a synthase enzyme.

4 The malonyl group is transferred from CoA to ACP.

5 Claisen-like condensation of malonyl ACP with acetyl synthase occurs, followed by decarboxylation to yield acetoacetyl ACP, a β-keto thioester.

6 Reduction of the ketone by NADPH yields the corresponding β-hydroxy thioester.

7 Dehydration of β-hydroxybutyryl ACP gives crotonyl ACP, an α,β-unsaturated thioester.

8 Reduction of the double bond yields the saturated, chain-elongated butyryl ACP.

Figure 29.5 MECHANISM: The pathway for fatty-acid biosynthesis from the two-carbon precursor, acetyl CoA. Individual steps are explained in the text.

Biochemists use this kind of format commonly, and we'll also use it on occasion in the remainder of this chapter.

Tetrahedral intermediate

In bacteria, ACP is a small protein of 77 residues that transports an acyl group from enzyme to enzyme. In vertebrates, however, ACP appears to be a long arm on a multienzyme synthase complex, whose apparent function is to shepherd an acyl group from site to site within the complex. As in acetyl CoA, the acyl group in acetyl ACP is linked by a thioester bond to the sulfur atom of phosphopantetheine. The phosphopantetheine is in turn linked to ACP through the side-chain $-OH$ group of a serine residue in the enzyme.

Phosphopantetheine

Acetyl ACP

Step 2, another priming reaction, involves a further exchange of thioester linkages by another nucleophilic acyl substitution and results in covalent bonding of the acetyl group to a cysteine residue in the synthase complex that will catalyze the upcoming condensation step.

Steps 3–4 of Figure 29.5: Carboxylation and Acyl Transfer   The third step is a *loading* reaction in which acetyl CoA is carboxylated by reaction with HCO_3^- and ATP to yield malonyl CoA plus ADP. This step requires the coenzyme biotin, which is bonded to the lysine residue of acetyl CoA carboxylase and acts as a carrier of CO_2. Biotin first reacts with bicarbonate ion to give *N*-carboxybiotin, which then reacts with the enolate ion of acetyl CoA and transfers the CO_2 group. Thus, biotin acts as a carrier of CO_2, binding it in one step and releasing it in another.

The mechanism of the CO_2 transfer reaction with acetyl CoA to give malonyl CoA is thought to involve CO_2 as the reactive species. One proposal is that loss of CO_2 is favored by hydrogen-bond formation between the *N*-carboxybiotin carbonyl group and a nearby acidic site in the enzyme. Simultaneous deprotonation of acetyl CoA by a basic site in the enzyme gives a thioester enolate ion that can react with CO_2 as it is formed (Figure 29.6).

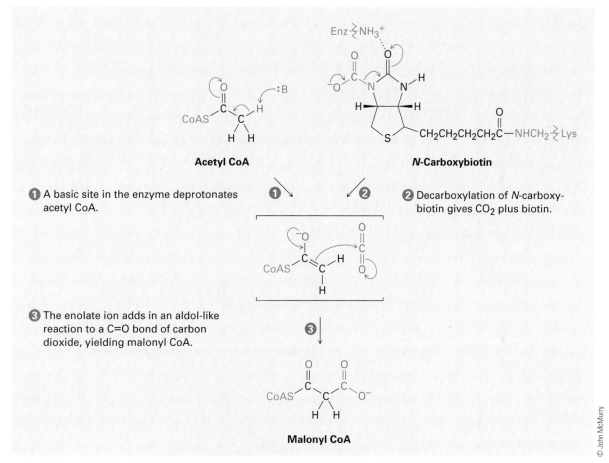

Acetyl CoA

N-Carboxybiotin

1 A basic site in the enzyme deprotonates acetyl CoA.

2 Decarboxylation of *N*-carboxybiotin gives CO_2 plus biotin.

3 The enolate ion adds in an aldol-like reaction to a C=O bond of carbon dioxide, yielding malonyl CoA.

Malonyl CoA

© John McMurry

Figure 29.6 MECHANISM: Mechanism of step 3 in Figure 29.5, the biotin-dependent carboxylation of acetyl CoA to yield malonyl CoA.

Following the formation of malonyl CoA, another nucleophilic acyl substitution reaction occurs in step 4 to form the more reactive malonyl ACP, thereby binding the malonyl group to an ACP arm of the multienzyme synthase. At this point, both acetyl and malonyl groups are bound to the enzyme, and the stage is set for their condensation.

Step 5 of Figure 29.5: Condensation   The key carbon–carbon bond-forming reaction that builds the fatty-acid chain occurs in step 5. This step is simply a Claisen condensation between acetyl synthase as the electrophilic acceptor and malonyl ACP as the nucleophilic donor. The mechanism of the condensation is thought to involve decarboxylation of malonyl ACP to give an enolate ion, followed by immediate addition of the enolate ion to the carbonyl group of acetyl synthase. Breakdown of the tetrahedral intermediate gives the four-carbon condensation product acetoacetyl ACP and frees the synthase binding site for attachment of the chain-elongated acyl group at the end of the sequence.

Malonyl ACP   Acetoacetyl ACP

Steps 6–8 of Figure 29.5: Reduction and Dehydration   The ketone carbonyl group in acetoacetyl ACP is next reduced to the alcohol β-hydroxybutyryl ACP by β-keto thioester reductase and NADPH, a reducing coenzyme closely related to NADH. R Stereochemistry results at the newly formed chirality center in the β-hydroxy thioester product. (Note that the systematic name of a butyryl group is *butanoyl*.)

Acetoacetyl ACP   NADPH β-Hydroxybutyryl ACP   NADP+

Subsequent dehydration of β-hydroxybutyryl ACP by an E1cB reaction in step 7 yields *trans*-crotonyl ACP, and the carbon–carbon double bond of crotonyl ACP is reduced by NADPH in step 8 to yield butyryl ACP. The double-bond reduction occurs by conjugate addition of a hydride ion from NADPH to the β carbon of *trans*-crotonyl ACP. In vertebrates, the reduction occurs by an overall syn addition, but other organisms carry out similar chemistry with different stereochemistry.

Crotonyl ACP   Butyryl ACP

The net effect of the eight steps in the fatty-acid biosynthesis pathway is to take two 2-carbon acetyl groups and combine them into a 4-carbon butyryl group. Further condensation of the butyryl group with another malonyl ACP yields a 6-carbon unit, and still further repetitions of the pathway add two more carbon atoms to the chain each time until the 16-carbon palmitoyl ACP is reached.

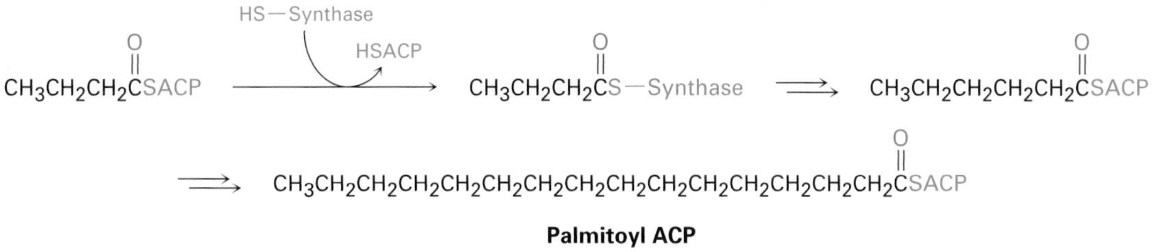

Palmitoyl ACP

Further chain elongation of palmitic acid occurs by reactions similar to those just described, but CoA rather than ACP is the carrier group, and separate enzymes are needed for each step rather than a multienzyme synthase complex.

Problem 29.4 | Write a mechanism for the dehydration reaction of β-hydroxybutyryl ACP to yield crotonyl ACP in step 7 of fatty-acid synthesis.

Problem 29.5 | Evidence for the role of acetate in fatty-acid biosynthesis comes from isotope-labeling experiments. If acetate labeled with ^{13}C in the methyl group ($^{13}CH_3CO_2H$) were incorporated into fatty acids, at what positions in the fatty-acid chain would you expect the ^{13}C label to appear?

Problem 29.6 | Does the reduction of acetoacetyl ACP in step 6 occur on the *Re* face or the *Si* face of the molecule?

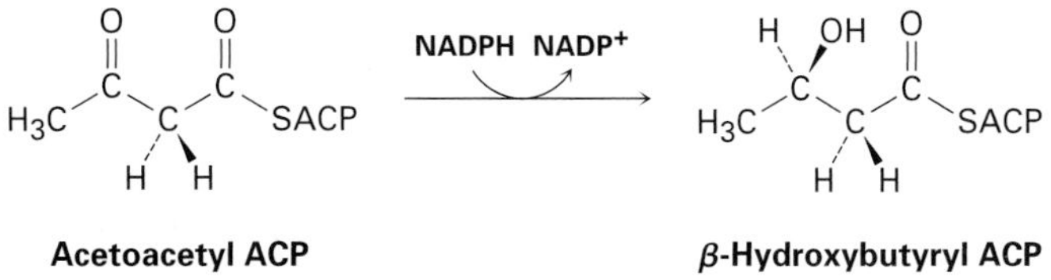

Acetoacetyl ACP **β-Hydroxybutyryl ACP**

29.5 | Catabolism of Carbohydrates: Glycolysis

Glucose is the body's primary short-term energy source. Its catabolism begins with **glycolysis**, a series of ten enzyme-catalyzed reactions that break down glucose into 2 equivalents of pyruvate, $CH_3COCO_2^-$. The steps of glycolysis, also called the *Embden–Meyerhoff pathway* after its discoverers, are summarized in Figure 29.7.

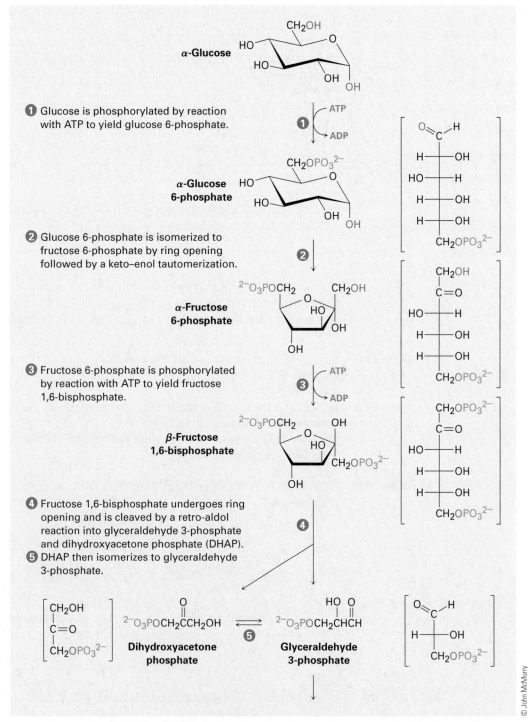

α-Glucose

1 Glucose is phosphorylated by reaction with ATP to yield glucose 6-phosphate.

α-Glucose 6-phosphate

2 Glucose 6-phosphate is isomerized to fructose 6-phosphate by ring opening followed by a keto–enol tautomerization.

α-Fructose 6-phosphate

3 Fructose 6-phosphate is phosphorylated by reaction with ATP to yield fructose 1,6-bisphosphate.

β-Fructose 1,6-bisphosphate

4 Fructose 1,6-bisphosphate undergoes ring opening and is cleaved by a retro-aldol reaction into glyceraldehyde 3-phosphate and dihydroxyacetone phosphate (DHAP).
5 DHAP then isomerizes to glyceraldehyde 3-phosphate.

Dihydroxyacetone phosphate

Glyceraldehyde 3-phosphate

© John McMurry

Active Figure 29.7 MECHANISM: The 10-step glycolysis pathway for catabolizing glucose to two molecules of pyruvate. Individual steps are described in the text. *Sign in at* **www.cengage.com/login** *to see a simulation based on this figure and to take a short quiz.*

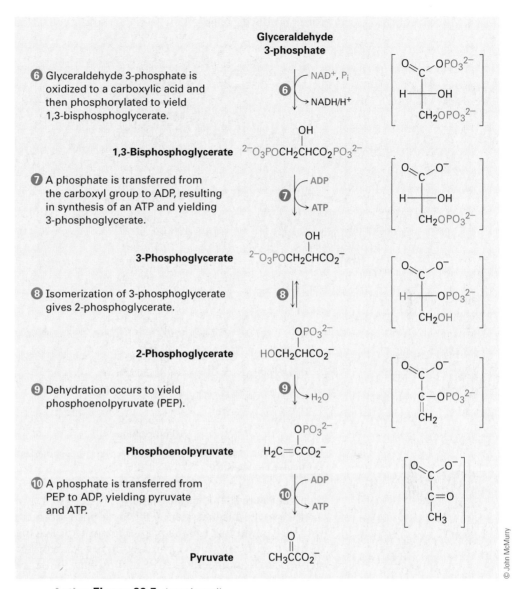

⑥ Glyceraldehyde 3-phosphate is oxidized to a carboxylic acid and then phosphorylated to yield 1,3-bisphosphoglycerate.

Glyceraldehyde 3-phosphate

1,3-Bisphosphoglycerate $^{2-}O_3POCH_2CHCO_2PO_3{}^{2-}$

⑦ A phosphate is transferred from the carboxyl group to ADP, resulting in synthesis of an ATP and yielding 3-phosphoglycerate.

3-Phosphoglycerate $^{2-}O_3POCH_2CHCO_2{}^-$

⑧ Isomerization of 3-phosphoglycerate gives 2-phosphoglycerate.

2-Phosphoglycerate $HOCH_2CHCO_2{}^-$

⑨ Dehydration occurs to yield phosphoenolpyruvate (PEP).

Phosphoenolpyruvate $H_2C{=}CCO_2{}^-$

⑩ A phosphate is transferred from PEP to ADP, yielding pyruvate and ATP.

Pyruvate $CH_3CCO_2{}^-$

Active Figure 29.7 *(continued)*

Steps 1–2 of Figure 29.7: Phosphorylation and Isomerization Glucose, produced by the digestion of dietary carbohydrates, is first phosphorylated at the C6 hydroxyl group by reaction with ATP in a process catalyzed by hexokinase. As noted in Section 29.1, the reaction requires Mg^{2+} as a cofactor to complex with the negatively charged phosphate oxygens. The glucose 6-phosphate that results is isomerized in step 2 by glucose 6-phosphate isomerase to give fructose 6-phosphate. The isomerization takes place by initial opening of the glucose hemiacetal ring to the open-chain form, followed by keto–enol tautomerization to a cis enediol, $HO{-}C{=}C{-}OH$. But because glucose and fructose share a common enediol, further tautomerization to a different keto form produces open-chain fructose, and cyclization completes the process (Figure 29.8).

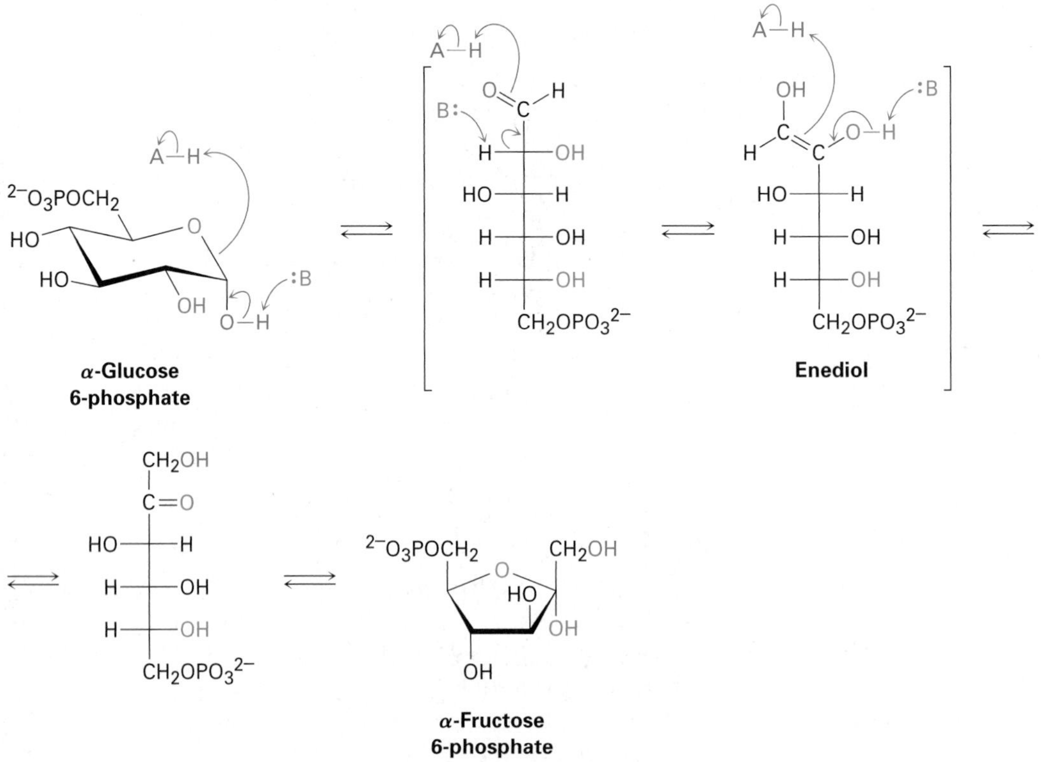

Figure 29.8 Mechanism of step 2 in glycolysis, the isomerization of glucose 6-phosphate to fructose 6-phosphate.

Step 3 of Figure 29.7: Phosphorylation Fructose 6-phosphate is converted in step 3 to fructose 1,6-bisphosphate (FBP) by a phosphofructokinase-catalyzed reaction with ATP (recall that the prefix *bis-* means two). The mechanism is similar to that in step 1, with Mg^{2+} ion again required as cofactor. Interestingly, the product of step 2 is the α anomer of fructose 6-phosphate, but it is the β anomer that is phosphorylated in step 3, implying that the two anomers equilibrate rapidly through the open-chain form. The result of step 3 is a molecule ready to be split into the two three-carbon intermediates that will ultimately become two molecules of pyruvate.

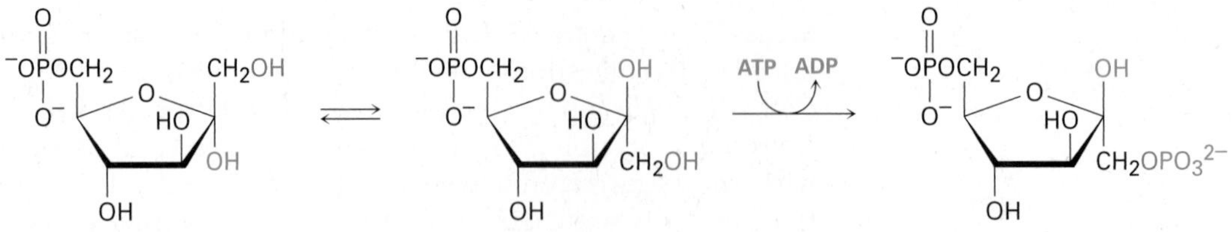

α-Fructose 6-phosphate | β-Fructose 6-phosphate | β-Fructose 1,6-bisphosphate (FBP)

Step 4 of Figure 29.7: Cleavage Fructose 1,6-bisphosphate is cleaved in step 4 into two 3-carbon pieces, dihydroxyacetone phosphate (DHAP) and glyceraldehyde 3-phosphate (GAP). The bond between C3 and C4 of fructose 1,6-bisphosphate breaks, and a C=O group is formed at C4. Mechanistically, the cleavage is the reverse of an aldol reaction (Section 23.1) and is catalyzed by an aldolase. A forward aldol reaction joins two aldehydes or ketones to give a β-hydroxy carbonyl compound, while a retro aldol reaction such as that occurring here cleaves a β-hydroxy carbonyl compound into two aldehydes or ketones.

Fructose 1,6-bisphosphate → **Glyceraldehyde 3-phosphate (GAP)** + **Dihydroxyacetone phosphate (DHAP)**

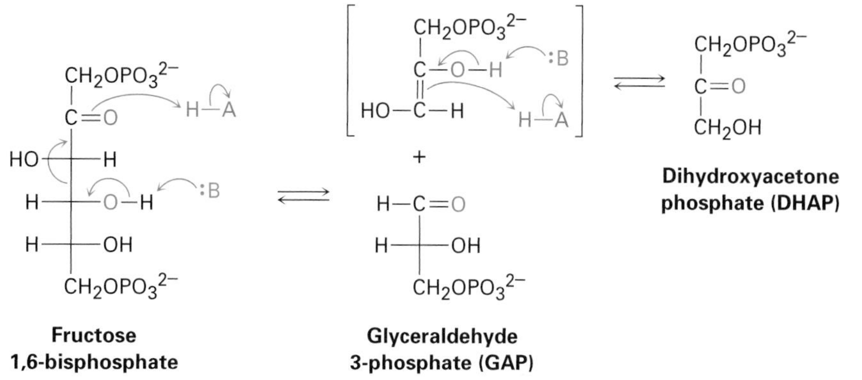

Two classes of aldolases are used by organisms for catalysis of the retro-aldol reaction. In fungi, algae, and some bacteria, the retro-aldol reaction is catalyzed by class II aldolases, which function by coordination of the fructose carbonyl group with Zn^{2+} as Lewis acid. In plants and animals, however, the reaction is catalyzed by class I aldolases and does not take place on the free ketone. Instead, fructose 1,6-bisphosphate undergoes reaction with the side-chain $-NH_2$ group of a lysine residue on the aldolase to yield a protonated enzyme-bound imine (Section 19.8), often called a **Schiff base** in biochemistry. Because of its positive charge, the iminium ion is a better electron acceptor than a ketone carbonyl group. Retro-aldol reaction ensues, giving glyceraldehyde 3-phosphate and an enamine, which is protonated to give another iminium ion that is hydrolyzed to yield dihydroxyacetone phosphate (Figure 29.9 on page 1148).

Step 5 of Figure 29.7: Isomerization Dihydroxyacetone phosphate is isomerized in step 5 by triose phosphate isomerase to form a second equivalent of glyceraldehyde 3-phosphate. As in the conversion of glucose 6-phosphate to fructose 6-phosphate in step 2, the isomerization takes place by keto–enol tautomerization through a common enediol intermediate. A base deprotonates of C1 and then reprotonates C2 using the same hydrogen. The net result of steps 4 and 5 is the production of two glyceraldehyde 3-phosphate molecules, both of which pass down the rest of the pathway. Thus, each of the remaining five steps of glycolysis takes place twice for every glucose molecule that enters at step 1.

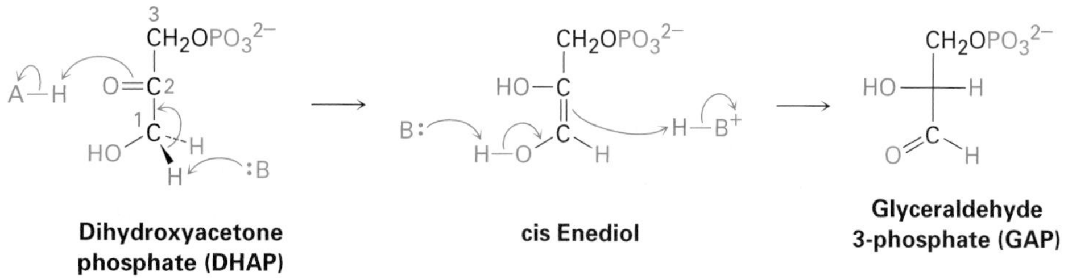

Dihydroxyacetone phosphate (DHAP) → **cis Enediol** → **Glyceraldehyde 3-phosphate (GAP)**

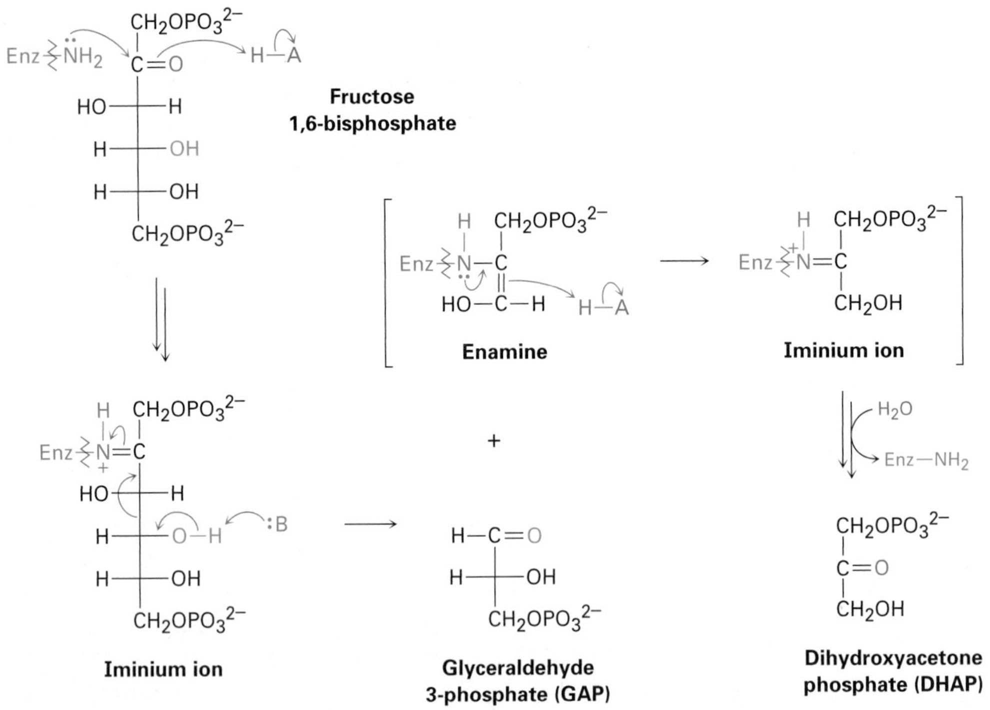

Figure 29.9 Mechanism of step 4 in Figure 29.7, the cleavage of fructose 1,6-bisphosphate to yield glyceraldehyde 3-phosphate and dihydroxyacetone phosphate.

Steps 6–7 of Figure 29.7: Oxidation, Phosphorylation, and Dephosphorylation

Glyceraldehyde 3-phosphate is oxidized and phosphorylated in step 6 to give 1,3-bisphosphoglycerate (Figure 29.10). The reaction is catalyzed by glyceraldehyde 3-phosphate dehydrogenase and begins by nucleophilic addition of the —SH group of a cysteine residue in the enzyme to the aldehyde carbonyl group to yield a *hemithioacetal,* the sulfur analog of a hemiacetal. Oxidation of the hemithioacetal —OH group by NAD⁺ then yields a thioester, which reacts with phosphate ion in a nucleophilic acyl substitution step to yield 1,3-bisphospho-glycerate, a mixed anhydride between a carboxylic acid and phosphoric acid.

Like all anhydrides (Section 21.5), the mixed carboxylic–phosphoric anhy-dride is a reactive substrate in nucleophilic acyl (or phosphoryl) substitution reactions. Reaction of 1,3-bisphosphoglycerate with ADP occurs in step 7 by substitution on phosphorus, resulting in transfer of a phosphate group to ADP and giving ATP plus 3-phosphoglycerate. The process is catalyzed by phospho-glycerate kinase and requires Mg^{2+} as cofactor. Together, steps 6 and 7 accom-plish the oxidation of an aldehyde to a carboxylic acid.

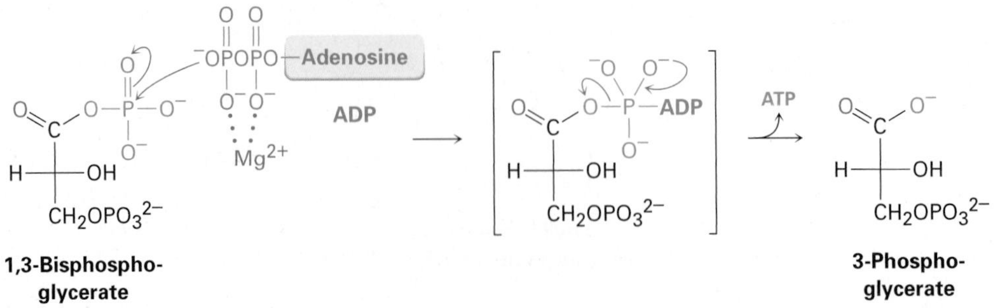

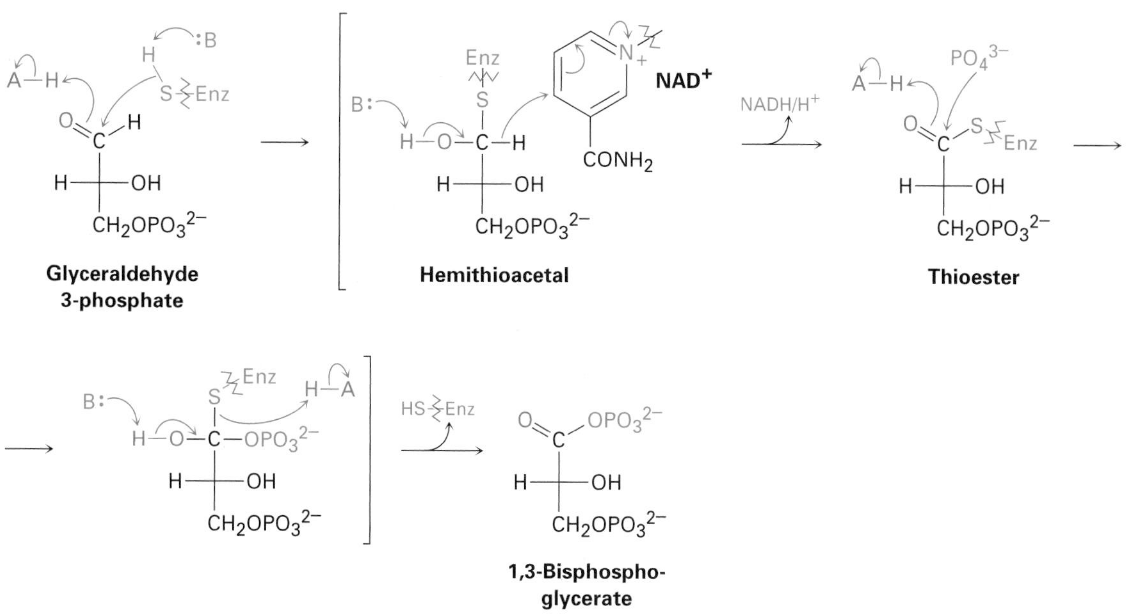

Figure 29.10 Mechanism of step 6 in Figure 29.7, the oxidation and phosphorylation of glyceraldehyde 3-phosphate to give 1,3-bisphosphoglycerate.

Step 8 of Figure 29.7: Isomerization 3-Phosphoglycerate isomerizes to 2-phosphoglycerate in a step catalyzed by phosphoglycerate mutase. In plants, 3-phosphoglycerate transfers its phosphoryl group from its C3 oxygen to a histidine residue on the enzyme in one step and then accepts the same phosphoryl group back onto the C2 oxygen in a second step. In animals and yeast, however, the enzyme contains a phosphorylated histidine, which transfers its phosphoryl group to the C2 oxygen of 3-phosphoglycerate and forms 2,3-bisphosphoglycerate as intermediate. The same histidine then accepts a phosphoryl group from the C3 oxygen to yield the isomerized product plus regenerated enzyme.

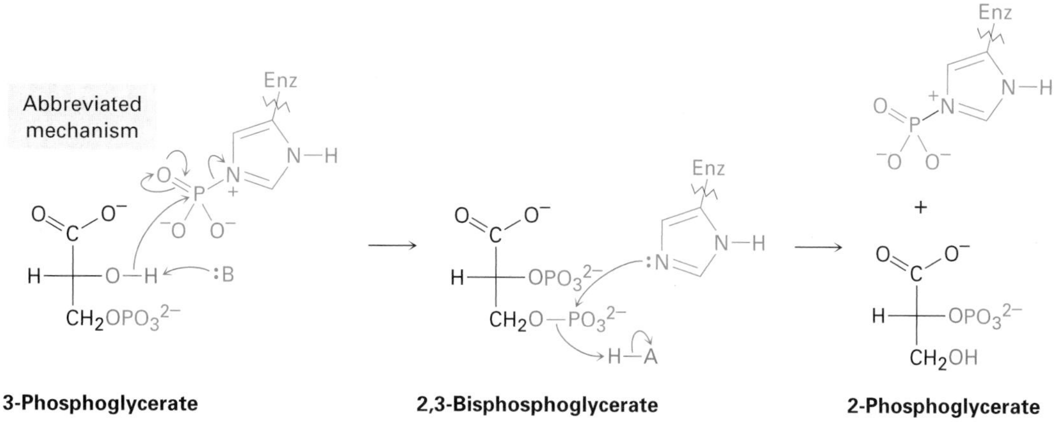

Steps 9–10 of Figure 29.7: Dehydration and Dephosphorylation Like most β-hydroxy carbonyl compounds produced in aldol reactions, 2-phosphoglycerate undergoes a ready dehydration in step 9 by an E1cB mechanism (Section 23.3). The process is catalyzed by enolase, and the product is

phosphoenolpyruvate, abbreviated PEP. Two Mg^{2+} ions are associated with the 2-phosphoglycerate to neutralize the negative charges.

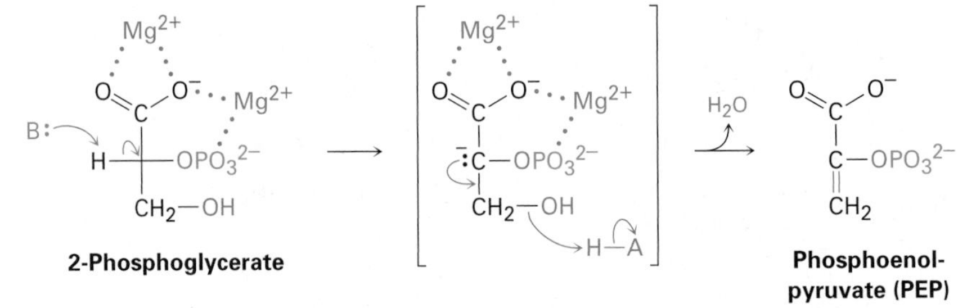

2-Phosphoglycerate → **Phosphoenol-pyruvate (PEP)**

Transfer of the phosphoryl group to ADP in step 10 then generates ATP and gives enolpyruvate, which undergoes tautomerization to pyruvate. The reaction is catalyzed by pyruvate kinase and requires that a molecule of fructose 1,6-bis-phosphate also be present, as well as 2 equivalents of Mg^{2+}. One Mg^{2+} ion coordinates to ADP, and the other increases the acidity of a water molecule necessary for protonation of the enolate ion.

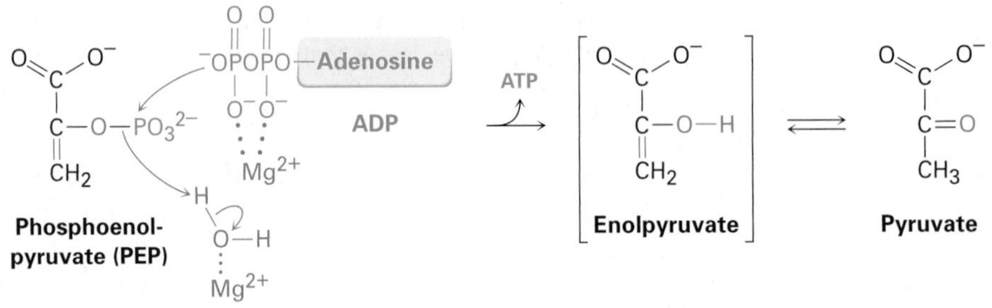

Phosphoenol-pyruvate (PEP) **Enolpyruvate** **Pyruvate**

The overall result of glycolysis can be summarized by the following equation:

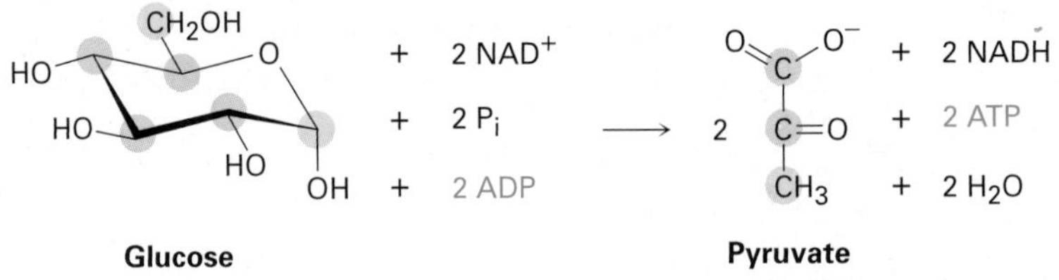

Glucose **Pyruvate**

Problem 29.7 Identify the two steps in glycolysis in which ATP is produced.

Problem 29.8 Look at the entire glycolysis pathway and make a list of the kinds of organic reactions that take place—nucleophilic acyl substitutions, aldol reactions, E1cB reactions, and so forth.

29.6 | Conversion of Pyruvate to Acetyl CoA

Pyruvate, produced by catabolism of glucose (and by degradation of several amino acids), can undergo several further transformations depending on the conditions and on the organism. In the absence of oxygen, pyruvate can be either reduced by NADH to yield lactate [$CH_3CH(OH)CO_2^-$] or, in yeast, fermented to give ethanol. Under typical aerobic conditions in mammals, how-ever, pyruvate is converted by a process called *oxidative decarboxylation* to give acetyl CoA plus CO_2. (*Oxidative* because the oxidation state of the carbonyl carbon rises from that of a ketone to that of a thioester.)

The conversion occurs through a multistep sequence of reactions catalyzed by a complex of enzymes and cofactors called the *pyruvate dehydrogenase complex*. The process occurs in three stages, each catalyzed by one of the enzymes in the complex, as outlined in Figure 29.11 on page 1152. Acetyl CoA, the ultimate product, then acts as fuel for the final stage of catabolism, the citric acid cycle. All the steps have laboratory analogies.

Step 1 of Figure 29.11: Addition of Thiamin Diphosphate   The conversion of pyruvate to acetyl CoA begins by reaction of pyruvate with thiamin diphosphate, a deriva-tive of vitamin B_1. Formerly called thiamin *pyro*phosphate, thiamin diphosphate is usually abbreviated as TPP. The spelling *thiamine* is also correct and frequently used.

The key structural feature in thiamin diphosphate is the presence of a thia-zolium ring—a five-membered, unsaturated heterocycle containing a sulfur atom and a positively charged nitrogen atom. The thiazolium ring is weakly acidic, with a pK_a of approximately 18 for the ring hydrogen between N and S. Bases can therefore deprotonate thiamin diphosphate, leading to formation of a nucleophilic ylide much like the phosphonium ylides used in Wittig reactions (Section 19.11). As in the Wittig reaction, the TPP ylide is a nucleophile and adds to the ketone carbonyl group of pyruvate to yield an alcohol addition product.

Thiamin diphosphate (TPP)   **Thiamin diphosphate ylide (adjacent + and − charges)**

Pyruvate   **Thiamin diphosphate ylide**

Step 2 of Figure 29.11: Decarboxylation   The TPP addition product, which con-tains an iminium ion β to a carboxylate anion, undergoes decarboxylation in much the same way that a β-keto acid decarboxylates in the acetoacetic ester synthesis (Section 22.7). The C=N$^+$ bond of the pyruvate addition product acts

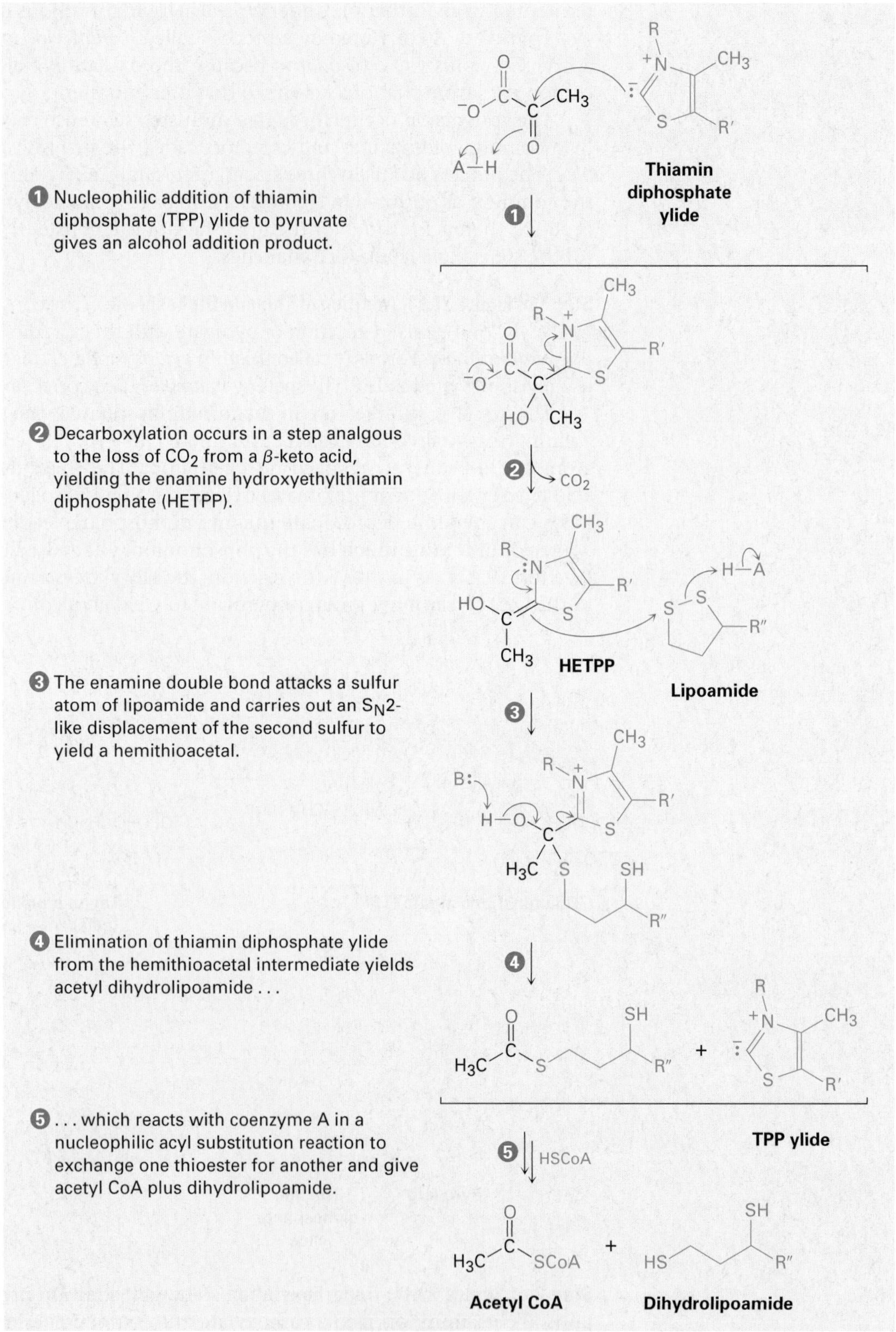
Thiamin diphosphate ylide
1 Nucleophilic addition of thiamin diphosphate (TPP) ylide to pyruvate gives an alcohol addition product.
2 Decarboxylation occurs in a step analgous to the loss of CO_2 from a β-keto acid, yielding the enamine hydroxyethylthiamin diphosphate (HETPP).
CO_2
HETPP
Lipoamide
3 The enamine double bond attacks a sulfur atom of lipoamide and carries out an S_N2-like displacement of the second sulfur to yield a hemithioacetal.
4 Elimination of thiamin diphosphate ylide from the hemithioacetal intermediate yields acetyl dihydrolipoamide . . .
TPP ylide
5 . . . which reacts with coenzyme A in a nucleophilic acyl substitution reaction to exchange one thioester for another and give acetyl CoA plus dihydrolipoamide.
HSCoA
Acetyl CoA
Dihydrolipoamide

Figure 29.11 MECHANISM: Mechanism of the conversion of pyruvate to acetyl CoA through a multistep sequence of reactions that requires three different enzymes and four different coenzymes. The individual steps are explained in the text.

like the C=O bond of a β-keto acid to accept electrons as CO_2 leaves, giving hydroxyethylthiamin diphosphate (HETPP).

Thiamin addition product   **Hydroxyethylthiamin diphosphate (HETTP)**   + CO_2

Step 3 of Figure 29.11: Reaction with Lipoamide Hydroxyethylthiamin diphosphate is an enamine ($R_2N—C=C$), which, like all enamines, is nucleophilic (Section 23.11). It therefore reacts with the enzyme-bound disulfide lipoamide by nucleophilic attack on a sulfur atom, displacing the second sulfur in an S_N2-like process.

Lipoic acid   **Lysine**

Lipoamide: Lipoic acid is linked through an amide bond to a lysine residue in the enzyme

HETPP   **Lipoamide**

Step 4 of Figure 29.11: Elimination of Thiamin Diphosphate The product of the HETPP reaction with lipoamide is a hemithioacetal, which eliminates thiamin diphosphate ylide. This elimination is the reverse of the ketone addition in step 1 and generates acetyl dihydrolipoamide.

Acetyl dihydrolipoamide   **TPP ylide**

Step 5 of Figure 29.11: Acyl Transfer Acetyl dihydrolipoamide, a thioester, undergoes a nucleophilic acyl substitution reaction with coenzyme A to yield acetyl CoA plus dihydrolipoamide. The dihydrolipoamide is then oxidized back to lipoamide by FAD (Section 29.3), and the $FADH_2$ that results is in turn oxidized back to FAD by NAD^+, completing the catalytic cycle.

Acetyl CoA

+

Dihydrolipoamide

Lipoamide

Problem 29.9 | Which carbon atoms in glucose end up as $-CH_3$ carbons in acetyl CoA? Which carbons end up as CO_2?

29.7 | The Citric Acid Cycle

Sir Hans Adolf Krebs

Sir Hans Adolf Krebs (1900–1981) was born in Hildesheim, Germany, and received an M.D. in 1925 from the University of Hamburg. In 1933, his appointment in Germany was terminated by the government, so he moved to England, first at the University of Cambridge, then at the University of Sheffield (1935–1954), and finally at the University of Oxford (1954–1967). He received the 1953 Nobel Prize in medicine for his work on elucidating pathways in intermediary metabolism.

The initial stages of catabolism result in the conversion of both fats and carbohydrates into acetyl groups that are bonded through a thioester link to coenzyme A. Acetyl CoA then enters the next stage of catabolism—the **citric acid cycle**, also called the *tricarboxylic acid (TCA) cycle,* or *Krebs cycle,* after Hans Krebs, who unraveled its complexities in 1937. The overall result of the cycle is the conversion of an acetyl group into two molecules of CO_2 plus reduced coenzymes by the eight-step sequence of reactions shown in Figure 29.12.

As its name implies, the citric acid *cycle* is a closed loop of reactions in which the product of the final step (oxaloacetate) is a reactant in the first step. The intermediates are constantly regenerated and flow continuously through the cycle, which operates as long as the oxidizing coenzymes NAD^+ and FAD are available. To meet this condition, the reduced coenzymes NADH and $FADH_2$ must be reoxidized via the electron-transport chain, which in turn relies on oxygen as the ultimate electron acceptor. Thus, the cycle is dependent on the availability of oxygen and on the operation of the electron-transport chain.

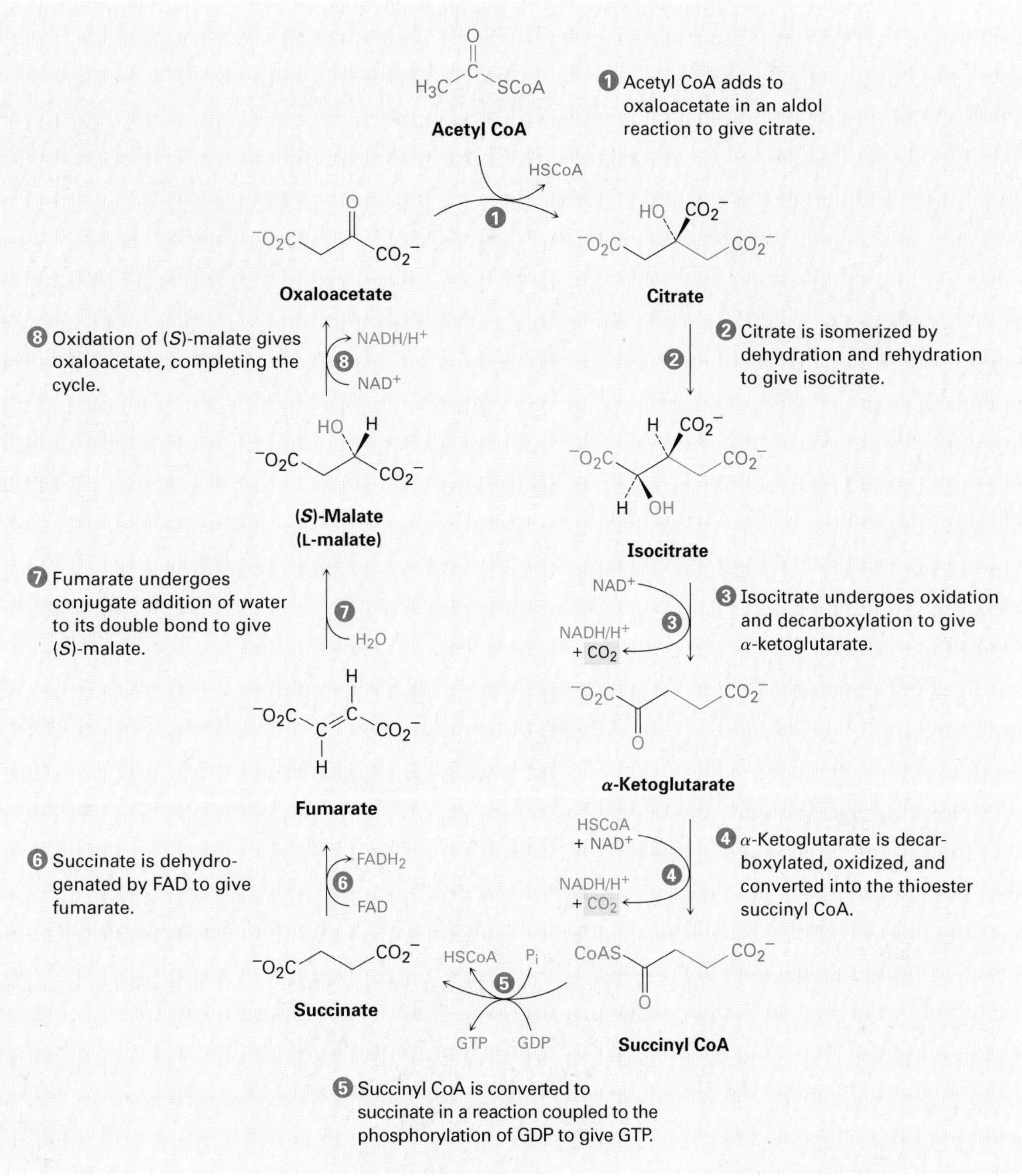

Figure 29.12 MECHANISM: The citric acid cycle is an eight-step series of reactions that results in the conversion of an acetyl group into two molecules of CO_2 plus reduced coenzymes. Individual steps are explained in the text.

The following labels appear around the cycle diagram:

Acetyl CoA

❶ Acetyl CoA adds to oxaloacetate in an aldol reaction to give citrate.

HSCoA

Oxaloacetate

Citrate

❽ Oxidation of (S)-malate gives oxaloacetate, completing the cycle.

NADH/H⁺
NAD⁺

❷ Citrate is isomerized by dehydration and rehydration to give isocitrate.

(S)-Malate (L-malate)

Isocitrate

❼ Fumarate undergoes conjugate addition of water to its double bond to give (S)-malate.

H_2O

NAD⁺
NADH/H⁺ + CO_2

❸ Isocitrate undergoes oxidation and decarboxylation to give α-ketoglutarate.

Fumarate

α-Ketoglutarate

❻ Succinate is dehydrogenated by FAD to give fumarate.

FADH₂
FAD

HSCoA + NAD⁺
NADH/H⁺ + CO_2

❹ α-Ketoglutarate is decarboxylated, oxidized, and converted into the thioester succinyl CoA.

Succinate

HSCoA Pᵢ
GTP GDP

Succinyl CoA

❺ Succinyl CoA is converted to succinate in a reaction coupled to the phosphorylation of GDP to give GTP.

Step 1 of Figure 29.12: Addition to Oxaloacetate   Acetyl CoA enters the citric acid cycle in step 1 by nucleophilic addition to the oxaloacetate carbonyl group, to give (S)-citryl CoA. The addition is an aldol reaction and is catalyzed by citrate synthase, as discussed in Section 26.11. (S)-Citryl CoA is then hydrolyzed to citrate by a typical nucleophilic acyl substitution reaction, catalyzed by the same citrate synthase enzyme.

   Note that the hydroxyl-bearing carbon of citrate is a prochirality center and contains two identical "arms." Because the initial aldol reaction of acetyl CoA to oxaloacetate occurs specifically from the *Si* face of the ketone carbonyl group, the *pro-S* arm of citrate is derived from acetyl CoA and the *pro-R* arm is derived from oxaloacetate.

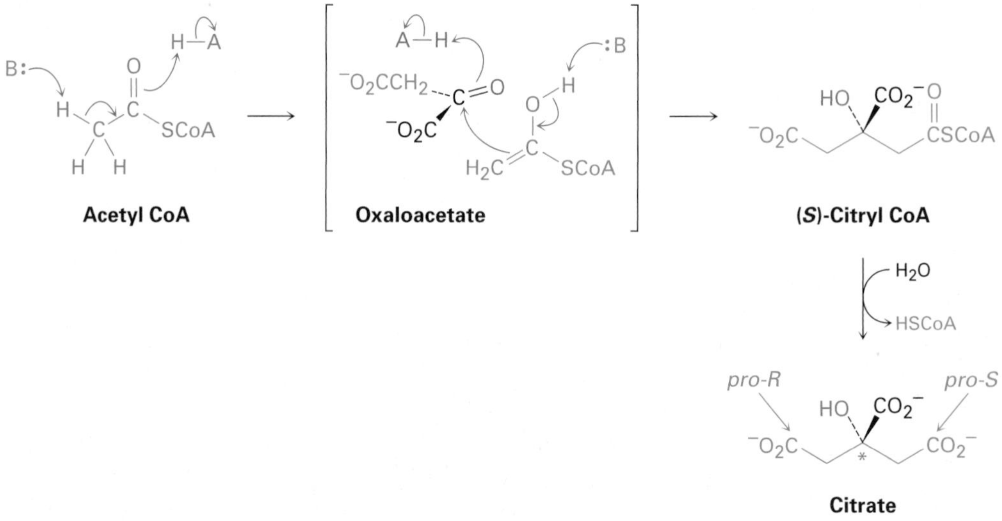

Step 2 of Figure 29.12: Isomerization   Citrate, a prochiral tertiary alcohol, is next converted into its isomer, (2R,3S)-isocitrate, a chiral secondary alcohol. The isomerization occurs in two steps, both of which are catalyzed by the same aconitase enzyme. The initial step is an E1cB dehydration of a β-hydroxy acid to give *cis*-aconitate, the same sort of reaction that occurs in step 9 of glycolysis (Figure 29.7). The second step is a conjugate nucleophilic addition of water to the C=C bond (Section 19.13). The dehydration of citrate takes place specifically on the *pro-R* arm—the one derived from oxaloacetate—rather than on the *pro-S* arm derived from acetyl CoA.

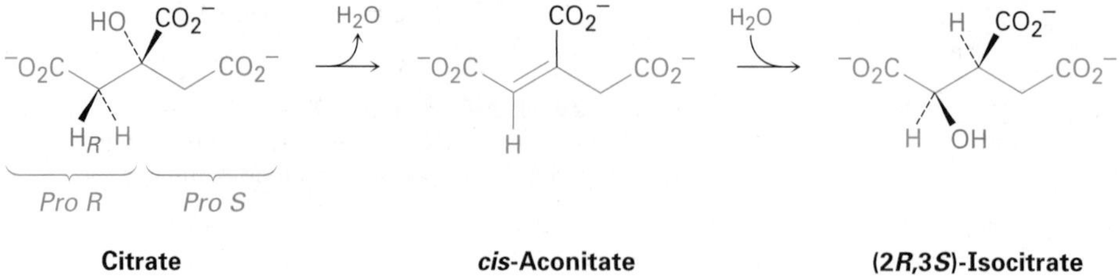

Step 3 of Figure 29.12: Oxidation and Decarboxylation (2*R*,3*S*)-Isocitrate, a secondary alcohol, is oxidized by NAD^+ in step 3 to give the ketone oxalosuccinate, which loses CO_2 to give α-ketoglutarate. Catalyzed by isocitrate dehydrogenase, the decarboxylation is a typical reaction of a β-keto acid, just like that in the acetoacetic ester synthesis (Section 22.7). The enzyme requires a divalent cation as cofactor, presumably to polarize the ketone carbonyl group.

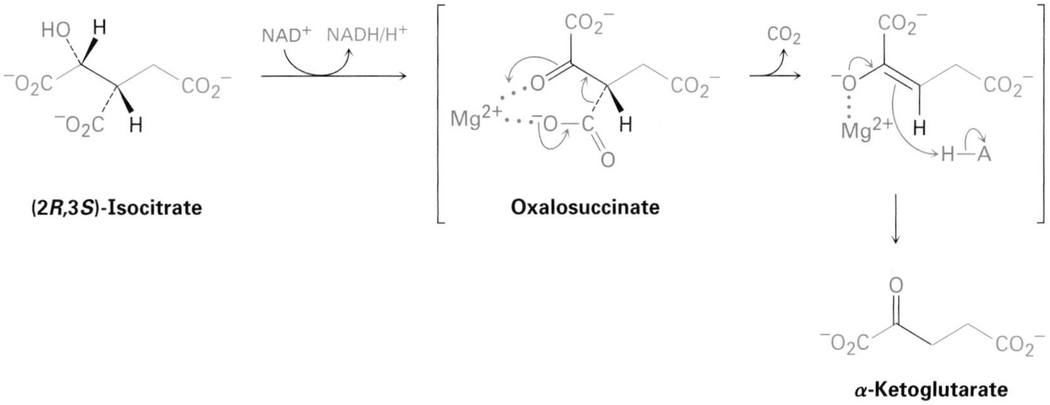

(2*R*,3*S*)-Isocitrate   **Oxalosuccinate**

α-Ketoglutarate

Step 4 of Figure 29.12: Oxidative Decarboxylation The transformation of α-ketoglutarate to succinyl CoA in step 4 is a multistep process just like the transformation of pyruvate to acetyl CoA that we saw in Figure 29.11. In both cases, an α-keto acid loses CO_2 and is oxidized to a thioester in a series of steps catalyzed by a multienzyme dehydrogenase complex. As in the conversion of pyruvate to acetyl CoA, the reaction involves an initial nucleophilic addition reaction to α-ketoglutarate by thiamin diphosphate ylide, followed by decarboxylation, reaction with lipoamide, elimination of TPP ylide, and finally a transesterification of the dihydrolipoamide thioester with coenzyme A.

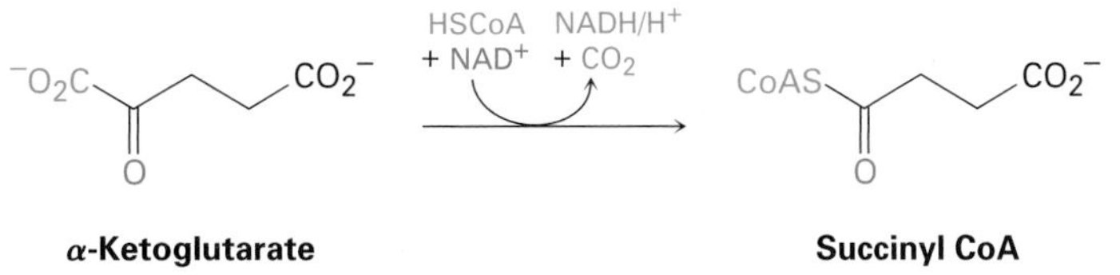

α-Ketoglutarate   **Succinyl CoA**

Step 5 of Figure 29.12: Acyl CoA Cleavage Succinyl CoA is converted to succinate in step 5. The reaction is catalyzed by succinyl CoA synthetase and is coupled with phosphorylation of guanosine diphosphate (GDP) to give guanosine triphosphate (GTP). The overall transformation is similar to that of steps 6 through 8 in glycolysis (Figure 29.7), in which a thioester is converted into an acyl phosphate and a phosphate group is then transferred to ADP.

The overall result is a "hydrolysis" of the thioester group without involvement of water.

Succinyl CoA   **Acyl phosphate**

Step 6 of Figure 29.12: Dehydrogenation   Succinate is dehydrogenated in step 6 by the FAD-dependent succinate dehydrogenase to give fumarate. The process is analogous to what occurs during the β-oxidation pathway of fatty-acid catabolism (Section 29.3). The reaction is stereospecific, removing the *pro-S* hydrogen from one carbon and the *pro-R* hydrogen from the other.

Succinate   **Fumarate**

Steps 7–8 of Figure 29.12: Hydration and Oxidation   The final two steps in the citric acid cycle are the conjugate nucleophilic addition of water to fumarate to yield (*S*)-malate (L-malate) and the oxidation of (*S*)-malate by NAD$^+$ to give oxaloacetate. The addition is catalyzed by fumarase and is mechanistically similar to the addition of water to *cis*-aconitate in step 2. The reaction occurs through an enolate-ion intermediate, which is protonated on the side opposite the OH, leading to a net anti addition.

Fumarate   **(*S*)-Malate**

The final step is the oxidation of (S)-malate by NAD^+ to give oxaloacetate, a reaction catalyzed by malate dehydrogenase. The citric acid cycle has now returned to its starting point, ready to revolve again. The overall result of the cycle is

Acetyl CoA $+$ $3 NAD^+$ $+$ FAD $+$ GDP $+$ P_i $+$ $2 H_2O$

$\longrightarrow$ $2 CO_2$ $+$ HSCoA $+$ 3 NADH $+$ $2 H^+$ $+$ $FADH_2$ $+$ GTP

Problem 29.10   Which of the substances in the citric acid cycle are tricarboxylic acids, thus giving the cycle its alternative name?

Problem 29.11   Write mechanisms for step 2 of the citric acid cycle, the dehydration of citrate and the addition of water to aconitate.

Problem 29.12   Is the *pro-R* or *pro-S* hydrogen removed from citrate during the dehydration in step 2 of the citric acid cycle? Does the elimination reaction occur with syn or anti geometry?

Citrate   ***cis*-Aconitate**

29.8 | Carbohydrate Biosynthesis: Gluconeogenesis

Glucose is the body's primary fuel when food is plentiful, but in times of fasting or prolonged exercise, glucose stores can become depleted. Most tissues then begin metabolizing fats as their source of acetyl CoA, but the brain is different. The brain relies almost entirely on glucose for fuel and is dependent on receiving a continuous supply in the blood. When the supply of glucose fails for even a brief time, irreversible damage can occur. Thus, a pathway for synthesizing glucose from simple precursors is needed.

Higher organisms are not able to synthesize glucose from acetyl CoA but must instead use one of the three-carbon precursors lactate, glycerol, or alanine, all of which are readily converted into pyruvate.

(S)-Lactate   **Alanine**   **Glycerol**

Pyruvate   *Gluconeogenesis* →   **Glucose**

Figure 29.13
MECHANISM: The gluconeogenesis pathway for the biosynthesis of glucose from pyruvate. Individual steps are explained in the text.

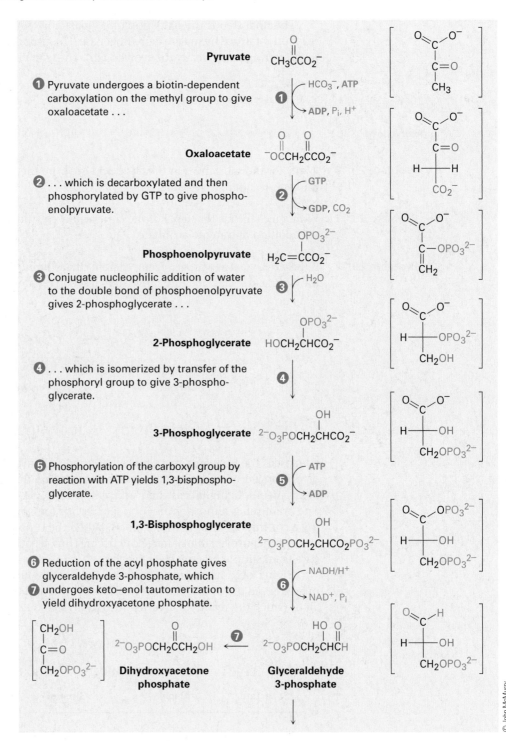

Pyruvate

1 Pyruvate undergoes a biotin-dependent carboxylation on the methyl group to give oxaloacetate . . .

Oxaloacetate

2 . . . which is decarboxylated and then phosphorylated by GTP to give phosphoenolpyruvate.

Phosphoenolpyruvate

3 Conjugate nucleophilic addition of water to the double bond of phosphoenolpyruvate gives 2-phosphoglycerate . . .

2-Phosphoglycerate

4 . . . which is isomerized by transfer of the phosphoryl group to give 3-phosphoglycerate.

3-Phosphoglycerate

5 Phosphorylation of the carboxyl group by reaction with ATP yields 1,3-bisphosphoglycerate.

1,3-Bisphosphoglycerate

6 Reduction of the acyl phosphate gives glyceraldehyde 3-phosphate, which

7 undergoes keto–enol tautomerization to yield dihydroxyacetone phosphate.

Dihydroxyacetone phosphate

Glyceraldehyde 3-phosphate

© John McMurry

Pyruvate then becomes the starting point for **gluconeogenesis**, the 11-step biosynthetic pathway by which organisms make glucose (Figure 29.13). The gluconeogenesis pathway by which glucose is made, however, is not the reverse of the glycolysis pathway by which it is degraded. As with the catabolic and anabolic pathways for fatty acids (Sections 29.3 and 29.4), the catabolic and anabolic pathways for carbohydrates differ in some details so that both are energetically favorable.

Figure 29.13 *(continued)*

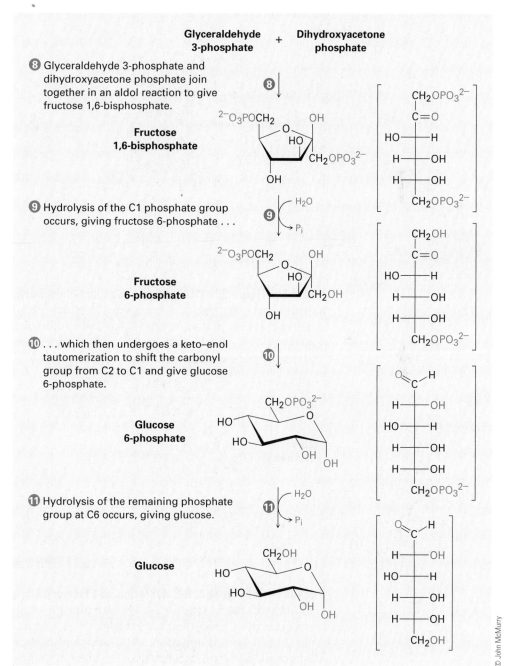

⑧ Glyceraldehyde 3-phosphate and dihydroxyacetone phosphate join together in an aldol reaction to give fructose 1,6-bisphosphate.

⑨ Hydrolysis of the C1 phosphate group occurs, giving fructose 6-phosphate . . .

⑩ . . . which then undergoes a keto–enol tautomerization to shift the carbonyl group from C2 to C1 and give glucose 6-phosphate.

⑪ Hydrolysis of the remaining phosphate group at C6 occurs, giving glucose.

Step 1 of Figure 29.13: Carboxylation Gluconeogenesis begins with the carboxylation of pyruvate to yield oxaloacetate. The reaction is catalyzed by pyruvate carboxylase and requires ATP, bicarbonate ion, and the coenzyme biotin, which acts as a carrier to transport CO_2 to the enzyme active site. The mechanism is analogous to that of step 3 in fatty-acid biosynthesis (Figure 29.6), in which acetyl CoA is carboxylated to yield malonyl CoA.

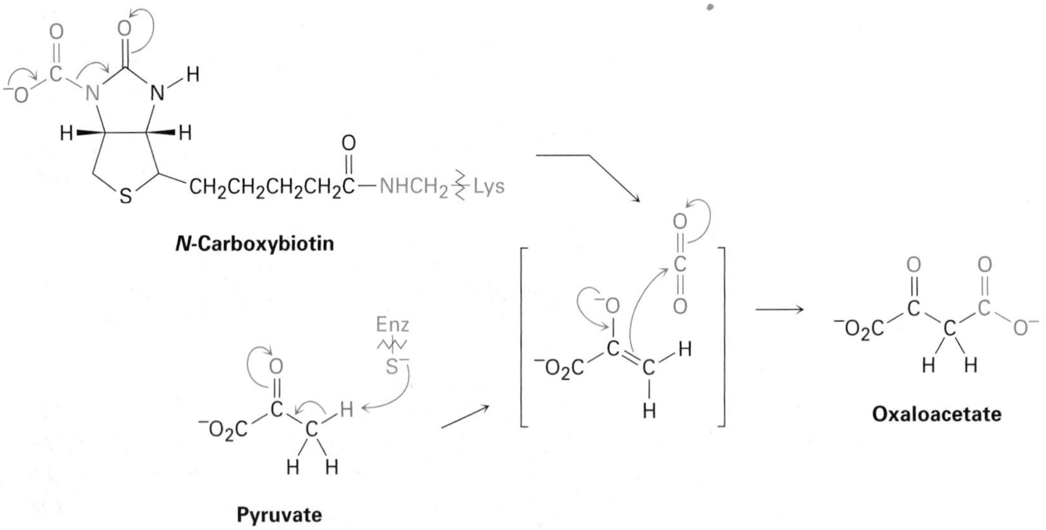

Step 2 of Figure 29.13: Decarboxylation and Phosphorylation Decarboxylation of oxaloacetate, a β-keto acid, occurs by the typical retro-aldol mechanism like that in step 3 in the citric acid cycle (Figure 29.12), and phosphorylation of the resultant pyruvate enolate ion by GTP occurs concurrently to give phosphoenolpyruvate. The reaction is catalyzed by phosphoenolpyruvate carboxykinase.

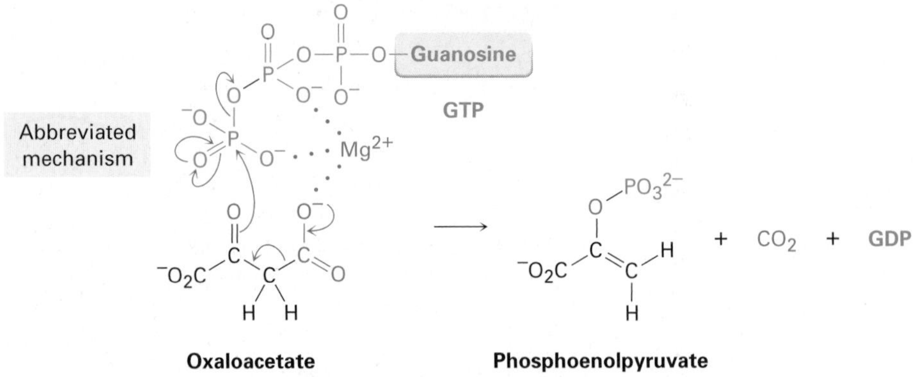

Steps 3–4 of Figure 29.13: Hydration and Isomerization Conjugate nucleophilic addition of water to the double bond of phosphoenolpyruvate gives 2-phosphoglycerate by a process similar to that of step 7 in the citric acid cycle (Figure 29.12). Phosphorylation of C3 and dephosphorylation of C2 then yields 3-phosphoglycerate. Mechanistically, these steps are the reverse of steps 9 and 8 in glycolysis (Figure 29.7), which have equilibrium constants near 1 so that substantial amounts of reactant and product are both present.

Steps 5–7 of Figure 29.13: Phosphorylation, Reduction, and Tautomerization Reaction of 3-phosphoglycerate with ATP generates the corresponding acyl phosphate, 1,3-bisphosphoglycerate, which binds to the glyceraldehyde 3-phosphate dehydrogenase by a thioester bond to a cysteine residue. Reduction by NADH/H$^+$ yields the aldehyde, and keto–enol tautomerization of the aldehyde gives dihydroxyacetone phosphate. All three steps are mechanistically the reverse of the corresponding steps 7, 6, and 5 of glycolysis and have equilibrium constants near 1.

Step 8 of Figure 29.13: Aldol Reaction Dihydroxyacetone phosphate and glyceraldehyde 3-phosphate, the two 3-carbon units produced in step 7, join by an aldol reaction to give fructose 1,6-bisphosphate, the reverse of step 4 in glycolysis. As in glycolysis (Figure 29.9), the reaction is catalyzed in plants and animals by a class I aldolase and takes place on an iminium ion formed by reaction of dihydroxyacetone phosphate with a side-chain lysine $-NH_2$ group on the enzyme. Loss of a proton from the neighboring carbon then generates an enamine, an aldol-like reaction ensues, and the product is hydrolyzed.

Steps 9–10 of Figure 29.13: Hydrolysis and Isomerization   Hydrolysis of the phosphate group at C1 of fructose 1,6-bisphosphate gives fructose 6-phosphate. Although the result of the reaction is the exact opposite of step 3 in glycolysis, the mechanism is not. In glycolysis, the phosphorylation is accomplished by reaction of the fructose with ATP. The reverse of that process, however—the reaction of fructose 1,6-bisphosphate with ADP to give fructose 6-phosphate and ATP—is energetically unfavorable because ATP is too high in energy. Thus, an alternative pathway is used in which the C1 phosphate group is removed by a direct hydrolysis reaction, catalyzed by fructose 1,6-bisphosphatase.

Following hydrolysis, keto–enol tautomerization of the carbonyl group from C2 to C1 gives glucose 6-phosphate. The isomerization is the reverse of step 2 in glycolysis.

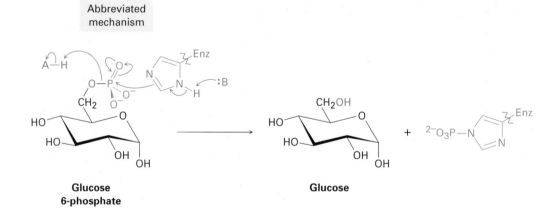

Fructose **1,6-bisphosphate**	**Fructose** **6-phosphate**	**Glucose** **6-phosphate**

Step 11 of Figure 29.13: Hydrolysis   The final step in gluconeogenesis is the conversion of glucose 6-phosphate to glucose by another phosphatase-catalyzed hydrolysis reaction. As just discussed for the hydrolysis of fructose 1,6-bisphosphate in step 9, and for the same energetic reasons, the mechanism of the glucose 6-phosphate hydrolysis is not the exact opposite of the corresponding step 1 in glycolysis.

Interestingly, however, the mechanisms of the two phosphate hydrolysis reactions in steps 9 and 11 are not the same. In step 9, water is the nucleophile, but in the glucose 6-phosphate reaction of step 11, a histidine residue on the enzyme attacks phosphorus, giving a phosphoryl enzyme intermediate that subsequently reacts with water.

Abbreviated
mechanism

Glucose
6-phosphate

Glucose

The overall result of gluconeogenesis is summarized by the following equation:

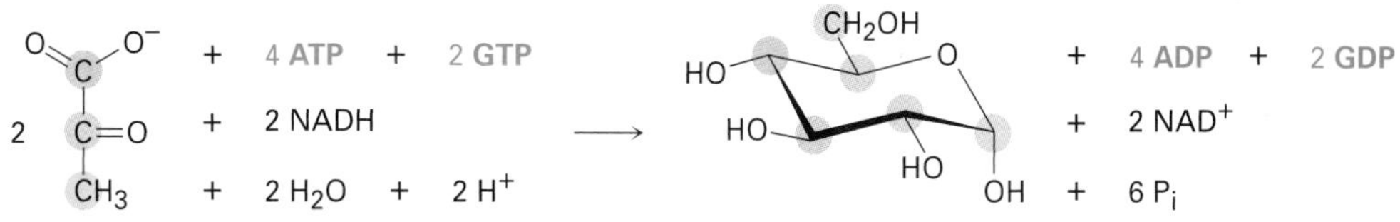

Pyruvate   **Glucose**

Problem 29.13   Write a mechanism for step 6 of gluconeogenesis, the reduction of 3-phosphoglyceryl phosphate with $NADH/H^+$ to yield glyceraldehyde 3-phosphate.

29.9 | Catabolism of Proteins: Transamination

The catabolism of proteins is much more complex than that of fats and carbohydrates because each of the 20 amino acids is degraded through its own unique pathway. The general idea, however, is that the amino nitrogen atom is removed and the substance that remains is converted into a compound that enters the citric acid cycle.

Most amino acids lose their nitrogen atom by a **transamination** reaction in which the $-NH_2$ group of the amino acid changes places with the keto group of α-ketoglutarate. The products are a new α-keto acid plus glutamate. The overall process occurs in two parts, is catalyzed by aminotransferase enzymes, and involves participation of the coenzyme pyridoxal phosphate (PLP), a derivative of pyridoxine (vitamin B_6). Different aminotransferases differ in their specificity for amino acids, but the mechanism remains the same.

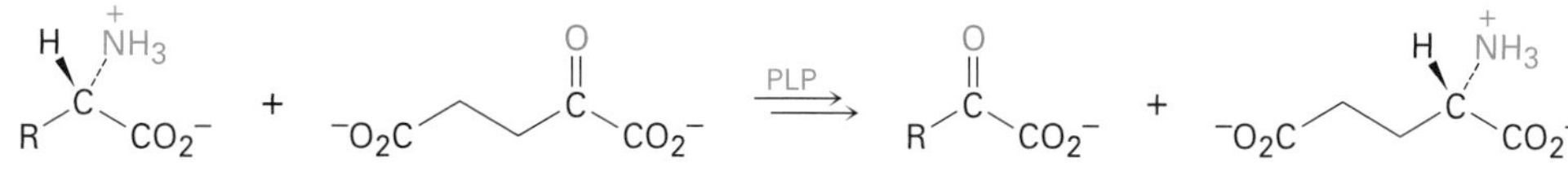

An α-amino acid   **α-Ketoglutarate**   **An α-keto acid**   **Glutamate**

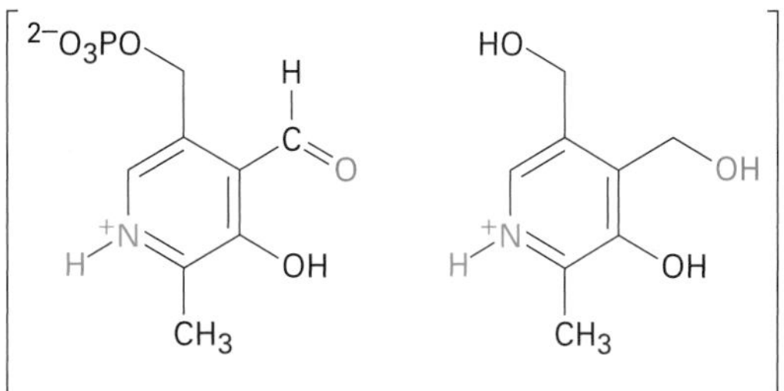

Pyridoxal phosphate (PLP)   **Pyridoxine (vitamin B_6)**

The mechanism of the first part of transamination is shown in Figure 29.14. The process begins with reaction between the α-amino acid and pyridoxal phosphate, which is covalently bonded to the aminotransferase by an imine linkage between the side-chain $-NH_2$ group of a lysine residue and the PLP aldehyde group. Deprotonation/reprotonation of the PLP–amino acid imine in steps 2 and 3 effects tautomerization of the imine C=N bond, and hydrolysis of the tautomerized imine in step 4 gives an α-keto acid plus pyridoxamine phosphate (PMP).

Step 1 of Figure 29.14: Transimination The first step in trans*amination* is trans*imination*—the reaction of the PLP–enzyme imine with an α-amino acid to give a PLP–amino acid imine plus expelled enzyme as the leaving group. The reaction occurs by nucleophilic addition of the amino acid $-NH_2$ group to the C=N bond of the PLP imine, much as an amine adds to the C=O bond of a ketone or aldehyde in a nucleophilic addition reaction (Section 19.8). The protonated diamine intermediate undergoes a proton transfer and expels the lysine amino group in the enzyme to complete the step.

Steps 2–4 of Figure 29.14: Tautomerization and Hydrolysis Following formation of the PLP–amino acid imine in step 1, a tautomerization of the C=N bond occurs in step 2. The basic lysine residue in the enzyme that was expelled as a leaving group during transimination deprotonates the acidic α position of the amino acid, with the protonated pyridine ring of PLP acting as the electron acceptor as shown in step 2 of Figure 29.2. Reprotonation occurs on the carbon atom next to the ring (step 3), generating a tautomeric product that is the imine of an α-keto acid with pyridoxamine phosphate, abbreviated PMP (Figure 29.15).

Hydrolysis of this PMP–α-keto acid imine in step 4 then completes the first part of the transamination reaction. The hydrolysis is the mechanistic reverse of

① An amino acid reacts with the enzyme-bound PLP imine by nucleophilic addition of its –NH$_2$ group to the C=N bond of the imine, giving a PLP–amino acid imine and releasing the enzyme amino group.

PLP–amino acid imine (Schiff base)

② Deprotonation of the acidic α carbon of the amino acid gives an intermediate α-keto acid imine . . .

α-Keto acid imine

③ . . . that is reprotonated on the PLP carbon. The net result of this deprotonation/reprotonation sequence is tautomerization of the imine C=N bond.

α-Keto acid imine tautomer

④ Hydrolysis of the α-keto acid imine by nucleophilic addition of water to the C=N bond gives the transamination products pyridoxamine phosphate (PMP) and α-keto acid.

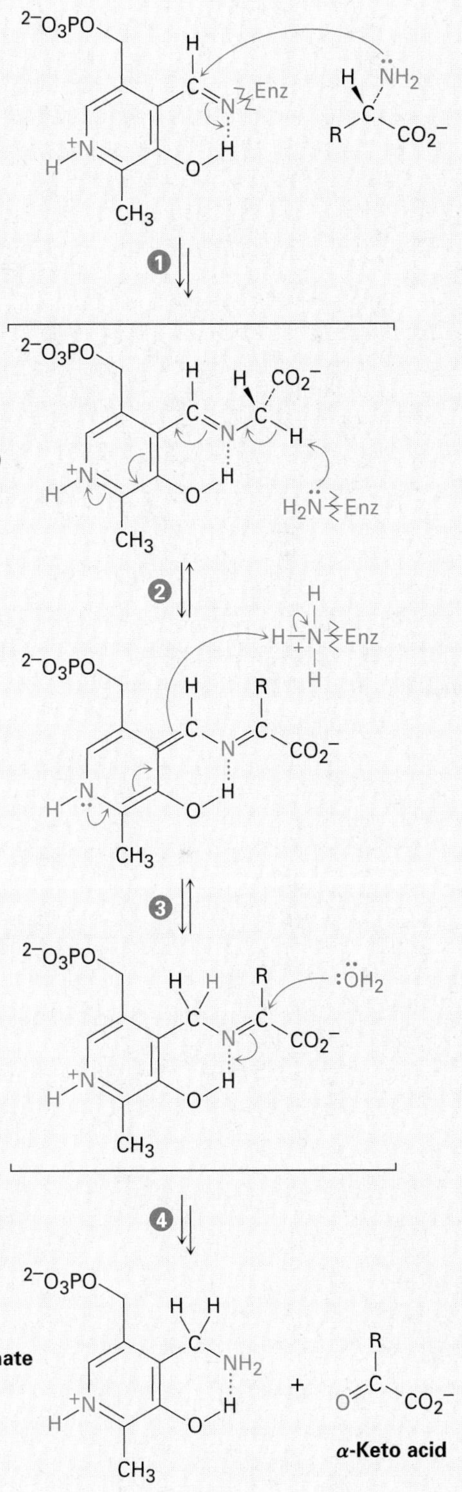

Pyridoxamine phosphate (PMP)

α-Keto acid

© John McMurry

Figure 29.14 MECHANISM: Mechanism of the enzyme-catalyzed, PLP-dependent transamination of an α-amino acid to give an α-keto acid. Individual steps are explained in the text.

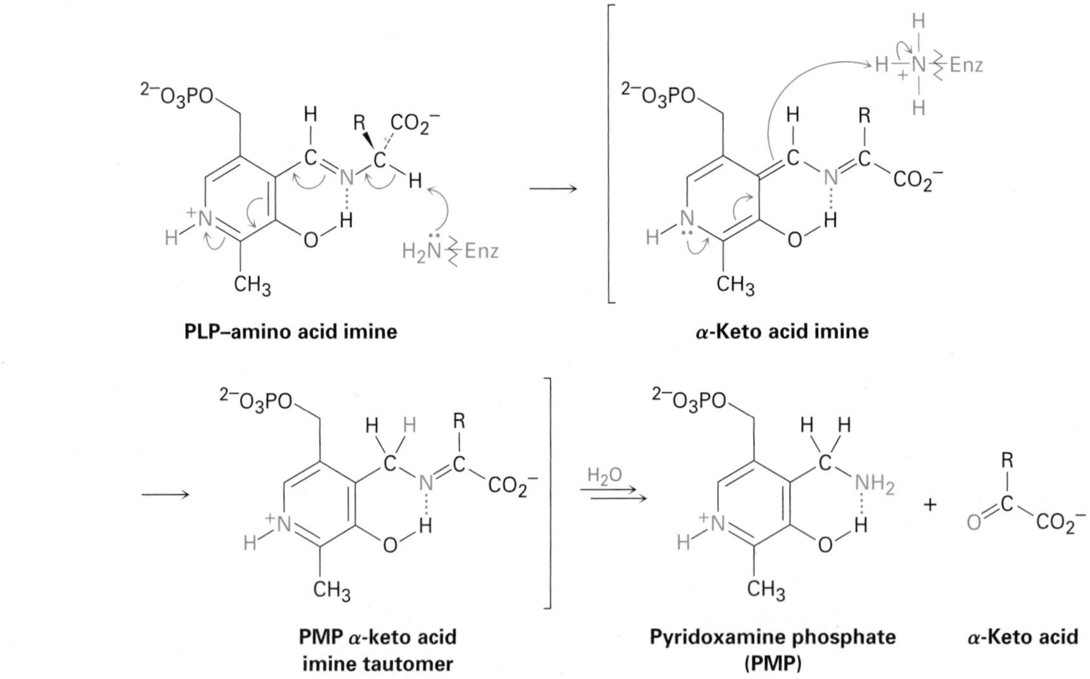

PLP–amino acid imine

α-Keto acid imine

PMP α-keto acid imine tautomer

Pyridoxamine phosphate (PMP)

α-Keto acid

Figure 29.15 Mechanism of steps 2–4 of amino acid transamination, the conversion of a PLP–amino acid imine to PMP and an α-keto acid.

imine formation and occurs by nucleophilic addition of water to the imine, followed by proton transfer and expulsion of PMP as leaving group.

With PLP plus the α-amino acid now converted into PMP plus an α-keto acid, PMP must be transformed back into PLP to complete the catalytic cycle. The conversion occurs by another transamination reaction, this one between PMP and an α-keto acid, usually α-ketoglutarate. PLP plus glutamate are the products, and the mechanism of the process is the reverse of that shown in Figure 29.14. That is, PMP and α-ketoglutarate give an imine; the PMP–ketoglutarate imine undergoes tautomerization of the C=N bond to give a PLP–glutamate imine; and the PLP–glutamate imine reacts with a lysine residue on the enzyme in a transimination process to yield PLP–enzyme imine plus glutamate.

PMP

α-Ketoglutarate

PLP–enzyme imine

Glutamate

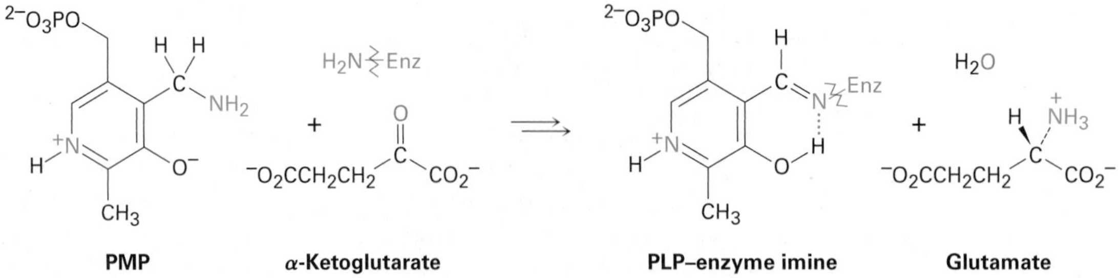

Problem 29.14 Write all the steps in the transamination reaction of PMP with α-ketoglutarate plus a lysine residue in the enzyme to give the PLP–enzyme imine plus glutamate.

Problem 29.15 What α-keto acid is formed on transamination of leucine?

Problem 29.16 From what amino acid is the following α-keto acid derived?

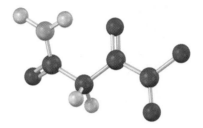

29.10 | Some Conclusions about Biological Chemistry

As promised in the chapter introduction, the past few sections have been a fast-paced tour of a large number of reactions. Following it all undoubtedly required a lot of work and a lot of page turning to look at earlier sections.

After examining the various metabolic pathways, perhaps the main conclusion about biological chemistry is the remarkable similarity between the mechanisms of biological reactions and the mechanisms of laboratory reactions. In all the pathways described in this chapter, terms like *imine formation, aldol reaction, nucleophilic acyl substitution reaction, E1cB reaction,* and *Claisen reaction* appear constantly. Biological reactions are not mysterious—the vitalistic theory described on page 1 died long ago. There are clear, understandable reasons for the reactions carried out within living organisms. Biological chemistry *is* organic chemistry.

Focus On . . .

Basal Metabolism

The minimum amount of energy per unit time an organism must expend to stay alive is called the organism's *basal metabolic rate (BMR).* This rate is measured by monitoring respiration and finding the rate of oxygen consumption, which is proportional to the amount of energy used. Assuming an average dietary mix of fats, carbohydrates, and proteins, approximately 4.82 kcal are required for each liter of oxygen consumed.

The average basal metabolic rate for humans is about 65 kcal/h, or 1600 kcal/day. Obviously, the rate varies for different people depending on sex, age, weight, and physical condition. As a rule, the BMR is lower for older people than for younger people, is lower for females than for males, and is lower for people in good physical condition than for those who are out of shape and overweight. A BMR substantially above the expected value indicates an unusually rapid metabolism, perhaps caused by a fever or some biochemical abnormality.

(continued)

Endurance trail runners can use up to 10,000 kcal to fuel their prodigious energy needs in runs of over 100 miles.

The total number of calories a person needs each day is the sum of the basal requirement plus the energy used for physical activities, as shown in Table 29.1. A relatively inactive person needs about 30% above basal requirements per day, a lightly active person needs about 50% above basal, and a very active person such as an athlete or construction worker may need 100% above basal requirements. Some endurance athletes in ultradistance events can use as many as 10,000 kcal/day above the basal level. Each day that your caloric intake is above what you use, fat is stored in your body and your weight rises. Each day that your caloric intake is below what you use, fat in your body is metabolized and your weight drops.

Table 29.1	Energy Cost of Various Activities[a]	
Activity		**Kcal/min**
Sleeping		1.2
Sitting, reading		1.6
Standing still		1.8
Walking		3–6
Tennis		7–9
Basketball		9–10
Walking up stairs		10–18
Running		9–22

[a]For a 70 kg man.

SUMMARY AND KEY WORDS

anabolism, 1126

β-oxidation pathway, 1133

catabolism, 1126

citric acid cycle, 1154

gluconeogenesis, 1161

glycolysis, 1143

metabolism, 1126

Schiff base, 1147

transamination, 1165

Metabolism is the sum of all chemical reactions in the body. Reactions that break down large molecules into smaller fragments are called **catabolism**; reactions that build up large molecules from small pieces are called **anabolism**. Although the details of specific biochemical pathways are sometimes complex, all the reactions that occur follow the normal rules of organic chemical reactivity.

The catabolism of fats begins with digestion, in which ester bonds are hydrolyzed to give glycerol and fatty acids. The fatty acids are degraded in the four-step **β-oxidation pathway** by removal of two carbons at a time, yielding acetyl CoA. Catabolism of carbohydrates begins with the hydrolysis of glycoside bonds to give glucose, which is degraded in the ten-step **glycolysis** pathway. Pyruvate, the initial product of glycolysis, is then converted into acetyl CoA. Acetyl CoA next enters the eight-step **citric acid cycle**, where it is further degraded into CO_2. The cycle is a closed loop of reactions in which the product of the final step (oxaloacetate) is a reactant in the first step. The intermediates are constantly regenerated and flow continuously through the cycle, which operates as long as the oxidizing coenzymes NAD^+ and FAD are available.

Catabolism of proteins is more complex than that of fats or carbohydrates because each of the 20 different amino acids is degraded by its own unique pathway. In general, though, the amino nitrogen atoms are removed and the substances that remain are converted into compounds that enter the citric acid cycle. Most amino acids lose their nitrogen atom by **transamination**, a reaction in which the $-NH_2$ group of the amino acid changes places with the keto group of an α-keto acid such as α-ketoglutarate. The products are a new α-keto acid and glutamate.

The energy released in catabolic pathways is used in the *electron-transport chain* to make molecules of adenosine triphosphate, ATP. ATP, the final result of food catabolism, couples to and drives many otherwise unfavorable reactions.

Biomolecules are synthesized as well as degraded, but the pathways for anabolism and catabolism are not the exact reverse of one another. Fatty acids are biosynthesized from acetate by an 8-step pathway, and carbohydrates are made from pyruvate by the 11-step **gluconeogenesis** pathway.

EXERCISES

Organic KNOWLEDGE TOOLS

CENGAGENOW™ Sign in at **www.cengage.com/login** to assess your knowledge of this chapter's topics by taking a pre-test. The pre-test will link you to interactive organic chemistry resources based on your score in each concept area.

ⓦWL Online homework for this chapter may be assigned in Organic OWL.

■ indicates problems assignable in Organic OWL.

VISUALIZING CHEMISTRY

(Problems 29.1–29.16 appear within the chapter.)

29.17 ■ Identify the amino acid that is a catabolic precursor of each of the following α-keto acids:

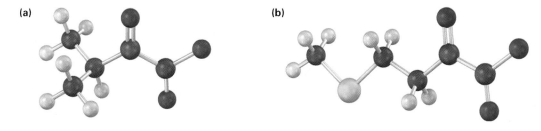

(a)   (b)

29.18 ■ Identify the following intermediate in the citric acid cycle, and tell whether it has *R* or *S* stereochemistry:

29.19 The following compound is an intermediate in the biosynthesis of one of the twenty common α-amino acids. Which one is it likely to be, and what kind of chemical change must take place to complete the biosynthesis?

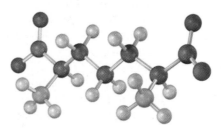

29.20 The following compound is an intermediate in the pentose phosphate pathway, an alternative route for glucose metabolism. Identify the sugar it is derived from.

ADDITIONAL PROBLEMS

29.21 What chemical events occur during the digestion of food?

29.22 What is the difference between digestion and metabolism?

29.23 What is the difference between anabolism and catabolism?

29.24 ■ Draw the structure of adenosine 5′-monophosphate (AMP), an intermediate in some biochemical pathways.

29.25 ■ Cyclic adenosine monophosphate (cyclic AMP), a modulator of hormone action, is related to AMP (Problem 29.24) but has its phosphate group linked to *two* hydroxyl groups at C3′ and C5′ of the sugar. Draw the structure of cyclic AMP.

29.26 ■ What general kind of reaction does ATP carry out?

29.27 ■ What general kind of reaction does NAD^+ carry out?

29.28 ■ What general kind of reaction does FAD carry out?

29.29 Why aren't the glycolysis and gluconeogenesis pathways the exact reverse of one another?

29.30 ■ Lactate, a product of glucose catabolism in oxygen-starved muscles, can be converted into pyruvate by oxidation. What coenzyme do you think is needed? Write the equation in the normal biochemical format using a curved arrow.

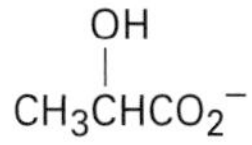

$$\underset{\displaystyle CH_3CHCO_2^-}{\overset{\displaystyle OH}{|}} \quad \textbf{Lactate}$$

29.31 ■ How many moles of acetyl CoA are produced by catabolism of the following substances?
(a) 1.0 mol glucose   (b) 1.0 mol palmitic acid   (c) 1.0 mol maltose

29.32 ■ How many grams of acetyl CoA (MW = 809.6 amu) are produced by catabolism of the following substances? Which substances is the most efficient precursor of acetyl CoA on a weight basis?
(a) 100.0 g glucose   (b) 100.0 g palmitic acid   (c) 100.0 g maltose

29.33 Write the equation for the final step in the β-oxidation pathway of any fatty acid with an even number of carbon atoms.

29.34 ■ Show the products of each of the following reactions:

(a)

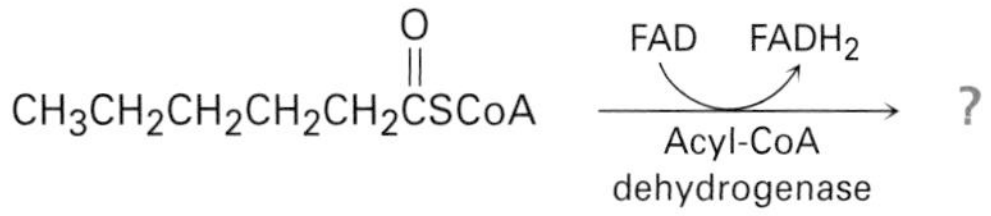

$$CH_3CH_2CH_2CH_2CH_2\overset{\displaystyle O}{\overset{\displaystyle ||}{C}}SCoA \xrightarrow[\substack{Acyl\text{-}CoA \\ dehydrogenase}]{FAD \quad FADH_2} \quad ?$$

(b)

$$\text{Product of (a)} \quad + \quad H_2O \xrightarrow[]{\substack{Enoyl\text{-}CoA \\ hydratase}} \quad ?$$

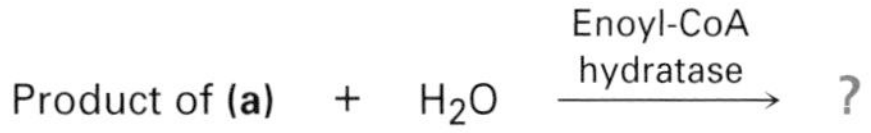

(c)

$$\text{Product of (b)} \xrightarrow[\substack{\beta\text{-Hydroxyacyl-CoA} \\ dehydrogenase}]{NAD^+ \quad NADH/H^+} \quad ?$$

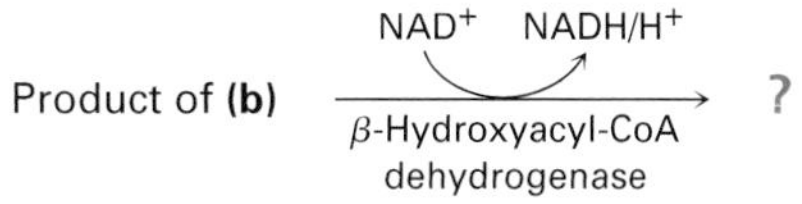

29.35 ■ What is the structure of the α-keto acid formed by transamination of each of the following amino acids?
(a) Threonine   (b) Phenylalanine   (c) Asparagine

29.36 ■ What enzyme cofactor is associated with each of the following kinds of reactions?
(a) Transamination   (b) Carboxylation of a ketone
(c) Decarboxylation of an α-keto acid

29.37 The glycolysis pathway shown in Figure 29.7 has a number of intermediates that contain phosphate groups. Why can 3-phosphoglyceryl phosphate and phosphoenolpyruvate transfer a phosphate group to ADP while glucose 6-phosphate cannot?

29.38 ■ In the *pentose phosphate* pathway for degrading sugars, ribulose 5-phosphate is converted to ribose 5-phosphate. Propose a mechanism for the isomerization.

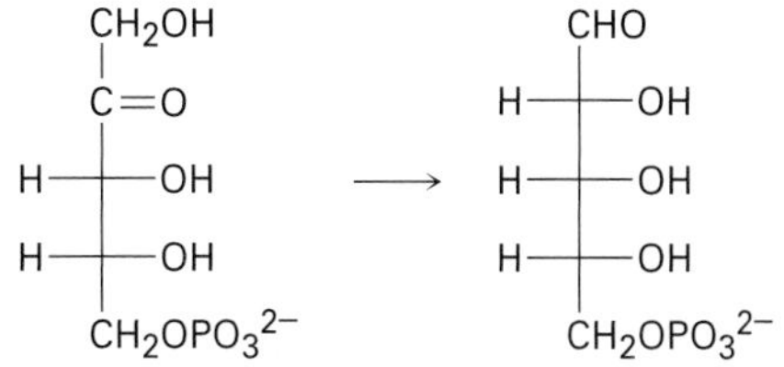

Ribulose 5-phosphate   **Ribose 5-phosphate**

29.39 ■ Another step in the pentose phosphate pathway for degrading sugars (see Problem 29.38) is the conversion of ribose 5-phosphate to glyceraldehyde 3-phosphate. What kind of organic process is occurring? Propose a mechanism for the conversion.

Ribose 5-phosphate → **Glyceraldehyde 3-phosphate** + (CHO, $CH_2OPO_3{}^{2-}$)

29.40 Write a mechanism for the conversion of α-ketoglutarate to succinyl CoA in step 4 of the citric acid cycle (Figure 29.12).

29.41 ■ In step 2 of the citric acid cycle (Figure 29.12), *cis*-aconitate reacts with water to give (2R,3S)-isocitrate. Does −OH add from the *Re* face of the double bond or from the *Si* face? What about −H? Does the addition of water occur with syn or anti geometry?

cis-**Aconitate** → (2R,3S)-**Isocitrate**

29.42 ■ The primary fate of acetyl CoA under normal metabolic conditions is degradation in the citric acid cycle to yield CO_2. When the body is stressed by prolonged starvation, however, acetyl CoA is converted into compounds called *ketone bodies,* which can be used by the brain as a temporary fuel. Fill in the missing information indicated by the four question marks in the following biochemical pathway for the synthesis of ketone bodies from acetyl CoA:

Acetyl CoA → **Acetoacetyl CoA** → **Acetoacetate** → **Acetone**, **3-Hydroxybutyrate**

Ketone bodies

29.43 ■ The initial reaction in Problem 29.42, conversion of two molecules of acetyl CoA to one molecule of acetoacetyl CoA, is a Claisen reaction. Assuming that there is a base present, show the mechanism of the reaction.

29.44 In step 6 of fatty-acid biosynthesis (Figure 29.5), acetoacetyl ACP is reduced stereospecifically by NADPH to yield an alcohol. Does hydride ion add to the *Si* face or the *Re* face of acetoacetyl ACP?

Acetoacetyl ACP   **β-Hydroxybutyryl ACP**

29.45 In step 7 of fatty-acid biosynthesis (Figure 29.5), dehydration of a β-hydroxy thioester occurs to give *trans*-crotonyl ACP. Is the dehydration a syn elimination or an anti elimination?

***trans*-Crotonyl ACP**

29.46 ■ In step 8 of fatty-acid biosynthesis (Figure 29.5), reduction of *trans*-crotonyl ACP gives butyryl ACP. A hydride from NADPH adds to C3 of the crotonyl group from the *Re* face, and protonation on C2 occurs on the *Si* face. Is the reduction a syn addition or an anti addition?

Crotonyl ACP   **Butyryl ACP**

29.47 One of the steps in the pentose phosphate pathway for glucose catabolism is the reaction of sedoheptulose 7-phosphate with glyceraldehyde 3-phosphate in the presence of a transaldolase to yield erythrose 4-phosphate and fructose 6-phosphate.

Sedoheptulose 7-phosphate   **Glyceraldehyde 3-phosphate**   **Erythrose 4-phosphate**   **Fructose 6-phosphate**

(a) The first part of the reaction is formation of a protonated Schiff base of sedoheptulose 7-phosphate with a lysine residue in the enzyme followed by a retro-aldol cleavage to give an enamine plus erythrose 4-phosphate. Show the structure of the enamine and the mechanism by which it is formed.

(b) The second part of the reaction is nucleophilic addition of the enamine to glyceraldehyde 3-phosphate followed by hydrolysis of the Schiff base to give fructose 6-phosphate. Show the mechanism.

29.48 One of the steps in the pentose phosphate pathway for glucose catabolism is the reaction of xylulose 5-phosphate with ribose 5-phosphate in the presence of a transketolase to give glyceraldehyde 3-phosphate and sedoheptulose 7-phosphate.

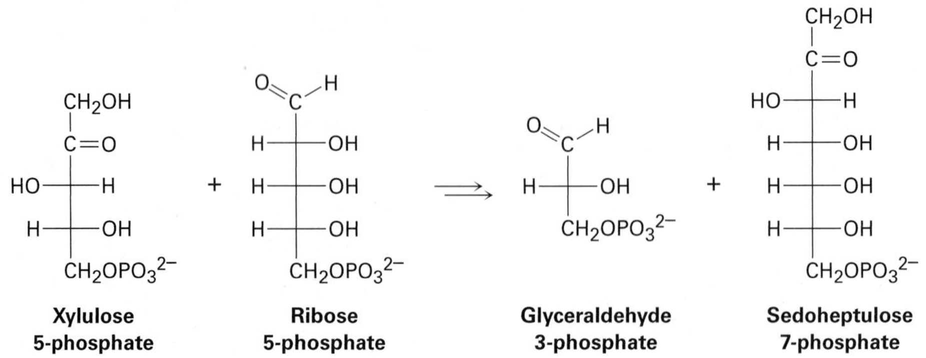

(a) The first part of the reaction is nucleophilic addition of thiamin diphosphate (TPP) ylide to xylulose 5-phosphate, followed by a retro-aldol cleavage to give glyceraldehyde 3-phosphate and a TPP-containing enamine. Show the structure of the enamine and the mechanism by which it is formed.

(b) The second part of the reaction is addition of the enamine to ribose 5-phosphate followed by loss of TPP ylide to give sedoheptulose 7-phosphate. Show the mechanism.

29.49 The amino acid tyrosine is biologically degraded by a series of steps that include the following transformations:

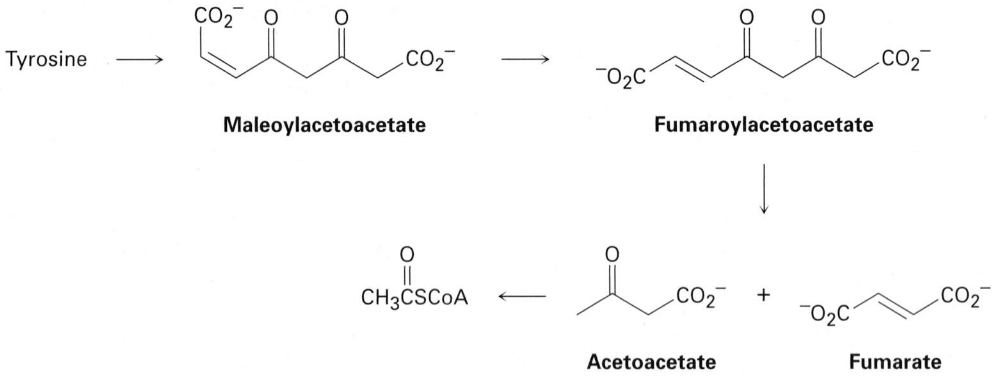

The double-bond isomerization of maleoylacetoacetate to fumaroyl acetoacetate is catalyzed by practically any nucleophile, :Nu⁻. Propose a mechanism.

29.50 Propose a mechanism for the conversion of fumaroylacetoacetate to fumarate plus acetoacetate (Problem 29.49).

29.51 Propose a mechanism for the conversion of acetoacetate to acetyl CoA (Problem 29.49).

29.52 Design your own degradative pathway. You know the rules (organic mechanisms), and you've seen the kinds of reactions that occur in the biological degradation of fats and carbohydrates into acetyl CoA. If you were Mother Nature, what series of steps would you use to degrade the amino acid serine into acetyl CoA?

Serine   **Acetyl CoA**

29.53 The amino acid serine is biosynthesized by a route that involves reaction of 3-phosphohydroxypyruvate with glutamate. Propose a mechanism.

3-Phosphohydroxypyruvate   **3-Phosphoserine**

29.54 The amino acid leucine is biosynthesized from α-ketoisocaproate, which is itself prepared from α-ketoisovalerate by a multistep route that involves (1) reaction with acetyl CoA, (2) hydrolysis, (3) dehydration, (4) hydration, (5) oxidation, and (6) decarboxylation. Show the steps in the transformation, and propose a mechanism for each.

α-**Ketoisovalerate** α-**Ketoisocaproate**

29.55 The amino acid cysteine, $C_3H_7NO_2S$, is biosynthesized from a substance called cystathionine by a multistep pathway.

Cystathionine

(a) The first step is a transamination. What is the product?
(b) The second step is an E1cB reaction. Show the products and the mechanism of the reaction.
(c) The final step is a double-bond reduction. What organic cofactor is required for this reaction, and what is the product represented by the question mark in the equation?

■ Assignable in OWL

30

Orbitals and Organic Chemistry: Pericyclic Reactions

Most organic reactions take place by polar mechanisms, in which a nucleophile donates two electrons to an electrophile in forming a new bond. Other reactions take place by radical mechanisms, in which each of two reactants donates one electron in forming a new bond. Both kinds of reactions occur frequently in the laboratory and in living organisms. Less common, however, is the third major class of organic reaction mechanisms—*pericyclic reactions*.

A **pericyclic reaction** is one that occurs by a concerted process through a cyclic transition state. The word *concerted* means that all bonding changes occur at the same time and in a single step; no intermediates are involved. Rather than try to expand this definition now, we'll begin by briefly reviewing some of the ideas of molecular orbital theory introduced in Chapters 1 and 14 and then looking individually at the three main classes of pericyclic reactions: *electrocyclic reactions, cycloadditions,* and *sigmatropic rearrangements.*

WHY THIS CHAPTER?

The broad outlines of both polar and radical reactions have been known for nearly a century, but our understanding of pericyclic reactions emerged more recently. Prior to the mid-1960s, in fact, they were even referred to on occasion as "no-mechanism reactions." They occur largely in laboratory rather than biological processes, but a knowledge of them is necessary, both for completeness in studying organic chemistry and in understanding those biological pathways where they do occur.

30.1 | Molecular Orbitals and Pericyclic Reactions of Conjugated Pi Systems

A conjugated polyene, as we saw in Section 14.1, is one with alternating double and single bonds. According to molecular orbital (MO) theory, the *p* orbitals on the sp^2-hybridized carbons of a conjugated polyene interact to form a set of

Sean Duggan

π molecular orbitals whose energies depend on the number of nodes they have between nuclei. Those molecular orbitals with fewer nodes are lower in energy than the isolated p atomic orbitals and are *bonding MOs;* those molecular orbitals with more nodes are higher in energy than the isolated p orbitals and are *antibonding MOs*. Pi molecular orbitals of ethylene and 1,3-butadiene are shown in Figure 30.1.

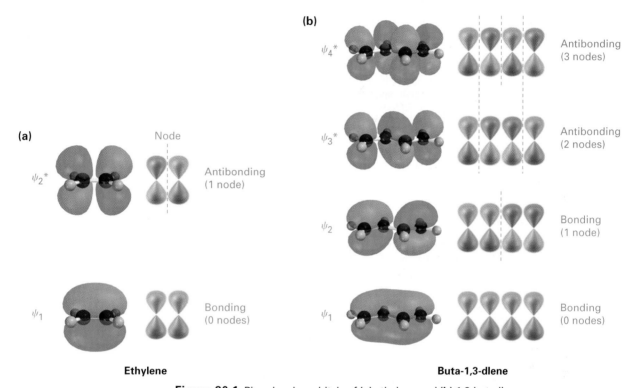

Figure 30.1 Pi molecular orbitals of **(a)** ethylene and **(b)** 1,3-butadiene.

A similar sort of molecular orbital description can be derived for any conjugated π electron system. 1,3,5-Hexatriene, for example, has three double bonds and six π MOs, as shown in Figure 30.2. In the ground state, only the three bonding orbitals, ψ_1, ψ_2, and ψ_3, are filled. On irradiation with ultraviolet light, however, an electron is promoted from the highest-energy filled orbital (ψ_3) to the lowest-energy unfilled orbital ($\psi_4{}^\star$) to give an excited state (Section 14.7), in which ψ_3 and $\psi_4{}^\star$ are each half-filled. (An asterisk denotes an antibonding orbital.)

What do molecular orbitals and their nodes have to do with pericyclic reactions? The answer is, *everything*. According to a series of rules formulated in the mid-1960s by R. B. Woodward and Roald Hoffmann, a pericyclic reaction can take place only if the symmetries of the reactant MOs are the same as the symmetries of the product MOs. In other words, *the lobes of reactant MOs must be of the correct algebraic sign for bonding to occur in the transition state leading to product.*

If the symmetries of reactant and product orbitals match up, or correlate, the reaction is said to be **symmetry-allowed**. If the symmetries of reactant and product orbitals don't correlate, the reaction is **symmetry-disallowed**.

Robert Burns Woodward	Roald Hoffmann	Kenichi Fukui
Robert Burns Woodward (1917–1979) was born in Boston, Massachusetts. He entered the Massachusetts Institute of Technology at age 16, was expelled, reentered, obtained a B.S. degree at age 19, and received a Ph.D. at age 20. He then moved to Harvard University, where he joined the faculty in 1940 at age 23 and remained as professor until his death. His vast scientific contributions included determining the structure of penicillin, pioneering the use of spectroscopic tools for structure elucidation, and turning the field of synthetic organic chemistry into an art form. He received the 1965 Nobel Prize for his work in organic synthesis.	**Roald Hoffmann** (1937–) was born in Zloczow, Poland, just prior to World War II. As a boy, he survived the Holocaust by hiding in the attic of a village schoolhouse. In 1949, he immigrated to the United States, where he received an undergraduate degree at Columbia University and a Ph.D. at Harvard University in 1962. During a further 3-year stay at Harvard as Junior Fellow, he began the collaboration with R. B. Woodward that led to the development of the Woodward–Hoffmann rules for pericyclic reactions. In 1965, he moved to Cornell University, where he remains as professor. He received the 1981 Nobel Prize in chemistry.	**Kenichi Fukui** (1918–1998) was born in Nara Prefecture, Japan, and received a Ph.D. in 1948 from Kyoto Imperial University. He remained at Kyoto University as professor of chemistry until 1982 and then became president of that institution from 1982 to 1988. He received the 1981 Nobel Prize in chemistry, the first Japanese scientist to be thus honored.

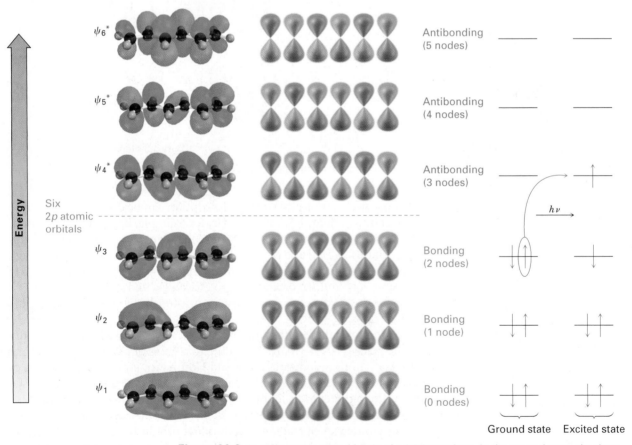

Figure 30.2 The six π molecular orbitals of 1,3,5-hexatriene. In the ground state, the three bonding MOs are filled. In the excited state, ψ_3 and ψ_4^* each have one electron.

Symmetry-allowed reactions often occur under relatively mild conditions, but symmetry-disallowed reactions can't occur by concerted paths. Either they take place by nonconcerted, high-energy pathways, or they don't take place at all.

The Woodward–Hoffmann rules for pericyclic reactions require an analysis of all reactant and product molecular orbitals, but Kenichi Fukui at Kyoto Imperial University in Japan introduced a simplified version. According to Fukui, we need to consider only two molecular orbitals, called the **frontier orbitals**. These frontier orbitals are the **highest occupied molecular orbital (HOMO)** and the **lowest unoccupied molecular orbital (LUMO)**. In ground-state 1,3,5-hexatriene, for example, ψ_3 is the HOMO and ψ_4^* is the LUMO (Figure 30.2). In excited-state 1,3,5-hexatriene, however, ψ_4^* is the HOMO and ψ_5^* is the LUMO.

Problem 30.1 | Look at Figure 30.1, and tell which molecular orbital is the HOMO and which is the LUMO for both ground and excited states of ethylene and 1,3-butadiene.

30.2 | Electrocyclic Reactions

The best way to understand how orbital symmetry affects pericyclic reactions is to look at some examples. Let's look first at a group of polyene rearrangements called *electrocyclic reactions*. An **electrocyclic reaction** is a pericyclic process that involves the cyclization of a conjugated polyene. One π bond is broken, the other π bonds change position, a new σ bond is formed, and a cyclic compound results. For example, a conjugated triene can be converted into a cyclohexadiene, and a conjugated diene can be converted into a cyclobutene.

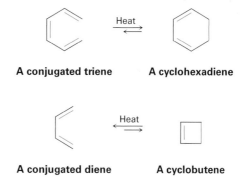

A conjugated triene   **A cyclohexadiene**

A conjugated diene   **A cyclobutene**

Both reactions are reversible, and the position of the equilibrium depends on the specific case. In general, the triene ⇌ cyclohexadiene equilibrium favors the cyclic product, whereas the diene ⇌ cyclobutene equilibrium favors the unstrained open-chain product.

The most striking feature of electrocyclic reactions is their stereochemistry. For example, (2E,4Z,6E)-2,4,6-octatriene yields only *cis*-5,6-dimethyl-1,3-cyclohexadiene when heated, and (2E,4Z,6Z)-2,4,6-octatriene yields only *trans*-5,6-dimethyl-1,3-cyclohexadiene. Remarkably, however, the stereochemical results change completely when the reactions are carried out under what are called **photochemical**, rather than thermal, conditions. Irradiation, or *photolysis,* of (2*E*,4*Z*,6*E*)-2,4,6-octatriene with ultraviolet light yields *trans*-5,6-dimethyl-1,3-cyclohexadiene (Figure 30.3).

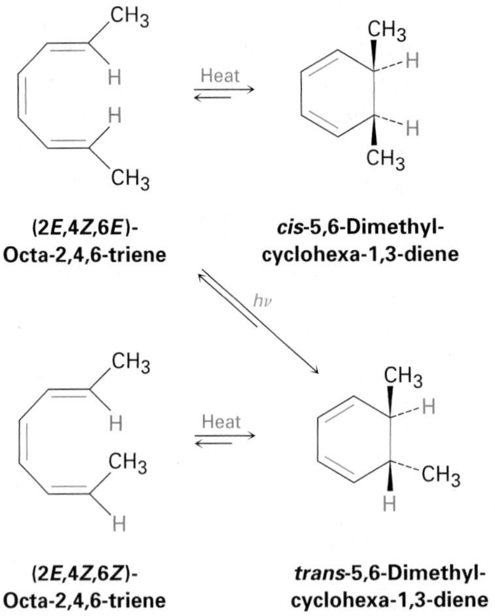

Figure 30.3 Electrocyclic interconversions of 2,4,6-octatriene isomers and 5,6-dimethyl-1,3-cyclohexadiene isomers.

(2*E*,4*Z*,6*E*)-
Octa-2,4,6-triene

cis-5,6-Dimethyl-
cyclohexa-1,3-diene

hν

(2*E*,4*Z*,6*Z*)-
Octa-2,4,6-triene

trans-5,6-Dimethyl-
cyclohexa-1,3-diene

A similar result is obtained for the thermal electrocyclic ring-opening of 3,4-dimethylcyclobutene. The trans isomer yields only (2*E*,4*E*)-2,4-hexadiene when heated, and the cis isomer yields only (2*E*,4*Z*)-2,4-hexadiene. On UV irradiation, however, the results are opposite. Cyclization of the 2*E*,4*E* isomer under photochemical conditions yields cis product (Figure 30.4).

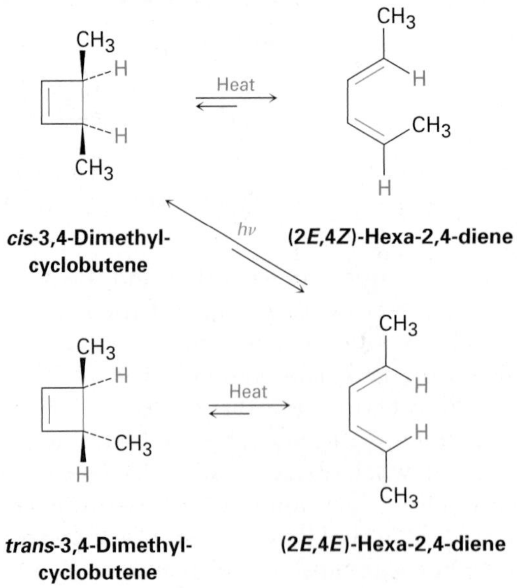

Figure 30.4 Electrocyclic interconversions of 2,4-hexadiene isomers and 3,4-dimethylcyclobutene isomers.

cis-3,4-Dimethyl-
cyclobutene

hν

(2*E*,4*Z*)-Hexa-2,4-diene

trans-3,4-Dimethyl-
cyclobutene

(2*E*,4*E*)-Hexa-2,4-diene

To account for these results, we need to look at the two outermost lobes of the polyene MOs—the lobes that interact when bonding occurs. There are two possibilities: the lobes of like sign can be either on the same side or on opposite sides of the molecule.

Like lobes on same side   or   **Like lobes on opposite side**

For a bond to form, the outermost π lobes must rotate so that favorable bonding interaction is achieved—a positive lobe with a positive lobe or a negative lobe with a negative lobe. If two lobes of like sign are on the *same* side of the molecule, the two orbitals must rotate in *opposite* directions—one clockwise and one counterclockwise. This kind of motion is referred to as **disrotatory**.

Clockwise   Counterclockwise

Conversely, if lobes of like sign are on *opposite* sides of the molecule, both orbitals must rotate in the *same* direction, either both clockwise or both counterclockwise. This kind of motion is called **conrotatory**.

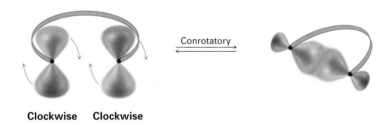

Clockwise   Clockwise

30.3 | Stereochemistry of Thermal Electrocyclic Reactions

How can we predict whether conrotatory or disrotatory motion will occur in a given case? According to frontier orbital theory, *the stereochemistry of an electrocyclic reaction is determined by the symmetry of the polyene HOMO.* The electrons in the HOMO are the highest-energy, most loosely held electrons, and are therefore most easily moved during reaction. For thermal reactions, the ground-state electronic configuration is used to identify the HOMO; for photochemical reactions, the excited-state electronic configuration is used.

Let's look again at the thermal ring closure of conjugated trienes. According to Figure 30.2, the HOMO of a conjugated triene in its ground state has lobes of like sign on the same side of the molecule, a symmetry that predicts disrotatory ring closure. This disrotatory cyclization is exactly what is observed in the thermal cyclization of 2,4,6-octatriene. The 2*E*,4*Z*,6*E* isomer yields cis product; the 2*E*,4*Z*,6*Z* isomer yields trans product (Figure 30.5).

Active Figure 30.5 Thermal cyclizations of 2,4,6-octatrienes occur by disrotatory ring closures. *Sign in at* **www.cengage.com/login** *to see a simulation based on this figure and to take a short quiz.*

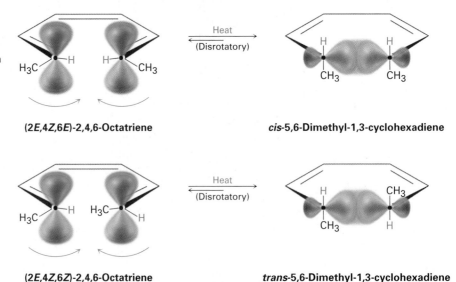

(2*E*,4*Z*,6*E*)-2,4,6-Octatriene   **cis-5,6-Dimethyl-1,3-cyclohexadiene**

(2*E*,4*Z*,6*Z*)-2,4,6-Octatriene   **trans-5,6-Dimethyl-1,3-cyclohexadiene**

In the same way, the ground-state HOMO of a conjugated diene (Figure 30.1) has a symmetry that predicts conrotatory ring closure. In practice, however, the conjugated diene reaction can be observed only in the reverse direction (cyclobutene → diene) because of the position of the equilibrium. We therefore find that the 3,4-dimethylcyclobutene ring *opens* in a conrotatory fashion. *cis*-3,4-Dimethylcyclobutene yields (2*E*,4*Z*)-2,4-hexadiene, and *trans*-3,4-dimethylcyclobutene yields (2*E*,4*E*)-2,4-hexadiene by conrotatory opening (Figure 30.6).

Figure 30.6 Thermal ringopenings of *cis*- and *trans*-dimethylcyclobutene occur by conrotatory paths.

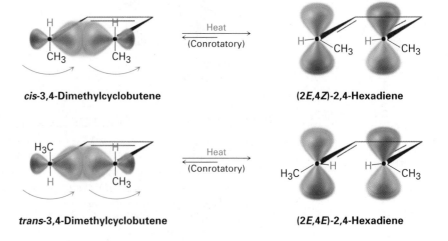

cis-3,4-Dimethylcyclobutene   **(2*E*,4*Z*)-2,4-Hexadiene**

trans-3,4-Dimethylcyclobutene   **(2*E*,4*E*)-2,4-Hexadiene**

Note that a conjugated diene and a conjugated triene react with opposite stereochemistry. The diene opens and closes by a conrotatory path, whereas the triene opens and closes by a disrotatory path. The difference is due to the different symmetries of the diene and triene HOMOs.

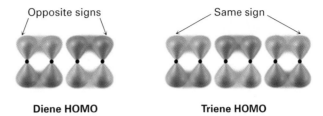

Diene HOMO **Triene HOMO**

It turns out that there is an alternating relationship between the number of electron pairs (double bonds) undergoing bond reorganization and the stereochemistry of ring opening or closure. Polyenes with an even number of electron pairs undergo thermal electrocyclic reactions in a conrotatory sense, whereas polyenes with an odd number of electron pairs undergo the same reactions in a disrotatory sense.

Problem 30.2 Draw the products you would expect from conrotatory and disrotatory cyclizations of (2Z,4Z,6Z)-2,4,6-octatriene. Which of the two paths would you expect the thermal reaction to follow?

Problem 30.3 *trans*-3,4-Dimethylcyclobutene can open by two conrotatory paths to give either (2E,4E)-2,4-hexadiene or (2Z,4Z)-2,4-hexadiene. Explain why both products are symmetry-allowed, and then account for the fact that only the 2E,4E isomer is obtained in practice.

30.4 Photochemical Electrocyclic Reactions

We noted previously that photochemical electrocyclic reactions take a different stereochemical course than their thermal counterparts, and we can now explain this difference. Ultraviolet irradiation of a polyene causes an excitation of one electron from the ground-state HOMO to the ground-state LUMO, thus changing their symmetries. But because electronic excitation changes the symmetries of HOMO and LUMO, it also changes the reaction stereochemistry. (2E,4E)-2,4-Hexadiene, for instance, undergoes photochemical cyclization by a disrotatory path, whereas the thermal reaction is conrotatory. Similarly, (2E,4Z,6E)-2,4,6-octatriene undergoes photochemical cyclization by a conrotatory path, whereas the thermal reaction is disrotatory (Figure 30.7).

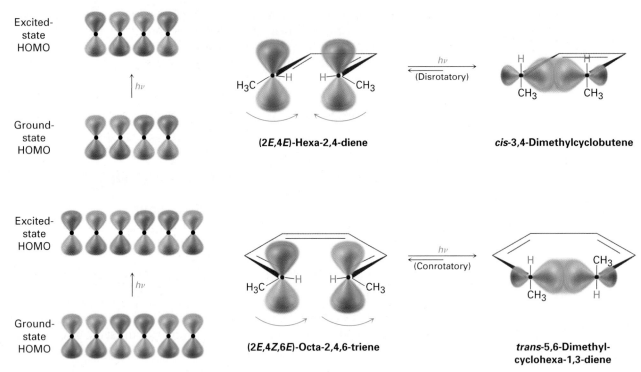

Excited-state HOMO

Ground-state HOMO

$h\nu$

$h\nu$ (Disrotatory)

(2E,4E)-Hexa-2,4-diene

cis-3,4-Dimethylcyclobutene

Excited-state HOMO

Ground-state HOMO

$h\nu$

$h\nu$ (Conrotatory)

(2E,4Z,6E)-Octa-2,4,6-triene

trans-5,6-Dimethyl-cyclohexa-1,3-diene

Figure 30.7 Photochemical cyclizations of conjugated dienes and trienes. The two processes occur with different stereochemistry because of their different orbital symmetries.

Thermal and photochemical electrocyclic reactions *always* take place with opposite stereochemistry because the symmetries of the frontier orbitals are always different. Table 30.1 gives some simple rules that make it possible to predict the stereochemistry of electrocyclic reactions.

Table 30.1	Stereochemical Rules for Electrocyclic Reactions	
Electron pairs (double bonds)	Thermal reaction	Photochemical reaction
Even number	Conrotatory	Disrotatory
Odd number	Disrotatory	Conrotatory

Problem 30.4 | What product would you expect to obtain from the photochemical cyclization of (2E,4Z,6E)-2,4,6-octatriene? Of (2E,4Z,6Z)-2,4,6-octatriene?

30.5 | Cycloaddition Reactions

A **cycloaddition reaction** is one in which two unsaturated molecules add to one another, yielding a cyclic product. As with electrocyclic reactions, cycloadditions are controlled by the orbital symmetry of the reactants. Symmetry-allowed processes often take place readily, but symmetry-disallowed processes take place with difficulty, if at all, and then only by nonconcerted pathways. Let's look at two examples to see how they differ.

The Diels–Alder cycloaddition reaction (Section 14.4) is a pericyclic process that takes place between a diene (four π electrons) and a dienophile (two π electrons) to yield a cyclohexene product. Many thousands of examples of Diels–Alder reactions are known. They often take place easily at room temperature or slightly above, and they are stereospecific with respect to substituents. For example, room-temperature reaction between 1,3-butadiene and diethyl maleate (cis) yields exclusively the cis-disubstituted cyclohexene product. A similar reaction between 1,3-butadiene and diethyl fumarate (trans) yields exclusively the trans-disubstituted product.

In contrast with the $[4 + 2]$-π-electron Diels–Alder reaction, the $[2 + 2]$ thermal cycloaddition between two alkenes does not occur. Only the photochemical $[2 + 2]$ cycloaddition takes place to yield cyclobutane products.

For a successful cycloaddition to take place, the terminal π lobes of the two reactants must have the correct symmetry for bonding to occur. This can happen in either of two ways, called *suprafacial* and *antarafacial*. **Suprafacial** cycloadditions take place when a bonding interaction occurs between lobes on the same face of one reactant and lobes on the same face of the other reactant. **Antarafacial** cycloadditions take place when a bonding interaction occurs between lobes on the same face of one reactant and lobes on *opposite* faces of the other reactant (Figure 30.8).

Figure 30.8 **(a)** Suprafacial cycloaddition occurs when there is bonding between lobes on the same face of one reactant and lobes on the same face of the other reactant. **(b)** Antarafacial cycloaddition occurs when there is bonding between lobes on the same face of one reactant and lobes on opposite faces of the other, which requires a twist in one π system.

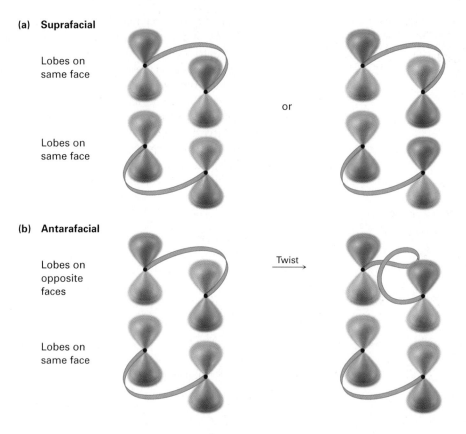

(a) **Suprafacial**

Lobes on same face

Lobes on same face

or

(b) **Antarafacial**

Lobes on opposite faces

Twist

Lobes on same face

Note that both suprafacial and antarafacial cycloadditions are symmetry-allowed. Geometric constraints often make antarafacial reactions difficult, however, because there must be a twisting of the π orbital system in one of the reactants. Thus, suprafacial cycloadditions are the most common for small π systems.

30.6 | Stereochemistry of Cycloadditions

How can we predict whether a given cycloaddition reaction will occur with suprafacial or with antarafacial geometry? According to frontier orbital theory, a cycloaddition reaction takes place when a bonding interaction occurs between the HOMO of one reactant and the LUMO of the other. An intuitive explanation of this rule is to imagine that one reactant donates electrons to the other. As with electrocyclic reactions, it's the electrons in the HOMO of the first reactant that are least tightly held and most likely to be donated. But when the second reactant accepts those electrons, they must go into a *vacant,* unoccupied orbital—the LUMO.

For a [4 + 2]-π-electron cycloaddition (Diels–Alder reaction), let's arbitrarily select the diene LUMO and the alkene HOMO. The symmetries of the two ground-state orbitals are such that bonding of the terminal lobes can occur with suprafacial geometry (Figure 30.9), so the Diels–Alder reaction takes place readily under thermal conditions. Note that, as with electrocyclic reactions, we need be concerned only with the *terminal* lobes. For purposes of prediction, interactions among the interior lobes need not be considered.

Figure 30.9 Interaction of diene LUMO and alkene HOMO in a suprafacial [4 + 2] cycloaddition reaction (Diels–Alder reaction).

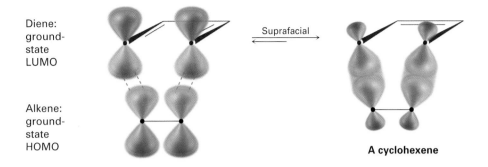

Diene: ground-state LUMO

Alkene: ground-state HOMO

Suprafacial

A cyclohexene

In contrast with the thermal [4 + 2] Diels–Alder reaction, the [2 + 2] cyclo-addition of two alkenes to yield a cyclobutane can only be observed photochemically. The explanation follows from orbital-symmetry arguments. Looking at the ground-state HOMO of one alkene and the LUMO of the second alkene, it's apparent that a thermal [2 + 2] cycloaddition must take place by an antarafacial pathway (Figure 30.10a). Geometric constraints make the antarafacial transition state difficult, however, and so concerted thermal [2 + 2] cycloadditions are not observed.

In contrast with the thermal process, photochemical [2 + 2] cycloadditions *are* observed. Irradiation of an alkene with UV light excites an electron from ψ_1, the ground-state HOMO, to ψ_2^*, which becomes the excited-state HOMO. Interaction between the excited-state HOMO of one alkene and the LUMO of the second alkene allows a photochemical [2 + 2] cycloaddition reaction to occur by a suprafacial pathway (Figure 30.10b).

Figure 30.10 **(a)** Interaction of a ground-state HOMO and a ground-state LUMO in a potential [2 + 2] cycloaddition does not occur thermally because the antarafacial geometry is too strained. **(b)** Interaction of an excited-state HOMO and a ground-state LUMO in a photochemical [2 + 2] cycloaddition reaction is less strained, however, and occurs with suprafacial geometry.

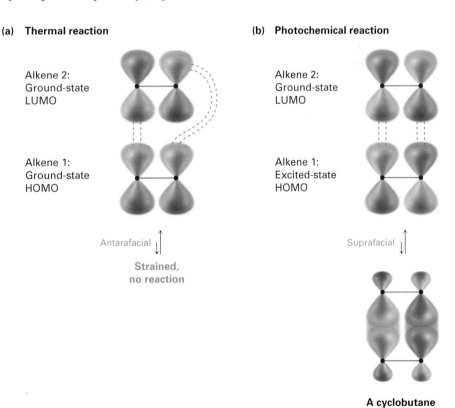

(a) **Thermal reaction**

Alkene 2: Ground-state LUMO

Alkene 1: Ground-state HOMO

Antarafacial

Strained, no reaction

(b) **Photochemical reaction**

Alkene 2: Ground-state LUMO

Alkene 1: Excited-state HOMO

Suprafacial

A cyclobutane

The photochemical [2 + 2] cycloaddition reaction occurs smoothly and represents one of the best methods known for synthesizing cyclobutane rings. For example:

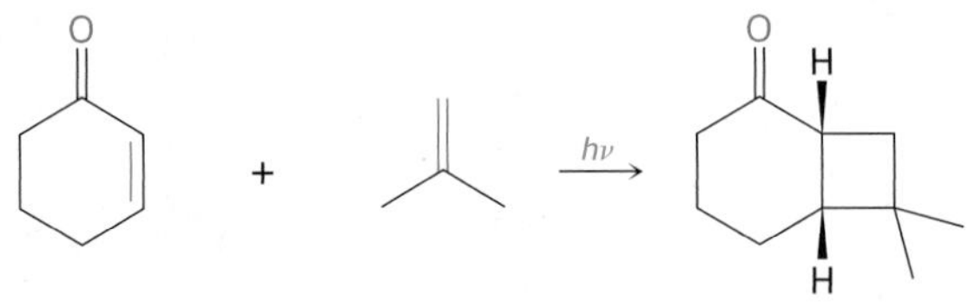

Cyclohex-2-enone   **2-Methylpropene**   **(40%)**

Thermal and photochemical cycloaddition reactions always take place with opposite stereochemistry. As with electrocyclic reactions, we can categorize cycloadditions according to the total number of electron pairs (double bonds) involved in the rearrangement. Thus, a thermal Diels–Alder [4 + 2] reaction between a diene and a dienophile involves an odd number (three) of electron pairs and takes place by a suprafacial pathway. A thermal [2 + 2] reaction between two alkenes involves an even number (two) of electron pairs and must take place by an antarafacial pathway. For photochemical cyclizations, these selectivities are reversed. The general rules are given in Table 30.2.

Table 30.2 | **Stereochemical Rules for Cycloaddition Reactions**

Electron pairs (double bonds)	Thermal reaction	Photochemical reaction
Even number	Antarafacial	Suprafacial
Odd number	Suprafacial	Antarafacial

Problem 30.5   What stereochemistry would you expect for the product of the Diels–Alder reaction between (2E,4E)-2,4-hexadiene and ethylene? What stereochemistry would you expect if (2E,4Z)-2,4-hexadiene were used instead?

Problem 30.6   1,3-Cyclopentadiene reacts with cycloheptatrienone to give the product shown. Tell what kind of reaction is involved, and explain the observed result. Is the reaction suprafacial or antarafacial?

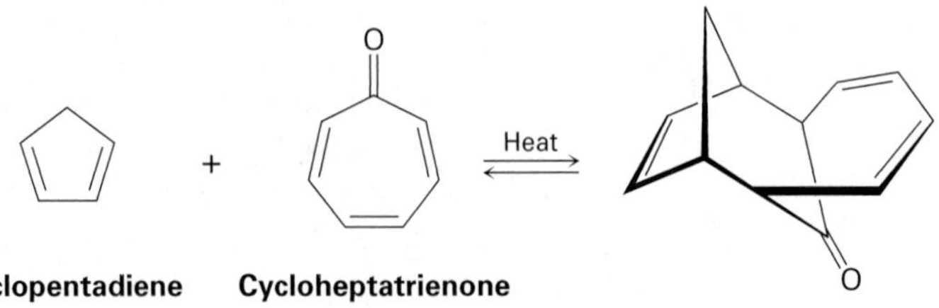

Cyclopentadiene   **Cycloheptatrienone**

30.7 Sigmatropic Rearrangements

CENGAGENOW™ Click *Organic Interactive* to **predict products from a variety of sigmatropic rearrangement reactions.**

A **sigmatropic rearrangement**, the third general kind of pericyclic reaction, is a process in which a σ-bonded substituent atom or group migrates across a π electron system from one position to another. A σ bond is broken in the reactant, the π bonds move, and a new σ bond is formed in the product. The σ-bonded group can be either at the end or in the middle of the π system, as the following [1,5] and [3,3] rearrangements illustrate:

A [1,5] sigmatropic rearrangement

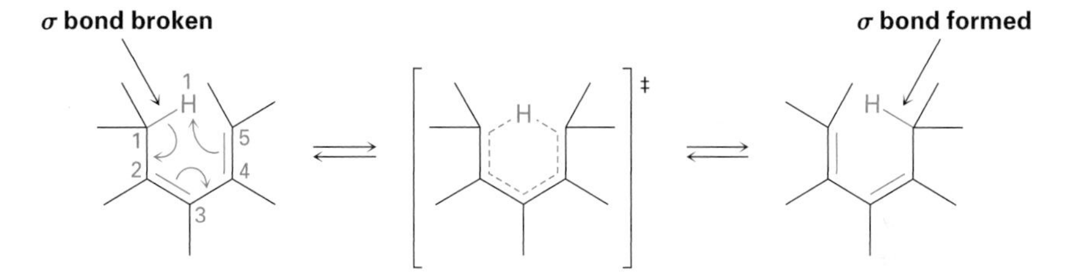

A 1,3-diene **Cyclic transition state** **A 1,3-diene**

A [3,3] sigmatropic rearrangement

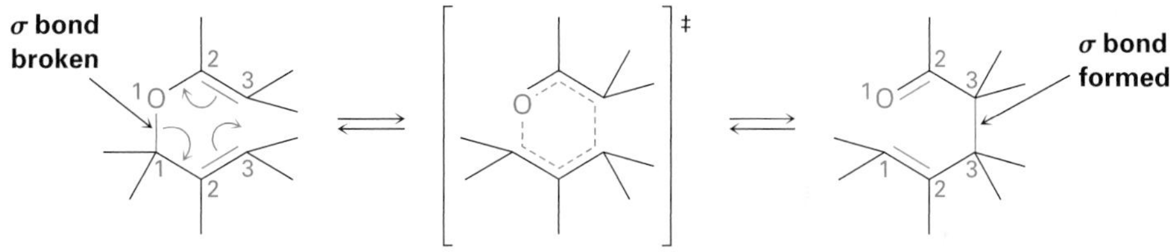

An allylic vinylic ether **Cyclic transition state** **An unsaturated ketone**

The notations [1,5] and [3,3] describe the kind of rearrangement that is occurring. The numbers refer to the two groups connected by the σ bond and designate the positions in those groups to which migration occurs. For example, in the [1,5] sigmatropic rearrangement of a diene, the two groups connected by the σ bond are a hydrogen atom and a pentadienyl group. Migration occurs to position 1 of the H group (the only possibility) and to position 5 of the pentadienyl group. In the [3,3] Claisen rearrangement (Section 18.4), the two groups connected by the σ bond are an allylic group and a vinylic ether group. Migration occurs to position 3 of the allylic group and also to position 3 of the vinylic ether.

Sigmatropic rearrangements, like electrocyclic reactions and cycloadditions, are controlled by orbital symmetries. There are two possible modes of reaction: migration of a group across the same face of the π system is called a *suprafacial* rearrangement, and migration of a group from one face of the π system to the other face is called an *antarafacial* rearrangement (Figure 30.11).

Figure 30.11 Suprafacial and antarafacial sigmatropic rearrangements.

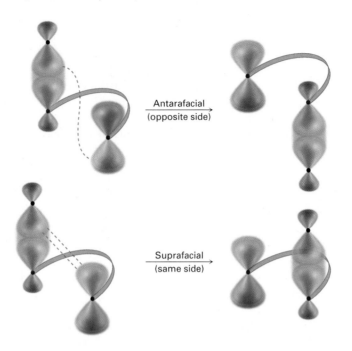

Both suprafacial and antarafacial sigmatropic rearrangements are symmetry-allowed, but suprafacial rearrangements are often easier for geometric reasons. The rules for sigmatropic rearrangements are identical to those for cycloaddition reactions (Table 30.3).

Table 30.3	Stereochemical Rules for Sigmatropic Rearrangements	
Electron pairs (double bonds)	Thermal reaction	Photochemical reaction
Even number	Antarafacial	Suprafacial
Odd number	Suprafacial	Antarafacial

Problem 30.7 | Classify the following sigmatropic reaction by order [x,y], and tell whether it will proceed with suprafacial or antarafacial stereochemistry:

30.8 | Some Examples of Sigmatropic Rearrangements

Because a [1,5] sigmatropic rearrangement involves three electron pairs (two π bonds and one σ bond), the orbital-symmetry rules in Table 30.3 predict a suprafacial reaction. In fact, the [1,5] suprafacial shift of a hydrogen atom across two double bonds of a π system is one of the most commonly observed of all sigmatropic rearrangements. For example, 5-methyl-1,3-cyclopentadiene rapidly rearranges at room temperature to yield a mixture of 1-methyl-, 2-methyl-, and 5-methyl-substituted products.

As another example, heating 5,5,5-trideuterio-(1,3Z)-1,3-pentadiene causes scrambling of deuterium between positions 1 and 5.

Both these [1,5] hydrogen shifts occur by a symmetry-allowed suprafacial rearrangement, as illustrated in Figure 30.12. In contrast with these thermal [1,5] sigmatropic hydrogen shifts, however, thermal [1,3] hydrogen shifts are unknown. Were they to occur, they would have to proceed by a strained antarafacial reaction pathway.

Figure 30.12 An orbital view of a suprafacial [1,5] hydrogen shift.

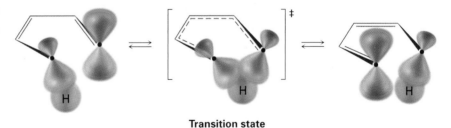

Transition state

Two other important sigmatropic reactions are the Claisen rearrangement of an allyl aryl ether discussed in Section 18.4 and the **Cope rearrangement** of a 1,5-hexadiene. These two, along with the Diels–Alder reaction, are the most useful pericyclic reactions for organic synthesis; many thousands of examples of all three are known. Note that the Claisen rearrangement occurs with both allylic *aryl* ethers and allylic *vinylic* ethers.

Claisen rearrangement

An allylic aryl ether

An o-allylphenol

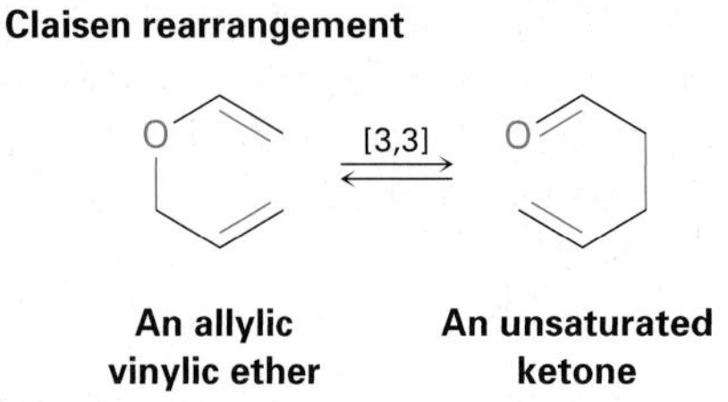
Claisen rearrangement
O
[3,3]
O
An allylic vinylic ether
An unsaturated ketone

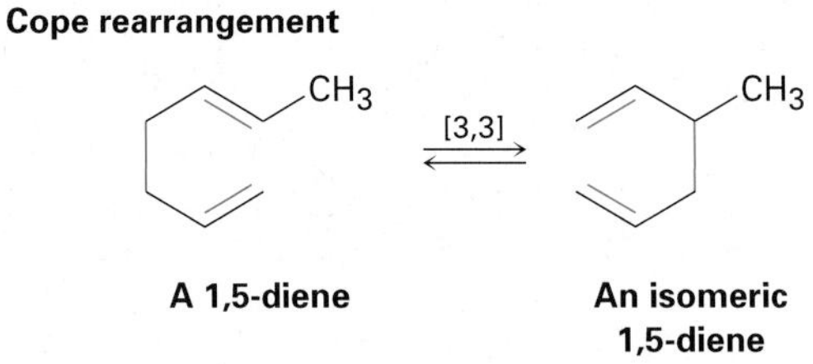
Cope rearrangement
CH3
[3,3]
CH3
A 1,5-diene
An isomeric 1,5-diene

Both Cope and Claisen rearrangements involve reorganization of an odd number of electron pairs (two π bonds and one σ bond), and both react by suprafacial pathways (Figure 30.13).

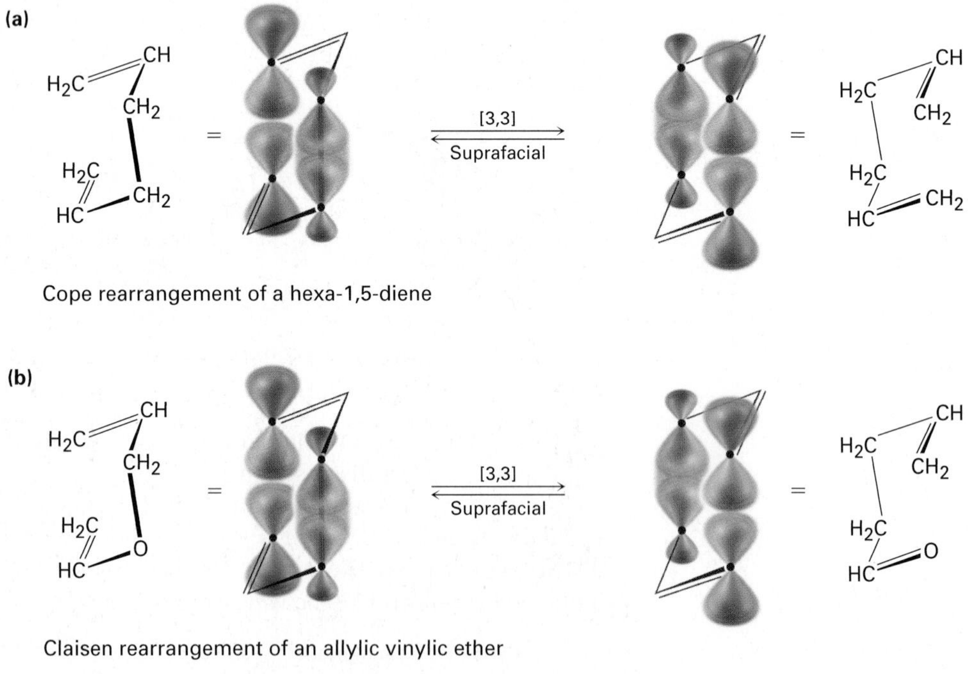
(a)
H2C
CH
CH2
H2C
HC
CH2
[3,3]
Suprafacial
H2C
CH
CH2
H2C
HC
CH2
Cope rearrangement of a hexa-1,5-diene
(b)
H2C
CH
CH2
H2C
HC
O
[3,3]
Suprafacial
H2C
CH
CH2
H2C
HC
O
Claisen rearrangement of an allylic vinylic ether

Figure 30.13 Suprafacial [3,3] **(a)** Cope and **(b)** Claisen rearrangements.

Biological examples of pericyclic reactions are relatively rare, although one much-studied example occurs during biosynthesis in bacteria of the essential amino acid phenylalanine. Phenylalanine arises from the precursor chorismate, through a Claisen rearrangement to prephenate, followed by decarboxylation to phenylpyruvate and reductive amination (Figure 30.14). You might note that the reductive amination of phenylpyruvate is the exact reverse of the transamination process discussed in Section 29.9, by which amino acids are deaminated. In addition, reductive amination of ketones is a standard method for preparing amines in the laboratory, as we saw in Section 24.6.

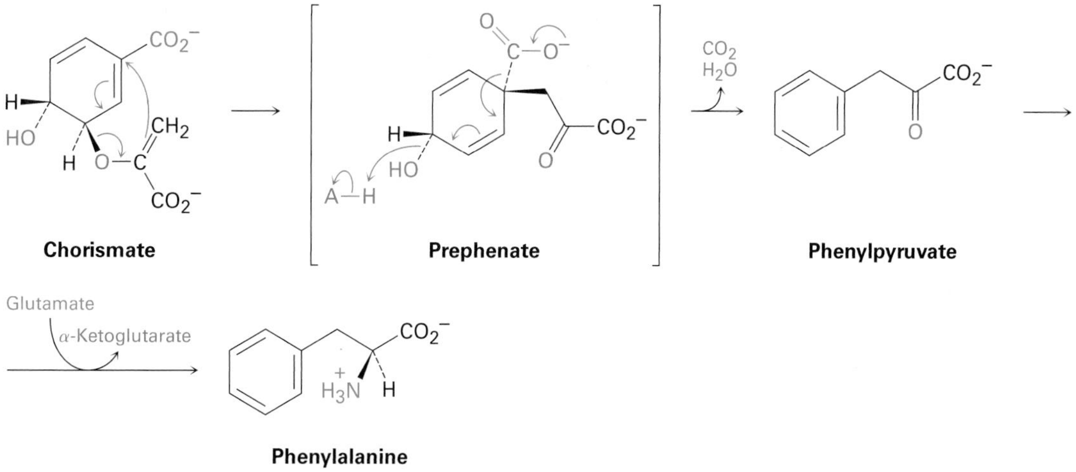

Figure 30.14 Pathway for the bacterial biosynthesis of phenylalanine from chorismate, involving a Claisen rearrangement.

Problem 30.8 | Propose a mechanism to account for the fact that heating 1-deuterioindene scrambles the isotope label to all three positions on the five-membered ring.

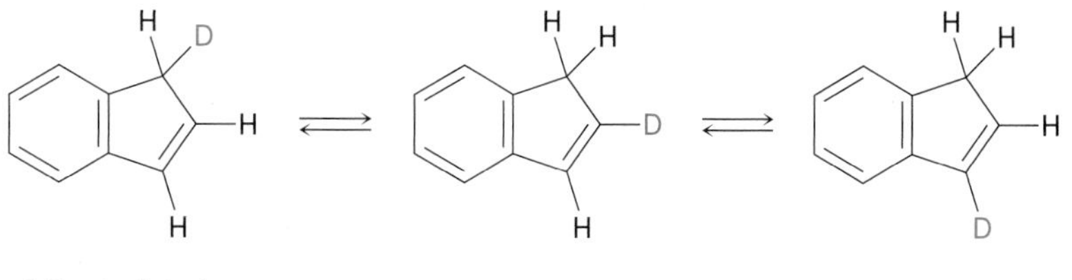

1-Deuterioindene

Problem 30.9 | When a 2,6-disubstituted allyl phenyl ether is heated in an attempted Claisen rearrangement, migration occurs to give the *p*-allyl product as the result of two sequential pericyclic reactions. Explain.

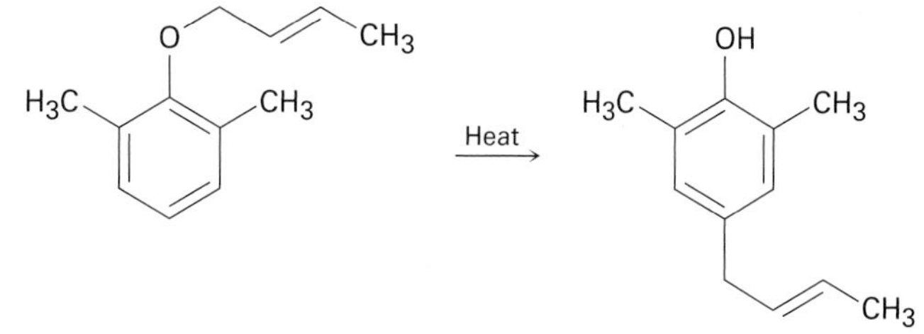

30.9 | A Summary of Rules for Pericyclic Reactions

How can you keep straight all the rules about pericyclic reactions? The summary information in Tables 30.1 to 30.3 can be distilled into one mnemonic phrase that provides an easy way to predict the stereochemical outcome of any pericyclic reaction:

The Electrons Circle Around (TECA)

*T*hermal reactions with an *E*ven number of electron
pairs are *C*onrotatory or *A*ntarafacial.

A change either from thermal to photochemical or from an even to an odd number of electron pairs changes the outcome from conrotatory/antarafacial to disrotatory/suprafacial. A change from both thermal and even to photochemical and odd causes no change because two negatives make a positive.

These selection rules are summarized in Table 30.4, thereby giving you the ability to predict the stereochemistry of literally thousands of pericyclic reactions.

Table 30.4 | **Stereochemical Rules for Pericyclic Reactions**

Electronic state	Electron pairs	Stereochemistry
Ground state (thermal)	Even number	Antara–con
	Odd number	Supra–dis
Excited state (photochemical)	Even number	Supra–dis
	Odd number	Antara–con

Problem 30.10 | Predict the stereochemistry of the following pericyclic reactions:
(a) The thermal cyclization of a conjugated tetraene
(b) The photochemical cyclization of a conjugated tetraene
(c) A photochemical [4 + 4] cycloaddition
(d) A thermal [2 + 6] cycloaddition
(e) A photochemical [3,5] sigmatropic rearrangement

Focus On . . .

Vitamin D, the Sunshine Vitamin

Synthesizing vitamin D takes dedication and hard work.

Vitamin D, discovered in 1918, is a general name for two related compounds, *cholecalciferol* (vitamin D_3) and *ergocalciferol* (vitamin D_2). Both are steroids (Section 27.6) and differ only in the nature of the hydrocarbon side chain attached to the five-membered ring. Cholecalciferol comes from dairy products and fish; ergocalciferol comes from some vegetables. Their function in the body is to control the calcification of bones by increasing intestinal absorption of calcium. When sufficient vitamin D is present, approximately 30% of ingested calcium is absorbed, but in the absence of vitamin D, calcium absorption falls to about 10%. A deficiency of vitamin D thus leads to poor bone growth and to the childhood disease known as *rickets*.

Actually, neither vitamin D_2 nor D_3 is present in foods. Rather, foods contain the precursor molecules 7-dehydrocholesterol and ergosterol. In the presence of sunlight, however, both precursors are converted under the skin to the active vitamins, hence the nickname for vitamin D, the "sunshine vitamin."

7-Dehydrocholesterol
Ergosterol

R = $CH(CH_3)CH_2CH_2CH_2CH(CH_3)_2$
R = $CH(CH_3)CH{=}CHCH(CH_3)CH(CH_3)_2$

Cholecalciferol
Ergocalciferol

Pericyclic reactions are unusual in living organisms, and the photochemical synthesis of vitamin D is one of only a few well-studied examples. The reaction takes place in two steps, an electrocyclic ring-opening of a cyclohexadiene to yield a hexatriene, followed by a sigmatropic [1,7] H shift to yield an isomeric hexatriene. Further metabolic processing in the liver and the kidney introduces several −OH groups to give the active form of the vitamin.

SUMMARY AND KEY WORDS

A **pericyclic reaction** is one that takes place in a single step through a cyclic transition state without intermediates. There are three major classes of pericyclic processes: electrocyclic reactions, cycloaddition reactions, and sigmatropic rearrangements. The stereochemistry of these reactions is controlled by the symmetry of the orbitals involved in bond reorganization.

Electrocyclic reactions involve the cyclization of conjugated polyenes. For example, 1,3,5-hexatriene cyclizes to 1,3-cyclohexadiene on heating. Electrocyclic reactions can occur by either **conrotatory** or **disrotatory** paths, depending on the symmetry of the terminal lobes of the π system. Conrotatory cyclization requires that both lobes rotate in the same direction, whereas disrotatory cyclization requires that the lobes rotate in opposite directions. The reaction course in a specific case can be found by looking at the symmetry of the **highest occupied molecular orbital (HOMO)**.

Cycloaddition reactions are those in which two unsaturated molecules add together to yield a cyclic product. For example, Diels–Alder reaction between a diene (four π electrons) and a dienophile (two π electrons) yields a cyclohexene. Cycloadditions can take place either by **suprafacial** or **antarafacial** pathways. Suprafacial cycloaddition involves interaction between lobes on the same face of one component and on the same face of the second component. Antarafacial cycloaddition involves interaction between lobes on the same face of one component and on opposite faces of the other component. The reaction course in a specific case can be found by looking at the symmetry of the HOMO of one component and the **lowest unoccupied molecular orbital (LUMO)** of the other component.

Sigmatropic rearrangements involve the migration of a σ-bonded group across a π electron system. For example, Claisen rearrangement of an allylic vinylic ether yields an unsaturated carbonyl compound, and **Cope rearrangement** of a 1,5-hexadiene yields an isomeric 1,5-hexadiene. Sigmatropic rearrangements can occur with either suprafacial or antarafacial stereochemistry; the selection rules for a given case are the same as those for cycloaddition reactions.

The stereochemistry of any pericyclic reaction can be predicted by counting the total number of electron pairs (bonds) involved in bond reorganization and then applying the mnemonic "The Electrons Circle Around." That is, **thermal** (ground-state) reactions involving an even number of electron pairs occur with either conrotatory or antarafacial stereochemistry. Exactly the opposite rules apply to **photochemical** (excited-state) **reactions**.

EXERCISES

VISUALIZING CHEMISTRY

(Problems 30.1–30.10 appear within the chapter.)

30.11 ■ Predict the product obtained when the following substance is heated:

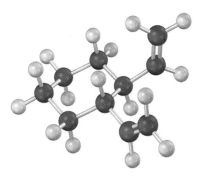

30.12 The ^{13}C NMR spectrum of homotropilidene taken at room temperature shows only three peaks. Explain.

ADDITIONAL PROBLEMS

30.13 ■ Have the following reactions taken place in a conrotatory or disrotatory manner? Under what conditions, thermal or photochemical, would you carry out each reaction?

(a)

(b)

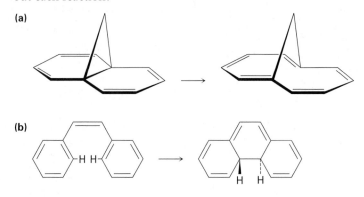

30.14 ■ What stereochemistry—antarafacial or suprafacial—would you expect to observe in the following reactions?
(a) A photochemical [1,5] sigmatropic rearrangement
(b) A thermal [4 + 6] cycloaddition
(c) A thermal [1,7] sigmatropic rearrangement
(d) A photochemical [2 + 6] cycloaddition

30.15 The following thermal isomerization occurs under relatively mild conditions. Identify the pericyclic reactions involved, and show how the rearrangement occurs.

30.16 ■ Would you expect the following reaction to proceed in a conrotatory or disrotatory manner? Show the stereochemistry of the cyclobutene product, and explain your answer.

30.17 Heating (1Z,3Z,5Z)-1,3,5-cyclononatriene to 100 °C causes cyclization and formation of a bicyclic product. Is the reaction conrotatory or disrotatory? What is the stereochemical relationship of the two hydrogens at the ring junctions, cis or trans?

(1Z,3Z,5Z)-Cyclonona-1,3,5-triene

30.18 (2E,4Z,6Z,8E)-2,4,6,8-Decatetraene has been cyclized to give 7,8-dimethyl-1,3,5-cyclooctatriene. Predict the manner of ring closure—conrotatory or disrotatory—for both thermal and photochemical reactions, and predict the stereochemistry of the product in each case.

30.19 Answer Problem 30.18 for the thermal and photochemical cyclizations of (2E,4Z,6Z,8Z)-2,4,6,8-decatetraene.

30.20 The cyclohexadecaoctaene shown isomerizes to two different isomers, depending on reaction conditions. Explain the observed results, and indicate whether each reaction is conrotatory or disrotatory.

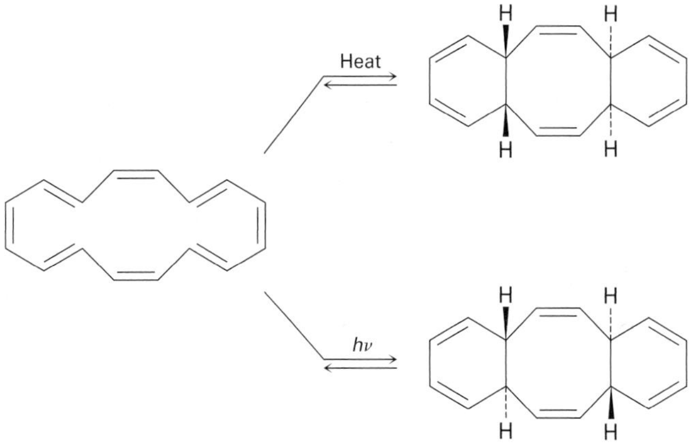

30.21 ■ Which of the following reactions is more likely to occur? Explain.

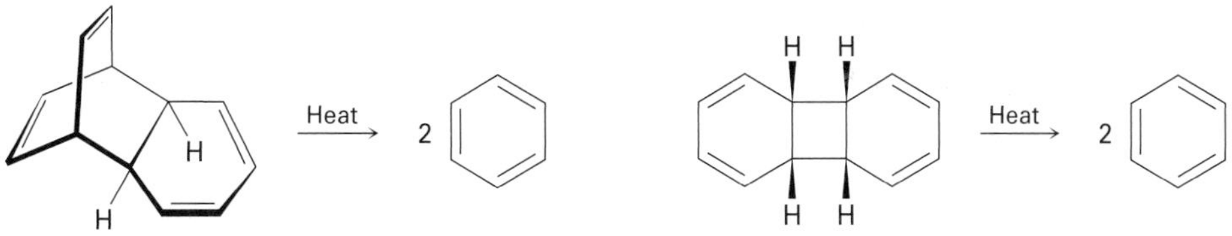

30.22 Bicyclohexadiene, also known as *Dewar benzene,* is extremely stable despite the fact that its rearrangement to benzene is energetically favored. Explain why the rearrangement is so slow.

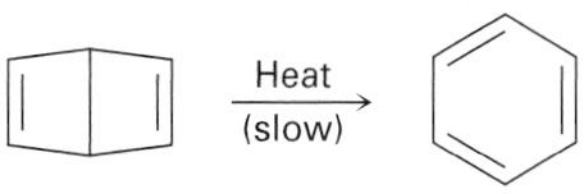

30.23 ■ The following thermal rearrangement involves two pericyclic reactions in sequence. Identify them, and propose a mechanism to account for the observed result.

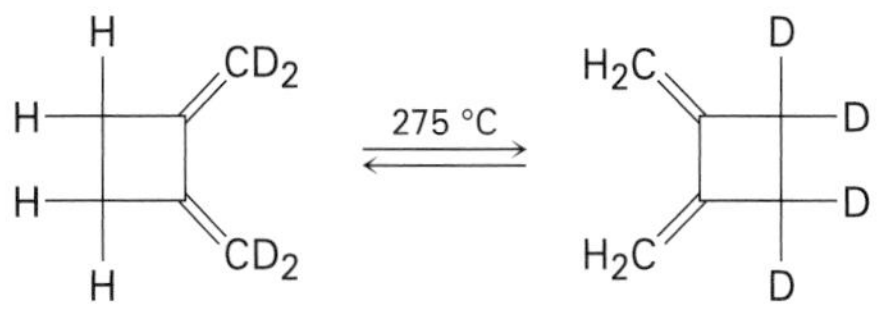

30.24 ■ Predict the product of the following pericyclic reaction. Is this [5,5] shift a suprafacial or an antarafacial process?

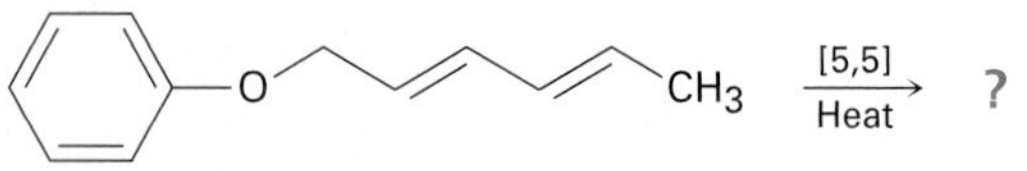

30.25 Ring-opening of the *trans*-cyclobutene isomer shown takes place at much lower temperature than a similar ring-opening of the *cis*-cyclobutene isomer. Explain the temperature effect, and identify the stereochemistry of each reaction as either conrotatory or disrotatory.

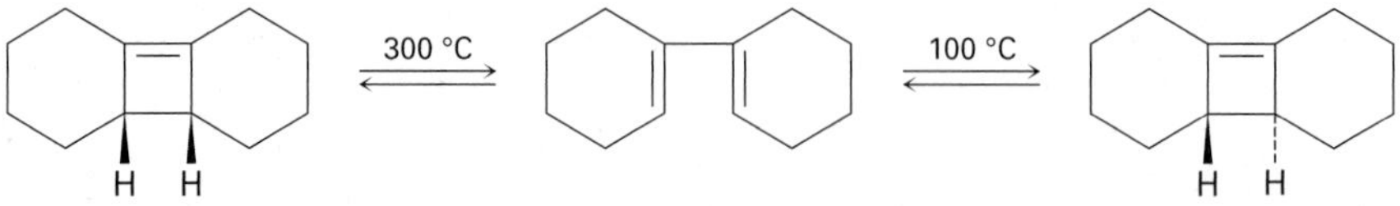

30.26 ■ Photolysis of the *cis*-cyclobutene isomer in Problem 30.25 yields *cis*-cyclo-dodecaen-7-yne, but photolysis of the trans isomer yields *trans*-cyclododecaen-7-yne. Explain these results, and identify the type and stereochemistry of the pericyclic reaction.

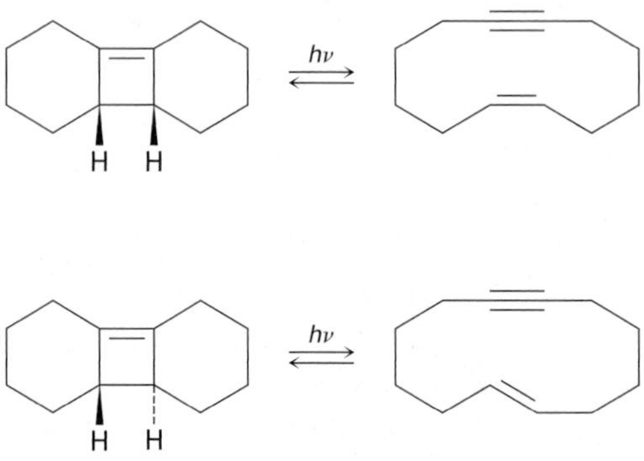

30.27 Propose a pericyclic mechanism to account for the following transformation:

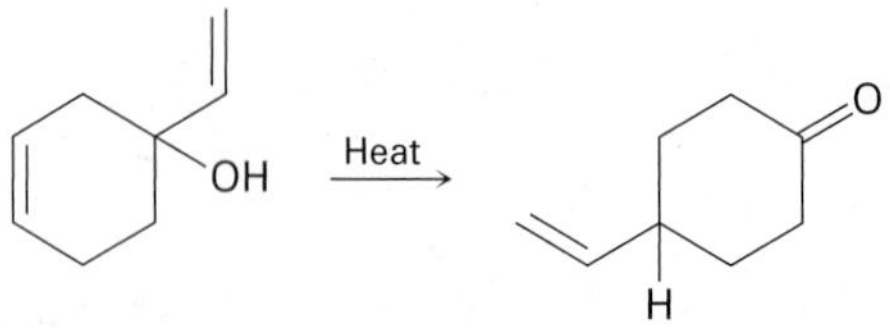

30.28 Vinyl-substituted cyclopropanes undergo thermal rearrangement to yield cyclopentenes. Propose a mechanism for the reaction, and identify the pericyclic process involved.

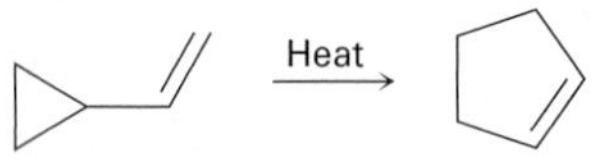

Vinylcyclopropane   **Cyclopentene**

30.29 ■ The following reaction takes place in two steps, one of which is a cycloaddition and the other of which is a *reverse* cycloaddition. Identify the two pericyclic reactions, and show how they occur.

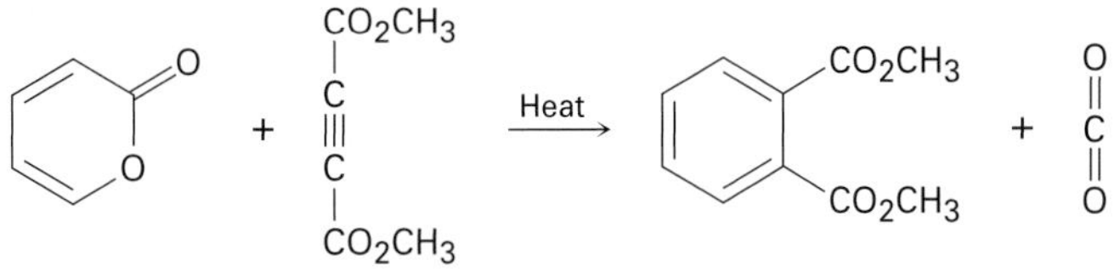

30.30 Two sequential pericyclic reactions are involved in the following furan synthesis. Identify them, and propose a mechanism for the transformation.

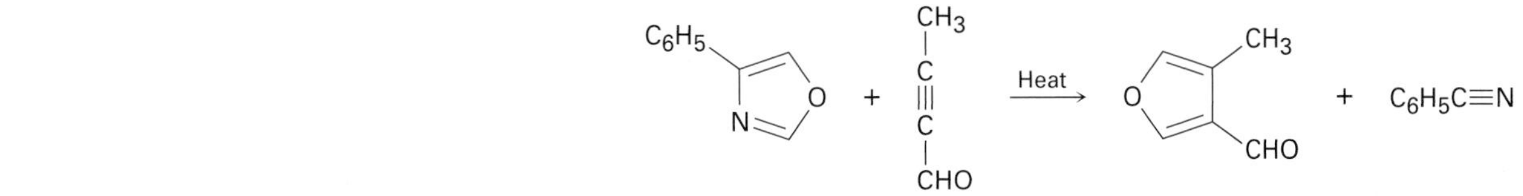

30.31 ■ The following synthesis of dienones occurs readily. Propose a mechanism to account for the results, and identify the kind of pericyclic reaction involved.

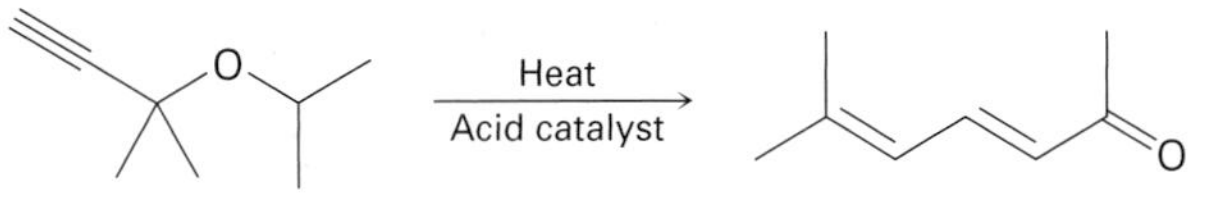

30.32 Karahanaenone, a terpenoid isolated from oil of hops, has been synthesized by the thermal reaction shown. Identify the kind of pericyclic reaction, and explain how karahanaenone is formed.

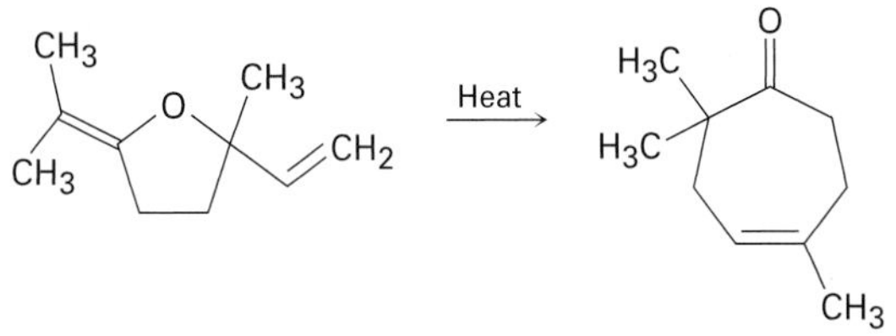

Karahanaenone

30.33 The ^{1}H NMR spectrum of bullvalene at 100 °C consists only of a single peak at 4.22 δ. Explain.

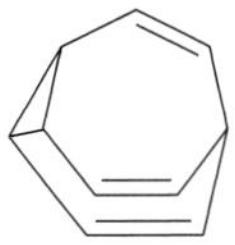

Bullvalene

30.34 The following rearrangement was devised and carried out to prove the stereo-chemistry of [1,5] sigmatropic hydrogen shifts. Explain how the observed result confirms the predictions of orbital symmetry.

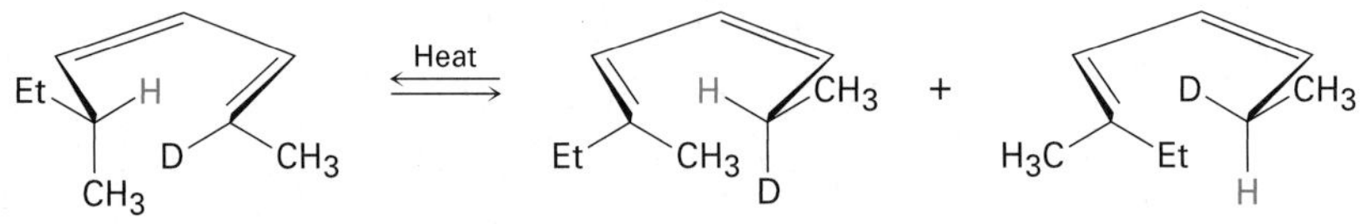

30.35 The following reaction is an example of a [2,3] sigmatropic rearrangement. Would you expect the reaction to be suprafacial or antarafacial? Explain.

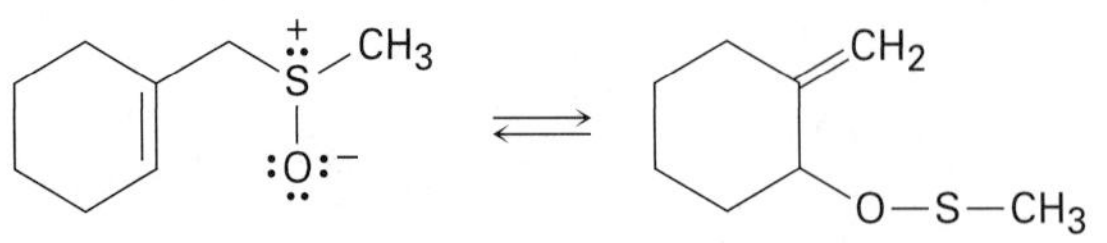

30.36 When the compound having a cyclobutene fused to a five-membered ring is heated, (1Z,3Z)-1,3-cycloheptadiene is formed. When the related compound having a cyclobutene fused to an eight-membered ring is heated, however, (1E,3Z)-1,3-cyclodecadiene is formed. Explain these results, and suggest a reason why opening of the eight-membered ring occurs at a lower temperature.

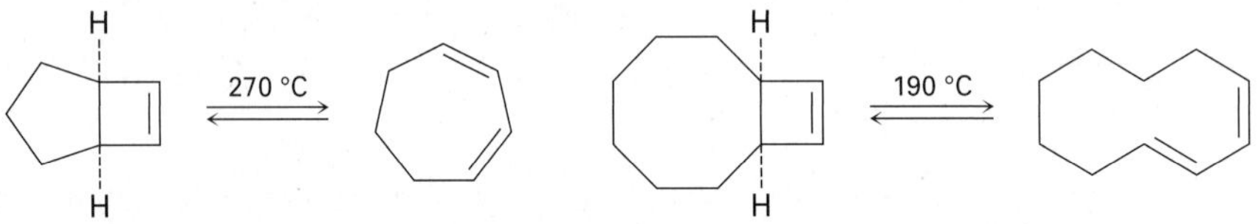

30.37 In light of your answer to Problem 30.36, explain why a mixture of products occurs in the following reaction:

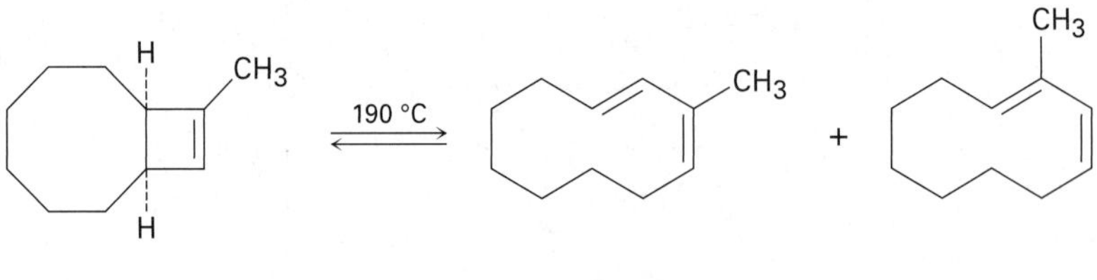

30.38 ■ The sex hormone estrone has been synthesized by a route that involves the following step. Identify the pericyclic reactions involved, and propose a mechanism.

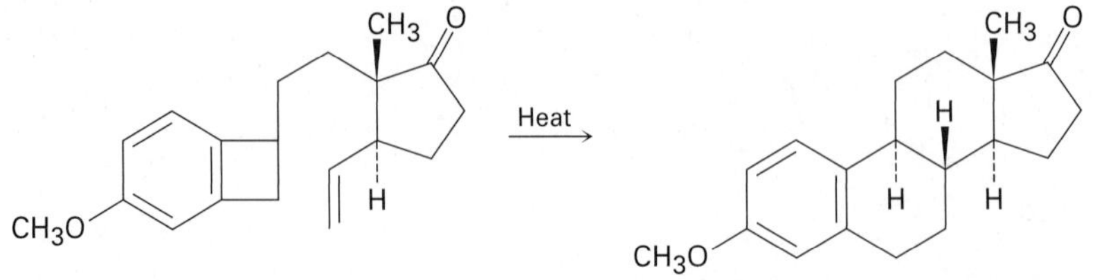

Estrone methyl ether

30.39 ■ Coronafacic acid, a bacterial toxin, was synthesized using a key step that involves three sequential pericyclic reactions. Identify them, and propose a mechanism for the overall transformation. How would you complete the synthesis?

Coronafacic acid

30.40 The following rearrangement of *N*-allyl-*N,N*-dimethylanilinium ion has been observed. Propose a mechanism.

N-Allyl-N,N-dimethylanilinium ion   **o-Allyl-N,N-dimethylanilinium ion**

31

Synthetic Polymers

Polymers are a fundamental part of the modern world, showing up in everything from coffee cups to cars to clothing. In medicine, too, their importance is growing for purposes as diverse as cardiac pacemakers, artificial heart valves, and biodegradable sutures.

We've seen on several occasions in previous chapters that a polymer, whether synthetic or biological, is a large molecule built up by repetitive bonding together of many smaller units, or monomers. Polyethylene, for instance, is a synthetic polymer made from ethylene (Section 7.10), nylon is a synthetic polyamide made from a diacid and a diamine (Section 21.9), and proteins are biological polyamides made from amino acids. Note that polymers are often drawn by indicating their repeating unit in parentheses. The repeat unit in polystyrene, for example, comes from the monomer styrene.

$$H_2C=CH \quad \longrightarrow \quad \left(CH_2-CH\right)_n$$

Styrene   **Polystyrene**

WHY THIS CHAPTER?

Our treatment of polymers has thus far been dispersed over several chapters, but it's now time to take a more comprehensive view. In the present chapter, we'll look further at how polymers are made, and we'll see how polymer structure correlates with physical properties. No course in organic chemistry would be complete without a look at polymers.

Sean Duggan

31.1 | Chain-Growth Polymers

Synthetic polymers are classified by their method of synthesis as either *chain-growth* or *step-growth*. The categories are somewhat imprecise but nevertheless provide a useful distinction. **Chain-growth polymers** are produced by chain-reaction polymerization in which an initiator adds to a carbon–carbon double bond of an unsaturated substrate (a *vinyl monomer*) to yield a reactive intermediate. This intermediate reacts with a second molecule of monomer to yield a new intermediate, which reacts with a third monomer unit, and so on.

The initiator can be a radical, an acid, or a base. Historically, as we saw in Section 7.10, radical polymerization was the most common method because it can be carried out with practically any vinyl monomer. Acid-catalyzed (cationic) polymerization, by contrast, is effective only with vinyl monomers that contain an electron-donating group (EDG) capable of stabilizing the chain-carrying carbocation intermediate. Thus, isobutylene (2-methylpropene) polymerizes rapidly under cationic conditions, but ethylene, vinyl chloride, and acrylonitrile do not. Isobutylene polymerization is carried out commercially at $-80\ °C$, using BF_3 and a small amount of water to generate $BF_3OH^-\ H^+$ catalyst. The product is used in the manufacture of truck and bicycle inner tubes.

where $BzO\cdot$ =   Benzoyloxy, $PhCO_2\cdot$

where   EDG = an electron-donating group

Vinyl monomers with electron-withdrawing substituents (EWG) can be polymerized by basic (anionic) catalysts. The chain-carrying step is conjugate nucleophilic addition of an anion to the unsaturated monomer (Section 19.13).

where   EWG = an electron-withdrawing group

Acrylonitrile ($H_2C=CHCN$), methyl methacrylate [$H_2C=C(CH_3)CO_2CH_3$], and styrene ($H_2C=CHC_6H_5$) can all be polymerized anionically. The polystyrene used in foam coffee cups, for example, is prepared by anionic polymerization of styrene using butyllithium as catalyst.

Styrene

Polystyrene

An interesting example of anionic polymerization accounts for the remarkable properties of "super glue," one drop of which can support up to 2000 lb. Super glue is simply a solution of pure methyl α-cyanoacrylate, which has two electron-withdrawing groups that make anionic addition particularly easy. Trace amounts of water or bases on the surface of an object are sufficient to initiate polymerization of the cyanoacrylate and bind articles together. Skin is a good source of the necessary basic initiators, and many people have found their fingers stuck together after inadvertently touching super glue. So good is super glue at binding tissues together that related cyanoacrylate esters such as Dermabond are used in hospitals in place of sutures to close wounds.

Methyl α-cyanoacrylate

Dermabond
(2-ethylhexyl α-cyanoacrylate)

Problem 31.1 Order the following monomers with respect to their expected reactivity toward cationic polymerization, and explain your answer:

$$H_2C=CHCH_3, \ H_2C=CHCl, \ H_2C=CH-C_6H_5, \ H_2C=CHCO_2CH_3$$

Problem 31.2 Order the following monomers with respect to their expected reactivity toward anionic polymerization, and explain your answer:

$$H_2C=CHCH_3, \ H_2C=CHC\equiv N, \ H_2C=CHC_6H_5$$

Problem 31.3 Polystyrene is produced commercially by reaction of styrene with butyllithium as an anionic initiator. Using resonance structures, explain how the chain-carrying intermediate is stabilized.

31.2 Stereochemistry of Polymerization: Ziegler–Natta Catalysts

Although we didn't point it out previously, the polymerization of a substituted vinyl monomer can lead to a polymer with numerous chirality centers in its chain. For example, propylene might polymerize with any of the three stereochemical outcomes shown in Figure 31.1. The polymer having all methyl groups on the same side of the zigzag backbone is called **isotactic**, the one in which the methyl groups alternate regularly on opposite sides of the backbone is called **syndiotactic**, and the one having the methyl groups randomly oriented is called **atactic**.

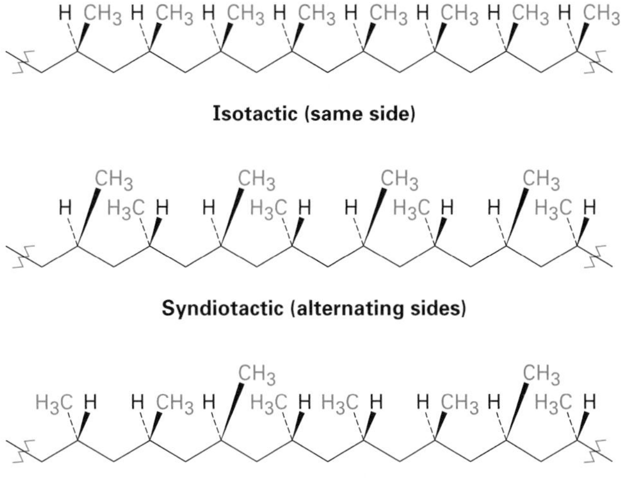

Figure 31.1 Isotactic, syndiotactic, and atactic forms of polypropylene.

Isotactic (same side)

Syndiotactic (alternating sides)

Atactic (random)

The three different stereochemical forms of polypropylene all have somewhat different properties, and all can be made by using the right polymerization catalyst. Propylene polymerization using radical initiators does not work well, but polymerization using *Ziegler–Natta catalysts* allows preparation of isotactic, syndiotactic, and atactic polypropylene.

Ziegler–Natta catalysts—there are many different formulations—are organometallic transition-metal complexes prepared by treatment of an alkylaluminum with a titanium compound. Triethylaluminum and titanium tetrachloride form a typical preparation.

$$(CH_3CH_2)_3Al + TiCl_4 \longrightarrow \text{A Ziegler–Natta catalyst}$$

Following their introduction in 1953, Ziegler–Natta catalysts revolutionized the field of polymer chemistry because of two advantages: the resultant polymers are linear, with practically no chain branching, and they are stereochemically controllable. Isotactic, syndiotactic, and atactic forms can all be produced, depending on the catalyst system used.

The active form of a Ziegler–Natta catalyst is an alkyltitanium intermediate with a vacant coordination site on the metal. Coordination of alkene monomer to the titanium occurs, and the coordinated alkene then inserts into the carbon–titanium bond to extend the alkyl chain. A new coordination site opens up during the insertion step, so the process repeats indefinitely.

The linear polyethylene produced by the Ziegler–Natta process, called *high-density polyethylene,* is a highly crystalline polymer with 4000 to 7000 ethylene units per chain and molecular weights in the range 100,000 to 200,000 amu. High-density polyethylene has greater strength and heat resistance than the branched product of radical-induced polymerization, called *low-density polyethylene,* and is used to produce plastic squeeze bottles and molded housewares.

Polyethylenes of even higher molecular weights are produced for specialty applications. So-called high-molecular-weight (HMW) polyethylene contains 10,000 to 18,000 monomer units per chain (MW = 300,000–500,000 amu) and is used for pipes and large containers. Ultrahigh-molecular-weight (UHMW) polyethylene contains more than 100,000 monomer units per chain and has molecular weights ranging from 3,000,000 to 6,000,000 amu. It is used in bearings, conveyor belts, and bulletproof vests among other applications requiring unusual wear resistance.

Problem 31.4 | Vinylidene chloride, $H_2C=CCl_2$, does not polymerize in isotactic, syndiotactic, and atactic forms. Explain.

Problem 31.5 | Polymers such as polypropylene contain a large number of chirality centers. Would you therefore expect samples of isotactic, syndiotactic, or atactic polypropylene to rotate plane-polarized light? Explain.

31.3 | Copolymers

Up to this point we've discussed only **homopolymers**—polymers that are made up of identical repeating units. In practice, however, *copolymers* are more important commercially. **Copolymers** are obtained when two or more different monomers are allowed to polymerize together. For example, copolymerization of vinyl chloride with vinylidene chloride (1,1-dichloroethylene) in a 1:4 ratio leads to the polymer Saran.

Copolymerization of monomer mixtures often leads to materials with properties quite different from those of either corresponding homopolymer, giving the polymer chemist a vast amount of flexibility for devising new materials. Table 31.1 lists some common copolymers and their commercial applications.

Table 31.1 | Some Common Copolymers and Their Uses

Monomers	Structures	Trade name	Uses
Vinyl chloride Vinylidene chloride		Saran	Fibers, food packaging
Styrene 1,3-Butadiene		SBR (styrene–butadiene rubber)	Tires, rubber articles
Hexafluoropropene Vinylidene fluoride		Viton	Gaskets, seals
Acrylonitrile 1,3-Butadiene		Nitrile rubber	Adhesives, hoses
Isobutylene Isoprene		Butyl rubber	Inner tubes
Acrylonitrile 1,3-Butadiene Styrene		ABS (monomer initials)	Pipes, high-impact applications

Several different types of copolymers can be defined, depending on the distribution of monomer units in the chain. If monomer A is copolymerized with monomer B, for instance, the resultant product might have a random distribution of the two units throughout the chain, or it might have an alternating distribution.

$$+A—A—A—B—A—B—B—A—B—A—A—A—B—B—B+$$

Random copolymer

$$+A—B—A—B—A—B—A—B—A—B—A—B—A—B—A+$$

Alternating copolymer

The exact distribution of monomer units depends on the initial proportions of the two reactant monomers and their relative reactivities. In practice, neither perfectly random nor perfectly alternating copolymers are usually found. Most copolymers have many random imperfections.

Two other forms of copolymers that can be prepared under certain conditions are called *block copolymers* and *graft copolymers*. **Block copolymers** are those in which different blocks of identical monomer units alternate with each other; **graft copolymers** are those in which homopolymer branches of one monomer unit are "grafted" onto a homopolymer chain of another monomer unit.

Block copolymers are prepared by initiating the polymerization of one monomer as if growing a homopolymer chain and then adding an excess of the second monomer to the still-active reaction mix. Graft copolymers are made by gamma irradiation of a completed homopolymer chain in the presence of the second monomer. The high-energy irradiation knocks hydrogen atoms off the homopolymer chain at random points, thus generating radical sites that can initiate polymerization of the added monomer.

Problem 31.6 | Draw the structure of an alternating segment of butyl rubber, a copolymer of isoprene (2-methyl-1,3-butadiene) and isobutylene (2-methylpropene) prepared using a cationic initiator.

Problem 31.7 | Irradiation of poly(-1,3-butadiene), followed by addition of styrene, yields a graft copolymer that is used to make rubber soles for shoes. Draw the structure of a representative segment of this styrene–butadiene graft copolymer.

31.4 | Step-Growth Polymers

Step-growth polymers are produced by reactions in which each bond in the polymer is formed stepwise, independently of the others. Like the polyamides (nylons) and polyesters that we saw in Section 21.9, most step-growth polymers

CENGAGENOW™ Click *Organic Interactive* to **predict products from simple polymerization reactions**.

are produced by reaction between two difunctional reactants. Nylon 66, for instance, is made by reaction between the six-carbon adipic acid and the six-carbon hexamethylenediamine (1,6-hexanediamine). Alternatively, a single reactant with two different functional groups can polymerize. Nylon 6 is made by polymerization of the six-carbon caprolactam. The reaction is initiated by addition of a small amount of water, which hydrolyzes some caprolactam to 6-aminohexanoic acid. Nucleophilic addition of the amino group to caprolactam then propagates the polymerization.

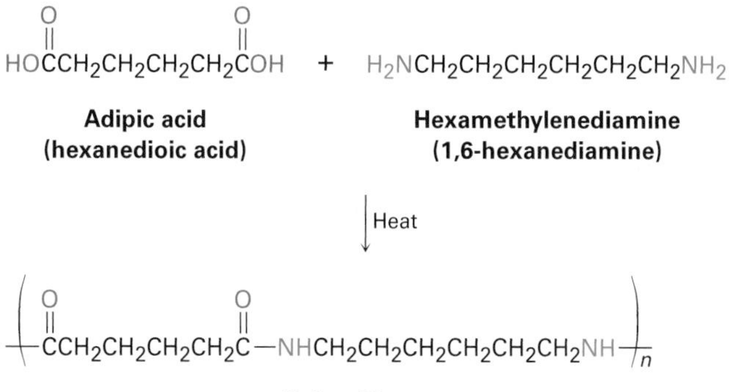

Adipic acid
(hexanedioic acid)   **Hexamethylenediamine**
                              (1,6-hexanediamine)

Heat

Nylon 66

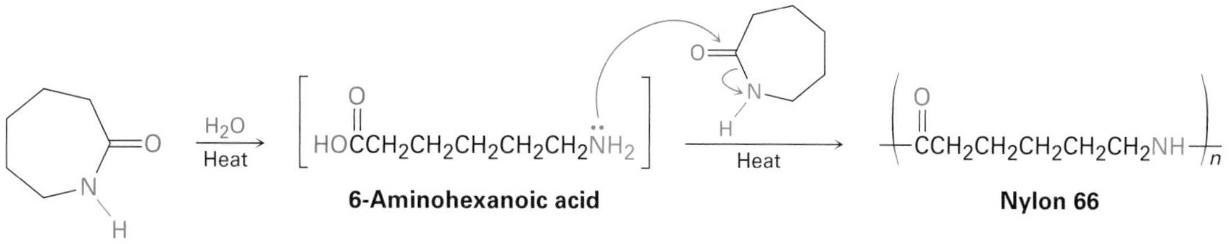

Caprolactam   **6-Aminohexanoic acid**   **Nylon 66**

Polycarbonates

Polycarbonates are like polyesters, but their carbonyl group is linked to two —OR groups, $[O=C(OR)_2]$. Lexan, for instance, is a polycarbonate prepared from diphenyl carbonate and a diphenol called bisphenol A. Lexan has an unusually high impact strength, making it valuable for use in machinery housings, telephones, bicycle safety helmets, and bulletproof glass.

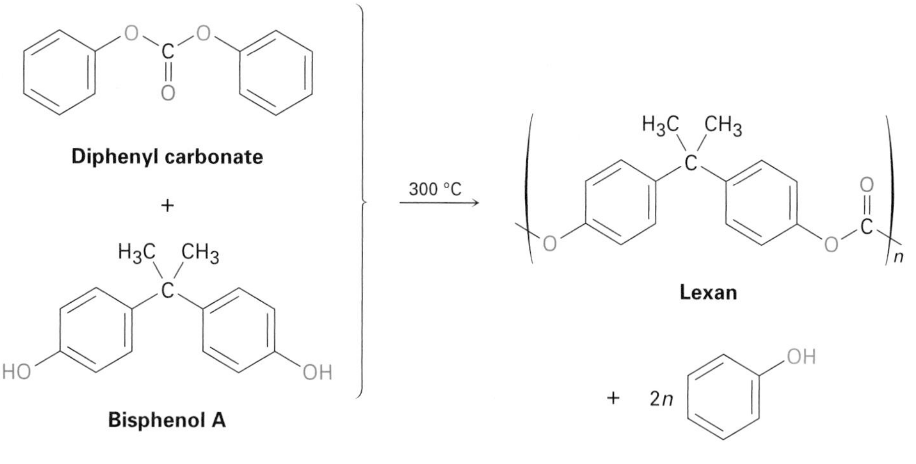

Diphenyl carbonate

+

Bisphenol A

300 °C

Lexan

+ 2n

Polyurethanes

A *urethane* is a carbonyl-containing functional group in which the carbonyl carbon is bonded to both an −OR group and an −NR$_2$ group. As such, a urethane is halfway between a carbonate and a urea.

A carbonate   **A urethane**   **A urea**

A urethane is typically prepared by nucleophilic addition reaction between an alcohol and an isocyanate (R−N=C=O), so a **polyurethane** is prepared by reaction between a diol and a diisocyanate. The diol is usually a low-molecular-weight polymer (MW ≈ 1000 amu) with hydroxyl end-groups; the diisocyanate is often toluene-2,4-diisocyanate.

Toluene-2,4-diisocyanate   **A polyurethane**

Several different kinds of polyurethanes are produced, depending on the nature of the polymeric alcohol used. One major use of polyurethane is in the stretchable spandex fibers used for bathing suits and athletic gear. These polyurethanes have a fairly low degree of cross-linking, so the resultant polymer is soft and elastic. A second major use of polyurethanes is in the foams used for insulation. Foaming occurs when a small amount of water is added during polymerization, giving a carbamic acid intermediate that spontaneously loses bubbles of CO_2.

A carbamic acid

Polyurethane foams are generally made using a *poly*alcohol rather than a diol as the monomer, so the polymer has a high amount of three-dimensional cross-linking. The result is a rigid but very light foam suitable for use as thermal insulation in building construction and portable ice chests.

Problem 31.8   Poly(ethylene terephthalate), or PET, is a polyester used to make soft-drink bottles. It is prepared by reaction of ethylene glycol with 1,4-benzenedicarboxylic acid (terephthalic acid). Draw the structure of PET.

Problem 31.9   Show the mechanism of the nucleophilic addition reaction of an alcohol with an isocyanate to yield a urethane.

31.5 | Polymer Structure and Physical Properties

Polymers aren't really that different from other organic molecules. They're much larger, of course, but their chemistry is similar to that of analogous small molecules. Thus, the alkane chains of polyethylene undergo radical-initiated halogenation, the aromatic rings of polystyrene undergo typical electrophilic aromatic substitution reactions, and the amide linkages of a nylon are hydrolyzed by aqueous base.

The major difference between small and large organic molecules is in their physical properties. For instance, their large size means that polymers experience substantially larger van der Waals forces than do small molecules (Section 2.13). But because van der Waals forces operate only at close distances, they are strongest in polymers like high-density polyethylene, in which chains can pack together closely in a regular way. Many polymers, in fact, have regions that are essentially crystalline. These regions, called **crystallites**, consist of highly ordered portions in which the zigzag polymer chains are held together by van der Waals forces (Figure 31.2).

Figure 31.2 Crystallites in linear polyethylene. The long polymer chains are arranged in parallel lines in the crystallite regions.

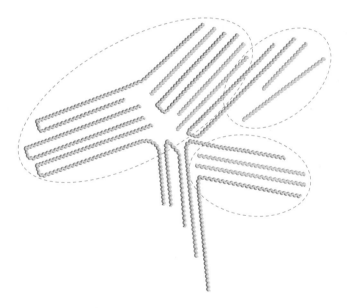

As you might expect, polymer crystallinity is strongly affected by the steric requirements of substituent groups on the chains. Linear polyethylene is highly crystalline, but poly(methyl methacrylate) is noncrystalline because the chains can't pack closely together in a regular way. Polymers with a high degree of crystallinity are generally hard and durable. When heated, the crystalline regions melt at the **melt transition temperature**, T_m, to give an amorphous material.

Noncrystalline, amorphous polymers like poly(methyl methacrylate), sold under the trade name Plexiglas, have little or no long-range ordering among chains but can nevertheless be very hard at room temperature. When heated, the hard amorphous polymer becomes soft and flexible at a point called the **glass transition temperature**, T_g. Much of the art in polymer synthesis lies in finding methods for controlling the degree of crystallinity and the glass transition temperature, thereby imparting useful properties to the polymer.

In general, polymers can be divided into four major categories, depending on their physical behavior: *thermoplastics, fibers, elastomers,* and *thermosetting resins.* **Thermoplastics** are the polymers most people think of when the word *plastic* is mentioned. These polymers have a high T_g and are therefore hard at room temperature but become soft and viscous when heated. As a result, they can be molded into toys, beads, telephone housings, or any of a thousand other items. Because thermoplastics have little or no cross-linking, the individual chains can slip past one another in the melt. Some thermoplastic polymers, such as poly(methyl methacrylate) and polystyrene, are amorphous and non-crystalline; others, such as polyethylene and nylon, are partially crystalline. Among the better-known thermoplastics is poly(ethylene terephthalate), or PET, used for making plastic soft-drink bottles.

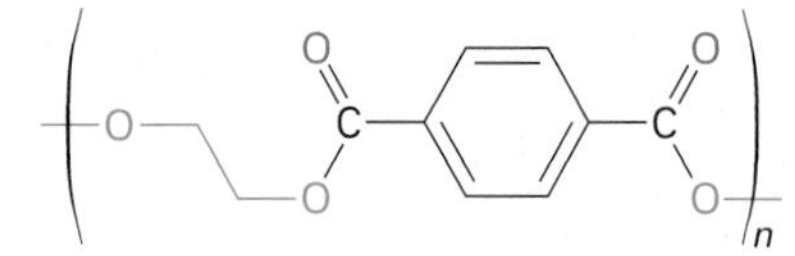

Poly(ethylene terephthalate)

Plasticizers—small organic molecules that act as lubricants between chains—are usually added to thermoplastics to keep them from becoming brittle at room temperature. An example is poly(vinyl chloride), which is brittle when pure but becomes supple and pliable when a plasticizer is added. In fact, most drip bags used in hospitals to deliver intravenous saline solutions are made of poly(vinyl chloride), although replacements are appearing. Dialkyl phthalates such as di(2-ethylhexyl) phthalate (generally called dioctyl phthalate) are commonly used as plasticizers, although questions about their safety have been raised. The U.S. Food and Drug Administration (FDA) has advised the use of alternative materials in compromised patients and infants but has found no evidence of toxicity for most patients.

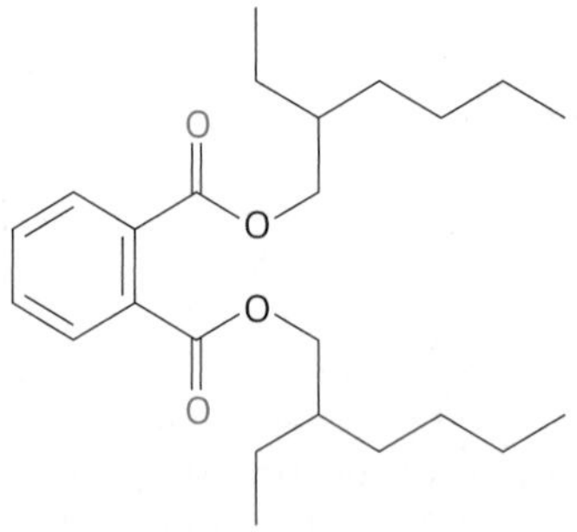

Di(2-ethylhexyl) phthalate (or dioctyl phthalate), a plasticizer

Fibers are thin threads produced by extruding a molten polymer through small holes in a die, or spinneret. The fibers are then cooled and drawn out, which orients the crystallite regions along the axis of the fiber and adds considerable tensile strength (Figure 31.3). Nylon, Dacron, and polyethylene all have the semicrystalline structure necessary for drawing into oriented fibers.

Elastomers are amorphous polymers that have the ability to stretch out and spring back to their original shapes. These polymers must have low T_g values and a small amount of cross-linking to prevent the chains from slipping over one another. In addition, the chains must have an irregular shape to prevent crystallite

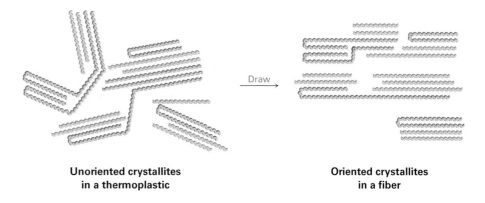

**Unoriented crystallites
in a thermoplastic**

Draw

**Oriented crystallites
in a fiber**

formation. When stretched, the randomly coiled chains straighten out and orient along the direction of the pull. Van der Waals forces are too weak and too few to maintain this orientation, however, and the elastomer therefore reverts to its random coiled state when the stretching force is released (Figure 31.4).

Figure 31.4 Unstretched and stretched forms of an elastomer.

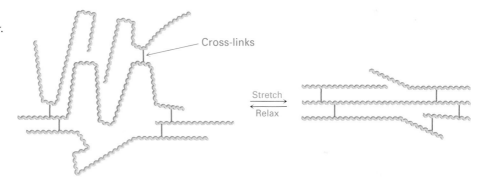

Cross-links

Stretch
Relax

Natural rubber (Chapter 7 *Focus On*) is the most common example of an elastomer. Rubber has the long chains and occasional cross-links needed for elasticity, but its irregular geometry prevents close packing of the chains into crystallites. Gutta-percha, by contrast, is highly crystalline and is not an elastomer (Figure 31.5).

Figure 31.5 **(a)** Natural rubber is elastic and noncrystalline because of its cis double-bond geometry, but **(b)** gutta-percha is nonelastic and crystalline because its geometry allows for better packing together of chains.

(a)

(b)

Thermosetting resins are polymers that become highly cross-linked and solidify into a hard, insoluble mass when heated. *Bakelite,* a thermosetting resin first produced in 1907, has been in commercial use longer than any other synthetic polymer. It is widely used for molded parts, adhesives, coatings, and even high-temperature applications such as missile nose cones.

Chemically, Bakelite is a *phenolic resin,* produced by reaction of phenol and formaldehyde. On heating, water is eliminated, many cross-links form, and the polymer sets into a rocklike mass. The cross-linking in Bakelite and other thermosetting resins is three-dimensional and is so extensive that we can't really speak of polymer "chains." A piece of Bakelite is essentially one large molecule.

Bakelite

Problem 31.10 What product would you expect to obtain from catalytic hydrogenation of natural rubber? Would the product be syndiotactic, atactic, or isotactic?

Problem 31.11 Propose a mechanism to account for the formation of Bakelite from acid-catalyzed polymerization of phenol and formaldehyde.

Focus On . . .

Biodegradable Polymers

The high chemical stability of many polymers is both a blessing and a curse. Heat resistance, wear resistance, and long life are valuable characteristics of clothing fibers, bicycle helmets, underground pipes, food wrappers, and many other items. Yet when those items outlive their usefulness, disposal becomes a problem.

Recycling of unwanted polymers is the best solution, and six types of plastics in common use are frequently stamped with identifying codes assigned by the Society of the Plastics Industry (Table 31.2). After being sorted by type, the

(continued)

What happens to the plastics that end up here?

items to be recycled are shredded into small chips, washed, dried, and melted for reuse. Soft-drink bottles, for instance, are made from recycled poly(ethylene terephthalate), trash bags are made from recycled low-density polyethylene, and garden furniture is made from recycled polypropylene and mixed plastics.

Table 31.2 | Recyclable Plastics

Polymer	Recycling code	Use
Poly(ethylene terephthalate)	1—PET	Soft-drink bottles
High-density polyethylene	2—HDPE	Bottles
Poly(vinyl chloride)	3—V	Floor mats
Low-density polyethylene	4—DPE	Grocery bags
Polypropylene	5—PP	Furniture
Polystyrene	6—PS	Molded articles
Mixed plastics	7	Benches, plastic lumber

Frequently, however, plastics are simply thrown away rather than recycled, and much work has therefore been carried out on developing *biodegradable* polymers, which can be broken down rapidly by soil microorganisms. Among the most common biodegradable polymers are polyglycolic acid (PGA), polylactic acid (PLA), and polyhydroxybutyrate (PHB). All are polyesters and are therefore susceptible to hydrolysis of their ester links. Copolymers of PGA with PLA have found a particularly wide range of uses. A 90/10 copolymer of polyglycolic acid with polylactic acid is used to make absorbable sutures, for instance. The sutures are entirely degraded and absorbed by the body within 90 days after surgery.

$$\underset{\textbf{Glycolic acid}}{\text{HOCH}_2\overset{\overset{\displaystyle O}{\|}}{\text{C}}\text{OH}} \qquad \underset{\textbf{Lactic acid}}{\text{HOCHC}\overset{\overset{\displaystyle O}{\|}}{\text{C}}\text{OH}} \qquad \underset{\textbf{3-Hydroxybutyric acid}}{\text{HOCHCH}_2\overset{\overset{\displaystyle O}{\|}}{\text{C}}\text{OH}}$$

Glycolic acid   Lactic acid (CH_3)   3-Hydroxybutyric acid (CH_3)

↓ Heat   ↓ Heat   ↓ Heat

$$\left(\!\text{OCH}_2\overset{\overset{\displaystyle O}{\|}}{\text{C}}\!\right)_n \qquad \left(\!\text{OCHC}\overset{\overset{\displaystyle O}{\|}}{}\!\right)_n \qquad \left(\!\text{OCHCH}_2\overset{\overset{\displaystyle O}{\|}}{\text{C}}\!\right)_n$$

Poly(glycolic acid)   **Poly(lactic acid)**   **Poly(hydroxybutyrate)**

SUMMARY AND KEY WORDS

Synthetic polymers can be classified as either chain-growth polymers or step-growth polymers. **Chain-growth polymers** are prepared by chain-reaction polymerization of *vinyl monomers* in the presence of a radical, an anion, or a cation initiator. Radical polymerization is sometimes used, but alkenes such as 2-methylpropene that have electron-donating substituents on the double bond polymerize easily by a cationic route through carbocation intermediates. Similarly, monomers such as methyl α-cyanoacrylate that have electron-withdrawing substituents on the double bond polymerize by an anionic, conjugate addition pathway.

Copolymerization of two monomers gives a product with properties different from those of either homopolymer. **Graft copolymers** and **block copolymers** are two examples.

Alkene polymerization can be carried out in a controlled manner using a **Ziegler–Natta catalyst**. Ziegler–Natta polymerization minimizes the amount of chain branching in the polymer and leads to stereoregular chains—either **isotactic** (substituents on the same side of the chain) or **syndiotactic** (substituents on alternate sides of the chain), rather than **atactic** (substituents randomly disposed).

Step-growth polymers, the second major class of polymers, are prepared by reactions between difunctional molecules, with the individual bonds in the polymer formed independently of one another. **Polycarbonates** are formed from a diester and a diol, and **polyurethanes** are formed from a diisocyanate and a diol.

The chemistry of synthetic polymers is similar to the chemistry of small molecules with the same functional groups, but the physical properties of polymers are greatly affected by size. Polymers can be classified by physical property into four groups: **thermoplastics**, **fibers**, **elastomers**, and **thermosetting resins**. The properties of each group can be accounted for by the structure, the degree of crystallinity, and the amount of cross-linking they contain.

EXERCISES

Organic KNOWLEDGE TOOLS

CENGAGENOW™ Sign in at **www.cengage.com/login** to assess your knowledge of this chapter's topics by taking a pre-test. The pre-test will link you to interactive organic chemistry resources based on your score in each concept area.

OWL Online homework for this chapter may be assigned in Organic OWL.

■ indicates problems assignable in Organic OWL.

VISUALIZING CHEMISTRY

(Problems 31.1–31.11 appear within the chapter.)

31.12 Identify the structural class to which the following polymer belongs, and show the structure of the monomer units used to make it:

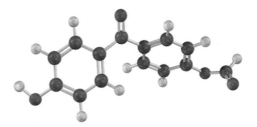

31.13 ■ Show the structures of the polymers that could be made from the following monomers (yellow-green = Cl):

(a)

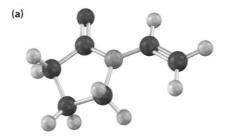

(b)

ADDITIONAL PROBLEMS

31.14 ■ Identify the monomer units from which each of the following polymers is made, and tell whether each is a chain-growth or a step-growth polymer.

(a) $+CH_2-O+_n$ **(b)** $+CF_2-CFCl+_n$ **(c)** $\left(-NHCH_2CH_2CH_2\overset{\overset{O}{\|}}{C}-\right)_n$

(d) $\left(-O-\underset{}{\bigcirc}-\overset{\overset{O}{\|}}{C}-\right)_n$ **(e)** $\left(-O-\underset{}{\bigcirc}-O-\overset{\overset{O}{\|}}{C}-\right)_n$

31.15 Draw a three-dimensional representation of segments of the following polymers:
(a) Syndiotactic polyacrylonitrile (b) Atactic poly(methyl methacrylate)
(c) Isotactic poly(vinyl chloride)

31.16 ■ Draw the structure of Kodel, a polyester prepared by heating dimethyl 1,4-benzenedicarboxylate with 1,4-bis(hydroxymethyl)cyclohexane.

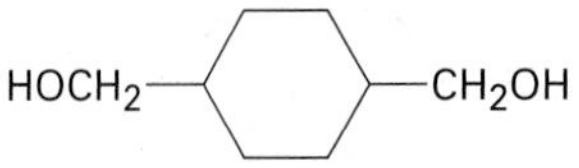

HOCH₂—⬡—CH₂OH **1,4-Bis(hydroxymethyl)cyclohexane**

31.17 ■ Show the structure of the polymer that results from heating the following diepoxide and diamine:

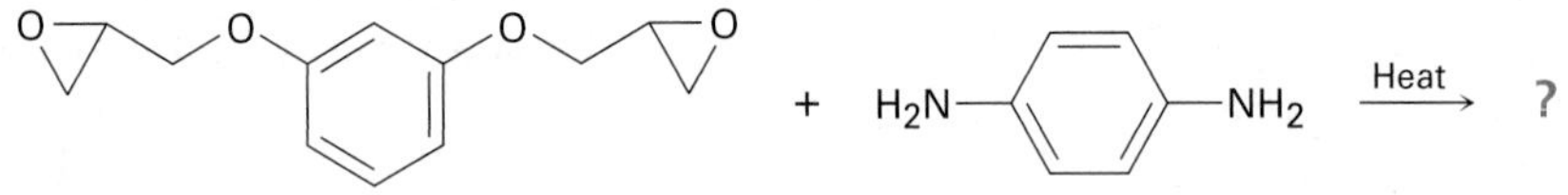

31.18 ■ Nomex, a polyamide used in such applications as fire-retardant clothing, is prepared by reaction of 1,3-benzenediamine with 1,3-benzenedicarbonyl chloride. Show the structure of Nomex.

31.19 ■ Nylon 10,10 is an extremely tough, strong polymer used to make reinforcing rods for concrete. Draw a segment of nylon 10,10, and show its monomer units.

31.20 1,3-Cyclopentadiene undergoes thermal polymerization to yield a polymer that has no double bonds in the chain. On strong heating, the polymer breaks down to regenerate cyclopentadiene. Propose a structure for the polymer.

31.21 When styrene, $C_6H_5CH=CH_2$, is copolymerized in the presence of a few percent p-divinylbenzene, a hard, insoluble, cross-linked polymer is obtained. Show how this cross-linking of polystyrene chains occurs.

31.22 ■ Poly(ethylene glycol), or Carbowax, is made by anionic polymerization of ethylene oxide using NaOH as catalyst. Propose a mechanism.

$+O-CH_2CH_2\frac{}{}_n$ **Poly(ethylene glycol)**

31.23 ■ Nitroethylene, $H_2C=CHNO_2$, is a sensitive compound that must be prepared with great care. Attempted purification of nitroethylene by distillation often results in low recovery of product and a white coating on the inner walls of the distillation apparatus. Explain.

31.24 Poly(vinyl butyral) is used as the plastic laminate in the preparation of automobile windshield safety glass. How would you synthesize this polymer?

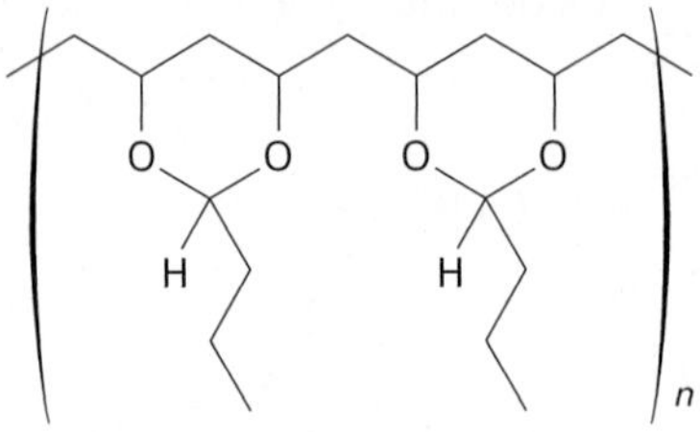

Poly(vinyl butyral)

31.25 ■ What is the structure of the polymer produced by anionic polymerization of β-propiolactone using NaOH as catalyst?

β-Propiolactone

31.26 Glyptal is a highly cross-linked thermosetting resin produced by heating glycerol and phthalic anhydride (1,2-benzenedicarboxylic acid anhydride). Show the structure of a representative segment of glyptal.

31.27 Melmac, a thermosetting resin often used to make plastic dishes, is prepared by heating melamine with formaldehyde. Look at the structure of Bakelite shown in Section 31.5, and then propose a structure for Melmac.

Melamine

31.28 Epoxy adhesives are cross-linked resins prepared in two steps. The first step involves S_N2 reaction of the disodium salt of bisphenol A with epichlorohydrin to form a low-molecular-weight prepolymer. This prepolymer is then "cured" into a cross-linked resin by treatment with a triamine such as $H_2NCH_2CH_2NHCH_2CH_2NH_2$.

Bisphenol A   **Epichlorohydrin**

(a) What is the structure of the prepolymer?
(b) How does addition of the triamine to the prepolymer result in cross-linking?

31.29 The polyurethane foam used for home insulation uses methanediphenyl-diisocyanate (MDI) as monomer. The MDI is prepared by acid-catalyzed reaction of aniline with formaldehyde, followed by treatment with phosgene, $COCl_2$. Propose mechanisms for both steps.

MDI

■ Assignable in OWL

31.30 Write the structure of a representative segment of polyurethane prepared by reaction of ethylene glycol with MDI (Problem 31.29).

31.31 The smoking salons of the Hindenburg and other hydrogen-filled dirigibles of the 1930s were insulated with urea–formaldehyde polymer foams. The structure of this polymer is highly cross-linked, like that of Bakelite (Section 31.5). Propose a structure.

31.32 The polymeric resin used for Merrifield solid-phase peptide synthesis (Section 26.8) is prepared by treating polystyrene with *N*-(hydroxymethyl) phthalimide and trifluoromethanesulfonic acid, followed by reaction with hydrazine. Propose a mechanism for both steps.

31.33 2-Ethyl-1-hexanol, used in the synthesis of di(2-ethylhexyl) phthalate plasticizer, is made commercially from butanal. Show the likely synthesis route.

Nomenclature of Polyfunctional Organic Compounds

With more than 30 million organic compounds now known and thousands more being created daily, naming them all is a real problem. Part of the problem is due to the sheer complexity of organic structures, but part is also due to the fact that chemical names have more than one purpose. For Chemical Abstracts Service (CAS), which catalogs and indexes the worldwide chemical literature, each compound must have only one correct name. It would be chaos if half the entries for CH_3Br were indexed under "M" for methyl bromide and half under "B" for bromomethane. Furthermore, a CAS name must be strictly systematic so that it can be assigned and interpreted by computers; common names are not allowed.

People, however, have different requirements than computers. For people—which is to say chemists in their spoken and written communications—it's best that a chemical name be pronounceable and that it be as easy as possible to assign and interpret. Furthermore, it's convenient if names follow historical precedents, even if that means a particularly well-known compound might have more than one name. People can readily understand that bromomethane and methyl bromide both refer to CH_3Br.

As noted in the text, chemists overwhelmingly use the nomenclature system devised and maintained by the International Union of Pure and Applied Chemistry, or IUPAC. Rules for naming monofunctional compounds were given throughout the text as each new functional group was introduced, and a list of where these rules can be found is given in Table A.1.

Table A.1   Nomenclature Rules for Functional Groups

Functional group	Text section	Functional group	Text section
Acid anhydrides	21.1	Aromatic compounds	15.1
Acid halides	21.1	Carboxylic acids	20.1
Acyl phosphates	21.1	Cycloalkanes	4.1
Alcohols	17.1	Esters	21.1
Aldehydes	19.1	Ethers	18.1
Alkanes	3.4	Ketones	19.1
Alkenes	6.3	Nitriles	20.1
Alkyl halides	10.1	Phenols	17.1
Alkynes	8.1	Sulfides	18.8
Amides	21.1	Thioesters	21.1
Amines	24.1	Thiols	18.8

Naming a monofunctional compound is reasonably straightforward, but even experienced chemists often encounter problems when faced with naming a complex polyfunctional compound. Take the following compound, for instance. It has three functional groups, ester, ketone, and C=C, but how should it be named? As an ester with an *-oate* ending, a ketone with an *-one* ending, or an alkene with an *-ene* ending? It's actually named methyl 3-(2-oxocyclohex-6-enyl)propanoate.

Methyl 3-(2-oxo**cylohex-6-enyl**)propanoate

The name of a polyfunctional organic molecule has four parts—suffix, parent, prefixes, and locants—which must be identified and expressed in the proper order and format. Let's look at each of the four.

Name Part 1. The Suffix: Functional-Group Precedence

Although a polyfunctional organic molecule might contain several different functional groups, we must choose just one suffix for nomenclature purposes. It's not correct to use two suffixes. Thus, keto ester **1** must be named either as a ketone with an *-one* suffix or as an ester with an *-oate* suffix but can't be named as an *-onoate*. Similarly, amino alcohol **2** must be named either as an alcohol (*-ol*) or as an amine (*-amine*) but can't be named as an *-olamine* or *-aminol*.

1. $CH_3CCH_2CH_2COCH_3$ (with two C=O)

2. $CH_3CHCH_2CH_2CH_2NH_2$ (with OH)

The only exception to the rule requiring a single suffix is when naming compounds that have double or triple bonds. Thus, the unsaturated acid $H_2C=CHCH_2CO_2H$ is but-3-enoic acid, and the acetylenic alcohol $HC\equiv CCH_2CH_2CH_2OH$ is pent-5-yn-1-ol.

How do we choose which suffix to use? Functional groups are divided into two classes, **principal groups** and **subordinate groups**, as shown in Table A.2. Principal groups can be cited either as prefixes or as suffixes, while subordinate groups are cited only as prefixes. Within the principal groups, an order of priority has been established, with the proper suffix for a given compound determined by choosing the principal group of highest priority. For example, Table A.2 indicates that keto ester **1** should be named as an ester rather than as a ketone because an ester functional group is higher in priority than a ketone. Similarly, amino alcohol **2** should be named as an alcohol rather than as an amine.

Table A.2 | Classification of Functional Groups[a]

Functional group	Name as suffix	Name as prefix
Principal groups		
Carboxylic acids	-oic acid -carboxylic acid	carboxy
Acid anhydrides	-oic anhydride -carboxylic anhydride	—
Esters	-oate -carboxylate	alkoxycarbonyl
Thioesters	-thioate -carbothioate	alkylthiocarbonyl
Acid halides	-oyl halide -carbonyl halide	halocarbonyl
Amides	-amide -carboxamide	carbamoyl
Nitriles	-nitrile -carbonitrile	cyano
Aldehydes	-al -carbaldehyde	oxo
Ketones	-one	oxo
Alcohols	-ol	hydroxy
Phenols	-ol	hydroxy
Thiols	-thiol	mercapto
Amines	-amine	amino
Imines	-imine	imino
Ethers	ether	alkoxy
Sulfides	sulfide	alkylthio
Disulfides	disulfide	—
Alkenes	-ene	—
Alkynes	-yne	—
Alkanes	-ane	—
Subordinate groups		
Azides	—	azido
Halides	—	halo
Nitro compounds	—	nitro

[a]Principal groups are listed in order of decreasing priority; subordinate groups have no priority order.

Thus, the name of **1** is methyl 4-oxopentanoate, and the name of **2** is 5-aminopentan-2-ol. Further examples are shown:

$$\underset{\substack{O\\||}}{CH_3C}CH_2CH_2\underset{\substack{O\\||}}{C}OCH_3$$

1. Methyl 4-oxopentanoate
(an ester with a ketone group)

$$CH_3\underset{\substack{|\\OH}}{C}HCH_2CH_2CH_2NH_2$$

2. 5-Aminopentan-2-ol
(an alcohol with an amine group)

$$CH_3\underset{\substack{|\\CHO}}{C}HCH_2CH_2CH_2\underset{\substack{O\\||}}{C}OCH_3$$

3. Methyl 5-methyl-6-oxohexanoate
(an ester with an aldehyde group)

$$H_2N\underset{\substack{O\\||}}{C}CH_2\underset{\substack{OH\\|}}{C}HCH_2CH_2\underset{\substack{O\\||}}{C}OH$$

4. 5-Carbamoyl-4-hydroxypentanoic acid
(a carboxylic acid with amide and alcohol groups)

5. 3-Oxocyclohexanecarbaldehyde
(an aldehyde with a ketone group)

Name Part 2. The Parent: Selecting the Main Chain or Ring

The parent, or base, name of a polyfunctional organic compound is usually easy to identify. If the principal group of highest priority is part of an open chain, the parent name is that of the longest chain containing the largest number of principal groups. For example, compounds **6** and **7** are isomeric aldehydo amides, which must be named as amides rather than as aldehydes according to Table A.2. The longest chain in compound **6** has six carbons, and the substance is therefore named 5-methyl-6-oxohexanamide. Compound **7** also has a chain of six carbons, but the longest chain that contains both principal functional groups has only four carbons. The correct name of **7** is 4-oxo-3-propylbutanamide.

$$\underset{\substack{O\\||}}{HC}\underset{\substack{|\\CH_3}}{C}HCH_2CH_2CH_2\underset{\substack{O\\||}}{C}NH_2$$

6. 5-Methyl-6-oxohexanamide

$$CH_3CH_2CH_2\underset{\substack{CHO\\|}}{C}HCH_2\underset{\substack{O\\||}}{C}NH_2$$

7. 4-Oxo-3-propylbutanamide

If the highest-priority principal group is attached to a ring, the parent name is that of the ring system. Compounds **8** and **9**, for instance, are isomeric keto nitriles and must both be named as nitriles according to Table A.2. Substance **8** is named as a benzonitrile because the −CN functional group is a substituent on the aromatic ring, but substance **9** is named as an acetonitrile because the −CN functional group is on an open chain. The correct names are 2-acetyl-(4-bromomethyl)benzonitrile (**8**) and (2-acetyl-4-bromophenyl)acetonitrile (**9**).

As further examples, compounds **10** and **11** are both keto acids and must be named as acids, but the parent name in **10** is that of a ring system (cyclohexanecarboxylic acid) and the parent name in **11** is that of an open chain (propanoic acid). The full names are *trans*-2-(3-oxopropyl)cyclohexanecarboxylic acid (**10**) and 3-(2-oxocyclohexyl)propanoic acid (**11**).

8. 2-Acetyl-(4-bromomethyl)**benzonitrile**

9. (2-Acetyl-4-bromophenyl)**acetonitrile**

10. *trans*-2-(3-oxopropyl)**cyclo-hexanecarboxylic acid**

11. 3-(2-Oxocyclohexyl)**propanoic acid**

Name Parts 3 and 4. The Prefixes and Locants

With parent name and suffix established, the next step is to identify and give numbers, or *locants*, to all substituents on the parent chain or ring. These substituents include all alkyl groups and all functional groups other than the one cited in the suffix. For example, compound **12** contains three different functional groups (carboxyl, keto, and double bond). Because the carboxyl group is highest in priority and because the longest chain containing the functional groups has seven carbons, compound **12** is a heptenoic acid. In addition, the main chain has a keto (oxo) substituent and three methyl groups. Numbering from the end nearer the highest-priority functional group, compound **12** is named (*E*)-2,5,5-trimethyl-4-oxohept-2-enoic acid. Look back at some of the other compounds we've named to see other examples of how prefixes and locants are assigned.

12. (*E*)-2,5,5-Trimethyl-4-oxo**hept-2-enoic acid**

Writing the Name

With the name parts established, the entire name is then written out. Several additional rules apply:

1. **Order of prefixes.** When the substituents have been identified, the main chain has been numbered, and the proper multipliers such as *di*- and *tri*- have been assigned, the name is written with the substituents listed in alphabetical,

rather than numerical, order. Multipliers such as *di-* and *tri-* are not used for alphabetization purposes, but the prefix *iso-* is used.

$$H_2NCH_2CH_2\overset{\overset{\displaystyle OH}{|}}{C}H\overset{\overset{\displaystyle}{|}}{C}HCH_3$$
$$CH_3$$

13. 5-Amino-3-methyl**pentan-2-ol**

2. **Use of hyphens; single- and multiple-word names.** The general rule is to determine whether the parent is itself an element or compound. If so, then the name is written as a single word; if not, then the name is written as multiple words. Methylbenzene is written as one word, for instance, because the parent—benzene—is itself a compound. Diethyl ether, however, is written as two words because the parent—ether—is a class name rather than a compound name. Some further examples follow:

$$H_3C-Mg-CH_3$$

14. Dimethylmagnesium
(one word, because
magnesium is an element)

$$HOCH_2CH_2\overset{\overset{\displaystyle O}{\|}}{C}O\overset{\overset{\displaystyle}{|}}{C}HCH_3$$
$$CH_3$$

15. Isopropyl 3-hydroxypropanoate
(two words, because "propanoate"
is not a compound)

16. 4-(Dimethylamino)**pyridine**
(one word, because pyridine
is a compound)

17. Methyl **cyclopentanecarbothioate**
(two words, because "cyclopentanecarbothioate" is not a compound)

3. **Parentheses.** Parentheses are used to denote complex substituents when ambiguity would otherwise arise. For example, chloromethylbenzene has two substituents on a benzene ring, but (chloromethyl)benzene has only one complex substituent. Note that the expression in parentheses is not set off by hyphens from the rest of the name.

18. *p*-Chloromethyl**benzene**

19. (Chloromethyl)**benzene**

$$HO\overset{\overset{\displaystyle O}{\|}}{C}\overset{\overset{\displaystyle}{|}}{C}HCH_2CH_2\overset{\overset{\displaystyle O}{\|}}{C}OH$$
$$CH_3CHCH_2CH_3$$

20. 2-(1-Methylpropyl)**pentanedioic acid**

Additional Reading

Further explanations of the rules of organic nomenclature can be found online at http://www.acdlabs.com/iupac/nomenclature/ and in the following references:

1. "A Guide to IUPAC Nomenclature of Organic Compounds," CRC Press, Boca Raton, FL, 1993.
2. "Nomenclature of Organic Chemistry, Sections A, B, C, D, E, F, and H," International Union of Pure and Applied Chemistry, Pergamon Press, Oxford, 1979.

Acidity Constants for Some Organic Compounds

Compound	pK_a	Compound	pK_a	Compound	pK_a
CH_3SO_3H	−1.8	CH_2ClCO_2H	2.8	(3-chlorobenzoic acid)	3.8
$CH(NO_2)_3$	0.1	$HO_2CCH_2CO_2H$	2.8; 5.6		
(2,4,6-trinitrophenol)	0.3	CH_2BrCO_2H	2.9	(4-chlorobenzoic acid)	4.0
CCl_3CO_2H	0.5	(2-chlorobenzoic acid)	3.0	$CH_2BrCH_2CO_2H$	4.0
CF_3CO_2H	0.5	(salicylic acid)	3.0	(2,6-dinitrophenol)	4.1
CBr_3CO_2H	0.7	CH_2ICO_2H	3.2		
$HO_2CC{\equiv}CCO_2H$	1.2; 2.5	$CHOCO_2H$	3.2	(benzoic acid)	4.2
HO_2CCO_2H	1.2; 3.7	(4-nitrobenzoic acid)	3.4		
$CHCl_2CO_2H$	1.3			$H_2C{=}CHCO_2H$	4.2
$CH_2(NO_2)CO_2H$	1.3	(2,4-dinitrobenzoic acid)	3.5	$HO_2CCH_2CH_2CO_2H$	4.2; 5.7
$HC{\equiv}CCO_2H$	1.9			$HO_2CCH_2CH_2CH_2CO_2H$	4.3; 5.4
$Z\ HO_2CCH{=}CHCO_2H$	1.9; 6.3	$HSCH_2CO_2H$	3.5; 10.2	(pentachlorophenol)	4.5
(2-nitrobenzoic acid)	2.4	$CH_2(NO_2)_2$	3.6		
		$CH_3OCH_2CO_2H$	3.6		
CH_3COCO_2H	2.4	$CH_3COCH_2CO_2H$	3.6	$H_2C{=}C(CH_3)CO_2H$	4.7
$NCCH_2CO_2H$	2.5	$HOCH_2CO_2H$	3.7	CH_3CO_2H	4.8
$CH_3C{\equiv}CCO_2H$	2.6	HCO_2H	3.7		
CH_2FCO_2H	2.7				

Compound	pK_a	Compound	pK_a	Compound	pK_a
$CH_3CH_2CO_2H$	4.8	$CH_3COCH_2COCH_3$	9.0	[benzyl]—CH_2OH	15.4
$(CH_3)_3CCO_2H$	5.0	[resorcinol, HO—C6H4—OH]	9.3; 11.1	CH_3OH	15.5
$CH_3COCH_2NO_2$	5.1			$H_2C{=}CHCH_2OH$	15.5
[1,3-cyclohexanedione]	5.3	[catechol, OH OH]	9.3; 12.6	CH_3CH_2OH	16.0
$O_2NCH_2CO_2CH_3$	5.8	[benzyl CH_2SH]	9.4	$CH_3CH_2CH_2OH$	16.1
[2-oxocyclopentane-CHO]	5.8			CH_3COCH_2Br	16.1
[2,4,6-trichlorophenol]	6.2	[hydroquinone, OH HO]	9.9; 11.5	[cyclohexanone]	16.7
[benzenethiol, —SH]	6.6	[phenol, OH]	9.9	CH_3CHO	17
HCO_3H	7.1	$CH_3COCH_2SOCH_3$	10.0	$(CH_3)_2CHCHO$	17
[2-nitrophenol]	7.2	[o-cresol, OH CH3]	10.3	$(CH_3)_2CHOH$	17.1
$(CH_3)_2CHNO_2$	7.7	CH_3NO_2	10.3	$(CH_3)_3COH$	18.0
[2,4-dichlorophenol]	7.8	CH_3SH	10.3	CH_3COCH_3	19.3
CH_3CO_3H	8.2	$CH_3COCH_2CO_2CH_3$	10.6	[fluorene]	23
[2-chlorophenol]	8.5	CH_3COCHO	11.0	$CH_3CO_2CH_2CH_3$	25
$CH_3CH_2NO_2$	8.5	$CH_2(CN)_2$	11.2	$HC{\equiv}CH$	25
F_3C—[phenol]—OH	8.7	CCl_3CH_2OH	12.2	CH_3CN	25
		Glucose	12.3	$CH_3SO_2CH_3$	28
		$(CH_3)_2C{=}NOH$	12.4	$(C_6H_5)_3CH$	32
		$CH_2(CO_2CH_3)_2$	12.9	$(C_6H_5)_2CH_2$	34
		$CHCl_2CH_2OH$	12.9	CH_3SOCH_3	35
		$CH_2(OH)_2$	13.3	NH_3	36
		$HOCH_2CH(OH)CH_2OH$	14.1	$CH_3CH_2NH_2$	36
		CH_2ClCH_2OH	14.3	$(CH_3CH_2)_2NH$	40
		[cyclopentadiene]	15.0	[toluene, CH3]	41
				[benzene]	43
				$H_2C{=}CH_2$	44
				CH_4	~60

An acidity list covering more than 5000 organic compounds has been published: E.P. Serjeant and B. Dempsey (eds.), "Ionization Constants of Organic Acids in Aqueous Solution," IUPAC Chemical Data Series No. 23, Pergamon Press, Oxford, 1979.

Glossary

Absolute configuration (Section 9.5): The exact three-dimensional structure of a chiral molecule. Absolute configurations are specified verbally by the Cahn–Ingold–Prelog *R,S* convention and are represented on paper by Fischer projections.

Absorbance (Section 14.7): In optical spectroscopy, the logarithm of the intensity of the incident light divided by the intensity of the light transmitted through a sample; $A = \log I_0/I$.

Absorption spectrum (Section 12.5): A plot of wavelength of incident light versus amount of light absorbed. Organic molecules show absorption spectra in both the infrared and the ultraviolet regions of the electromagnetic spectrum.

Acetal (Section 19.10): A functional group consisting of two −OR groups bonded to the same carbon, $R_2C(OR')_2$. Acetals are often used as protecting groups for ketones and aldehydes.

Acetoacetic ester synthesis (Section 22.7): The synthesis of a methyl ketone by alkylation of an alkyl halide, followed by hydrolysis and decarboxylation.

Acetyl group (Section 19.1): The CH_3CO- group.

Acetylide anion (Section 8.7): The anion formed by removal of a proton from a terminal alkyne.

Achiral (Section 9.2): Having a lack of handedness. A molecule is achiral if it has a plane of symmetry and is thus superimposable on its mirror image.

Acid anhydride (Section 21.1): A functional group with two acyl groups bonded to a common oxygen atom, RCO_2COR'.

Acid halide (Section 21.1): A functional group with an acyl group bonded to a halogen atom, RCOX.

Acidity constant, K_a (Section 2.8): A measure of acid strength. For any acid HA, the acidity constant is given by the expression $K_a = K_{eq} [H_2O] = \dfrac{[H_3O^+] [A^-]}{[HA]}$.

Activating group (Section 16.4): An electron-donating group such as hydroxyl (−OH) or amino (−NH₂) that increases the reactivity of an aromatic ring toward electrophilic aromatic substitution.

Activation energy (Section 5.9): The difference in energy between ground state and transition state in a reaction. The amount of activation energy determines the rate at which the reaction proceeds. Most organic reactions have activation energies of 40–100 kJ/mol.

Acyl group (Sections 16.3, 19.1): A −COR group.

Acyl phosphate (Section 21.8): A functional group with an acyl group bonded to a phosphate, $RCO_2PO_3{}^{2-}$ or $RCO_2PO_3R'^-$.

Acylation (Sections 16.3, 21.4): The introduction of an acyl group, −COR, onto a molecule. For example, acylation of an alcohol yields an ester, acylation of an amine yields an amide, and acylation of an aromatic ring yields an alkyl aryl ketone.

Acylium ion (Section 16.3): A resonance-stabilized carbocation in which the positive charge is located at a carbonyl-group carbon, $R-\overset{+}{C}{=}O \leftrightarrow R-C{\equiv}O^+$. Acylium ions are strongly electrophilic and are involved as intermediates in Friedel–Crafts acylation reactions.

Adams catalyst (Section 7.7): The PtO_2 catalyst used for hydrogenations.

1,2-Addition (Sections 14.2, 19.13): The addition of a reactant to the two ends of a double bond.

1,4-Addition (Sections 14.2, 19.13): Addition of a reactant to the ends of a conjugated π system. Conjugated dienes yield 1,4 adducts when treated with electrophiles such as HCl. Conjugated enones yield 1,4 adducts when treated with nucleophiles such as cyanide ion.

Addition reaction (Section 5.1): The reaction that occurs when two reactants add together to form a single new product with no atoms "left over."

Adrenocortical hormone (Section 27.6): A steroid hormone secreted by the adrenal glands. There are two types of adrenocortical hormones: mineralocorticoids and glucocorticoids.

Alcohol (Chapter 17 introduction): A compound with an −OH group bonded to a saturated, alkane-like carbon, ROH.

Aldaric acid (Section 25.6): The dicarboxylic acid resulting from oxidation of an aldose.

Aldehyde (Chapter 19 introduction): A compound containing the −CHO functional group.

Alditol (Section 25.6): The polyalcohol resulting from reduction of the carbonyl group of a sugar.

Aldol reaction (Section 23.1): The carbonyl condensation reaction of an aldehyde or ketone to give a β-hydroxy carbonyl compound.

Aldonic acid (Section 25.6): The monocarboxylic acid resulting from mild oxidation of the −CHO group of an aldose.

Aldose (Section 25.1): A carbohydrate with an aldehyde functional group.

Alicyclic (Section 4.1): An aliphatic cyclic hydrocarbon such as a cycloalkane or cycloalkene.

Aliphatic (Section 3.2): A nonaromatic hydrocarbon such as a simple alkane, alkene, or alkyne.

Alkaloid (Chapter 2 *Focus On*): A naturally occurring organic base, such as morphine.

Alkane (Section 3.2): A compound of carbon and hydrogen that contains only single bonds.

Alkene (Chapter 6 introduction): A hydrocarbon that contains a carbon–carbon double bond, $R_2C{=}CR_2$.

Alkoxide ion (Section 17.2): The anion RO^- formed by deprotonation of an alcohol.

Alkoxymercuration reaction (Section 18.2): A method for synthesizing ethers by mercuric-ion catalyzed addition of an alcohol to an alkene.

Alkyl group (Section 3.3): The partial structure that remains when a hydrogen atom is removed from an alkane.

Alkylamine (Section 24.1): An amino-substituted alkane.

Alkylation (Sections 8.8, 16.3, 18.2, 22.7): Introduction of an alkyl group onto a molecule. For example, aromatic rings can be alkylated to yield arenes, and enolate anions can be alkylated to yield α-substituted carbonyl compounds.

Alkyne (Chapter 8 introduction): A hydrocarbon that contains a carbon–carbon triple bond, $RC{\equiv}CR$.

Allyl group (Section 6.3): A $H_2C{=}CHCH_2-$ substituent.

Allylic (Section 10.5): The position next to a double bond. For example, $H_2C{=}CHCH_2Br$ is an allylic bromide.

α-Amino acid (Section 26.1): A difunctional compound with an amino group on the carbon atom next to a carboxyl group, $RCH(NH_2)CO_2H$.

α Anomer (Section 25.5): The cyclic hemiacetal form of a sugar that has the hemiacetal −OH group on the side of the ring opposite the terminal −CH_2OH.

α Helix (Section 26.9): The coiled secondary structure of a protein.

α Position (Chapter 22 introduction): The position next to a carbonyl group.

α-Substitution reaction (Section 22.2): The substitution of the α hydrogen atom of a carbonyl compound by reaction with an electrophile.

Amide (Chapter 21 introduction): A compound containing the −$CONR_2$ functional group.

Amidomalonate synthesis (Section 26.3): A method for preparing α-amino acids by alkylation of diethyl amidomalonate with an alkyl halide.

Amine (Chapter 24 introduction): A compound containing one or more organic substituents bonded to a nitrogen atom, RNH_2, R_2NH, or R_3N.

Amino acid (*See* α-Amino acid; Section 26.1)

Amino sugar (Section 25.7): A sugar with one of its −OH groups replaced by −NH_2.

Amphiprotic (Section 26.1): Capable of acting either as an acid or as a base. Amino acids are amphiprotic.

Amplitude (Section 12.5): The height of a wave measured from the midpoint to the maximum. The intensity of radiant energy is proportional to the square of the wave's amplitude.

Anabolism (Section 29.1): The group of metabolic pathways that build up larger molecules from smaller ones.

Androgen (Section 27.6): A male steroid sex hormone.

Angle strain (Section 4.3): The strain introduced into a molecule when a bond angle is deformed from its ideal value. Angle strain is particularly important in small-ring cycloalkanes, where it results from compression of bond angles to less than their ideal tetrahedral values.

Annulation (Section 23.12): The building of a new ring onto an existing molecule.

Anomers (Section 25.5): Cyclic stereoisomers of sugars that differ only in their configuration at the hemiacetal (anomeric) carbon.

Antarafacial (Section 30.6): A pericyclic reaction that takes place on opposite faces of the two ends of a π electron system.

Anti conformation (Section 3.7): The geometric arrangement around a carbon–carbon single bond in which the two largest substituents are 180° apart as viewed in a Newman projection.

Anti periplanar (Section 11.8): Describing a stereochemical relationship whereby two bonds on adjacent carbons lie in the same plane at an angle of 180°.

Anti stereochemistry (Section 7.2): The opposite of syn. An anti addition reaction is one in which the two ends of the double bond are attacked from different sides. An anti elimination reaction is one in which the two groups leave from opposite sides of the molecule.

Antiaromatic (Section 15.3): Referring to a planar, conjugated molecule with $4n$ π electrons. Delocalization of the π electrons leads to an increase in energy.

Antibonding MO (Section 1.11): A molecular orbital that is higher in energy than the atomic orbitals from which it is formed.

Anticodon (Section 28.5): A sequence of three bases on tRNA that reads the codons on mRNA and brings the correct amino acids into position for protein synthesis.

Arene (Section 15.1): An alkyl-substituted benzene.

Arenediazonium salt (Section 24.7): An aromatic compound $Ar-\overset{+}{N}\equiv N\ X^-$; used in the Sandmeyer reaction.

Aromaticity (Chapter 15 introduction): The special characteristics of cyclic conjugated molecules. These characteristics include unusual stability, the presence of a ring current in the ^1H NMR spectrum, and a tendency to undergo substitution reactions rather than addition reactions on treatment with electrophiles. Aromatic molecules are planar, cyclic, conjugated species that have $4n + 2$ π electrons.

Arylamine (Section 24.1): An amino-substituted aromatic compound, $Ar-NH_2$.

Atactic (Section 31.2): A chain-growth polymer in which the substituents are randomly oriented along the backbone.

Atomic mass (Section 1.1): The weighted average mass of an element's naturally occurring isotopes.

Atomic number, Z (Section 1.1): The number of protons in the nucleus of an atom.

ATZ Derivative (Section 26.6): An anilinothiazolinone, formed from an amino acid during Edman degradation of a peptide.

Aufbau principle (Section 1.3): The rules for determining the electron configuration of an atom.

Axial bond (Section 4.6): A bond to chair cyclohexane that lies along the ring axis perpendicular to the rough plane of the ring.

Azide synthesis (Section 24.6): A method for preparing amines by S_N2 reaction of an alkyl halide with azide ion, followed by reduction.

Azo compound (Section 24.8): A compound with the general structure $R-N=N-R'$.

Backbone (Section 26.4): The continuous chain of atoms running the length of a polymer.

Base peak (Section 12.1): The most intense peak in a mass spectrum.

Basicity constant, K_b (Section 24.3): A measure of base strength. For any base B, the basicity constant is given by the expression

$$B + H_2O \rightleftharpoons BH^+ + OH^-$$

$$K_b = \frac{[BH^+]\ [OH^-]}{[B]}$$

Bent bonds (Section 4.4): The bonds in small rings such as cyclopropane that bend away from the internuclear line and overlap at a slight angle, rather than head-on. Bent bonds are highly strained and highly reactive.

Benzoyl group (Section 19.1): The C_6H_5CO- group.

Benzyl group (Section 15.1): The $C_6H_5CH_2-$ group.

Benzylic (Section 11.5): The position next to an aromatic ring.

Benzyne (Section 16.8): An unstable compound having a triple bond in a benzene ring.

β Anomer (Section 25.5): The cyclic hemiacetal form of a sugar that has the hemiacetal $-OH$ group on the same side of the ring as the terminal $-CH_2OH$.

β-Diketone (Section 22.5): A 1,3-diketone.

β-Keto ester (Section 22.5): A 3-oxoester.

β-Oxidation pathway (Section 29.3): The metabolic pathway for degrading fatty acids.

β-Pleated sheet (Section 26.9): A type of secondary structure of a protein.

Betaine (Section 19.11): A neutral dipolar molecule with nonadjacent positive and negative charges. For example, the adduct of a Wittig reagent with a carbonyl compound is a betaine.

Bicycloalkane (Section 4.9): A cycloalkane that contains two rings.

Bimolecular reaction (Section 11.2): A reaction whose rate-limiting step occurs between two reactants.

Block copolymer (Section 31.3): A polymer in which different blocks of identical monomer units alternate with one another.

Boat cyclohexane (Section 4.5): A conformation of cyclohexane that bears a slight resemblance to a boat. Boat cyclohexane has no angle strain but has a large number of eclipsing interactions that make it less stable than chair cyclohexane.

Boc derivative (Section 26.7): A butyloxycarbonyl amide protected amino acid.

Bond angle (Section 1.6): The angle formed between two adjacent bonds.

Bond dissociation energy, D (Section 5.8): The amount of energy needed to break a bond homolytically and produce two radical fragments.

Bond length (Section 1.5): The equilibrium distance between the nuclei of two atoms that are bonded to each other.

Bond strength (Section 1.5): An alternative name for bond dissociation energy.

Bonding MO (Section 1.11): A molecular orbital that is lower in energy than the atomic orbitals from which it is formed.

Branched-chain alkane (Section 3.2): An alkane that contains a branching connection of carbons as opposed to a straight-chain alkane.

Bridgehead atom (Section 4.9): An atom that is shared by more than one ring in a polycyclic molecule.

Bromohydrin (Section 7.3): A 1,2-disubstituted bromo-alcohol; obtained by addition of HOBr to an alkene.

Bromonium ion (Section 7.2): A species with a divalent, positively charged bromine, R_2Br^+.

Brønsted–Lowry acid (Section 2.7): A substance that donates a hydrogen ion (proton; H^+) to a base.

Brønsted–Lowry base (Section 2.7): A substance that accepts H^+ from an acid.

C-terminal amino acid (Section 26.4): The amino acid with a free $-CO_2H$ group at the end of a protein chain.

Cahn–Ingold–Prelog sequence rules (Sections 6.5, 9.5): A series of rules for assigning relative priorities to substituent groups on a double-bond carbon atom or on a chirality center.

Cannizzaro reaction (Section 19.12): The disproportionation reaction of an aldehyde to yield an alcohol and a carboxylic acid on treatment with base.

Carbanion (Section 19.7): A carbon anion, or substance that contains a trivalent, negatively charged carbon atom ($R_3C:^-$). Carbanions are sp^3-hybridized and have eight electrons in the outer shell of the negatively charged carbon.

Carbene (Section 7.6): A neutral substance that contains a divalent carbon atom having only six electrons in its outer shell ($R_2C:$).

Carbinolamine (Section 19.8): A molecule that contains the $R_2C(OH)NH_2$ functional group. Carbinolamines are produced as intermediates during the nucleophilic addition of amines to carbonyl compounds.

Carbocation (Sections 5.5, 6.9): A carbon cation, or substance that contains a trivalent, positively charged carbon atom having six electrons in its outer shell (R_3C^+).

Carbohydrate (Section 25.1): A polyhydroxy aldehyde or ketone. Carbohydrates can be either simple sugars, such as glucose, or complex sugars, such as cellulose.

Carbonyl condensation reaction (Section 23.1): A reaction that joins two carbonyl compounds together by a combination of α-substitution and nucleophilic addition reactions.

Carbonyl group (Section 2.1): The C=O functional group.

Carboxyl group (Section 20.1): The $-CO_2H$ functional group.

Carboxylation (Section 20.5): The addition of CO_2 to a molecule.

Carboxylic acid (Chapter 20 introduction): A compound containing the $-CO_2H$ functional group.

Carboxylic acid derivative (Chapter 21 introduction): A compound in which an acyl group is bonded to an electronegative atom or substituent Y that can act as a leaving group in a substitution reaction, RCOY.

Catabolism (Section 29.1): The group of metabolic pathways that break down larger molecules into smaller ones.

Cation radical (Section 12.1): A reactive species formed by loss of an electron from a neutral molecule.

Chain-growth polymer (Section 31.1): A polymer whose bonds are produced by chain reactions. Polyethylene and other alkene polymers are examples.

Chain reaction (Section 5.3): A reaction that, once initiated, sustains itself in an endlessly repeating cycle of propagation steps. The radical chlorination of alkanes is an example of a chain reaction that is initiated by irradiation with light and then continues in a series of propagation steps.

Chair cyclohexane (Section 4.5): A three-dimensional conformation of cyclohexane that resembles the rough shape of a chair. The chair form of cyclohexane is the lowest-energy conformation of the molecule.

Chemical shift (Section 13.3): The position on the NMR chart where a nucleus absorbs. By convention, the chemical shift of tetramethylsilane (TMS) is set at zero, and all other absorptions usually occur downfield (to the left on the chart). Chemical shifts are expressed in delta units, δ, where 1 δ equals 1 ppm of the spectrometer operating frequency.

Chiral (Section 9.2): Having handedness. Chiral molecules are those that do not have a plane of symmetry and are therefore not superimposable on their mirror image. A chiral molecule thus exists in two forms, one right-handed and one left-handed. The most common cause of chirality in a molecule is the presence of a carbon atom that is bonded to four different substituents.

Chiral environment (Section 9.14): Chiral surroundings or conditions in which a molecule resides.

Chirality center (Section 9.2): An atom (usually carbon) that is bonded to four different groups.

Chlorohydrin (Section 7.3): A 1,2-disubstituted chloroalcohol; obtained by addition of HOCl to an alkene.

Chromatography (Chapter 12 *Focus On,* Section 26.7): A technique for separating a mixture of compounds into pure components. Different compounds adsorb to a stationary support phase and are then carried along it at different rates by a mobile phase.

Cis–trans isomers (Sections 4.2, 6.4): Stereoisomers that differ in their stereochemistry about a double bond or ring.

Citric acid cycle (Section 29.7): The metabolic pathway by which acetyl CoA is degraded to CO_2.

Claisen condensation reaction (Section 23.7): The carbonyl condensation reaction of an ester to give a β-keto ester product.

Claisen rearrangement reaction (Sections 18.4, 30.8): The pericyclic conversion of an allyl phenyl ether to an *o*-allylphenol by heating.

Coding strand (Section 28.4): The strand of double-helical DNA that contains the gene.

Codon (Section 28.5): A three-base sequence on a messenger RNA chain that encodes the genetic information necessary to cause a specific amino acid to be incorporated into a protein. Codons on mRNA are read by complementary anticodons on tRNA.

Coenzyme (Section 26.10): A small organic molecule that acts as a cofactor.

Cofactor (Section 26.10): A small nonprotein part of an enzyme that is necessary for biological activity.

Combinatorial chemistry (Chapter 16 *Focus On*): A procedure in which anywhere from a few dozen to several hundred thousand substances are prepared simultaneously.

Complex carbohydrate (Section 25.1): A carbohydrate that is made of two or more simple sugars linked together.

Concerted (Section 30.1): A reaction that takes place in a single step without intermediates. For example, the Diels–Alder cycloaddition reaction is a concerted process.

Condensed structure (Section 1.12): A shorthand way of writing structures in which carbon–hydrogen and carbon–carbon bonds are understood rather than shown explicitly. Propane, for example, has the condensed structure $CH_3CH_2CH_3$.

Configuration (Section 9.5): The three-dimensional arrangement of atoms bonded to a chirality center.

Conformation (Section 3.6): The three-dimensional shape of a molecule at any given instant, assuming that rotation around single bonds is frozen.

Conformational analysis (Section 4.8): A means of assessing the energy of a substituted cycloalkane by totaling the steric interactions present in the molecule.

Conformer (Section 3.6): A conformational isomer.

Conjugate acid (Section 2.7): The product that results from protonation of a Brønsted–Lowry base.

Conjugate addition (Section 19.13): Addition of a nucleophile to the β carbon atom of an α,β-unsaturated carbonyl compound.

Conjugate base (Section 2.7): The product that results from deprotonation of a Brønsted–Lowry acid.

Conjugation (Chapter 14 introduction): A series of overlapping *p* orbitals, usually in alternating single and multiple bonds. For example, buta-1,3-diene is a conjugated diene, but-3-en-2-one is a conjugated enone, and benzene is a cyclic conjugated triene.

Conrotatory (Section 30.2): A term used to indicate that *p* orbitals must rotate in the same direction during electrocyclic ring-opening or ring closure.

Constitutional isomers (Sections 3.2, 9.9): Isomers that have their atoms connected in a different order. For example, butane and 2-methylpropane are constitutional isomers.

Cope rearrangement (Section 30.8): The sigmatropic rearrangement of a hexa-1,5-diene.

Copolymer (Section 31.3): A polymer obtained when two or more different monomers are allowed to polymerize together.

Coupling constant, *J* (Section 13.11): The magnitude (expressed in hertz) of the interaction between nuclei whose spins are coupled.

Covalent bond (Section 1.5): A bond formed by sharing electrons between atoms.

Cracking (Chapter 3 *Focus On*): A process used in petroleum refining in which large alkanes are thermally cracked into smaller fragments.

Crown ether (Section 18.7): A large-ring polyether; used as a phase-transfer catalyst.

Crystallite (Section 31.5): A highly ordered crystal-like region within a long polymer chain.

Curtius rearrangement (Section 24.6): The conversion of an acid chloride into an amine by reaction with azide ion, followed by heating with water.

Cyanohydrin (Section 19.6): A compound with an −OH group and a −CN group bonded to the same carbon atom; formed by addition of HCN to an aldehyde or ketone.

Cycloaddition reaction (Sections 14.4, 30.6): A pericyclic reaction in which two reactants add together in a single step to yield a cyclic product. The Diels–Alder reaction between a diene and a dienophile to give a cyclohexene is an example.

Cycloalkane (Section 4.1): An alkane that contains a ring of carbons.

D Sugar (Section 25.3): A sugar whose hydroxyl group at the chirality center farthest from the carbonyl group points to the right when drawn in Fischer projection.

***d,l* form** (Section 9.8): The racemic modification of a compound.

Deactivating group (Section 16.4): An electron-withdrawing substituent that decreases the reactivity of an aromatic ring toward electrophilic aromatic substitution.

Debye, D (Section 2.2): A unit for measuring dipole moments; $1 \text{ D} = 3.336 \times 10^{-30}$ coulomb meter (C · m).

Decarboxylation (Section 22.7): The loss of carbon dioxide from a molecule. β-Keto acids decarboxylate readily on heating.

Degenerate orbitals (Section 15.2): Two or more orbitals that have the same energy level.

Degree of unsaturation (Section 6.2): The number of rings and/or multiple bonds in a molecule.

Dehydration (Sections 7.1, 11.10, 17.6): The loss of water from an alcohol. Alcohols can be dehydrated to yield alkenes.

Dehydrohalogenation (Sections 7.1, 11.8): The loss of HX from an alkyl halide. Alkyl halides undergo dehydrohalogenation to yield alkenes on treatment with strong base.

Delocalization (Section 10.5): A spreading out of electron density over a conjugated π electron system. For example, allylic cations and allylic anions are delocalized because their charges are spread out over the entire π electron system.

Delta scale (Section 13.3): An arbitrary scale used to calibrate NMR charts. One delta unit (δ) is equal to 1 part per million (ppm) of the spectrometer operating frequency.

Denaturation (Section 26.9): The physical changes that occur in a protein when secondary and tertiary structures are disrupted.

Deoxy sugar (Section 25.7): A sugar with one of its −OH groups replaced by an −H.

Deoxyribonucleic acid (DNA) (Section 28.1): The biopolymer consisting of deoxyribonucleotide units linked together through phosphate–sugar bonds. Found in the nucleus of cells, DNA contains an organism's genetic information.

DEPT-NMR (Section 13.6): An NMR method for distinguishing among signals due to CH_3, CH_2, CH, and quaternary carbons. That is, the number of hydrogens attached to each carbon can be determined.

Deshielding (Section 13.2): An effect observed in NMR that causes a nucleus to absorb downfield (to the left) of tetramethylsilane (TMS) standard. Deshielding is caused by a withdrawal of electron density from the nucleus.

Deuterium isotope effect (Section 11.8): A tool used in mechanistic investigations to establish whether a C−H bond is broken in the rate-limiting step of a reaction.

Dextrorotatory (Section 9.3): A word used to describe an optically active substance that rotates the plane of polarization of plane-polarized light in a right-handed (clockwise) direction.

Diastereomers (Section 9.6): Non–mirror-image stereoisomers; diastereomers have the same configuration at one or more chirality centers but differ at other chirality centers.

Diastereotopic (Section 13.8): Two hydrogens in a molecule whose replacement by some other group leads to different diastereomers.

1,3-Diaxial interaction (Section 4.8): The strain energy caused by a steric interaction between axial groups three carbon atoms apart in chair cyclohexane.

Diazonium salt (Section 24.8): A compound with the general structure $RN_2^+ X^-$.

Diazotization (Section 24.8): The conversion of a primary amine, RNH_2, into a diazonium ion, RN_2^+, by treatment with nitrous acid.

Dideoxy DNA sequencing (Section 28.6): A biochemical method for sequencing DNA strands.

Dieckmann cyclization reaction (Section 23.9): An intramolecular Claisen condensation reaction to give a cyclic β-keto ester.

Diels–Alder reaction (Sections 14.4, 30.6): The cyclo-addition reaction of a diene with a dienophile to yield a cyclohexene.

Dienophile (Section 14.5): A compound containing a double bond that can take part in the Diels–Alder cycloaddition reaction. The most reactive dienophiles are those that have electron-withdrawing groups on the double bond.

Digestion (Section 29.1): The first stage of catabolism, in which food is broken down by hydrolysis of ester, glycoside (acetal), and peptide (amide) bonds to yield fatty acids, simple sugars, and amino acids.

Dipole moment, μ (Section 2.2): A measure of the net polarity of a molecule. A dipole moment arises when the centers of mass of positive and negative charges within a molecule do not coincide.

Dipole–dipole force (Section 2.13): A noncovalent electro-static interaction between dipolar molecules.

Disaccharide (Section 25.8): A carbohydrate formed by linking two simple sugars through an acetal bond.

Dispersion force (Section 2.13): A noncovalent interaction between molecules that arises because of constantly changing electron distributions within the molecules.

Disrotatory (Section 30.2): A term used to indicate that p orbitals rotate in opposite directions during electrocyclic ring-opening or ring closing.

Disulfide (Section 18.8): A compound of the general structure RSSR′.

DNA (*See* Deoxyribonucleic acid; Section 28.1)

Double helix (Section 28.2): The structure of DNA in which two polynucleotide strands coil around each other.

Doublet (Section 13.11): A two-line NMR absorption caused by spin–spin splitting when the spin of the nucleus under observation couples with the spin of a neighboring magnetic nucleus.

Downfield (Section 13.3): Referring to the left-hand portion of the NMR chart.

E geometry (Section 6.5): A term used to describe the stereochemistry of a carbon–carbon double bond. The two groups on each carbon are assigned priorities according to the Cahn–Ingold–Prelog sequence rules, and the two carbons are compared. If the high-priority groups on each carbon are on opposite sides of the double bond, the bond has E geometry.

E1 reaction (Section 11.10): A unimolecular elimination reaction in which the substrate spontaneously dissociates to give a carbocation intermediate, which loses a proton in a separate step.

E1cB reaction (Section 11.10): A unimolecular elimination reaction in which a proton is first removed to give a carbanion intermediate, which then expels the leaving group in a separate step.

E2 reaction (Section 11.8): A bimolecular elimination reaction in which both the hydrogen and the leaving group are lost in the same step.

Eclipsed conformation (Section 3.6): The geometric arrangement around a carbon–carbon single bond in which the bonds to substituents on one carbon are parallel to the bonds to substituents on the neighboring carbon as viewed in a Newman projection.

Eclipsing strain (Section 3.6): The strain energy in a molecule caused by electron repulsions between eclipsed bonds. Eclipsing strain is also called torsional strain.

Edman degradation (Section 26.6): A method for N-terminal sequencing of peptide chains by treatment with *N*-phenylisothiocyanate.

Eicosanoid (Section 27.4): A lipid derived biologically from eicosa-5,8,11,14-tetraenoic acid, or arachidonic acid. Prostaglandins, thromboxanes and leukotrienes are examples.

Elastomer (Section 31.5): An amorphous polymer that has the ability to stretch out and spring back to its original shape.

Electrocyclic reaction (Section 30.3): A unimolecular peri-cyclic reaction in which a ring is formed or broken by a concerted reorganization of electrons through a cyclic transition state. For example, the cyclization of hexa-1,3,5-triene to yield cyclohexa-1,3-diene is an electrocyclic reaction.

Electromagnetic spectrum (Section 12.5): The range of electromagnetic energy, including infrared, ultraviolet, and visible radiation.

Electron configuration (Section 1.3): A list of the orbitals occupied by electrons in an atom.

Electron-dot structure (Section 1.4): A representation of a molecule showing valence electrons as dots.

Electron-transport chain (Section 29.1): The final stage of catabolism in which ATP is produced.

Electronegativity (Section 2.1): The ability of an atom to attract electrons in a covalent bond. Electronegativity increases across the periodic table from right to left and from bottom to top.

Electrophile (Section 5.4): An "electron-lover," or substance that accepts an electron pair from a nucleophile in a polar bond-forming reaction.

Electrophilic addition reaction (Section 6.7): The addition of an electrophile to a carbon–carbon double bond to yield a saturated product.

Electrophilic aromatic substitution (Chapter 16 introduction): A reaction in which an electrophile (E^+) reacts with an aromatic ring and substitutes for one of the ring hydrogens.

Electrophoresis (Section 26.2): A technique used for separating charged organic molecules, particularly proteins and amino acids. The mixture to be separated is placed on a buffered gel or paper, and an electric potential is applied across the ends of the apparatus. Negatively charged molecules migrate toward the positive electrode, and positively charged molecules migrate toward the negative electrode.

Electrostatic potential map (Section 2.1): A molecular representation that uses color to indicate the charge distribution in the molecule as derived from quantum-mechanical calculations.

Elimination reaction (Section 5.1): What occurs when a single reactant splits into two products.

Elution (Chapter 12 *Focus On*): The removal of a substance from a chromatography column.

Embden–Meyerhof pathway (Section 29.5): An alternative name for glycolysis.

Enamine (Section 19.8): A compound with the $R_2N—CR=CR_2$ functional group.

Enantiomers (Section 9.1): Stereoisomers of a chiral substance that have a mirror-image relationship. Enantiomers must have opposite configurations at all chirality centers.

Enantioselective synthesis (Chapter 19 *Focus On*): A reaction method that yields only a single enantiomer of a chiral product starting from an achiral substrate.

Enantiotopic (Section 13.8): Two hydrogens in a molecule whose replacement by some other group leads to different enantiomers.

3′ End (Section 28.1): The end of a nucleic acid chain with a free hydroxyl group at C3′.

5′ End (Section 28.1): The end of a nucleic acid chain with a free hydroxyl group at C5′.

Endergonic (Section 5.7): A reaction that has a positive free-energy change and is therefore nonspontaneous. In a reaction energy diagram, the product of an endergonic reaction has a higher energy level than the reactants.

Endo (Section 14.5): A term indicating the stereochemistry of a substituent in a bridged bicycloalkane. An endo substituent is syn to the larger of the two bridges.

Endothermic (Section 5.7): A reaction that absorbs heat and therefore has a positive enthalpy change.

Energy diagram (Section 5.9): A representation of the course of a reaction, in which free energy is plotted as a function of reaction progress. Reactants, transition states, intermediates, and products are represented, and their appropriate energy levels are indicated.

Enol (Sections 8.4, 22.1): A vinylic alcohol that is in equilibrium with a carbonyl compound.

Enolate ion (Section 22.1): The anion of an enol.

Enthalpy change, ΔH (Section 5.7): The heat of reaction. The enthalpy change that occurs during a reaction is a measure of the difference in total bond energy between reactants and products.

Entropy change, ΔS (Section 5.7): The change in amount of molecular randomness. The entropy change that occurs during a reaction is a measure of the difference in randomness between reactants and products.

Enzyme (Section 26.10): A biological catalyst. Enzymes are large proteins that catalyze specific biochemical reactions.

Epoxide (Section 7.8): A three-membered-ring ether functional group.

Equatorial bond (Section 4.6): A bond to cyclohexane that lies along the rough equator of the ring.

ESI (Section 12.4): Electrospray ionization, a mild method for ionizing a molecule so that fragmentation is minimized during mass spectrometry.

Essential oil (Chapter 6 *Focus On*): The volatile oil obtained by steam distillation of a plant extract.

Ester (Chapter 21 introduction): A compound containing the $—CO_2R$ functional group.

Estrogen (Section 27.6): A female steroid sex hormone.

Ether (Chapter 18 introduction): A compound that has two organic substituents bonded to the same oxygen atom, ROR′.

Exergonic (Section 5.7): A reaction that has a negative free-energy change and is therefore spontaneous. On a reaction energy diagram, the product of an exergonic reaction has a lower energy level than that of the reactants.

Exo (Section 14.5): A term indicating the stereochemistry of a substituent in a bridged bicycloalkane. An exo substituent is anti to the larger of the two bridges.

Exon (Section 28.4): A section of DNA that contains genetic information.

Exothermic (Section 5.7): A reaction that releases heat and therefore has a negative enthalpy change.

Fat (Section 27.1): A solid triacylglycerol derived from an animal source.

Fatty acid (Section 27.1): A long, straight-chain carboxylic acid found in fats and oils.

Fiber (Section 31.5): A thin thread produced by extruding a molten polymer through small holes in a die.

Fibrous protein (Section 26.9): A protein that consists of polypeptide chains arranged side by side in long threads. Such proteins are tough, insoluble in water, and used in nature for structural materials such as hair, hooves, and fingernails.

Fingerprint region (Section 12.7): The complex region of the infrared spectrum from 1500 to 400 cm^{-1}.

First-order reaction (Section 11.4): A reaction whose rate-limiting step is unimolecular and whose kinetics therefore depend on the concentration of only one reactant.

Fischer esterification reaction (Section 21.3): The acid-catalyzed nucleophilic acyl substitution reaction of a carboxylic acid with an alcohol to yield an ester.

Fischer projection (Section 25.2): A means of depicting the absolute configuration of a chiral molecule on a flat page. A Fischer projection uses a cross to represent the chirality center. The horizontal arms of the cross represent bonds coming out of the plane of the page, and the vertical arms of the cross represent bonds going back into the plane of the page.

Fmoc derivative (Section 26.7): A fluorenylmethyloxy-carbonyl amide-protected amino acid.

Formal charge (Section 2.3): The difference in the number of electrons owned by an atom in a molecule and by the same atom in its elemental state.

Formyl group (Section 19.1): A −CHO group.

Frequency (Section 12.5): The number of electromagnetic wave cycles that travel past a fixed point in a given unit of time. Frequencies are expressed in units of cycles per second, or hertz.

Friedel–Crafts reaction (Section 16.3): An electrophilic aromatic substitution reaction to alkylate or acylate an aromatic ring.

Frontier orbitals (Section 30.1): The highest occupied (HOMO) and lowest unoccupied (LUMO) molecular orbitals.

FT-NMR (Section 13.4): Fourier-transform NMR; a rapid technique for recording NMR spectra in which all magnetic nuclei absorb at the same time.

Functional group (Section 3.1): An atom or group of atoms that is part of a larger molecule and that has a characteristic chemical reactivity.

Furanose (Section 25.5): The five-membered-ring form of a simple sugar.

Gabriel amine synthesis (Section 24.6): A method for preparing an amine by S_N2 reaction of an alkyl halide with potassium phthalimide, followed by hydrolysis.

Gauche conformation (Section 3.7): The conformation of butane in which the two methyl groups lie 60° apart as viewed in a Newman projection. This conformation has 3.8 kJ/mol steric strain.

Geminal (Section 19.5): Referring to two groups attached to the same carbon atom. For example, 1,1-dibromo-propane is a geminal dibromide.

Gibbs free-energy change, ΔG (Section 5.7): The free-energy change that occurs during a reaction, given by the equation $\Delta G = \Delta H - T\Delta S$. A reaction with a negative free-energy change is spontaneous, and a reaction with a positive free-energy change is nonspontaneous.

Gilman reagent (Section 10.8): A diorganocopper reagent, R_2CuLi.

Glass transition temperature, T_g (Section 31.5): The temperature at which a hard, amorphous polymer becomes soft and flexible.

Globular protein (Section 26.9): A protein that is coiled into a compact, nearly spherical shape. These proteins, which are generally water-soluble and mobile within the cell, are the structural class to which enzymes belong.

Gluconeogenesis (Section 29.8): The anabolic pathway by which organisms make glucose from simple precursors.

Glycal assembly method (Section 25.11): A method for linking monosaccharides together to synthesis polysaccharides.

Glycerophospholipid (Section 27.3): A lipid that contains a glycerol backbone linked to two fatty acids and a phosphoric acid.

Glycoconjugate (Section 25.6): A biological molecule in which a carbohydrate is linked through a glycoside bond to a lipid or protein.

Glycol (Section 7.8): A diol, such as ethylene glycol, $HOCH_2CH_2OH$.

Glycolipid (Section 25.6): A biological molecule in which a carbohydrate is linked through a glycoside bond to a lipid.

Glycoprotein (Section 25.6): A biological molecule in which a carbohydrate is linked through a glycoside bond to a protein.

Glycolysis (Section 29.5): A series of ten enzyme-catalyzed reactions that break down glucose into 2 equivalents of pyruvate, $CH_3COCO_2^-$.

Glycoside (Section 25.6): A cyclic acetal formed by reaction of a sugar with another alcohol.

Graft copolymer (Section 31.3): A copolymer in which homopolymer branches of one monomer unit are "grafted" onto a homopolymer chain of another monomer unit.

Green chemistry (Chapter 11 *Focus On*): The design and implementation of chemical products and processes that reduce waste and minimize or eliminate the generation of hazardous substances.

Grignard reagent (Section 10.7): An organomagnesium halide, RMgX.

Ground state (Section 1.3): The most stable, lowest-energy electron configuration of a molecule or atom.

Haloform reaction (Section 22.6): The reaction of a methyl ketone with halogen and base to yield a haloform (CHX_3) and a carboxylic acid.

Halohydrin (Section 7.3): A 1,2-disubstituted haloalcohol, such as that obtained on addition of HOBr to an alkene.

Halonium ion (Section 7.2): A species containing a positively charged, divalent halogen. Three-membered-ring bromonium ions are implicated as intermediates in the electrophilic addition of Br_2 to alkenes.

Hammond postulate (Section 6.10): A postulate stating that we can get a picture of what a given transition state looks like by looking at the structure of the nearest stable species. Exergonic reactions have transition states that resemble reactant; endergonic reactions have transition states that resemble product.

Heat of hydrogenation (Section 6.6): The amount of heat released when a carbon–carbon double bond is hydrogenated.

Heat of reaction (Section 5.7): An alternative name for the enthalpy change in a reaction, ΔH.

Hell–Volhard–Zelinskii (HVZ) reaction (Section 22.4): The reaction of a carboxylic acid with Br_2 and phosphorus to give an α-bromo carboxylic acid.

Hemiacetal (Section 19.10): A functional group consisting of one $-OR$ and one $-OH$ group bonded to the same carbon.

Henderson–Hasselbalch equation (Sections 20.3, 24.5, 26.2): An equation for determining the extent of deprotonation of a weak acid at various pH values.

Heterocycle (Sections 15.5, 24.9): A cyclic molecule whose ring contains more than one kind of atom. For example, pyridine is a heterocycle that contains five carbon atoms and one nitrogen atom in its ring.

Heterolytic bond breakage (Section 5.2): The kind of bond-breaking that occurs in polar reactions when one fragment leaves with both of the bonding electrons: A:B → A$^+$ + B:$^-$.

Hofmann elimination (Section 24.7): The elimination reaction of an amine to yield an alkene by reaction with iodomethane, followed by heating with Ag_2O.

Hofmann rearrangement (Section 24.6): The conversion of an amide into an amine by reaction with Br_2 and base.

HOMO (Sections 14.7, 30.2): An acronym for highest occupied molecular orbital. The symmetries of the HOMO and LUMO are important in pericyclic reactions.

Homolytic bond breakage (Section 5.2): The kind of bond-breaking that occurs in radical reactions when each fragment leaves with one bonding electron: A:B → A· + B·.

Homopolymer (Section 31.3): A polymer made up of identical repeating units.

Homotopic (Section 13.8): Hydrogens that give the identical structure on replacement by X and thus show identical NMR absorptions.

Hormone (Section 27.6): A chemical messenger that is secreted by an endocrine gland and carried through the bloodstream to a target tissue.

Hückel's rule (Section 15.3): A rule stating that monocyclic conjugated molecules having $4n + 2$ π electrons (n = an integer) are aromatic.

Hund's rule (Section 1.3): If two or more empty orbitals of equal energy are available, one electron occupies each, with their spins parallel, until all are half-full.

Hybrid orbital (Section 1.6): An orbital derived from a combination of atomic orbitals. Hybrid orbitals, such as the sp^3, sp^2, and sp hybrids of carbon, are strongly directed and form stronger bonds than atomic orbitals do.

Hydration (Section 7.4): Addition of water to a molecule, such as occurs when alkenes are treated with aqueous sulfuric acid to give alcohols.

Hydride shift (Section 6.11): The shift of a hydrogen atom and its electron pair to a nearby cationic center.

Hydroboration (Section 7.5): Addition of borane (BH_3) or an alkylborane to an alkene. The resultant trialkylborane products are useful synthetic intermediates that can be oxidized to yield alcohols.

Hydrocarbon (Section 3.2): A compound that contains only carbon and hydrogen.

Hydrogen bond (Section 2.13): A weak attraction between a hydrogen atom bonded to an electronegative atom and an electron lone pair on another electronegative atom.

Hydrogenation (Section 7.7): Addition of hydrogen to a double or triple bond to yield a saturated product.

Hydrogenolysis (Section 26.7): Cleavage of a bond by reaction with hydrogen. Benzylic ethers and esters, for instance, are cleaved by hydrogenolysis.

Hydrophilic (Section 2.13): Water-loving; attracted to water.

Hydrophobic (Section 2.13): Water-fearing; repelled by water.

Hydroquinone (Section 17.10): A 1,4-dihydroxybenzene.

Hydroxylation (Section 7.8): Addition of two −OH groups to a double bond.

Hyperconjugation (Sections 6.6, 6.9): An interaction that results from overlap of a vacant p orbital on one atom with a neighboring C−H σ bond. Hyperconjugation is important in stabilizing carbocations and in stabilizing substituted alkenes.

Imide (Section 24.6): A compound with the −CONHCO− functional group.

Imine (Section 19.8): A compound with the $R_2C{=}NR$ functional group.

Inductive effect (Sections 2.1, 6.9, 16.4): The electron-attracting or electron-withdrawing effect transmitted through σ bonds. Electronegative elements have an electron-withdrawing inductive effect.

Infrared (IR) spectroscopy (Section 12.6): A kind of optical spectroscopy that uses infrared energy. IR spectroscopy is particularly useful in organic chemistry for determining the kinds of functional groups present in molecules.

Initiator (Section 5.3): A substance with an easily broken bond that is used to initiate a radical chain reaction. For example, radical chlorination of alkanes is initiated when light energy breaks the weak Cl−Cl bond to form Cl· radicals.

Integration (Section 13.10): A technique for measuring the area under an NMR peak to determine the relative number of each kind of proton in a molecule. Integrated peak areas are superimposed over the spectrum as a "stair-step" line, with the height of each step proportional to the area underneath the peak.

Intermediate (Section 5.10): A species that is formed during the course of a multistep reaction but is not the final product. Intermediates are more stable than transition states but may or may not be stable enough to isolate.

Intramolecular, intermolecular (Section 23.6): A reaction that occurs within the same molecule is intramolecular; a reaction that occurs between two molecules is intermolecular.

Intron (Section 28.4): A section of DNA that does not contain genetic information.

Ion pair (Section 11.5): A loose complex between two ions in solution. Ion pairs are implicated as intermediates in S_N1 reactions to account for the partial retention of stereochemistry that is often observed.

Isoelectric point, pI (Section 26.2): The pH at which the number of positive charges and the number of negative charges on a protein or an amino acid are equal.

Isomers (Sections 3.2, 9.9): Compounds that have the same molecular formula but different structures.

Isoprene rule (Chapter 6 *Focus On*): An observation to the effect that terpenoids appear to be made up of isoprene (2-methylbuta-1,3-diene) units connected head-to-tail.

Isotactic (Section 31.2): A chain-growth polymer in which the substituents are regularly oriented on the same side of the backbone.

Isotopes (Section 1.1): Atoms of the same element that have different mass numbers.

IUPAC system of nomenclature (Section 3.4): Rules for naming compounds, devised by the International Union of Pure and Applied Chemistry.

Kekulé structure (Section 1.4): A method of representing molecules in which a line between atoms indicates a bond.

Keto–enol tautomerism (Sections 8.4, 22.1): The rapid equilibration between a carbonyl form and vinylic alcohol form of a molecule.

Ketone (Chapter 19 introduction): A compound with two organic substituents bonded to a carbonyl group, $R_2C{=}O$.

Ketose (Section 25.1): A carbohydrate with a ketone functional group.

Kiliani–Fischer synthesis (Section 25.6): A method for lengthening the chain of an aldose sugar.

Kinetic control (Section 14.3): A reaction that follows the lowest activation energy pathway is said to be kinetically controlled. The product is the most rapidly formed but is not necessarily the most stable.

Kinetics (Section 11.2): Referring to reaction rates. Kinetic measurements are useful for helping to determine reaction mechanisms.

Koenigs–Knorr reaction (Section 25.6): A method for the synthesis of glycosides by reaction of an alcohol with a pyranosyl bromide.

Krebs cycle (Section 29.7): An alternative name for the citric acid cycle, by which acetyl CoA is degraded to CO_2.

ʟ Sugar (Section 25.3): A sugar whose hydroxyl group at the chirality center farthest from the carbonyl group points to the left when drawn in Fischer projection.

Lactam (Section 21.7): A cyclic amide.

Lactone (Section 21.6): A cyclic ester.

Leaving group (Section 11.2): The group that is replaced in a substitution reaction.

Levorotatory (Section 9.3): An optically active substance that rotates the plane of polarization of plane-polarized light in a left-handed (counterclockwise) direction.

Lewis acid (Section 2.11): A substance with a vacant low-energy orbital that can accept an electron pair from a base. All electrophiles are Lewis acids.

Lewis base (Section 2.11): A substance that donates an electron lone pair to an acid. All nucleophiles are Lewis bases.

Lewis structure (Section 1.5): A representation of a molecule showing valence electrons as dots.

Lindlar catalyst (Section 8.5): A hydrogenation catalyst used to convert alkynes to cis alkenes.

Line-bond structure (Section 1.5): A representation of a molecule showing covalent bonds as lines between atoms.

1→4 Link (Section 25.8): An acetal link between the C1 −OH group of one sugar and the C4 −OH group of another sugar.

Lipid (Section 27.1): A naturally occurring substance isolated from cells and tissues by extraction with a nonpolar solvent. Lipids belong to many different structural classes, including fats, terpenes, prostaglandins, and steroids.

Lipid bilayer (Section 27.3): The ordered lipid structure that forms a cell membrane.

Lipoprotein (Chapter 27 *Focus On*): A complex molecule with both lipid and protein parts that transports lipids through the body.

Lone-pair electrons (Section 1.4): Nonbonding valence-shell electron pairs. Lone-pair electrons are used by nucleophiles in their reactions with electrophiles.

LUMO (Sections 14.4, 30.2): An acronym for lowest unoccupied molecular orbital. The symmetries of the LUMO and the HOMO are important in determining the stereochemistry of pericyclic reactions.

Magnetic resonance imaging, MRI (Chapter 13 *Focus On*): A medical diagnostic technique based on nuclear magnetic resonance.

MALDI (Section 12.4): Matrix-assisted laser desorption ionization; a mild method for ionizing a molecule so that fragmentation is minimized during mass spectrometry.

Malonic ester synthesis (Section 22.7): The synthesis of a carboxylic acid by alkylation of an alkyl halide, followed by hydrolysis and decarboxylation.

Markovnikov's rule (Section 6.8): A guide for determining the regiochemistry (orientation) of electrophilic addition reactions. In the addition of HX to an alkene, the hydrogen atom bonds to the alkene carbon that has fewer alkyl substituents.

Mass number, A (Section 1.1): The total of protons plus neutrons in an atom.

Mass spectrometry (Section 12.1): A technique for measuring the mass, and therefore the molecular weight (MW), of ions.

McLafferty rearrangement (Section 12.3): A mass-spectral fragmentation pathway for carbonyl compounds.

Mechanism (Section 5.2): A complete description of how a reaction occurs. A mechanism must account for all starting materials and all products and must describe the details of each individual step in the overall reaction process.

Meisenheimer complex (Section 16.7): An intermediate formed by addition of a nucleophile to a halo-substituted aromatic ring.

Melt transition temperature, T_m (Section 31.5): The temperature at which crystalline regions of a polymer melt to give an amorphous material.

Mercapto group (Section 18.8): An alternative name for the thiol group, −SH.

Meso compound (Section 9.7): A compound that contains chirality centers but is nevertheless achiral by virtue of a symmetry plane.

Messenger RNA (Section 28.4): A kind of RNA formed by transcription of DNA and used to carry genetic messages from DNA to ribosomes.

Meta- (Section 15.1): A naming prefix used for 1,3-disubstituted benzenes.

Metabolism (Section 29.1): A collective name for the many reactions that go on in the cells of living organisms.

Methylene group (Section 6.3): A −CH_2− or =CH_2 group.

Micelle (Section 27.2): A spherical cluster of soaplike molecules that aggregate in aqueous solution. The ionic heads of the molecules lie on the outside, where they are solvated by water, and the organic tails bunch together on the inside of the micelle.

Michael reaction (Section 23.10): The conjugate addition reaction of an enolate ion to an unsaturated carbonyl compound.

Molar absorptivity (Section 14.7): A quantitative measure of the amount of UV light absorbed by a sample.

Molecular ion (Section 12.1): The cation produced in the mass spectrometer by loss of an electron from the parent molecule. The mass of the molecular ion corresponds to the molecular weight of the sample.

Molecular mechanics (Chapter 4 *Focus On*): A computer-based method for calculating the minimum-energy conformation of a molecule.

Molecular orbital (MO) theory (Section 1.11): A description of covalent bond formation as resulting from a mathematical combination of atomic orbitals (wave functions) to form molecular orbitals.

Molecule (Section 1.5): A neutral collection of atoms held together by covalent bonds.

Molozonide (Section 7.9): The initial addition product of ozone with an alkene.

Monomer (Section 7.10, Chapter 31 introduction): The simple starting unit from which a polymer is made.

Monosaccharide (Section 25.1): A simple sugar.

Monoterpenoid (Chapter 6 *Focus On*, Section 27.5): A ten-carbon lipid.

Multiplet (Section 13.11): A pattern of peaks in an NMR spectrum that arises by spin–spin splitting of a single absorption because of coupling between neighboring magnetic nuclei.

Mutarotation (Section 25.5): The change in optical rotation observed when a pure anomer of a sugar is dissolved in water. Mutarotation is caused by the reversible opening and closing of the acetal linkage, which yields an equilibrium mixture of anomers.

$n + 1$ rule (Section 13.11): A hydrogen with n other hydrogens on neighboring carbons shows $n + 1$ peaks in its ^{1}H NMR spectrum.

N-terminal amino acid (Section 26.4): The amino acid with a free $-NH_2$ group at the end of a protein chain.

Newman projection (Section 3.6): A means of indicating stereochemical relationships between substituent groups on neighboring carbons. The carbon–carbon bond is viewed end-on, and the carbons are indicated by a circle. Bonds radiating from the center of the circle are attached to the front carbon, and bonds radiating from the edge of the circle are attached to the rear carbon.

Nitrile (Section 20.1): A compound containing the $C\equiv N$ functional group.

Nitrogen rule (Section 24.10): A compound with an odd number of nitrogen atoms has an odd-numbered molecular weight.

Node (Section 1.2): A surface of zero electron density within an orbital. For example, a p orbital has a nodal plane passing through the center of the nucleus, perpendicular to the axis of the orbital.

Nonbonding electrons (Section 1.4): Valence electrons that are not used in forming covalent bonds.

Noncovalent interaction (Section 2.13): An interaction between molecules, commonly called intermolecular forces or van der Waals forces. Hydrogen bonds, dipole–dipole forces, and dispersion forces are examples.

Normal alkane (Section 3.2): A straight-chain alkane, as opposed to a branched alkane. Normal alkanes are denoted by the suffix n, as in n-C_4H_{10} (n-butane).

NSAID (Chapter 15 *Focus On*): A nonsteroidal anti-inflammatory drug, such as aspirin or ibuprofen.

Nuclear magnetic resonance, NMR (Chapter 13 introduction): A spectroscopic technique that provides information about the carbon–hydrogen framework of a molecule. NMR works by detecting the energy absorptions accompanying the transitions between nuclear spin states that occur when a molecule is placed in a strong magnetic field and irradiated with radiofrequency waves.

Nucleophile (Section 5.4): A "nucleus-lover," or species that donates an electron pair to an electrophile in a polar bond-forming reaction. Nucleophiles are also Lewis bases.

Nucleophilic acyl substitution reaction (Section 21.2): A reaction in which a nucleophile attacks a carbonyl compound and substitutes for a leaving group bonded to the carbonyl carbon.

Nucleophilic addition reaction (Section 19.4): A reaction in which a nucleophile adds to the electrophilic carbonyl group of a ketone or aldehyde to give an alcohol.

Nucleophilic aromatic substitution reaction (Section 16.7): The substitution reaction of an aryl halide by a nucleophile.

Nucleophilic substitution reaction (Section 11.1): A reaction in which one nucleophile replaces another attached to a saturated carbon atom.

Nucleophilicity (Section 11.3): The ability of a substance to act as a nucleophile in an S_N2 reaction.

Nucleoside (Section 28.1): A nucleic acid constituent, consisting of a sugar residue bonded to a heterocyclic purine or pyrimidine base.

Nucleotide (Section 28.1): A nucleic acid constituent, consisting of a sugar residue bonded both to a heterocyclic purine or pyrimidine base and to a phosphoric acid.

Nucleotides are the monomer units from which DNA and RNA are constructed.

Nylon (Section 21.9): A synthetic polyamide step-growth polymer.

Olefin (Chapter 6 introduction): An alternative name for an alkene.

Optical isomers (Section 9.4): An alternative name for enantiomers. Optical isomers are isomers that have a mirror-image relationship.

Optically active (Section 9.3): A substance that rotates the plane of polarization of plane-polarized light.

Orbital (Section 1.2): A wave function, which describes the volume of space around a nucleus in which an electron is most likely to be found.

Organic chemistry (Chapter 1 introduction): The study of carbon compounds.

Ortho- (Section 15.1): A naming prefix used for 1,2-disubstituted benzenes.

Oxidation (Sections 7.8, 10.9): A reaction that causes a decrease in electron ownership by carbon, either by bond formation between carbon and a more electronegative atom (usually oxygen, nitrogen, or a halogen) or by bond-breaking between carbon and a less electronegative atom (usually hydrogen).

Oxime (Section 19.8): A compound with the $R_2C{=}NOH$ functional group.

Oxirane (Section 7.8): An alternative name for an epoxide.

Oxymercuration (Section 7.4): A method for double-bond hydration using aqueous mercuric acetate as the reagent.

Ozonide (Section 7.9): The product formed by addition of ozone to a carbon–carbon double bond. Ozonides are usually treated with a reducing agent, such as zinc in acetic acid, to produce carbonyl compounds.

Para- (Section 15.1): A naming prefix used for 1,4-disubstituted benzenes.

Paraffin (Section 3.5): A common name for alkanes.

Parent peak (Section 12.1): The peak in a mass spectrum corresponding to the molecular ion. The mass of the parent peak therefore represents the molecular weight of the compound.

Pauli exclusion principle (Section 1.3): No more than two electrons can occupy the same orbital, and those two must have spins of opposite sign.

Peptide (Section 26.4): A short amino acid polymer in which the individual amino acid residues are linked by amide bonds.

Peptide bond (Section 26.4): An amide bond in a peptide chain.

Pericyclic reaction (Chapter 30 introduction): A reaction that occurs by a concerted reorganization of bonding electrons in a cyclic transition state.

Periplanar (Section 11.8): A conformation in which bonds to neighboring atoms have a parallel arrangement. In an eclipsed conformation, the neighboring bonds are syn periplanar; in a staggered conformation, the bonds are anti periplanar.

Peroxide (Section 18.1): A molecule containing an oxygen–oxygen bond functional group, ROOR′ or ROOH.

Peroxyacid (Section 7.8): A compound with the $-CO_3H$ functional group. Peroxyacids react with alkenes to give epoxides.

Phenol (Chapter 17 introduction): A compound with an $-OH$ group directly bonded to an aromatic ring, ArOH.

Phenyl (Section 15.1): The name for the $-C_6H_5$ unit when the benzene ring is considered as a substituent. A phenyl group is abbreviated as $-Ph$.

Phospholipid (Section 27.3): A lipid that contains a phosphate residue. For example, glycerophospholipids contain a glycerol backbone linked to two fatty acids and a phosphoric acid.

Phosphoric acid anhydride (Section 29.1): A substance that contains PO_2PO link, analogous to the CO_2CO link in carboxylic acid anhydrides.

Photochemical reaction (Section 30.3): A reaction carried out by irradiating the reactants with light.

Pi (π) bond (Section 1.8): The covalent bond formed by sideways overlap of atomic orbitals. For example, carbon–carbon double bonds contain a π bond formed by sideways overlap of two p orbitals.

PITC (Section 26.6): Phenylisothiocyanate; used in the Edman degradation.

Plane of symmetry (Section 9.2): A plane that bisects a molecule such that one half of the molecule is the mirror image of the other half. Molecules containing a plane of symmetry are achiral.

Plane-polarized light (Section 9.3): Ordinary light that has its electromagnetic waves oscillating in a single plane rather than in random planes. The plane of polarization is rotated when the light is passed through a solution of a chiral substance.

Plasticizer (Section 31.5): A small organic molecule added to polymers to act as a lubricant between polymer chains.

Polar aprotic solvent (Section 11.3): A polar solvent that can't function as a hydrogen ion donor. Polar aprotic solvents such as dimethyl sulfoxide (DMSO) and dimethylformamide (DMF) are particularly useful in S_N2 reactions because of their ability to solvate cations.

Polar covalent bond (Section 2.1): A covalent bond in which the electron distribution between atoms is unsymmetrical.

Polar reaction (Section 5.2): A reaction in which bonds are made when a nucleophile donates two electrons to an electrophile and in which bonds are broken when one fragment leaves with both electrons from the bond.

Polarity (Section 2.1): The unsymmetrical distribution of electrons in a molecule that results when one atom attracts electrons more strongly than another.

Polarizability (Section 5.4): The measure of the change in a molecule's electron distribution in response to changing electric interactions with solvents or ionic reagents.

Polycarbonate (Section 31.4): A polyester in which the carbonyl groups are linked to two −OR groups, $[O=C(OR)_2]$.

Polycyclic (Section 4.9): A compound that contains more than one ring.

Polycyclic aromatic compound (Section 15.7): A compound with two or more benzene-like aromatic rings fused together.

Polymer (Sections 7.10, 21.9, Chapter 31 introduction): A large molecule made up of repeating smaller units. For example, polyethylene is a synthetic polymer made from repeating ethylene units, and DNA is a biopolymer made of repeating deoxyribonucleotide units.

Polymerase chain reaction, PCR (Section 28.8): A method for amplifying small amounts of DNA to produce larger amounts.

Polysaccharide (Section 25.1): A carbohydrate that is made of many simple sugars linked together by acetal bonds.

Polyunsaturated fatty acid (Section 27.1): A fatty acid that contains more than one double bond.

Polyurethane (Section 31.4): A step-growth polymer prepared by reaction between a diol and a diisocyanate.

Primary, secondary, tertiary, quaternary (Section 3.3): Terms used to describe the substitution pattern at a specific site. A primary site has one organic substituent attached to it, a secondary site has two organic substituents, a tertiary site has three, and a quaternary site has four.

	Carbon	Carbocation	Hydrogen	Alcohol	Amine
Primary	RCH_3	RCH_2^+	RCH_3	RCH_2OH	RNH_2
Secondary	R_2CH_2	R_2CH^+	R_2CH_2	R_2CHOH	R_2NH
Tertiary	R_3CH	R_3C^+	R_3CH	R_3COH	R_3N
Quaternary	R_4C				

Primary structure (Section 26.9): The amino acid sequence in a protein.

pro-R (Section 9.13): One of two identical atoms in a compound, whose replacement leads to an R chirality center.

pro-S (Section 9.13): One of two identical atoms in a compound whose replacement leads to an S chirality center.

Prochiral (Section 9.13): A molecule that can be converted from achiral to chiral in a single chemical step.

Prochirality center (Section 9.13): An atom in a compound that can be converted into a chirality center by changing one of its attached substituents.

Propagation step (Section 5.3): The step or series of steps in a radical chain reaction that carry on the chain. The propagation steps must yield both product and a reactive intermediate.

Prostaglandin (Section 27.4): A lipid derived from arachidonic acid. Prostaglandins are present in nearly all body tissues and fluids, where they serve many important hormonal functions.

Protecting group (Sections 17.8, 19.10, 26.7): A group that is introduced to protect a sensitive functional group toward reaction elsewhere in the molecule. After serving its protective function, the group is removed.

Protein (Section 26.4): A large peptide containing 50 or more amino acid residues. Proteins serve both as structural materials and as enzymes that control an organism's chemistry.

Protic solvent (Section 11.3): A solvent such as water or alcohol that can act as a proton donor.

Pyramidal inversion (Section 24.2): The rapid stereochemical inversion of a trivalent nitrogen compound.

Pyranose (Section 25.5): The six-membered-ring form of a simple sugar.

Quartet (Section 13.11): A set of four peaks in an NMR spectrum, caused by spin–spin splitting of a signal by three adjacent nuclear spins.

Quaternary (*See* Primary)

Quaternary ammonium salt (Section 24.1): An ionic compound containing a positively charged nitrogen atom with four attached groups, $R_4N^+ X^-$.

Quaternary structure (Section 26.9): The highest level of protein structure, involving a specific aggregation of individual proteins into a larger cluster.

Quinone (Section 17.10): A cyclohexa-2,5-diene-1,4-dione.

R group (Section 3.3): A generalized abbreviation for an organic partial structure.

R,S convention (Section 9.5): A method for defining the absolute configuration at chirality centers using the Cahn–Ingold–Prelog sequence rules.

Racemic mixture (Section 9.8): A mixture consisting of equal parts (+) and (−) enantiomers of a chiral substance.

Radical (Section 5.2): A species that has an odd number of electrons, such as the chlorine radical, Cl·.

Radical reaction (Section 5.2): A reaction in which bonds are made by donation of one electron from each of two reactants and in which bonds are broken when each fragment leaves with one electron.

Rate constant (Section 11.2): The constant k in a rate equation.

Rate equation (Section 11.2): An equation that expresses the dependence of a reaction's rate on the concentration of reactants.

Rate-limiting step (Section 11.4): The slowest step in a multistep reaction sequence. The rate-limiting step acts as a kind of bottleneck in multistep reactions.

Re face (Section 9.13): One of two faces of a planar, sp^2-hybridized atom.

Rearrangement reaction (Section 5.1): What occurs when a single reactant undergoes a reorganization of bonds and atoms to yield an isomeric product.

Reducing sugar (Section 25.6): A sugar that reduces silver ion in the Tollens test or cupric ion in the Fehling or Benedict tests.

Reduction (Sections 7.7, 10.9): A reaction that causes an increase of electron ownership by carbon, either by bond-breaking between carbon and a more electronegative atom or by bond formation between carbon and a less electronegative atom.

Reductive amination (Sections 24.6, 26.3): A method for preparing an amine by reaction of an aldehyde or ketone with ammonia and a reducing agent.

Refining (Chapter 3 *Focus On*): The process by which petroleum is converted into gasoline and other useful products.

Regiochemistry (Section 6.8): A term describing the orientation of a reaction that occurs on an unsymmetrical substrate.

Regiospecific (Section 6.8): A term describing a reaction that occurs with a specific regiochemistry to give a single product rather than a mixture of products.

Replication (Section 28.3): The process by which double-stranded DNA uncoils and is replicated to produce two new copies.

Replication fork (Section 28.3): The point of unraveling in a DNA chain where replication occurs.

Residue (Section 26.4): An amino acid in a protein chain.

Resolution (Section 9.8): The process by which a racemic mixture is separated into its two pure enantiomers.

Resonance effect (Section 16.4): The donation or withdrawal of electrons through orbital overlap with neighboring π bonds. For example, an oxygen or nitrogen substituent donates electrons to an aromatic ring by overlap of the O or N orbital with the aromatic ring p orbitals.

Resonance form (Section 2.4): An individual Lewis structure of a resonance hybrid.

Resonance hybrid (Section 2.4): A molecule, such as benzene, that can't be represented adequately by a single Kekulé structure but must instead be considered as an average of two or more resonance structures. The resonance structures themselves differ only in the positions of their electrons, not their nuclei.

Restriction endonuclease (Section 28.6): An enzyme that is able to cleave a DNA molecule at points in the chain where a specific base sequence occurs.

Retrosynthetic (Sections 8.9, 16.11): A strategy for planning organic syntheses by working backward from the final product to the starting material.

Ribonucleic acid (RNA) (Section 28.1): The biopolymer found in cells that serves to transcribe the genetic information found in DNA and uses that information to direct the synthesis of proteins.

Ribosomal RNA (Section 28.4): A kind of RNA used in the physical makeup of ribosomes.

Ring current (Section 15.8): The circulation of π electrons induced in aromatic rings by an external magnetic field. This effect accounts for the downfield shift of aromatic ring protons in the ^1H NMR spectrum.

Ring-flip (Section 4.6): A molecular motion that converts one chair conformation of cyclohexane into another chair conformation. The effect of a ring-flip is to convert an axial substituent into an equatorial substituent.

RNA (*See* Ribonucleic acid; Section 28.1)

Robinson annulation reaction (Section 23.12): A synthesis of cyclohexenones by sequential Michael reaction and intramolecular aldol reaction.

s-cis conformation (Section 14.5): The conformation of a conjugated diene that is cis-like around the single bond.

Saccharide (Section 25.1): A sugar.

Salt bridge (Section 26.9): The ionic attraction between two oppositely charged groups in a protein chain.

Sandmeyer reaction (Section 24.8): The nucleophilic substitution reaction of an arenediazonium salt with a cuprous halide to yield an aryl halide.

Sanger dideoxy method (Section 2.6): The most commonly used method of DNA sequencing.

Saponification (Section 21.6): An old term for the base-induced hydrolysis of an ester to yield a carboxylic acid salt.

Saturated (Section 3.2): A molecule that has only single bonds and thus can't undergo addition reactions. Alkanes are saturated, but alkenes are unsaturated.

Sawhorse structure (Section 3.6): A manner of representing stereochemistry that uses a stick drawing and gives a perspective view of the conformation around a single bond.

Schiff base (Sections 19.8, 29.5): An alternative name for an imine, $R_2C=NR'$, used primarily in biochemistry.

Second-order reaction (Section 11.2): A reaction whose rate-limiting step is bimolecular and whose kinetics are therefore dependent on the concentration of two reactants.

Secondary (*See* Primary)

Secondary structure (Section 26.9): The level of protein substructure that involves organization of chain sections into ordered arrangements such as β-pleated sheets or α helices.

Semiconservative replication (Section 28.3): The process by which DNA molecules are made containing one strand of old DNA and one strand of new DNA.

Sequence rules (Sections 6.5, 9.5): A series of rules for assigning relative priorities to substituent groups on a double-bond carbon atom or on a chirality center.

Sesquiterpenoid (Section 27.5): A 15-carbon lipid.

Shell (electron) (Section 1.2): A group of an atom's electrons with the same principal quantum number.

Shielding (Section 13.2): An effect observed in NMR that causes a nucleus to absorb toward the right (upfield) side of the chart. Shielding is caused by donation of electron density to the nucleus.

***Si* face** (Section 9.13): One of two faces of a planar, sp^2-hybridized atom.

Sialic acid (Section 25.7): A group of more than 300 carbohydrates based on acetylneuramic acid.

Side chain (Section 26.1): The substituent attached to the α carbon of an amino acid.

Sigma (σ) bond (Section 1.6): A covalent bond formed by head-on overlap of atomic orbitals.

Sigmatropic reaction (Section 30.8): A pericyclic reaction that involves the migration of a group from one end of a π electron system to the other.

Simmons–Smith reaction (Section 7.6): The reaction of an alkene with CH_2I_2 and Zn–Cu to yield a cyclopropane.

Simple sugar (Section 25.1): A carbohydrate that cannot be broken down into smaller sugars by hydrolysis.

Skeletal structure (Section 1.12): A shorthand way of writing structures in which carbon atoms are assumed to be at each intersection of two lines (bonds) and at the end of each line.

S_N1 reaction (Section 11.4): A unimolecular nucleophilic substitution reaction.

S_N2 reaction (Section 11.2): A bimolecular nucleophilic substitution reaction.

Solid-phase synthesis (Section 26.8): A technique of synthesis whereby the starting material is covalently bound to a solid polymer bead and reactions are carried out on the bound substrate. After the desired transformations have been effected, the product is cleaved from the polymer.

Solvation (Sections 11.3): The clustering of solvent molecules around a solute particle to stabilize it.

***sp* Orbital** (Section 1.9): A hybrid orbital derived from the combination of an s and a p atomic orbital. The two sp orbitals that result from hybridization are oriented at an angle of 180° to each other.

***sp²* Orbital** (Section 1.8): A hybrid orbital derived by combination of an s atomic orbital with two p atomic orbitals. The three sp^2 hybrid orbitals that result lie in a plane at angles of 120° to each other.

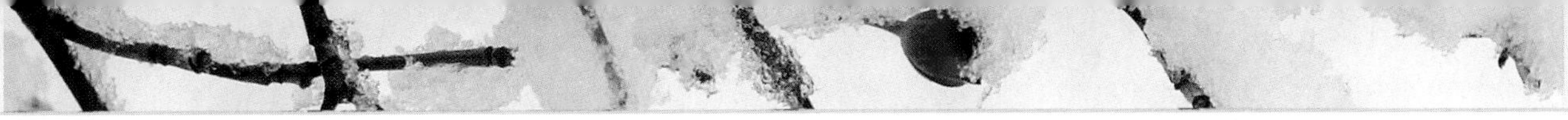

sp^3 Orbital (Section 1.6): A hybrid orbital derived by combination of an *s* atomic orbital with three *p* atomic orbitals. The four sp^3 hybrid orbitals that result are directed toward the corners of a regular tetrahedron at angles of 109° to each other.

Specific rotation, $[\alpha]_D$ (Section 9.3): The optical rotation of a chiral compound under standard conditions.

Sphingomyelin (Section 27.3): A phospholipid that has sphingosine as its backbone.

Spin–spin splitting (Section 13.11): The splitting of an NMR signal into a multiplet because of an interaction between nearby magnetic nuclei whose spins are coupled. The magnitude of spin–spin splitting is given by the coupling constant, *J*.

Staggered conformation (Section 3.4): The three-dimensional arrangement of atoms around a carbon–carbon single bond in which the bonds on one carbon bisect the bond angles on the second carbon as viewed end-on.

Step-growth polymer (Sections 21.9, 31.4): A polymer in which each bond is formed independently of the others. Polyesters and polyamides (nylons) are examples.

Stereochemistry (Chapters 3, 4, 9): The branch of chemistry concerned with the three-dimensional arrangement of atoms in molecules.

Stereoisomers (Section 4.2): Isomers that have their atoms connected in the same order but have different three-dimensional arrangements. The term *stereoisomer* includes both enantiomers and diastereomers.

Stereospecific (Section 7.6): A term indicating that only a single stereoisomer is produced in a given reaction rather than a mixture.

Steric strain (Sections 3.7): The strain imposed on a molecule when two groups are too close together and try to occupy the same space. Steric strain is responsible both for the greater stability of trans versus cis alkenes and for the greater stability of equatorially substituted versus axially substituted cyclohexanes.

Steroid (Section 27.6): A lipid whose structure is based on a tetracyclic carbon skeleton with three 6-membered and one 5-membered ring. Steroids occur in both plants and animals and have a variety of important hormonal functions.

Stork reaction (Section 23.11): A carbonyl condensation between an enamine and an α,β-unsaturated acceptor in a Michael-like reaction to yield a 1,5-dicarbonyl product.

Straight-chain alkane (Section 3.2): An alkane whose carbon atoms are connected without branching.

Substitution reaction (Section 5.1): What occurs when two reactants exchange parts to give two new products. S_N1 and S_N2 reactions are examples.

Sulfide (Section 18.8): A compound that has two organic substituents bonded to the same sulfur atom, RSR′.

Sulfone (Section 18.8): A compound of the general structure RSO_2R'.

Sulfonium ion (Section 18.8): A species containing a positively charged, trivalent sulfur atom, R_3S^+.

Sulfoxide (Section 18.8): A compound of the general structure RSOR′.

Suprafacial (Section 30.6): A word used to describe the geometry of pericyclic reactions. Suprafacial reactions take place on the same side of the two ends of a π electron system.

Symmetry-allowed, symmetry-disallowed (Section 30.2): A symmetry-allowed reaction is a pericyclic process that has a favorable orbital symmetry for reaction through a concerted pathway. A symmetry-disallowed reaction is one that does not have favorable orbital symmetry for reaction through a concerted pathway.

Symmetry plane (Section 9.2): A plane that bisects a molecule such that one half of the molecule is the mirror image of the other half. Molecules containing a plane of symmetry are achiral.

Syn periplanar (Section 11.8): Describing a stereochemical relationship in which two bonds on adjacent carbons lie in the same plane and are eclipsed.

Syn stereochemistry (Section 7.5): The opposite of anti. A syn addition reaction is one in which the two ends of the double bond react from the same side. A syn elimination is one in which the two groups leave from the same side of the molecule.

Syndiotactic (Section 31.2): A chain-growth polymer in which the substituents regularly alternate on opposite sides of the backbone.

Tautomers (Sections 8.4, 22.1): Isomers that are rapidly interconverted.

Template strand (Section 28.4): The strand of double-helical DNA that does not contain the gene.

Terpenoid (Chapter 6 *Focus On*, Section 27.5): A lipid that is formally derived by head-to-tail polymerization of isoprene units.

Tertiary (*See* Primary)

Tertiary structure (Section 26.9): The level of protein structure that involves the manner in which the entire protein chain is folded into a specific three-dimensional arrangement.

Thermodynamic control (Section 14.3): An equilibrium reaction that yields the lowest-energy, most stable product is said to be thermodynamically controlled.

Thermoplastic (Section 31.5): A polymer that has a high T_g and is therefore hard at room temperature, but becomes soft and viscous when heated.

Thermosetting resin (Section 31.5): A polymer that becomes highly cross-linked and solidifies into a hard, insoluble mass when heated.

Thioester (Section 21.8): A compound with the RCOSR′ functional group.

Thiol (Section 18.8): A compound containing the −SH functional group.

Thiolate ion (Section 18.8): The anion of a thiol, RS⁻.

TMS (Section 13.3): Tetramethylsilane; used as an NMR calibration standard.

TOF (Section 12.4): Time-of flight mass spectrometry; a sensitive method of mass detection accurate to about 3 ppm.

Tollens' reagent (Section 19.3): A solution of Ag_2O in aqueous ammonia; used to oxidize aldehydes to carboxylic acids.

Torsional strain (Section 3.6): The strain in a molecule caused by electron repulsion between eclipsed bonds. Torsional strain is also called eclipsing strain.

Tosylate (Section 11.1): A *p*-toluenesulfonate ester; useful as a leaving group in nucleophilic substitution reactions.

Transamination (Section 29.9): The exchange of an amino group and a keto group between reactants.

Transcription (Section 28.4): The process by which the genetic information encoded in DNA is read and used to synthesize RNA in the nucleus of the cell. A small portion of double-stranded DNA uncoils, and complementary ribonucleotides line up in the correct sequence for RNA synthesis.

Transfer RNA (Section 28.4): A kind of RNA that transports amino acids to the ribosomes, where they are joined together to make proteins.

Transition state (Section 5.9): An activated complex between reactants, representing the highest energy point on a reaction curve. Transition states are unstable complexes that can't be isolated.

Translation (Section 28.5): The process by which the genetic information transcribed from DNA onto mRNA is read by tRNA and used to direct protein synthesis.

Tree diagram (Section 13.12): A diagram used in NMR to sort out the complicated splitting patterns that can arise from multiple couplings.

Triacylglycerol (Section 27.1): A lipid, such as those found in animal fat and vegetable oil, that is, a triester of glycerol with long-chain fatty acids.

Tricarboxylic acid cycle (Section 29.7): An alternative name for the citric acid cycle by which acetyl CoA is degraded to CO_2.

Triplet (Section 13.11): A symmetrical three-line splitting pattern observed in the ^1H NMR spectrum when a proton has two equivalent neighbor protons.

Turnover number (Section 26.10): The number of substrate molecules acted on by an enzyme per unit time.

Twist-boat conformation (Section 4.5): A conformation of cyclohexane that is somewhat more stable than a pure boat conformation.

Ultraviolet (UV) spectroscopy (Section 14.7): An optical spectroscopy employing ultraviolet irradiation. UV spectroscopy provides structural information about the extent of π electron conjugation in organic molecules.

Unimolecular reaction (Section 11.4): A reaction that occurs by spontaneous transformation of the starting material without the intervention of other reactants. For example, the dissociation of a tertiary alkyl halide in the S_N1 reaction is a unimolecular process.

Unsaturated (Section 6.2): A molecule that has one or more multiple bonds.

Upfield (Section 13.3): The right-hand portion of the NMR chart.

Urethane (Section 31.4): A functional group in which a carbonyl group is bonded to both an −OR group and an −NR₂ group.

Uronic acid (Section 25.6): The monocarboxylic acid resulting from enzymatic oxidation of the −CH₂OH group of an aldose.

Valence bond theory (Section 1.5): A bonding theory that describes a covalent bond as resulting from the overlap of two atomic orbitals.

Valence shell (Section 1.4): The outermost electron shell of an atom.

Van der Waals forces (Section 2.13): Intermolecular forces that are responsible for holding molecules together in the liquid and solid states.

Vicinal (Section 8.2): A term used to refer to a 1,2-disubstitution pattern. For example, 1,2-dibromoethane is a vicinal dibromide.

Vinyl group (Section 6.3): An $H_2C=CH-$ substituent.

Vinyl monomer (Sections 7.10, 31.1): A substituted alkene monomer used to make chain-growth polymers.

Vinylic (Section 8.3): A term that refers to a substituent at a double-bond carbon atom. For example, chloroethylene is a vinylic chloride, and enols are vinylic alcohols.

Vitamin (Section 26.10): A small organic molecule that must be obtained in the diet and is required in trace amounts for proper growth and function.

Vulcanization (Section 14.6): A technique for cross-linking and hardening a diene polymer by heating with a few percent by weight of sulfur.

Walden inversion (Section 11.1): The inversion of configuration at a chirality center that accompanies an S_N2 reaction.

Wave equation (Section 1.2): A mathematical expression that defines the behavior of an electron in an atom.

Wave function (Section 1.2): A solution to the wave equation for defining the behavior of an electron in an atom. The square of the wave function defines the shape of an orbital.

Wavelength, λ (Section 12.5): The length of a wave from peak to peak. The wavelength of electromagnetic radiation is inversely proportional to frequency and inversely proportional to energy.

Wavenumber, $\tilde{\nu}$ (Section 12.6): The reciprocal of the wavelength in centimeters.

Wax (Section 27.1): A mixture of esters of long-chain carboxylic acids with long-chain alcohols.

Williamson ether synthesis (Section 18.2): A method for synthesizing ethers by S_N2 reaction of an alkyl halide with an alkoxide ion.

Wittig reaction (Section 19.11): The reaction of a phosphorus ylide with a ketone or aldehyde to yield an alkene.

Wohl degradation (Section 25.6): A method for shortening the chain of an aldose sugar.

Wolff–Kishner reaction (Section 19.9): The conversion of an aldehyde or ketone into an alkane by reaction with hydrazine and base.

Wood alcohol (Chapter 17 introduction): An old name for methanol.

Ylide (Section 19.11): A neutral dipolar molecule with adjacent positive and negative charges. The phosphoranes used in Wittig reactions are ylides.

Z geometry (Section 6.5): A term used to describe the stereochemistry of a carbon–carbon double bond. The two groups on each carbon are assigned priorities according to the Cahn–Ingold–Prelog sequence rules, and the two carbons are compared. If the high-priority groups on each carbon are on the same side of the double bond, the bond has Z geometry.

Zaitsev's rule (Section 11.7): A rule stating that E2 elimination reactions normally yield the more highly substituted alkene as major product.

Ziegler–Natta catalyst (Section 31.2): A catalyst of an alkylaluminum and a titanium compound used for preparing alkene polymers.

Zwitterion (Section 26.1): A neutral dipolar molecule in which the positive and negative charges are not adjacent. For example, amino acids exist as zwitterions, $H_3\overset{+}{N}-CHR-CO_2{}^-$.

Answers to In-Text Problems

The following answers are meant only as a quick check while you study. Full answers for all problems are provided in the accompanying *Study Guide and Solutions Manual*.

CHAPTER 16

16.1 *o*-, *m*-, and *p*-Bromotoluene

16.2 *o*-Xylene: 2; *p*-xylene: 1; *m*-xylene: 3

16.3 D$^+$ does electrophilic substitutions on the ring.

16.4 No rearrangement: (a), (b), (e)

16.5 *tert*-Butylbenzene

16.6 (a) $(CH_3)_2CHCOCl$ (b) $PhCOCl$

16.7

and others

16.8

and others

16.9 (a) *o*- and *p*-Bromonitrobenzene
(b) *m*-Bromonitrobenzene
(c) *o*- and *p*-Chlorophenol
(d) *o*- and *p*-Bromoaniline

16.10 (a) Phenol > Toluene > Benzene > Nitrobenzene
(b) Phenol > Benzene > Chlorobenzene > Benzoic acid
(c) Aniline > Benzene > Bromobenzene > Benzaldehyde

16.11 Alkylbenzenes are more reactive than benzene itself, but acylbenzenes are less reactive.

16.12 Toluene is more reactive; the trifluoromethyl group is electron-withdrawing.

16.13 The nitrogen electrons are donated to the nearby carbonyl group and are less available to the ring.

16.14 The meta intermediate is most favored.

16.15 (a) Ortho and para to $-OCH_3$
(b) Ortho and para to $-NH_2$
(c) Ortho and para to $-Cl$

16.16 (a) Reaction occurs ortho and para to the $-CH_3$ group.
(b) Reaction occurs ortho and para to the $-OCH_3$ group.

16.17 The phenol is deprotonated by KOH to give an anion that carries out a nucleophilic acyl substitution reaction on the fluoronitrobenzene.

16.18 Only one benzyne intermediate can form from *p*-bromotoluene; two different benzyne intermediates can form from *m*-bromotoluene.

16.19 (a) *m*-Nitrobenzoic acid
(b) *p-tert*-Butylbenzoic acid

16.20 A benzyl radical is more stable than a primary alkyl radical by 52 kJ/mol and is similar in stability to an allyl radical.

16.21 1. CH_3CH_2Cl, $AlCl_3$; 2. NBS; 3. KOH, ethanol

16.22 1. $PhCOCl$, $AlCl_3$; 2. H_2/Pd

16.23 (a) 1. HNO_3, H_2SO_4; 2. Cl_2, $FeCl_3$
(b) 1. CH_3COCl, $AlCl_3$; 2. Cl_2, $FeCl_3$; 3. H_2/Pd
(c) 1. CH_3CH_2COCl, $AlCl_3$; 2. Cl_2, $FeCl_3$; 3. H_2/Pd; 4. HNO_3, H_2SO_4
(d) 1. CH_3Cl, $AlCl_3$; 2. Br_2, $FeBr_3$; 3. SO_3, H_2SO_4

16.24 (a) Friedel–Crafts acylation does not occur on a deactivated ring.
(b) Rearrangement occurs during Friedel–Crafts alkylation with primary halides; chlorination occurs ortho to the alkyl group.

CHAPTER 17

17.1 (a) 5-Methyl-2,4-hexanediol
(b) 2-Methyl-4-phenyl-2-butanol
(c) 4,4-Dimethylcyclohexanol
(d) *trans*-2-Bromocyclopentanol
(e) 4-Bromo-3-methylphenol
(f) 2-Cyclopenten-1-ol

17.2 (a)

$$H_3C\text{-}\overset{\displaystyle H}{\underset{\displaystyle }{C}}=C\overset{\displaystyle CH_2OH}{\underset{\displaystyle CH_2CH_3}{}}$$

(b) cyclohexenol structure with OH

(c) cycloheptane with OH, H and Cl, H substituents

(d) $CH_3CHCH_2CH_2CH_2OH$ with OH

(e) 2,6-dimethylphenol: H_3C and CH_3 with OH

(f) phenol with CH_2CH_2OH and OH

17.3 Hydrogen-bonding is more difficult in hindered alcohols.

17.4 (a) $HC\equiv CH < (CH_3)_2CHOH < CH_3OH < (CF_3)_2CHOH$
(b) *p*-Methylphenol < Phenol < *p*-(Trifluoromethyl)phenol
(c) Benzyl alcohol < Phenol < *p*-Hydroxybenzoic acid

17.5 The electron-withdrawing nitro group stabilizes an alkoxide ion, but the electron-donating methoxyl group destabilizes the anion.

17.6 (a) 2-Methyl-3-pentanol
(b) 2-Methyl-4-phenyl-2-butanol
(c) *meso*-5,6-Decanediol

17.7 (a) $NaBH_4$ (b) $LiAlH_4$ (c) $LiAlH_4$

17.8 (a) Benzaldehyde or benzoic acid (or ester)
(b) Acetophenone
(c) Cyclohexanone
(d) 2-Methylpropanal or 2-methylpropanoic acid (or ester)

17.9 (a) 1-Methylcyclopentanol
(b) 1,1-Diphenylethanol
(c) 3-Methyl-3-hexanol

17.10 (a) Acetone + CH_3MgBr, or ethyl acetate + 2 CH_3MgBr
(b) Cyclohexanone + CH_3MgBr
(c) 3-Pentanone + CH_3MgBr, or 2-butanone + CH_3CH_2MgBr, or ethyl acetate + 2 CH_3CH_2MgBr
(d) 2-Butanone + $PhMgBr$, or ethyl phenyl ketone + CH_3MgBr, or acetophenone + CH_3CH_2MgBr
(e) Formaldehyde + $PhMgBr$
(f) Formaldehyde + $(CH_3)_2CHCH_2MgBr$

17.11 Cyclohexanone + CH_3CH_2MgBr

17.12 1. *p*-TosCl, pyridine; 2. NaCN

17.13 (a) 2-Methyl-2-pentene
(b) 3-Methylcyclohexene
(c) 1-Methylcyclohexene
(d) 2,3-Dimethyl-2-pentene
(e) 2-Methyl-2-pentene

17.14 (a) 1-Phenylethanol (b) 2-Methyl-1-propanol
(c) Cyclopentanol

17.15 (a) Hexanoic acid, hexanal (b) 2-Hexanone
(c) Hexanoic acid, no reaction

17.16 S_N2 reaction of F^- on silicon with displacement of alkoxide ion.

17.17 Protonation of 2-methylpropene gives the *tert*-butyl cation, which carries out an electrophilic aromatic substitution reaction.

17.18 Disappearance of $-OH$ absorption; appearance of $C=O$

17.19 (a) Singlet (b) Doublet (c) Triplet
(d) Doublet (e) Doublet (f) Singlet

CHAPTER 18

18.1 (a) Diisopropyl ether
(b) Cyclopentyl propyl ether
(c) *p*-Bromoanisole or 4-bromo-1-methoxybenzene
(d) 1-Methoxycyclohexene
(e) Ethyl isobutyl ether
(f) Allyl vinyl ether

18.2 A mixture of diethyl ether, dipropyl ether, and ethyl propyl ether is formed in a $1:1:2$ ratio.

18.3 (a) $CH_3CH_2CH_2O^- + CH_3Br$
(b) $PhO^- + CH_3Br$
(c) $(CH_3)_2CHO^- + PhCH_2Br$
(d) $(CH_3)_3CCH_2O^- + CH_3CH_2Br$

18.4

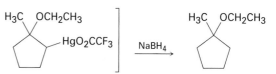

18.5 (a) Either method (b) Williamson
(c) Alkoxymercuration (d) Williamson

18.6 (a) Bromoethane > 2-Bromopropane > Bromobenzene

(b) Bromoethane > Chloroethane > 1-Iodopropene

18.7 (a)

2-phenyl-2-bromopropane + CH₃OH

(b)

$$CH_3CH_2\overset{\underset{\displaystyle CH_3}{|}}{C}HOH \quad + \quad CH_3CH_2CH_2Br$$

18.8 Protonation of the oxygen atom, followed by E1 reaction

18.9 Br^- and I^- are better nucleophiles than Cl^-.

18.10 *o*-(1-Methylallyl)phenol

18.11 Epoxidation of *cis*-2-butene yields *cis*-2,3-epoxybutane, while epoxidation of *trans*-2-butene yields *trans*-2,3-epoxybutane.

18.12 (a)

(b)

18.13 (a) 1-Methylcyclohexene + OsO_4; then $NaHSO_3$

(b) 1-Methylcyclohexene + *m*-chloroperoxybenzoic acid, then H_3O^+

18.14 (a)

$$CH_3CH_2\overset{\underset{\displaystyle CH_3}{|}}{C}-\overset{\displaystyle HO \ \ ^{*}OH}{\overset{|}{C}H_2}$$

(b)

$$CH_3CH_2\overset{\underset{\displaystyle CH_3}{|}}{C}-\overset{\displaystyle HO^{*} \ OH}{\overset{|}{C}H_2}$$

(c)

18.16 (a) 2-Butanethiol

(b) 2,2,6-Trimethyl-4-heptanethiol

(c) 2-Cyclopentene-1-thiol

(d) Ethyl isopropyl sulfide

(e) *o*-Di(methylthio)benzene

(f) 3-(Ethylthio)cyclohexanone

18.17 (a) 1. $LiAlH_4$; 2. PBr_3; 3. $(H_2N)_2C{=}S$; 4. H_2O, NaOH

(b) 1. HBr; 2. $(H_2N)_2C{=}S$; 3. H_2O, NaOH

18.18 1,2-Epoxybutane

PREVIEW OF CARBONYL CHEMISTRY

1. Acetyl chloride is more electrophilic than acetone.

2.

3. (a) Nucleophilic acyl substitution

(b) Nucleophilic addition

(c) Carbonyl condensation

CHAPTER 19

19.1 (a) 2-Methyl-3-pentanone

(b) 3-Phenylpropanal

(c) 2,6-Octanedione

(d) *trans*-2-Methylcyclohexanecarbaldehyde

(e) Pentanedial

(f) *cis*-2,5-Dimethylcyclohexanone

19.2

(a)

$$CH_3\overset{\underset{\displaystyle CH_3}{|}}{C}HCH_2CHO$$

(b)

$$CH_3\overset{\underset{\displaystyle Cl}{|}}{C}HCH_2\overset{\underset{\displaystyle O}{\|}}{C}CH_3$$

(c)

(d)

(e)

$$H_2C{=}\overset{\underset{\displaystyle CH_3}{|}}{C}CH_2CHO$$

(f)

$$CH_3CH_2\overset{\underset{\displaystyle CH_3}{|}}{C}HCH_2CH_2\overset{\underset{\displaystyle CH_3CHCl}{|}}{C}HCHO$$

19.3 (a) PCC   (b) 1. O_3; 2. Zn   (c) DIBAH

19.4 (a) $Hg(OAc)_2$, H_3O^+

(b) 1. CH_3COCl, $AlCl_3$; 2. Br_2, $FeBr_3$

(c) 1. Mg; 2. CH_3CHO; 3. H_3O^+; 4. PCC

(d) 1. BH_3; 2. H_2O_2, NaOH; 3. PCC

19.5

19.6 The electron-withdrawing nitro group in *p*-nitrobenzaldehyde polarizes the carbonyl group.

19.7 $CCl_3CH(OH)_2$

19.8 Labeled water adds reversibly to the carbonyl group.

19.9 The equilibrium is unfavorable for sterically hindered ketones.

19.10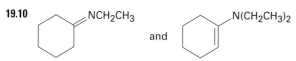
and

19.11 The steps are the exact reverse of the forward reaction.

19.12

$$\text{(cyclopentanone)} =O \ + \ (CH_3CH_2)_2NH \ \longrightarrow$$

$$\text{(cyclopentene)}-N(CH_2CH_3)_2$$

19.13 (a) H_2/Pd (b) N_2H_4, KOH
(c) 1. H_2/Pd; 2. N_2H_4, KOH

19.14 The mechanism is identical to that between a ketone and 2 equivalents of a monoalcohol (text Figure 19.12).

19.15

$$CH_3O_2C-\text{(aryl)}-\overset{CH_3}{\underset{}{CH}}-CHO \ + \ CH_3OH$$

19.16 (a) Cyclohexanone + $(Ph)_3P{=}CHCH_3$
(b) 2-Cyclohexenone + $(Ph)_3P{=}CH_2$
(c) Acetone + $(Ph)_3P{=}CHCH_2CH_2CH_3$
(d) Acetone + $(Ph)_3P{=}CHPh$
(e) $PhCOCH_3$ + $(Ph)_3P{=}CHPh$
(f) 2-Cyclohexenone + $(Ph)_3P{=}CH_2$

19.17

β-Carotene

19.18 Intramolecular Cannizzaro reaction

19.19 Addition of the *pro-R* hydrogen of NADH takes place on the *Re* face of pyruvate.

19.20 The $-OH$ group adds to the *Re* face at C2, and $-H$ adds to the *Re* face at C3, to yield (2*R*,3*S*)-isocitrate.

19.21

19.22 (a) 3-Buten-2-one + $(CH_3CH_2CH_2)_2CuLi$
(b) 3-Methyl-2-cyclohexenone + $(CH_3)_2CuLi$
(c) 4-*tert*-Butyl-2-cyclohexenone + $(CH_3CH_2)_2CuLi$
(d) Unsaturated ketone + $(H_2C{=}CH)_2CuLi$

19.23 Look for appearance of either an alcohol or a saturated ketone in the product.

19.24 (a) $1715 \ cm^{-1}$ (b) $1685 \ cm^{-1}$
(c) $1750 \ cm^{-1}$ (d) $1705 \ cm^{-1}$
(e) $1715 \ cm^{-1}$ (f) $1705 \ cm^{-1}$

19.25 (a) Different peaks due to McLafferty rearrangement
(b) Different peaks due to α cleavage and McLafferty rearrangement
(c) Different peaks due to McLafferty rearrangement

19.26 IR: $1750 \ cm^{-1}$; MS: 140, 84

CHAPTER 20

20.1 (a) 3-Methylbutanoic acid
(b) 4-Bromopentanoic acid
(c) 2-Ethylpentanoic acid
(d) *cis*-4-Hexenoic acid
(e) 2,4-Dimethylpentanenitrile
(f) *cis*-1,3-Cyclopentanedicarboxylic acid

20.2

(a) $CH_3CH_2CH_2\overset{H_3C}{\underset{}{CH}}\overset{CH_3}{\underset{}{CH}}CO_2H$
(b) $CH_3\overset{CH_3}{\underset{}{CH}}CH_2CH_2CO_2H$

(c)
(d)

(e)

(f) $CH_3CH_2CH{=}CHCN$

20.3 Dissolve the mixture in ether, extract with aqueous NaOH, separate and acidify the aqueous layer, and extract with ether.

20.4 43%

20.5 (a) 82% dissociation (b) 73% dissociation

20.6 Lactic acid is stronger because of the inductive effect of the $-OH$ group.

20.7 The dianion is destabilized by repulsion between charges.

20.8 More reactive

20.9 (a) *p*-Methylbenzoic acid < Benzoic acid < *p*-Chlorobenzoic acid

 (b) Acetic acid < Benzoic acid < *p*-Nitrobenzoic acid

20.10 (a) 1. Mg; 2. CO_2; 3. H_3O^+

 (b) 1. Mg; 2. CO_2; 3. H_3O^+ or 1. NaCN; 2. H_3O^+

20.11 1. NaCN; 2. H_3O^+; 3. $LiAlH_4$

20.12 1. PBr_3; 2. NaCN; 3. H_3O^+; 4. $LiAlH_4$

20.13 (a) Propanenitrile + CH_3CH_2MgBr, then H_3O^+

 (b) *p*-Nitrobenzonitrile + CH_3MgBr, then H_3O^+

20.14 1. NaCN; 2. CH_3CH_2MgBr, then H_3O^+

20.15 A carboxylic acid has a very broad −OH absorption at 2500–3300 cm^{-1}.

20.16 4-Hydroxycyclohexanone: **H**−C−O absorption near 4 δ in ^1H spectrum and **C**=O absorption near 210 δ in ^{13}C spectrum. Cyclopentanecarboxylic acid: $−CO_2$**H** absorption near 12 δ in ^1H spectrum and $−$**C**O_2H absorption near 170 δ in ^{13}C spectrum.

CHAPTER 21

21.1 (a) 4-Methylpentanoyl chloride

 (b) Cyclohexylacetamide

 (c) Isopropyl 2-methylpropanoate

 (d) Benzoic anhydride

 (e) Isopropyl cyclopentanecarboxylate

 (f) Cyclopentyl 2-methylpropanoate

 (g) *N*-Methyl-4-pentenamide

 (h) (*R*)-2-Hydroxypropanoyl phosphate

 (i) Ethyl 2,3-Dimethyl-2-butenethioate

21.2

(a) $C_6H_5CO_2C_6H_5$ (b) $CH_3CH_2CH_2CON(CH_3)CH_2CH_3$

(c) $(CH_3)_2CHCH_2CH(CH_3)COCl$ (d)

(e)

$CH_3CH_2CCH_2COCH_2CH_3$

(f)

(g)

(h)

21.3

21.4 (a) Acetyl chloride > Methyl acetate > Acetamide

 (b) Hexafluoroisopropyl acetate > 2,2,2-Trichloroethyl acetate > Methyl acetate

21.5 (a) $CH_3CO_2^- Na^+$ (b) CH_3CONH_2

 (c) $CH_3CO_2CH_3 + CH_3CO_2^- Na^+$

 (d) $CH_3CONHCH_3$

21.6

21.7 (a) Acetic acid + 1-butanol

 (b) Butanoic acid + methanol

 (c) Cyclopentanecarboxylic acid + isopropyl alcohol

21.8

21.9 (a) Propanoyl chloride + methanol

 (b) Acetyl chloride + ethanol

 (c) Benzoyl chloride + ethanol

21.10 Benzoyl chloride + cyclohexanol

21.11 This is a typical nucleophilic acyl substitution reaction, with morpholine as the nucleophile and chloride as the leaving group.

21.12 (a) Propanoyl chloride + methylamine

 (b) Benzoyl chloride + diethylamine

 (c) Propanoyl chloride + ammonia

21.13 (a) Benzoyl chloride + $[(CH_3)_2CH]_2CuLi$, or 2-methylpropanoyl chloride + Ph_2CuLi

 (b) 2-Propenoyl chloride + $(CH_3CH_2CH_2)_2CuLi$, or butanoyl chloride + $(H_2C=CH)_2CuLi$

21.14 This is a typical nucleophilic acyl substitution reaction, with *p*-hydroxyaniline as the nucleophile and acetate ion as the leaving group.

21.15 Monomethyl ester of benzene-1,2-dicarboxylic acid

21.16 Reaction of a carboxylic acid with an alkoxide ion gives the carboxylate ion.

21.17 $HOCH_2CH_2CH_2CHO$

21.18 (a) $CH_3CH_2CH_2CH(CH_3)CH_2OH$
(b) $PhOH + PhCH_2OH$

21.19 (a) Ethyl benzoate + 2 CH_3MgBr
(b) Ethyl acetate + 2 $PhMgBr$
(c) Ethyl pentanoate + 2 CH_3CH_2MgBr

21.20 (a) H_2O, NaOH
(b) Benzoic acid + BH_3
(c) $LiAlH_4$

21.21 1. Mg; 2. CO_2, then H_3O^+; 3. $SOCl_2$; 4. $(CH_3)_2NH$; 5. $LiAlH_4$

21.22

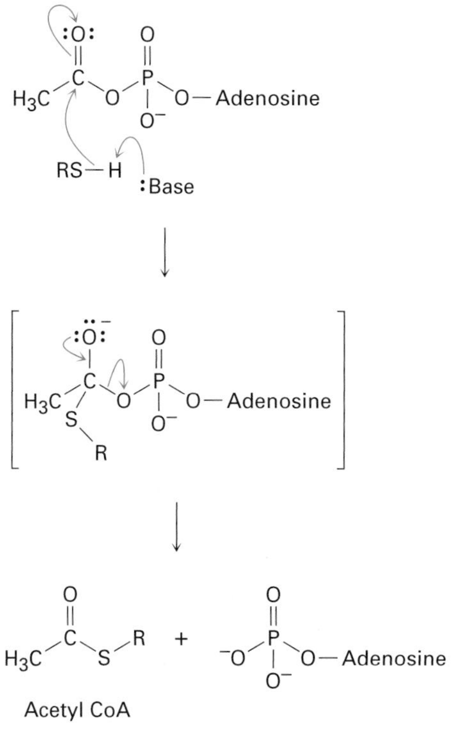

Acetyl CoA

21.23 (a)

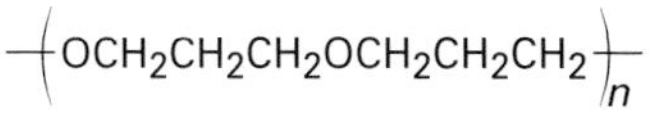

(b)

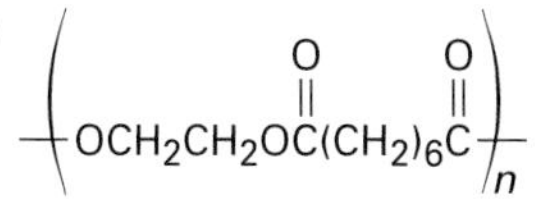

(c)

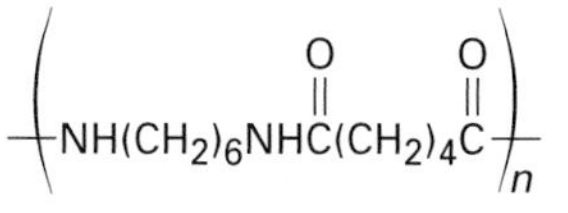

21.24

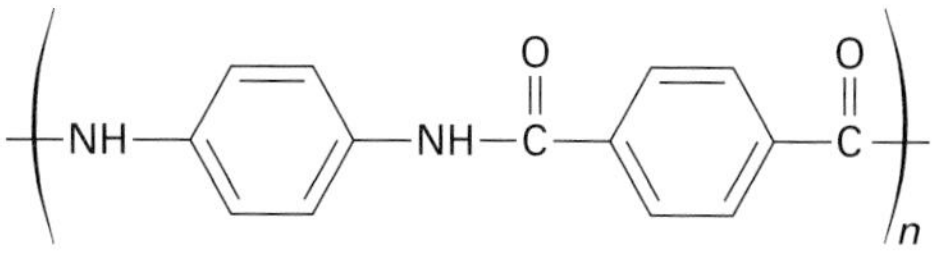

21.25 The product has a large amount of cross-linking.

21.26 (a) Ester   (b) Acid chloride
(c) Carboxylic acid
(d) Aliphatic ketone or cyclohexanone

21.27 (a) $CH_3CH_2CH_2CO_2CH_2CH_3$ and other possibilities
(b) $CH_3CON(CH_3)_2$
(c) $CH_3CH=CHCOCl$ or $H_2C=C(CH_3)COCl$

CHAPTER 22

22.1 (a)

(b)

$H_2C=CSCH_3$ with OH

(c)

$H_2C=COCH_2CH_3$ with OH

(d) $CH_3CH=CHOH$

(e)

$CH_3CH=COH$ with OH

(f)

$PhCH=CCH_3$ or $PhCH_2C=CH_2$ (with OH)

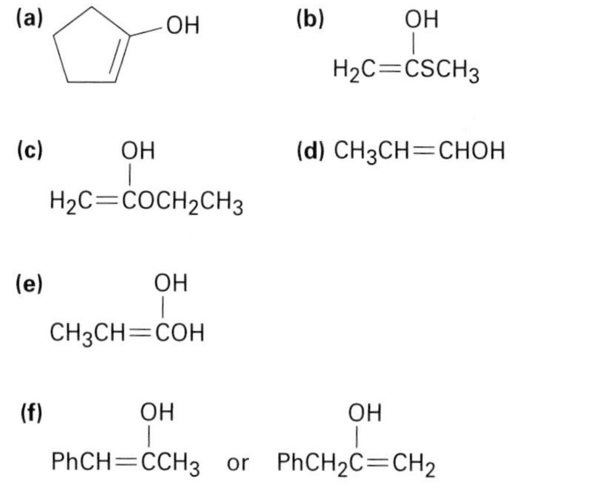

22.2 (a) 4   (b) 3   (c) 3   (d) 2   (e) 4   (f) 5

22.3

Equivalent; more stable

Equivalent; less stable

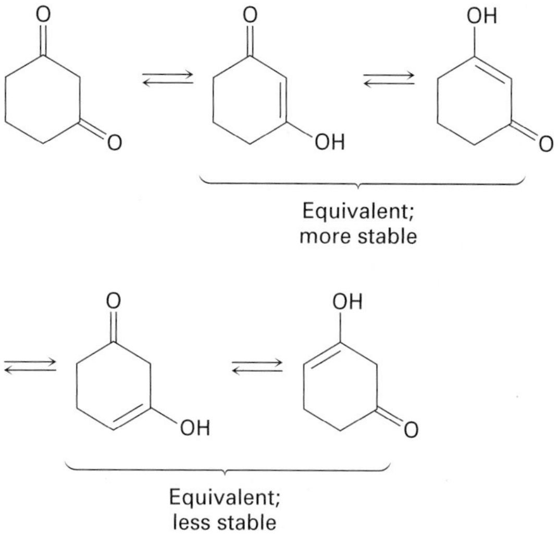

22.4 Acid-catalyzed formation of an enol is followed by deuteronation of the enol double bond and dedeuteronation of oxygen.

22.5 1. Br_2; 2. Pyridine, heat

22.6 The intermediate α-bromo acid bromide undergoes a nucleophilic acyl substitution reaction with methanol to give an α-bromo ester.

22.7 (a) $CH_3\underline{CH}_2CHO$ (b) $(CH_3)_3CCOC\underline{H}_3$
(c) $CH_3CO_2\underline{H}$ (d) $PhCON\underline{H}_2$
(e) $CH_3CH_2C\underline{H}_2CN$ (f) $C\underline{H}_3CON(CH_3)_2$

22.8 $^-\!:CH_2C\!\equiv\!N\!: \longleftrightarrow H_2C\!=\!C\!=\!\ddot{N}\!:^-$

22.9 Acid is regenerated, but base is used stoichiometrically.

22.10 (a) 1. $Na^+\,^-OEt$; 2. $PhCH_2Br$; 3. H_3O^+
(b) 1. $Na^+\,^-OEt$; 2. $CH_3CH_2CH_2Br$; 3. $Na^+\,^-OEt$;
    4. CH_3Br; 5. H_3O^+
(c) 1. $Na^+\,^-OEt$; 2. $(CH_3)_2CHCH_2Br$; 3. H_3O^+

22.11 Malonic ester has only two acidic hydrogens to be replaced.

22.12 1. $Na^+\,^-OEt$; 2. $(CH_3)_2CHCH_2Br$; 3. $Na^+\,^-OEt$;
4. CH_3Br; 5. H_3O^+

22.13 (a) $(CH_3)_2CHCH_2Br$ (b) $PhCH_2CH_2Br$

22.14 None can be prepared.

22.15 1. $2\,Na^+\,^-OEt$; 2. $BrCH_2CH_2CH_2CH_2Br$; 3. H_3O^+

22.16 (a) Alkylate phenylacetone with CH_3I
(b) Alkylate pentanenitrile with CH_3CH_2I
(c) Alkylate cyclohexanone with $H_2C\!=\!CHCH_2Br$
(d) Alkylate cyclohexanone with excess CH_3I
(e) Alkylate $C_6H_5COCH_2CH_3$ with CH_3I
(f) Alkylate methyl 3-methylbutanoate with CH_3CH_2I

CHAPTER 23

23.1 (a)

$$CH_3CH_2CH_2\underset{\underset{\textstyle CH_2CH_3}{|}}{\overset{\overset{\textstyle OH}{|}}{C}H}\overset{\overset{\textstyle O}{||}}{C}H$$

(b)

(c)

23.2 The reverse reaction is the exact opposite of the forward reaction.

23.3

(a)

(b)

(c)

$$(CH_3)_2CHCH_2CH\!=\!\underset{\underset{\textstyle CH(CH_3)_2}{|}}{C}\overset{\overset{\textstyle O}{||}}{C}H$$

23.4

and

23.5 (a) Not an aldol product   (b) 3-Pentanone

23.6 1. NaOH; 2. $LiAlH_4$; 3. H_2/Pd

23.7

23.8 (a) $C_6H_5CHO + CH_3COCH_3$
(b), (c) Not easily prepared

23.9 The CH_2 position between the two carbonyl groups is so acidic that it is completely deprotonated to give a stable enolate ion.

23.10

23.11 (a)

$$CH_3\underset{\underset{\textstyle CH(CH_3)_2}{|}}{\overset{\overset{\textstyle CH_3}{|}}{CH}CH_2}\overset{\overset{\textstyle O}{||}}{C}\overset{\overset{\textstyle O}{||}}{C}HCOEt$$

(b)

$$PhCH_2\overset{\overset{\textstyle O}{||}}{C}\underset{\underset{\textstyle Ph}{|}}{\overset{\overset{\textstyle O}{||}}{C}}HCOEt$$

(c)

$$C_6H_{11}CH_2\overset{\overset{\textstyle O}{||}}{C}\underset{\underset{\textstyle C_6H_{11}}{|}}{\overset{\overset{\textstyle O}{||}}{C}}HCOEt$$

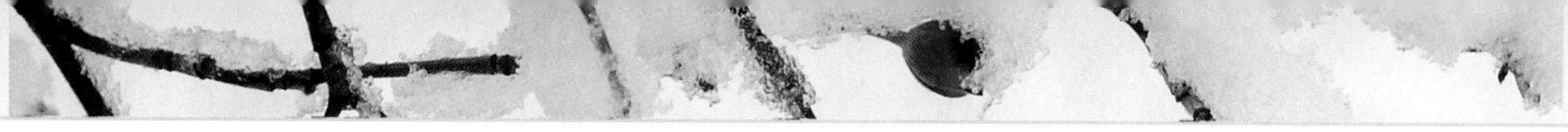

23.12 The cleavage reaction is the exact reverse of the forward reaction.

23.13

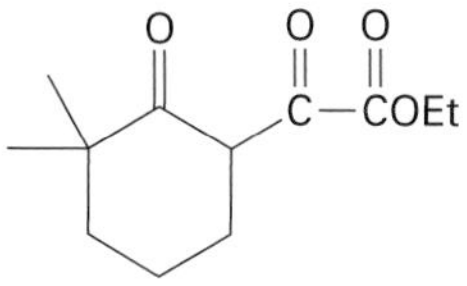

23.14

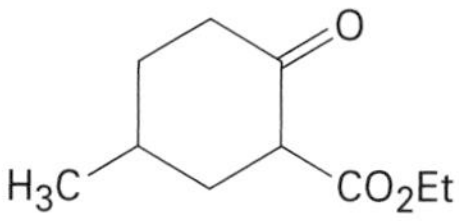

23.15

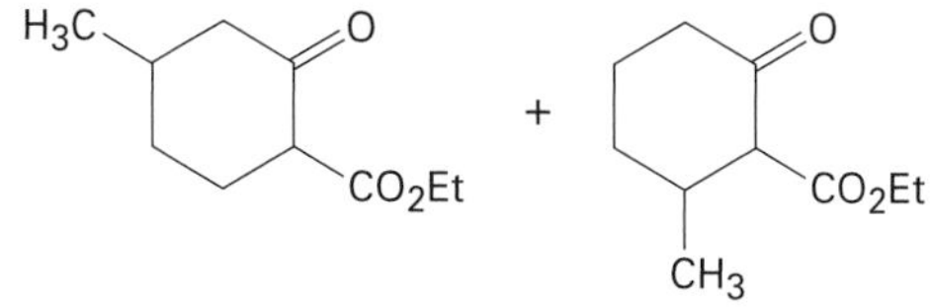

23.16 (a)

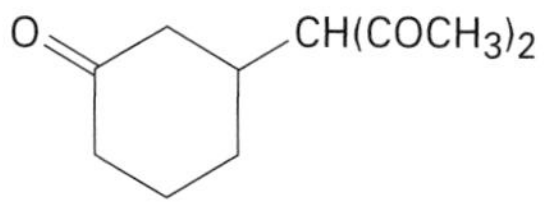

(b) $(CH_3CO)_2CHCH_2CH_2CN$

(c)

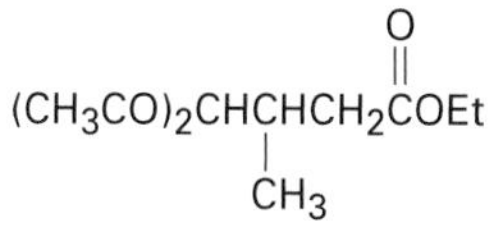

$(CH_3CO)_2CHCHCH_2\overset{\displaystyle O}{\overset{\|}{C}}OEt$
　　　　　$|$
　　　　　CH_3

23.17

(a)

$(EtO_2C)_2CHCH_2CH_2\overset{\displaystyle O}{\overset{\|}{C}}CH_3$

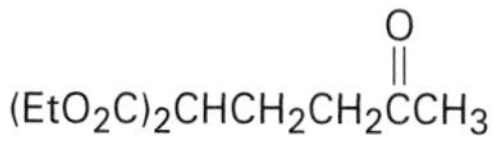

(b)

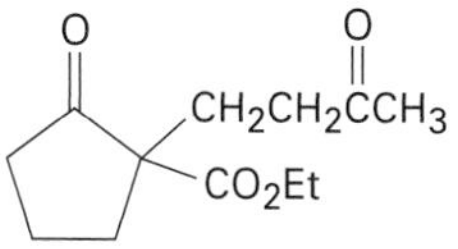

23.18 $CH_3CH_2COCH=CH_2 + CH_3CH_2NO_2$

23.19

(a)

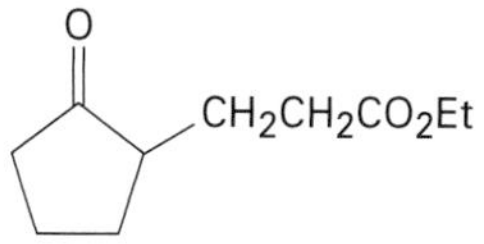

(b)

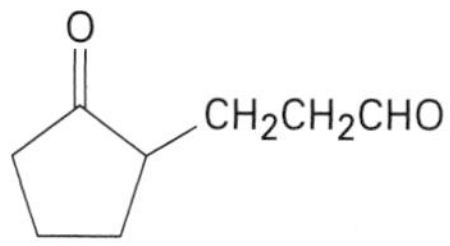

(c)

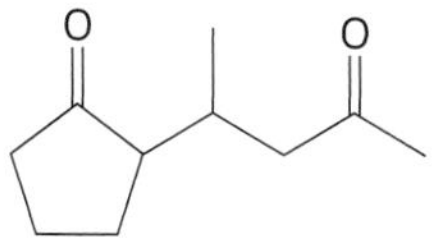

23.20 (a) Cyclopentanone enamine + propenenitrile
(b) Cyclohexanone enamine + methyl propenoate

23.21

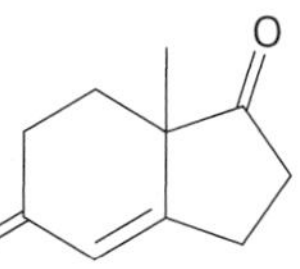

23.22 2,5,5-Trimethyl-1,3-cyclohexanedione + 1-penten-3-one

CHAPTER 24

24.1 (a) *N*-Methylethylamine
(b) Tricyclohexylamine
(c) *N*-Methyl-*N*-propylcyclohexylamine
(d) *N*-Methylpyrrolidine
(e) Diisopropylamine
(f) 1,3-Butanediamine

24.2

(a) $[(CH_3)_2CH]_3N$

(b) $(H_2C=CHCH_2)_2NH$

(c)

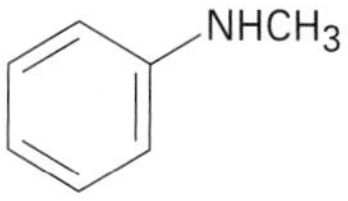

(d)

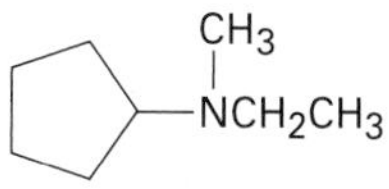

(e)

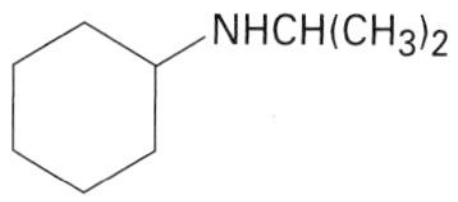

(f)

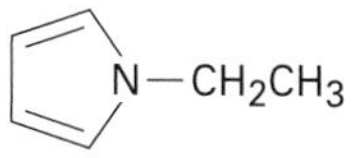

24.3

(a)

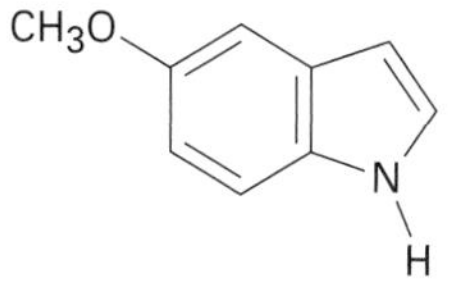

(b)

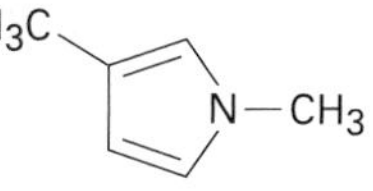

(c)

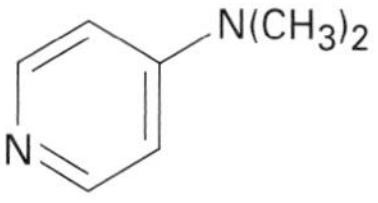

(d)

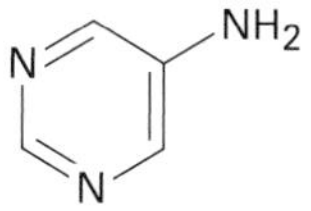

24.4 (a) $CH_3CH_2NH_2$ (b) NaOH
(c) CH_3NHCH_3

24.5 Propylamine is stronger; benzylamine $pK_b = 4.67$; propylamine $pK_b = 3.29$

24.6 (a) *p*-Nitroaniline < *p*-Aminobenzaldehyde < *p*-Bromoaniline
(b) *p*-Aminoacetophenone < *p*-Chloroaniline < *p*-Methylaniline
(c) *p*-(Trifluoromethyl)aniline < *p*-(Fluoromethyl)aniline < *p*-Methylaniline

24.7 Pyrimidine is essentially 100% neutral (unprotonated).

24.8 (a) Propanenitrile or propanamide
(b) *N*-Propylpropanamide
(c) Benzonitrile or benzamide
(d) *N*-Phenylacetamide

24.9 The reaction takes place by two nucleophilic acyl substitution reactions.

24.10

or

24.11 (a) Ethylamine + acetone, or isopropylamine + acetaldehyde
(b) Aniline + acetaldehyde
(c) Cyclopentylamine + formaldehyde, or methylamine + cyclopentanone

24.12

24.13 (a) 4,4-Dimethylpentanamide or 4,4-dimethylpentanoyl azide
(b) *p*-Methylbenzamide or *p*-methylbenzoyl azide

24.14 (a) 3-Octene and 4-octene
(b) Cyclohexene
(c) 3-Heptene
(d) Ethylene and cyclohexene

24.15 $H_2C{=}CHCH_2CH_2CH_2N(CH_3)_2$

24.16 1. HNO_3, H_2SO_4; 2. H_2/PtO_2; 3. $(CH_3CO)_2O$; 4. $HOSO_2Cl$; 5. aminothiazole; 6. H_2O, NaOH

24.17 (a) 1. HNO_3, H_2SO_4; 2. H_2/PtO_2; 3. 2 CH_3Br
(b) 1. HNO_3, H_2SO_4; 2. H_2/PtO_2; 3. $(CH_3CO)_2O$; 4. Cl_2; 5. H_2O, NaOH
(c) 1. HNO_3, H_2SO_4; 2. Cl_2, $FeCl_3$; 3. $SnCl_2$
(d) 1. HNO_3, H_2SO_4; 2. H_2/PtO_2; 3. $(CH_3CO)_2O$; 4. 2 CH_3Cl, $AlCl_3$; 5. H_2O, NaOH

24.18 (a) 1. CH_3Cl, $AlCl_3$; 2. HNO_3, H_2SO_4; 3. $SnCl_2$; 4. $NaNO_2$, H_2SO_4; 5. CuBr; 6. $KMnO_4$, H_2O
(b) 1. HNO_3, H_2SO_4; 2. Br_2, $FeBr_3$; 3. $SnCl_2$, H_3O^+; 4. $NaNO_2$, H_2SO_4; 5. CuCN; 6. H_3O^+
(c) 1. HNO_3, H_2SO_4; 2. Cl_2, $FeCl_3$; 3. $SnCl_2$; 4. $NaNO_2$, H_2SO_4; 5. CuBr
(d) 1. CH_3Cl, $AlCl_3$; 2. HNO_3, H_2SO_4; 3. $SnCl_2$; 4. $NaNO_2$, H_2SO_4; 5. CuCN; 6. H_3O^+
(e) 1. HNO_3, H_2SO_4; 2. H_2/PtO_2; 3. $(CH_3CO)_2O$; 4. 2 Br_2; 5. H_2O, NaOH; 6. $NaNO_2$, H_2SO_4; 7. CuBr

24.19 1. HNO_3, H_2SO_4; 2. $SnCl_2$; 3a. 2 equiv. CH_3I; 3b. $NaNO_2$, H_2SO_4; 4. product of 3a + product of 3b

24.20

24.21 4.1% protonated

24.22

Attack at C2:

Unfavorable

Attack at C3:

Attack at C4:

Unfavorable

24.23 The side-chain nitrogen is more basic than the ring nitrogen.

24.24 Reaction at C2 is disfavored because the aromaticity of the benzene ring is lost.

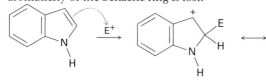

24.25 $(CH_3)_3CCOCH_3 \longrightarrow (CH_3)_3CCH(NH_2)CH_3$

CHAPTER 25

25.1 (a) Aldotetrose
(b) Ketopentose
(c) Ketohexose
(d) Aldopentose

25.2 (a) *S*   (b) *R*   (c) *S*

25.3 A, B, and C are the same.

25.4

$$HOCH_2 \overset{\overset{H}{|}}{\underset{\underset{Cl}{|}}{C}} CH_3 \quad R$$

25.5

CHO
H——OH   *R*
H——OH   *R*
CH₂OH

25.6 (a) L-Erythrose; 2*S*,3*S*
(b) D-Xylose; 2*R*,3*S*,4*R*
(c) D-Xylulose; 3*S*,4*R*

25.7

CHO
H——OH
HO——H   L-(+)-Arabinose
HO——H
CH₂OH

25.8

(a)
CHO
HO——H
H——OH
HO——H
CH₂OH

(b)
CHO
HO——H
H——OH
H——OH
HO——H
CH₂OH

(c)
CHO
HO——H
H——OH
HO——H
HO——H
CH₂OH

25.9 16 D and 16 L aldoheptoses

25.10

CHO
H——OH
H——OH   D-Ribose
H——OH
CH₂OH

25.11

25.12

25.13

β-D-Galactopyranose   β-D-Mannopyranose

25.14

25.15 α-D-Allopyranose

25.16

25.17 D-Galactitol has a plane of symmetry and is a meso compound, whereas D-glucitol is chiral.

25.18 The −CHO end of L-gulose corresponds to the −CH_2OH end of D-glucose after reduction.

25.19 D-Allaric acid has a symmetry plane and is a meso compound, but D-glucaric acid is chiral.

25.20 D-Allose and D-galactose yield meso aldaric acids; the other six D-hexoses yield optically active aldaric acids.

25.21 D-Allose + D-altrose

25.22 L-Xylose

25.23 D-Xylose and D-lyxose

25.24

25.25 (a) The hemiacetal ring is reduced.
(b) The hemiacetal ring is oxidized.
(c) All hydroxyl groups are acetylated.

CHAPTER 26

26.1 Aromatic: Phe, Tyr, Trp, His; sulfur-containing: Cys, Met; alcohols: Ser, Thr; hydrocarbon side chains: Ala, Ile, Leu, Val, Phe

26.2 The sulfur atom in the −CH_2SH group of cysteine makes the side chain higher in priority than the −CO_2H group.

26.3

L-Threonine   Diastereomers of L-threonine

26.4 Net positive at pH = 5.3; net negative at pH = 7.3

26.5 (a) Start with 3-phenylpropanoic acid:
1. Br_2, PBr_3; 2. NH_3
(b) Start with 3-methylbutanoic acid:
1. Br_2, PBr_3; 2. NH_3

26.6 (a) $(CH_3)_2CHCH_2Br$ (b)

(c)

(d) $CH_3SCH_2CH_2Br$

26.7

26.8 Val-Tyr-Gly (VYG), Tyr-Gly-Val (YGV), Gly-Val-Tyr (GVY), Val-Gly-Tyr (VGY), Tyr-Val-Gly (YVG), Gly-Tyr-Val (GYV)

26.9

26.10

26.11

+ $(CH_3)_2CHCHO$ + CO_2

26.12 Trypsin: Asp-Arg + Val-Tyr-Ile-His-Pro-Phe

Chymotrypsin: Asp-Arg-Val-Tyr + Ile-His-Pro-Phe

26.13 Methionine

26.14

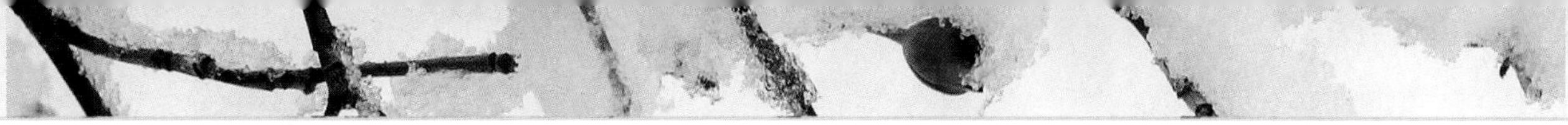

26.15 (a) Arg-Pro-Leu-Gly-Ile-Val

(b) Val-Met-Trp-Asp-Val-Leu (VMWNVL)

26.16 This is a typical nucleophilic acyl substitution reaction, with the amine of the amino acid as the nucleophile and *tert*-butyl carbonate as the leaving group. The *tert*-butyl carbonate then loses CO_2 and gives *tert*-butoxide, which is protonated.

26.17 (1) Protect the amino group of leucine.
(2) Protect the carboxylic acid group of alanine.
(3) Couple the protected amino acids with DCC.
(4) Remove the leucine protecting group.
(5) Remove the alanine protecting group.

26.18 (a) Lyase   (b) Hydrolase   (c) Oxidoreductase

CHAPTER 27

27.1 $CH_3(CH_2)_{18}CO_2CH_2(CH_2)_{30}CH_3$

27.2 Glyceryl tripalmitate is higher melting.

27.3 $[CH_3(CH_2)_7CH{=}CH(CH_2)_7CO_2^-]_2\,Mg^{2+}$

27.4 Glyceryl dioleate monopalmitate → glycerol + 2 sodium oleate + sodium palmitate

27.5

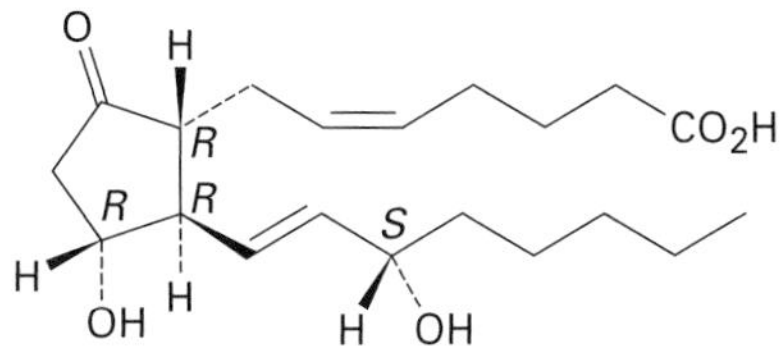

27.6 The *pro-S* hydrogen is cis to the $-CH_3$ group; the *pro-R* hydrogen is trans.

27.7

(a)

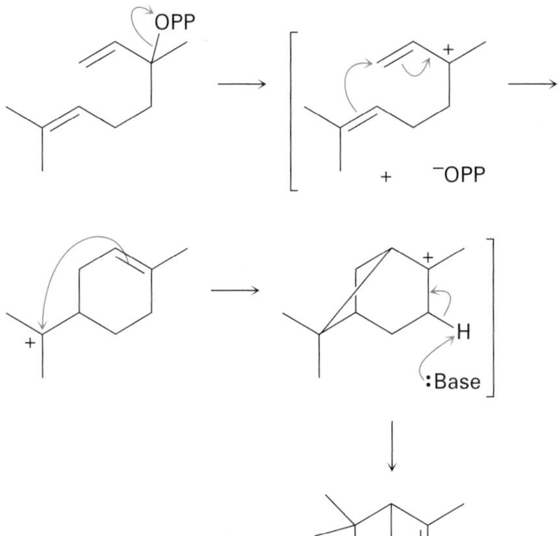

α-Pinene

(b)

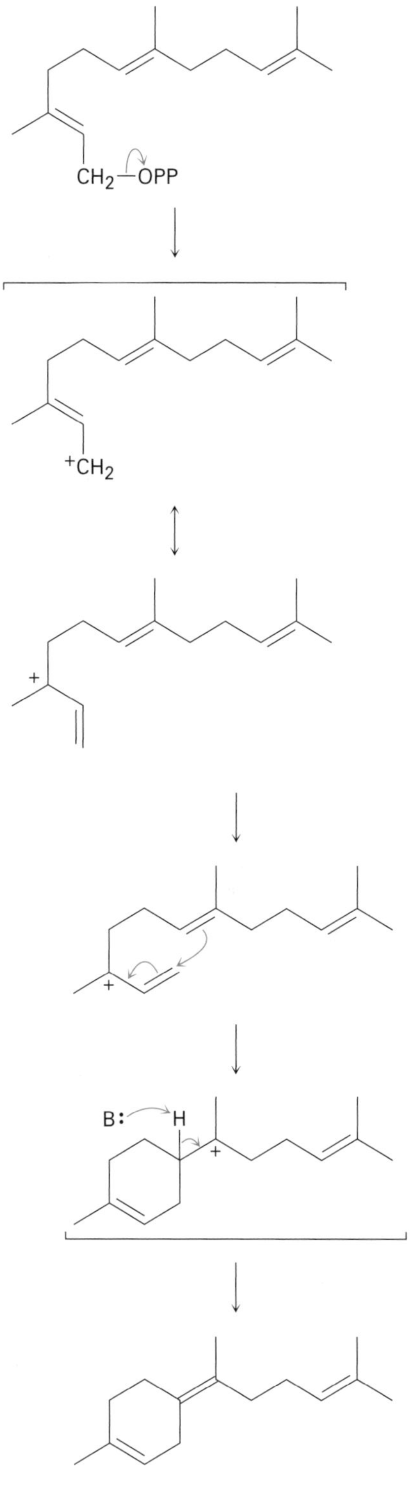

γ-Bisabolene

27.8

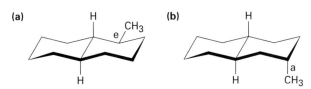

(a) H CH₃ (e)

(b) H CH₃ (a)

27.9

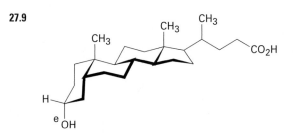

27.10 Three methyl groups are removed, the side-chain double bond is reduced, and the double bond in the B ring is migrated.

CHAPTER 28

28.3 (5′) ACGGATTAGCC (3′)

28.4

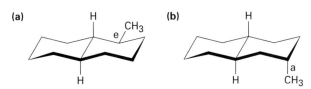

28.5 (3′) CUAAUGGCAU (5′)

28.6 (5′) ACTCTGCGAA (3′)

28.7 (a) GCU, GCC, GCA, GCG
(b) UUU, UUC
(c) UUA, UUG, CUU, CUC, CUA, CUG
(d) UAU, UAC

28.8 (a) AGC, GGC, UGC, CGC
(b) AAA, GAA
(c) UAA, CAA, GAA, GAG, UAG, CAG
(d) AUA, GUA

28.9 Leu-Met-Ala-Trp-Pro-Stop

28.10 (5′) TTA-GGG-CCA-AGC-CAT-AAG (3′)

28.11 The cleavage is an S$_N$1 reaction that occurs by protonation of the oxygen atom followed by loss of the stable triarylmethyl carbocation.

28.12

RO—P(=O)(OR′)—O—CH₂—CHC≡N   :NH₃   E2 reaction

CHAPTER 29

29.1 HOCH₂CH(OH)CH₂OH + ATP ⟶
HOCH₂CH(OH)CH₂OPO₃²⁻ + ADP

29.2 Caprylyl CoA ⟶ Hexanoyl CoA ⟶
Butyryl CoA ⟶ 2 Acetyl CoA

29.3 (a) 8 acetyl CoA; 7 passages
(b) 10 acetyl CoA; 9 passages

29.4 The dehydration is an E1cB reaction.

29.5 At C2, C4, C6, C8, and so forth

29.6 The *Si* face

29.7 Steps 7 and 10

29.8 Steps 1, 3: Phosphate transfers; steps 2, 5, 8: isomerizations; step 4: retro-aldol reaction; step 5: oxidation and nucleophilic acyl substitution; steps 7, 10: phosphate transfers; step 9: E2 dehydration

29.9 C1 and C6 of glucose become −CH₃ groups; C3 and C4 become CO₂.

29.10 Citrate and isocitrate

29.11 E2 elimination of water, followed by conjugate addition

29.12 *pro-R*; anti geometry

29.13 The reaction occurs by two sequential nucleophilic acyl substitutions, the first by a cysteine residue in the enzyme, with phosphate as leaving group, and the second by hydride donation from NADH, with the cysteine residue as leaving group.

29.14 Initial imine formation between PMP and α-ketoglutarate is followed by double-bond rearrangement to an isomeric imine and hydrolysis.

29.15 (CH₃)₂CHCH₂COCO₂⁻

29.16 Asparagine

CHAPTER 30

30.1 Ethylene: ψ_1 is the HOMO and ψ_2^* is the LUMO in the ground state; ψ_2^* is the HOMO and there is no LUMO in the excited state. 1,3-Butadiene: ψ_2 is the HOMO and ψ_3^* is the LUMO in the ground state; ψ_3^* is the HOMO and ψ_4^* is the LUMO in the excited state.

30.2 Disrotatory: *cis*-5,6-dimethyl-1,3-cyclohexadiene; conrotatory: *trans*-5,6-dimethyl-1,3-cyclohexadiene. Disrotatory closure occurs.

30.3 The more stable of two allowed products is formed.

30.4 *trans*-5,6-Dimethyl-1,3-cyclohexadiene; *cis*-5,6-dimethyl-1,3-cyclohexadiene

30.5 *cis*-3,6-Dimethylcyclohexene; *trans*-3,6-dimethylcyclohexene

30.6 A [6 + 4] suprafacial cycloaddition

30.7 An antarafacial [1,7] sigmatropic rearrangement

30.8 A series of [1,5] hydrogen shifts occur.

30.9 Claisen rearrangement is followed by a Cope rearrangement.

30.10 (a) Conrotatory (b) Disrotatory
(c) Suprafacial (d) Antarafacial
(e) Suprafacial

CHAPTER 31

31.1 $H_2C{=}CHCO_2CH_3 < H_2C{=}CHCl < H_2C{=}CHCH_3 < H_2C{=}CH{-}C_6H_5$

31.2 $H_2C{=}CHCH_3 < H_2C{=}CHC_6H_5 < H_2C{=}CHC{\equiv}N$

31.3 The intermediate is a resonance-stabilized benzylic carbanion, Ph$-\overset{..}{\overline{C}}$HR.

31.4 The polymer has no chirality centers.

31.5 No, the polymers are racemic.

31.6

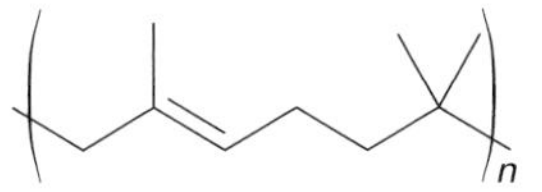

31.7 Polybutadiene chain / Polystyrene chain

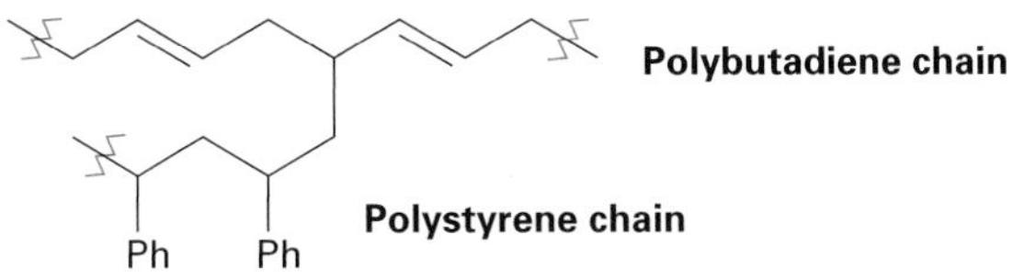

31.8

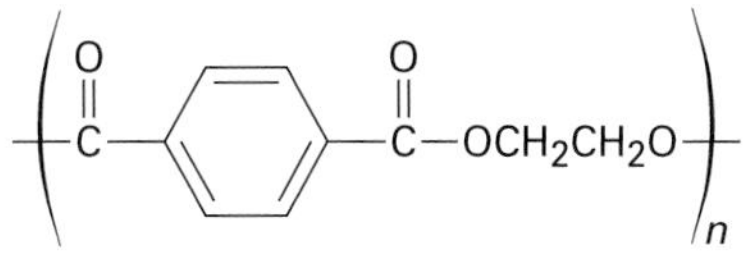

31.9

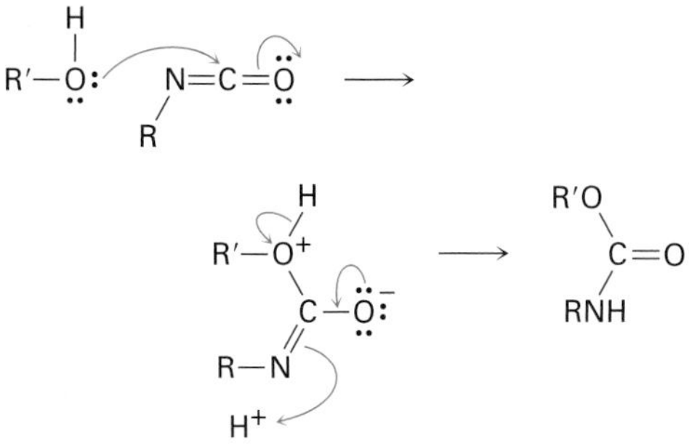

31.10 Atactic

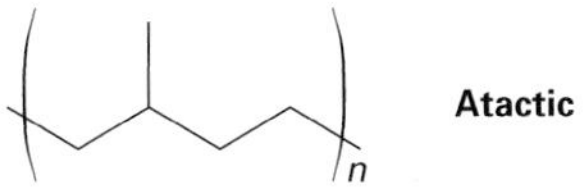

31.11

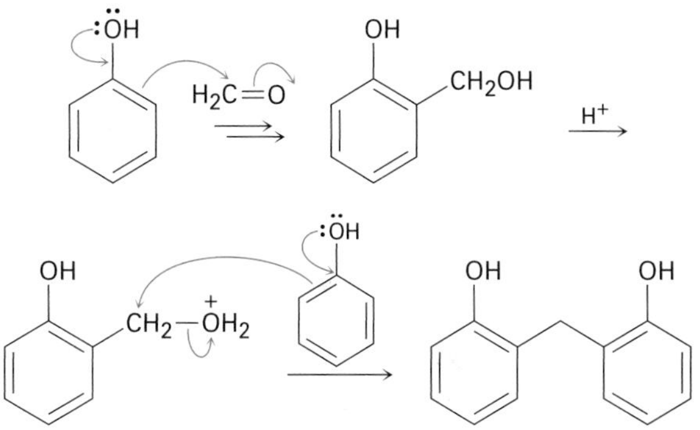

Index

The boldfaced references refer to pages where terms are defined.

Benzene, acylation of, 557–558
  alkylation of, 554–557
  bond lengths in, 521
  bromination of, 548–550
  chlorination of, 550
  discovery of, 518
  electrostatic potential map of, 44, 521, 565
  Friedel–Crafts reactions of, 554–558
  heat of hydrogenation of, 521
  Hückel $4n + 2$ rule and, 523–524
  iodination of, 551
  molecular orbitals of, 522, 531
  nitration of, 551–552
  ^{13}C NMR absorption of, 536
  reaction with Br_2, 548–550
  reaction with Cl_2, 550
  reaction with H_2SO_4/HNO_3, 552–553
  reaction with HNO_3, 551–552
  reaction with I_2, 551
  resonance in, 44, 521
  stability of, 520–522
  structure of, 520–522
  sulfonation of, 552–553
  toxicity of, 516
  UV absorption of, 503
Benzenediazonium ion, electrostatic potential map of, 945
Benzenesulfonic acid, synthesis of, 552
Benzodiazepine, combinatorial library of, 586
Benzoic acid, pK_a of, 756
  ^{13}C NMR absorptions in, 771
  substituent effects on acidity of, 759–761
Benzophenone, structure of, 697
Benzoquinone, electrostatic potential map of, 631
Benzoyl group, **697**
Benzoyl peroxide, ethylene polymerization and, 240
Benzo[a]pyrene, carcinogenicity of, 532
  structure of, 532
Benzyl ester, hydrogenolysis of, 1034
Benzyl group, **518**
Benzylic, **377**
Benzylic acid rearrangement, 836
Benzylic carbocation, electrostatic potential map of, 377
  resonance in, 377
  S_N1 reaction and, 376–377

Benzylic halide, S_N1 reaction and, 377
  S_N2 reaction and, 377–378
Benzylic radical, resonance in, 578
  spin-density surface of, 578
Benzylpenicillin, discovery of, 824
  structure of, 1
Benzyne, **575**
  Diels–Alder reaction of, 575
  electrostatic potential map of, 576
  evidence for, 575
  structure of, 576
Bergman, Torbern, 2
Bergström, Sune K., **1068**
Beta anomer, **984**
Beta-carotene, structure of, 172
  industrial synthesis of, 722
  UV spectrum of, 504
Beta-diketone, **851**
  Michael reactions and, 895
Beta-keto ester, **851**
  alkylation of, 859–860
  cyclic, 892–893
  decarboxylation of, 857, 860
  Michael reactions and, 895
  pK_a of, 852
  synthesis of, 892–893
Beta-lactam antibiotics, 824–825
Beta oxidation pathway, **1133**–1137
  mechanism of, 1133–1136
Beta-pleated sheet (protein), **1038**
  molecular model of, 1039
  secondary protein structure and, 1038–1039
Betaine, **720**
Bextra, structure of, 544
BHA, synthesis of, 629
BHT, synthesis of, 629
Bicycloalkane, **129**
Bijvoet, J. M., 299
Bimolecular, **363**
Biodegradable polymers, 821, 1219
Biological acids,
  Henderson–Hasselbalch equation and, 758–759
Biological mass spectrometry, 417–418
Biological oxidation, NAD^+ and, 625–626
Biological reaction(s), aldol reaction, 901–902
  alkene halogenation, 218
  alkylation, 863–864
  aromatic hydroxylation, 553–554

carboxylation, 764
Claisen condensation reaction, 901
Claisen rearrangement, 1194–1195
dehydration, 622
elimination reactions, 393
oxidation, 625–626
radical reactions, 243–244
characteristics of, 162–164
comparison with laboratory reactions, 162–164
conventions for writing, 162, 190
energy diagram of, 161
reduction, 723–725
reductive amination, 932
substitution reactions, 381–383
Biological reduction, NADH and, 610–611
Biological substitution reactions, diphosphate leaving group in, 381–382
Biomass, carbohydrates and, 973
Biosynthesis, fatty acids, 1138–1143
Biot, Jean Baptiste, **295**
Biotin, fatty acid biosynthesis and, 1141
  stereochemistry of, 325
  structure of, 1045
Bisphenol A, epoxy resins from, 673
  polymers from, 821
Bloch, Konrad Emil, **1084**
Block copolymer, **1212**
  synthesis of, 1212
Blood groups, antigenic determinants in, 1004
  compatibility of, 1004
  types of, 1004
Boat conformation (cyclohexane), steric strain in, 118
Boc (*tert*-butoxycarbonyl amide), 1034
  amino acid derivatives of, 1034
Bond, covalent, **11**–12
  molecular orbital description of, 21–22, 485–486,
  pi, **16**
  sigma, **10**
  valence bond description of, 10–12
Bond angle, **13**
Bond dissociation energy (D), **155**
  table of, 156
Bond length, **12**
Bond strength, **11**

Diterpenoid, **203**, **1071**
DMAPP, *see* Dimethylallyl
    diphosphate
DMF, *see* Dimethylformamide
DMSO, *see* Dimethyl sulfoxide
DMT (dimethoxytrityl ether), DNA
    synthesis and, 1114
DNA, *see* Deoxyribonucleic acid
DNA fingerprinting, 1118–1119
    reliability of, 1119
    STR loci and, 1118
Dopamine, molecular model of, 930
Double bond, electronic structure of,
    16
    length of, 16
    molecular orbitals in, 22
    see also Alkene
    strength of, 16
Double helix (DNA), **1103**–1105
Doublet (NMR), 462
Downfield (NMR), **445**
Drugs, approval procedure for, 165
    chiral, 320–322
    origin of, 164

E configuration, **180**
    assignment of, 180–183
E1 reaction, **384**, 391–392
    carbocations and, 391–392
    deuterium isotope effect and, 392
    kinetics of, 392
    mechanism of, 391–392
    rate-limiting step in, 392
    stereochemistry of, 392
    Zaitsev's rule and, 392
E1cB reaction, **385**, 393
    carbanion intermediate in, 393
    mechanism of, 393
E2 reaction, **385**–391
    alcohol oxidation and, 625
    cyclohexane conformation and,
        389–391
    deuterium isotope effect and,
        386–387
    geometry of, 387–388
    kinetics of, 386
    mechanism of, 386
    menthyl chloride and, 390
    neomenthyl chloride and, 390
    rate law for, 386
    stereochemistry of, 387–388
    Zaitsev's rule and, 389–390
E85 ethanol, 600
Ebonite, structure of, 246

Eclipsed conformation, **94**
    molecular model of, 94
Edman, Pehr Victor, **1031**
Edman degradation, **1031**–1032
    mechanism of, 1032
Eicosanoid, **1067**–1070
    biosynthesis of, 1069–1070
    naming, 1069
Elaidic acid, from vegetable oil,
    1063
Elastomer, **1216**
    characteristics of, 1216–1217
    cross links in, 1217
Electrocyclic reaction, **1181**–1186
    conrotatory motion in, 1183
    disrotatory motion in, 1183
    examples of, 1181–1182
    HOMO and, 1183–1186
    photochemical, 1185–1186
    stereochemical rules for, 1186
    stereochemistry of, 1183–1186
    thermal, 1183–1185
Electromagnetic radiation, **418**–419
    amplitude of, 419
    characteristics of, 419–420
    energy of, 420
    frequency of, 419
    kinds of, 419
    wavelength of, 419
Electromagnetic spectrum, **419**
    regions in, 419
Electron, delocalization of, 341–342,
    486
    lone-pair, **9**
    nonbonding, **9**
Electron configuration, ground state,
    6
    rules for assigning, 6
    table of, 6
Electron movement, curved arrows
    and, 44–45, 57–58
Electron shell, **5**
Electron-dot structure, **9**
Electron-transport chain, **1127**
Electronegativity, **36**
    inductive effects and, 37
    polar covalent bonds and, 36–37
    table of, 36
Electrophile, **145**
    characteristics of, 149–151
    curved arrows and, 149–151
    electrostatic potential maps of,
        145
    examples of, 145

Electrophilic addition reaction,
    188–190
    carbocation rearrangements in,
        200–201
    energy diagram of, 158, 160–161
    Hammond postulate and, 197–199
    intermediate in, 160
    Markovnikov's rule and, 191–193
    mechanism of, 147–148, 188–189
    regiospecificity of, 191–193
    stereochemistry and, 311–313
Electrophilic aromatic substitution
    reaction, **547**
    arylamines and, 939–940
    biological example of, 551
    inductive effects in, 562
    kinds of, 547
    mechanism of, 548–549
    orientation in, 560–561
    pyridine and, 949
    pyrrole and, 947–948
    resonance effects in, 562–563
    substituent effects in, 560–563
Electrophoresis, **1025**
    DNA sequencing and, 1113
Electrospray ionization (ESI) mass
    spectrometry, 417–418
Electrostatic potential map, **37**
    acetaldehyde, 688
    acetamide, 791, 922
    acetate ion, 43, 53, 56, 757
    acetic acid, 53, 55
    acetic acid dimer, 755
    acetic anhydride, 791
    acetone, 55, 56, 78
    acetone anion, 56
    acetyl azide, 830
    acetyl chloride, 791
    acetylene, 262
    acetylide anion, 271
    acid anhydride, 791
    acid chloride, 791
    acyl cation, 558
    adenine, 1104
    alanine, 1017
    alanine zwitterion, 1017
    alcohol, 75
    alkene, 74, 147
    alkyl halide, 75
    alkyne, 74
    allyl carbocation, 377, 489
    amide, 791
    amine, 75
    amine hydrogen bonding, 920

Fragmentation (mass spectrum), 410–413
Free radical, **139**
Free-energy change (Δ*G*), **153**
Free-energy change (Δ*G*°), standard, **153**
Fremy's salt, **631**
Frequency (*ν*), **419**–420
Friedel, Charles, **555**
Friedel–Crafts acylation reaction, 557–558
   acyl cations in, 557–558
   arylamines and, 939–940
   mechanism of, 557–558
Friedel–Crafts alkylation reaction, **554**–557
   arylamines and, 939–940
   biological example of, 558–559
   limitations of, 555–556
   mechanism of, 554–555
   polyalkylation in, 556
   rearrangements in, 556–557
Frontier orbitals, **1181**
Fructose, anomers of, 985–986
   furanose form of, 985–986
   sweetness of, 1005
Fructose-1,6-bisphosphate aldolase, X-ray crystal structure of, 865
L-Fucose, biosynthesis of, 1015
   structure of, 996
Fukui, Kenichi, **1180**
Fumarate, hydration of, 221–222
   malate from, 221–222
Functional group, **73**–77
   carbonyl compounds and, 75
   importance of, 73–74
   IR spectroscopy of, 425–429
   multiple bonds in, 74
   polarity patterns of, 143
   table of, 76–77
Furan, industrial synthesis of, 946
Furanose, **985**–986
   fructose and, 985–986

γ, see Gamma
Gabriel, Siegmund, **929**
Gabriel amine synthesis, **929**
Galactose, biosynthesis of, 1011
   configuration of, 982
   Wohl degradation of, 995
γ-aminobutyric acid, structure of, 1020
γ rays, electromagnetic spectrum and, 419

Gasoline, manufacture of, 99–100
   octane number of, 100
Gatterman–Koch reaction, 596
Gauche conformation, **95**
   butane and, 95–96
   steric strain in, 96
Gel electrophoresis, DNA sequencing and, 1113
Gem, *see* Geminal, 705
Geminal (gem), **705**
Genome, size of in humans, 1107
Gentamicin, structure of, 1002
Geraniol, biosynthesis of, 382
Geranyl diphosphate, biosynthesis of, 1077–1078
   monoterpenoids from, 1077–1078
Gibbs free-energy change (Δ*G*), **153**
Gibbs free-energy change (Δ*G*°), standard, **153**
   equilibrium constant and, 154
Gilman, Henry, **347**
Gilman reagent, **347**
   conjugate carbonyl addition reactions of, 728–729
   organometallic coupling reactions of, 346–347
   reaction with acid chlorides, 805
   reaction with alkyl halides, 346–347
   reaction with enones, 728–729
Glass transition temperature (polymers), **1215**
Globo H hexasaccharide, function of, 1004
   structure of, 1005
Glucocorticoid, **1083**
Gluconeogenesis, **1159**–1165
   overall result of, 1165
   steps in, 1160–1161
Glucosamine, biosynthesis of, 1012
   structure of, 1002
Glucose, *α* anomer of, 985
   anabolism of, 1159–1165
   anomers of, 984–985
   β anomer of, 985
   biosynthesis of, 1159–1165
   catabolism of, 1143–1150
   chair conformation of, 119
   configuration of, 982
   Fischer projection of, 978
   from pyruvate, 1159–1165
   glycosides of, 989–990
   keto-enol tautomerization of, 1145–1146

Koenigs–Knorr reaction of, 990
   molecular model of, 119, 126, 985
   mutarotation of, 985–986
   pentaacetyl ester of, 988
   pentamethyl ether of, 988
   pyranose form of, 984–985
   pyruvate from, 1143–1150
   reaction with acetic anhydride, 988
   reaction with ATP, 1129
   reaction with iodomethane, 988
   sweetness of, 1005
   Williamson ether synthesis with, 988
Glutamic acid, structure and properties of, 1019
Glutamine, structure and properties of, 1018
Glutaric acid, structure of, 753
Glutathione, function of, 668
   prostaglandin biosynthesis and, 1070
   structure of, 668
Glycal, **1002**
Glycal assembly method, **1002**
(+)-Glyceraldehyde, absolute configuration of, 980
(−)-Glyceraldehyde, configuration of, 300
(*R*)-Glyceraldehyde, Fischer projection of, 976
   molecular model of, 976, 977
Glyceric acid, structure of, 753
Glycerol, catabolism of, 1132–1133
sn-Glycerol 3-phosphate, naming of, 1132
Glycerophospholipid, **1066**
Glycine, structure and properties of, 1018
Glycoconjugate, **991**
Glycogen, function of, 1001
   structure of, 1001
Glycol, **234**, **662**
Glycolic acid, p*K*a of, 756
   structure of, 753
Glycolipid, **991**
Glycolysis, 903–904, **1143**–1150
   overall result of, 1150
   steps in, 1143–1145
Glycoprotein, **991**
   biosynthesis of, 991
Glycoside, **989**
   Koenigs–Knorr reaction and, 990
   occurrence of, 989
   synthesis of, 990

Glyptal, structure of, 1223
Goodyear, Charles, 499
GPP, *see* Geranyl diphosphate
Graft copolymer, **1212**
    synthesis of, 1212
Grain alcohol, 599
Green chemistry, **395**–396
    ibuprofen synthesis and, 396
    ionic liquids and, 956–957
    principles of, 395–396
Grignard, François Auguste Victor,
    345
Grignard reaction, aldehydes and,
    614
    carboxylic acids and, 614
    esters and, 614
    formaldehyde and, 614
    ketones and, 614
    limitations of, 615
    mechanism of, 708–709
    strategy for, 616
Grignard reagent, **345**
    alkanes from, 346
    carboxylation of, 763
    carboxylic acids from, 763
    electrostatic potential map of, 345,
        708
    from alkyl halides, 345
    reaction with acids, 346
    reaction with aldehydes, 614,
        708–709
    reaction with carboxylic acids, 614
    reaction with CO_2, 763
    reaction with epoxides, 665
    reaction with esters, 614, 813
    reaction with formaldehyde, 614
    reaction with ketones, 614,
        708–709
    reaction with nitriles, 769
    reaction with oxetanes, 680
Guanine, electrostatic potential map
    of, 1104
    protection of, 1114–1115
    structure of, 1101
Gulose, configuration of, 982
Guncotton, 1000
Gutta-percha, structure of, 499

Hagemann's ester, synthesis of, 912
Halo group, directing effect of,
        567–568
    inductive effect of, 562
    orienting effect of, 561
    resonance effect of, 563

Haloalkane, *see* Alkyl halide
Haloform reaction, **854**–855
Halogen, inductive effect of, 562
    resonance effect of, 563
Halogenation, aldehydes and,
        846–848
    alkenes and, 215–218
    alkynes and, 262–263
    aromatic compounds and,
        548–551
    carboxylic acids and, 849
    ketones and, 846–848
Halohydrin, **218**
    epoxides from, 234, 661
    reaction with base, 234, 661
Hammond, George Simms, **197**
Hammond postulate, **197**–199
    carbocation stability and, 197–199
    endergonic reactions and,
        197–198
    exergonic reactions and, 197–198
    Markovnikov's rule and, 198–199
    radical stability and, 338
    S_N1 reaction and, 376
Handedness, molecular, 290–293
HDL, heart disease and, 1090–1091
Heart disease, cholesterol and,
        1090–1091
Heat of combustion, **113**
Heat of hydrogenation, **186**
    table of, 187
Heat of reaction, **154**
Helicase, DNA replication and, 1106
Hell–Volhard–Zelinskii reaction, **849**
    amino acid synthesis and, 1025
    mechanism of, 849
Heme, biosynthesis of, 966
    structure of, 946
Hemiacetal, **717**
Hemiketal, **717**
Hemithioacetal, **1148**
Henderson–Hasselbalch equation,
        biological amines and, 925–926
    amino acids and, 1022–1023
    biological acids and, **758**–759
Hertz (Hz), **419**
Heterocycle, **528**, **945**
    aromatic, 528–529
    polycyclic, 950–951
Heterocyclic amine, **918**
    basicity of, 922–923
    names for, 918
Heterolytic, **139**
Hevea brasieliensis, rubber from, 245

Hexachlorophene, synthesis of, 595,
        629
Hexamethylphosphoramide, S_N2
    reaction and, 371
Hexane, IR spectrum of, 424
    mass spectrum of, 413
Hexa-1,3,5-triene, molecular orbitals
    of, 1180
    UV absorption of, 503
Hex-1-ene, IR spectrum of, 424
Hex-2-ene, mass spectrum of, 415
Hexokinase, active site in, 163
    molecular model of, 163
Hex-1-yne, IR spectrum of, 424
High-density polyethylene, synthesis
    of, 1210
High-molecular-weight polyethylene,
    uses of, 1210
High-pressure liquid
    chromatography, **432**
    amino acid analysis and, 1030
Highest occupied molecular orbital
    (HOMO), **500**, **1181**
    cycloaddition reactions and,
        1188–1189
    electrocyclic reactions and,
        1183–1186
    UV spectroscopy and, 500
Histamine, structure of, 965
Histidine, electrostatic potential map
    of, 1021
    structure and properties of, 1019
HMPA, *see*,
        Hexamethylphosphoramide
Hoffmann, Roald, **1180**
Hoffmann-La Roche Co., vitamin C
    synthesis and, 773
von Hofmann, August Wilhelm, **933**
Hofmann elimination reaction,
        936–938
    biological example of, 937
    molecular model of, 937
    mechanism of, 937
    regiochemistry of, 937
    Zaitsev's rule and, 937
Hofmann rearrangement, **933**–934
    mechanism of, 933–934
HOMO, *see* Highest occupied
    molecular orbital
Homocysteine, structure of, 1020
Homolytic, **139**
Homopolymer, **1210**
Homotopic (NMR), **455**
Honey, sugars in, 999

polarizability of, 144
reaction with alkyl halides, 668–669
reaction with Br$_2$, 668
reaction with NaH, 668
sulfides from, 668–669
thiolate ions from, 668
Thiolate ion, **668**
Thionyl chloride, reaction with alcohols, 344, 618
reaction with amides, 766–767
reaction with carboxylic acids, 794–795
Thiophene, aromaticity of, 530
Thiourea, reaction with alkyl halides, 667
Threonine, stereoisomers of, 302–303
structure and properties of, 1019
Threose, configuration of, 982
molecular model of, 294
Thromboxane B$_2$, structure of, 1068
Thymine, electrostatic potential map of, 1104
structure of, 1101
Thyroxine, biosynthesis of, 551
structure of, 1020
TIme-of-flight (TOF) mass spectrometry, 417–418
Titration curve, alanine, 1023
TMS, *see* Tetramethylsilane
see Trimethylsilyl ether
Tollens' reagent, **701**
Tollens' test, 992
Toluene, electrostatic potential map of, 565
IR spectrum of, 534
^{13}C NMR absorptions of, 536
^{1}H NMR spectrum of, 465
Toluene-2,4-diisocyanate, polyurethanes from, 1214
p-Toluenesulfonyl chloride, reaction with alcohols, 618–619
Torsional strain, **94**
Tosylate, **360–361**
from alcohols, 618–619
S$_N$2 reactions and, 369, 619
uses of, 619
Toxicity, chemicals and, 25–26
Trans fatty acid, from hydrogenation of fats, 232–233
from vegetable oils, 1063
Transamination, **1165–1168**
mechanism of, 1167

Transcription (DNA), **1108–1109**
coding strand in, 1108
primer strand in, 1108
promoter sites in, 1108
template strand in, 1108
Transfer RNA, **1108**
anticodons in, 1109–1111
function of, 1109–1111
molecular model of, 1111
shape of, 1111
Transferase, 1041–1042
Transition state, **158**
Hammond postulate and, 197–199
Translation (RNA), **1109**–1111
Tranylcypromine, synthesis of, 935
Tree diagram (NMR), 466
Triacylglycerol, **1061**
catabolism of, 1130–1137
Trialkylsulfonium ions, alkylations with, 669
chirality of, 315
Tributyltin hydride, reaction with alkyl halides, 358
Tricarboxylic acid cycle, *see* Citric acid cycle
Trichloroacetic acid, pK_a of, 759
Trifluoroacetic acid, pK_a of, 756
Trifluoromethylbenzene, electrostatic potential map of, 565
Triglyceride, *see* Triacylglycerol, 1061
Trimethylamine, bond angles in, 919
bond lengths in, 919
electrostatic potential map of, 921
molecular model of, 919
Trimethylammonium chloride, IR spectrum of, 953
Trimethylsilyl ether, cleavage of, 627–628
from alcohols, 626–628
synthesis of, 626–627
Trimetozine, synthesis of, 804
2,4,6-Trinitrochlorobenzene, electrostatic potential map of, 572
Triphenylphosphine, reaction with alkyl halides, 721
Triple bond, electronic structure of, 18
length of, 18
see also Alkyne
strength of, 18
Triplet (NMR), 460
Trisubstituted aromatic compound, synthesis of, 581–584

Trisubstituted cyclohexane, naming, 663
Triterpenoid, **1071**
tRNA, *see* Transfer RNA
Trypsin, peptide cleavage with, 1033
Tryptophan, pK_a of, 52
structure and properties of, 1019
Tswett, Mikhail, 431
Turnover number (enzyme), **1041**
Twist-boat conformation (cyclohexane), **118**
steric strain in, 118
molecular model of, 118
Tyrosine, biosynthesis of, 622
catabolism of, 1176
iodination of, 551
structure and properties of, 1019

Ubiquinones, structure and function of, 632
Ultrahigh-molecular-weight polyethylene, uses of, 1210
Ultraviolet light, electromagnetic spectrum and, 419
wavelength of, 500
Ultraviolet spectroscopy, **500–503**
absorbance and, 501
aromatic compounds, 534
conjugation and, 502–503
HOMO−LUMO transition in, 500–502
molar absorptivity and, 502
Ultraviolet spectrum, benzene, 503
β-carotene, 504
buta-1,3-diene, 501
but-3-en-2-one, 503
cyclohexa-1,3-diene, 503
hexa-1,3,5-triene, 503
ergosterol, 514
isoprene, 503
Unimolecular, **373**
Unsaturated, **174**
Unsaturated aldehyde, conjugate addition reactions of, 725–729
Unsaturated ketone, conjugate addition reactions of, 725–729
Unsaturation, degree of, **174**
Upfield, (NMR), **445**
Uracil, structure of, 1101
Urea, from ammonium cyanate, 2
Urethane, **1214**
Uric acid, pK_a of, 778

Aromaticity

Aromaticity

Aromaticity is a special stabilization in certain arrays of π systems that leads to reactivity that is much different from that of normal alkenes.

A classic example is benzene. As revealed by attempts to hydrogenate its double bonds, compared to other systems, approximately 151 kJ/mol (36 kcal/mol) of additional energy is needed to form cyclohexane.

Normal systems:

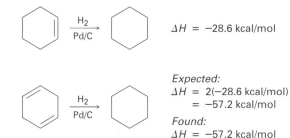

$\Delta H = -28.6$ kcal/mol

Expected:
$\Delta H = 2(-28.6$ kcal/mol$)$
$= -57.2$ kcal/mol

Found:
$\Delta H = -57.2$ kcal/mol

Special systems:

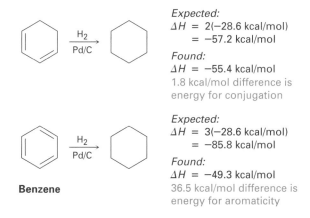

Benzene

Expected:
$\Delta H = 2(-28.6$ kcal/mol$)$
$= -57.2$ kcal/mol

Found:
$\Delta H = -55.4$ kcal/mol
1.8 kcal/mol difference is energy for conjugation

Expected:
$\Delta H = 3(-28.6$ kcal/mol$)$
$= -85.8$ kcal/mol

Found:
$\Delta H = -49.3$ kcal/mol
36.5 kcal/mol difference is energy for aromaticity

To be aromatic, three criteria must be met:

1. Molecule must be flat.
2. Molecule must possess a cyclic array of atoms.
3. Molecule must have a *p* orbital on every atom in that ring.

If all of these rules are satisfied, then there is one final step to determine whether the molecule is aromatic, nonaromatic, or antiaromatic.

Examples of molecules that are not aromatic:

Follows Rules 2 and 3, but does not adopt a flat shape.

Follows Rules 1 and 2, but does not have a *p* orbital on every atom.

Hückel's Rule

1. Count the number of π electrons.
2. If there are $4n + 2$ π electrons, then the molecule is aromatic.
3. If there are $4n$ π electrons, then the molecule is antiaromatic *if* there are 6 or fewer carbon atoms in the ring.
4. If anything else, the molecule is nonaromatic.

Because antiaromaticity is so destabilizing, a molecule with more carbon atoms, like cyclooctatetraene, will adopt a non-planar shape to avoid it.

Understanding aromaticity allows a comprehension of reactivity, particularly for aromatic rings possessing heteroatoms.

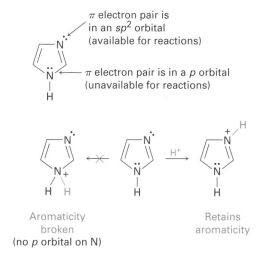

π electron pair is in an sp^2 orbital (available for reactions)

π electron pair is in a *p* orbital (unavailable for reactions)

Aromaticity broken
(no *p* orbital on N)

Retains aromaticity

Aromatic Substitution Reactions

Critical Positional Relationships

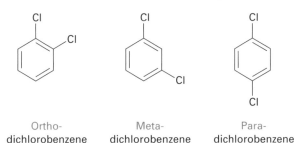

Ortho-
dichlorobenzene

Meta-
dichlorobenzene

Para-
dichlorobenzene

Electrophilic Aromatic Substitution (EAS)

Overall reaction

$$\text{benzene} \xrightarrow[\substack{MX_3 \\ (\text{catalyst})}]{X_2} \text{benzene–X}$$

General mechanism

$$:\overset{..}{\underset{..}{X}}-\overset{..}{\underset{..}{X}}: \ \ MX_3 \longrightarrow :\overset{..}{\underset{..}{X}}-\overset{..}{\underset{..}{X}}\overset{+}{-}\bar{M}X_3$$

Catalyst contains an open valence site; resultant complex is the new, highly reactive, electrophile.

$$:\overset{..}{\underset{..}{X}}-\overset{..}{\underset{..}{X}}\overset{+}{-}\bar{M}X_3$$

$$\overset{..}{\underset{..}{X}}: \quad MX_3$$

$$\overset{+}{:X}-MX_3$$

$$\longrightarrow \text{benzene–X} \quad H-X$$

$$MX_3$$

Determining Selectivity of Electrophile Addition When Multiple Groups Are Present

Mutually reinforcing

$$\text{EWG ... EDG} \xrightarrow{E^+} \text{EWG ... E ... EDG}$$

Competition

$$\text{EDG}_1 \ \text{EDG}_2 \xrightarrow{E^+} \text{EDG}_1 \ \text{EDG}_2 \ \text{E} \quad +$$

EDG$_1$ is a better director than EDG$_2$

Minor product or not observed

EWG: Electron-withdrawing group; cannot provide a lone pair of electrons to the benzene ring through resonance but can accept a lone pair from the ring. These groups are *meta*-directors.

EDG: Electron-donating group; can provide a lone pair of electrons to the benzene ring through resonance. These groups are *ortho*- or *para*-directors.

> Electrophile goes to the site that is controlled by the strongest directing group (that is, the most electron-rich) on the ring at the least hindered site.

Some Important Rules/Exceptions for EAS Reactions

1. Friedel–Crafts alkylations or acylations cannot be performed on a ring possessing an EWG, because the carbon-based electrophiles are not reactive enough to react with an electron-deficient aromatic ring.
2. Friedel–Crafts alkylations are often uncontrolled because alkyl electrophiles can undergo hydride shifts prior to reaction to form a more stable cation, and overalkylation is often a problem.

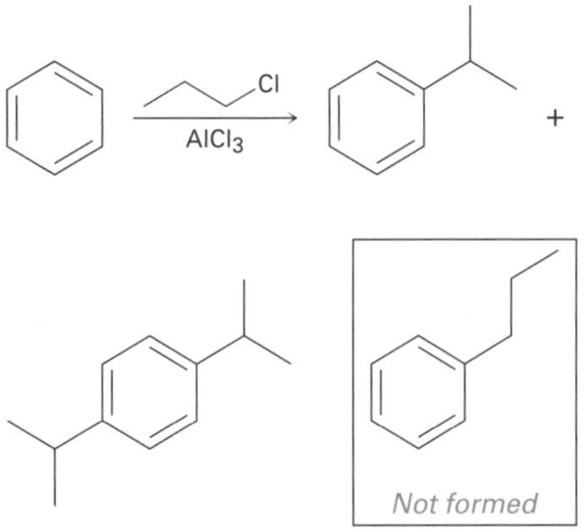

Not formed

Nucleophilic Aromatic Substitution (NAS)

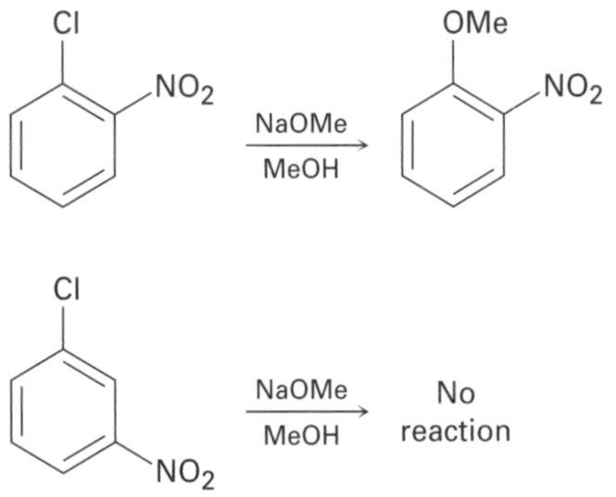

1. Ring must be electron-deficient (has an EWG attached).
2. EWG must be *ortho* or *para* to the leaving group.

Nucleophile Strength

$$R\overset{\bar{}}{N}H \;>\; RO^- \text{ or } HO^- \;>\; Br^- \;>\; NR_3 \;>\; Cl^- \;>$$

$$F^- \;>\; ROH/H_2O \;>\; C{=}C \;>\; \bigcirc$$

Organic Oxidation

* Decrease in the number of bonds to hydrogen and/or
* Increase in the number of bonds to oxygen

Organic Reduction

* Increase in the number of bonds to hydrogen and/or
* Decrease in the number of bonds to oxygen

Important pK_a Values

The lower the pK_a value, the more acidic the proton (more easily removed).

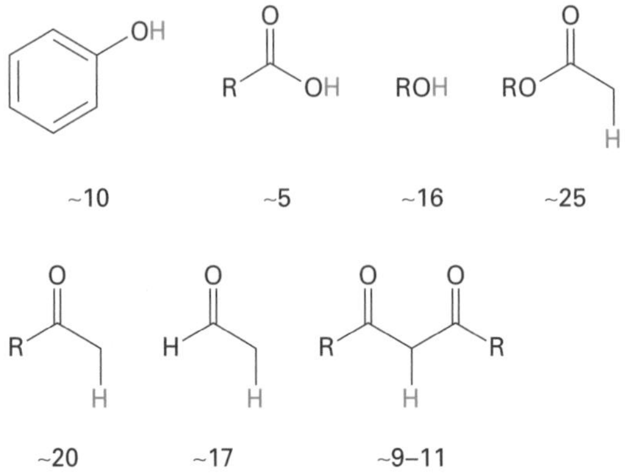

~10   ~5   ~16   ~25

~20   ~17   ~9–11

Alcohols, Ethers, and Carbonyls

Alcohols

Alcohols have the ability to do many things:

1. Can behave as weak acids and weak bases
2. Can also serve as both nucleophile and electrophile

Reactions of alcohols follow a common mechanism:

activated species

Ethers

Ethers are generally unreactive except upon exposure to strong acids at elevated temperatures.

The exception is epoxides. Epoxides can be ruptured with nucleophiles under mildly acidic or basic conditions. In general, epoxides are attacked by nucleophiles at the less hindered position via an S_N2 mechanism **unless** the opening involves an epoxide with a tertiary center under acidic conditions, initiating an S_N1-like process.

Carbonyls

Typical modes of reactivity:

Example is formation of imines

Typical chemistry for reduction reactions and nucleophilic acyl substitution

Nucleophilic Acyl Substitution and Reduction Chemistry

(Reagents that affect the reaction are shown inside the table.)

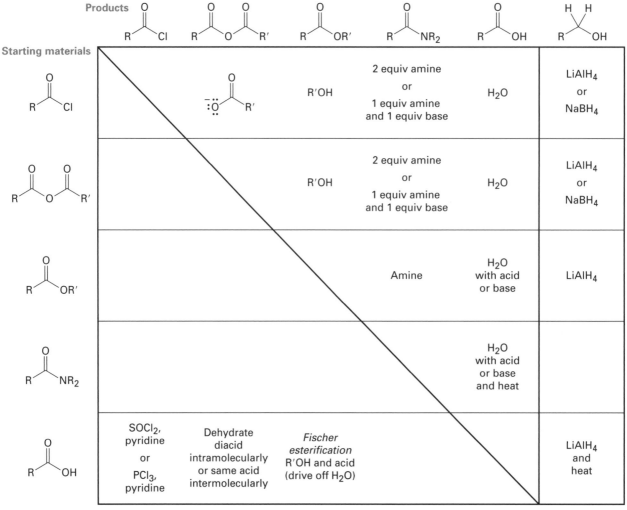

The carbonyl compounds are listed in decreasing order of reactivity toward nucleophilic acyl substitution in the starting material row. This ordering means that a given starting material can form new carbonyl products through substitution reactions only if the resultant products are less reactive (that is, they have a worse leaving group attached).

Reaction pathways with different mechanisms, like Fischer esterification, are needed to move up the chart (acid to ester).

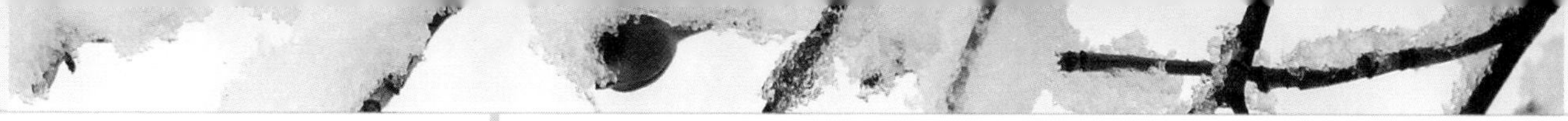

Carbonyls, Enolates, and Amines

Equilibria-Based Reactions of Carbonyls

Reactions That Typically Favor One Side of the Equation

The following reaction has little synthetic utility because the reaction favors the starting material. However, it explains why Cr(VI) oxidations under aqueous conditions provide carboxylic acids; the hydrated aldehyde is oxidized further.

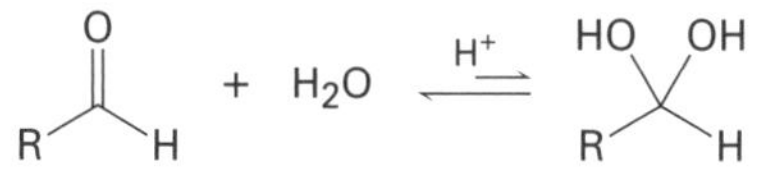

Following is the general exception to the trend of carbonyls in the presence of acids, as the equilibrium with HCN lies on the side of the products.

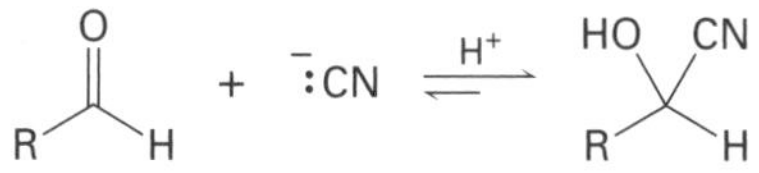

Reactions That Can Be Controlled to Go to Either Side through Le Châtelier's Principle

Use excess nucleophile and drive off water to favor products; add lots of water to favor starting materials.

Uses

Reaction to protect carbonyls or convert acetals into carbonyls:

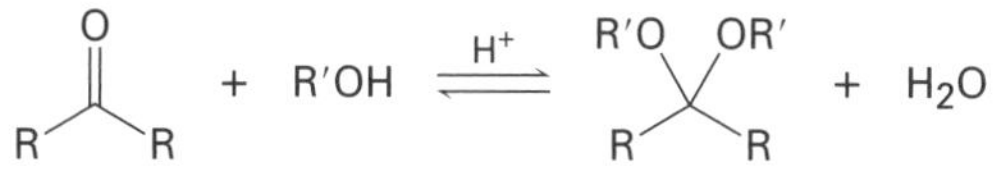

Uses

Way to make imines with primary amines or enamines with secondary amines; intermediate imine hydrolysis to form aldehydes from DIBAH reductions of nitriles uses this reaction:

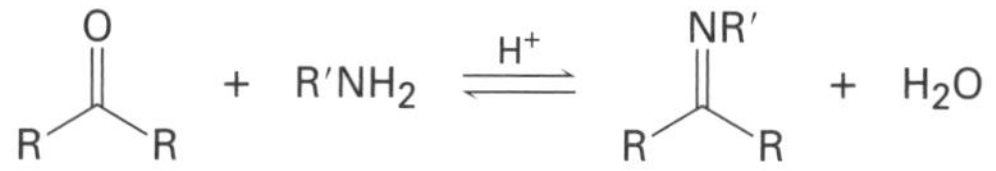

Uses

Way to convert carboxylic acids into a more reactive carbonyl:

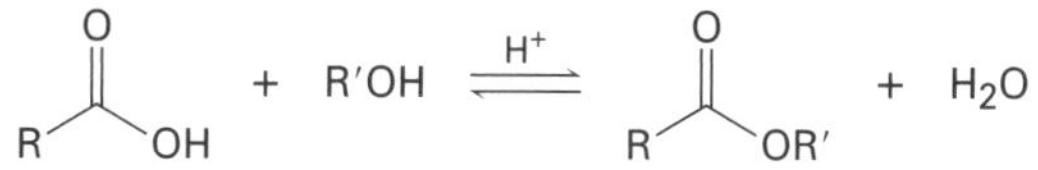

Enolate Chemistry

Although there are two potential nucleophiles (carbon and oxygen) in the following, the carbon version is usually where reactions occur, be they alkylations or condensations.

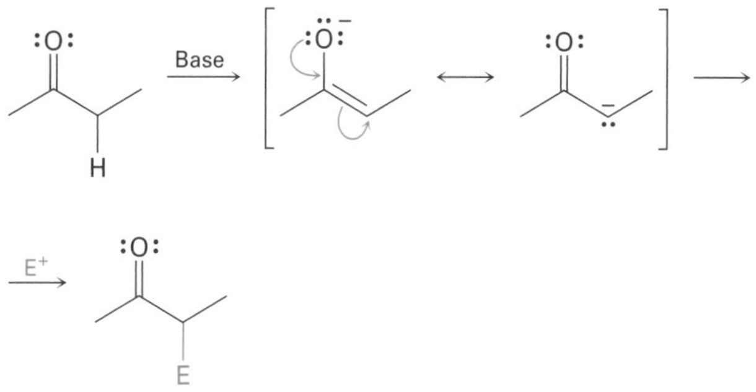

Two Types of Base Conditions

1. NaOR/ROH: Equilibrium (thermodynamic) conditions because both base and proton source are present at the same time.
2. LDA: Kinetic conditions where one can selectively remove most acidic proton (or most accessible) with control.

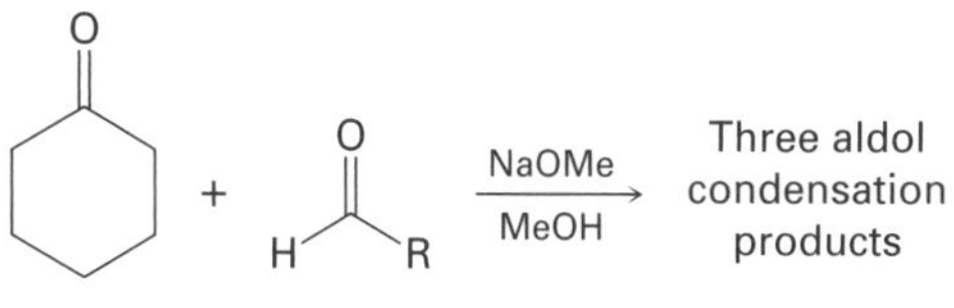

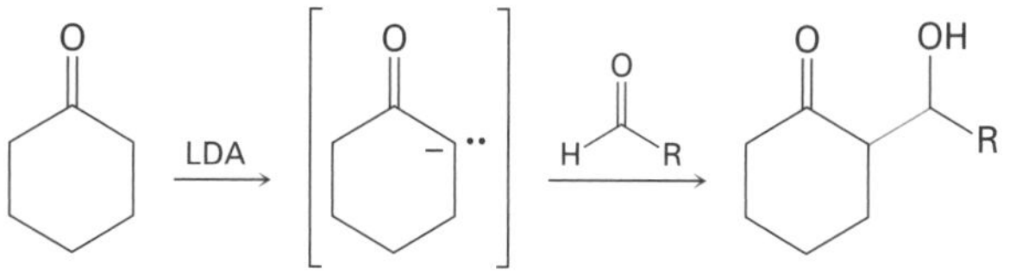

Names of Typical Enolate Reactions

1. Aldol condensation: Intermolecular or intramolecular reaction between aldehydes and ketones
2. Claisen condensation: Intermolecular reaction between two esters
3. Dieckmann condensation: Intramolecular reaction between two esters
4. Michael reaction: Electrophile is an α,β-unsaturated carbonyl
5. Robinson annulation: Michael reaction and intramolecular aldol condensation

Typical Functional Group Patterns to Recognize Condensation Reactions

Enamines Are Enolate Surrogates

The best way to get controlled alkylations, especially with aldehydes that easily undergo self-aldol condensation.

Amines

Amines are best synthesized via reductive amination; this reaction is the most general synthesis and allows for the preparation of 1°, 2°, and 3° amines. Gabriel amine synthesis makes only 1° amines.

Removal of Amines from Molecules (Hofmann Elimination)

Non-Zaitsev product

Pericyclic Reactions

Pericyclic Reactions

A pericyclic reaction is a concerted reaction mechanism, meaning all electrons move at once and there are no reaction intermediates. All are reversible processes, although in many cases, one side of the equilibrium is heavily favored (as denoted below by a single reaction arrow).

(a) Diels–Alder reaction ($4\pi + 2\pi$ cycloaddition)

Diene   Dienophile

The electronic interaction is between the HOMO of the diene and the LUMO of the dienophile.

Size of ΔE determines how much energy is needed to promote the reaction. Diene should be electron-rich; dienophile should be electron-poor.

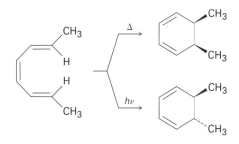

Endo product

Exo T.S.   Endo T.S. (favored)
            π orbital overlap

(b) Cope rearrangement

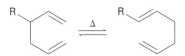

Generally uncontrolled equilibrium unless one side is more stable. Here, the compound on the right is favored because a more substituted alkene results.

(c) Claisen rearrangement

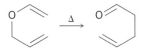

Formation of a carbonyl drives the reaction to this side.

(d) Electrocyclic ring closures

e^-	Thermal	Photochemical
2	Disrotatory	Conrotatory
4	Conrotatory	Disrotatory
6	Disrotatory	Conrotatory
8	Conrotatory	Disrotatory

Easy mnemonic for how to make the table on an exam: "CDs are read with light," so 2 π electrons are conrotatory, while 4 π electrons are disrotatory (the first letters of the words in compact disc). Heat does the opposite of light.

Kinetic and Thermodynamic Products

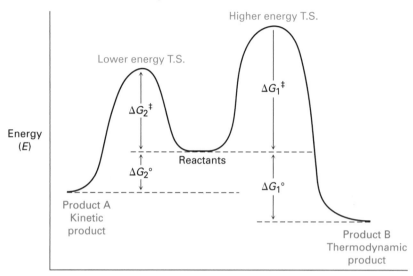

The Diels–Alder reaction is a prime example of this concept: an *endo* transition state is lower in energy than an *exo* transition state, leading to faster formation of the *endo* product. If you compare steric strain in the possible products, however, the *endo* is less stable than the *exo* in most cases, but it is still formed selectively because the reaction is under kinetic control. If enough heat were applied to the product, it could eventually be converted to the thermodynamic product. Usually, at least for Diels–Alder reactions, such a reversion is hard to achieve.

Biomolecules

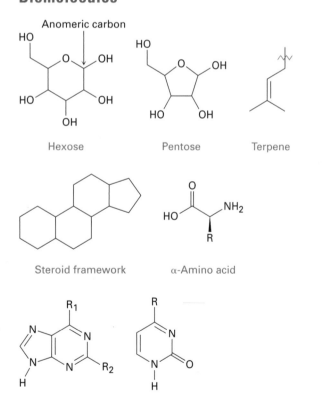